ELECTRIC MACHINERY FUNDAMENTALS

ELECTRIC MACHINERY FUNDAMENTALS

FIFTH EDITION

Stephen J. Chapman

BAE Systems Australia

Mc Graw Hill Connect Learn Succeed™

The McGraw-Hill Companies

ELECTRIC MACHINERY FUNDAMENTALS, FIFTH EDITION

Published by McGraw-Hill, a business unit of The McGraw-Hill Companies, Inc., 1221 Avenue of the Americas, New York, NY 10020.

This book is printed on acid-free paper containing 10% postconsumer waste.

1 2 3 4 5 6 7 8 9 0 DOC/DOC 1 0 9 8 7 6 5 4 3 2 1

ISBN 978-0-07-352954-7
MHID 0-07-352954-0

Vice President & Editor-in-Chief: *Marty Lange*
Vice President EDP/Central Publishing Services: *Kimberly Meriwether David*
Global Publisher: *Raghothaman Srinivasan*
Senior Sponsoring Editor: *Peter E. Massar*
Senior Marketing Manager: *Curt Reynolds*
Development Editor: *Darlene M. Schueller*
Senior Project Manager: *Joyce Watters*
Design Coordinator: *Brenda A. Rolwes*
Cover Designer: *Studio Montage, St. Louis, Missouri*
Cover Image: © *Stockbyte/Punchstock Images RF*
Buyer: *Laura Fuller*
Compositor: *Laserwords Private Limited*
Typeface: *10/12 Times Roman*
Printer: *R R Donnelley*

Library of Congress Cataloging-in-Publication Data

Chapman, Stephen J.
Electric machinery fundamentals / Stephen J. Chapman. —5th ed.
p. cm.
ISBN-13: 978-0-07-352954-7 (alk. paper)
ISBN-10: 0-07-352954-0 (alk. paper)
1. Electric machinery. I. Title.

TK2000.C46 2012
621.31'042—dc22 2010050474

www.mhhe.com

FOR MY DAUGHTER SARAH RIVKAH CHAPMAN,
WHO WILL LIKELY USE THIS BOOK IN HER
STUDIES AT SWINBURNE UNIVERSITY,
MELBOURNE.

ABOUT THE AUTHOR

Stephen J. Chapman received a B.S. in Electrical Engineering from Louisiana State University (1975) and an M.S.E. in Electrical Engineering from the University of Central Florida (1979), and pursued further graduate studies at Rice University.

From 1975 to 1980, he served as an officer in the U.S. Navy, assigned to teach electrical engineering at the U.S. Naval Nuclear Power School in Orlando, Florida. From 1980 to 1982, he was affiliated with the University of Houston, where he ran the power systems program in the College of Technology.

From 1982 to 1988 and from 1991 to 1995, he served as a member of the technical staff of the Massachusetts Institute of Technology's Lincoln Laboratory, both at the main facility in Lexington, Massachusetts, and at the field site on Kwajalein Atoll in the Republic of the Marshall Islands. While there, he did research in radar signal processing systems. He ultimately became the leader of four large operational range instrumentation radars at the Kwajalein field site (TRADEX, ALTAIR, ALCOR, and MMW).

From 1988 to 1991, Mr. Chapman was a research engineer for Shell Development Company in Houston, Texas, where he did seismic signal processing research. He was also affiliated with the University of Houston, where he continued to teach on a part-time basis.

Mr. Chapman is currently manager of systems modeling and operational analysis for BAE Systems Australia, in Melbourne.

Mr. Chapman is a senior member of the Institute of Electrical and Electronic Engineers (and several of its component societies). He is also a member of Engineers Australia.

BRIEF CONTENTS

Chapter 1	Introduction to Machinery Principles	1
Chapter 2	Transformers	65
Chapter 3	AC Machinery Fundamentals	152
Chapter 4	Synchronous Generators	191
Chapter 5	Synchronous Motors	271
Chapter 6	Induction Motors	307
Chapter 7	DC Machinery Fundamentals	404
Chapter 8	DC Motors and Generators	464
Chapter 9	Single-Phase and Special-Purpose Motors	565
Appendix A	Three-Phase Circuits	613
Appendix B	Coil Pitch and Distributed Windings	639
Appendix C	Salient-Pole Theory of Synchronous Machines	659
Appendix D	Tables of Constants and Conversion Factors	669

TABLE OF CONTENTS

Chapter 1 Introduction to Machinery Principles 1

1.1 Electrical Machines, Transformers, and Daily Life 1
1.2 A Note on Units and Notation 2

Notation

1.3 Rotational Motion, Newton's Law, and Power Relationships 3

Angular Position θ / Angular Velocity ω / Angular Acceleration α / Torque τ / Newton's Law of Rotation / Work W / Power P

1.4 The Magnetic Field 8

Production of a Magnetic Field / Magnetic Circuits / Magnetic Behavior of Ferromagnetic Materials / Energy Losses in a Ferromagnetic Core

1.5 Faraday's Law—Induced Voltage from a Time-Changing Magnetic Field 28
1.6 Production of Induced Force on a Wire 33
1.7 Induced Voltage on a Conductor Moving in a Magnetic Field 34
1.8 The Linear DC Machine—A Simple Example 36

Starting the Linear DC Machine / The Linear DC Machine as a Motor / The Linear DC Machine as a Generator / Starting Problems with the Linear Machine

1.9 Real, Reactive, and Apparent Power in Single-Phase AC Circuits 47

Alternative Forms of the Power Equations / Complex Power / The Relationships between Impedance Angle, Current Angle, and Power / The Power Triangle

1.10 Summary 53

Questions 54
Problems 55
References 64

Chapter 2 Transformers 65

2.1 Why Transformers Are Important to Modern Life 66
2.2 Types and Construction of Transformers 67
2.3 The Ideal Transformer 69
Power in an Ideal Transformer / Impedance Transformation through a Transformer / Analysis of Circuits Containing Ideal Transformers
2.4 Theory of Operation of Real Single-Phase Transformers 77
The Voltage Ratio across a Transformer / The Magnetization Current in a Real Transformer / The Current Ratio on a Transformer and the Dot Convention
2.5 The Equivalent Circuit of a Transformer 86
The Exact Equivalent Circuit of a Real Transformer / Approximate Equivalent Circuits of a Transformer / Determining the Values of Components in the Transformer Model
2.6 The Per-Unit System of Measurements 94
2.7 Transformer Voltage Regulation and Efficiency 99
The Transformer Phasor Diagram / Transformer Efficiency
2.8 Transformer Taps and Voltage Regulation 108
2.9 The Autotransformer 109
Voltage and Current Relationships in an Autotransformer / The Apparent Power Rating Advantage of Autotransformers / The Internal Impedance of an Autotransformer
2.10 Three-Phase Transformers 116
Three-Phase Transformer Connections / The Per-Unit System for Three-Phase Transformers
2.11 Three-Phase Transformation Using Two Transformers 126
The Open-Δ (or V-V) Connection / The Open-Wye–Open-Delta Connection / The Scott-T Connection / The Three-Phase T Connection
2.12 Transformer Ratings and Related Problems 134
The Voltage and Frequency Ratings of a Transformer / The Apparent Power Rating of a Transformer / The Problem of Current Inrush / The Transformer Nameplate
2.13 Instrument Transformers 140
2.14 Summary 142
Questions 143
Problems 144
References 151

Chapter 3 AC Machinery Fundamentals 152

3.1 A Simple Loop in a Uniform Magnetic Field 153

The Voltage Induced in a Simple Rotating Loop / The Torque Induced in a Current-Carrying Loop

3.2 The Rotating Magnetic Field 160

Proof of the Rotating Magnetic Field Concept / The Relationship between Electrical Frequency and the Speed of Magnetic Field Rotation / Reversing the Direction of Magnetic Field Rotation

3.3 Magnetomotive Force and Flux Distribution on AC Machines 169

3.4 Induced Voltage in AC Machines 172

The Induced Voltage in a Coil on a Two-Pole Stator / The Induced Voltage in a Three-Phase Set of Coils / The RMS Voltage in a Three-Phase Stator

3.5 Induced Torque in an AC Machine 178

3.6 Winding Insulation in an AC Machine 182

3.7 AC Machine Power Flows and Losses 182

The Losses in AC Machines / The Power-Flow Diagram

3.8 Voltage Regulation and Speed Regulation 186

3.9 Summary 187

Questions 187

Problems 188

References 190

Chapter 4 Synchronous Generators 191

4.1 Synchronous Generator Construction 192

4.2 The Speed of Rotation of a Synchronous Generator 197

4.3 The Internal Generated Voltage of a Synchronous Generator 197

4.4 The Equivalent Circuit of a Synchronous Generator 198

4.5 The Phasor Diagram of a Synchronous Generator 202

4.6 Power and Torque in Synchronous Generators 205

4.7 Measuring Synchronous Generator Model Parameters 208

The Short-Circuit Ratio

4.8 The Synchronous Generator Operating Alone 213

The Effect of Load Changes on a Synchronous Generator Operating Alone / Example Problems

4.9 Parallel Operation of AC Generators 224

The Conditions Required for Paralleling / The General Procedure for Paralleling Generators / Frequency–Power and Voltage–Reactive Power Characteristics of a Synchronous Generator / Operation of Generators in Parallel with Large Power Systems / Operation of Generators in Parallel with Other Generators of the Same Size

4.10 Synchronous Generator Transients 244

Transient Stability of Synchronous Generators / Short-Circuit Transients in Synchronous Generators

4.11 Synchronous Generator Ratings 251

The Voltage, Speed, and Frequency Ratings / Apparent Power and Power-Factor Ratings / Synchronous Generator Capability Curves / Short-Time Operation and Service Factor

4.12 Summary 261

Questions 262

Problems 263

References 270

Chapter 5 Synchronous Motors 271

5.1 Basic Principles of Motor Operation 271

The Equivalent Circuit of a Synchronous Motor / The Synchronous Motor from a Magnetic Field Perspective

5.2 Steady-State Synchronous Motor Operation 275

The Synchronous Motor Torque-Speed Characteristic Curve / The Effect of Load Changes on a Synchronous Motor / The Effect of Field Current Changes on a Synchronous Motor / The Synchronous Motor and Power-Factor Correction / The Synchronous Capacitor or Synchronous Condenser

5.3 Starting Synchronous Motors 290

Motor Starting by Reducing Electrical Frequency / Motor Starting with an External Prime Mover / Motor Starting by Using Amortisseur Windings / The Effect of Amortisseur Windings on Motor Stability

5.4 Synchronous Generators and Synchronous Motors 297

5.5 Synchronous Motor Ratings 298

5.6 Summary 298

Questions 300

Problems 300

References 306

Chapter 6 Induction Motors 307

6.1 Induction Motor Construction 309

6.2 Basic Induction Motor Concepts 311

The Development of Induced Torque in an Induction Motor / The Concept of Rotor Slip / The Electrical Frequency on the Rotor

6.3 The Equivalent Circuit of an Induction Motor 315

The Transformer Model of an Induction Motor / The Rotor Circuit Model / The Final Equivalent Circuit

6.4 Power and Torque in Induction Motors 321

Losses and the Power-Flow Diagram / Power and Torque in an Induction Motor / Separating the Rotor Copper Losses and the Power Converted in an Induction Motor's Equivalent Circuit

6.5 Induction Motor Torque–Speed Characteristics 328

Induced Torque from a Physical Standpoint / The Derivation of the Induction Motor Induced-Torque Equation / Comments on the Induction Motor Torque–Speed Curve / Maximum (Pullout) Torque in an Induction Motor

6.6 Variations in Induction Motor Torque–Speed Characteristics 343

Control of Motor Characteristics by Cage Rotor Design / Deep-Bar and Double-Cage Rotor Designs / Induction Motor Design Classes

6.7 Trends in Induction Motor Design 353

6.8 Starting Induction Motors 357

Induction Motor Starting Circuits

6.9 Speed Control of Induction Motors 363

Induction Motor Speed Control by Pole Changing / Speed Control by Changing the Line Frequency / Speed Control by Changing the Line Voltage / Speed Control by Changing the Rotor Resistance

6.10 Solid-State Induction Motor Drives 372

Frequency (Speed) Adjustment / A Choice of Voltage and Frequency Patterns / Independently Adjustable Acceleration and Deceleration Ramps / Motor Protection

6.11 Determining Circuit Model Parameters 380

The No-Load Test / The DC Test for Stator Resistance / The Locked-Rotor Test

6.12 The Induction Generator 388

The Induction Generator Operating Alone / Induction Generator Applications

6.13 Induction Motor Ratings 393

6.14 Summary 394

Questions 396

Problems 397

References 402

Chapter 7 DC Machinery Fundamentals 404

7.1 A Simple Rotating Loop between Curved Pole Faces 404

The Voltage Induced in a Rotating Loop / Getting DC Voltage Out of the Rotating Loop / The Induced Torque in the Rotating Loop

7.2 Commutation in a Simple Four-Loop DC Machine 416

7.3 Commutation and Armature Construction in Real DC Machines 421

The Rotor Coils / Connections to the Commutator Segments / The Lap Winding / The Wave Winding / The Frog-Leg Winding

7.4 Problems with Commutation in Real Machines 433

Armature Reaction / L di/dt *Voltages / Solutions to the Problems with Commutation*

7.5 The Internal Generated Voltage and Induced Torque Equations of Real DC Machines 445

7.6 The Construction of DC Machines 449

Pole and Frame Construction / Rotor or Armature Construction / Commutator and Brushes / Winding Insulation

7.7 Power Flow and Losses in DC Machines 455

The Losses in DC Machines / The Power Flow Diagram

7.8 Summary 458

Questions 458

Problems 458

References 461

Chapter 8 DC Motors and Generators 464

8.1 Introduction to DC Motors 465

8.2 The Equivalent Circuit of a DC Motor 467

8.3 The Magnetization Curve of a DC Machine 468

8.4 Separately Excited and Shunt DC Motors 469

The Terminal Characteristic of a Shunt DC Motor / Nonlinear Analysis of a Shunt DC Motor / Speed Control of Shunt DC Motors / The Effect of an Open Field Circuit

8.5 The Permanent-Magnet DC Motor 491

8.6 The Series DC Motor 493

Induced Torque in a Series DC Motor / The Terminal Characteristic of a Series DC Motor / Speed Control of Series DC Motors

8.7 The Compounded DC Motor 500

The Torque–Speed Characteristic of a Cumulatively Compounded DC Motor / The Torque–Speed Characteristic of a Differentially Compounded DC Motor / The Nonlinear Analysis of Compounded DC Motors / Speed Control in the Cumulatively Compounded DC Motor

8.8 DC Motor Starters 505

DC Motor Problems on Starting / DC Motor Starting Circuits

8.9 The Ward-Leonard System and Solid-State Speed Controllers 514

Protection Circuit Section / Start/Stop Circuit Section / High-Power Electronics Section / Low-Power Electronics Section

8.10 DC Motor Efficiency Calculations 524

8.11 Introduction to DC Generators 526

8.12 The Separately Excited Generator 528

The Terminal Characteristic of a Separately Excited DC Generator / Control of Terminal Voltage / Nonlinear Analysis of a Separately Excited DC Generator

8.13 The Shunt DC Generator 534

Voltage Buildup in a Shunt Generator / The Terminal Characteristic of a Shunt DC Generator / Voltage Control for a Shunt DC Generator / The Analysis of Shunt DC Generators

8.14 The Series DC Generator 540

The Terminal Characteristic of a Series Generator

8.15 The Cumulatively Compounded DC Generator 543

The Terminal Characteristic of a Cumulatively Compounded DC Generator / Voltage Control of Cumulatively Compounded DC Generators / Analysis of Cumulatively Compounded DC Generators

8.16 The Differentially Compounded DC Generator 547

The Terminal Characteristic of a Differentially Compounded DC Generator / Voltage Control of Differentially Compounded DC Generators / Graphical Analysis of a Differentially Compounded DC Generator

8.17 Summary 551

Questions 552

Problems 553

References 564

Chapter 9 Single-Phase and Special-Purpose Motors 565

9.1 The Universal Motor 566

Applications of Universal Motors / Speed Control of Universal Motors

9.2 Introduction to Single-Phase Induction Motors 569

The Double-Revolving-Field Theory of Single-Phase Induction Motors / The Cross-Field Theory of Single-Phase Induction Motors

9.3 Starting Single-Phase Induction Motors 578

Split-Phase Windings / Capacitor-Start Motors / Permanent Split-Capacitor and Capacitor-Start, Capacitor-Run Motors / Shaded-Pole Motors / Comparison of Single-Phase Induction Motors

9.4 Speed Control of Single-Phase Induction Motors 588

9.5 The Circuit Model of a Single-Phase Induction Motor 590

Circuit Analysis with the Single-Phase Induction Motor Equivalent Circuit

9.6 Other Types of Motors 597

Reluctance Motors / Hysteresis Motors / Stepper Motors / Brushless DC Motors

9.7 Summary 609

Questions 610

Problems 611

References 612

Appendix A Three-Phase Circuits 613

A.1 Generation of Three-Phase Voltages and Currents 613

A.2 Voltages and Currents in a Three-Phase Circuit 617

Voltages and Currents in the Wye (Y) Connection / Voltages and Currents in the Delta (Δ) Connection

A.3 Power Relationships in Three-Phase Circuits 622

Three-Phase Power Equations Involving Phase Quantities / Three-Phase Power Equations Involving Line Quantities

A.4 Analysis of Balanced Three-Phase Systems 625

A.5 One-Line Diagrams 632

A.6 Using the Power Triangle 632

Questions 635

Problems 636

Reference 638

Appendix B Coil Pitch and Distributed Windings 639

B.1 The Effect of Coil Pitch on AC Machines 639

The Pitch of a Coil / The Induced Voltage of a Fractional-Pitch Coil / Harmonic Problems and Fractional-Pitch Windings

B.2 Distributed Windings in AC Machines 648

The Breadth or Distribution Factor / The Generated Voltage Including Distribution Effects / Tooth or Slot Harmonics

B.3 Summary 656

Questions 657

Problems 657

References 658

Appendix C Salient-Pole Theory of Synchronous Machines 659

C.1 Development of the Equivalent Circuit of a Salient-Pole Synchronous Generator 660

C.2 Torque and Power Equations of a Salient-Pole Machines 666

Problems 667

Appendix D Tables of Constants and Conversion Factors 669

PREFACE

In the years since the first edition of *Electric Machinery Fundamentals* was published, there has been rapid advance in the development of larger and more sophisticated solid-state motor drive packages. The first edition of this book stated that dc motors were the method of choice for demanding variable-speed applications. That statement is no longer true today. Now, the system of choice for speed control applications is most often an ac induction motor with a solid-state motor drive. DC motors have been largely relegated to special-purpose applications where a dc power source is readily available, such as in automotive electrical systems.

The third edition of the book was extensively restructured to reflect these changes. The material on ac motors and generators is now covered in Chapters 3 through 6, before the material on dc machines. In addition, the dc machinery coverage was reduced compared to earlier editions. This edition continues with this same basic structure.

In addition, the former Chapter 3 on solid-state electronics has been deleted from the fifth edition. Feedback from users has indicated that that material was too detailed for a quick overview, and not detailed enough for a solid-state electronics course. Since very few instructors were using this material, it has been removed from this edition and added as a supplement on the book's website. Any instructor or student wishing to continue using the material in this chapter can freely download it.

Learning objectives have been added to the beginning of each chapter to enhance student learning.

Chapter 1 provides an introduction to basic machinery concepts, and concludes by applying those concepts to a linear dc machine, which is the simplest possible example of a machine. Chapter 2 covers transformers, which are not rotating machines, but which share many similar analysis techniques.

After Chapter 2, an instructor may choose to teach either dc or ac machinery first. Chapters 3 through 6 cover ac machinery, and Chapters 7 and 8 cover dc machinery. These chapter sequences have been made completely independent of each other, so that an instructor can cover the material in the order which best suits his or her needs. For example, a one-semester course with a primary concentration in ac machinery might consist of parts of Chapters 1, 2, 3, 4, 5, and 6, with any remaining time devoted to dc machinery. A one-semester course with a primary concentration in dc machinery might consist of parts of Chapters 1, 2, 7, and 8, with any remaining time devoted to ac machinery. Chapter 9 is devoted to single-phase and special-purpose motors, such as universal motors, stepper motors, brushless dc motors, and shaded-pole motors.

The homework problems and the ends of chapters have been revised and corrected, and more than 70% of the problems are either new or modified since the last edition.

In recent years, there have been major changes in the methods used to teach machinery to electrical engineering and electrical technology students. Excellent analytical tools such as MATLAB® have become widely available in university engineering curricula. These tools make very complex calculations simple to perform, and they allow students to explore the behavior of problems interactively. This edition of *Electric Machinery Fundamentals* makes selected use of MATLAB to enhance a student's learning experience where appropriate. For example, students use MATLAB in Chapter 6 to calculate the torque–speed characteristics of induction motors, and to explore the properties of double-cage induction motors.

This text does not teach MATLAB; it assumes that the student is familiar with it through previous work. Also, the book does *not* depend on a student having MATLAB. MATLAB provides an enhancement to the learning experience if it is available, but if it is not, the examples involving MATLAB can simply be skipped, and the remainder of the text still makes sense.

This book would never have been possible without the help of dozens of people over the past 25 years. It is gratifying for me to see the book still popular after all that time, and much of that is due to the excellent feedback provided by reviewers. For this edition, I would especially like to thank:

Ashoka K.S. Bhat
University of Victoria

William Butuk
Lakehead University

Shaahin Filizadeh
University of Manitoba

Jesús Fraile-Ardanuy
Universidad Politécnica de Madrid

Riadh Habash
University of Ottawa

Floyd Henderson
Michigan Technological University

Rajesh Kavasseri
North Dakota State University

Ali Keyhani
The Ohio State University

Andrew Knight
University of Alberta

Xiaomin Kou
University of Wisconsin–Platteville

Ahmad Nafisi
California Polytechnic State University, San Luis Obispo

Subhasis Nandi
University of Victoria

M. Hashem Nehrir
Montana State University–Bozeman

Ali Shaban
California Polytechnic State University, San Luis Obispo

Kuang Sheng
Rutgers University

Barna Szabados
McMaster University

Tristan J. Tayag
Texas Christian University

Rajiv K. Varma
The University of Western Ontario

Stephen J. Chapman
Melbourne, Victoria, Australia

MCGRAW-HILL DIGITAL OFFERINGS INCLUDE:

Supplemental materials supporting the book are available from the book's website at www.mhhe.com/chapman. The materials available at that address include MATLAB source code, the supplement "Introduction to Power Electronics," pointers to sites of interest to machinery students, a list of errata in the text, some supplemental topics which are not covered in the main text, and supplemental MATLAB tools.

CHAPTER 1

INTRODUCTION TO MACHINERY PRINCIPLES

OBJECTIVES

- Learn the basics of rotational mechanics: angular velocity, angular acceleration, torque, and Newton's law for rotation.
- Learn how to produce a magnetic field.
- Understand magnetic circuits.
- Understand the behavior of ferromagnetic materials.
- Understand hysteresis in ferromagnetic materials.
- Understand Faraday's law.
- Understand how to produce an induced force on a wire.
- Understand how to produce an induced voltage across a wire.
- Understand the operation of a simple linear machine.
- Be able to work with real, reactive, and apparent powers.

1.1 ELECTRICAL MACHINES, TRANSFORMERS, AND DAILY LIFE

An **electrical machine** is a device that can convert either mechanical energy to electrical energy or electrical energy to mechanical energy. When such a device is used to convert mechanical energy to electrical energy, it is called a *generator*. When it converts electrical energy to mechanical energy, it is called a *motor*. Since any given electrical machine can convert power in either direction, any machine can be used as either a generator or a motor. Almost all practical motors and generators convert energy from one form to another through the action of a magnetic field, and only machines using magnetic fields to perform such conversions are considered in this book.

The *transformer* is an electrical device that is closely related to electrical machines. It converts ac electrical energy at one voltage level to ac electrical energy at another voltage level. Since transformers operate on the same principles as generators and motors, depending on the action of a magnetic field to accomplish the change in voltage level, they are usually studied together with generators and motors.

These three types of electric devices are ubiquitous in modern daily life. Electric motors in the home run refrigerators, freezers, vacuum cleaners, blenders, air conditioners, fans, and many similar appliances. In the workplace, motors provide the motive power for almost all tools. Of course, generators are necessary to supply the power used by all these motors.

Why are electric motors and generators so common? The answer is very simple: Electric power is a clean and efficient energy source that is easy to transmit over long distances, and easy to control. An electric motor does not require constant ventilation and fuel the way that an internal-combustion engine does, so the motor is very well suited for use in environments where the pollutants associated with combustion are not desirable. Instead, heat or mechanical energy can be converted to electrical form at a distant location, the energy can be transmitted over long distances to the place where it is to be used, and it can be used cleanly in any home, office, or factory. Transformers aid this process by reducing the energy loss between the point of electric power generation and the point of its use.

1.2 A NOTE ON UNITS AND NOTATION

The design and study of electric machines and power systems are among the oldest areas of electrical engineering. Study began in the latter part of the nineteenth century. At that time, electrical units were being standardized internationally, and these units came to be universally used by engineers. Volts, amperes, ohms, watts, and similar units, which are part of the metric system of units, have long been used to describe electrical quantities in machines.

In English-speaking countries, though, mechanical quantities had long been measured with the English system of units (inches, feet, pounds, etc.). This practice was followed in the study of machines. Therefore, for many years the electrical and mechanical quantities of machines have been measured with different systems of units.

In 1954, a comprehensive system of units based on the metric system was adopted as an international standard. This system of units became known as the *Système International* (SI) and has been adopted throughout most of the world. The United States is practically the sole holdout—even Britain and Canada have switched over to SI.

The SI units will inevitably become standard in the United States as time goes by, and professional societies such as the Institute of Electrical and Electronics Engineers (IEEE) have standardized on metric units for all work. However, many people have grown up using English units, and this system will remain in daily use for a long time. Engineering students and working engineers in the United States today must be familiar with both sets of units, since they will encounter both throughout their professional lives. Therefore, this book includes problems and examples using both SI and English units. The emphasis in the examples is on SI units, but the older system is not entirely neglected.

Notation

In this book, vectors, electrical phasors, and other complex values are shown in bold face (e.g., **F**), while scalars are shown in italic face (e.g., *R*). In addition, a special font is used to represent magnetic quantities such as magnetomotive force (e.g., $\mathcal{F}$).

1.3 ROTATIONAL MOTION, NEWTON'S LAW, AND POWER RELATIONSHIPS

Almost all electric machines rotate about an axis, called the *shaft* of the machine. Because of the rotational nature of machinery, it is important to have a basic understanding of rotational motion. This section contains a brief review of the concepts of distance, velocity, acceleration, Newton's law, and power as they apply to rotating machinery. For a more detailed discussion of the concepts of rotational dynamics, see References 2, 4, and 5.

In general, a three-dimensional vector is required to completely describe the rotation of an object in space. However, machines normally turn on a fixed shaft, so their rotation is restricted to one angular dimension. Relative to a given end of the machine's shaft, the direction of rotation can be described as either *clockwise* (CW) or *counterclockwise* (CCW). For the purpose of this volume, a counterclockwise angle of rotation is assumed to be positive, and a clockwise one is assumed to be negative. For rotation about a fixed shaft, all the concepts in this section reduce to scalars.

Each major concept of rotational motion is defined below and is related to the corresponding idea from linear motion.

Angular Position θ

The angular position θ of an object is the angle at which it is oriented, measured from some arbitrary reference point. Angular position is usually measured in radians or degrees. It corresponds to the linear concept of distance along a line.

Angular Velocity ω

Angular velocity (or speed) is the rate of change in angular position with respect to time. It is assumed positive if the rotation is in a counterclockwise direction. Angular velocity is the rotational analog of the concept of velocity on a line. One-dimensional linear velocity along a line is defined as the rate of change of the displacement along the line (r) with respect to time.

$$v = \frac{dr}{dt} \tag{1–1}$$

Similarly, angular velocity ω is defined as the rate of change of the angular displacement θ with respect to time.

$$\omega = \frac{d\theta}{dt} \tag{1–2}$$

If the units of angular position are radians, then angular velocity is measured in radians per second.

In dealing with ordinary electric machines, engineers often use units other than radians per second to describe shaft speed. Frequently, the speed is given in revolutions per second or revolutions per minute. Because speed is such an important quantity in the study of machines, it is customary to use different symbols for speed when it is expressed in different units. By using these different symbols, any possible confusion as to the units intended is minimized. The following symbols are used in this book to describe angular velocity:

ω_m	angular velocity expressed in radians per second
f_m	angular velocity expressed in revolutions per second
n_m	angular velocity expressed in revolutions per minute

The subscript m on these symbols indicates a mechanical quantity, as opposed to an electrical quantity. If there is no possibility of confusion between mechanical and electrical quantities, the subscript is often left out.

These measures of shaft speed are related to each other by the following equations:

$$n_m = 60f_m \tag{1–3a}$$

$$f_m = \frac{\omega_m}{2\pi} \tag{1–3b}$$

Angular Acceleration α

Angular acceleration is the rate of change in angular velocity with respect to time. It is assumed positive if the angular velocity is increasing in an algebraic sense. Angular acceleration is the rotational analog of the concept of acceleration on a line. Just as one-dimensional linear acceleration is defined by the equation

$$a = \frac{dv}{dt} \tag{1–4}$$

angular acceleration is defined by

$$\alpha = \frac{d\omega}{dt} \tag{1–5}$$

If the units of angular velocity are radians per second, then angular acceleration is measured in radians per second squared.

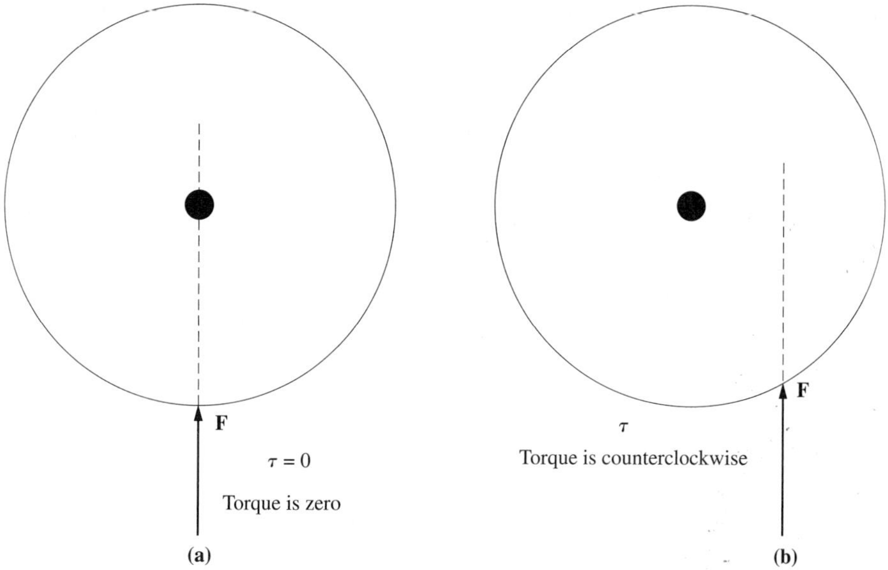

FIGURE 1–1
(a) A force applied to a cylinder so that it passes through the axis of rotation. $\tau = 0$.
(b) A force applied to a cylinder so that its line of action misses the axis of rotation. Here τ is counterclockwise.

Torque τ

In linear motion, a *force* applied to an object causes its velocity to change. In the absence of a net force on the object, its velocity is constant. The greater the force applied to the object, the more rapidly its velocity changes.

There exists a similar concept for rotation. When an object is rotating, its angular velocity is constant unless a *torque* is present on it. The greater the torque on the object, the more rapidly the angular velocity of the object changes.

What is torque? It can loosely be called the "twisting force" on an object. Intuitively, torque is fairly easy to understand. Imagine a cylinder that is free to rotate about its axis. If a force is applied to the cylinder in such a way that its line of action passes through the axis (Figure 1–1a), then the cylinder will not rotate. However, if the same force is placed so that its line of action passes to the right of the axis (Figure 1–1b), then the cylinder will tend to rotate in a counterclockwise direction. The torque or twisting action on the cylinder depends on (1) the magnitude of the applied force and (2) the distance between the axis of rotation and the line of action of the force.

The torque on an object is defined as the product of the force applied to the object and the smallest distance between the line of action of the force and the object's axis of rotation. If **r** is a vector pointing from the axis of rotation to the point

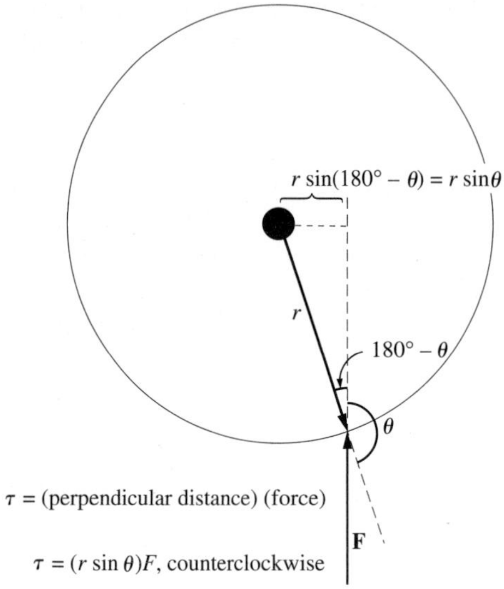

FIGURE 1–2
Derivation of the equation for the torque on an object.

of application of the force, and if **F** is the applied force, then the torque can be described as

$$\begin{aligned}\tau &= \text{(force applied)(perpendicular distance)} \\ &= (F)\,(r \sin \theta) \\ &= rF \sin \theta \end{aligned} \tag{1–6}$$

where θ is the angle between the vector **r** and the vector **F**. The direction of the torque is clockwise if it would tend to cause a clockwise rotation and counterclockwise if it would tend to cause a counterclockwise rotation (Figure 1–2).

The units of torque are newton-meters in SI units and pound-feet in the English system.

Newton's Law of Rotation

Newton's law for objects moving along a straight line describes the relationship between the force applied to an object and its resulting acceleration. This relationship is given by the equation

$$F = ma \tag{1–7}$$

where

F = net force applied to an object
m = mass of the object
a = resulting acceleration

In SI units, force is measured in newtons, mass in kilograms, and acceleration in meters per second squared. In the English system, force is measured in pounds, mass in slugs, and acceleration in feet per second squared.

A similar equation describes the relationship between the torque applied to an object and its resulting angular acceleration. This relationship, called *Newton's law of rotation,* is given by the equation

$$\tau = J\alpha \tag{1–8}$$

where τ is the net applied torque in newton-meters or pound-feet and α is the resulting angular acceleration in radians per second squared. The term J serves the same purpose as an object's mass in linear motion. It is called the *moment of inertia* of the object and is measured in kilogram-meters squared or slug-feet squared. Calculation of the moment of inertia of an object is beyond the scope of this book. For information about it see Ref. 2.

Work W

For linear motion, work is defined as the application of a *force* through a *distance.* In equation form,

$$W = \int F\,dr \tag{1–9}$$

where it is assumed that the force is collinear with the direction of motion. For the special case of a constant force applied collinearly with the direction of motion, this equation becomes just

$$W = Fr \tag{1–10}$$

The units of work are joules in SI and foot-pounds in the English system.

For rotational motion, work is the application of a *torque* through an *angle.* Here the equation for work is

$$W = \int \tau\,d\theta \tag{1–11}$$

and if the torque is constant,

$$W = \tau\theta \tag{1–12}$$

Power P

Power is the rate of doing work, or the increase in work per unit time. The equation for power is

$$P = \frac{dW}{dt} \tag{1–13}$$

It is usually measured in joules per second (watts), but also can be measured in foot-pounds per second or in horsepower.

By this definition, and assuming that force is constant and collinear with the direction of motion, power is given by

$$P = \frac{dW}{dt} = \frac{d}{dt}(Fr) = F\left(\frac{dr}{dt}\right) = Fv \tag{1–14}$$

Similarly, assuming constant torque, power in rotational motion is given by

$$P = \frac{dW}{dt} = \frac{d}{dt}(\tau\theta) = \tau\left(\frac{d\theta}{dt}\right) = \tau\omega$$

$$P = \tau\omega \tag{1–15}$$

Equation (1–15) is very important in the study of electric machinery, because it can describe the mechanical power on the shaft of a motor or generator.

Equation (1–15) is the correct relationship among power, torque, and speed if power is measured in watts, torque in newton-meters, and speed in radians per second. If other units are used to measure any of the above quantities, then a constant must be introduced into the equation for unit conversion factors. It is still common in U.S. engineering practice to measure torque in pound-feet, speed in revolutions per minute, and power in either watts or horsepower. If the appropriate conversion factors are included in each term, then Equation (1–15) becomes

$$P\ (\text{watts}) = \frac{\tau\ (\text{lb-ft})\ n\ (\text{r/min})}{7.04} \tag{1–16}$$

$$P\ (\text{horsepower}) = \frac{\tau\ (\text{lb-ft})\ n\ (\text{r/min})}{5252} \tag{1–17}$$

where torque is measured in pound-feet and speed is measured in revolutions per minute.

1.4 THE MAGNETIC FIELD

As previously stated, magnetic fields are the fundamental mechanism by which energy is converted from one form to another in motors, generators, and transformers. Four basic principles describe how magnetic fields are used in these devices:

1. A current-carrying wire produces a magnetic field in the area around it.
2. A time-changing magnetic field induces a voltage in a coil of wire if it passes through that coil. (This is the basis of *transformer action.*)
3. A current-carrying wire in the presence of a magnetic field has a force induced on it. (This is the basis of *motor action.*)
4. A moving wire in the presence of a magnetic field has a voltage induced in it. (This is the basis of *generator action.*)

This section describes and elaborates on the production of a magnetic field by a current-carrying wire, while later sections of this chapter explain the remaining three principles.

Production of a Magnetic Field

The basic law governing the production of a magnetic field by a current is Ampere's law:

$$\oint \mathbf{H} \cdot d\mathbf{l} = I_{\text{net}} \tag{1–18}$$

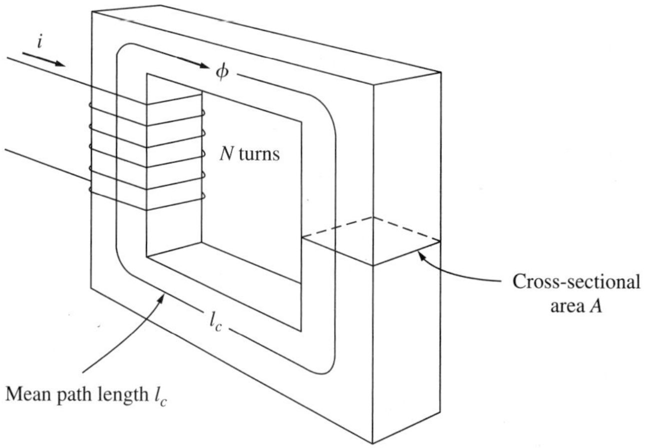

FIGURE 1–3
A simple magnetic core.

where **H** is the magnetic field intensity produced by the current I_{net}, and dl is a differential element of length along the path of integration. In SI units, I is measured in amperes and H is measured in ampere-turns per meter. To better understand the meaning of this equation, it is helpful to apply it to the simple example in Figure 1–3. Figure 1–3 shows a rectangular core with a winding of N turns of wire wrapped about one leg of the core. If the core is composed of iron or certain other similar metals (collectively called *ferromagnetic materials*), essentially all the magnetic field produced by the current will remain inside the core, so the path of integration in Ampere's law is the mean path length of the core l_c. The current passing within the path of integration I_{net} is then Ni, since the coil of wire cuts the path of integration N times while carrying current i. Ampere's law thus becomes

$$Hl_c = Ni \tag{1–19}$$

Here H is the magnitude of the magnetic field intensity vector **H**. Therefore, the magnitude of the magnetic field intensity in the core due to the applied current is

$$H = \frac{Ni}{l_c} \tag{1–20}$$

The magnetic field intensity **H** is in a sense a measure of the "effort" that a current is putting into the establishment of a magnetic field. The strength of the magnetic field flux produced in the core also depends on the material of the core. The relationship between the magnetic field intensity **H** and the resulting magnetic flux density **B** produced within a material is given by

$$\mathbf{B} = \mu\mathbf{H} \tag{1–21}$$

where

$\mathbf{H}$ = magnetic field intensity
μ = magnetic *permeability* of material
$\mathbf{B}$ = resulting magnetic flux density produced

The actual magnetic flux density produced in a piece of material is thus given by a product of two terms:

H, representing the effort exerted by the current to establish a magnetic field
μ, representing the relative ease of establishing a magnetic field in a given material

The units of magnetic field intensity are ampere-turns per meter, the units of permeability are henrys per meter, and the units of the resulting flux density are webers per square meter, known as teslas (T).

The permeability of free space is called μ_0, and its value is

$$\mu_0 = 4\pi \times 10^{-7} \text{ H/m} \tag{1–22}$$

The permeability of any other material compared to the permeability of free space is called its *relative permeability:*

$$\mu_r = \frac{\mu}{\mu_0} \tag{1–23}$$

Relative permeability is a convenient way to compare the magnetizability of materials. For example, the steels used in modern machines have relative permeabilities of 2000 to 6000 or even more. This means that, for a given amount of current, 2000 to 6000 times more flux is established in a piece of steel than in a corresponding area of air. (The permeability of air is essentially the same as the permeability of free space.) Obviously, the metals in a transformer or motor core play an extremely important part in increasing and concentrating the magnetic flux in the device.

Also, because the permeability of iron is so much higher than that of air, the great majority of the flux in an iron core like that in Figure 1–3 remains inside the core instead of traveling through the surrounding air, which has much lower permeability. The small leakage flux that does leave the iron core is very important in determining the flux linkages between coils and the self-inductances of coils in transformers and motors.

In a core such as the one shown in Figure 1–3, the magnitude of the flux density is given by

$$B = \mu H = \frac{\mu N i}{l_c} \tag{1–24}$$

Now the total flux in a given area is given by

$$\phi = \int_A \mathbf{B} \cdot d\mathbf{A} \tag{1–25a}$$

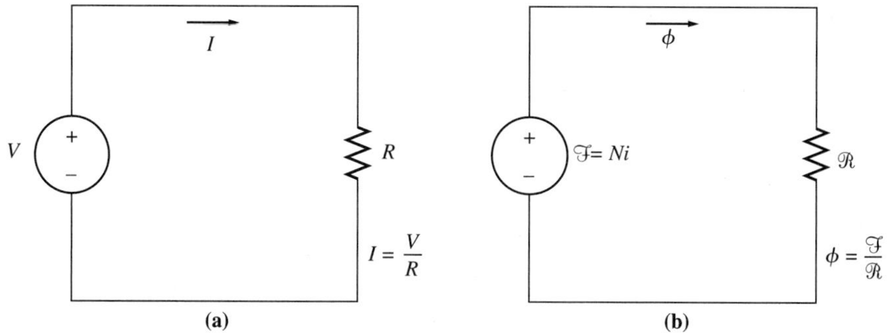

FIGURE 1–4
(a) A simple electric circuit. (b) The magnetic circuit analog to a transformer core.

where $d\mathbf{A}$ is the differential unit of area. If the flux density vector is perpendicular to a plane of area A, and if the flux density is constant throughout the area, then this equation reduces to

$$\phi = BA \tag{1–25b}$$

Thus, the total flux in the core in Figure 1–3 due to the current i in the winding is

$$\phi = BA = \frac{\mu NiA}{l_c} \tag{1–26}$$

where A is the cross-sectional area of the core.

Magnetic Circuits

In Equation (1–26) we see that the *current* in a coil of wire wrapped around a core produces a magnetic flux in the core. This is in some sense analogous to a voltage in an electric circuit producing a current flow. It is possible to define a "magnetic circuit" whose behavior is governed by equations analogous to those for an electric circuit. The magnetic circuit model of magnetic behavior is often used in the design of electric machines and transformers to simplify the otherwise quite complex design process.

In a simple electric circuit such as the one shown in Figure 1–4a, the voltage source V drives a current I around the circuit through a resistance R. The relationship between these quantities is given by Ohm's law:

$$V = IR$$

In the electric circuit, it is the voltage or electromotive force that drives the current flow. By analogy, the corresponding quantity in the magnetic circuit is called the *magnetomotive force* (mmf). The magnetomotive force of the magnetic circuit is equal to the effective current flow applied to the core, or

$$\mathcal{F} = Ni \tag{1–27}$$

where $\mathcal{F}$ is the symbol for magnetomotive force, measured in ampere-turns.

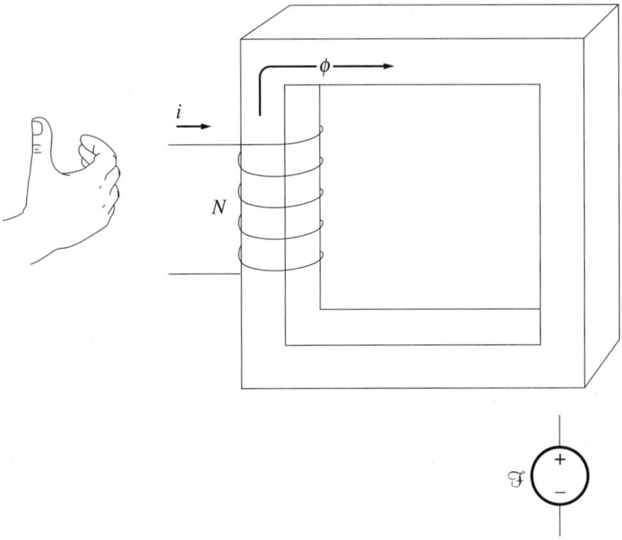

FIGURE 1–5
Determining the polarity of a magnetomotive force source in a magnetic circuit.

Like the voltage source in the electric circuit, the magnetomotive force in the magnetic circuit has a polarity associated with it. The *positive* end of the mmf source is the end from which the flux exits, and the *negative* end of the mmf source is the end at which the flux reenters. The polarity of the mmf from a coil of wire can be determined from a modification of the right-hand rule: If the fingers of the right hand curl in the direction of the current flow in a coil of wire, then the thumb will point in the direction of the positive mmf (see Figure 1–5).

In an electric circuit, the applied voltage causes a current I to flow. Similarly, in a magnetic circuit, the applied magnetomotive force causes flux ϕ to be produced. The relationship between voltage and current in an electric circuit is Ohm's law ($V = IR$); similarly, the relationship between magnetomotive force and flux is

$$\boxed{\mathcal{F} = \phi \mathcal{R}} \tag{1–28}$$

where

$\mathcal{F}$ = magnetomotive force of circuit
ϕ = flux of circuit
$\mathcal{R}$ = *reluctance* of circuit

The reluctance of a magnetic circuit is the counterpart of electrical resistance, and its units are ampere-turns per weber.

There is also a magnetic analog of conductance. Just as the conductance of an electric circuit is the reciprocal of its resistance, the *permeance* $\mathcal{P}$ of a magnetic circuit is the reciprocal of its reluctance:

$$\mathcal{P} = \frac{1}{\mathcal{R}} \tag{1–29}$$

The relationship between magnetomotive force and flux can thus be expressed as

$$\phi = \mathcal{F}\mathcal{P} \tag{1–30}$$

Under some circumstances, it is easier to work with the permeance of a magnetic circuit than with its reluctance.

What is the reluctance of the core in Figure 1–3? The resulting flux in this core is given by Equation (1–26):

$$\phi = BA = \frac{\mu NiA}{l_c} \tag{1–26}$$

$$= Ni\left(\frac{\mu A}{l_c}\right)$$

$$\phi = \mathcal{F}\left(\frac{\mu A}{l_c}\right) \tag{1–31}$$

By comparing Equation (1–31) with Equation (1–28), we see that the reluctance of the core is

$$\mathcal{R} = \frac{l_c}{\mu A} \tag{1–32}$$

Reluctances in a magnetic circuit obey the same rules as resistances in an electric circuit. The equivalent reluctance of a number of reluctances in series is just the sum of the individual reluctances:

$$\mathcal{R}_{\text{eq}} = \mathcal{R}_1 + \mathcal{R}_2 + \mathcal{R}_3 + \cdots \tag{1–33}$$

Similarly, reluctances in parallel combine according to the equation

$$\frac{1}{\mathcal{R}_{\text{eq}}} = \frac{1}{\mathcal{R}_1} + \frac{1}{\mathcal{R}_2} + \frac{1}{\mathcal{R}_3} + \cdots \tag{1–34}$$

Permeances in series and parallel obey the same rules as electrical conductances.

Calculations of the flux in a core performed by using the magnetic circuit concepts are *always* approximations—at best, they are accurate to within about 5 percent of the real answer. There are a number of reasons for this inherent inaccuracy:

1. The magnetic circuit concept assumes that all flux is confined within a magnetic core. Unfortunately, this is not quite true. The permeability of a ferromagnetic core is 2000 to 6000 times that of air, but a small fraction of the flux escapes from the core into the surrounding low-permeability air. This flux

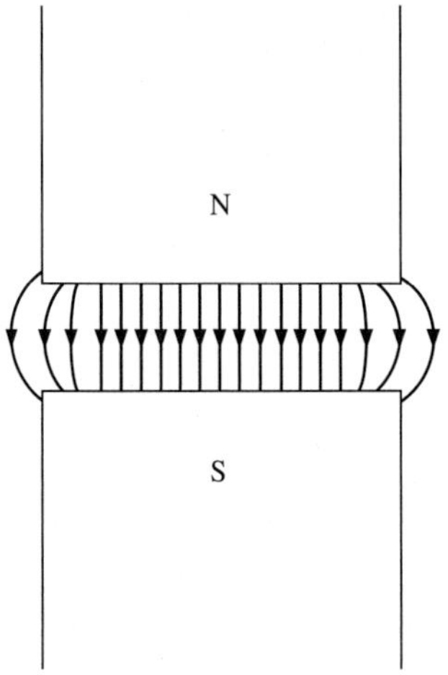

FIGURE 1–6
The fringing effect of a magnetic field at an air gap. Note the increased cross-sectional area of the air gap compared with the cross-sectional area of the metal.

outside the core is called *leakage flux,* and it plays a very important role in electric machine design.

2. The calculation of reluctance assumes a certain mean path length and cross-sectional area for the core. These assumptions are not really very good, especially at corners.
3. In ferromagnetic materials, the permeability varies with the amount of flux already in the material. This nonlinear effect is described in detail. It adds yet another source of error to magnetic circuit analysis, since the reluctances used in magnetic circuit calculations depend on the permeability of the material.
4. If there are air gaps in the flux path in a core, the effective cross-sectional area of the air gap will be larger than the cross-sectional area of the iron core on either side. The extra effective area is caused by the "fringing effect" of the magnetic field at the air gap (Figure 1–6).

It is possible to partially offset these inherent sources of error by using a "corrected" or "effective" mean path length and cross-sectional area instead of the actual physical length and area in the calculations.

There are many inherent limitations to the concept of a magnetic circuit, but it is still the easiest design tool available for calculating fluxes in practical machinery design. Exact calculations using Maxwell's equations are just too difficult, and they are not needed anyway, since satisfactory results may be achieved with this approximate method.

The following examples illustrate basic magnetic circuit calculations. Note that in these examples the answers are given to three significant digits.

Example 1–1. A ferromagnetic core is shown in Figure 1–7a. Three sides of this core are of uniform width, while the fourth side is somewhat thinner. The depth of the core (into the page) is 10 cm, and the other dimensions are shown in the figure. There is a 200-turn coil wrapped around the left side of the core. Assuming relative permeability μ_r of 2500, how much flux will be produced by a 1-A input current?

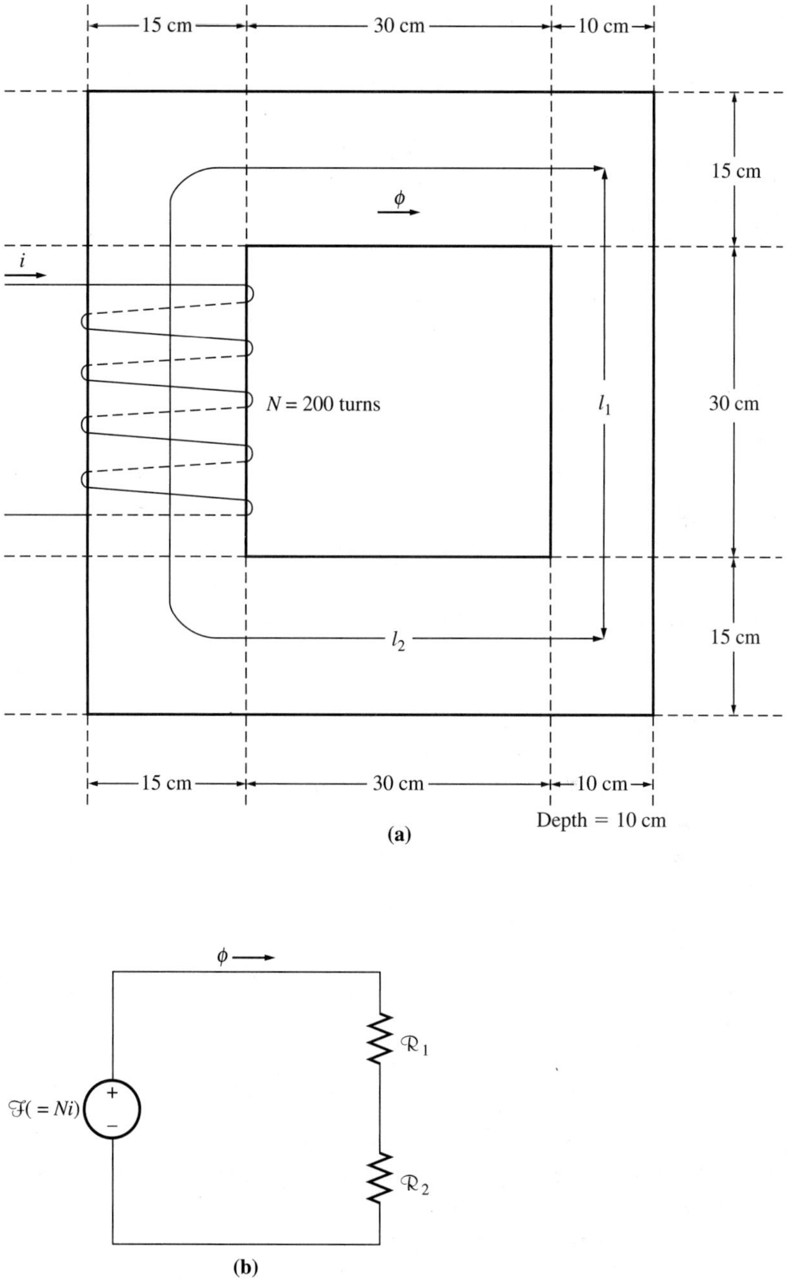

FIGURE 1–7
(a) The ferromagnetic core of Example 1–1. (b) The magnetic circuit corresponding to (a).

Solution

We will solve this problem twice, once by hand and once by a MATLAB program, and show that both approaches yield the same answer.

Three sides of the core have the same cross-sectional areas, while the fourth side has a different area. Thus, the core can be divided into two regions: (1) the single thinner side and (2) the other three sides taken together. The magnetic circuit corresponding to this core is shown in Figure 1–7b.

The mean path length of region 1 is 45 cm, and the cross-sectional area is 10 × 10 cm = 100 cm^2. Therefore, the reluctance in the first region is

$$\mathcal{R}_1 = \frac{l_1}{\mu A_1} = \frac{l_1}{\mu_r \mu_0 A_1} \tag{1–32}$$

$$= \frac{0.45 \text{ m}}{(2500)(4\pi \times 10^{-7})(0.01 \text{ m}^2)}$$

$$= 14{,}300 \text{ A} \bullet \text{turns/Wb}$$

The mean path length of region 2 is 130 cm, and the cross-sectional area is 15 × 10 cm = 150 cm^2. Therefore, the reluctance in the second region is

$$\mathcal{R}_2 = \frac{l_2}{\mu A_2} = \frac{l_2}{\mu_r \mu_0 A_2} \tag{1–32}$$

$$= \frac{1.3 \text{ m}}{(2500)(4\pi \times 10^{-7})(0.015 \text{ m}^2)}$$

$$= 27{,}600 \text{ A} \bullet \text{turns/Wb}$$

Therefore, the total reluctance in the core is

$$\mathcal{R}_{\text{eq}} = \mathcal{R}_1 + \mathcal{R}_2$$

$$= 14{,}300 \text{ A} \bullet \text{turns/Wb} + 27{,}600 \text{ A} \bullet \text{turns/Wb}$$

$$= 41{,}900 \text{ A} \bullet \text{turns/Wb}$$

The total magnetomotive force is

$$\mathcal{F} = Ni = (200 \text{ turns})(1.0 \text{ A}) = 200 \text{ A} \bullet \text{turns}$$

The total flux in the core is given by

$$\phi = \frac{\mathcal{F}}{\mathcal{R}} = \frac{200 \text{ A} \bullet \text{turns}}{41{,}900 \text{ A} \bullet \text{turns/ Wb}}$$

$$= 0.0048 \text{ Wb}$$

This calculation can be performed by using a MATLAB script file, if desired. A simple script to calculate the flux in the core is shown below.

```
% M-file: ex1_1.m
% M-file to calculate the flux in Example 1-1.
l1 = 0.45;                        % Length of region 1
l2 = 1.3;                         % Length of region 2
a1 = 0.01;                        % Area of region 1
a2 = 0.015;                       % Area of region 2
ur = 2500;                        % Relative permeability
```

```
u0 = 4*pi*1E-7;                 % Permeability of free space
n = 200;                        % Number of turns on core
i = 1;                          % Current in amps

% Calculate the first reluctance
r1 = l1 / (ur * u0 * a1);
disp (['r1 = ' num2str(r1)]);

% Calculate the second reluctance
r2 = l2 / (ur * u0 * a2);
disp (['r2 = ' num2str(r2)]);

% Calculate the total reluctance
rtot = r1 + r2;

% Calculate the mmf
mmf = n * i;

% Finally, get the flux in the core
flux = mmf / rtot;

% Display result
disp (['Flux = ' num2str(flux)]);
```

When this program is executed, the results are:

```
» ex1_1
r1 = 14323.9449
r2 = 27586.8568
Flux = 0.004772
```

This program produces the same answer as our hand calculations to the number of significant digits in the problem.

Example 1–2. Figure 1–8a shows a ferromagnetic core whose mean path length is 40 cm. There is a small gap of 0.05 cm in the structure of the otherwise whole core. The cross-sectional area of the core is 12 cm^2, the relative permeability of the core is 4000, and the coil of wire on the core has 400 turns. Assume that fringing in the air gap increases the effective cross-sectional area of the air gap by 5 percent. Given this information, find *(a)* the total reluctance of the flux path (iron plus air gap) and *(b)* the current required to produce a flux density of 0.5 T in the air gap.

Solution

The magnetic circuit corresponding to this core is shown in Figure 1–8b.

(a) The reluctance of the core is

$$\mathcal{R}_c = \frac{l_c}{\mu A_c} = \frac{l_c}{\mu_r \mu_0 A_c} \tag{1–32}$$

$$= \frac{0.4 \text{ m}}{(4000)(4\pi \times 10^{-7})(0.002 \text{ m}^2)}$$

$$= 66{,}300 \text{ A} \bullet \text{turns/Wb}$$

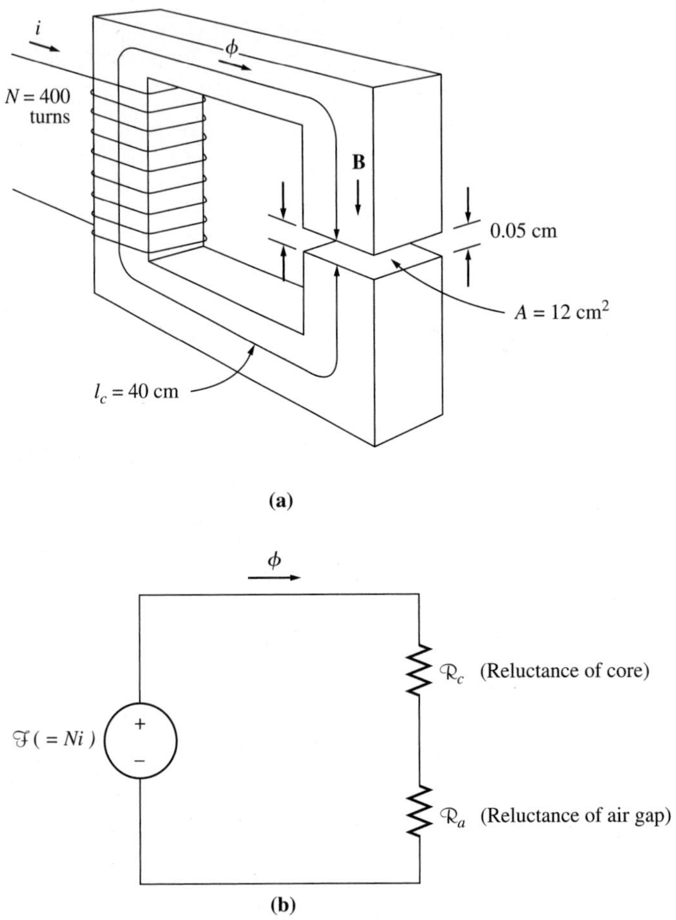

FIGURE 1–8
(a) The ferromagnetic core of Example 1–2. (b) The magnetic circuit corresponding to (a).

The effective area of the air gap is $1.05 \times 12\ \text{cm}^2 = 12.6\ \text{cm}^2$, so the reluctance of the air gap is

$$\mathcal{R}_a = \frac{l_a}{\mu_0 A_a} \qquad (1\text{–}32)$$

$$= \frac{0.0005\ \text{m}}{(4\pi \times 10^{-7})(0.00126\ \text{m}^2)}$$

$$= 316{,}000\ \text{A} \bullet \text{turns/Wb}$$

Therefore, the total reluctance of the flux path is

$$\mathcal{R}_{\text{eq}} = \mathcal{R}_c + \mathcal{R}_a$$

$$= 66{,}300\ \text{A} \bullet \text{turns/Wb} + 316{,}000\ \text{A} \bullet \text{turns/Wb}$$

$$= 382{,}300\ \text{A} \bullet \text{turns/Wb}$$

Note that the air gap contributes most of the reluctance even though it is 800 times shorter than the core.

(b) Equation (1–28) states that

$$\mathcal{F} = \phi \mathcal{R} \tag{1–28}$$

Since the flux $\phi = BA$ and $\mathcal{F} = Ni$, this equation becomes

$$Ni = BA\mathcal{R}$$

so

$$\begin{aligned} i &= \frac{BA\mathcal{R}}{N} \\ &= \frac{(0.5 \text{ T})(0.00126 \text{ m}^2)(383{,}200 \text{ A} \cdot \text{turns/Wb})}{400 \text{ turns}} \\ &= 0.602 \text{ A} \end{aligned}$$

Notice that, since the *air-gap* flux was required, the effective air-gap area was used in the above equation.

Example 1–3. Figure 1–9a shows a simplified rotor and stator for a dc motor. The mean path length of the stator is 50 cm, and its cross-sectional area is 12 cm^2. The mean path length of the rotor is 5 cm, and its cross-sectional area also may be assumed to be 12 cm^2. Each air gap between the rotor and the stator is 0.05 cm wide, and the cross-sectional area of each air gap (including fringing) is 14 cm^2. The iron of the core has a relative permeability of 2000, and there are 200 turns of wire on the core. If the current in the wire is adjusted to be 1 A, what will the resulting flux density in the air gaps be?

Solution

To determine the flux density in the air gap, it is necessary to first calculate the magnetomotive force applied to the core and the total reluctance of the flux path. With this information, the total flux in the core can be found. Finally, knowing the cross-sectional area of the air gaps enables the flux density to be calculated.

The reluctance of the stator is

$$\begin{aligned} \mathcal{R}_s &= \frac{l_s}{\mu_r \mu_0 A_s} \\ &= \frac{0.5 \text{ m}}{(2000)(4\pi \times 10^{-7})(0.0012 \text{ m}^2)} \\ &= 166{,}000 \text{ A} \cdot \text{turns/Wb} \end{aligned}$$

The reluctance of the rotor is

$$\begin{aligned} \mathcal{R}_r &= \frac{l_r}{\mu_r \mu_0 A_r} \\ &= \frac{0.05 \text{ m}}{(2000)(4\pi \times 10^{-7})(0.0012 \text{ m}^2)} \\ &= 16{,}600 \text{ A} \cdot \text{turns/Wb} \end{aligned}$$

The reluctance of the air gaps is

$$\begin{aligned} \mathcal{R}_a &= \frac{l_a}{\mu_r \mu_0 A_a} \\ &= \frac{0.0005 \text{ m}}{(1)(4\pi \times 10^{-7})(0.0014 \text{ m}^2)} \\ &= 284{,}000 \text{ A} \cdot \text{turns/Wb} \end{aligned}$$

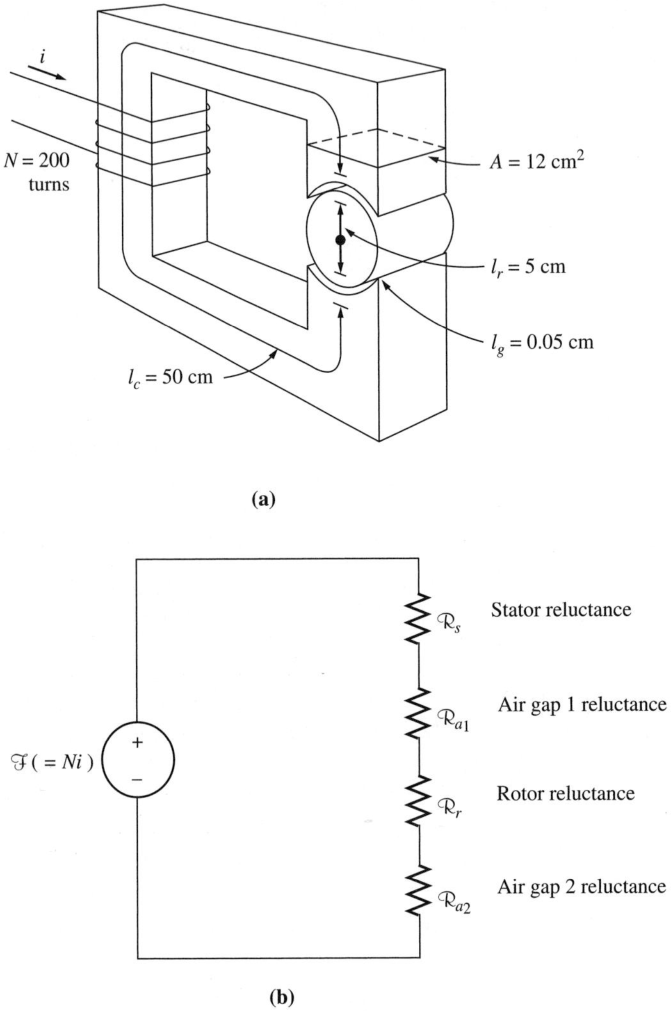

FIGURE 1–9
(a) A simplified diagram of a rotor and stator for a dc motor. (b) The magnetic circuit corresponding to (a).

The magnetic circuit corresponding to this machine is shown in Figure 1–9b. The total reluctance of the flux path is thus

$$\begin{aligned}\mathcal{R}_{\text{eq}} &= \mathcal{R}_s + \mathcal{R}_{a1} + \mathcal{R}_r + \mathcal{R}_{a2} \\ &= 166{,}000 + 284{,}000 + 16{,}600 + 284{,}000 \text{ A} \bullet \text{turns/Wb} \\ &= 751{,}000 \text{ A} \bullet \text{turns/Wb}\end{aligned}$$

The net magnetomotive force applied to the core is

$$\mathcal{F} = Ni = (200 \text{ turns})(1.0 \text{ A}) = 200 \text{ A} \bullet \text{turns}$$

Therefore, the total flux in the core is

$$\phi = \frac{\mathcal{F}}{\mathcal{R}} = \frac{200 \text{ A} \cdot \text{turns}}{751{,}000 \text{ A} \cdot \text{turns/Wb}}$$
$$= 0.00266 \text{ Wb}$$

Finally, the magnetic flux density in the motor's air gap is

$$B = \frac{\phi}{A} = \frac{0.000266 \text{ Wb}}{0.0014 \text{ m}^2} = 0.19 \text{ T}$$

Magnetic Behavior of Ferromagnetic Materials

Earlier in this section, magnetic permeability was defined by the equation

$$\mathbf{B} = \mu\mathbf{H} \tag{1–21}$$

It was explained that the permeability of ferromagnetic materials is very high, up to 6000 times the permeability of free space. In that discussion and in the examples that followed, the permeability was assumed to be constant regardless of the magnetomotive force applied to the material. Although permeability is constant in free space, this most certainly is *not* true for iron and other ferromagnetic materials.

To illustrate the behavior of magnetic permeability in a ferromagnetic material, apply a direct current to the core shown in Figure 1–3, starting with 0 A and slowly working up to the maximum permissible current. When the flux produced in the core is plotted versus the magnetomotive force producing it, the resulting plot looks like Figure 1–10a. This type of plot is called a *saturation curve* or a *magnetization curve.* At first, a small increase in the magnetomotive force produces a huge increase in the resulting flux. After a certain point, though, further increases in the magnetomotive force produce relatively smaller increases in the flux. Finally, an increase in the magnetomotive force produces almost no change at all. The region of this figure in which the curve flattens out is called the *saturation region,* and the core is said to be *saturated.* In contrast, the region where the flux changes very rapidly is called the *unsaturated region* of the curve, and the core is said to be *unsaturated.* The transition region between the unsaturated region and the saturated region is sometimes called the *knee* of the curve. Note that the flux produced in the core is linearly related to the applied magnetomotive force in the unsaturated region, and approaches a constant value regardless of magnetomotive force in the saturated region.

Another closely related plot is shown in Figure 1–10b. Figure 1–10b is a plot of magnetic flux density **B** versus magnetizing intensity **H**. From Equations (1–20) and (1–25b),

$$H = \frac{Ni}{l_c} = \frac{\mathcal{F}}{l_c} \tag{1–20}$$

$$\phi = BA \tag{1–25b}$$

it is easy to see that *magnetizing intensity is directly proportional to magnetomotive force* and *magnetic flux density is directly proportional to flux* for any given core. Therefore, the relationship between B and H has the same shape as the relationship

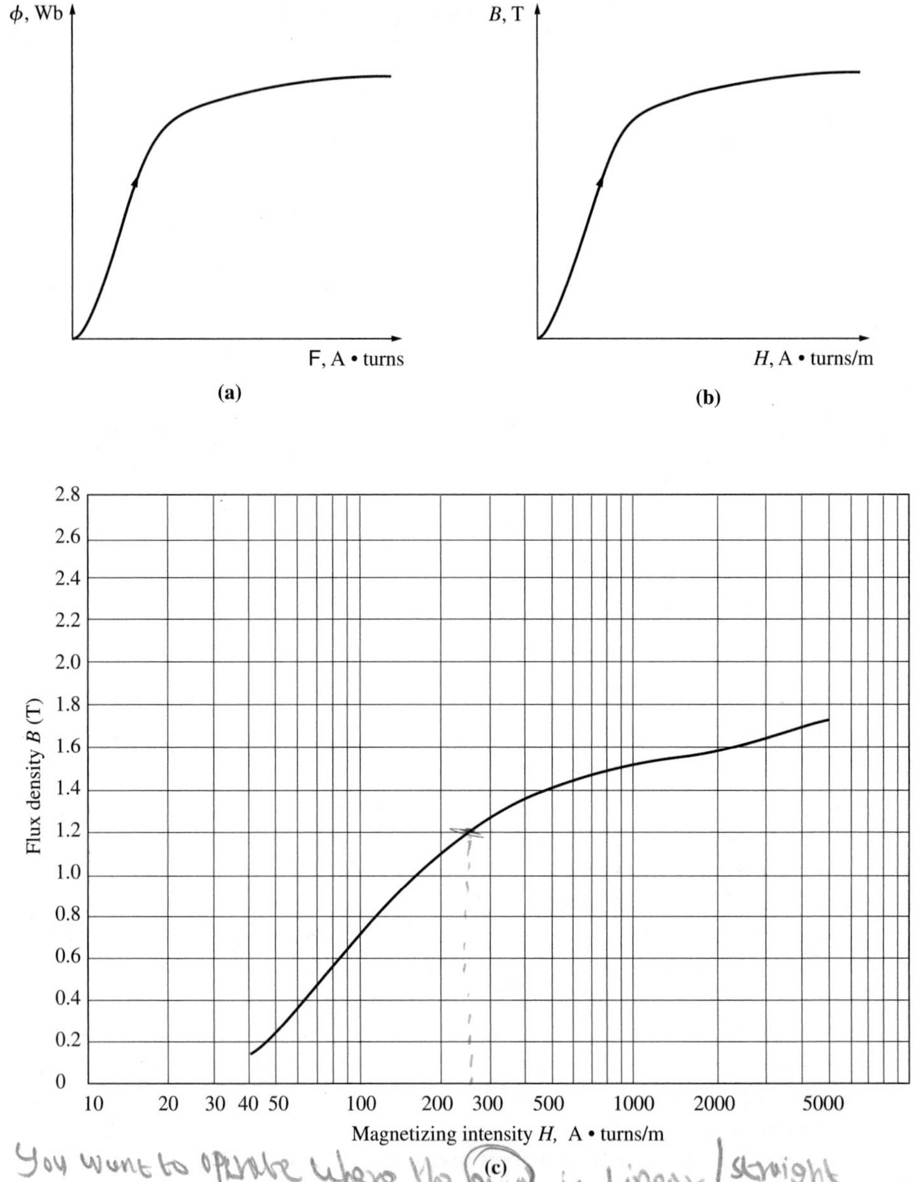

You want to operate where the graph is Linear / straight Line

FIGURE 1–10
(a) Sketch of a dc magnetization curve for a ferromagnetic core. (b) The magnetization curve expressed in terms of flux density and magnetizing intensity. (c) A detailed magnetization curve for a typical piece of steel. (d) A plot of relative permeability μ_r as a function of magnetizing intensity H for a typical piece of steel.

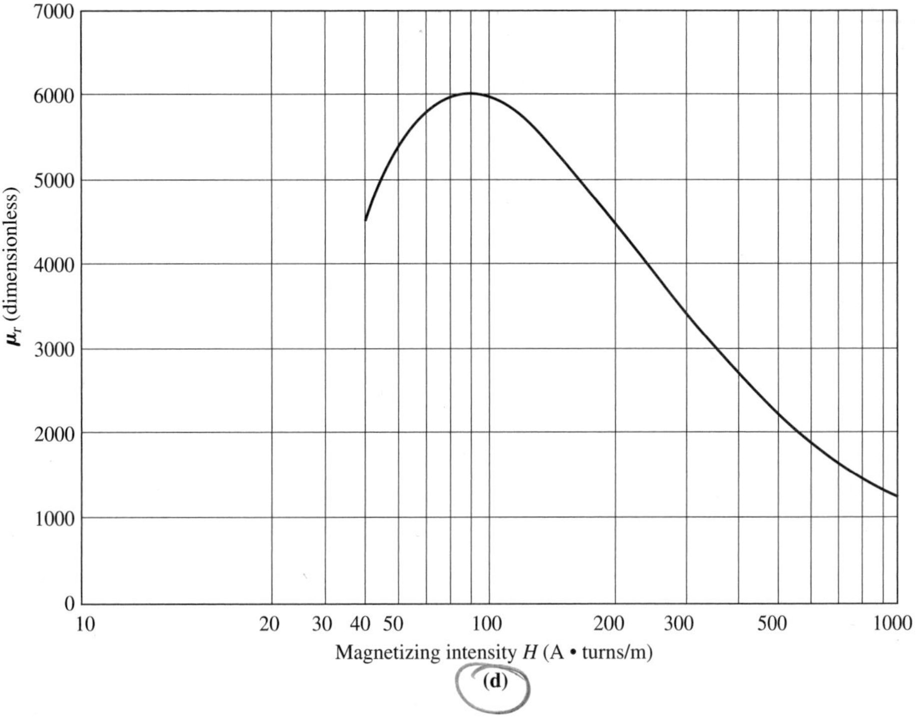

FIGURE 1–10
(continued)

between flux and magnetomotive force. The slope of the curve of flux density versus magnetizing intensity at any value of H in Figure 1–10b is by definition the permeability of the core at that magnetizing intensity. The curve shows that the permeability is large and relatively constant in the unsaturated region and then gradually drops to a very low value as the core becomes heavily saturated.

Figure 1–10c is a magnetization curve for a typical piece of steel shown in more detail and with the magnetizing intensity on a logarithmic scale. Only with the magnetizing intensity shown logarithmically can the huge saturation region of the curve fit onto the graph.

The advantage of using a ferromagnetic material for cores in electric machines and transformers is that one gets many times more flux for a given magnetomotive force with iron than with air. However, if the resulting flux has to be proportional, or nearly so, to the applied magnetomotive force, then the core *must* be operated in the unsaturated region of the magnetization curve.

Since real generators and motors depend on magnetic flux to produce voltage and torque, they are designed to produce as much flux as possible. As a result, most real machines operate near the knee of the magnetization curve, and the flux in their cores is not linearly related to the magnetomotive force producing it. This nonlinearity accounts for many of the peculiar behaviors of machines that will be explained in future chapters. We will use MATLAB to calculate solutions to problems involving the nonlinear behavior of real machines.

Example 1–4. Find the relative permeability of the typical ferromagnetic material whose magnetization curve is shown in Figure 1–10c at (a) $H = 50$, (b) $H = 100$, (c) $H = 500$, and (d) $H = 1000$ A • turns/m.

Solution

The permeability of a material is given by

$$\mu = \frac{B}{H}$$

and the relative permeability is given by

$$\mu_r = \frac{\mu}{\mu_0} \tag{1–23}$$

Thus, it is easy to determine the permeability at any given magnetizing intensity.

(a) At $H = 50$ A • turns/m, $B = 0.25$ T, so

$$\mu = \frac{B}{H} = \frac{0.25 \text{ T}}{50 \text{ A} \bullet \text{turns/m}} = 0.0050 \text{ H/m}$$

and

$$\mu_r = \frac{\mu}{\mu_0} = \frac{0.0050 \text{ H/m}}{4\pi \times 10^{-7} \text{ H/m}} = 3980$$

(b) At $H = 100$ A • turns/m, $B = 0.72$ T, so

$$\mu = \frac{B}{H} = \frac{0.72 \text{ T}}{100 \text{ A} \bullet \text{turns/m}} = 0.0072 \text{ H/m}$$

and

$$\mu_r = \frac{\mu}{\mu_0} = \frac{0.0072 \text{ H/m}}{4\pi \times 10^{-7} \text{ H/m}} = 5730$$

(c) At $H = 500$ A • turns/m, $B = 1.40$ T, so

$$\mu = \frac{B}{H} = \frac{1.40 \text{ T}}{500 \text{ A} \bullet \text{turns/m}} = 0.0028 \text{ H/m}$$

and

$$\mu_r = \frac{\mu}{\mu_0} = \frac{0.0028 \text{ H/m}}{4\pi \times 10^{-7} \text{ H/m}} = 2230$$

(d) At $H = 1000$ A • turns/m, $B = 1.51$ T, so

$$\mu = \frac{B}{H} = \frac{1.51 \text{ T}}{1000 \text{ A} \bullet \text{turns/m}} = 0.00151 \text{ H/m}$$

and

$$\mu_r = \frac{\mu}{\mu_0} = \frac{0.00151 \text{ H/m}}{4\pi \times 10^{-7} \text{ H/m}} = 1200$$

Notice that as the magnetizing intensity is increased, the relative permeability first increases and then starts to drop off. The relative permeability of a typical ferromagnetic material as a function of the magnetizing intensity is shown in Figure 1–10d. This shape is fairly typical of all ferromagnetic materials. It can easily be seen from the curve for μ_r versus H that the assumption of constant relative permeability made in Examples 1–1 to 1–3 is valid only over a relatively narrow range of magnetizing intensities (or magnetomotive forces).

In the following example, the relative permeability is not assumed to be constant. Instead, the relationship between B and H is given by a graph.

Example 1–5. A square magnetic core has a mean path length of 55 cm and a cross-sectional area of 150 cm^2. A 200-turn coil of wire is wrapped around one leg of the core. The core is made of a material having the magnetization curve shown in Figure 1–10c.

(a) How much current is required to produce 0.012 Wb of flux in the core?
(b) What is the core's relative permeability at that current level?
(c) What is its reluctance?

Solution

(a) The required flux density in the core is

$$B = \frac{\phi}{A} = \frac{1.012 \text{ Wb}}{0.015 \text{ m}^2} = 0.8 \text{ T}$$

From Figure 1–10c, the required magnetizing intensity is

$$H = 115 \text{ A} \bullet \text{turns/m}$$

From Equation (1–20), the magnetomotive force needed to produce this magnetizing intensity is

$$\begin{aligned} \mathcal{F} &= Ni = Hl_c \\ &= (115 \text{ A} \bullet \text{turns/m})(0.55 \text{ m}) = 63.25 \text{ A} \bullet \text{turns} \end{aligned}$$

so the required current is

$$i = \frac{\mathcal{F}}{N} = \frac{63.25 \text{ A} \bullet \text{turns}}{200 \text{ turns}} = 0.316 \text{ A}$$

(b) The core's permeability at this current is

$$\mu = \frac{B}{H} = \frac{0.8 \text{ T}}{115 \text{ A} \bullet \text{turns/m}} = 0.00696 \text{ H/m}$$

Therefore, the relative permeability is

$$\mu_r = \frac{\mu}{\mu_0} = \frac{0.00696 \text{ H/m}}{4\pi \times 10^{-7} \text{ H/m}} = 5540$$

(c) The reluctance of the core is

$$\mathcal{R} = \frac{\mathcal{F}}{\phi} = \frac{63.25 \text{ A} \bullet \text{turns}}{0.012 \text{ Wb}} = 5270 \text{ A} \bullet \text{turns/Wb}$$

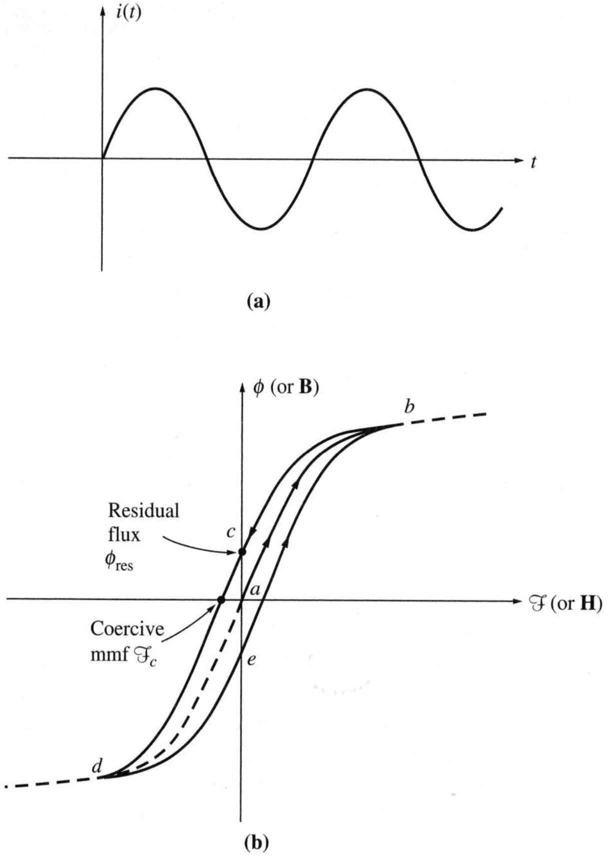

FIGURE 1–11
The hysteresis loop traced out by the flux in a core when the current *i(t)* is applied to it.

Energy Losses in a Ferromagnetic Core

Instead of applying a direct current to the windings on the core, let us now apply an alternating current and observe what happens. The current to be applied is shown in Figure 1–11a. Assume that the flux in the core is initially zero. As the current increases for the first time, the flux in the core traces out path *ab* in Figure 1–11b. This is basically the saturation curve shown in Figure 1–10. However, when the current falls again, *the flux traces out a different path from the one it followed when the current increased.* As the current decreases, the flux in the core traces out path *bcd*, and later when the current increases again, the flux traces out path *deb*. Notice that the amount of flux present in the core depends not only on the amount of current applied to the windings of the core, but also on the previous history of the flux in the core. This dependence on the preceding flux history and the resulting failure to retrace flux paths is called *hysteresis*. Path *bcdeb* traced out in Figure 1–11b as the applied current changes is called a *hysteresis loop*.

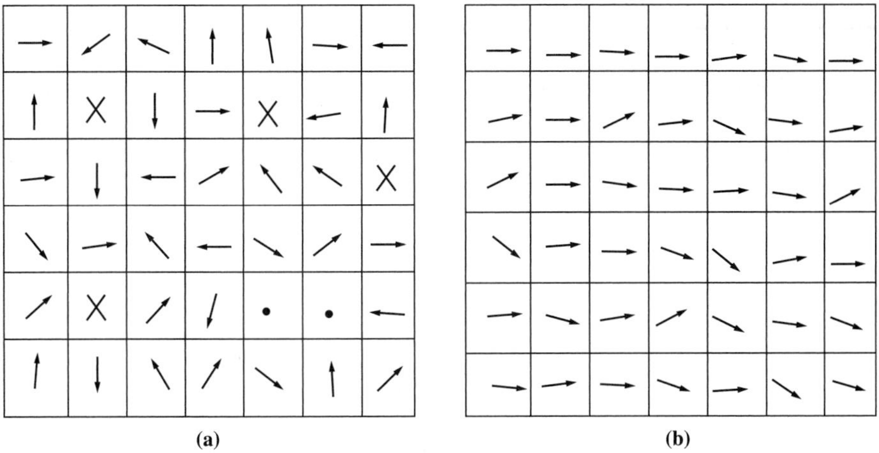

FIGURE 1–12
(a) Magnetic domains oriented randomly. (b) Magnetic domains lined up in the presence of an external magnetic field.

Notice that if a large magnetomotive force is first applied to the core and then removed, the flux path in the core will be *abc*. When the magnetomotive force is removed, the flux in the core *does not* go to zero. Instead, a magnetic field is left in the core. This magnetic field is called the *residual flux* in the core. It is in precisely this manner that permanent magnets are produced. To force the flux to zero, an amount of magnetomotive force known as the *coercive magnetomotive force* $\mathcal{F}_c$ must be applied to the core in the opposite direction.

Why does hysteresis occur? To understand the behavior of ferromagnetic materials, it is necessary to know something about their structure. The atoms of iron and similar metals (cobalt, nickel, and some of their alloys) tend to have their magnetic fields closely aligned with each other. Within the metal, there are many small regions called *domains*. In each domain, all the atoms are aligned with their magnetic fields pointing in the same direction, so each domain within the material acts as a small permanent magnet. The reason that a whole block of iron can appear to have no flux is that these numerous tiny domains are oriented randomly within the material. An example of the domain structure within a piece of iron is shown in Figure 1–12.

When an external magnetic field is applied to this block of iron, it causes domains that happen to point in the direction of the field to grow at the expense of domains pointed in other directions. Domains pointing in the direction of the magnetic field grow because the atoms at their boundaries physically switch orientation to align themselves with the applied magnetic field. The extra atoms aligned with the field increase the magnetic flux in the iron, which in turn causes more atoms to switch orientation, further increasing the strength of the magnetic field. It is this positive feedback effect that causes iron to have a permeability much higher than air.

As the strength of the external magnetic field continues to increase, whole domains that are aligned in the wrong direction eventually reorient themselves as a unit to line up with the field. Finally, when nearly all the atoms and domains in the iron are lined up with the external field, any further increase in the magnetomotive force can cause only the same flux increase that it would in free space. (Once everything is aligned, there can be no more feedback effect to strengthen the field.) At this point, the iron is *saturated* with flux. This is the situation in the saturated region of the magnetization curve in Figure 1–10.

The key to hysteresis is that when the external magnetic field is removed, the domains do not completely randomize again. Why do the domains remain lined up? Because turning the atoms in them requires *energy.* Originally, energy was provided by the external magnetic field to accomplish the alignment; when the field is removed, there is no source of energy to cause all the domains to rotate back. The piece of iron is now a permanent magnet.

Once the domains are aligned, some of them will remain aligned until a source of external energy is supplied to change them. Examples of sources of external energy that can change the boundaries between domains and/or the alignment of domains are magnetomotive force applied in another direction, a large mechanical shock, and heating. Any of these events can impart energy to the domains and enable them to change alignment. (It is for this reason that a permanent magnet can lose its magnetism if it is dropped, hit with a hammer, or heated.)

The fact that turning domains in the iron requires energy leads to a common type of energy loss in all machines and transformers. The *hysteresis loss* in an iron core is the energy required to accomplish the reorientation of domains during each cycle of the alternating current applied to the core. It can be shown that the area enclosed in the hysteresis loop formed by applying an alternating current to the core is directly proportional to the energy lost in a given ac cycle. The smaller the applied magnetomotive force excursions on the core, the smaller the area of the resulting hysteresis loop and so the smaller the resulting losses. Figure 1–13 illustrates this point.

Another type of loss should be mentioned at this point, since it is also caused by varying magnetic fields in an iron core. This loss is the *eddy current* loss. The mechanism of eddy current losses is explained later after Faraday's law has been introduced. Both hysteresis and eddy current losses cause heating in the core material, and both losses must be considered in the design of any machine or transformer. Since both losses occur within the metal of the core, they are usually lumped together and called *core losses.*

1.5 FARADAY'S LAW—INDUCED VOLTAGE FROM A TIME-CHANGING MAGNETIC FIELD

So far, attention has been focused on the production of a magnetic field and on its properties. It is now time to examine the various ways in which an existing magnetic field can affect its surroundings.

The first major effect to be considered is called *Faraday's law.* It is the basis of transformer operation. Faraday's law states that if a flux passes through a turn of a coil of wire, a voltage will be induced in the turn of wire that is directly

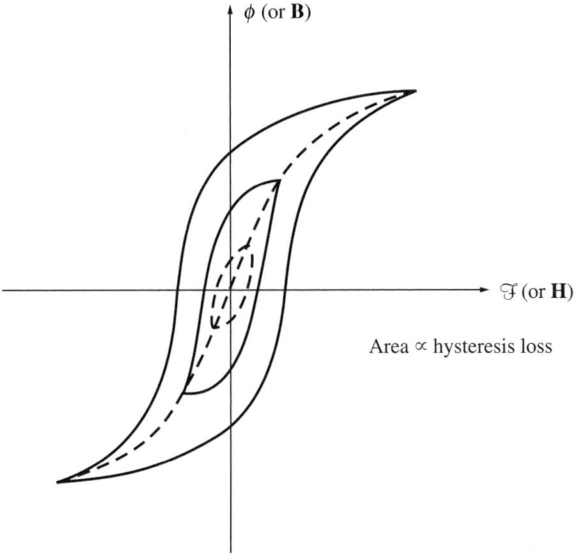

FIGURE 1–13
The effect of the size of magnetomotive force excursions on the magnitude of the hysteresis loss.

proportional to the *rate of change* in the flux with respect to time. In equation form,

$$e_{\text{ind}} = -\frac{d\phi}{dt} \tag{1–35}$$

where e_{ind} is the voltage induced in the turn of the coil and ϕ is the flux passing through the turn. If a coil has N turns and if the same flux passes through all of them, then the voltage induced across the whole coil is given by

$$\boxed{e_{\text{ind}} = -N\frac{d\phi}{dt}} \tag{1–36}$$

where

e_{ind} = voltage induced in the coil
N = number of turns of wire in coil
ϕ = flux passing through coil

The minus sign in the equations is an expression of *Lenz's law.* Lenz's law states that the direction of the voltage buildup in the coil is such that if the coil ends were short circuited, it would produce current that would cause a flux *opposing* the original flux change. Since the induced voltage opposes the change that causes it, a minus sign is included in Equation (1–36). To understand this concept clearly, examine

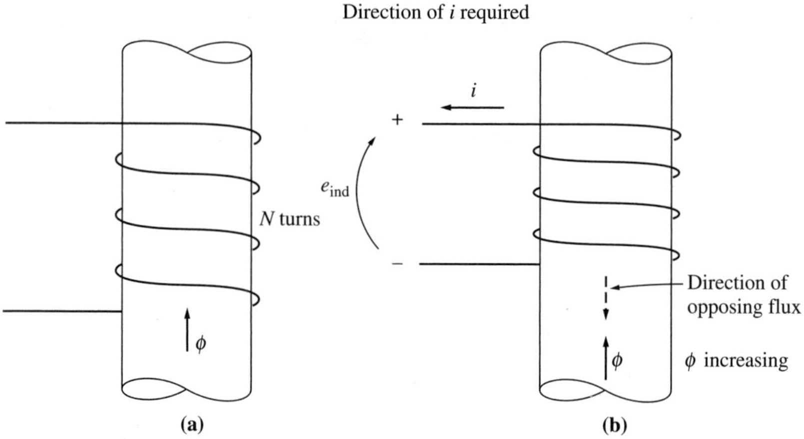

FIGURE 1–14
The meaning of Lenz's law: (a) A coil enclosing an increasing magnetic flux; (b) determining the resulting voltage polarity.

Figure 1–14. If the flux shown in the figure is *increasing* in strength, then the voltage built up in the coil will tend to establish a flux that will oppose the increase. A current flowing as shown in Figure 1–14b would produce a flux opposing the increase, so the voltage on the coil must be built up with the polarity required to drive that current through the external circuit. Therefore, the voltage must be built up with the polarity shown in the figure. Since the polarity of the resulting voltage can be determined from physical considerations, the minus sign in Equations (1–35) and (1–36) is often left out. It is left out of Faraday's law in the remainder of this book.

There is one major difficulty involved in using Equation (1–36) in practical problems. That equation assumes that exactly the same flux is present in each turn of the coil. Unfortunately, the flux leaking out of the core into the surrounding air prevents this from being true. If the windings are tightly coupled, so that the vast majority of the flux passing through one turn of the coil does indeed pass through all of them, then Equation (1–36) will give valid answers. But if leakage is quite high or if extreme accuracy is required, a different expression that does not make that assumption will be needed. The magnitude of the voltage in the *i*th turn of the coil is always given by

$$e_i = \frac{d(\phi_i)}{dt} \tag{1–37}$$

If there are N turns in the coil of wire, the total voltage on the coil is

$$e_{ind} = \sum_{i=1}^{N} e_i \tag{1–38}$$

$$= \sum_{i=1}^{N} \frac{d(\phi_i)}{dt} \tag{1–39}$$

$$= \frac{d}{dt}\left(\sum_{i=1}^{N} \phi_i\right) \tag{1–40}$$

The term in parentheses in Equation (1–40) is called *the flux linkage* λ of the coil, and Faraday's law can be rewritten in terms of flux linkage as

$$\boxed{e_{\text{ind}} = \frac{d\lambda}{dt}} \tag{1–41}$$

where

$$\boxed{\lambda = \sum_{i=1}^{N} \phi_i} \tag{1–42}$$

The units of flux linkage are weber-turns.

Faraday's law is the fundamental property of magnetic fields involved in transformer operation. The effect of Lenz's law in transformers is to predict the polarity of the voltages induced in transformer windings.

Faraday's law also explains the eddy current losses mentioned previously. A time-changing flux induces voltage *within* a ferromagnetic core in just the same manner as it would in a wire wrapped around that core. These voltages cause swirls of current to flow within the core, much like the eddies seen at the edges of a river. It is the shape of these currents that gives rise to the name *eddy currents.* These eddy currents are flowing in a resistive material (the iron of the core), so energy is dissipated by them. The lost energy goes into heating the iron core.

The amount of energy lost due to eddy currents depends on the size of the current swirls and the resistivity of the material in which the current flows. The larger the size of the swirl, the greater the resulting induced voltage will be (due to the larger flux inside the swirl). The larger the induced voltage, the larger the current flow that results, and therefore the greater the $I^2 R$ losses will be. On the other hand, the greater the resistivity of the material containing the currents, the lower the current flow will be for a given induced voltage in the swirl.

These facts give us two possible approaches to reduce the eddy current losses in a transformer or an electric machine. If a ferromagnetic core that may be subject to alternating fluxes is broken up into many small strips, or *laminations*, then the maximum size of a current swirl will be reduced, resulting in a lower induced voltage, a lower current, and lower losses. This reduction is roughly proportional to the width of these laminations, so smaller laminations are better. The core is built up out of many of these laminations in parallel. An insulating resin is used between the strips, so that the current paths for eddy currents are limited to very small areas. Because the insulating layers are extremely thin, this action reduces eddy current losses with very little effect on the core's magnetic properties.

The second approach to reducing eddy current losses is to increase the resistivity of the core material. This is often done by adding some silicon to the steel of the core. If the resistance of the core is higher, the eddy currents will be smaller for a given flux, and the resulting $I^2 R$ losses will be smaller.

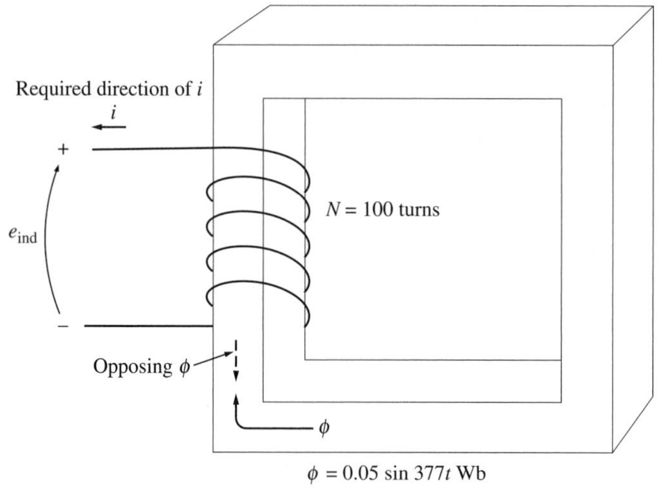

FIGURE 1–15
The core of Example 1–6. Determination of the voltage polarity at the terminals is shown.

Either laminations or high-resistivity materials can be used to control eddy currents. In many cases, both approaches are combined. Together, they can reduce the eddy current losses to the point where they are much smaller than the hysteresis losses in the core.

Example 1–6. Figure 1–15 shows a coil of wire wrapped around an iron core. The flux in the core is given by the equation

$$\phi = 0.05 \sin 377t \qquad \text{Wb}$$

If there are 100 turns on the core, what voltage is produced at the terminals of the coil? Of what polarity is the voltage during the time when flux is *increasing* in the reference direction shown in the figure? Assume that all the magnetic flux stays within the core (i.e., assume that the flux leakage is zero).

Solution
By the same reasoning as in the discussion on pages 29–30, the direction of the voltage while the flux is increasing in the reference direction must be positive to negative, as shown in Figure 1–15. The *magnitude* of the voltage is given by

$$e_{ind} = N\frac{d\phi}{dt}$$

$$= (100 \text{ turns}) \frac{d}{dt}(0.05 \sin 377t)$$

$$= 1885 \cos 377t$$

or alternatively,

$$e_{ind} = 1885 \sin(377t + 90°) \text{ V}$$

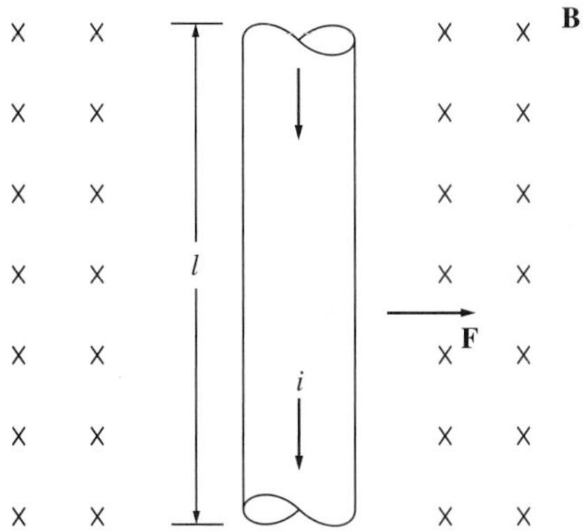

FIGURE 1–16
A current-carrying wire in the presence of a magnetic field.

1.6 PRODUCTION OF INDUCED FORCE ON A WIRE

A second major effect of a magnetic field on its surroundings is that it induces a force on a current-carrying wire within the field. The basic concept involved is illustrated in Figure 1–16. The figure shows a conductor present in a uniform magnetic field of flux density **B**, pointing into the page. The conductor itself is l meters long and contains a current of i amperes. The force induced on the conductor is given by

$$\mathbf{F} = i(\mathbf{l} \times \mathbf{B}) \tag{1–43}$$

where

i = magnitude of current in wire

$\mathbf{l}$ = length of wire, with direction of **l** defined to be in the direction of current flow

$\mathbf{B}$ = magnetic flux density vector

The direction of the force is given by the right-hand rule: If the index finger of the right hand points in the direction of the vector **l** and the middle finger points in the direction of the flux density vector **B**, then the thumb points in the direction of the resultant force on the wire. The magnitude of the force is given by the equation

$$F = ilB \sin\theta \tag{1–44}$$

where θ is the angle between the wire and the flux density vector.

Example 1–7. Figure 1–16 shows a wire carrying a current in the presence of a magnetic field. The magnetic flux density is 0.25 T, directed into the page. If the wire is 1.0 m long and carries 0.5 A of current in the direction from the top of the page to the bottom of the page, what are the magnitude and direction of the force induced on the wire?

Solution

The direction of the force is given by the right-hand rule as being to the right. The magnitude is given by

$$\begin{aligned} F &= ilB \sin\theta && (1–44) \\ &= (0.5\text{ A})(1.0\text{ m})(0.25\text{ T}) \sin 90^\circ = 0.125\text{ N} \end{aligned}$$

Therefore,

$$\mathbf{F} = 0.125 \text{ N, directed to the right}$$

The induction of a force in a wire by a current in the presence of a magnetic field is the basis of *motor action.* Almost every type of motor depends on this basic principle for the forces and torques which make it move.

1.7 INDUCED VOLTAGE ON A CONDUCTOR MOVING IN A MAGNETIC FIELD

There is a third major way in which a magnetic field interacts with its surroundings. If a wire with the proper orientation moves through a magnetic field, a voltage is induced in it. This idea is shown in Figure 1–17. The voltage induced in the wire is given by

$$e_{\text{ind}} = (\mathbf{v} \times \mathbf{B}) \cdot \mathbf{l} \tag{1–45}$$

where

$\mathbf{v}$ = velocity of the wire

$\mathbf{B}$ = magnetic flux density vector

$\mathbf{l}$ = length of conductor in the magnetic field

Vector $\mathbf{l}$ points along the direction of the wire toward the end making the smallest angle with respect to the vector $\mathbf{v} \times \mathbf{B}$. The voltage in the wire will be built up so that the positive end is in the direction of the vector $\mathbf{v} \times \mathbf{B}$. The following examples illustrate this concept.

Example 1–8. Figure 1–17 shows a conductor moving with a velocity of 5.0 m/s to the right in the presence of a magnetic field. The flux density is 0.5 T into the page, and the wire is 1.0 m in length, oriented as shown. What are the magnitude and polarity of the resulting induced voltage?

Solution

The direction of the quantity $\mathbf{v} \times \mathbf{B}$ in this example is up. Therefore, the voltage on the conductor will be built up positive at the top with respect to the bottom of the wire. The direction of vector $\mathbf{l}$ is up, so that it makes the smallest angle with respect to the vector $\mathbf{v} \times \mathbf{B}$.

Since $\mathbf{v}$ is perpendicular to $\mathbf{B}$ and since $\mathbf{v} \times \mathbf{B}$ is parallel to $\mathbf{l}$, the magnitude of the induced voltage reduces to

$$\begin{aligned} e_{\text{ind}} &= (\mathbf{v} \times \mathbf{B}) \cdot \mathbf{l} \qquad (1\text{–}45) \\ &= (vB \sin 90^\circ)\, l \cos 0^\circ \\ &= vBl \\ &= (5.0 \text{ m/s})(0.5 \text{ T})(1.0 \text{ m}) \\ &= 2.5 \text{ V} \end{aligned}$$

Thus the induced voltage is 2.5 V, positive at the top of the wire.

Example 1–9. Figure 1–18 shows a conductor moving with a velocity of 10 m/s to the right in a magnetic field. The flux density is 0.5 T, out of the page, and the wire is 1.0 m in length, oriented as shown. What are the magnitude and polarity of the resulting induced voltage?

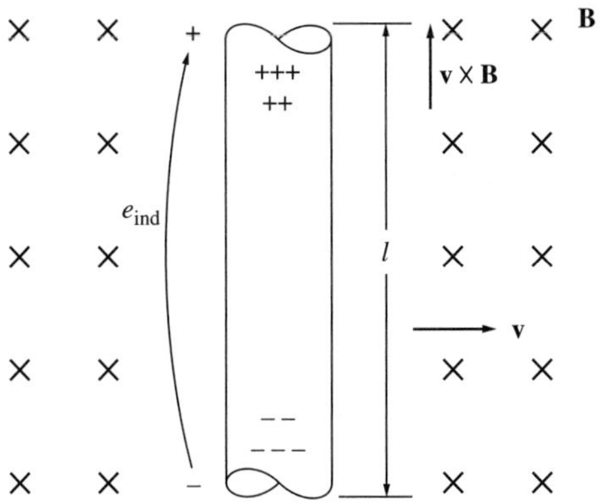

FIGURE 1–17
A conductor moving in the presence of a magnetic field.

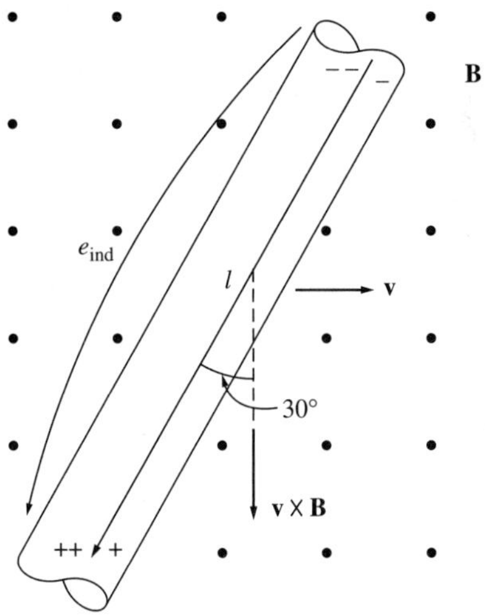

FIGURE 1–18
The conductor of Example 1–9.

Solution

The direction of the quantity $\mathbf{v} \times \mathbf{B}$ is down. The wire is not oriented on an up-down line, so choose the direction of $\mathbf{l}$ as shown to make the smallest possible angle with the direction of $\mathbf{v} \times \mathbf{B}$. The voltage is positive at the bottom of the wire with respect to the top of the wire. The magnitude of the voltage is

$$\begin{aligned} e_{ind} &= (\mathbf{v} \times \mathbf{B}) \cdot \mathbf{l} \qquad (1\text{–}45) \\ &= (vB \sin 90°)\, l \cos 30° \\ &= (10.0 \text{ m/s})(0.5 \text{ T})(1.0 \text{ m}) \cos 30° \\ &= 4.33 \text{ V} \end{aligned}$$

The induction of voltages in a wire moving in a magnetic field is fundamental to the operation of all types of generators. For this reason, it is called *generator action.*

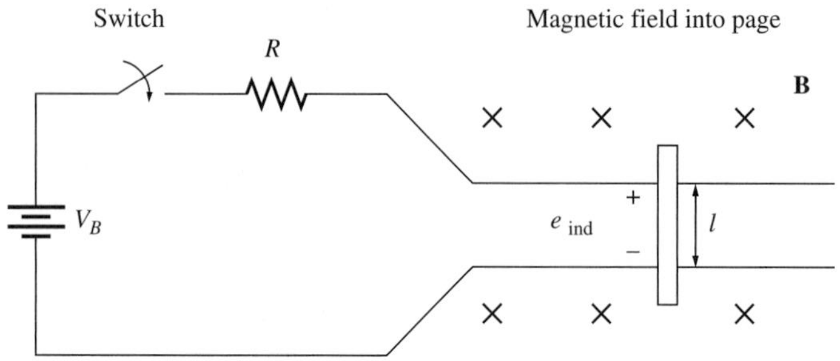

FIGURE 1–19
A linear dc machine. The magnetic field points into the page.

1.8 THE LINEAR DC MACHINE—A SIMPLE EXAMPLE

A *linear dc machine* is about the simplest and easiest-to-understand version of a dc machine, yet it operates according to the same principles and exhibits the same behavior as real generators and motors. It thus serves as a good starting point in the study of machines.

A linear dc machine is shown in Figure 1–19. It consists of a battery and a resistance connected through a switch to a pair of smooth, frictionless rails. Along the bed of this "railroad track" is a constant, uniform-density magnetic field directed into the page. A bar of conducting metal is lying across the tracks.

How does such a strange device behave? Its behavior can be determined from an application of four basic equations to the machine. These equations are

1. The equation for the force on a wire in the presence of a magnetic field:

$$\boxed{\mathbf{F} = i(\mathbf{l} \times \mathbf{B})} \tag{1–43}$$

where $\mathbf{F}$ = force on wire
i = magnitude of current in wire
$\mathbf{l}$ = length of wire, with direction of $\mathbf{l}$ defined to be in the direction of current flow
$\mathbf{B}$ = magnetic flux density vector

2. The equation for the voltage induced on a wire moving in a magnetic field:

$$\boxed{e_{\text{ind}} = (\mathbf{v} \times \mathbf{B}) \cdot \mathbf{l}} \tag{1–45}$$

where e_{ind} = voltage induced in wire
$\mathbf{v}$ = velocity of the wire
$\mathbf{B}$ = magnetic flux density vector
$\mathbf{l}$ = length of conductor in the magnetic field

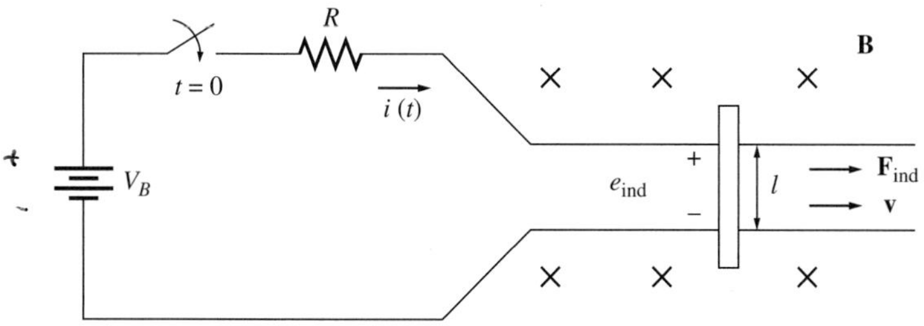

FIGURE 1–20
Starting a linear dc machine.

3. Kirchhoff's voltage law for this machine. From Figure 1–19 this law gives

$$V_B - iR - e_{\text{ind}} = 0$$

$$\boxed{V_B = e_{\text{ind}} + iR = 0} \tag{1–46}$$

4. Newton's law for the bar across the tracks:

$$\boxed{F_{\text{net}} = ma} \tag{1–7}$$

We will now explore the fundamental behavior of this simple dc machine using these four equations as tools.

Starting the Linear DC Machine

Figure 1–20 shows the linear dc machine under starting conditions. To start this machine, simply close the switch. Now a current flows in the bar, which is given by Kirchhoff's voltage law:

$$i = \frac{V_B - e_{ind}}{R} \tag{1–47}$$

Since the bar is initially at rest, $e_{\text{ind}} = 0$, so $i = V_B/R$. The current flows down through the bar across the tracks. But from Equation (1–43), a current flowing through a wire in the presence of a magnetic field induces a force on the wire. Because of the geometry of the machine, this force is

$$F_{\text{ind}} = ilB \qquad \text{to the right} \tag{1–48}$$

Therefore, the bar will accelerate to the right (by Newton's law). However, when the velocity of the bar begins to increase, a voltage appears across the bar. The voltage is given by Equation (1–45), which reduces for this geometry to

$$e_{\text{ind}} = vBl \qquad \text{positive upward} \tag{1–49}$$

The voltage now reduces the current flowing in the bar, since by Kirchhoff's voltage law

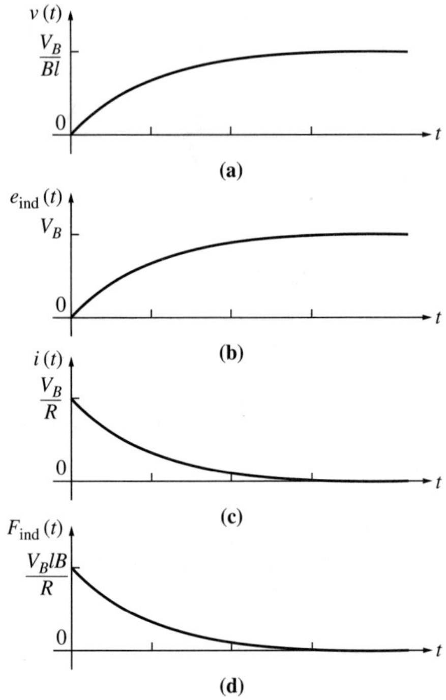

FIGURE 1–21
The linear dc machine on starting.
(a) Velocity $v(t)$ as a function of time; (b) induced voltage $e_{ind}(t)$; (c) current $i(t)$; (d) induced force $F_{ind}(t)$.

$$i\downarrow = \frac{V_B - e_{ind}\uparrow}{R} \tag{1–47}$$

As e_{ind} increases, the current i decreases.

The result of this action is that eventually the bar will reach a constant steady-state speed where the net force on the bar is zero. This will occur when e_{ind} has risen all the way up to equal the voltage V_B. At that time, the bar will be moving at a speed given by

$$V_B = e_{ind} = v_{ss}Bl$$
$$v_{ss} = \frac{V_B}{Bl} \tag{1–50}$$

The bar will continue to coast along at this no-load speed forever unless some external force disturbs it. When the motor is started, the velocity v, induced voltage e_{ind}, current i, and induced force F_{ind} are as sketched in Figure 1–21.

To summarize, at starting, the linear dc machine behaves as follows:

1. Closing the switch produces a current flow $i = V_B/R$.
2. The current flow produces a force on the bar given by $F = ilB$.

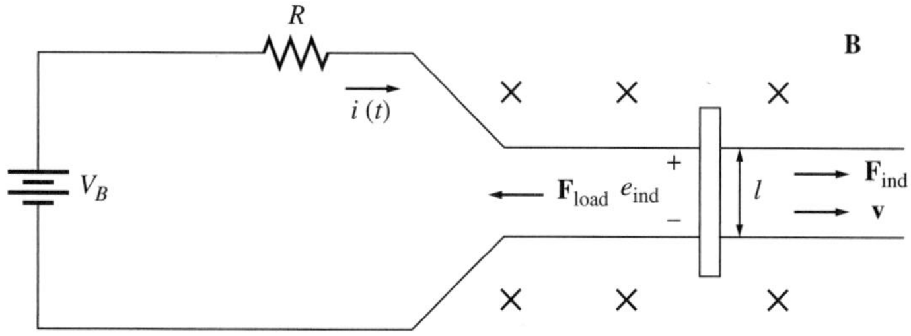

FIGURE 1–22
The linear dc machine as a motor.

3. The bar accelerates to the right, producing an induced voltage e_{ind} as it speeds up.
4. This induced voltage reduces the current flow $i = (V_B - e_{\text{ind}}\uparrow)/R$.
5. The induced force is thus decreased ($F = i\downarrow lB$) until eventually $F = 0$. At that point, $e_{\text{ind}} = V_B$, $i = 0$, and the bar moves at a constant no-load speed $v_{ss} = V_B/Bl$.

This is precisely the behavior observed in real motors on starting.

The Linear DC Machine as a Motor

Assume that the linear machine is initially running at the no-load steady-state conditions described above. What will happen to this machine if an external load is applied to it? To find out, let's examine Figure 1–22. Here, a force $\mathbf{F}_{\text{load}}$ is applied to the bar opposite the direction of motion. Since the bar was initially at steady state, application of the force $\mathbf{F}_{\text{load}}$ will result in a net force on the bar in the direction *opposite* the direction of motion ($\mathbf{F}_{\text{net}} = \mathbf{F}_{\text{load}} - \mathbf{F}_{\text{ind}}$). The effect of this force will be to slow the bar. But just as soon as the bar begins to slow down, the induced voltage on the bar drops ($e_{\text{ind}} = v\downarrow Bl$). As the induced voltage decreases, the current flow in the bar rises:

$$i\uparrow = \frac{V_B - e_{\text{ind}}\downarrow}{R} \tag{1–47}$$

Therefore, the induced force rises too ($F_{\text{ind}} = i\uparrow lB$). The overall result of this chain of events is that the induced force rises until it is equal and opposite to the load force, and the bar again travels in steady state, but at a lower speed. When a load is attached to the bar, the velocity v, induced voltage e_{ind}, current i, and induced force F_{ind} are as sketched in Figure 1–23.

There is now an induced force in the direction of motion of the bar, and power is being *converted from electrical form to mechanical form* to keep the bar moving. The power being converted is

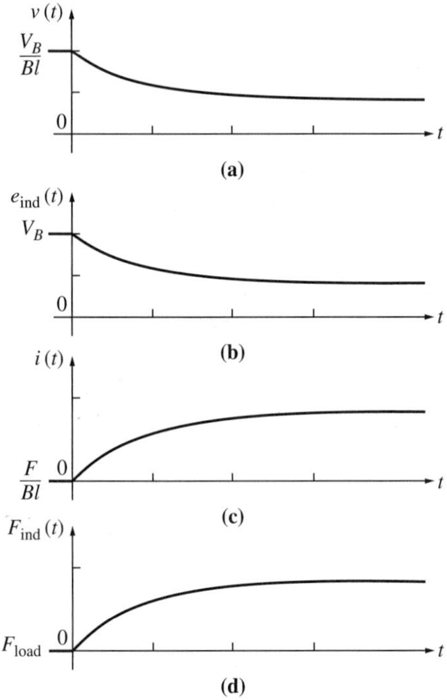

FIGURE 1–23
The linear dc machine operating at no-load conditions and then loaded as a motor. (a) Velocity $v(t)$ as a function of time; (b) induced voltage $e_{ind}(t)$; (c) current $i(t)$; (d) induced force $F_{ind}(t)$.

$$P_{conv} = e_{ind}i = F_{ind}v \tag{1–51}$$

An amount of electric power equal to $e_{ind}i$ is consumed in the bar and is replaced by mechanical power equal to $F_{ind}v$. Since power is converted from electrical to mechanical form, this bar is operating as a *motor*.

To summarize this behavior:

1. A force $\mathbf{F}_{load}$ is applied opposite to the direction of motion, which causes a net force $\mathbf{F}_{net}$ opposite to the direction of motion.
2. The resulting acceleration $a = F_{net}/m$ is negative, so the bar slows down ($v\downarrow$).
3. The voltage $e_{ind} = v\downarrow Bl$ falls, and so $i = (V_B - e_{ind}\downarrow)/R$ increases.
4. The induced force $F_{ind} = i\uparrow lB$ increases until $|\mathbf{F}_{ind}| = |\mathbf{F}_{load}|$ at a lower speed v.
5. An amount of electric power equal to $e_{ind}i$ is now being converted to mechanical power equal to $F_{ind}v$, and the machine is acting as a motor.

A real dc motor behaves in a precisely analogous fashion when it is loaded: As a load is added to its shaft, the motor begins to slow down, which reduces its internal voltage, increasing its current flow. The increased current flow increases its induced torque, and the induced torque will equal the load torque of the motor at a new, slower speed.

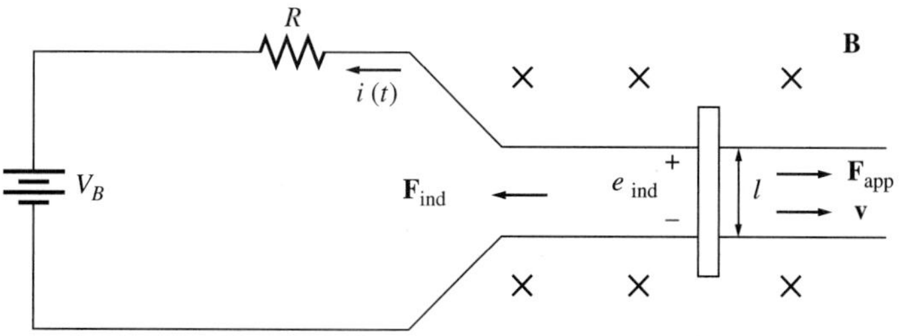

FIGURE 1–24
The linear dc machine as a generator.

Note that the power converted from electrical form to mechanical form by this linear motor was given by the equation $P_{\text{conv}} = F_{\text{ind}}v$. The power converted from electrical form to mechanical form in a real rotating motor is given by the equation

$$P_{\text{conv}} = \tau_{\text{ind}}\omega \tag{1–52}$$

where the induced torque τ_{ind} is the rotational analog of the induced force F_{ind}, and the angular velocity ω is the rotational analog of the linear velocity v.

The Linear DC Machine as a Generator

Suppose that the linear machine is again operating under no-load steady-state conditions. This time, apply a force *in the direction of motion* and see what happens.

Figure 1–24 shows the linear machine with an applied force $\mathbf{F}_{\text{app}}$ in the direction of motion. Now the applied force will cause the bar to accelerate in the direction of motion, and the velocity v of the bar will increase. As the velocity increases, $e_{\text{ind}} = v\uparrow Bl$ will increase and will be larger than the battery voltage V_B. With $e_{\text{ind}} > V_B$, the current reverses direction and is now given by the equation

$$i = \frac{e_{\text{ind}} - V_B}{R} \tag{1–53}$$

Since this current now flows *up* through the bar, it induces a force in the bar given by

$$F_{\text{ind}} = ilB \qquad \text{to the left} \tag{1–54}$$

The direction of the induced force is given by the right-hand rule. This induced force opposes the applied force on the bar.

Finally, the induced force will be equal and opposite to the applied force, and the bar will be moving at a *higher* speed than before. Notice that now *the battery is charging*. The linear machine is now serving as a generator, converting mechanical power $F_{\text{ind}}v$ into electric power $e_{\text{ind}}i$.

To summarize this behavior:

1. A force $\mathbf{F}_{app}$ is applied in the direction of motion; $\mathbf{F}_{net}$ is in the direction of motion.
2. Acceleration $a = F_{net}/m$ is positive, so the bar speeds up ($v\uparrow$).
3. The voltage $e_{ind} = v\uparrow Bl$ increases, and so $i = (e_{ind}\uparrow - V_B)/R$ increases.
4. The induced force $F_{ind} = i\uparrow lB$ increases until $|\mathbf{F}_{ind}| = |\mathbf{F}_{load}|$ at a higher speed v.
5. An amount of mechanical power equal to $F_{ind}v$ is now being converted to electric power $e_{ind}i$, and the machine is acting as a generator.

Again, a real dc generator behaves in precisely this manner: A torque is applied to the shaft *in the direction of motion,* the speed of the shaft increases, the internal voltage increases, and current flows out of the generator to the loads. The amount of mechanical power converted to electrical form in the real rotating generator is again given by Equation (1–52):

$$P_{conv} = \tau_{ind}\omega \tag{1–52}$$

It is interesting that the same machine acts as *both motor and generator.* The only difference between the two is whether the externally applied forces are in the direction of motion (generator) or opposite to the direction of motion (motor). Electrically, when $e_{ind} > V_B$, the machine acts as a generator, and when $e_{ind} < V_B$, the machine acts as a motor. Whether the machine is a motor or a generator, both induced force (motor action) and induced voltage (generator action) are present at all times. This is generally true of all machines—both actions are present, and it is only the relative directions of the external forces with respect to the direction of motion that determine whether the overall machine behaves as a motor or as a generator.

Another very interesting fact should be noted: This machine was a generator when it moved rapidly and a motor when it moved more slowly, but whether it was a motor or a generator, it always moved in the same direction. Many beginning machinery students expect a machine to turn one way as a generator and the other way as a motor. *This does not occur.* Instead, there is merely a small change in operating speed and a reversal of current flow.

Starting Problems with the Linear Machine

A linear machine is shown in Figure 1–25. This machine is supplied by a 250-V dc source, and its internal resistance R is given as about 0.10 Ω. (The resistor R models the internal resistance of a real dc machine, and this is a fairly reasonable internal resistance for a medium-size dc motor.)

Providing actual numbers in this figure highlights a major problem with machines (and their simple linear model). At starting conditions, the speed of the bar is zero, so $e_{ind} = 0$. The current flow at starting is

$$i_{start} = \frac{V_B}{R} = \frac{250\ V}{0.1\ \Omega} = 2500\ \text{A}$$

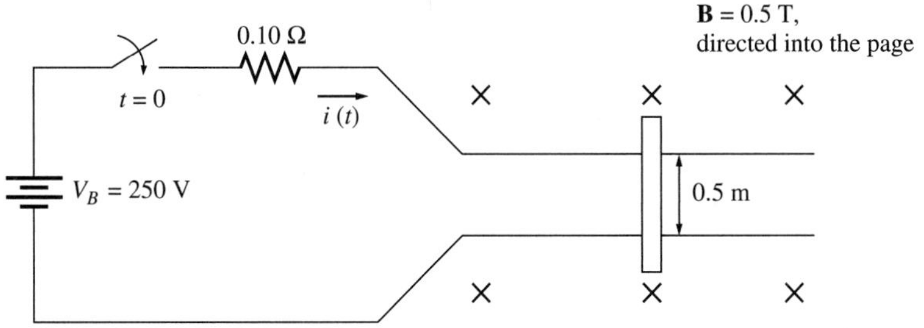

FIGURE 1–25
The linear dc machine with component values illustrating the problem of excessive starting current.

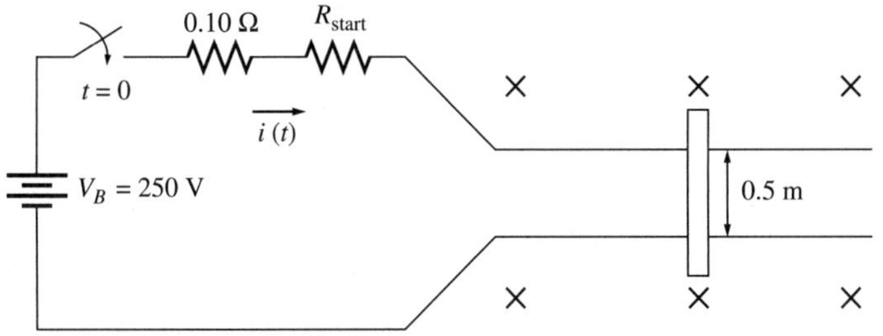

FIGURE 1–26
A linear dc machine with an extra series resistor inserted to control the starting current.

This current is very high, often in excess of 10 times the rated current of the machine. Such currents can cause severe damage to a motor. Both real ac and real dc machines suffer from similar high-current problems on starting.

How can such damage be prevented? The easiest method for this simple linear machine is to insert an extra resistance into the circuit during starting to limit the current flow until e_{ind} builds up enough to limit it. Figure 1–26 shows a starting resistance inserted into the machine circuitry.

The same problem exists in real dc machines, and it is handled in precisely the same fashion—a resistor is inserted into the motor armature circuit during starting. The control of high starting current in real ac machines is handled in a different fashion, which will be described in Chapter 6.

Example 1–10. The linear dc machine shown in Figure 1–27a has a battery voltage of 120 V, an internal resistance of 0.3 Ω, and a magnetic flux density of 0.1 T.

(*a*) What is this machine's maximum starting current? What is its steady-state velocity at no load?

(*b*) Suppose that a 30-N force pointing to the right were applied to the bar. What would the steady-state speed be? How much power would the bar be producing or consuming? How much power would the battery be producing or consuming?

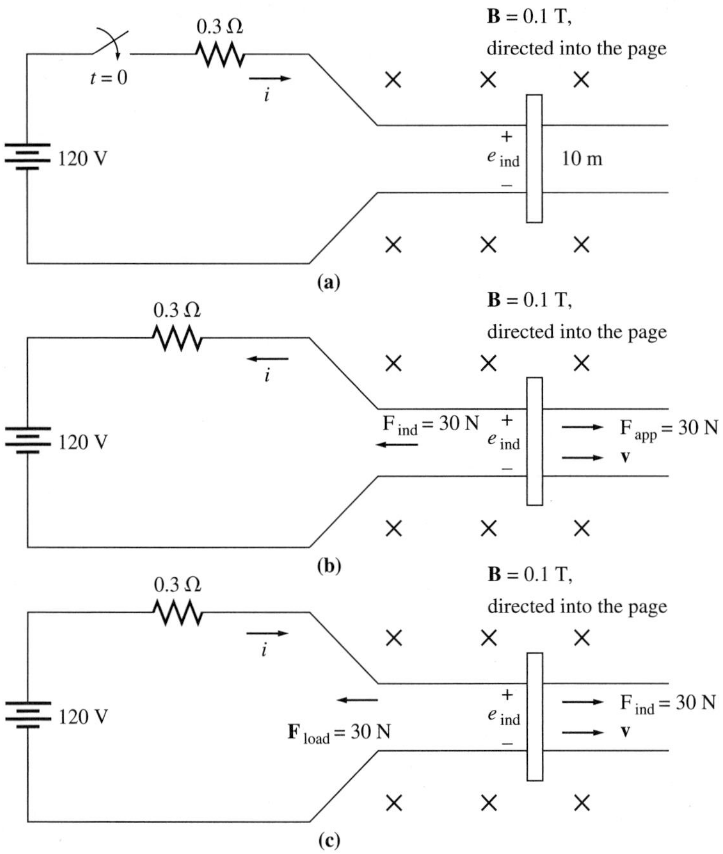

FIGURE 1–27
The linear dc machine of Example 1–10. (a) Starting conditions; (b) operating as a generator; (c) operating as a motor.

Explain the difference between these two figures. Is this machine acting as a motor or as a generator?

(c) Now suppose a 30-N force pointing to the left were applied to the bar. What would the new steady-state speed be? Is this machine a motor or a generator now?

(d) Assume that a force pointing to the left is applied to the bar. Calculate speed of the bar as a function of the force for values from 0 N to 50 N in 10-N steps. Plot the velocity of the bar versus the applied force.

(e) Assume that the bar is unloaded and that it suddenly runs into a region where the magnetic field is weakened to 0.08 T. How fast will the bar go now?

Solution

(a) At starting conditions, the velocity of the bar is 0, so $e_{\text{ind}} = 0$. Therefore,

$$i = \frac{V_B - e_{\text{ind}}}{R} = \frac{120\text{ V} - 0\text{ V}}{0.3\ \Omega} = 400\text{ A}$$

When the machine reaches steady state, $\mathbf{F}_{\text{ind}} = 0$ and $i = 0$. Therefore,

$$VB = e_{\text{ind}} = v_{ss}Bl$$

$$v_{\text{ss}} = \frac{V_B}{Bl}$$

$$= \frac{120\text{ V}}{(0.1\text{ T})(10\text{ m})} = 120\text{ m/s}$$

(b) Refer to Figure 1–27b. If a 30-N force to the right is applied to the bar, the final steady state will occur when the induced force $\mathbf{F}_{\text{ind}}$ is equal and opposite to the applied force $\mathbf{F}_{\text{app}}$, so that the net force on the bar is zero:

$$F_{\text{app}} = F_{\text{ind}} = ilB$$

Therefore,

$$i = \frac{F_{\text{ind}}}{lB} = \frac{30\text{ N}}{(10\text{ m})(0.1\text{ T})}$$

$$= 30\text{ A} \quad \text{flowing up through the bar}$$

The induced voltage e_{ind} on the bar must be

$$e_{\text{ind}} = V_B + iR$$

$$= 120\text{ V} + (30\text{A})(0.3\ \Omega) = 129\text{ V}$$

and the final steady-state speed must be

$$v_{\text{ss}} = \frac{e_{\text{ind}}}{Bl}$$

$$= \frac{129\text{ V}}{(0.1\text{ T})(10\text{ m})} = 129\text{ m/s}$$

The bar is *producing* $P = (129\text{ V})(30\text{ A}) = 3870\text{ W}$ of power, and the battery is *consuming* $P = (120\text{ V})(30\text{ A}) = 3600\text{ W}$. The difference between these two numbers is the 270 W of losses in the resistor. This machine is acting as a *generator.*

(c) Refer to Figure 1–25c. This time, the force is applied to the left, and the induced force is to the right. At steady state,

$$F_{\text{app}} = F_{\text{ind}} = ilB$$

$$i = \frac{F_{\text{ind}}}{lB} = \frac{30\text{ N}}{(10\text{ m})(0.1\text{ T})}$$

$$= 30\text{ A} \quad \text{flowing down through the bar}$$

The induced voltage e_{ind} on the bar must be

$$e_{\text{ind}} = V_B - iR$$

$$= 120\text{ V} - (30\text{ A})(0.3\ \Omega) = 111\text{ V}$$

and the final speed must be

$$v_{\text{ss}} = \frac{e_{\text{ind}}}{Bl}$$

$$= \frac{111\text{ V}}{(0.1\text{ T})(10\text{ m})} = 111\text{ m/s}$$

This machine is now acting as a *motor,* converting electric energy from the battery into mechanical energy of motion on the bar.

(*d*) This task is ideally suited for MATLAB. We can take advantage of MATLAB's vectorized calculations to determine the velocity of the bar for each value of force. The MATLAB code to perform this calculation is just a version of the steps that were performed by hand in part *c*. The program shown below calculates the current, induced voltage, and velocity in that order, and then plots the velocity versus the force on the bar.

```
% M-file: ex1_10.m
% M-file to calculate and plot the velocity of
% a linear motor as a function of load.
VB = 120;                        % Battery voltage (V)
r = 0.3;                         % Resistance (ohms)
l = 1;                           % Bar length (m)
B = 0.6;                         % Flux density (T)

% Select the forces to apply to the bar
F = 0:10:50;                     % Force (N)

% Calculate the currents flowing in the motor.
i = F ./ (l * B);                % Current (A)

% Calculate the induced voltages on the bar.
eind = VB - i .* r;              % Induced voltage (V)

% Calculate the velocities of the bar.
v_bar = eind ./ (l * B);         % Velocity (m/s)

% Plot the velocity of the bar versus force.
plot(F,v_bar);
title ('Plot of Velocity versus Applied Force');
xlabel ('Force (N)');
ylabel ('Velocity (m/s)');
axis ([0 50 0 200]);
```

The resulting plot is shown in Figure 1–28. Note that the bar slows down more and more as load increases.

(*e*) If the bar is initially unloaded, then $e_{\text{ind}} = V_B$. If the bar suddenly hits a region of weaker magnetic field, a transient will occur. Once the transient is over, though, e_{ind} will again equal V_B.

This fact can be used to determine the final speed of the bar. The *initial speed* was 120 m/s. The *final speed* is

$$VB = e_{\text{ind}} = v_{ss}Bl$$

$$v_{\text{ss}} = \frac{V_B}{Bl}$$

$$= \frac{120\text{ V}}{(0.08\text{ T})(10\text{ m})} = 150\text{ m/s}$$

Thus, when the flux in the linear motor weakens, the bar speeds up. The same behavior occurs in real dc motors: When the field flux of a dc motor weakens, it turns faster. Here, again, the linear machine behaves in much the same way as a real dc motor.

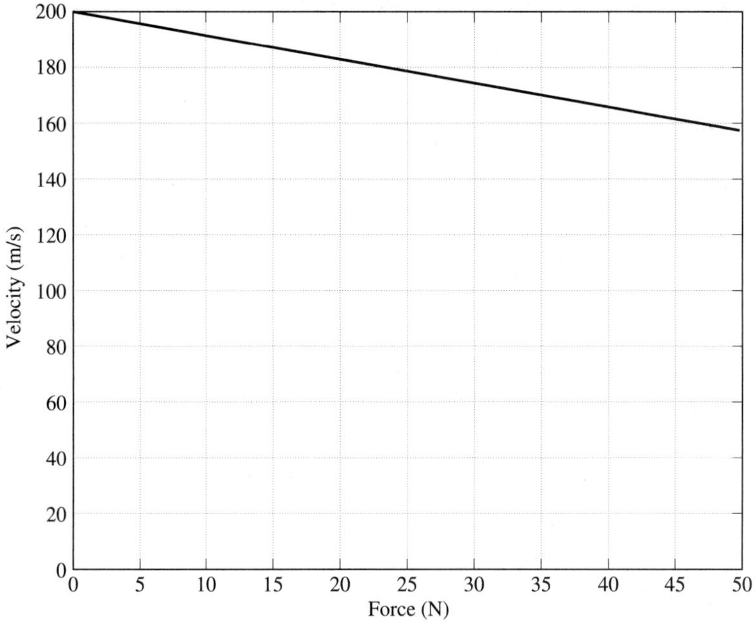

FIGURE 1–28
Plot of velocity versus force for a linear dc machine.

1.9 REAL, REACTIVE, AND APPARENT POWER IN SINGLE-PHASE AC CIRCUITS

This section describes the relationships among real, reactive, and apparent power in single-phase ac circuits. A similar discussion for three-phase ac circuits can be found in Appendix A.

In a dc circuit such as the one shown in Figure 1–29a, the power supplied to the dc load is simply the product of the voltage across the load and the current flowing through it.

$$P = VI \tag{1–55}$$

Unfortunately, the situation in sinusoidal ac circuits is more complex, because there can be a phase difference between the ac voltage and the ac current supplied to the load. The *instantaneous* power supplied to an ac load will still be the product of the instantaneous voltage and the instantaneous current, but the *average* power supplied to the load will be affected by the phase angle between the voltage and the current. We will now explore the effects of this phase difference on the average power supplied to an ac load.

Figure 1–29b shows a single-phase voltage source supplying power to a single-phase load with impedance $\mathbf{Z} = Z\angle\theta\ \Omega$. If we assume that the load is inductive, then the impedance angle θ of the load will be positive, and the current will lag the voltage by θ degrees.

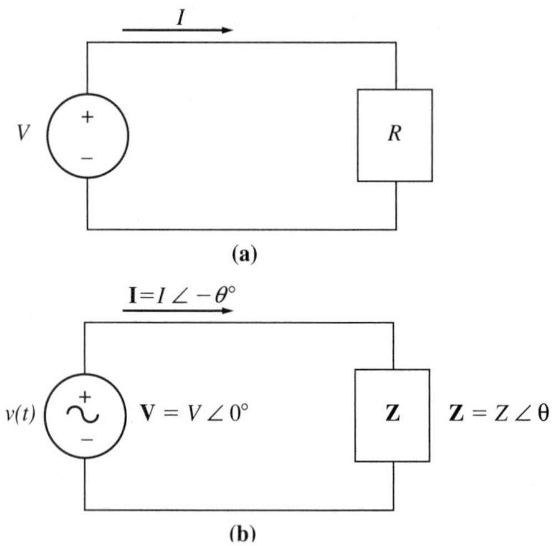

FIGURE 1–29
(a) A dc voltage source supplying a load with resistance R. (b) An ac voltage source supplying a load with impedance $\mathbf{Z} = Z \angle \theta\ \Omega$.

The voltage applied to this load is

$$v(t) = \sqrt{2}V \cos \omega t \tag{1–56}$$

where V is the rms value of the voltage applied to the load, and the resulting current flow is

$$i(t) = \sqrt{2}I \cos(\omega t - \theta) \tag{1–57}$$

where I is the rms value of the current flowing through the load.

The instantaneous power supplied to this load at any time t is

$$p(t) = v(t)i(t) = 2VI \cos \omega t \cos(\omega t - \theta) \tag{1–58}$$

The angle θ in this equation is the *impedance angle* of the load. For inductive loads, the impedance angle is positive, and the current waveform lags the voltage waveform by θ degrees.

If we apply trigonometric identities to Equation (1–58), it can be manipulated into an expression of the form

$$p(t) = VI \cos \theta\, (1 + \cos 2\omega t) + VI \sin \theta \sin 2\omega t \tag{1–59}$$

The first term of this equation represents the power supplied to the load by the component of current that is *in phase* with the voltage, while the second term represents the power supplied to the load by the component of current that is *90° out of phase* with the voltage. The components of this equation are plotted in Figure 1–30.

Note that the *first* term of the instantaneous power expression is always positive, but it produces pulses of power instead of a constant value. The average value of this term is

$$P = VI \cos \theta \tag{1–60}$$

which is the *average* or *real* power (P) supplied to the load by term 1 of the Equation (1–59). The units of real power are watts (W), where 1 W = 1 V × 1 A.

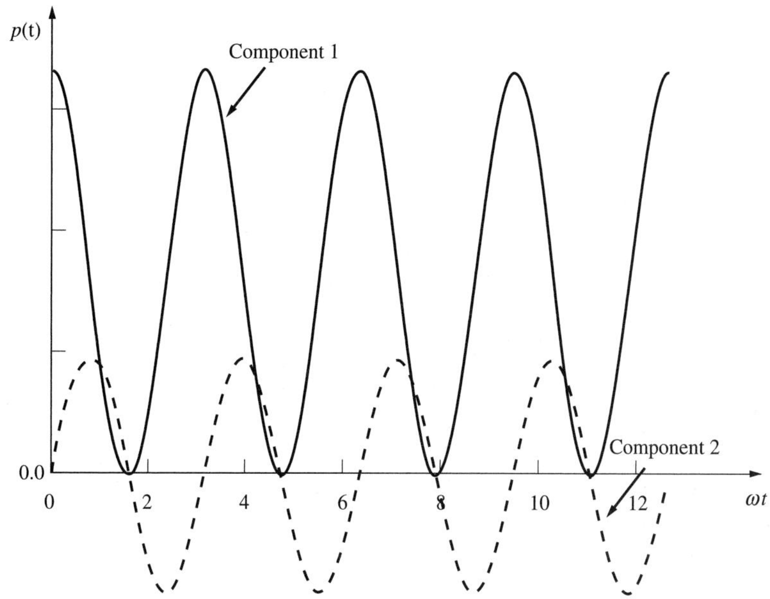

FIGURE 1–30
The components of power supplied to a single-phase load versus time. The first component represents the power supplied by the component of current *in phase* with the voltage, while the second term represents the power supplied by the component of current *90° out of phase* with the voltage.

Note that the *second* term of the instantaneous power expression is positive half of the time and negative half of the time, so that *the average power supplied by this term is zero.* This term represents power that is first transferred from the source to the load, and then returned from the load to the source. The power that continually bounces back and forth between the source and the load is known as *reactive power* (Q). Reactive power represents the energy that is first stored and then released in the magnetic field of an inductor, or in the electric field of a capacitor.

The reactive power of a load is given by

$$Q = VI \sin \theta \tag{1–61}$$

where θ is the impedance angle of the load. By convention, Q is positive for inductive loads and negative for capacitive loads, because the impedance angle θ is positive for inductive loads and negative for capacitive loads. The units of reactive power are volt-amperes reactive (var), where 1 var $= 1 \text{ V} \times 1 \text{ A}$. Even though the dimensional units are the same as for watts, reactive power is traditionally given a unique name to distinguish it from power actually supplied to a load.

The apparent power (S) supplied to a load is defined as the product of the voltage across the load and the current through the load. This is the power that "appears" to be supplied to the load if the phase angle differences between voltage and current are ignored. Therefore, the apparent power of a load is given by

$$S = VI \tag{1–62}$$

The units of apparent power are volt-amperes (VA), where 1 VA = 1 V × 1 A. As with reactive power, apparent power is given a distinctive set of units to avoid confusing it with real and reactive power.

Alternative Forms of the Power Equations

If a load has a constant impedance, then Ohm's law can be used to derive alternative expressions for the real, reactive, and apparent powers supplied to the load. Since the magnitude of the voltage across the load is given by

$$V = IZ \tag{1–63}$$

substituting Equation (1–63) into Equations (1–60) to (1–62) produces equations for real, reactive, and apparent power expressed in terms of current and impedance:

$$P = I^2Z \cos \theta \tag{1–64}$$

$$Q = I^2Z \sin \theta \tag{1–65}$$

$$S = I^2Z \tag{1–66}$$

where $|Z|$ is the magnitude of the load impedance Z.

Since the impedance of the load Z can be expressed as

$$Z = R + jX = |Z| \cos \theta + j|Z| \sin \theta$$

we see from this equation that $R = |Z| \cos \theta$ and $X = |Z| \sin \theta$, so the real and reactive powers of a load can also be expressed as

$$P = I^2R \tag{1–67}$$

$$Q = I^2X \tag{1–68}$$

where R is the resistance and X is the reactance of load Z.

Complex Power

For simplicity in computer calculations, real and reactive power are sometimes represented together as a *complex power* **S**, where

$$\mathbf{S} = P + jQ \tag{1–69}$$

The complex power **S** supplied to a load can be calculated from the equation

$$\mathbf{S} = \mathbf{VI}^* \tag{1–70}$$

where the asterisk represents the complex conjugate operator.

To understand this equation, let's suppose that the voltage applied to a load is $\mathbf{V} = V \angle \alpha$ and the current through the load is $\mathbf{I} = I \angle \beta$. Then the complex power supplied to the load is

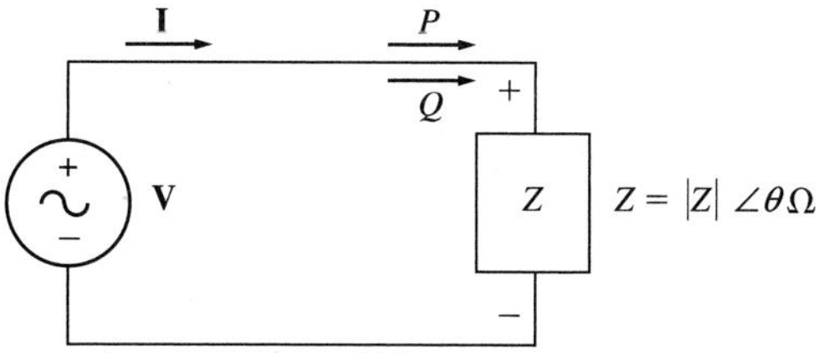

FIGURE 1–31
An inductive load has a *positive* impedance angle θ. This load produces a *lagging* current, and it consumes both real power P and reactive power Q from the source.

$$\begin{aligned}\mathbf{S} &= \mathbf{VI}^* = (V\angle\alpha)(I\angle-\beta) = VI\angle(\alpha-\beta) \\ &= VI\cos(\alpha-\beta) + jVI\sin(\alpha-\beta)\end{aligned}$$

The impedance angle θ is the difference between the angle of the voltage and the angle of the current ($\theta = \alpha - \beta$), so this equation reduces to

$$\begin{aligned}\mathbf{S} &= VI\cos\theta + jVI\sin\theta \\ &= P + jQ\end{aligned}$$

The Relationships between Impedance Angle, Current Angle, and Power

As we know from basic circuit theory, an inductive load (Figure 1–31) has a positive impedance angle θ, since the reactance of an inductor is positive. If the impedance angle θ of a load is positive, the phase angle of the current flowing through the load will *lag* the phase angle of the voltage across the load by θ.

$$\mathbf{I} = \frac{\mathbf{V}}{Z} = \frac{V\angle 0^\circ}{|Z|\angle\theta} = \frac{V}{|Z|}\angle-\theta$$

Also, if the impedance angle θ of a load is positive, the reactive power consumed by the load will be positive (Equation 1–65), and the load is said to be consuming both real and reactive power from the source.

In contrast, a capacitive load (Figure 1–32) has a negative impedance angle θ, since the reactance of a capacitor is negative. If the impedance angle θ of a load is negative, the phase angle of the current flowing through the load will *lead* the phase angle of the voltage across the load by θ. Also, if the impedance angle θ of a load is negative, the reactive power Q consumed by the load will be *negative* (Equation 1–65). In this case, we say that the load is consuming real power from the source and *supplying* reactive power to the source.

The Power Triangle

The real, reactive, and apparent powers supplied to a load are related by the *power triangle*. A power triangle is shown in Figure 1–33. The angle in the lower left

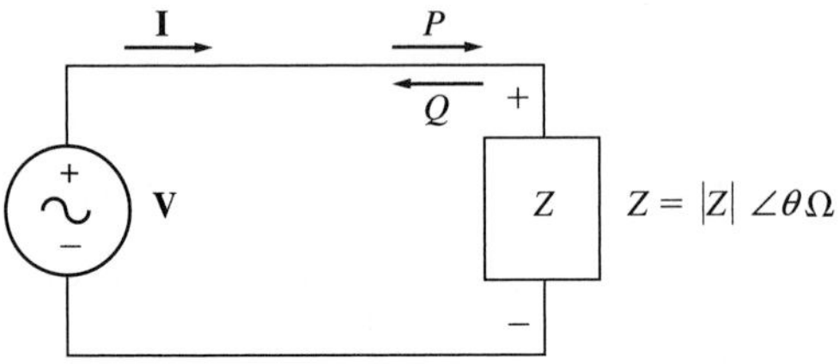

FIGURE 1–32
A capacitive load has a *negative* impedance angle θ. This load produces a *leading* current, and it consumes real power P from the source while supplying reactive power Q to the source.

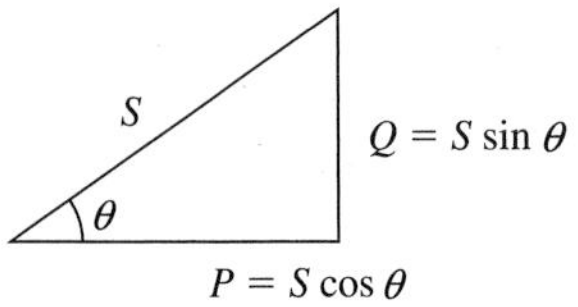

$$\cos\theta = \frac{P}{S}$$

$$\sin\theta = \frac{Q}{S}$$

$$\tan\theta = \frac{Q}{P}$$

FIGURE 1–33
The power triangle.

corner is the impedance angle θ. The adjacent side of this triangle is the real power P supplied to the load, the opposite side of the triangle is the reactive power Q supplied to the load, and the hypotenuse of the triangle is the apparent power S of the load.

The quantity $\cos\theta$ is usually known as the *power factor* of a load. The power factor is defined as the fraction of the apparent power S that is actually supplying real power to a load. Thus,

$$\text{PF} = \cos\theta \tag{1–71}$$

where θ is the impedance angle of the load.

Note that $\cos\theta = \cos(-\theta)$, so the power factor produced by an impedance angle of $+30°$ is exactly the same as the power factor produced by an impedance angle of $-30°$. Because we can't tell whether a load is inductive or capacitive from the power factor alone, it is customary to state whether the current is leading or lagging the voltage whenever a power factor is quoted.

The power triangle makes the relationships among real power, reactive power, apparent power, and the power factor clear, and provides a convenient way to calculate various power-related quantities if some of them are known.

Example 1–11. Figure 1–34 shows an ac voltage source supplying power to a load with impedance $Z = 20\angle -30°\ \Omega$. Calculate the current **I** supplied to the load, the power factor of the load, and the real, reactive, apparent, and complex power supplied to the load.

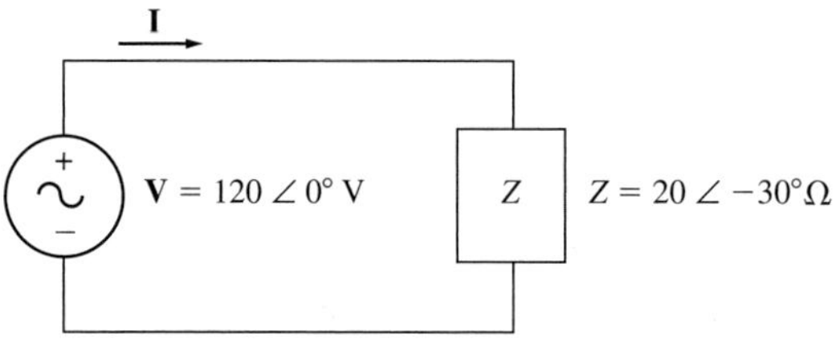

FIGURE 1–34
The circuit of Example 1–11.

Solution
The current supplied to this load is

$$\mathbf{I} = \frac{\mathbf{V}}{Z} = \frac{120\angle 0°\text{ V}}{20\angle -30°\ \Omega} = 6\angle 30°\text{ A}$$

The power factor of the load is

$$\text{PF} = \cos\theta = \cos(-30°) = 0.866\text{ leading} \qquad (1\text{–}71)$$

(Note that this is a capacitive load, so the impedance angle θ is negative, and the current *leads* the voltage.)

The real power supplied to the load is

$$P = VI\cos\theta \qquad (1\text{–}60)$$

$$P = (120\text{ V})(6\text{ A})\cos(-30°) = 623.5\text{ W}$$

The reactive power supplied to the load is

$$Q = VI\sin\theta \qquad (1\text{–}61)$$

$$Q = (120\text{ V})(6\text{ A})\sin(-30°) = -360\text{ VAR}$$

The apparent power supplied to the load is

$$S = VI \qquad (1\text{–}62)$$

$$Q = (120\text{ V})(6\text{ A}) = 720\text{ VA}$$

The complex power supplied to the load is

$$\begin{aligned}\mathbf{S} &= \mathbf{VI}^* \qquad (1\text{–}70)\\ &= (120\angle 0°\text{ V})(6\angle -30°\text{ A})^*\\ &= (120\angle 0°\text{ V})(6\angle 30°\text{ A}) = 720\angle 30°\text{ VA}\\ &= 623.5 - j360\text{ VA}\end{aligned}$$

1.10 SUMMARY

This chapter has reviewed briefly the mechanics of systems rotating about a single axis and introduced the sources and effects of magnetic fields important in the understanding of transformers, motors, and generators.

Historically, the English system of units has been used to measure the mechanical quantities associated with machines in English-speaking countries.

Recently, the SI units have superseded the English system almost everywhere in the world except in the United States, but rapid progress is being made even there. Since SI is becoming almost universal, most (but not all) of the examples in this book use this system of units for mechanical measurements. Electrical quantities are always measured in SI units.

In the section on mechanics, the concepts of angular position, angular velocity, angular acceleration, torque, Newton's law, work, and power were explained for the special case of rotation about a single axis. Some fundamental relationships (such as the power and speed equations) were given in both SI and English units.

The production of a magnetic field by a current was explained, and the special properties of ferromagnetic materials were explored in detail. The shape of the magnetization curve and the concept of hysteresis were explained in terms of the domain theory of ferromagnetic materials, and eddy current losses were discussed.

Faraday's law states that a voltage will be generated in a coil of wire that is proportional to the time rate of change in the flux passing through it. Faraday's law is the basis of transformer action, which is explored in detail in Chapter 3.

A current-carrying wire present in a magnetic field, if it is oriented properly, will have a force induced on it. This behavior is the basis of motor action in all real machines.

A wire moving through a magnetic field with the proper orientation will have a voltage induced in it. This behavior is the basis of generator action in all real machines.

A simple linear dc machine consisting of a bar moving in a magnetic field illustrates many of the features of real motors and generators. When a load is attached to it, it slows down and operates as a motor, converting electric energy into mechanical energy. When a force pulls the bar faster than its no-load steady-state speed, it acts as a generator, converting mechanical energy into electric energy.

In ac circuits, the real power P is the average power supplied by a source to a load. The reactive power Q is the component of power that is exchanged back and forth between a source and a load. By convention, positive reactive power is consumed by inductive loads ($+\theta$) and negative reactive power is consumed (or positive reactive power is supplied) by capacitive loads ($-\theta$). The apparent power S is the power that "appears" to be supplied to the load if only the magnitudes of the voltages and currents are considered.

QUESTIONS

1–1. What is torque? What role does torque play in the rotational motion of machines?

1–2. What is Ampere's law?

1–3. What is magnetizing intensity? What is magnetic flux density? How are they related?

1–4. How does the magnetic circuit concept aid in the design of transformer and machine cores?

1–5. What is reluctance?

1–6. What is a ferromagnetic material? Why is the permeability of ferromagnetic materials so high?

1–7. How does the relative permeability of a ferromagnetic material vary with magnetomotive force?

1–8. What is hysteresis? Explain hysteresis in terms of magnetic domain theory.

1–9. What are eddy current losses? What can be done to minimize eddy current losses in a core?

1–10. Why are all cores exposed to ac flux variations laminated?

1–11. What is Faraday's law?

1–12. What conditions are necessary for a magnetic field to produce a force on a wire?

1–13. What conditions are necessary for a magnetic field to produce a voltage in a wire?

1–14. Why is the linear machine a good example of the behavior observed in real dc machines?

1–15. The linear machine in Figure 1–19 is running at steady state. What would happen to the bar if the voltage in the battery were increased? Explain in detail.

1–16. Just how does a decrease in flux produce an increase in speed in a linear machine?

1–17. Will current be leading or lagging voltage in an inductive load? Will the reactive power of the load be positive or negative?

1–18. What are real, reactive, and apparent power? What units are they measured in? How are they related?

1–19. What is power factor?

PROBLEMS

1–1. A motor's shaft is spinning at a speed of 1800 r/min. What is the shaft speed in radians per second?

1–2. A flywheel with a moment of inertia of 4 kg • m^2 is initially at rest. If a torque of 6 N • m (counterclockwise) is suddenly applied to the flywheel, what will be the speed of the flywheel after 5 s? Express that speed in both radians per second and revolutions per minute.

1–3. A force of 10 N is applied to a cylinder of radius $r = 0.15$ m, as shown in Figure P1–1. The moment of inertia of this cylinder is $J = 4$ kg • m^2. What are the magnitude and direction of the torque produced on the cylinder? What is the angular acceleration α of the cylinder?

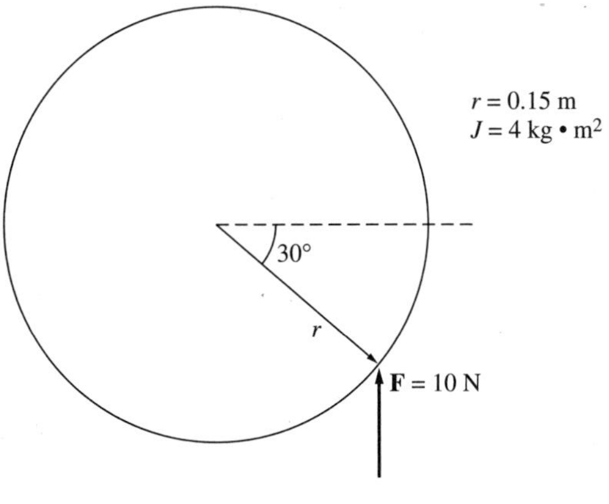

FIGURE P1–1
The cylinder of Problem 1–3.

1–4. A motor is supplying 50 N • m of torque to its load. If the motor s shaft is turning at 1500 r/min, what is the mechanical power supplied to the load in watts? In horsepower?

1–5. A ferromagnetic core is shown in Figure P1–2. The depth of the core is 5 cm. The other dimensions of the core are as shown in the figure. Find the value of the current that will produce a flux of 0.005 Wb. With this current, what is the flux density at the top of the core? What is the flux density at the right side of the core? Assume that the relative permeability of the core is 800.

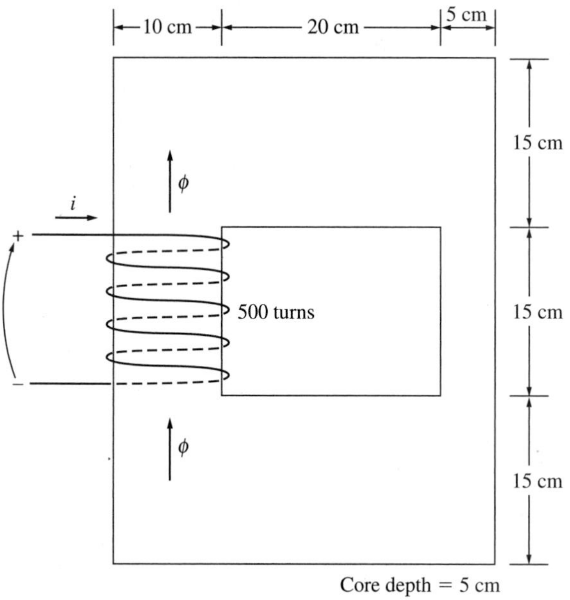

FIGURE P1–2
The core of Problems 1–5 and 1–16.

1–6. A ferromagnetic core with a relative permeability of 1500 is shown in Figure P1–3. The dimensions are as shown in the diagram, and the depth of the core is 5 cm. The air gaps on the left and right sides of the core are 0.050 and 0.070 cm, respectively. Because of fringing effects, the effective area of the air gaps is 5 percent larger than their physical size. If there are 300 turns in the coil wrapped around the center leg of the core and if the current in the coil is 1.0 A, what are the flux values for the left, center, and right legs of the core? What is the flux density in each air gap?

1–7. A two-legged core is shown in Figure P1–4. The winding on the left leg of the core (N_1) has 600 turns, and the winding on the right (N_2) has 200 turns. The coils are wound in the directions shown in the figure. If the dimensions are as shown, then what flux would be produced by currents $i_1 = 0.5$ A and $i_2 = 1.00$ A? Assume $\mu_r =$ 1200 and constant.

1–8. A core with three legs is shown in Figure P1–5. Its depth is 5 cm, and there are 100 turns on the leftmost leg. The relative permeability of the core can be assumed to be 2000 and constant. What flux exists in each of the three legs of the core? What is the flux density in each of the legs? Assume a 5 percent increase in the effective area of the air gap due to fringing effects.

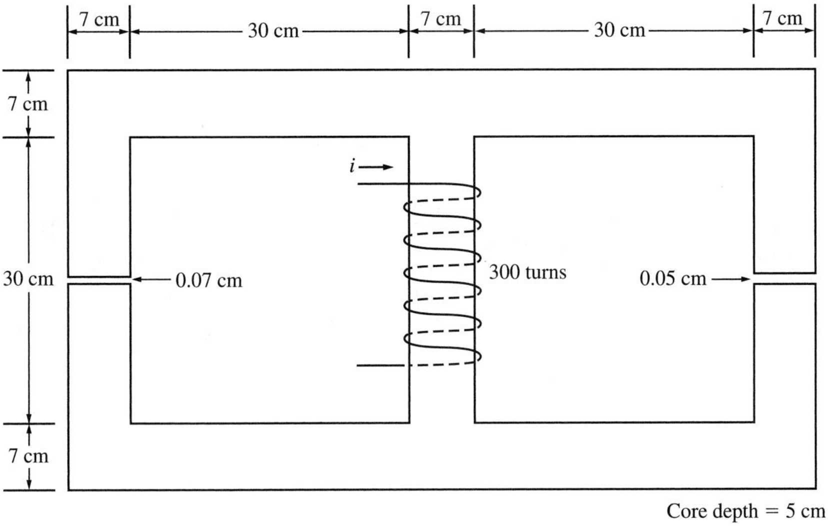

FIGURE P1–3
The core of Problem 1–6.

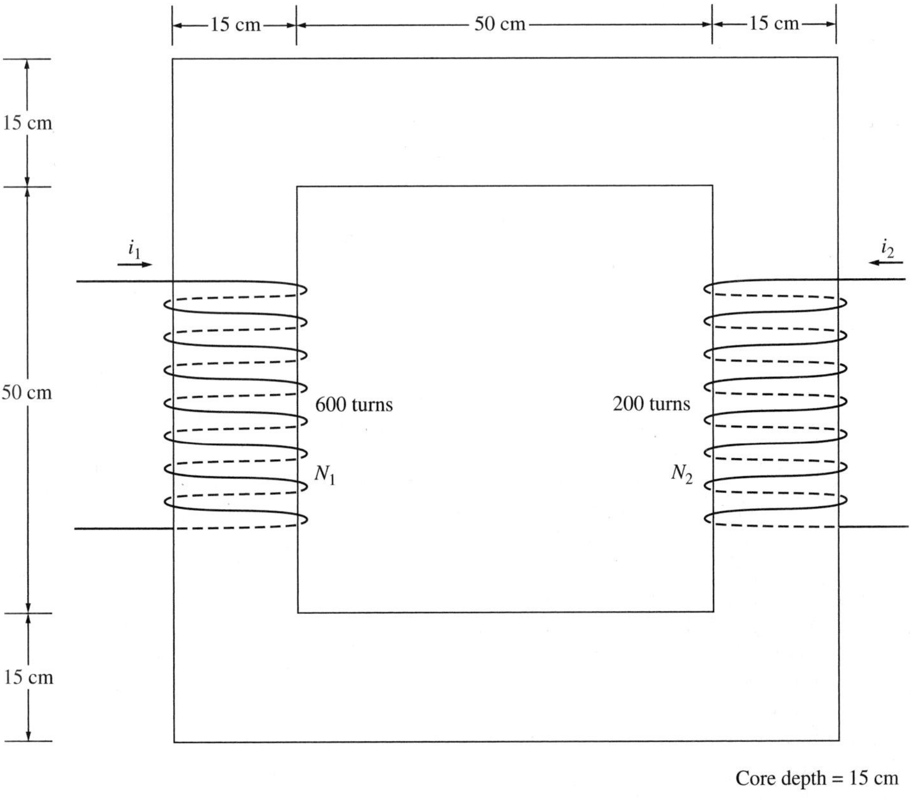

FIGURE P1–4
The core of Problems 1–7 and 1–12.

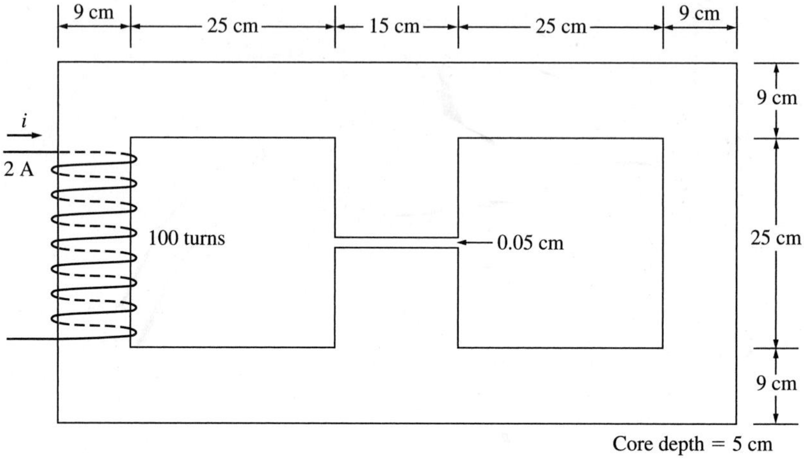

FIGURE P1–5
The core of Problem 1–8.

1–9. A wire is shown in Figure P1–6 that is carrying 2.0 A in the presence of a magnetic field. Calculate the magnitude and direction of the force induced on the wire.

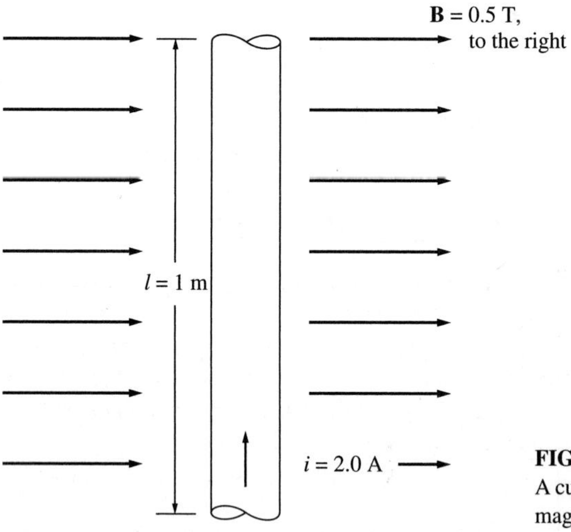

FIGURE P1–6
A current-carrying wire in a magnetic field (Problem 1–9).

1–10. A wire is shown in Figure P1–7 that is moving in the presence of a magnetic field. With the information given in the figure, determine the magnitude and direction of the induced voltage in the wire.

1–11. Repeat Problem 1–10 for the wire in Figure P1–8.

1–12. The core shown in Figure P1–4 is made of a steel whose magnetization curve is shown in Figure P1–9. Repeat Problem 1–7, but this time do *not* assume a constant value of μ_r. How much flux is produced in the core by the currents specified? What is the relative permeability of this core under these conditions? Was the assumption

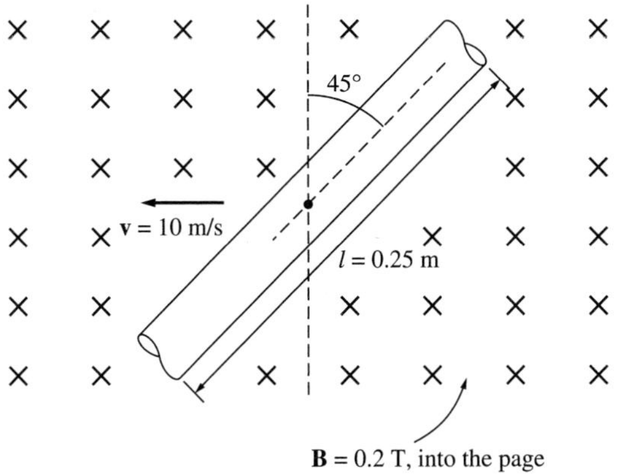

FIGURE P1–7
A wire moving in a magnetic field (Problem 1–10).

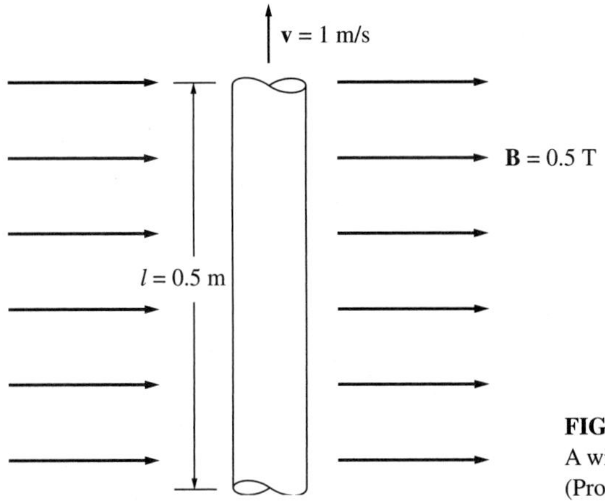

FIGURE P1–8
A wire moving in a magnetic field (Problem 1–11).

in Problem 1–7 that the relative permeability was equal to 1200 a good assumption for these conditions? Is it a good assumption in general?

1–13. A core with three legs is shown in Figure P1–10. Its depth is 5 cm, and there are 400 turns on the center leg. The remaining dimensions are shown in the figure. The core is composed of a steel having the magnetization curve shown in Figure 1–10*c*. Answer the following questions about this core:

(a) What current is required to produce a flux density of 0.5 T in the central leg of the core?

(b) What current is required to produce a flux density of 1.0 T in the central leg of the core? Is it twice the current in part *(a)*?

(c) What are the reluctances of the central and right legs of the core under the conditions in part *(a)*?

(d) What are the reluctances of the central and right legs of the core under the conditions in part *(b)*?

(e) What conclusion can you make about reluctances in real magnetic cores?

1–14. A two-legged magnetic core with an air gap is shown in Figure P1–11. The depth of the core is 5 cm, the length of the air gap in the core is 0.05 cm, and the number of turns on the coil is 1000. The magnetization curve of the core material is shown in

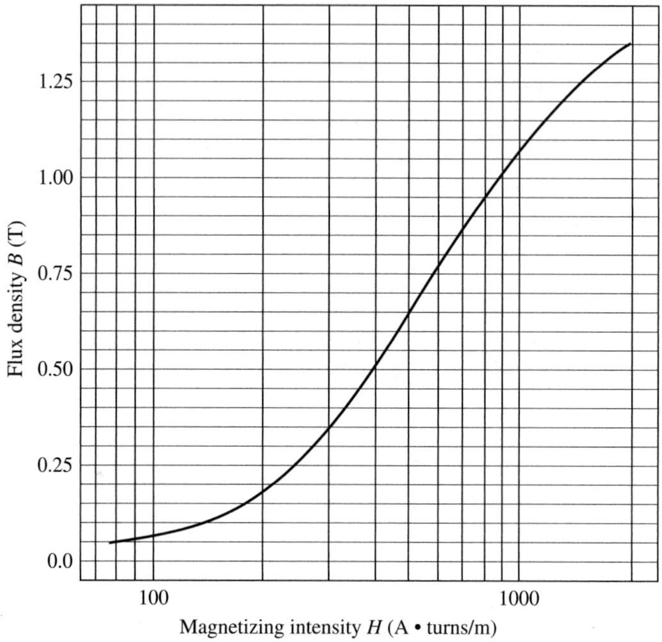

FIGURE P1–9
The magnetization curve for the core material of Problems 1–12 and 1–14.

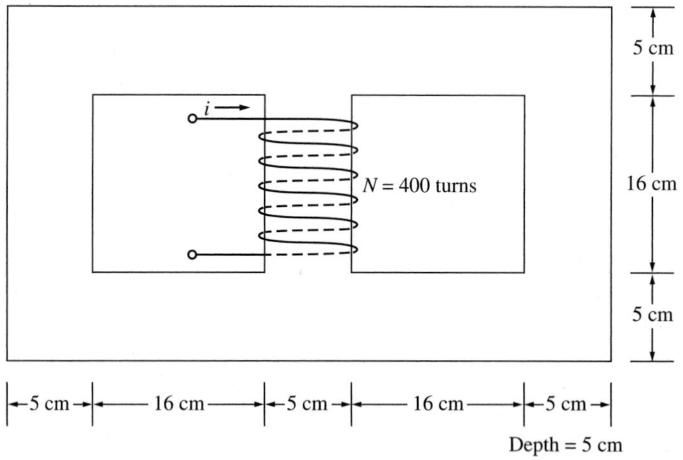

FIGURE P1–10
The core of Problem 1–13.

Figure P1–9. Assume a 5 percent increase in effective air-gap area to account for fringing. How much current is required to produce an air-gap flux density of 0.5 T? What are the flux densities of the four sides of the core at that current? What is the total flux present in the air gap?

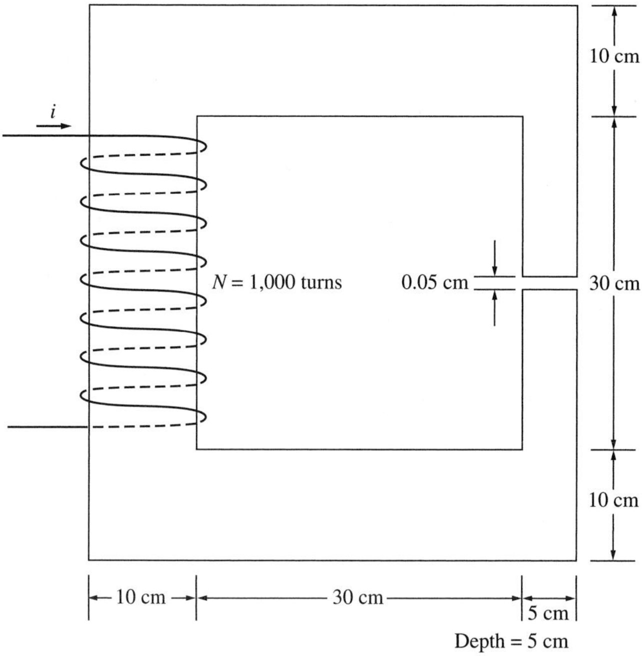

FIGURE P1–11
The core of Problem 1–14.

1–15. A transformer core with an effective mean path length of 6 in has a 200-turn coil wrapped around one leg. Its cross-sectional area is 0.25 in^2, and its magnetization curve is shown in Figure 1–10*c*. If current of 0.3 A is flowing in the coil, what is the total flux in the core? What is the flux density?

1–16. The core shown in Figure P1–2 has the flux ϕ shown in Figure P1–12. Sketch the voltage present at the terminals of the coil.

1–17. Figure P1–13 shows the core of a simple dc motor. The magnetization curve for the metal in this core is given by Figure 1–10*c* and *d*. Assume that the cross-sectional area of each air gap is 18 cm^2 and that the width of each air gap is 0.05 cm. The effective diameter of the rotor core is 5 cm.

(a) We wish to build a machine with as great a flux density as possible while avoiding excessive saturation in the core. What would be a reasonable maximum flux density for this core?

(b) What would be the total flux in the core at the flux density of part *(a)*?

(c) The maximum possible field current for this machine is 1 A. Select a reasonable number of turns of wire to provide the desired flux density while not exceeding the maximum available current.

1–18. Assume that the voltage applied to a load is $\mathbf{V} = 208\angle-30°$ V and the current flowing through the load is $\mathbf{I} = 2\angle20°$ A.

(a) Calculate the complex power **S** consumed by this load.

(b) Is this load inductive or capacitive?

(c) Calculate the power factor of this load.

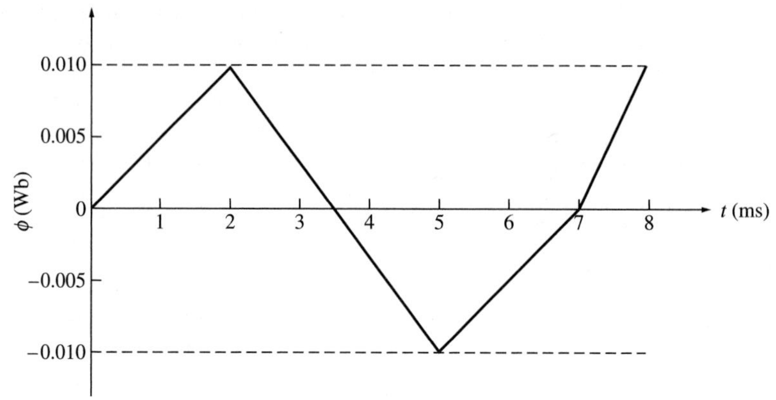

FIGURE P1–12
Plot of flux ϕ as a function of time for Problem 1–16.

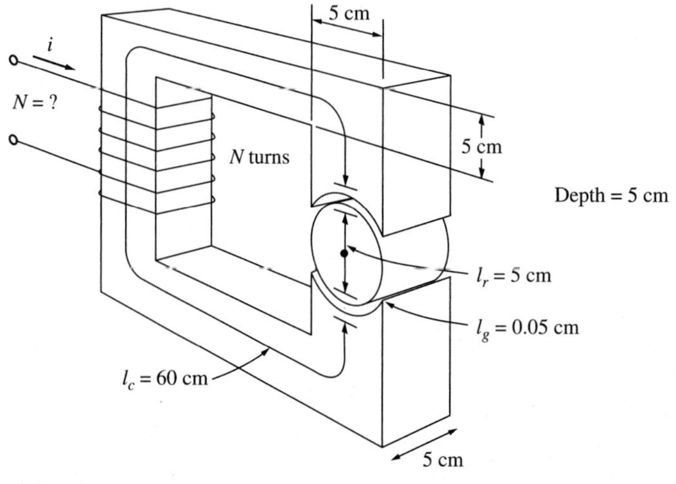

FIGURE P1–13
The core of Problem 1–17.

1–19. Figure P1–14 shows a simple single-phase ac power system with three loads. The voltage source is $\mathbf{V} = 240\angle 0°$ V, and the impedances of these three loads are

$$Z_1 = 10\angle 30°\ \Omega \qquad Z_2 = 10\angle 45°\ \Omega \qquad Z_3 = 10\angle -90°\ \Omega$$

Answer the following questions about this power system.

(a) Assume that the switch shown in the figure is initially open, and calculate the current **I**, the power factor, and the real, reactive, and apparent power being supplied by the source.

(b) How much real, reactive, and apparent power is being consumed by each load with the switch open?

(c) Assume that the switch shown in the figure is now closed, and calculate the current **I**, the power factor, and the real, reactive, and apparent power being supplied by the source.
(d) How much real, reactive, and apparent power is being consumed by each load with the switch closed?
(e) What happened to the current flowing from the source when the switch closed? Why?

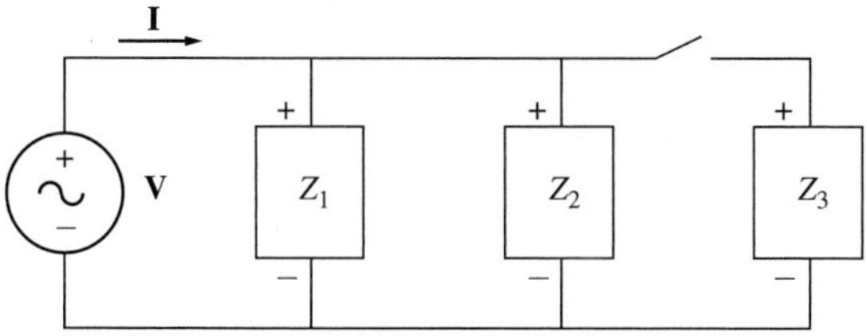

FIGURE P1–14
The circuit of Problem 1–19.

1–20. Demonstrate that Equation (1–59) can be derived from Equation (1–58) using simple trigonometric identities.

$$p(t) = v(t)i(t) = 2VI \cos \omega t \cos(\omega t - \theta) \qquad (1\text{–}58)$$

$$p(t) = VI \cos \theta\,(1 + \cos 2\omega t) + VI \sin \theta \sin 2\omega t \qquad (1\text{–}59)$$

Hint: The following identities will be useful:

$$\cos \alpha \cos \beta = \frac{1}{2}[\cos(\alpha - \beta) + \cos(\alpha + \beta)]$$
$$\cos(\alpha - \beta) = \cos \alpha \cos \beta + \sin \alpha \sin \beta$$

1–21. A linear machine shown in Figure P1–15 has a magnetic flux density of 0.5 T directed into the page, a resistance of 0.25 Ω, a bar length $l = 1.0$ m, and a battery voltage of 100 V.
(a) What is the initial force on the bar at starting? What is the initial current flow?
(b) What is the no-load steady-state speed of the bar?
(c) If the bar is loaded with a force of 25 N opposite to the direction of motion, what is the new steady-state speed? What is the efficiency of the machine under these circumstances?

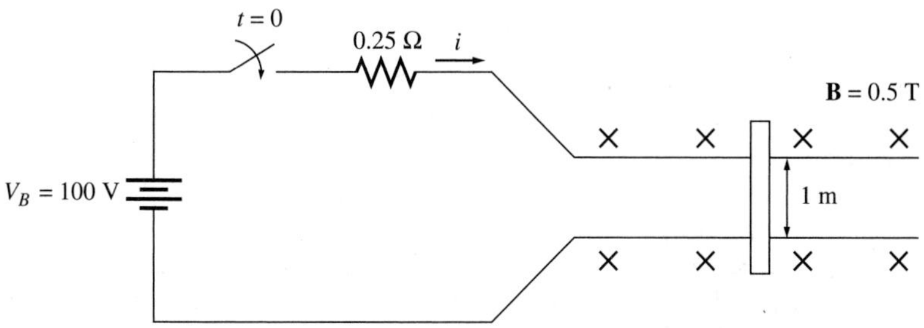

FIGURE P1–15
The linear machine in Problem 1–21.

1–22. A linear machine has the following characteristics:

$$\mathbf{B} = 0.5 \text{ T into page} \qquad R = 0.25\ \Omega$$
$$l = 0.5 \text{ m} \qquad V_B = 120 \text{ V}$$

(*a*) If this bar has a load of 20 N attached to it opposite to the direction of motion, what is the steady-state speed of the bar?

(*b*) If the bar runs off into a region where the flux density falls to 0.45 T, what happens to the bar? What is its final steady-state speed?

(*c*) Suppose V_B is now decreased to 100 V with everything else remaining as in part (*b*). What is the new steady-state speed of the bar?

(*d*) From the results for parts (*b*) and (*c*), what are two methods of controlling the speed of a linear machine (or a real dc motor)?

1–23. For the linear machine of Problem 1–22:

(*a*) When this machine is operating as a motor, calculate the speed of the bar for loads of 0 N to 30 N in 5 N steps. Plot the speed of the bar as a function of load.

(*b*) Assume that the motor is operation with a 30 N load, and calculate and plot the speed of the bar for magnetic flux densities of 0.3 T to 0.5 T in 0.05 T steps.

(*c*) Assume that the motor is running at no-load conditions with a flux density of 0.5 T. What is the speed of the bar? Now apply a 30 N load to the bar. What is the new speed of the bar? *What flux density would be required* to restore the loaded bar to the same speed that it had under no-load conditions?

REFERENCES

1. Alexander, Charles K., and Matthew N. O. Sadiiku: *Fundamentals of Electric Circuits*, 4th ed., McGraw-Hill, New York, 2008.
2. Beer, F., and E. Johnston, Jr.: *Vector Mechanics for Engineers: Dynamics*, 7th ed., McGraw-Hill, New York, 2004.
3. Hayt, William H.: *Engineering Electromagnetics*, 5th ed., McGraw-Hill, New York, 1989.
4. Mulligan, J. F.: *Introductory College Physics*, 2nd ed., McGraw-Hill, New York, 1991.
5. Sears, Francis W., Mark W. Zemansky, and Hugh D. Young: *University Physics*, Addison-Wesley, Reading, Mass., 1982.

CHAPTER
2

TRANSFORMERS

OBJECTIVES

- Understand the purpose of a transformer in a power system.
- Know the voltage, current, and impedance relationships across the windings of an ideal transformer.
- Understand how real transformers approximate the operation of an ideal transformer.
- Be able to explain how copper losses, leakage flux, hysteresis, and eddy currents are modeled in transformer equivalent circuits.
- Use a transformer equivalent circuit to find the voltage and current transformations across a transformer.
- Be able to calculate the losses and efficiency of a transformer.
- Be able to derive the equivalent circuit of a transformer from measurements.
- Understand the per-unit system of measurements.
- Be able to calculate the voltage regulation of a transformer.
- Understand the autotransformer.
- Understand three-phase transformers, including special cases where only two transformers are used.
- Understand transformer ratings.
- Understand instrument transformers—potential transformers and current transformers.

A *transformer* is a device that changes ac electric power at one frequency and voltage level to ac electric power at the same frequency and another voltage level through the action of a magnetic field. It consists of two or more coils of wire wrapped around a common ferromagnetic core. These coils are (usually) not directly connected. The only connection between the coils is the common magnetic flux present within the core.

One of the transformer windings is connected to a source of ac electric power, and the second (and perhaps third) transformer winding supplies electric power to loads. The transformer winding connected to the power source is called the *primary winding* or *input winding,* and the winding connected to the loads is called the *secondary winding* or *output winding.* If there is a third winding on the transformer, it is called the *tertiary winding.*

2.1 WHY TRANSFORMERS ARE IMPORTANT TO MODERN LIFE

The first power distribution system in the United States was a 120-V dc system invented by Thomas A. Edison to supply power for incandescent light bulbs. Edison's first central power station went into operation in New York City in September 1882. Unfortunately, his power system generated and transmitted power at such low voltages that very large currents were necessary to supply significant amounts of power. These high currents caused huge voltage drops and power losses in the transmission lines, severely restricting the service area of a generating station. In the 1880s, central power stations were located every few city blocks to overcome this problem. The fact that power could not be transmitted far with low-voltage dc

FIGURE 2–1
The first practical modern transformer, built by William Stanley in 1885. Note that the core is made up of individual sheets of metal (laminations). (*Courtesy of General Electric Company.*)

power systems meant that generating stations had to be small and localized and so were relatively inefficient.

The invention of the transformer and the concurrent development of ac power sources eliminated forever these restrictions on the range and power level of power systems. A transformer ideally changes one ac voltage level to another voltage level without affecting the actual power supplied. If a transformer steps up the voltage level of a circuit, it must decrease the current to keep the power into the device equal to the power out of it. Therefore, ac electric power can be generated at one central location, its voltage stepped up for transmission over long distances at very low losses, and its voltage stepped down again for final use. Since the transmission losses in the lines of a power system are proportional to the square of the current in the lines, raising the transmission voltage and reducing the resulting transmission currents by a factor of 10 with transformers reduces power transmission losses by a factor of 100. Without the transformer, it would simply not be possible to use electric power in many of the ways it is used today.

In a modern power system, electric power is generated at voltages of 12 to 25 kV. Transformers step up the voltage to between 110 kV and nearly 1000 kV for transmission over long distances at very low losses. Transformers then step down the voltage to the 12- to 34.5-kV range for local distribution and finally permit the power to be used safely in homes, offices, and factories at voltages as low as 120 V.

2.2 TYPES AND CONSTRUCTION OF TRANSFORMERS

The principal purpose of a transformer is to convert ac power at one voltage level to ac power of the same frequency at another voltage level. Transformers are also used for a variety of other purposes (e.g., voltage sampling, current sampling, and impedance transformation), but this chapter is primarily devoted to the power transformer.

Power transformers are constructed on one of two types of cores. One type of construction consists of a simple rectangular laminated piece of steel with the transformer windings wrapped around two sides of the rectangle. This type of construction is known as *core form* and is illustrated in Figure 2–2. The other type consists of a three-legged laminated core with the windings wrapped around the center leg. This type of construction is known as *shell form* and is illustrated in Figure 2–3. In either case, the core is constructed of thin laminations electrically isolated from each other in order to minimize eddy currents.

The primary and secondary windings in a physical transformer are wrapped one on top of the other with the low-voltage winding innermost. Such an arrangement serves two purposes:

1. It simplifies the problem of insulating the high-voltage winding from the core.
2. It results in much less leakage flux than would be the case if the two windings were separated by a distance on the core.

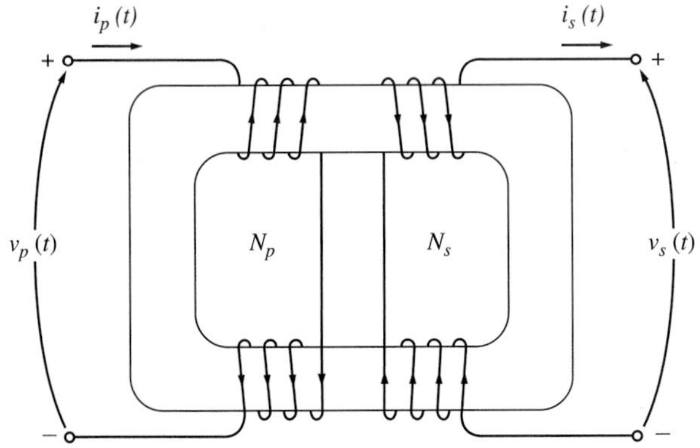

FIGURE 2–2
Core-form transformer construction.

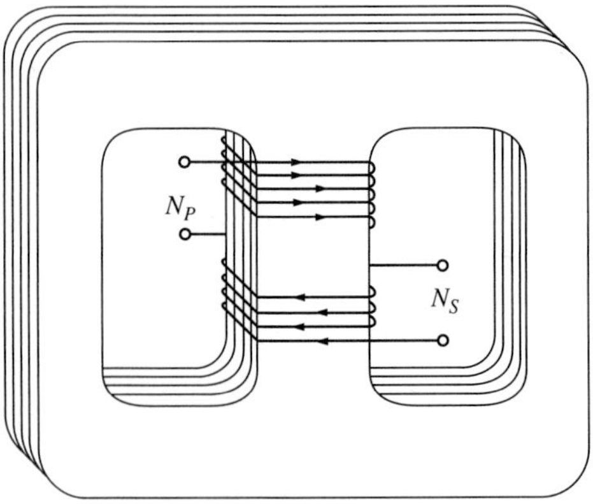

(a)

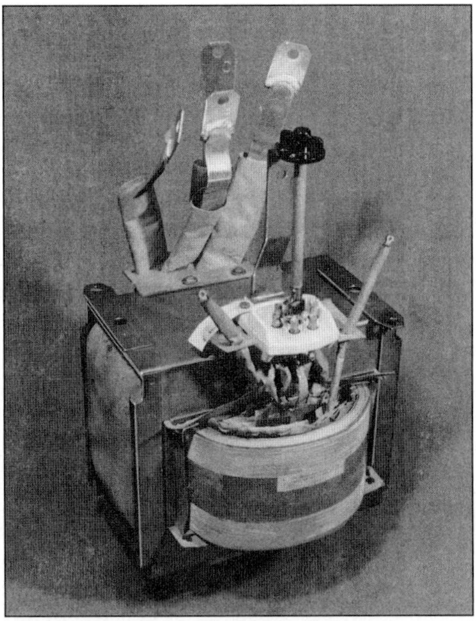

(b)

FIGURE 2–3
(a) Shell-form transformer construction. (b) A typical shell-form transformer. *(Courtesy of General Electric Company.)*

Power transformers are given a variety of different names, depending on their use in power systems. A transformer connected to the output of a generator and used to step its voltage up to transmission levels (110+ kV) is sometimes called a *unit transformer*. The transformer at the other end of the transmission line, which steps the voltage down from transmission levels to distribution levels (from 2.3 to 34.5 kV), is called a *substation transformer.* Finally, the transformer that takes the distribution voltage and steps it down to the final voltage at which the power is actually used (110, 208, 220 V, etc.) is called a *distribution transformer.* All these devices are essentially the same—the only difference among them is their intended use.

In addition to the various power transformers, two special-purpose transformers are used to measure voltage and current in electric machinery and power systems. The first of these special transformers is a device specially designed to sample a high voltage and produce a low secondary voltage directly proportional to it. Such a transformer is called a *potential transformer.* A power transformer also produces a secondary voltage directly proportional to its primary voltage; the difference between a potential transformer and a power transformer is that the potential transformer is designed to handle only a very small current. The second type of special transformer is a device designed to provide a secondary current much smaller than but directly proportional to its primary current. This device is called a *current transformer.* Both special-purpose transformers are discussed in a later section of this chapter.

2.3 THE IDEAL TRANSFORMER

An *ideal transformer* is a lossless device with an input winding and an output winding. The relationships between the input voltage and the output voltage, and between the input current and the output current, are given by two simple equations. Figure 2–4 shows an ideal transformer.

The transformer shown in Figure 2–4 has N_P turns of wire on its primary side and N_S turns of wire on its secondary side. The relationship between the voltage $v_P(t)$ applied to the primary side of the transformer and the voltage $v_S(t)$ produced on the secondary side is

$$\boxed{\frac{v_P(t)}{v_S(t)} = \frac{N_P}{N_S} = a} \qquad (2\text{–}1)$$

where a is defined to be the *turns ratio* of the transformer:

$$a = \frac{N_P}{N_S} \qquad (2\text{–}2)$$

The relationship between the current $i_P(t)$ flowing into the primary side of the transformer and the current $i_S(t)$ flowing out of the secondary side of the transformer is

$$\boxed{N_P i_P(t) = N_S i_S(t)} \qquad (2\text{–}3a)$$

or

$$\boxed{\frac{i_P(t)}{i_S(t)} = \frac{1}{a}} \qquad (2\text{–}3b)$$

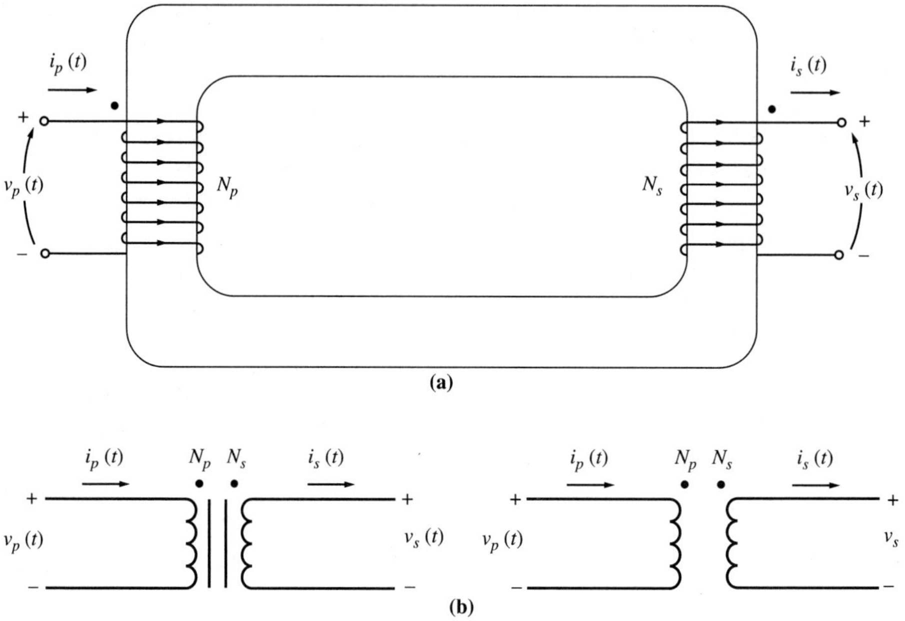

FIGURE 2–4
(a) Sketch of an ideal transformer. (b) Schematic symbols of a transformer. Sometimes the iron core is shown in the symbol, and sometimes not.

In terms of phasor quantities, these equations are

$$\boxed{\frac{\mathbf{V}_P}{\mathbf{V}_S} = a} \tag{2–4}$$

and

$$\boxed{\frac{\mathbf{I}_P}{\mathbf{I}_S} = \frac{1}{a}} \tag{2–5}$$

Notice that the phase angle of $\mathbf{V}_P$ is the same as the angle of $\mathbf{V}_S$ and the phase angle of $\mathbf{I}_P$ is the same as the phase angle of $\mathbf{I}_S$. The turns ratio of the ideal transformer affects the *magnitudes* of the voltages and currents, but not their *angles.*

Equations (2–1) to (2–5) describe the relationships between the magnitudes and angles of the voltages and currents on the primary and secondary sides of the transformer, but they leave one question unanswered: Given that the primary circuit's voltage is positive at a specific end of the coil, what would the *polarity* of the secondary circuit's voltage be? In real transformers, it would be possible to tell the secondary's polarity only if the transformer were opened and its windings examined. To avoid this necessity, transformers utilize the *dot convention.* The dots appearing at one end of each winding in Figure 2–4 tell the polarity of the voltage and current on the secondary side of the transformer. The relationship is as follows:

1. If the primary *voltage* is positive at the dotted end of the winding with respect to the undotted end, then the secondary voltage will be positive at the dotted end also. Voltage polarities are the same with respect to the dots on each side of the core.
2. If the primary *current* of the transformer flows *into* the dotted end of the primary winding, the secondary current will flow *out* of the dotted end of the secondary winding.

The physical meaning of the dot convention and the reason polarities work out this way will be explained in Section 2.4, which deals with the real transformer.

Power in an Ideal Transformer

The real power P_{in} supplied to the transformer by the primary circuit is given by the equation

$$\boxed{P_{\text{in}} = V_P I_P \cos \theta_P} \tag{2–6}$$

where θ_P is the angle between the primary voltage and the primary current. The real power P_{out} supplied by the transformer secondary circuit to its loads is given by the equation

$$\boxed{P_{\text{out}} = V_S I_S \cos \theta_S} \tag{2–7}$$

where θ_S is the angle between the secondary voltage and the secondary current. Since voltage and current angles are unaffected by an ideal transformer, $\theta_P = \theta_S = \theta$. The primary and secondary windings of an ideal transformer have the *same power factor.*

How does the power going into the primary circuit of the ideal transformer compare to the power coming out of the other side? It is possible to find out through a simple application of the voltage and current equations [Equations (2–4) and (2–5)]. The power out of a transformer is

$$P_{\text{out}} = V_S I_S \cos \theta \tag{2–8}$$

Applying the turns-ratio equations gives $V_S = V_P/a$ and $I_S = aI_P$, so

$$P_{\text{out}} = \frac{V_P}{a}(aI_P) \cos \theta$$

$$\boxed{P_{\text{out}} = V_P I_P \cos \theta = P_{\text{in}}} \tag{2–9}$$

Thus, *the output power of an ideal transformer is equal to its input power.*

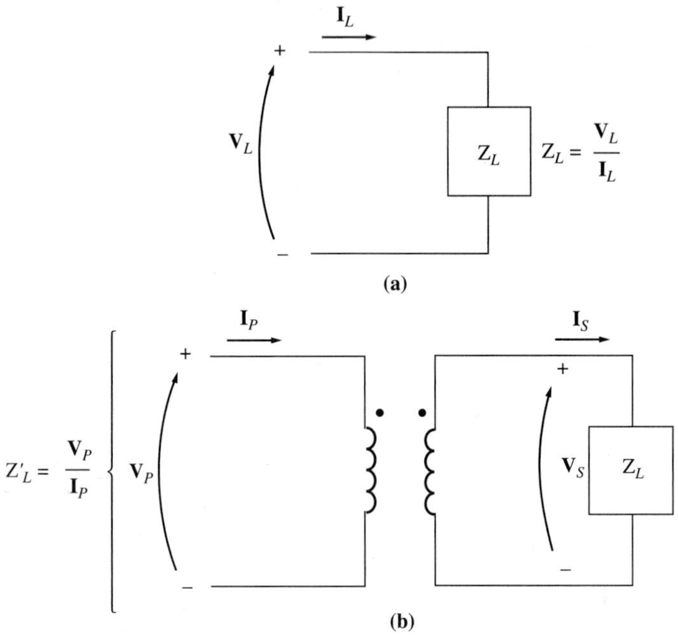

FIGURE 2–5
(a) Definition of impedance. (b) Impedance scaling through a transformer.

The same relationship applies to reactive power Q and apparent power S:

$$\boxed{Q_{\text{in}} = V_P I_P \sin\theta = V_S I_S \sin\theta = Q_{\text{out}}} \tag{2–10}$$

and

$$\boxed{S_{\text{in}} = V_P I_P = V_S I_S = S_{\text{out}}} \tag{2–11}$$

Impedance Transformation through a Transformer

The *impedance* of a device or an element is defined as the ratio of the phasor voltage across it to the phasor current flowing through it:

$$Z_L = \frac{\mathbf{V}_L}{\mathbf{I}_L} \tag{2–12}$$

One of the interesting properties of a transformer is that, since it changes voltage and current levels, it changes the *ratio* between voltage and current and hence the apparent impedance of an element. To understand this idea, refer to Figure 2–5. If the secondary current is called $\mathbf{I}_S$ and the secondary voltage $\mathbf{V}_S$, then the impedance of the load is given by

$$Z_L = \frac{\mathbf{V}_S}{\mathbf{I}_S} \tag{2–13}$$

The apparent impedance of the primary circuit of the transformer is

$$Z'_L = \frac{\mathbf{V}_P}{\mathbf{I}_P} \tag{2–14}$$

Since the primary voltage can be expressed as

$$\mathbf{V}_P = a\mathbf{V}_S$$

and the primary current can be expressed as

$$\mathbf{I}_P = \frac{\mathbf{I}_S}{a}$$

the apparent impedance of the primary is

$$Z'_L = \frac{\mathbf{V}_P}{\mathbf{I}_P} = \frac{a\mathbf{V}_S}{\mathbf{I}_S/a} = a^2\frac{\mathbf{V}_S}{\mathbf{I}_S}$$

$$\boxed{Z'_L = a^2 Z_L} \tag{2–15}$$

With a transformer, it is possible to match the magnitude of a load impedance to a source impedance simply by picking the proper turns ratio.

Analysis of Circuits Containing Ideal Transformers

If a circuit contains an ideal transformer, then the easiest way to analyze the circuit for its voltages and currents is to replace the portion of the circuit on one side of the transformer by an equivalent circuit with the same terminal characteristics. After the equivalent circuit has been substituted for one side, then the new circuit (without a transformer present) can be solved for its voltages and currents. In the portion of the circuit that was not replaced, the solutions obtained will be the correct values of voltage and current for the original circuit. Then the turns ratio of the transformer can be used to determine the voltages and currents on the other side of the transformer. The process of replacing one side of a transformer with its equivalent at the other side's voltage level is known as *referring* the first side of the transformer to the second side.

How is the equivalent circuit formed? Its shape is exactly the same as the shape of the original circuit. The values of voltages on the side being replaced are scaled by Equation (2–4), and the values of the impedances are scaled by Equation (2–15). The polarities of voltage sources in the equivalent circuit will be reversed from their direction in the original circuit if the dots on one side of the transformer windings are reversed compared to the dots on the other side of the transformer windings.

The solution for circuits containing ideal transformers is illustrated in the following example.

Example 2–1. A single-phase power system consists of a 480-V 60-Hz generator supplying a load $Z_{\text{load}} = 4 + j3\ \Omega$ through a transmission line of impedance $Z_{\text{line}} = 0.18 + j0.24\ \Omega$. Answer the following questions about this system.

(*a*) If the power system is exactly as described above (and shown in Figure 2–6a), what will the voltage at the load be? What will the transmission line losses be?

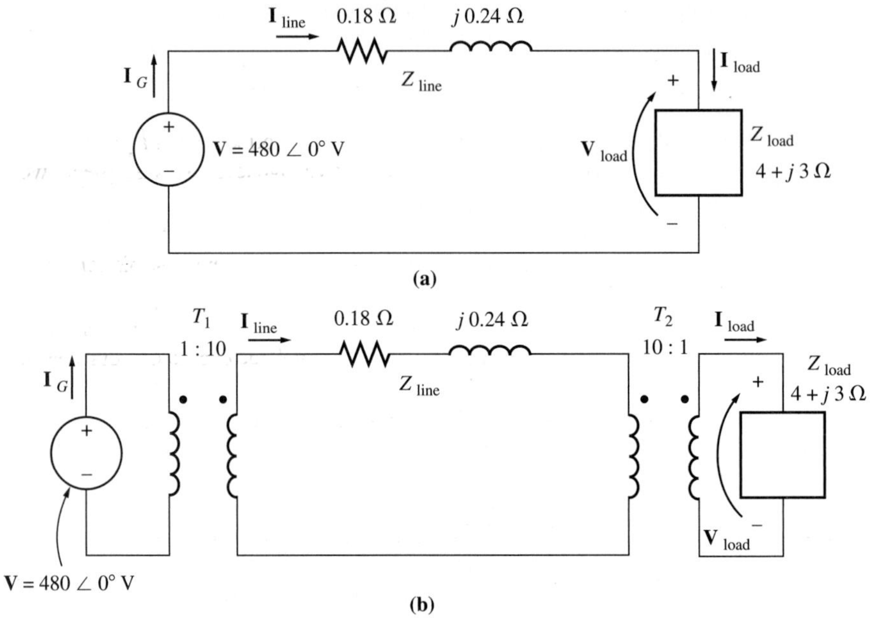

FIGURE 2–6
The power system of Example 2–1 (a) without and (b) with transformers at the ends of the transmission line.

(*b*) Suppose a 1:10 step-up transformer is placed at the generator end of the transmission line and a 10:1 step-down transformer is placed at the load end of the line (as shown in Figure 2–6b). What will the load voltage be now? What will the transmission line losses be now?

Solution

(*a*) Figure 2–6a shows the power system without transformers. Here $\mathbf{I}_G = \mathbf{I}_{\text{line}} = \mathbf{I}_{\text{load}}$. The line current in this system is given by

$$\begin{aligned}\mathbf{I}_{\text{line}} &= \frac{\mathbf{V}}{\mathbf{Z}_{\text{line}} + \mathbf{Z}_{\text{load}}} \\ &= \frac{480\angle 0^\circ \text{ V}}{(0.18\ \Omega + j0.24\ \Omega) + (4\ \Omega + j3\ \Omega)} \\ &= \frac{480\angle 0^\circ}{4.18 + j3.24} = \frac{480\angle 0^\circ}{5.29\angle 37.8^\circ} \\ &= 90.8\angle -37.8^\circ \text{ A}\end{aligned}$$

Therefore the load voltage is

$$\begin{aligned}\mathbf{V}_{\text{load}} &= \mathbf{I}_{\text{line}}\mathbf{Z}_{\text{load}} \\ &= (90.8\angle -37.8^\circ \text{ A})(4\ \Omega + j3\ \Omega) \\ &= (90.8\angle -37.8^\circ \text{ A})(5\angle 36.9^\circ\ \Omega) \\ &= 454\angle -0.9^\circ \text{ V}\end{aligned}$$

and the line losses are

$$
\begin{aligned}
P_{\text{loss}} &= (I_{\text{line}})^2 R_{\text{line}} \\
&= (90.8 \text{ A})^2 (0.18 \ \Omega) = 1484 \text{ W}
\end{aligned}
$$

(*b*) Figure 2–6b shows the power system with the transformers. To analyze this system, it is necessary to convert it to a common voltage level. This is done in two steps:

1. Eliminate transformer T_2 by referring the load over to the transmission line's voltage level.
2. Eliminate transformer T_1 by referring the transmission line's elements and the equivalent load at the transmission line's voltage over to the source side.

The value of the load's impedance when reflected to the transmission system's voltage is

$$
\begin{aligned}
\mathbf{Z}'_{\text{load}} &= \mathrm{a}^2 \mathbf{Z}_{\text{load}} \\
&= \left(\frac{10}{1}\right)^2 (4 \ \Omega + j3 \ \Omega) \\
&= 400 \ \Omega + j300 \ \Omega
\end{aligned}
$$

The total impedance at the transmission line level is now

$$
\begin{aligned}
\mathbf{Z}_{\text{eq}} &= \mathbf{Z}_{\text{line}} + \mathbf{Z}'_{\text{load}} \\
&= 400.18 + j300.24 \ \Omega = 500.3 \angle 36.88° \ \Omega
\end{aligned}
$$

This equivalent circuit is shown in Figure 2–7a. The total impedance at the transmission line level ($\mathbf{Z}_{\text{line}} + \mathbf{Z}'_{\text{load}}$) is now reflected across T_1 to the source's voltage level:

$$
\begin{aligned}
\mathbf{Z}'_{\text{eq}} &= a^2 \mathbf{Z}_{\text{eq}} \\
&= a^2(\mathbf{Z}_{\text{line}} + \mathbf{Z}'_{\text{load}}) \\
&= \left(\frac{1}{10}\right)^2 (0.18 \ \Omega + j0.24 \ \Omega + 400 \ \Omega + j300 \ \Omega) \\
&= (0.0018 \ \Omega + j0.0024 \ \Omega + 4 \ \Omega + j3 \ \Omega) \\
&= 5.003 \angle 36.88° \ \Omega
\end{aligned}
$$

Notice that $\mathbf{Z}''_{\text{load}} = 4 + j3 \ \Omega$ and $\mathbf{Z}'_{\text{line}} = 0.0018 + j0.0024 \ \Omega$. The resulting equivalent circuit is shown in Figure 2–7b. The generator's current is

$$
\mathbf{I}_G = \frac{480 \angle 0° \text{ V}}{5.003 \angle 36.88° \ \Omega} = 95.94 \angle -36.88° \text{ A}
$$

Knowing the current $\mathbf{I}_G$, we can now work back and find $\mathbf{I}_{\text{line}}$ and $\mathbf{I}_{\text{load}}$. Working back through T_1, we get

$$
\begin{aligned}
N_{P1}\mathbf{I}_G &= N_{S1}\mathbf{I}_{\text{line}} \\
\mathbf{I}_{\text{line}} &= \frac{N_{P1}}{N_{S1}} \mathbf{I}_G \\
&= \frac{1}{10}(95.94 \angle -36.88° \text{ A}) = 9.594 \angle -36.88° \text{ A}
\end{aligned}
$$

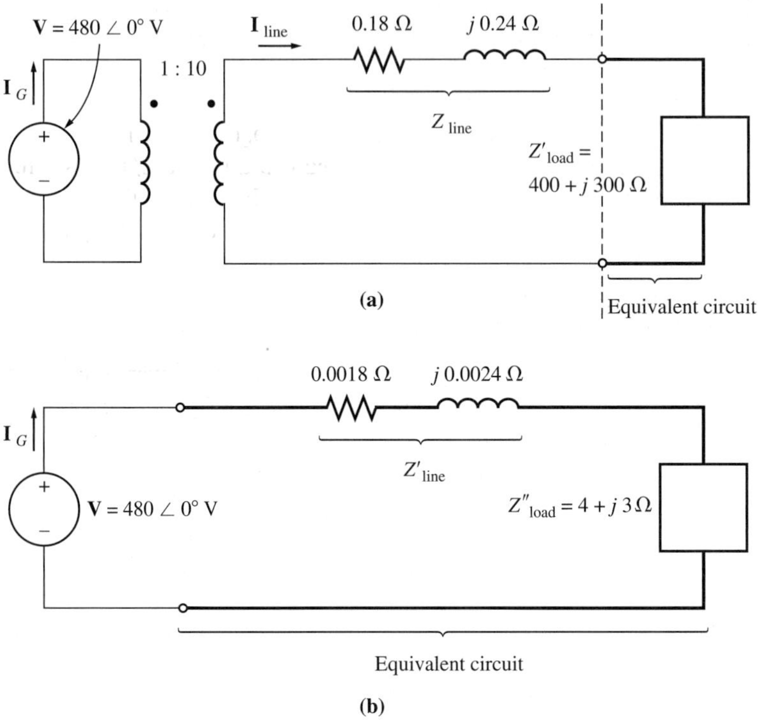

FIGURE 2–7
(a) System with the load referred to the transmission system voltage level. (b) System with the load and transmission line referred to the generator's voltage level.

Working back through T_2 gives

$$N_{P2}\mathbf{I}_{\text{line}} = N_{S2}\mathbf{I}_{\text{load}}$$

$$\mathbf{I}_{\text{load}} = \frac{N_{P2}}{N_{S2}}\mathbf{I}_{\text{line}}$$

$$= \frac{10}{1}(9.594 \angle -36.88° \text{ A}) = 95.94 \angle -36.88° \text{ A}$$

It is now possible to answer the questions originally asked. The load voltage is given by

$$\mathbf{V}_{\text{load}} = \mathbf{I}_{\text{load}} Z_{\text{load}}$$

$$= (95.94 \angle -36.88° \text{ A})(5 \angle 36.87° \ \Omega)$$

$$= 479.7 \angle -0.01° \text{ V}$$

and the line losses are given by

$$P_{\text{loss}} = (I_{\text{line}})^2 R_{\text{line}}$$

$$= (9.594 \text{ A})^2 \, (0.18 \ \Omega) = 16.7 \text{ W}$$

Notice that raising the transmission voltage of the power system reduced transmission losses by a factor of nearly 90! Also, the voltage at the load dropped much less in the system with transformers compared to the system without transformers. This simple example dramatically illustrates the advantages of using higher-voltage transmission lines as well as the extreme importance of transformers in modern power systems.

Real power systems generate electric power at voltages in the range of 4 to 30 kV. They then use *step-up transformers* to raise the voltage to a much higher level (say 500 kV) for transmission over long distances, and *step-down transformers* to reduce the voltage to a reasonable level for distribution and final use. As we have seen in Example 2.1, this can greatly decrease transmission losses in the power system.

2.4 THEORY OF OPERATION OF REAL SINGLE-PHASE TRANSFORMERS

The ideal transformers described in Section 2.3 can of course never actually be made. What can be produced are real transformers—two or more coils of wire physically wrapped around a ferromagnetic core. The characteristics of a real transformer approximate the characteristics of an ideal transformer, but only to a degree. This section deals with the behavior of real transformers.

To understand the operation of a real transformer, refer to Figure 2–8. Figure 2–8 shows a transformer consisting of two coils of wire wrapped around a transformer core. The primary of the transformer is connected to an ac power source, and the secondary winding is open-circuited. The hysteresis curve of the transformer is shown in Figure 2–9.

The basis of transformer operation can be derived from Faraday's law:

$$e_{\text{ind}} = \frac{d\lambda}{dt} \tag{1–41}$$

where λ is the flux linkage in the coil across which the voltage is being induced. The flux linkage λ is the sum of the flux passing through each turn in the coil added over all the turns of the coil:

$$\lambda = \sum_{i=1}^{N} \phi_i \tag{1–42}$$

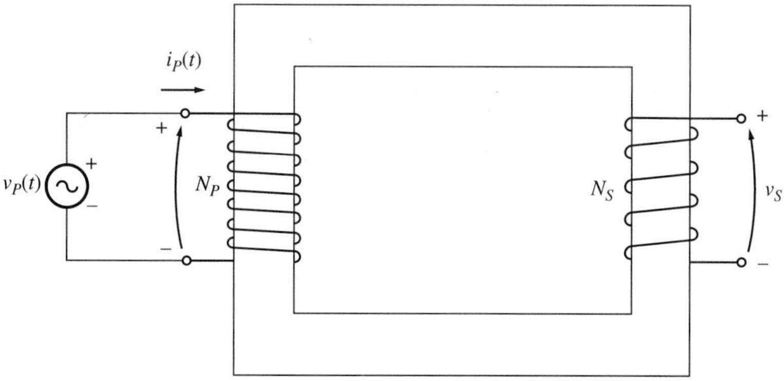

FIGURE 2–8
Sketch of a real transformer with no load attached to its secondary.

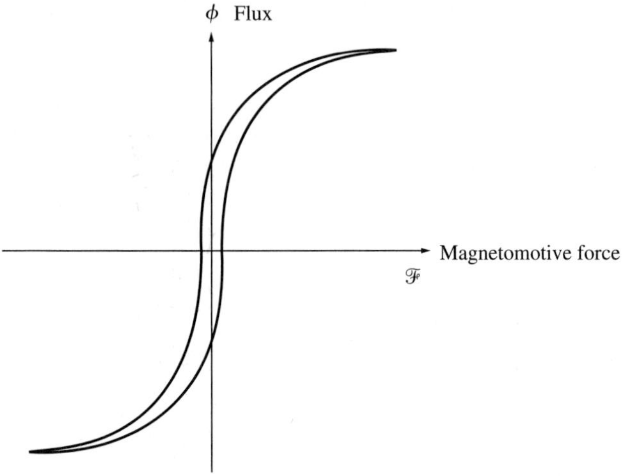

FIGURE 2–9
The hysteresis curve of the transformer.

The total flux linkage through a coil is not just $N\phi$, where N is the number of turns in the coil, because the flux passing through each turn of a coil is slightly different from the flux in the other turns, depending on the position of the turn within the coil.

However, it is possible to define an *average* flux per turn in a coil. If the total flux linkage in all the turns of the coils is λ and if there are N turns, then the *average flux per turn* is given by

$$\overline{\phi} = \frac{\lambda}{N} \tag{2–16}$$

and Faraday's law can be written as

$$e_{\text{ind}} = N\frac{d\overline{\phi}}{dt} \tag{2–17}$$

The Voltage Ratio across a Transformer

If the voltage of the source in Figure 2–8 is $v_P(t)$, then that voltage is placed directly across the coils of the primary winding of the transformer. How will the transformer react to this applied voltage? Faraday's law explains what will happen. When Equation (2–17) is solved for the average flux present in the primary winding of the transformer, and the winding resistance is ignored, the result is

$$\overline{\phi}_P = \frac{1}{N_P}\int v_P(t)\,dt \tag{2–18}$$

This equation states that the average flux in the winding is proportional to the integral of the voltage applied to the winding, and the constant of proportionality is the reciprocal of the number of turns in the primary winding $1/N_P$.

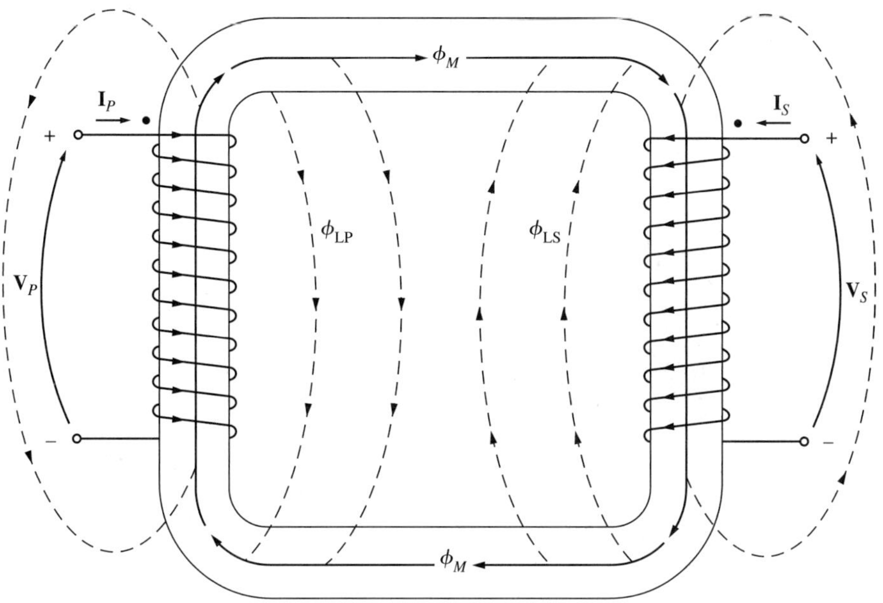

FIGURE 2–10
Mutual and leakage fluxes in a transformer core.

This flux is present in the *primary coil* of the transformer. What effect does it have on the secondary coil of the transformer? The effect depends on how much of the flux reaches the secondary coil. Not all the flux produced in the primary coil also passes through the secondary coil—some of the flux lines leave the iron core and pass through the air instead (see Figure 2–10). The portion of the flux that goes through one of the transformer coils but not the other one is called *leakage flux*. The flux in the primary coil of the transformer can thus be divided into two components: a *mutual flux*, which remains in the core and links both windings, and a small *leakage flux*, which passes through the primary winding but returns through the air, bypassing the secondary winding:

$$\boxed{\overline{\phi}_P = \phi_M + \phi_{LP}} \tag{2–19}$$

where $\overline{\phi}_P$ = total average primary flux

ϕ_M = flux component linking both primary and secondary coils

ϕ_{LP} = primary leakage flux

There is a similar division of flux in the secondary winding between mutual flux and leakage flux which passes through the secondary winding but returns through the air, bypassing the primary winding:

$$\boxed{\overline{\phi}_S = \phi_M + \phi_{LS}} \tag{2–20}$$

where $\bar{\phi}_S$ = total average secondary flux
ϕ_M = flux component linking both primary and secondary coils
ϕ_{LS} = secondary leakage flux

With the division of the average primary flux into mutual and leakage components, Faraday's law for the primary circuit can be reexpressed as

$$\begin{aligned} v_P(t) &= N_P \frac{d\bar{\phi}_P}{dt} \\ &= N_P \frac{d\phi_M}{dt} + N_P \frac{d\phi_{LP}}{dt} \qquad (2\text{–}21) \end{aligned}$$

The first term of this expression can be called $e_P(t)$, and the second term can be called $e_{LP}(t)$. If this is done, then Equation (2–21) can be rewritten as

$$v_P(t) = e_P(t) + e_{LP}(t) \qquad (2\text{–}22)$$

The voltage on the secondary coil of the transformer can also be expressed in terms of Faraday's law as

$$\begin{aligned} v_S(t) &= N_S \frac{d\bar{\phi}_S}{dt} \\ &= N_S \frac{d\phi_M}{dt} + N_S \frac{d\phi_{LS}}{dt} \qquad (2\text{–}23) \\ &= e_S(t) + e_{LS}(t) \qquad (2\text{–}24) \end{aligned}$$

The primary voltage *due to the mutual flux* is given by

$$e_P(t) = N_P \frac{d\phi_M}{dt} \qquad (2\text{–}25)$$

and the secondary voltage *due to the mutual flux* is given by

$$e_S(t) = N_S \frac{d\phi_M}{dt} \qquad (2\text{–}26)$$

Notice from these two relationships that

$$\frac{e_P(t)}{N_P} = \frac{d\phi_M}{dt} = \frac{e_S(t)}{N_S}$$

Therefore,

$$\boxed{\frac{e_P(t)}{e_S(t)} = \frac{N_P}{N_S} = a} \qquad (2\text{–}27)$$

This equation means that *the ratio of the primary voltage caused by the mutual flux to the secondary voltage caused by the mutual flux is equal to the turns ratio of the transformer.* Since in a well-designed transformer $\phi_M >> \phi_{LP}$ and $\phi_M >> \phi_{LS}$, the ratio of the total voltage on the primary of a transformer to the total voltage on the secondary of a transformer is approximately

$$\frac{v_P(t)}{v_S(t)} = \frac{N_P}{N_S} = a \tag{2–28}$$

The smaller the leakage fluxes of the transformer are, the closer the total transformer voltage ratio approximates that of the ideal transformer discussed in Section 2.3.

The Magnetization Current in a Real Transformer

When an ac power source is connected to a transformer as shown in Figure 2–8, a current flows in its primary circuit, *even when the secondary circuit is open-circuited.* This current is the current required to produce flux in a real ferromagnetic core, as explained in Chapter 1. It consists of two components:

1. The *magnetization current* i_M, which is the current required to produce the flux in the transformer core, and
2. The *core-loss current* i_{h+e}, which is the current required to make up for hysteresis and eddy current losses in the core.

Figure 2–11 shows the magnetization curve of a typical transformer core. If the flux in the transformer core is known, then the magnitude of the magnetization current can be found directly from Figure 2–11.

Ignoring for the moment the effects of leakage flux, we see that the average flux in the core is given by

$$\bar{\phi}_P = \frac{1}{N_P} \int v_P(t)dt \tag{2–18}$$

If the primary voltage is given by the expression $v_P(t) = V_M \cos \omega t$ V, then the resulting flux must be

$$\begin{aligned}\bar{\phi}_P &= \frac{1}{N_P} \int V_M \cos \omega t \, dt \\ &= \frac{V_M}{\omega N_P} \sin \omega t \quad \text{Wb}\end{aligned} \tag{2–29}$$

If the values of current required to produce a given flux (Figure 2–11a) are compared to the flux in the core at different times, it is possible to construct a sketch of the magnetization current in the winding on the core. Such a sketch is shown in Figure 2–11b. Notice the following points about the magnetization current:

1. The magnetization current in the transformer is not sinusoidal. The higher-frequency components in the magnetization current are due to magnetic saturation in the transformer core.
2. Once the peak flux reaches the saturation point in the core, a small increase in peak flux requires a very large increase in the peak magnetization current.
3. The fundamental component of the magnetization current lags the voltage applied to the core by 90°.
4. The higher-frequency components in the magnetization current can be quite large compared to the fundamental component. In general, the further a transformer core is driven into saturation, the larger the harmonic components will become.

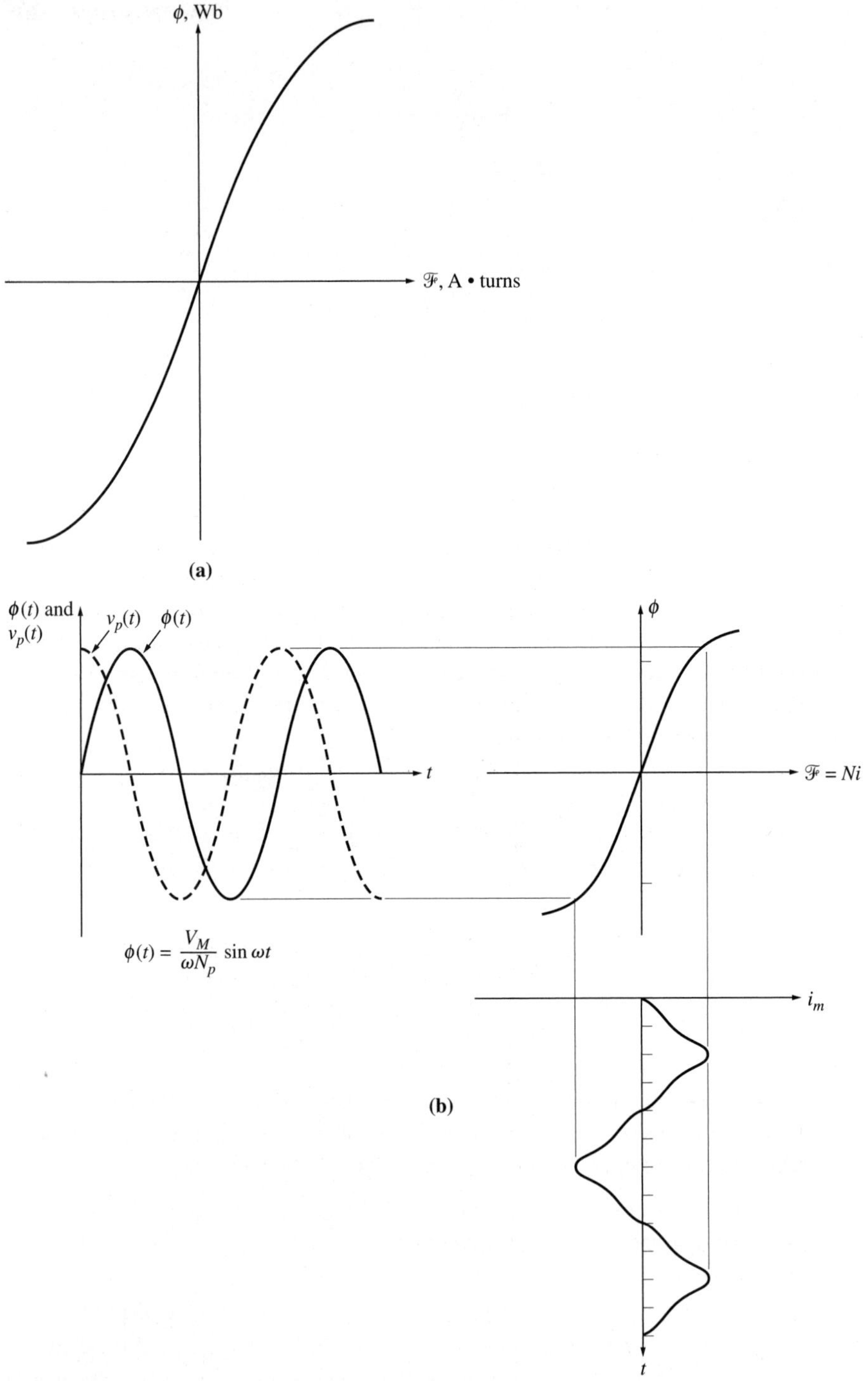

FIGURE 2–11
(a) The magnetization curve of the transformer core. (b) The magnetization current caused by the flux in the transformer core.

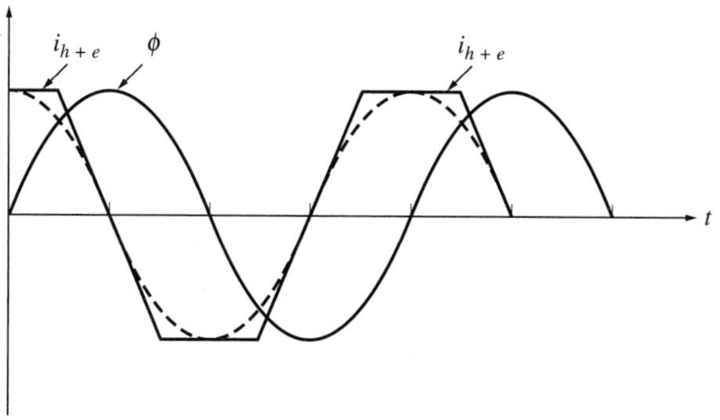

FIGURE 2–12
The core-loss current in a transformer.

The other component of the no-load current in the transformer is the current required to supply power to make up the hysteresis and eddy current losses in the core. This is the core-loss current. Assume that the flux in the core is sinusoidal. Since the eddy currents in the core are proportional to $d\phi/dt$, the eddy currents are largest when the flux in the core is passing through 0 Wb. Therefore, the core-loss current is greatest as the flux passes through zero. The total current required to make up for core losses is shown in Figure 2–12.

Notice the following points about the core-loss current:

1. The core-loss current is nonlinear because of the nonlinear effects of hysteresis.
2. The fundamental component of the core-loss current is in phase with the voltage applied to the core.

The total no-load current in the core is called the *excitation current* of the transformer. It is just the sum of the magnetization current and the core-loss current in the core:

$$i_{ex} = i_m + i_{h+e} \tag{2–30}$$

The total excitation current in a typical transformer core is shown in Figure 2–13. In a well-designed power transformer, the excitation current is much smaller than the full-load current of the transformer.

The Current Ratio on a Transformer and the Dot Convention

Now suppose that a load is connected to the secondary of the transformer. The resulting circuit is shown in Figure 2–14. Notice the dots on the windings of the transformer. As in the ideal transformer previously described, the dots help determine the polarity of the voltages and currents in the core without having to physically examine its windings. The physical significance of the dot convention is that *a current flowing into the dotted end of a winding produces a positive magnetomotive force* $\mathcal{F}$,

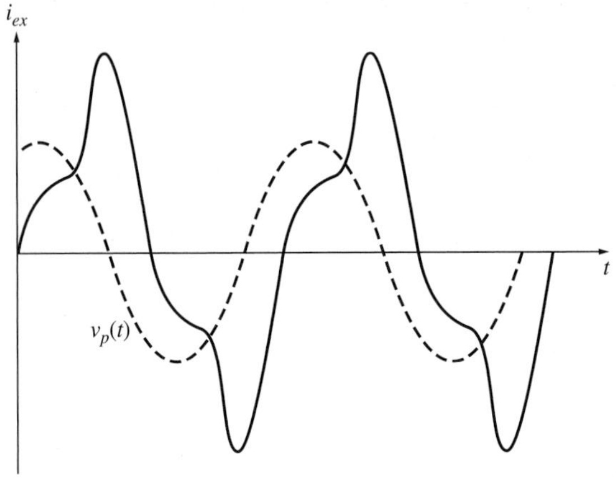

FIGURE 2–13
The total excitation current in a transformer.

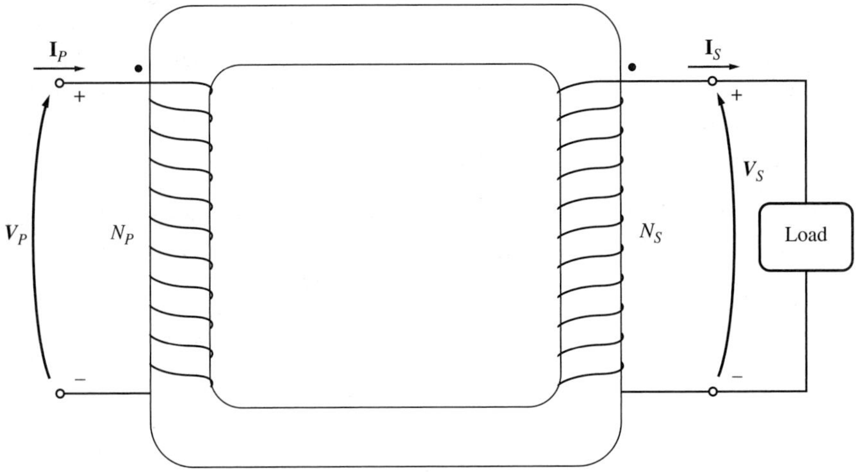

FIGURE 2–14
A real transformer with a load connected to its secondary.

while a current flowing into the undotted end of a winding produces a negative magnetomotive force. Therefore, two currents flowing into the dotted ends of their respective windings produce magnetomotive forces that add. If one current flows into a dotted end of a winding and one flows out of a dotted end, then the magnetomotive forces will subtract from each other.

In the situation shown in Figure 2–14, the primary current produces a positive magnetomotive force $\mathscr{F}_P = N_P i_P$, and the secondary current produces a negative magnetomotive force $\mathscr{F}_S = -N_S i_S$. Therefore, the net magnetomotive force on the core must be

$$\mathscr{F}_{\text{net}} = N_P i_P - N_S i_S \tag{2–31}$$

This net magnetomotive force must produce the net flux in the core, so the net magnetomotive force must be equal to

$$\boxed{\mathscr{F}_{\text{net}} = N_P i_P - N_S i_S = \phi \mathscr{R}} \tag{2–32}$$

where $\mathscr{R}$ is the reluctance of the transformer core. Because the reluctance of a well-designed transformer core is very small (nearly zero) until the core is saturated, the relationship between the primary and secondary currents is approximately

$$\mathscr{F}_{\text{net}} = N_P i_P - N_S i_S \approx 0 \tag{2–33}$$

as long as the core is unsaturated. Therefore,

$$\boxed{N_P i_P \approx N_S i_S} \tag{2–34}$$

or

$$\boxed{\frac{i_P}{i_S} \approx \frac{N_S}{N_P} = \frac{1}{a}} \tag{2–35}$$

It is the fact that the magnetomotive force in the core is nearly zero which gives the dot convention the meaning in Section 2.3. In order for the magnetomotive force to be nearly zero, current must flow *into one dotted end* and *out of the other dotted end.* The voltages must be built up in the same way with respect to the dots on each winding in order to drive the currents in the direction required. (The polarity of the voltages can also be determined by Lenz's law if the construction of the transformer coils is visible.)

What assumptions are required to convert a real transformer into the ideal transformer described previously? They are as follows:

1. The core must have no hysteresis or eddy currents.
2. The magnetization curve must have the shape shown in Figure 2–15. Notice that for an unsaturated core the net magnetomotive force $\mathscr{F}_{\text{net}} = 0$, implying that $N_P i_P = N_S i_S$.

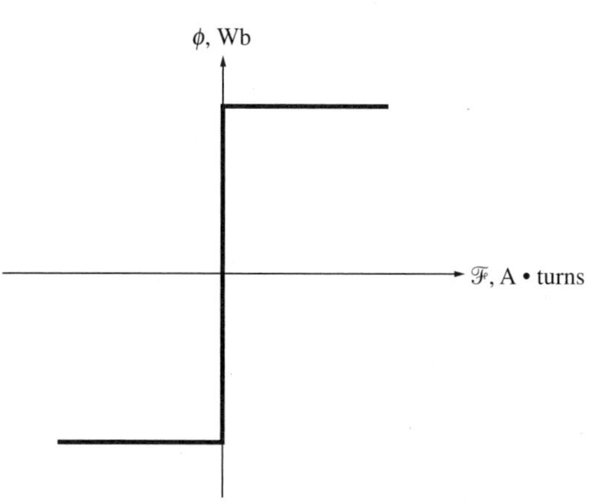

FIGURE 2–15
The magnetization curve of an ideal transformer.

3. The leakage flux in the core must be zero, implying that all the flux in the core couples both windings.
4. The resistance of the transformer windings must be zero.

While these conditions are never exactly met, well-designed power transformers can come quite close.

2.5 THE EQUIVALENT CIRCUIT OF A TRANSFORMER

The losses that occur in real transformers have to be accounted for in any accurate model of transformer behavior. The major items to be considered in the construction of such a model are

1. *Copper (I^2R) losses.* Copper losses are the resistive heating losses *in the primary and secondary windings* of the transformer. They are proportional to the square of the current in the windings.
2. *Eddy current losses.* Eddy current losses are resistive heating losses *in the core* of the transformer. They are proportional to the square of the voltage applied to the transformer.
3. *Hysteresis losses.* Hysteresis losses are associated with the rearrangement of the magnetic domains in the core during each half-cycle, as explained in Chapter 1. They are a complex, nonlinear function of the voltage applied to the transformer.
4. *Leakage flux.* The fluxes ϕ_{LP} and ϕ_{LS} which escape the core and pass through only one of the transformer windings are leakage fluxes. These escaped fluxes produce a *leakage inductance* in the primary and secondary coils, and the effects of this inductance must be accounted for.

The Exact Equivalent Circuit of a Real Transformer

It is possible to construct an equivalent circuit that takes into account all the major imperfections in real transformers. Each major imperfection is considered in turn, and its effect is included in the transformer model.

The easiest effect to model is the copper losses. Copper losses are resistive losses in the primary and secondary windings of the transformer core. They are modeled by placing a resistor R_P in the primary circuit of the transformer and a resistor R_S in the secondary circuit.

As explained in Section 2.4, the leakage flux in the primary windings ϕ_{LP} produces a voltage ϕ_{LP} given by

$$e_{LP}(t) = N_P \frac{d\phi_{LP}}{dt} \tag{2–36a}$$

and the leakage flux in the secondary windings ϕ_{LS} produces a voltage e_{LS} given by

$$e_{LS}(t) = N_S \frac{d\phi_{LS}}{dt} \tag{2–36b}$$

Since much of the leakage flux path is through air, and since air has a constant reluctance much higher than the core reluctance, the flux ϕ_{LP} is directly proportional to the primary circuit current i_P, and the flux ϕ_{LS} is directly proportional to the secondary current i_S:

$$\phi_{\mathrm{LP}} = (\mathcal{P}N_P)\,i_P \tag{2–37a}$$

$$\phi_{\mathrm{LS}} = (\mathcal{P}N_S)\,i_S \tag{2–37b}$$

where $\mathcal{P}$ = permeance of flux path

N_P = number of turns on primary coil

N_S = number of turns on secondary coil

Substitute Equations (2–37) into Equations (2–36). The result is

$$e_{\mathrm{LP}}(t) = N_P\frac{d}{dt}(\mathcal{P}N_P)i_P = N_P^2\mathcal{P}\frac{di_P}{dt} \tag{2–38a}$$

$$e_{\mathrm{LS}}(t) = N_S\frac{d}{dt}(\mathcal{P}N_S)i_S = N_S^2\mathcal{P}\frac{di_S}{dt} \tag{2–38b}$$

The constants in these equations can be lumped together. Then

$$\boxed{e_{\mathrm{LP}}(t) = L_P\frac{di_P}{dt}} \tag{2–39a}$$

$$\boxed{e_{\mathrm{LS}}(t) = L_S\frac{di_S}{dt}} \tag{2–39b}$$

where $L_P = N_P^2\mathcal{P}$ is the leakage inductance of the primary coil and $L_S = N_S^2\mathcal{P}$ is the leakage inductance of the secondary coil. Therefore, the leakage flux will be modeled by primary and secondary inductors.

How can the core excitation effects be modeled? The magnetization current i_m is a current proportional (in the unsaturated region) to the voltage applied to the core and *lagging the applied voltage by* 90°, so it can be modeled by a reactance X_M connected across the primary voltage source. The core-loss current i_{h+e} is a current proportional to the voltage applied to the core that is *in phase with the applied voltage,* so it can be modeled by a resistance R_C connected across the primary voltage source. (Remember that both these currents are really nonlinear, so the inductance X_M and the resistance R_C are, at best, approximations of the real excitation effects.)

The resulting equivalent circuit is shown in Figure 2–16. In this circuit, R_P is the resistance of the primary winding, X_P $(= \omega L_P)$ is the reactance due to the primary leakage inductance, R_S is the resistance of the secondary winding, and X_S $(= \omega L_s)$ is the reactance due to the secondary leakage inductance. The excitation branch is modeled by the resistance R_C (hysteresis and core losses) in parallel with the reactance X_M (the magnetization current).

Notice that the elements forming the excitation branch are placed inside the primary resistance R_P and reactance X_P. This is because the voltage actually applied to the core is really the input voltage less the internal voltage drops of the winding.

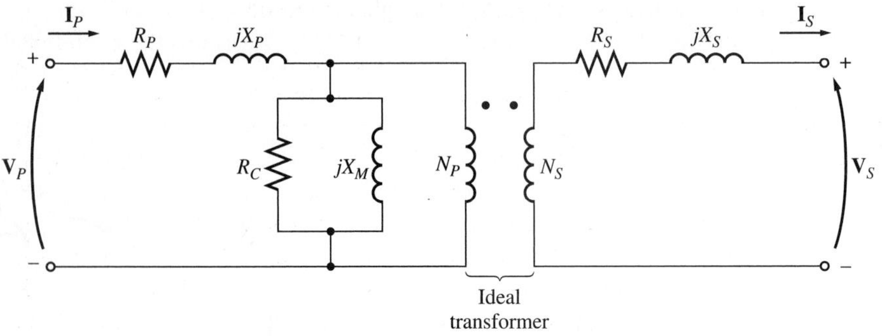

FIGURE 2–16
The model of a real transformer.

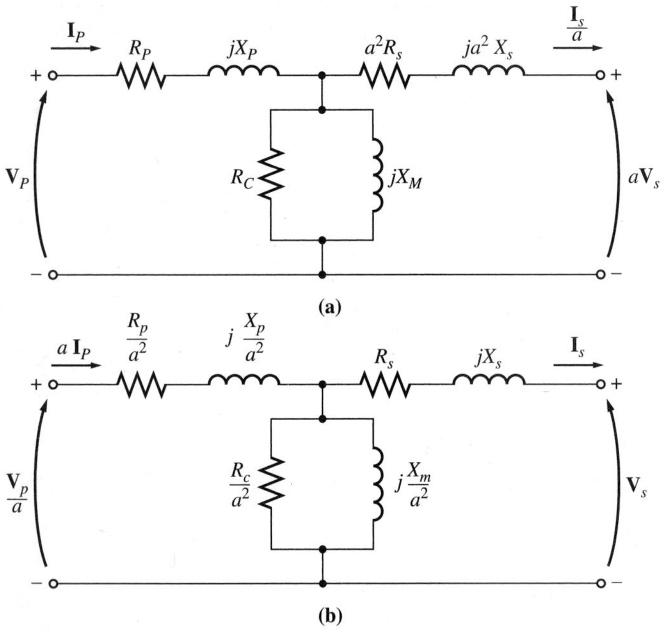

FIGURE 2–17
(a) The transformer model referred to its primary voltage level. (b) The transformer model referred to its secondary voltage level.

Although Figure 2–16 is an accurate model of a transformer, it is not a very useful one. To analyze practical circuits containing transformers, it is normally necessary to convert the entire circuit to an equivalent circuit at a single voltage level. (Such a conversion was done in Example 2–1.) Therefore, the equivalent circuit must be referred either to its primary side or to its secondary side in problem solutions. Figure 2–17a is the equivalent circuit of the transformer referred to its primary side, and Figure 2–17b is the equivalent circuit referred to its secondary side.

Approximate Equivalent Circuits of a Transformer

The transformer models shown before are often more complex than necessary in order to get good results in practical engineering applications. One of the principal complaints about them is that the excitation branch of the model adds another node to the circuit being analyzed, making the circuit solution more complex than necessary. The excitation branch has a very small current compared to the load current of the transformers. In fact, the excitation current is only about 2–3% of the full load current for typical power transformers. Because this is true, a simplified equivalent circuit can be produced that works almost as well as the original model. The excitation branch is simply moved to the front of the transformer, and the primary and secondary impedances are left in series with each other. These impedances are just added, creating the approximate equivalent circuits in Figure 2–18a and b.

In some applications, the excitation branch may be neglected entirely without causing serious error. In these cases, the equivalent circuit of the transformer reduces to the simple circuits in Figure 2–18c and d.

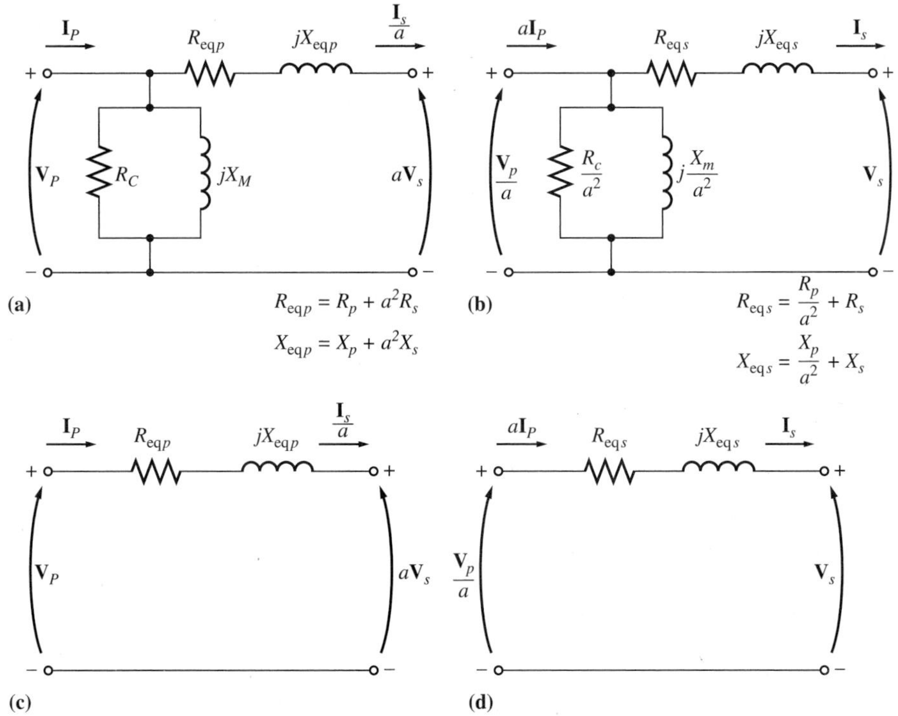

FIGURE 2–18
Approximate transformer models. (a) Referred to the primary side; (b) referred to the secondary side; (c) with no excitation branch, referred to the primary side; (d) with no excitation branch, referred to the secondary side.

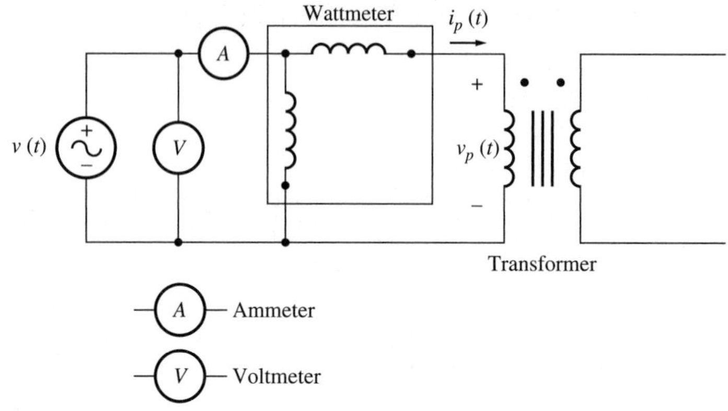

FIGURE 2–19
Connection for transformer open-circuit test.

Determining the Values of Components in the Transformer Model

It is possible to experimentally determine the values of the inductances and resistances in the transformer model. An adequate approximation of these values can be obtained with only two tests, the open-circuit test and the short-circuit test.

In the *open-circuit test*, one transformer winding is open-circuited, and the other winding is connected to full rated line voltage. Look at the equivalent circuit in Figure 2–17. Under the conditions described, all the input current must be flowing through the excitation branch of the transformer. The series elements, R_P and X_P are too small in comparison to R_C and X_M to cause a significant voltage drop, so essentially all the input voltage is dropped across the excitation branch.

The open-circuit test connections are shown in Figure 2–19. Full line voltage is applied to one side of the transformer, and the input voltage, input current, and input power to the transformer are measured. (This measurement is normally done on the *low-voltage* side of the transformer, since lower voltages are easier to work with.) From this information, it is possible to determine the power factor of the input current and therefore both the *magnitude* and the *angle* of the excitation impedance.

The easiest way to calculate the values of R_C and X_M is to look first at the *admittance* of the excitation branch. The conductance of the core-loss resistor is given by

$$G_C = \frac{1}{R_C} \tag{2–40}$$

and the susceptance of the magnetizing inductor is given by

$$B_M = \frac{1}{X_M} \tag{2–41}$$

Since these two elements are in parallel, their admittances add, and the total excitation admittance is

$$Y_E = G_C - jB_M \tag{2–42}$$

$$Y_E = \frac{1}{R_C} - j\frac{1}{X_M} \tag{2–43}$$

The *magnitude* of the excitation admittance (referred to the side of the transformer used for the measurement) can be found from the open-circuit test voltage and current:

$$|Y_E| = \frac{I_{OC}}{V_{OC}} \tag{2–44}$$

The *angle* of the admittance can be found from a knowledge of the circuit power factor. The open-circuit power factor (PF) is given by

$$\text{PF} = \cos\theta = \frac{P_{OC}}{V_{OC}I_{OC}} \tag{2–45}$$

and the power-factor angle θ is given by

$$\theta = \cos^{-1}\frac{P_{OC}}{V_{OC}I_{OC}} \tag{2–46}$$

The power factor is always lagging for a real transformer, so the angle of the current always lags the angle of the voltage by θ degrees. Therefore, the admittance Y_E is

$$Y_E = \frac{I_{OC}}{V_{OC}}\angle-\theta$$

$$Y_E = \frac{I_{OC}}{V_{OC}}\angle-\cos^{-1}\text{PF} \tag{2–47}$$

By comparing Equations (2–43) and (2–47), it is possible to determine the values of R_C and X_M referred to the low-voltage side directly from the open-circuit test data.

In the *short-circuit test*, the low-voltage terminals of the transformer are short-circuited, and the high-voltage terminals are connected to a variable voltage source, as shown in Figure 2–20. (This measurement is normally done on the *high-voltage* side of the transformer, since currents will be lower on that side, and lower currents are easier to work with.) The input voltage is adjusted until the current in the short-circuited windings is equal to its rated value. (Be sure to keep the primary voltage at a safe level. It would not be a good idea to burn out the transformer's windings while trying to test it.) The input voltage, current, and power are again measured.

Since the input voltage is so low during the short-circuit test, negligible current flows through the excitation branch. If the excitation current is ignored, then all the voltage drop in the transformer can be attributed to the series elements in the circuit. The magnitude of the series impedances referred to the primary side of the transformer is

$$|Z_{SE}| = \frac{V_{SC}}{I_{SC}} \tag{2–48}$$

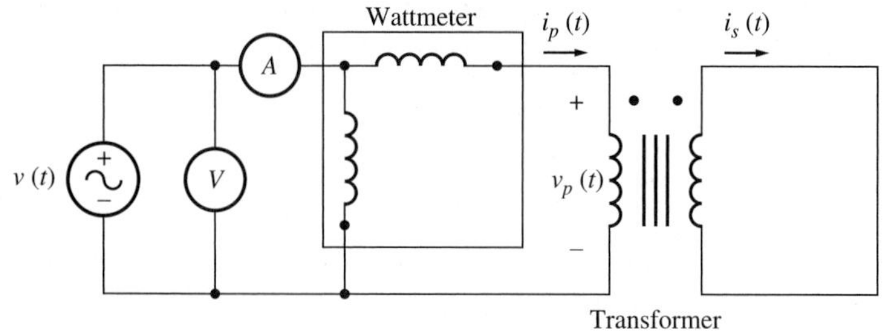

FIGURE 2–20
Connection for transformer short-circuit test.

The power factor of the current is given by

$$\text{PF} = \cos\theta = \frac{P_{\text{SC}}}{V_{\text{SC}} I_{\text{SC}}} \tag{2–49}$$

and is lagging. The current angle is thus negative, and the overall impedance angle θ is positive:

$$\theta = \cos^{-1}\frac{P_{\text{SC}}}{V_{\text{SC}} I_{\text{SC}}} \tag{2–50}$$

Therefore,

$$Z_{\text{SE}} = \frac{V_{\text{SC}}\angle 0°}{I_{\text{SC}}\angle -\theta°} = \frac{V_{\text{SC}}}{I_{\text{SC}}}\angle\theta° \tag{2–51}$$

The series impedance Z_{SE} is equal to

$$Z_{\text{SE}} = R_{\text{eq}} + jX_{\text{eq}}$$

$$Z_{\text{SE}} = (R_P + a^2R_S) + j(X_P + a^2X_S) \tag{2–52}$$

It is possible to determine the total series impedance referred to the high-voltage side by using this technique, but there is no easy way to split the series impedance into primary and secondary components. Fortunately, such separation is not necessary to solve normal problems.

Note that the open-circuit test is usually performed on the low-voltage side of the transformer, and the short-circuit test is usually performed on the high-voltage side of the transformer, so R_C and X_M are usually found referred to the low-voltage side, and R_{eq} and X_{eq} are usually found referred to the high-voltage side. All of the elements must be referred to the same side (either high or low) to create the final equivalent circuit.

Example 2–2. The equivalent circuit impedances of a 20-kVA, 8000/240 V, 60-Hz transformer are to be determined. The open-circuit test was performed on the secondary side of the transformer (to reduce the maximum voltage to be measured) and the short-circuit test were performed on the primary side of the transformer (to reduce the maximum current to be measured). The following data were taken:

Open-circuit test (on secondary)	**Short-circuit test (on primary)**
$V_{OC} = 240$ V	$V_{SC} = 489$ V
$I_{OC} = 7.133$ A	$I_{SC} = 2.5$ A
$V_{OC} = 400$ W	$P_{SC} = 240$ W

Find the impedances of the approximate equivalent circuit referred to the primary side, and sketch that circuit.

Solution

The turns ratio of this transformer is $a = 8000/240 = 33.3333$. The power factor during the *open-circuit* test is

$$\text{PF} = \cos\theta = \frac{P_{OC}}{V_{OC}I_{OC}} \tag{2–45}$$

$$\text{PF} = \cos\theta = \frac{400 \text{ W}}{(240 \text{ V})(7.133 \text{ A})}$$

$$\text{PF} = 0.234 \text{ lagging}$$

The excitation admittance is given by

$$Y_E = \frac{I_{OC}}{V_{OC}} \angle -\cos^{-1} \text{PF} \tag{2–47}$$

$$Y_E = \frac{7.133 \text{ A}}{240 \text{ V}} \angle -\cos^{-1} 0.234$$

$$Y_E = 0.0297 \angle -76.5° \text{ S}$$

$$Y_E = 0.00693 - j\,0.02888 = \frac{1}{R_C} - j\frac{1}{X_M}$$

Therefore, the values of the excitation branch *referred to the low-voltage (secondary) side* are

$$R_C = \frac{1}{0.00693} = 144\ \Omega$$

$$X_M = \frac{1}{0.02888} = 34.63\ \Omega$$

The power factor during the *short-circuit* test is

$$\text{PF} = \cos\theta = \frac{P_{SC}}{V_{SC}I_{SC}} \tag{2–49}$$

$$\text{PF} = \cos\theta = \frac{240 \text{ W}}{(489 \text{ V})(2.5 \text{ A})} = 0.196 \text{ lagging}$$

The series impedance is given by

$$Z_{SE} = \frac{V_{SC}}{I_{SC}} \angle \cos^{-1} \text{PF}$$

$$Z_{SE} = \frac{489 \text{ V}}{2.5 \text{ A}} \angle 78.7°$$

$$Z_{SE} = 195.6 \angle 78.7° = 38.4 + j192\ \Omega$$

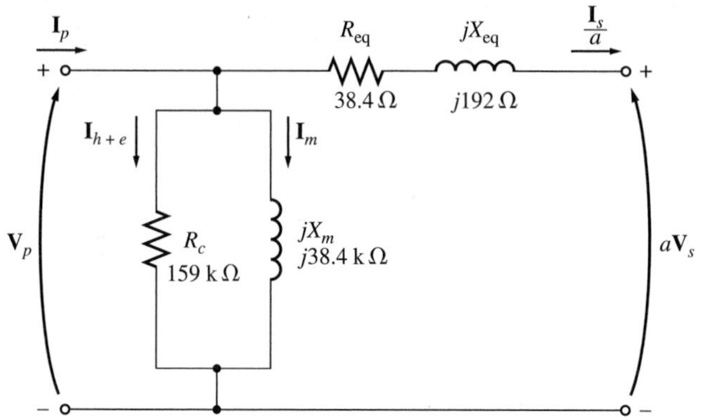

FIGURE 2–21
The equivalent circuit of Example 2–2.

Therefore, the equivalent resistance and reactance *referred to the high-voltage (primary) side* are

$$R_{eq} = 38.4\ \Omega \qquad X_{eq} = 192\ \Omega$$

The resulting simplified equivalent circuit referred to the high-voltage (primary) side can be found by converting the excitation branch values to the high-voltage side.

$$R_{C,P} = a^2 R_{C,S} = (33.333)^2\,(144\ \Omega) = 159\ \text{k}\Omega$$

$$X_{M,P} = a^2 X_{M,S} = (33.333)^2\,(34.63\ \Omega) = 38.4\ \text{k}\Omega$$

The resulting equivalent circuit is shown in Figure 2–21.

2.6 THE PER-UNIT SYSTEM OF MEASUREMENTS

As the relatively simple Example 2–1 showed, solving circuits containing transformers can be quite a tedious operation because of the need to refer all the different voltage levels on different sides of the transformers in the system to a common level. Only after this step has been taken can the system be solved for its voltages and currents.

There is another approach to solving circuits containing transformers which eliminates the need for explicit voltage-level conversions at every transformer in the system. Instead, the required conversions are handled automatically by the method itself, without ever requiring the user to worry about impedance transformations. Because such impedance transformations can be avoided, circuits containing many transformers can be solved easily with less chance of error. This method of calculation is known as the *per-unit (pu) system* of measurements.

There is yet another advantage to the per-unit system that is quite significant for electric machinery and transformers. As the size of a machine or transformer varies, its internal impedances vary widely. Thus, a primary circuit reactance of 0.1 Ω might be an atrociously high number for one transformer and a ridiculously low number for another—it all depends on the device's voltage and power ratings.

However, it turns out that in a per-unit system related to the device's ratings, *machine and transformer impedances fall within fairly narrow ranges* for each type and construction of device. This fact can serve as a useful check in problem solutions.

In the per-unit system, the voltages, currents, powers, impedances, and other electrical quantities are not measured in their usual SI units (volts, amperes, watts, ohms, etc.). Instead, *each electrical quantity is measured as a decimal fraction* of some base level. Any quantity can be expressed on a per-unit basis by the equation

$$\text{Quantity per unit} = \frac{\text{Actual value}}{\text{base value of quantity}} \tag{2–53}$$

where "actual value" is a value in volts, amperes, ohms, etc.

It is customary to select two base quantities to define a given per-unit system. The ones usually selected are voltage and power (or apparent power). Once these base quantities have been selected, all the other base values are related to them by the usual electrical laws. In a single-phase system, these relationships are

$$P_{\text{base}}, Q_{\text{base}}, \text{ or } S_{\text{base}} = V_{\text{base}} I_{\text{base}} \tag{2–54}$$

$$R_{\text{base}}, X_{\text{base}}, \text{ or } Z_{\text{base}} = \frac{V_{\text{base}}}{I_{\text{base}}} \tag{2–55}$$

$$Y_{\text{base}} = \frac{I_{\text{base}}}{V_{\text{base}}} \tag{2–56}$$

and

$$Z_{\text{base}} = \frac{(V_{\text{base}})^2}{S_{\text{base}}} \tag{2–57}$$

Once the base values of S (or P) and V have been selected, all other base values can be computed easily from Equations (2–54) to (2–57).

In a power system, a base apparent power and voltage are selected *at a specific point in the system*. A transformer has no effect on the base apparent power of the system, since the apparent power into a transformer equals the apparent power out of the transformer [Equation (2–11)]. On the other hand, voltage changes when it goes through a transformer, so the value of V_{base} changes at every transformer in the system according to its turns ratio. Because the *base quantities* change in passing through a transformer, the process of referring quantities to a common voltage level is automatically taken care of during per-unit conversion.

Example 2–3. A simple power system is shown in Figure 2–22. This system contains a 480-V generator connected to an ideal 1:10 step-up transformer, a transmission line, an ideal 20:1 step-down transformer, and a load. The impedance of the transmission line is $20 + j60\ \Omega$, and the impedance of the load is $10\angle 30°\Omega$. The base values for this system are chosen to be 480 V and 10 kVA at the generator.

(a) Find the base voltage, current, impedance, and apparent power at every point in the power system.
(b) Convert this system to its per-unit equivalent circuit.
(c) Find the power supplied to the load in this system.
(d) Find the power lost in the transmission line.

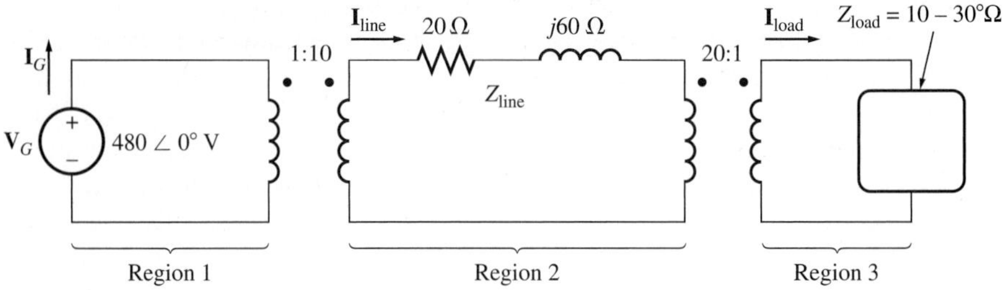

FIGURE 2–22
The power system of Example 2–3.

Solution

(a) In the generator region, $V_{base} = 480$ V and $S_{base} = 10$ kVA, so

$$I_{base\ 1} = \frac{S_{base}}{V_{base\ 1}} = \frac{10{,}000\text{ VA}}{480\text{ V}} = 20.83\text{ A}$$

$$Z_{base\ 1} = \frac{V_{base\ 1}}{I_{base\ 1}} = \frac{480\text{ V}}{20.83\text{ A}} = 23.04\ \Omega$$

The turns ratio of transformer T_1 is $a = 1/10 = 0.1$, so the base voltage *in the transmission line region* is

$$V_{base\ 2} = \frac{V_{base\ 1}}{a} = \frac{480\text{ V}}{0.1} = 4800\text{ V}$$

The other base quantities are

$$S_{base\ 2} = 10\text{ kVA}$$

$$I_{base\ 2} = \frac{10{,}000\text{ VA}}{4800\text{ V}} = 2.083\text{ A}$$

$$Z_{base\ 2} = \frac{4800\text{ V}}{2.083\text{ A}} = 2304\ \Omega$$

The turns ratio of transformer T_2 is $a = 20/1 = 20$, so the base voltage *in the load region* is

$$V_{base\ 3} = \frac{V_{base\ 2}}{a} = \frac{4800\text{ V}}{20} = 240\text{ V}$$

The other base quantities are

$$S_{base\ 3} = 10\text{ kVA}$$

$$I_{base\ 3} = \frac{10{,}000\text{ VA}}{240\text{ V}} = 41.67\text{ A}$$

$$Z_{base\ 3} = \frac{240\text{ V}}{41.67\text{ A}} = 5.76\ \Omega$$

(b) To convert a power system to a per-unit system, each component must be divided by its base value in its region of the system. The *generator's* per-unit voltage is its actual value divided by its base value:

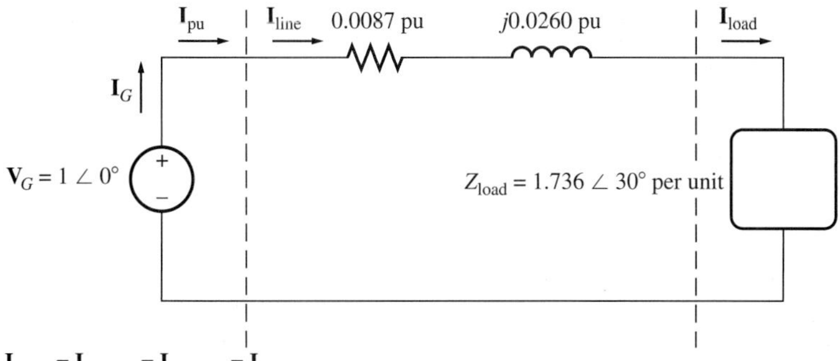

FIGURE 2–23
The per-unit equivalent circuit for Example 2–3.

$$V_{G,pu} = \frac{480 \angle 0° \text{ V}}{480 \text{ V}} = 1.0 \angle 0° \text{ pu}$$

The *transmission line's* per-unit impedance is its actual value divided by its base value:

$$Z_{line,pu} = \frac{20 + j60 \ \Omega}{2304 \ \Omega} = 0.0087 + j0.0260 \text{ pu}$$

The *load's* per-unit impedance is also given by actual value divided by base value:

$$Z_{load,pu} = \frac{10 \angle 30° \ \Omega}{5.76 \ \Omega} = 1.736 \angle 30° \text{ pu}$$

The per-unit equivalent circuit of the power system is shown in Figure 2–23.

(c) The current flowing in this per-unit power system is

$$\begin{aligned}
\mathbf{I}_{pu} &= \frac{\mathbf{V}_{pu}}{Z_{tot,pu}} \\
&= \frac{1 \angle 0°}{(0.0087 + j0.0260) + (1.736 \angle 30°)} \\
&= \frac{1 \angle 0°}{(0.0087 + j0.0260) + (1.503 + j0.868)} \\
&= \frac{1 \angle 0°}{1.512 + j0.894} = \frac{1 \angle 0°}{1.757 \angle 30.6°} \\
&= 0.569 \angle -30.6° \text{ pu}
\end{aligned}$$

Therefore, the per-unit power of the load is

$$P_{load,pu} = I_{pu}^2 R_{pu} = (0.569)^2(1.503) = 0.487$$

and the actual power supplied to the load is

$$\begin{aligned}
P_{load} &= P_{load,pu} S_{base} = (0.487)(10{,}000 \text{ VA}) \\
&= 4870 \text{ W}
\end{aligned}$$

(a)

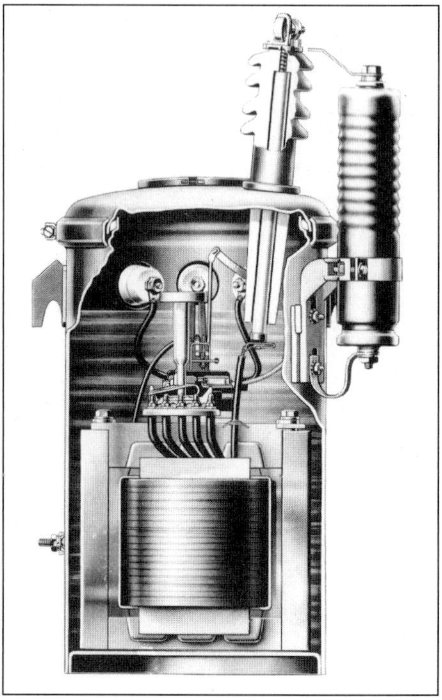

(b)

FIGURE 2–24
(a) A typical 13.2–kV to 120/240-V distribution transformer. (*Courtesy of General Electric Company.*) (b) A cutaway view of the distribution transformer showing the shell-form transformer inside it. (*Courtesy of General Electric Company.*)

(*d*) The per-unit power lost in the transmission line is

$$P_{\text{line,pu}} = I_{\text{pu}}^2 R_{\text{line,pu}} = (0.569)^2(0.0087) = 0.00282$$

and the actual power lost in the transmission line is

$$P_{\text{line}} = P_{\text{line,pu}} S_{\text{base}} = (0.00282)(10{,}000 \text{ VA})$$
$$= 28.2 \text{ W}$$

When only one device (transformer or motor) is being analyzed, its own ratings are usually used as the base for the per-unit system. If a per-unit system based on the transformer's own ratings is used, a power or distribution transformer's characteristics will not vary much over a wide range of voltage and power ratings. For example, the series resistance of a transformer is usually about 0.01 per unit, and the series reactance is usually between 0.02 and 0.10 per unit. In general, the larger the transformer, the smaller the series impedances. The magnetizing reactance is usually between about 10 and 40 per unit, while the core-loss resistance is usually between about 50 and 200 per unit. Because per-unit values provide a convenient and meaningful way to compare transformer characteristics when they are of different sizes, transformer impedances are normally given in per-unit or as a percentage on the transformer's nameplate (see Figure 2–45, later in this chapter).

The same idea applies to synchronous and induction machines as well: Their per-unit impedances fall within relatively narrow ranges over quite large size ranges.

If more than one machine and one transformer are included in a single power system, the system base voltage and power may be chosen arbitrarily, but the *entire system must have the same base.* One common procedure is to choose the system base quantities to be equal to the base of the largest component in the system. Per-unit values given to another base can be converted to the new base by converting them to their actual values (volts, amperes, ohms, etc.) as an in-between step. Alternatively, they can be converted directly by the equations

$$(P, Q, S)_{\text{pu on base 2}} = (P, Q, S)_{\text{pu on base 1}} \frac{S_{\text{base 1}}}{S_{\text{base 2}}} \tag{2–58}$$

$$V_{\text{pu on base 2}} = V_{\text{pu on base 1}} \frac{V_{\text{base 1}}}{V_{\text{base 2}}} \tag{2–59}$$

$$(R, X, Z)_{\text{pu on base 2}} = (R, X, Z)_{\text{pu on base 1}} \frac{(V_{\text{base 1}})^2(S_{\text{base 2}})}{(V_{\text{base 2}})^2(S_{\text{base 1}})} \tag{2–60}$$

Example 2–4. Sketch the approximate per-unit equivalent circuit for the transformer in Example 2–2. Use the transformer's ratings as the system base.

Solution

The transformer in Example 2–2 is rated at 20 kVA, 8000/240 V. The approximate equivalent circuit (Figure 2–21) developed in the example was referred to the high-voltage side of the transformer, so to convert it to per-unit, the primary circuit base impedance must be found. On the primary,

$$V_{\text{base 1}} = 8000 \text{ V}$$

$$S_{\text{base 1}} = 20{,}000 \text{ VA}$$

$$Z_{\text{base 1}} = \frac{(V_{\text{base 1}})^2}{S_{\text{base 1}}} = \frac{(8000 \text{ V})^2}{20{,}000 \text{ VA}} = 3200 \ \Omega$$

Therefore,

$$Z_{SE,\text{pu}} = \frac{38.4 + j192 \ \Omega}{3200 \ \Omega} = 0.012 + j0.06 \text{ pu}$$

$$R_{C,\text{pu}} = \frac{159 \text{ k}\Omega}{3200 \ \Omega} = 49.7 \text{ pu}$$

$$Z_{M,\text{pu}} = \frac{38.4 \text{ k}\Omega}{3200 \ \Omega} = 12 \text{ pu}$$

The per-unit approximate equivalent circuit, expressed to the transformer's own base, is shown in Figure 2–25.

2.7 TRANSFORMER VOLTAGE REGULATION AND EFFICIENCY

Because a real transformer has series impedances within it, the output voltage of a transformer varies with the load even if the input voltage remains constant. To

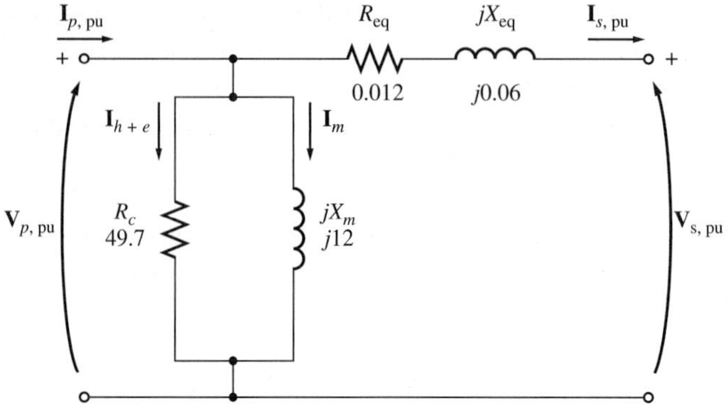

FIGURE 2–25
The per-unit equivalent circuit of Example 2–4.

conveniently compare transformers in this respect, it is customary to define a quantity called *voltage regulation* (VR). *Full-load voltage regulation* is a quantity that compares the output voltage of the transformer at no load with the output voltage at full load. It is defined by the equation

$$\boxed{\mathrm{VR} = \frac{V_{S,\mathrm{nl}} - V_{S,\mathrm{fl}}}{V_{S,\mathrm{fl}}} \times 100\%} \tag{2–61}$$

Since at no load, $V_S = V_P / a$, the voltage regulation can also be expressed as

$$\boxed{\mathrm{VR} = \frac{V_P / a - V_{S,\mathrm{fl}}}{V_{S,\mathrm{fl}}} \times 100\%} \tag{2–62}$$

If the transformer equivalent circuit is in the per-unit system, then voltage regulation can be expressed as

$$\boxed{\mathrm{VR} = \frac{V_{P,\mathrm{pu}} - V_{S,\mathrm{fl,pu}}}{V_{S,\mathrm{fl,pu}}} \times 100\%} \tag{2–63}$$

Usually it is a good practice to have as small a voltage regulation as possible. For an ideal transformer, VR = 0 percent. It is not always a good idea to have a low-voltage regulation, though—sometimes high-impedance and high-voltage regulation transformers are deliberately used to reduce the fault currents in a circuit.

How can the voltage regulation of a transformer be determined?

The Transformer Phasor Diagram

To determine the voltage regulation of a transformer, it is necessary to understand the voltage drops within it. Consider the simplified transformer equivalent circuit in Figure 2–18b. The effects of the excitation branch on transformer voltage regulation can be ignored, so only the series impedances need be considered. The voltage regulation of a transformer depends both on the magnitude of these series impedances and on the phase angle of the current flowing through the transformer. The easiest way to determine the effect of the impedances and the current phase angles on the transformer voltage regulation is to examine a *phasor diagram*, a sketch of the phasor voltages and currents in the transformer.

In all the following phasor diagrams, the phasor voltage $\mathbf{V}_S$ is assumed to be at an angle of 0°, and all other voltages and currents are compared to that reference. By applying Kirchhoff's voltage law to the equivalent circuit in Figure 2–18b, the primary voltage can be found as

$$\frac{\mathbf{V}_P}{a} = \mathbf{V}_S + R_{\text{eq}}\mathbf{I}_S + jX_{\text{eq}}\mathbf{I}_S \tag{2–64}$$

A transformer phasor diagram is just a visual representation of this equation.

Figure 2–26 shows a phasor diagram of a transformer operating at a lagging power factor. It is easy to see that $V_P/a > V_S$ for lagging loads, so the voltage regulation of a transformer with lagging loads must be greater than zero.

A phasor diagram at unity power factor is shown in Figure 2–27a. Here again, the voltage at the secondary is lower than the voltage at the primary, so VR $>$ 0.

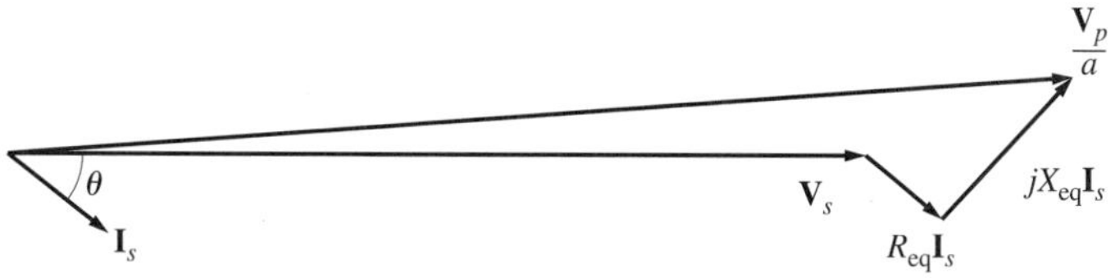

FIGURE 2–26
Phasor diagram of a transformer operating at a lagging power factor.

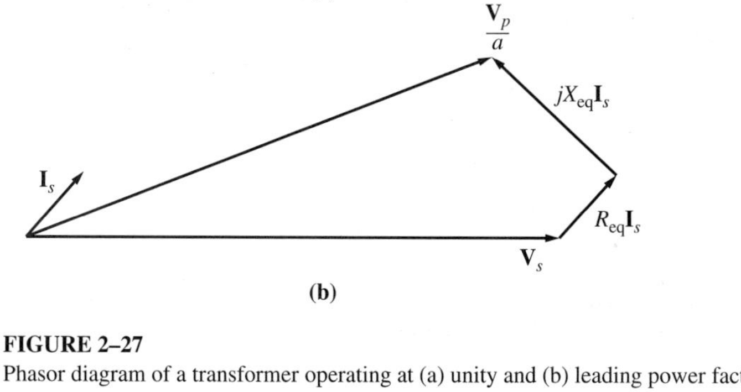

FIGURE 2–27
Phasor diagram of a transformer operating at (a) unity and (b) leading power factor.

However, this time the voltage regulation is a smaller number than it was with a lagging current. If the secondary current is leading, the secondary voltage can actually be *higher* than the referred primary voltage. If this happens, the transformer actually has a *negative* voltage regulation (see Figure 2–27b).

Transformer Efficiency

Transformers are also compared and judged on their efficiencies. The efficiency of a device is defined by the equation

$$\boxed{\eta = \frac{P_{\text{out}}}{P_{\text{in}}} \times 100\%} \tag{2–65}$$

$$\boxed{\eta = \frac{P_{\text{out}}}{P_{\text{out}} + P_{\text{loss}}} \times 100\%} \tag{2–66}$$

These equations apply to motors and generators as well as to transformers.

The transformer equivalent circuits make efficiency calculations easy. There are three types of losses present in transformers:

1. *Copper (I^2R) losses.* These losses are accounted for by the series resistance in the equivalent circuit.
2. *Hysteresis losses.* These losses were explained in Chapter 1. These losses are included in resistor R_C.
3. *Eddy current losses.* These losses were explained in Chapter 1. These losses are included in resistor R_C.

To calculate the efficiency of a transformer at a given load, just add the losses from each resistor and apply Equation (2–67). Since the output power is given by

$$P_{\text{out}} = V_S I_S \cos \theta_S \tag{2–7}$$

the efficiency of the transformer can be expressed by

$$\boxed{\eta = \frac{V_S I_S \cos \theta}{P_{\text{Cu}} + P_{\text{core}} + V_S I_S \cos \theta} \times 100\%} \tag{2–67}$$

Example 2–5. A 15-kVA, 2300/230-V transformer is to be tested to determine its excitation branch components, its series impedances, and its voltage regulation. The following test data have been taken from the transformer:

Open-circuit test (low voltage side)	**Short-circuit test (high voltage side)**
$V_{OC} = 230$ V	$V_{SC} = 47$ V
$I_{OC} = 2.1$ A	$I_{SC} = 6.0$ A
$V_{OC} = 50$ W	$P_{SC} = 160$ W

The data have been taken by using the connections shown in Figures 2–19 and 2–20.

(a) Find the equivalent circuit of this transformer referred to the high-voltage side.
(b) Find the equivalent circuit of this transformer referred to the low-voltage side.
(c) Calculate the full-load voltage regulation at 0.8 lagging power factor, 1.0 power factor, and at 0.8 leading power factor using the exact equation for $\mathbf{V}_P$.
(d) Plot the voltage regulation as load is increased from no load to full load at power factors of 0.8 lagging, 1.0, and 0.8 leading.
(e) What is the efficiency of the transformer at full load with a power factor of 0.8 lagging?

Solution

(a) The turns ratio of this transformer is $a = 2300/230 = 10$. The excitation branch values of the transformer equivalent circuit referred to the secondary (low voltage) side can be calculated from the *open-circuit test* data, and the series elements referred to the primary (high voltage) side can be calculated from the *short-circuit test* data. From the open-circuit test data, the open-circuit impedance angle is

$$\theta_{\mathrm{OC}} = \cos^{-1}\frac{P_{\mathrm{OC}}}{V_{\mathrm{OC}}I_{\mathrm{OC}}}$$

$$\theta_{\mathrm{OC}} = \cos^{-1}\frac{50 \text{ W}}{(230 \text{ V})(2.1 \text{ A})} = 84^\circ$$

The excitation admittance is thus

$$Y_E = \frac{I_{\mathrm{OC}}}{V_{\mathrm{OC}}} \angle -84^\circ$$

$$Y_E = \frac{2.1 \text{ A}}{230 \text{ V}} \angle -84^\circ \text{ S}$$

$$Y_E = 0.00913 \angle -84^\circ \text{ S} = 0.000954 - j0.00908 \text{ S}$$

The elements of the excitation branch referred to the secondary are

$$R_{C,S} = \frac{1}{0.000954} = 1050 \ \Omega$$

$$X_{M,S} = \frac{1}{0.00908} = 110 \ \Omega$$

From the short-circuit test data, the short-circuit impedance angle is

$$\theta_{\mathrm{SC}} = \cos^{-1}\frac{P_{\mathrm{SC}}}{V_{\mathrm{SC}}I_{\mathrm{SC}}}$$

$$\theta_{\mathrm{SC}} = \cos^{-1}\frac{160 \text{ W}}{(47 \text{ V})(6 \text{ A})} = 55.4^\circ$$

The equivalent series impedance is thus

$$Z_{\mathrm{SE}} = \frac{V_{\mathrm{SC}}}{I_{\mathrm{SC}}} \angle \theta_{\mathrm{SC}}$$

$$Z_{\mathrm{SE}} = \frac{47 \text{ V}}{6 \text{ A}} \angle 55.4^\circ \ \Omega$$

$$Z_{\mathrm{SE}} = 7.833 \angle 55.4^\circ = 4.45 + j6.45 \ \Omega$$

The series elements referred to the primary side are

$$R_{\text{eq,P}} = 4.45\ \Omega \qquad X_{\text{eq,P}} = 6.45\ \Omega$$

The resulting simplified equivalent circuit referred to the primary side can be found by converting the excitation branch values to the primary side.

$$R_{C,P} = a^2 R_{C,S} = (10)^2\,(1050\ \Omega) = 105\ \text{k}\Omega$$

$$X_{M,P} = a^2 X_{M,S} = (10)^2\,(110\ \Omega) = 11\ \text{k}\Omega$$

This equivalent circuit is shown in Figure 2–28*a*.

(b) To find the equivalent circuit referred to the low-voltage side, it is simply necessary to divide the impedance by a^2. Since $a = N_P/N_S = 10$, the resulting values are

$$R_C = 1050\ \Omega \qquad R_{\text{eq}} = 0.0445\ \Omega$$

$$X_M = 110\ \Omega \qquad X_{\text{eq}} = 0.0645\ \Omega$$

The resulting equivalent circuit is shown in Figure 2–28b.

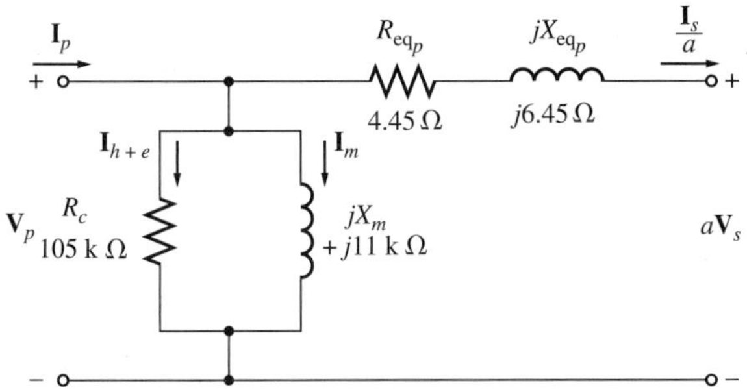

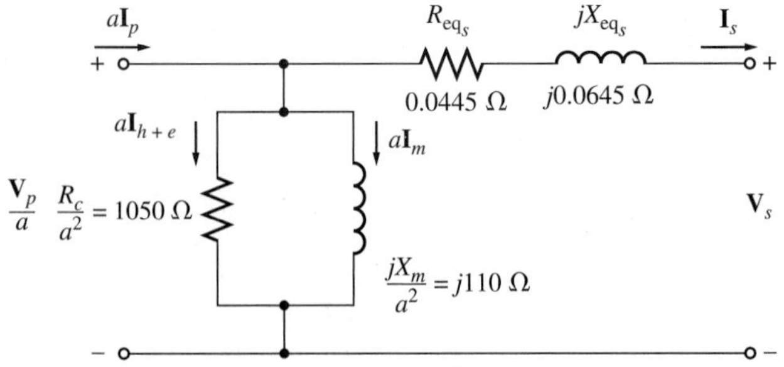

FIGURE 2–28
The transfer equivalent circuit for Example 2–5 referred to (a) its primary side and (b) its secondary side.

(c) The full-load current on the secondary side of this transformer is

$$I_{S,\text{rated}} = \frac{S_{\text{rated}}}{V_{S,\text{rated}}} = \frac{15{,}000\text{ VA}}{230\text{ V}} = 65.2\text{ A}$$

To calculate V_P/a, use Equation (2–64):

$$\frac{\mathbf{V}_P}{a} = \mathbf{V}_S + R_{\text{eq}}\mathbf{I}_S + jX_{\text{eq}}\mathbf{I}_S \tag{2–64}$$

At PF = 0.8 lagging, current $\mathbf{I}_S = 65.2\angle{-36.9°}$ A. Therefore,

$$\frac{\mathbf{V}_P}{a} = 230\angle 0°\text{ V} + (0.0445\ \Omega)(65.2\angle{-36.9°}\text{ A}) + j(0.0645\ \Omega)(65.2\angle{-36.9°}\text{ A})$$

$$= 230\angle 0°\text{ V} + 2.90\angle{-36.9°}\text{ V} + 4.21\angle 53.1°\text{ V}$$

$$= 230 + 2.32 - j1.74 + 2.52 + j3.36$$

$$= 234.84 + j1.62 = 234.85\angle 0.40°\text{ V}$$

The resulting voltage regulation is

$$\text{VR} = \frac{V_P/a - V_{S,\text{fl}}}{V_{S,\text{fl}}} \times 100\% \tag{2–62}$$

$$= \frac{234.85\text{ V} - 230\text{ V}}{230\text{ V}} \times 100\% = 2.1\%$$

At PF = 1.0, current $\mathbf{I}_S = 65.2\angle 0°$ A. Therefore,

$$\frac{\mathbf{V}_P}{a} = 230\angle 0°\text{ V} + (0.0445\ \Omega)(65.2\angle 0°\text{ A}) + j(0.0645\ \Omega)(65.2\angle 0°\text{ A})$$

$$= 230\angle 0°\text{ V} + 2.90\angle 0°\text{ V} + 4.21\angle 90°\text{ V}$$

$$= 230 + 2.90 + j4.21$$

$$= 232.9 + j4.21 = 232.94\angle 1.04°\text{ V}$$

The resulting voltage regulation is

$$\text{VR} = \frac{232.94\text{ V} - 230\text{ V}}{230\text{ V}} \times 100\% = 1.28\%$$

At PF = 0.8 leading, current $\mathbf{I}_S = 65.2\angle 36.9°$ A. Therefore,

$$\frac{\mathbf{V}_P}{a} = 230\angle 0°\text{ V} + (0.0445\ \Omega)(65.2\angle 36.9°\text{ A}) + j(0.0645\ \Omega)(65.2\angle 36.9°\text{ A})$$

$$= 230\angle 0°\text{ V} + 2.90\angle 36.9°\text{ V} + 4.21\angle 126.9°\text{ V}$$

$$= 230 + 2.32 + j1.74 - 2.52 + j3.36$$

$$= 229.80 + j5.10 = 229.85\angle 1.27°\text{ V}$$

The resulting voltage regulation is

$$\text{VR} = \frac{229.85\text{ V} - 230\text{ V}}{230\text{ V}} \times 100\% = -0.062\%$$

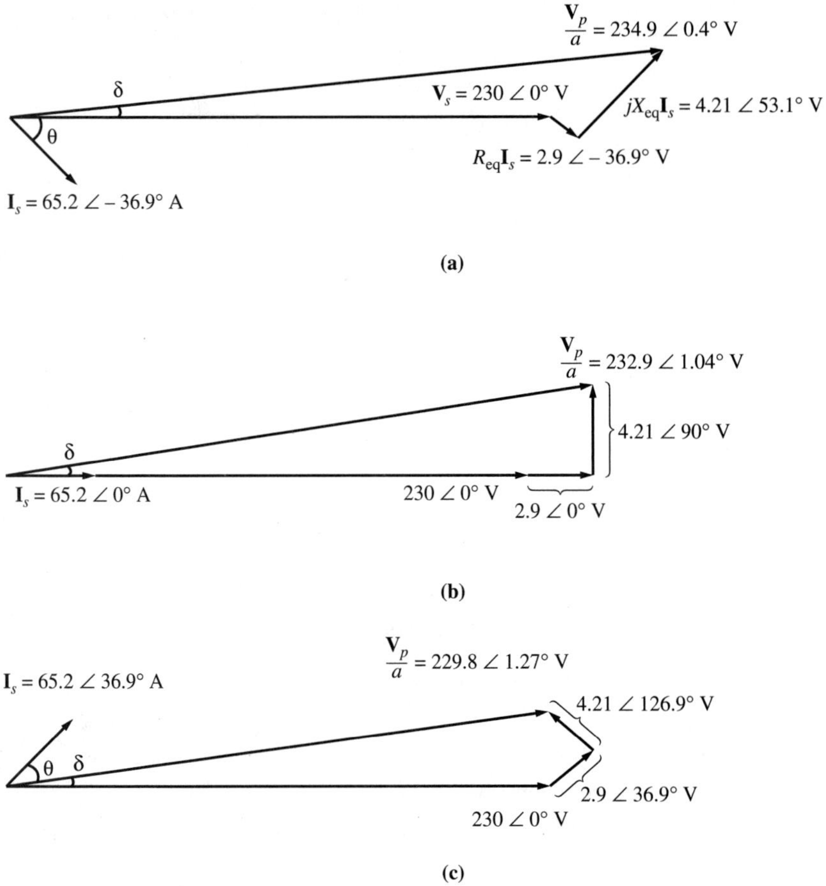

FIGURE 2–29
Transformer phasor diagrams for Example 2–5.

Each of these three phasor diagrams is shown in Figure 2–29.

(d) The best way to plot the voltage regulation as a function of load is to repeat the calculations in part *c* for many different loads using MATLAB. A program to do this is shown below.

```
% M-file: trans_vr.m
% M-file to calculate and plot the voltage regulation
% of a transformer as a function of load for power
% factors of 0.8 lagging, 1.0, and 0.8 leading.
VS = 230;                  % Secondary voltage (V)
amps = 0:6.52:65.2;        % Current values (A)
```

```
Req = 0.0445;                    % Equivalent R (ohms)
Xeq = 0.0645;                    % Equivalent X (ohms)

% Calculate the current values for the three
% power factors. The first row of I contains
% the lagging currents, the second row contains
% the unity currents, and the third row contains
% the leading currents.
I(1,:) = amps .* ( 0.8 - j*0.6);       % Lagging
I(2,:) = amps .* ( 1.0         );      % Unity
I(3,:) = amps .* ( 0.8 + j*0.6);       % Leading

% Calculate VP/a.
VPa = VS + Req.*I + j.*Xeq.*I;

% Calculate voltage regulation
VR = (abs(VPa) - VS) ./ VS .* 100;

% Plot the voltage regulation
plot(amps,VR(1,:),'b-');
hold on;
plot(amps,VR(2,:),'k–');
plot(amps,VR(3,:),'r-.');
title ('Voltage Regulation Versus Load');
xlabel ('Load (A)');
ylabel ('Voltage Regulation (%)');
legend('0.8 PF lagging','1.0 PF','0.8 PF leading');
hold off;
```

The plot produced by this program is shown in Figure 2–30.

(e) To find the efficiency of the transformer, first calculate its losses. The copper losses are

$$P_{\mathrm{Cu}} = (I_S)^2 R_{\mathrm{eq}} = (65.2\ \mathrm{A})^2(0.0445\ \Omega) = 189\ \mathrm{W}$$

The core losses are given by

$$P_{\mathrm{core}} = \frac{(V_P/a)^2}{R_C} = \frac{(234.85\ \mathrm{V})^2}{1050\ \Omega} = 52.5\ \mathrm{W}$$

The output power of the transformer at this power factor is

$$P_{\mathrm{out}} = V_S I_S \cos\theta$$

$$= (230\ \mathrm{V})(65.2\ \mathrm{A})\cos 36.9° = 12{,}000\ \mathrm{W}$$

Therefore, the efficiency of the transformer at this condition is

$$\eta = \frac{V_S I_S \cos\theta}{P_{\mathrm{Cu}} + P_{\mathrm{core}} + V_S I_S \cos\theta} \times 100\% \tag{2–68}$$

$$= \frac{12{,}000\ \mathrm{W}}{189\ \mathrm{W} + 52.5\ \mathrm{W} + 12{,}000\ \mathrm{W}} \times 100\%$$

$$= 98.03\%$$

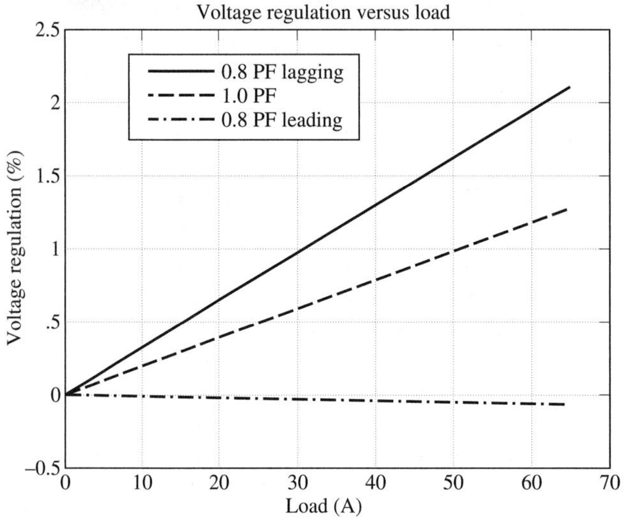

FIGURE 2–30
Plot of voltage regulation versus load for the transformer of Example 2–5.

2.8 TRANSFORMER TAPS AND VOLTAGE REGULATION

In previous sections of this chapter, transformers were described by their turns ratios or by their primary-to-secondary-voltage ratios. Throughout those sections, the turns ratio of a given transformer was treated as though it were completely fixed. In almost all real distribution transformers, this is not quite true. Distribution transformers have a series of *taps* in the windings to permit small changes in the turns ratio of the transformer after it has left the factory. A typical installation might have four taps in addition to the nominal setting with spacings of 2.5 percent of full-load voltage between them. Such an arrangement provides for adjustments up to 5 percent above or below the nominal voltage rating of the transformer.

Example 2–6. A 500-kVA, 13,200/480-V distribution transformer has four 2.5 percent taps on its primary winding. What are the voltage ratios of this transformer at each tap setting?

Solution
The five possible voltage ratings of this transformer are

+5.0% tap	13,860/480 V
+2.5% tap	13,530/480 V
Nominal rating	13,200/480 V
−2.5% tap	12,870/480 V
−5.0% tap	12,540/480 V

The taps on a transformer permit the transformer to be adjusted in the field to accommodate variations in local voltages. However, these taps normally cannot be changed while power is being applied to the transformer. They must be set once and left alone.

Sometimes a transformer is used on a power line whose voltage varies widely with the load. Such voltage variations might be due to a high line impedance between the generators on the power system and that particular load (perhaps it is located far out in the country). Normal loads need to be supplied an essentially constant voltage. How can a power company supply a controlled voltage through high-impedance lines to loads which are constantly changing?

One solution to this problem is to use a special transformer called a *tap changing under load* (TCUL) *transformer* or *voltage regulator*. Basically, a TCUL transformer is a transformer with the ability to change taps while power is connected to it. A voltage regulator is a TCUL transformer with built-in voltage sensing circuitry that automatically changes taps to keep the system voltage constant. Such special transformers are very common in modern power systems.

2.9 THE AUTOTRANSFORMER

On some occasions it is desirable to change voltage levels by only a small amount. For example, it may be necessary to increase a voltage from 110 to 120 V or from 13.2 to 13.8 kV. These small rises may be made necessary by voltage drops that occur in power systems a long way from the generators. In such circumstances, it is wasteful and excessively expensive to wind a transformer with two full windings, each rated at about the same voltage. A special-purpose transformer, called an *autotransformer*, is used instead.

A diagram of a step-up autotransformer is shown in Figure 2–31. In Figure 2–31a, the two coils of the transformer are shown in the conventional manner. In Figure 2–31b, the first winding is shown connected in an additive manner to the second winding. Now, the relationship between the voltage on the first winding and the voltage on the second winding is given by the turns ratio of the transformer. However, *the voltage at the output of the whole transformer is the sum of the voltage on the first winding and the voltage on the second winding*. The first winding here is called the *common winding*, because its voltage appears on both sides of the transformer. The smaller winding is called the *series winding*, because it is connected in series with the common winding.

A diagram of a step-down autotransformer is shown in Figure 2–32. Here the voltage at the input is the sum of the voltages on the series winding and the common winding, while the voltage at the output is just the voltage on the common winding.

Because the transformer coils are physically connected, a different terminology is used for the autotransformer than for other types of transformers. The voltage on the common coil is called the *common voltage* $\mathbf{V}_C$, and the current in that coil is called the *common current* $\mathbf{I}_C$. The voltage on the series coil is called the *series voltage* $\mathbf{V}_{SE}$, and the current in that coil is called the *series current* $\mathbf{I}_{SE}$.

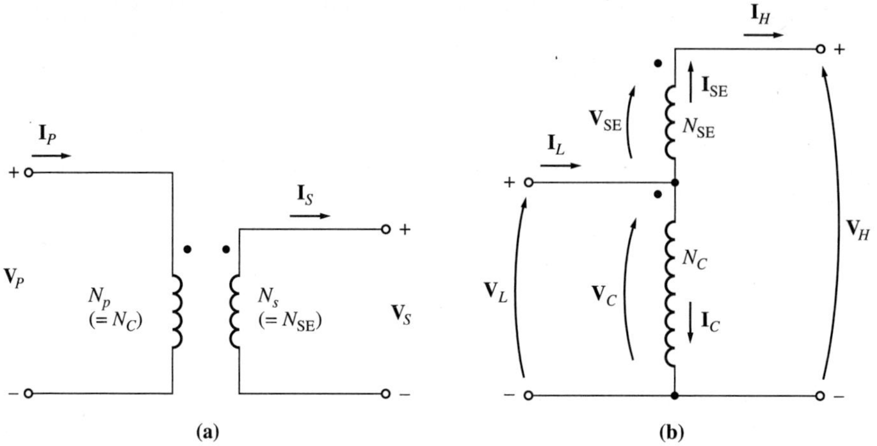

FIGURE 2–31
A transformer with its windings (a) connected in the conventional manner and (b) reconnected as an autotransformer.

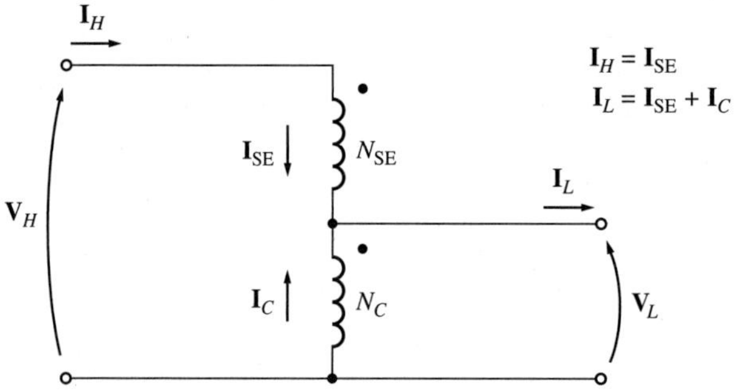

FIGURE 2–32
A step-down autotransformer connection.

The voltage and current on the low-voltage side of the transformer are called $\mathbf{V}_L$ and $\mathbf{I}_L$, respectively, while the corresponding quantities on the high-voltage side of the transformer are called $\mathbf{V}_H$ and $\mathbf{I}_H$. The primary side of the autotransformer (the side with power into it) can be either the high-voltage side or the low-voltage side, depending on whether the autotransformer is acting as a step-down or a step-up transformer. From Figure 2–31b the voltages and currents in the coils are related by the equations

$$\frac{\mathbf{V}_C}{\mathbf{V}_{SE}} = \frac{N_C}{N_{SE}} \tag{2–69}$$

$$N_C \mathbf{I}_C = N_{SE} \mathbf{I}_{SE} \tag{2–70}$$

The voltages in the coils are related to the voltages at the terminals by the equations

$$\mathbf{V}_L = \mathbf{V}_C \tag{2–71}$$

$$\mathbf{V}_H = \mathbf{V}_C + \mathbf{V}_{SE} \tag{2–72}$$

and the currents in the coils are related to the currents at the terminals by the equations

$$\mathbf{I}_L = \mathbf{I}_C + \mathbf{I}_{SE} \tag{2–73}$$

$$\mathbf{I}_H = \mathbf{I}_{SE} \tag{2–74}$$

Voltage and Current Relationships in an Autotransformer

What is the voltage relationship between the two sides of an autotransformer? It is quite easy to determine the relationship between $\mathbf{V}_H$ and $\mathbf{V}_L$. The voltage on the high side of the autotransformer is given by

$$\mathbf{V}_H = \mathbf{V}_C + \mathbf{V}_{SE} \tag{2–72}$$

But $\mathbf{V}_C/\mathbf{V}_{SE} = N_C/\mathrm{N}_{SE}$, so

$$\mathbf{V}_H = \mathbf{V}_C + \frac{N_{SE}}{N_C}\mathbf{V}_C \tag{2–75}$$

Finally, noting that $\mathbf{V}_L = V_C$, we get

$$\begin{aligned}\mathbf{V}_H &= \mathbf{V}_L + \frac{N_{SE}}{N_C}\mathbf{V}_L \\ &= \frac{N_{SE} + N_C}{N_C}\mathbf{V}_L\end{aligned} \tag{2–76}$$

or

$$\boxed{\frac{\mathbf{V}_L}{\mathbf{V}_H} = \frac{N_C}{N_{SE} + N_C}} \tag{2–77}$$

The current relationship between the two sides of the transformer can be found by noting that

$$\mathbf{I}_L = \mathbf{I}_C + \mathbf{I}_{SE} \tag{2–73}$$

From Equation (2–69), $\mathbf{I}_C = (N_{SE}/N_C)\mathbf{I}_{SE}$, so

$$\mathbf{I}_L = \frac{N_{SE}}{N_C}\mathbf{I}_{SE} + \mathbf{I}_{SE} \tag{2–78}$$

Finally, noting that $\mathbf{I}_H = \mathbf{I}_{SE}$, we find

$$\mathbf{I}_L = \frac{N_{SE}}{N_C}\mathbf{I}_H + \mathbf{I}_H$$

$$= \frac{N_{SE} + N_C}{N_C} \mathbf{I}_H \tag{2–79}$$

or

$$\boxed{\frac{\mathbf{I}_L}{\mathbf{I}_H} = \frac{N_{SE} + N_C}{N_C}} \tag{2–80}$$

The Apparent Power Rating Advantage of Autotransformers

It is interesting to note that not all the power traveling from the primary to the secondary in the autotransformer goes through the windings. As a result, if a conventional transformer is reconnected as an autotransformer, it can handle much more power than it was originally rated for.

To understand this idea, refer again to Figure 2–31b. Notice that the input apparent power to the autotransformer is given by

$$S_{in} = V_L I_L \tag{2–81}$$

and the output apparent power is given by

$$S_{out} = V_H I_H \tag{2–82}$$

It is easy to show, by using the voltage and current equations [Equations (2–77) and (2–80)], that the input apparent power is again equal to the output apparent power:

$$S_{in} = S_{out} = S_{IO} \tag{2–83}$$

where S_{IO} is defined to be the input and output apparent powers of the transformer. However, *the apparent power in the transformer windings* is

$$S_W = V_C I_C = V_{SE} I_{SE} \tag{2–84}$$

The relationship between the power going into the primary (and out the secondary) of the transformer and the power in the transformer's actual windings can be found as follows:

$$\begin{aligned} S_W &= V_C I_C \\ &= V_L(I_L - I_H) \\ &= V_L I_L - V_L I_H \end{aligned}$$

Using Equation (2–80), we get

$$S_W = V_L I_L - V_L I_L \frac{N_C}{N_{SE} + N_C}$$

$$= V_L I_L \frac{(N_{SE} + N_C) - N_C}{N_{SE} + N_C} \tag{2–85}$$

$$= S_{IO} \frac{N_{SE}}{N_{SE} + N_C} \tag{2–86}$$

Therefore, the ratio of the apparent power in the primary and secondary of the autotransformer to the apparent power actually traveling through its windings is

$$\boxed{\frac{S_{IO}}{S_W} = \frac{N_{SE} + N_C}{N_{SE}}} \qquad (2\text{–}87)$$

Equation (2–87) describes the *apparent power rating advantage* of an autotransformer over a conventional transformer. Here S_{IO} is the apparent power entering the primary and leaving the secondary of the transformer, while S_W is the apparent power actually traveling through the transformer's windings (the rest passes from primary to secondary without being coupled through the transformer's windings). Note that the smaller the series winding, the greater the advantage.

For example, a 5000-kVA autotransformer connecting a 110-kV system to a 138-kV system would have an N_C/N_{SE} turns ratio of 110:28. Such an autotransformer would actually have windings rated at

$$S_W = S_{IO}\frac{N_{SE}}{N_{SE} + N_C} \qquad (2\text{–}86)$$

$$= (5000\text{ kVA})\frac{28}{28 + 110} = 1015\text{ kVA}$$

The autotransformer would have windings rated at only about 1015 kVA, while a conventional transformer doing the same job would need windings rated at 5000 kVA. The autotransformer could be 5 times smaller than the conventional transformer and also would be much less expensive. For this reason, it is very advantageous to build transformers between two nearly equal voltages as autotransformers.

The following example illustrates autotransformer analysis and the rating advantage of autotransformers.

Example 2–7. A 100-VA, 120/12–V transformer is to be connected so as to form a step-up autotransformer (see Figure 2–33). A primary voltage of 120 V is applied to the transformer.

(*a*) What is the secondary voltage of the transformer?

(*b*) What is its maximum voltampere rating in this mode of operation?

(*c*) Calculate the rating advantage of this autotransformer connection over the transformer's rating in conventional 120/12–V operation.

Solution

To accomplish a step-up transformation with a 120-V primary, the ratio of the turns on the common winding N_C to the turns on the series winding N_{SE} in this transformer must be 120:12 (or 10:1).

(*a*) This transformer is being used as a step-up transformer. The secondary voltage is V_H, and from Equation (2–76),

$$\mathbf{V}_H = \frac{N_{SE} + N_C}{N_C}\mathbf{V}_L \qquad (2\text{–}76)$$

$$= \frac{12 + 120}{120}\,120\text{ V} = 132\text{ V}$$

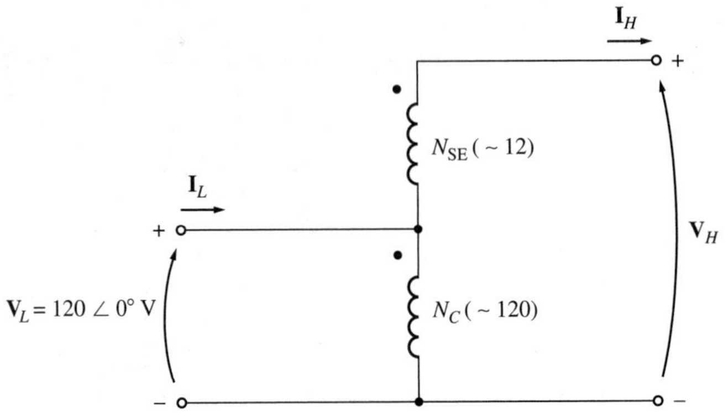

FIGURE 2–33
The autotransformer of Example 2–7.

(*b*) The maximum voltampere rating in either winding of this transformer is 100 VA. How much input or output apparent power can this provide? To find out, examine the series winding. The voltage V_{SE} on the winding is 12 V, and the voltampere rating of the winding is 100 VA. Therefore, the *maximum* series winding current is

$$I_{SE,max} = \frac{S_{max}}{V_{SE}} = \frac{100 \text{ VA}}{12\text{V}} = 8.33 \text{ A}$$

Since I_{SE} is equal to the secondary current I_S (or I_H) and since the secondary voltage $V_S = V_H = 132$ V, the secondary apparent power is

$$S_{out} = V_S I_S = V_H I_H$$
$$= (132 \text{ V})(8.33 \text{ A}) = 1100 \text{ VA} = S_{in}$$

(*c*) The rating advantage can be calculated from part (*b*) or separately from Equation (2–87). From part (*b*),

$$\frac{S_{IO}}{S_W} = \frac{1100 \text{ VA}}{100 \text{ VA}} = 11$$

From Equation (2–87),

$$\frac{S_{IO}}{S_W} = \frac{N_{SE} + N_C}{N_{SE}} \tag{2–87}$$

$$= \frac{12 + 120}{12} = \frac{132}{12} = 11$$

By either equation, the apparent power rating is increased by a factor of 11.

It is not normally possible to just reconnect an ordinary transformer as an autotransformer and use it in the manner of Example 2–7, because the insulation on the low-voltage side of the ordinary transformer may not be strong enough to withstand the full output voltage of the autotransformer connection. In transformers

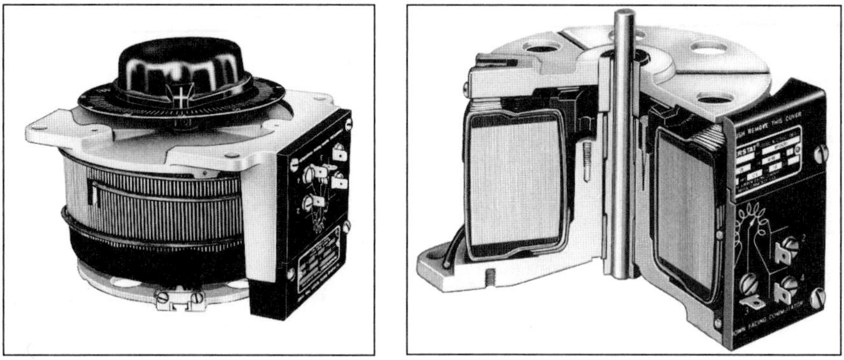

(a) **(b)**

FIGURE 2–34
(a) A variable-voltage autotransformer. (b) Cutaway view of the autotransformer. (*Courtesy of Superior Electric Company.*)

built specifically as autotransformers, the insulation on the smaller coil (the series winding) is made just as strong as the insulation on the larger coil.

It is common practice in power systems to use autotransformers whenever two voltages fairly close to each other in level need to be transformed, because the closer the two voltages are, the greater the autotransformer power advantage becomes. They are also used as variable transformers, where the low-voltage tap moves up and down the winding. This is a very convenient way to get a variable ac voltage. Such a variable autotransformer is shown in Figure 2–34.

The principal disadvantage of autotransformers is that, unlike ordinary transformers, *there is a direct physical connection between the primary and the secondary circuits,* so the *electrical isolation* of the two sides is lost. If a particular application does not require electrical isolation, then the autotransformer is a convenient and *inexpensive* way to tie nearly equal voltages together.

The Internal Impedance of an Autotransformer

Autotransformers have one additional disadvantage compared to conventional transformers. It turns out that, compared to a given transformer connected in the conventional manner, the effective per-unit impedance of an autotransformer is smaller by a factor equal to the reciprocal of the power advantage of the autotransformer connection.

The proof of this statement is left as a problem at the end of the chapter.

The reduced internal impedance of an autotransformer compared to a conventional two-winding transformer can be a serious problem in some applications where the series impedance is needed to limit current flows during power system faults (short circuits). The effect of the smaller internal impedance provided by an autotransformer must be taken into account in practical applications before autotransformers are selected.

Example 2–8. A transformer is rated at 1000 kVA, 12/1.2 kV, 60 Hz when it is operated as a conventional two-winding transformer. Under these conditions, its series resistance and reactance are given as 1 and 8 percent per unit, respectively. This transformer is to be used as a 13.2/12–kV step-down autotransformer in a power distribution system. In the autotransformer connection, (*a*) what is the transformer's rating when used in this manner and (*b*) what is the transformer's series impedance in per-unit?

Solution

(*a*) The N_C/N_{SE} turns ratio must be 12:1.2 or 10:1. The voltage rating of this transformer will be 13.2/12 kV, and the apparent power (voltampere) rating will be

$$S_{IO} = \frac{N_{SE} + N_C}{N_{SE}} S_W$$

$$= \frac{1 + 10}{1} 1000 \text{ kVA} = 11{,}000 \text{ kVA}$$

(*b*) The transformer's impedance in a per-unit system when connected in the conventional manner is

$$Z_{eq} = 0.01 + j0.08 \text{ pu} \qquad \text{separate windings}$$

The apparent power advantage of this autotransformer is 11, so the per-unit impedance of the autotransformer connected as described is

$$Z_{eq} = \frac{0.01 + j0.08}{11}$$

$$= 0.00091 + j0.00727 \text{ pu} \qquad \text{autotransformer}$$

2.10 THREE-PHASE TRANSFORMERS

Almost all the major power generation and distribution systems in the world today are three-phase ac systems. Since three-phase systems play such an important role in modern life, it is necessary to understand how transformers are used in them.

Transformers for three-phase circuits can be constructed in one of two ways. One approach is simply to take three single-phase transformers and connect them in a three-phase bank. An alternative approach is to make a three-phase transformer consisting of three sets of windings wrapped on a common core. These two possible types of transformer construction are shown in Figures 2–35 and 2–36. Both designs (three separate transformers and a single three-phase transformer) are in use today, and you are likely to run into both of them in practice. A single three-phase transformer is lighter, smaller, cheaper, and slightly more efficient, but using three separate single-phase transformers has the advantage that each unit in the bank could be replaced individually in the event of trouble. A utility would only need to stock a single spare single-phase transformer to back up all three phases, potentially saving money.

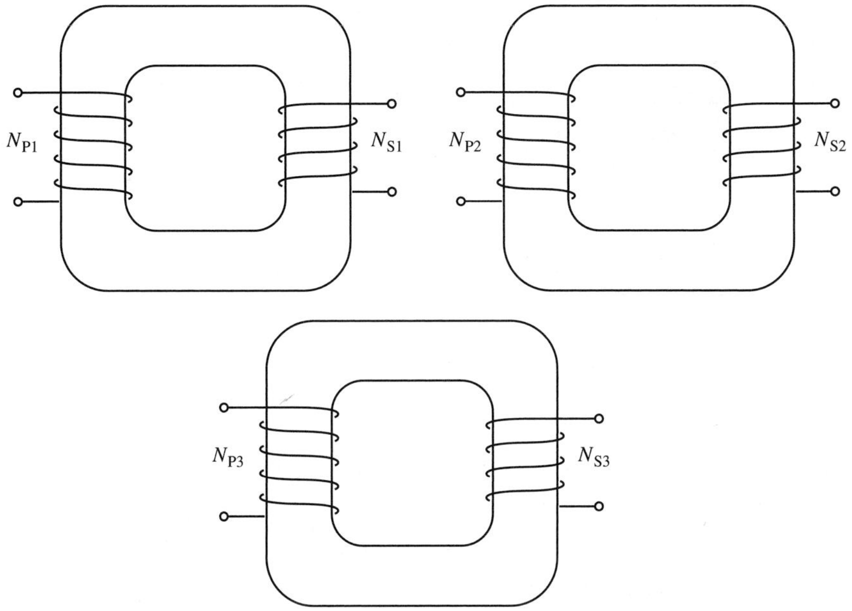

FIGURE 2–35
A three-phase transformer bank composed of independent transformers.

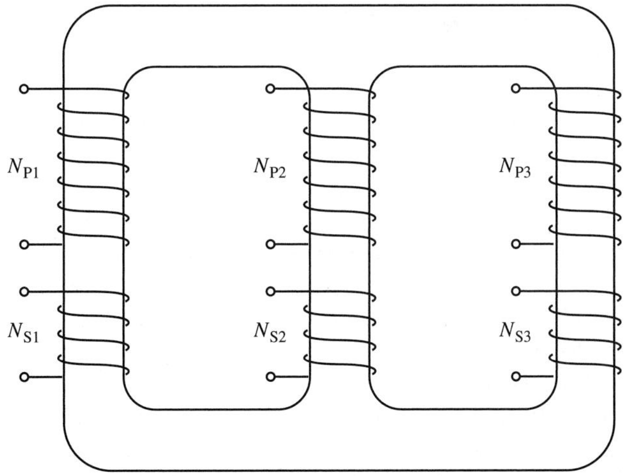

FIGURE 2–36
A three-phase transformer wound on a single three-legged core.

Three-Phase Transformer Connections

A three-phase transformer consists of three transformers, either separate or combined on one core. The primaries and secondaries of any three-phase transformer can be independently connected in either a wye (Y) or a delta (Δ). This gives a total of four possible connections for a three-phase transformer bank:

1. Wye–wye (Y–Y)
2. Wye–delta (Y–Δ)
3. Delta–wye (Δ–Y)
4. Delta–delta (Δ–Δ)

These connections are shown over the next several pages in Figure 2–37.

The key to analyzing any three-phase transformer bank is to look at a single transformer in the bank. *Any single transformer in the bank behaves exactly like the single-phase transformers already studied.* The impedance, voltage regulation, efficiency, and similar calculations for three-phase transformers are done on *a per-phase basis,* using exactly the same techniques already developed for single-phase transformers.

The advantages and disadvantages of each type of three-phase transformer connection are discussed below.

WYE–WYE CONNECTION. The Y–Y connection of three-phase transformers is shown in Figure 2–37a. In a Y–Y connection, the primary voltage on each phase of the transformer is given by $V_{\phi P} = V_{LP} / \sqrt{3}$. The primary-phase voltage is related to the secondary-phase voltage by the turns ratio of the transformer. The phase voltage on the secondary is then related to the line voltage on the secondary by $V_{LS} = \sqrt{3}V_{\phi S}$. Therefore, overall the voltage ratio on the transformer is

$$\boxed{\frac{V_{LP}}{V_{LS}} = \frac{\sqrt{3}V_{\phi P}}{\sqrt{3}V_{\phi S}} = a \qquad \text{Y–Y}} \tag{2–88}$$

The Y–Y connection has two very serious problems:

1. If loads on the transformer circuit are unbalanced, then the voltages on the phases of the transformer can become severely unbalanced.
2. Third-harmonic voltages can be large.

If a three-phase set of voltages is applied to a Y–Y transformer, the voltages in any phase will be 120° apart from the voltages in any other phase. However, *the third-harmonic components of each of the three phases will be in phase with each other,* since there are three cycles in the third harmonic for each cycle of the fundamental frequency. There are always some third-harmonic components in a transformer because of the nonlinearity of the core, and these components add up.

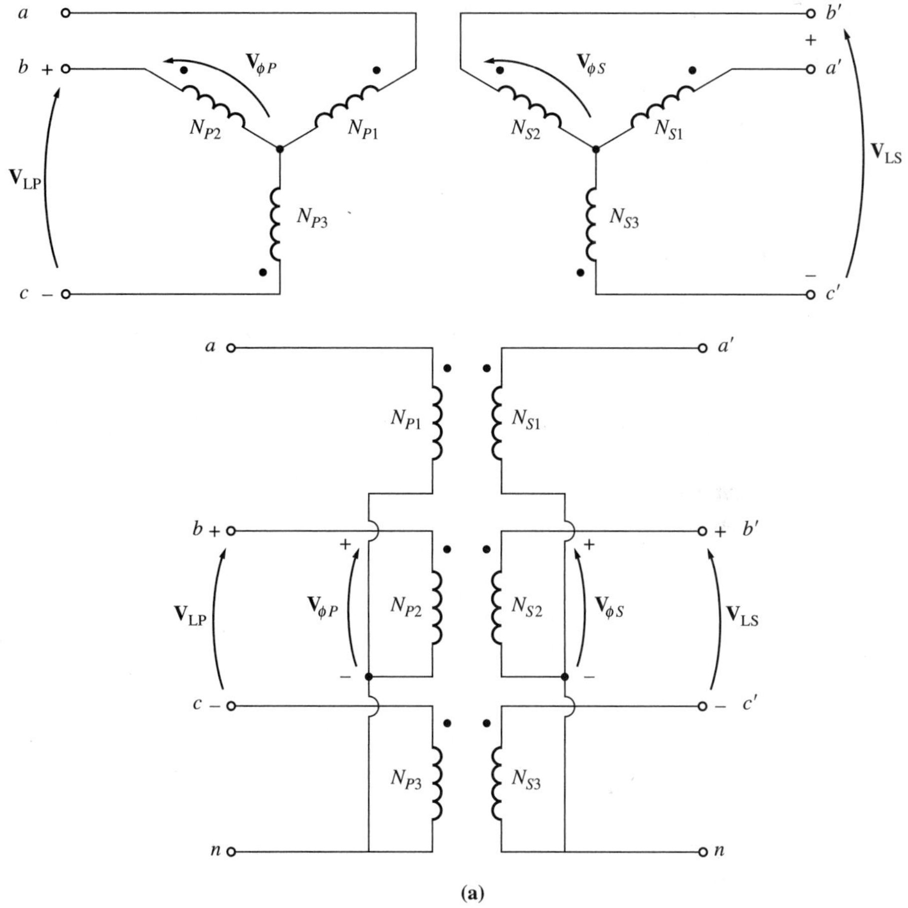

FIGURE 2–37
Three-phase transformer connections and wiring diagrams: (a) Y–Y; (b) Y–Δ; (c) Δ–Y; (d) Δ–Δ.

The result is a very large third-harmonic component of voltage on top of the 50- or 60-Hz fundamental voltage. This third-harmonic voltage can be larger than the fundamental voltage itself.

Both the unbalance problem and the third-harmonic problem can be solved using one of two techniques:

1. *Solidly ground the neutrals of the transformers,* especially the primary winding's neutral. This connection permits the additive third-harmonic components to cause a current flow in the neutral instead of building up large voltages. The neutral also provides a return path for any current imbalances in the load.
2. *Add a third (tertiary) winding connected in* Δ to the transformer bank. If a third Δ-connected winding is added to the transformer, then the third-harmonic

components of voltage in the Δ will add up, causing a circulating current flow within the winding. This suppresses the third-harmonic components of voltage in the same manner as grounding the transformer neutrals.

The Δ-connected tertiary windings need not even be brought out of the transformer case, but they often are used to supply lights and auxiliary power within the substation where it is located. The tertiary windings must be large enough to handle the circulating currents, so they are usually made about one-third the power rating of the two main windings.

One or the other of these correction techniques *must* be used any time a Y–Y transformer is installed. In practice, very few Y–Y transformers are used, since the same jobs can be done by one of the other types of three-phase transformers.

WYE–DELTA CONNECTION. The Y–Δ connection of three-phase transformers is shown in Figure 2–37b. In this connection, the primary line voltage is related to the primary phase voltage by $V_{\mathrm{LP}} = \sqrt{3}V_{\phi P}$, while the secondary line voltage is equal to the secondary phase voltage $V_{\mathrm{LS}} = V_{\phi S}$. The voltage ratio of each phase is

$$\frac{V_{\phi P}}{V_{\phi S}} = a$$

so the overall relationship between the line voltage on the primary side of the bank and the line voltage on the secondary side of the bank is

$$\frac{V_{\mathrm{LP}}}{V_{\mathrm{LS}}} = \frac{\sqrt{3}V_{\phi P}}{V_{\phi S}}$$

$$\boxed{\frac{V_{\mathrm{LP}}}{V_{\mathrm{LS}}} = \sqrt{3}a \qquad \text{Y–}\Delta} \tag{2–89}$$

The Y–Δ connection has no problem with third-harmonic components in its voltages, since they are consumed in a circulating current on the Δ side. This connection is also more stable with respect to unbalanced loads, since the Δ partially redistributes any imbalance that occurs.

This arrangement does have one problem, though. Because of the connection, the secondary voltage is shifted 30° relative to the primary voltage of the transformer. The fact that a phase shift has occurred can cause problems in paralleling the secondaries of two transformer banks together. The phase angles of transformer secondaries must be equal if they are to be paralleled, which means that attention must be paid to the direction of the 30° phase shift occurring in each transformer bank to be paralleled together.

In the United States, it is customary to make the secondary voltage lag the primary voltage by 30°. Although this is the standard, it has not always been observed, and older installations must be checked very carefully before a new transformer is paralleled with them, to make sure that their phase angles match.

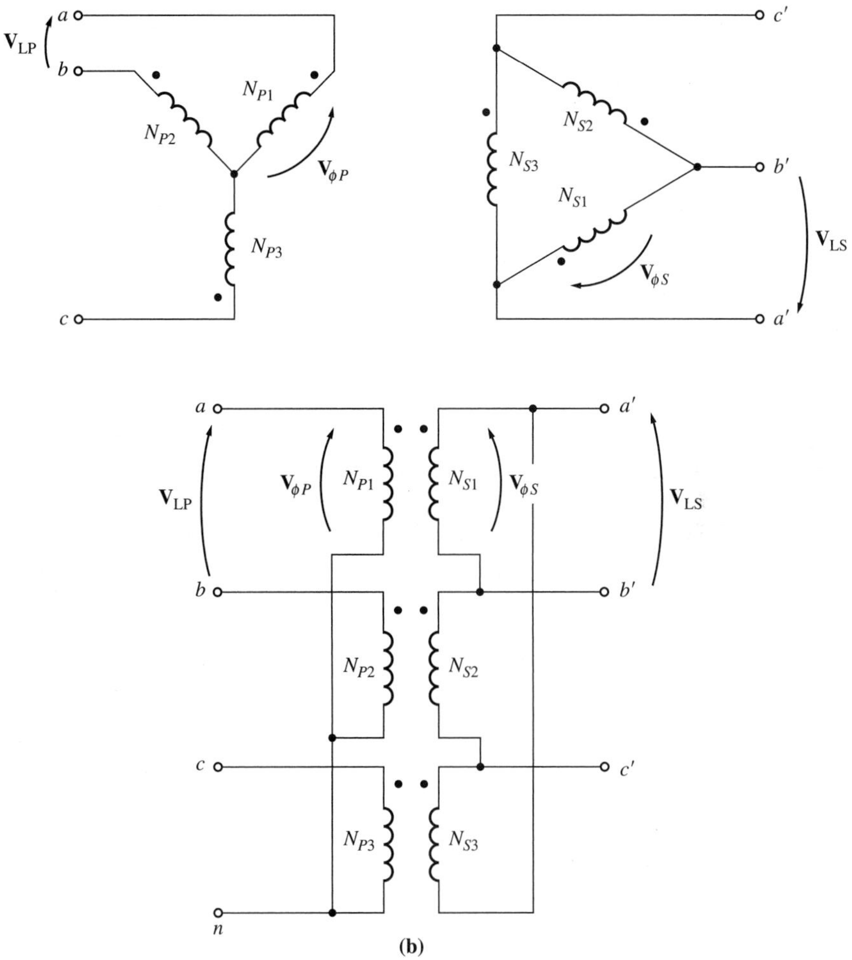

FIGURE 2–37
(b) Y–Δ *(continued)*

The connection shown in Figure 2–37b will cause the secondary voltage to be lagging if the system phase sequence is *abc*. If the system phase sequence is *acb*, then the connection shown in Figure 2–37b will cause the secondary voltage to be leading the primary voltage by 30°.

DELTA–WYE CONNECTION. A Δ–Y connection of three-phase transformers is shown in Figure 2–37c. In a Δ–Y connection, the primary line voltage is equal to the primary-phase voltage $V_{LP} = V_{\phi P}$, while the secondary voltages are related by $V_{LS} = \sqrt{3}V_{\phi S}$. Therefore, the line-to-line voltage ratio of this transformer connection is

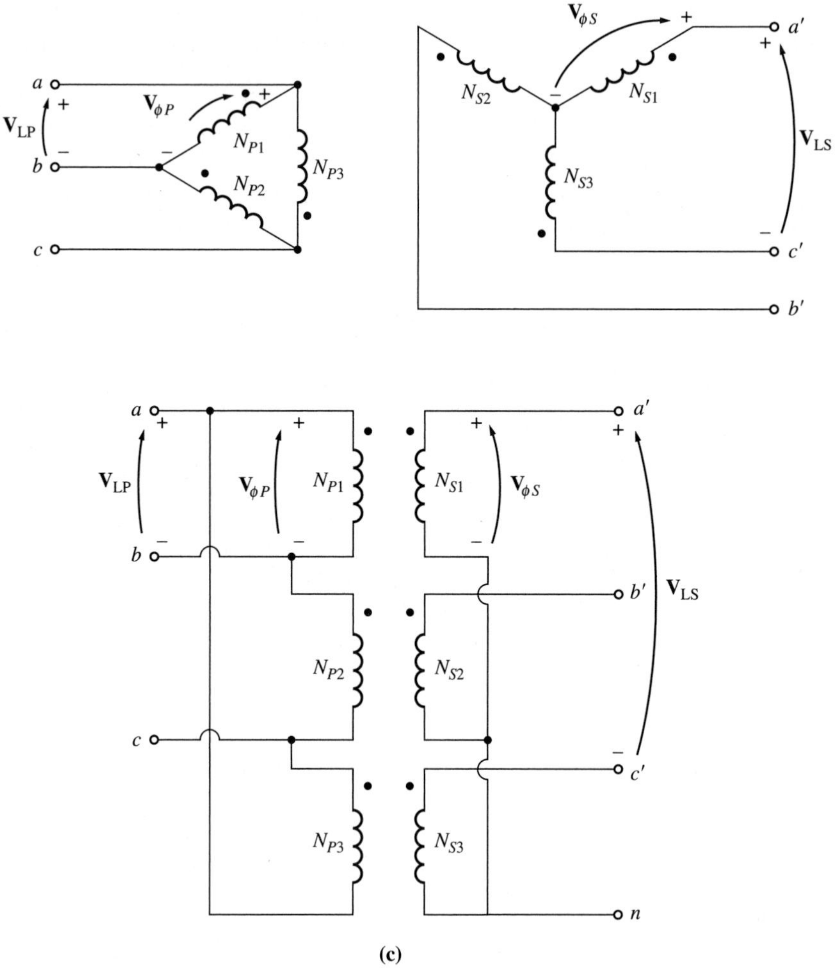

FIGURE 2–37
(c) Δ–Y *(continued)*

$$\frac{V_{LP}}{V_{LS}} = \frac{V_{\phi P}}{\sqrt{3}V_{\phi S}}$$

$$\boxed{\frac{V_{LP}}{V_{LS}} = \frac{a}{\sqrt{3}} \qquad \Delta\text{–Y}} \tag{2–90}$$

This connection has the same advantages and the same phase shift as the Y–Δ transformer. The connection shown in Figure 2–37c makes the secondary voltage lag the primary voltage by 30°, as before.

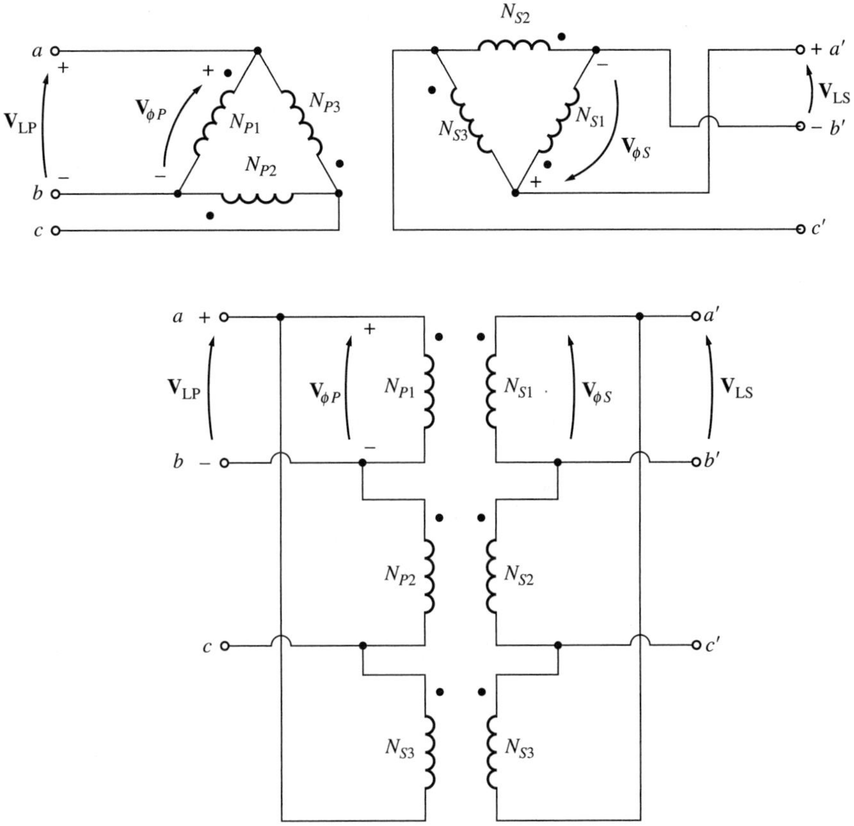

(d)

FIGURE 2–37
(d) Δ–Δ *(concluded)*

DELTA–DELTA CONNECTION. The Δ–Δ connection is shown in Figure 2–37d. In a Δ–Δ connection, $V_{LP} = V_{\phi P}$ and $V_{LS} = V_{\phi S}$, so the relationship between primary and secondary line voltages is

$$\boxed{\frac{V_{LP}}{V_{LS}} = \frac{V_{\phi P}}{V_{\phi S}} = a \qquad \Delta\text{–}\Delta} \tag{2–91}$$

This transformer has no phase shift associated with it and no problems with unbalanced loads or harmonics.

The Per-Unit System for Three-Phase Transformers

The per-unit system of measurements applies just as well to three-phase transformers as to single-phase transformers. The single-phase base equations (2–53) to (2–56) apply to three-phase systems on a *per-phase* basis. If the total base voltampere value of the transformer bank is called S_{base}, then the base voltampere value of one of the transformers $S_{1\phi,\text{base}}$ is

$$S_{1\phi,\text{base}} = \frac{S_{\text{base}}}{3} \tag{2–92}$$

and the base phase current and impedance of the transformer are

$$I_{\phi,\text{base}} = \frac{S_{1\phi,\text{base}}}{V_{\phi,\text{base}}} \tag{2–93a}$$

$$\boxed{I_{\phi,\text{base}} = \frac{S_{\text{base}}}{3\,V_{\phi,\text{base}}}} \tag{2–93b}$$

$$Z_{\text{base}} = \frac{(V_{\phi,\text{base}})^2}{S_{1\phi,\text{base}}} \tag{2–94a}$$

$$\boxed{Z_{\text{base}} = \frac{3(V_{\phi,\text{base}})^2}{S_{\text{base}}}} \tag{2–94b}$$

Line quantities on three-phase transformer banks can also be represented in the per-unit system. The relationship between the base line voltage and the base phase voltage of the transformer depends on the connection of windings. If the windings are connected in delta, $V_{\text{L,base}} = V_{\phi,\text{base}}$, while if the windings are connected in wye, $V_{L,\text{base}} = \sqrt{3}V_{\phi,\text{base}}$. The base line current in a three-phase transformer bank is given by

$$I_{L,\text{base}} = \frac{S_{\text{base}}}{\sqrt{3}V_{L,\text{base}}} \tag{2–95}$$

The application of the per-unit system to three-phase transformer problems is similar to its application in the single-phase examples already given.

Example 2–9. A 50-kVA, 13,800/208-V, Δ-Y distribution transformer has a resistance of 1 percent and a reactance of 7 percent per unit.

(a) What is the transformer's phase impedance referred to the high-voltage side?
(b) Calculate this transformer's voltage regulation at full load and 0.8 PF lagging, using the calculated high-side impedance.
(c) Calculate this transformer's voltage regulation under the same conditions, using the per-unit system.

Solution

(a) The high-voltage side of this transformer has a base line voltage of 13,800 V and a base apparent power of 50 kVA. Since the primary is Δ-connected, its phase voltage is equal to its line voltage. Therefore, its base impedance is

$$Z_{\text{base}} = \frac{3(V_{\phi,\,\text{base}})^2}{S_{\text{base}}} \tag{2–94b}$$

$$= \frac{3(13{,}800\ \text{V})^2}{50{,}000\ \text{VA}} = 11{,}426\ \Omega$$

The per-unit impedance of the transformer is

$$Z_{\text{eq}} = 0.01 + j0.07\ \text{pu}$$

so the high-side impedance in ohms is

$$\begin{aligned} Z_{\text{eq}} &= Z_{\text{eq,pu}} Z_{\text{base}} \\ &= (0.01 + j0.07\ \text{pu})(11{,}426\ \Omega) = 114.2 + j800\ \Omega \end{aligned}$$

(b) To calculate the voltage regulation of a three-phase transformer bank, determine the voltage regulation of any single transformer in the bank. The voltages on a single transformer are phase voltages, so

$$\text{VR} = \frac{V_{\phi P} - aV_{\phi S}}{aV_{\phi S}} \times 100\%$$

The rated transformer phase voltage on the primary is 13,800 V, so the rated phase current on the primary is given by

$$I_\phi = \frac{S}{3V_\phi}$$

The rated apparent power $S = 50$ kVA, so

$$I_\phi = \frac{50{,}000\ \text{VA}}{3(13{,}800\ \text{V})} = 1.208\ \text{A}$$

The rated phase voltage on the secondary of the transformer is $208\ \text{V}/\sqrt{3} = 120$ V. When referred to the high-voltage side of the transformer, this voltage becomes $V'_{\phi S} = aV_{\phi S} = 13{,}800$ V. Assume that the transformer secondary is operating at the rated voltage and current, and find the resulting primary phase voltage:

$$\begin{aligned} \mathbf{V}_{\phi P} &= a\mathbf{V}_{\phi S} + R_{\text{eq}}\mathbf{I}_\phi + jX_{\text{eq}}\mathbf{I}_\phi \\ &= 13{,}800\angle 0^\circ\ \text{V} + (114.2\ \Omega)(1.208\angle -36.87^\circ\ \text{A}) + (j800\ \Omega)(1.208\angle -36.87^\circ\ \text{A}) \\ &= 13{,}800 + 138\angle -36.87^\circ + 966.4\angle 53.13^\circ \\ &= 13{,}800 + 110.4 - j82.8 + 579.8 + j773.1 \\ &= 14{,}490 + j690.3 = 14{,}506\angle 2.73^\circ\ \text{V} \end{aligned}$$

Therefore,

$$\begin{aligned} \text{VR} &= \frac{V_{\phi P} - aV_{\phi S}}{aV_{\phi S}} \times 100\% \\ &= \frac{14{,}506 - 13{,}800}{13{,}800} \times 100\% = 5.1\% \end{aligned}$$

(c) In the per-unit system, the output voltage is $1\angle 0^\circ$, and the current is $1\angle -36.87^\circ$. Therefore, the input voltage is

$$\begin{aligned} V_P &= 1\angle 0^\circ + (0.01)(1\angle -36.87^\circ) + (j0.07)(1\angle -36.87^\circ) \\ &= 1 + 0.008 - j0.006 + 0.042 + j0.056 \\ &= 1.05 + j0.05 = 1.051\angle 2.73^\circ \end{aligned}$$

The voltage regulation is

$$\mathrm{VR} = \frac{1.051 - 1.0}{1.0} \times 100\% = 5.1\%$$

Of course, the voltage regulation of the transformer bank is the same whether the calculations are done in actual ohms or in the per-unit system.

2.11 THREE-PHASE TRANSFORMATION USING TWO TRANSFORMERS

In addition to the standard three-phase transformer connections, there are ways to perform three-phase transformation with only two transformers. These techniques are sometimes employed to create three-phase power at locations where not all three power lines are available. For example, in rural areas a power company will often run only one or two of the three phases on a distribution line, because the power requirements in the area do not justify the cost of running all three wires. If there is an isolated user of three-phase power along a route served by a distribution line with two of the three phases, these techniques can be used to create three-phase power for that local user.

All techniques that create three-phase power with only two transformers involve a reduction in the power-handling capability of the transformers, but they may be justified by certain economic situations.

Some of the more important two-transformer connections are

1. The open-Δ (or V–V) connection
2. The open-Y–open-Δ connection
3. The Scott-T connection
4. The three-phase T connection

Each of these transformer connections is described in this section.

The Open-Δ (or V–V) Connection

In some situations a full transformer bank may not be used to accomplish three-phase transformation. For example, suppose that a Δ–Δ transformer bank composed of separate transformers has a damaged phase that must be removed for repair. The resulting situation is shown in Figure 2–38. If the two remaining secondary voltages are $\mathbf{V}_A = V\angle 0°$ and $\mathbf{V}_A = V\angle 120°$ V, then the voltage across the gap where the third transformer used to be is given by

$$\begin{aligned}\mathbf{V}_C &= -\mathbf{V}_A - \mathbf{V}_B \\ &= -V\angle 0° - V\angle -120° \\ &= -V - (-0.5V - j0.866V) \\ &= -0.5V + j0.866V \\ &= V\angle 120° \qquad \mathrm{V}\end{aligned}$$

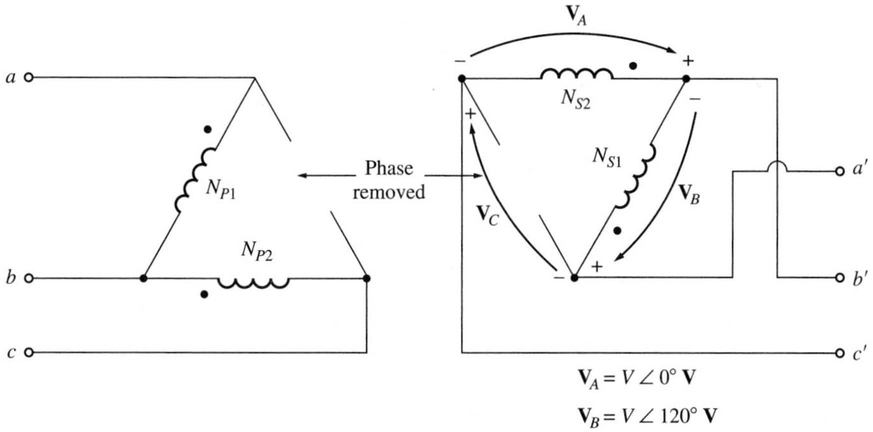

FIGURE 2–38
The open-Δ or V–V transformer connection.

This is exactly the same voltage that would be present if the third transformer were still there. Phase C is sometimes called a *ghost phase.* Thus, the open-delta connection lets a transformer bank get by with only two transformers, allowing some power flow to continue even with a damaged phase removed.

How much apparent power can the bank supply with one of its three transformers removed? At first, it seems that it could supply two-thirds of its rated apparent power, since two-thirds of the transformers are still present. Things are not quite that simple, though. To understand what happens when a transformer is removed, see Figure 2–39.

Figure 2–39a shows the transformer bank in normal operation connected to a resistive load. If the rated voltage of one transformer in the bank is V_ϕ and the rated current is I_ϕ, then the maximum power that can be supplied to the load is

$$P = 3V_\phi I_\phi \cos\theta$$

The angle between the voltage V_ϕ and the current I_ϕ in each phase is 0°, so the total power supplied by the transformer is

$$\begin{aligned} P &= 3V_\phi I_\phi \cos\theta \\ &= 3V_\phi I_\phi \end{aligned} \tag{2–96}$$

The open-delta transformer is shown in Figure 2–39b. It is important to note the angles on the voltages and currents in this transformer bank. Because one of the transformer phases is missing, the transmission line current is now equal to the phase current in each transformer, and the currents and voltages in the transformer bank differ in angle by 30°. Since the current and voltage angles differ in each of the two transformers, it is necessary to examine each transformer individually to determine the maximum power it can supply. For transformer 1, the voltage is at

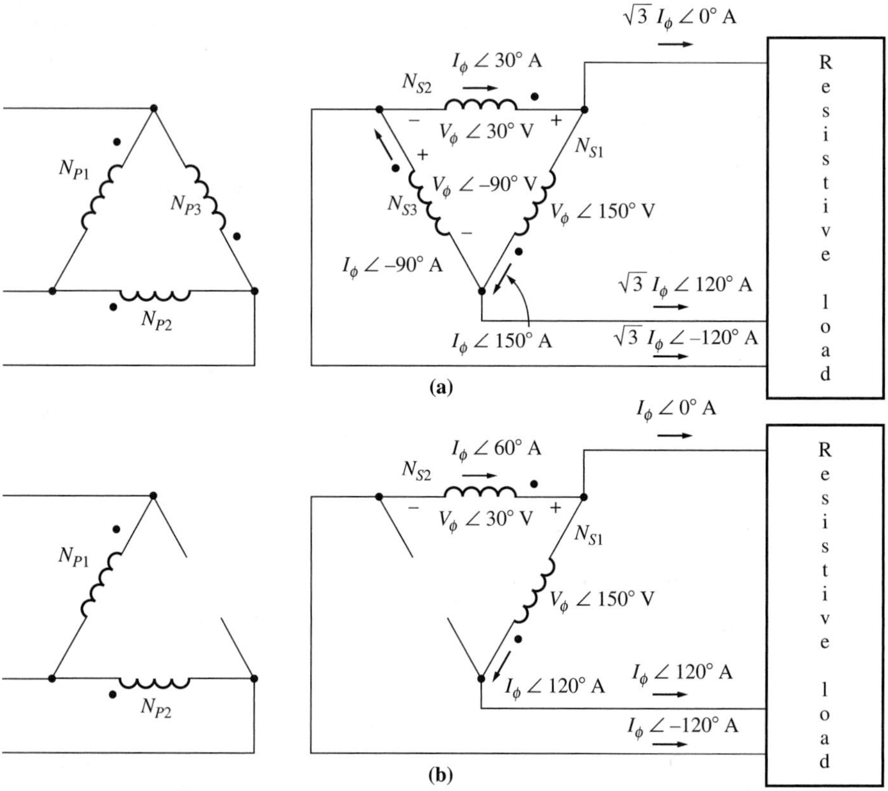

FIGURE 2–39
(a) Voltages and currents in a Δ–Δ transformer bank. (b) Voltages and currents in an open-Δ transformer bank.

an angle of 150° and the current is at an angle of 120°, so the expression for the maximum power in transformer 1 is

$$\begin{aligned} P_1 &= 3V_\phi I_\phi \cos(150° - 120°) \\ &= 3V_\phi I_\phi \cos 30° \\ &= \frac{\sqrt{3}}{2} V_\phi I_\phi \end{aligned} \tag{2–97}$$

For transformer 2, the voltage is at an angle of 30° and the current is at an angle of 60°, so its maximum power is

$$\begin{aligned} P_2 &= 3V_\phi I_\phi \cos(30° - 60°) \\ &= 3V_\phi I_\phi \cos(-30°) \\ &= \frac{\sqrt{3}}{2} V_\phi I_\phi \end{aligned} \tag{2–98}$$

Therefore, the total maximum power of the open-delta bank is given by

$$P = \sqrt{3} V_\phi I_\phi \tag{2–99}$$

The rated current is the same in each transformer whether there are two or three of them, and the voltage is the same on each transformer; so the ratio of the output power available from the open-delta bank to the output power available from the normal three-phase bank is

$$\frac{P_{\text{open }\Delta}}{P_{3\text{ phase}}} = \frac{\sqrt{3}V_\phi I_\phi}{3V_\phi I_\phi} = \frac{1}{\sqrt{3}} = 0.577 \tag{2–100}$$

The available power out of the open-delta bank is only 57.7 percent of the original bank's rating.

A good question that could be asked is: What happens to the rest of the open-delta bank's rating? After all, the total power that the two transformers together can produce is two-thirds that of the original bank's rating. To find out, examine the reactive power of the open-delta bank. The reactive power of transformer 1 is

$$\begin{aligned} Q_1 &= 3V_\phi I_\phi \sin\,(150° - 120°) \\ &= 3V_\phi I_\phi \sin 30° \\ &= \tfrac{1}{2}\, V_\phi I_\phi \end{aligned}$$

The reactive power of transformer 2 is

$$\begin{aligned} Q_2 &= 3V_\phi I_\phi \sin\,(30° - 60°) \\ &= 3V_\phi I_\phi \sin\,(-30°) \\ &= -\tfrac{1}{2}\, V_\phi I_\phi \end{aligned}$$

Thus one transformer is producing reactive power which the other one is consuming. It is this exchange of energy between the two transformers that limits the power output to 57.7 percent of the *original bank's rating* instead of the otherwise expected 66.7 percent.

An alternative way to look at the rating of the open-delta connection is that 86.6 percent of the rating *of the two remaining transformers* can be used.

Open-delta connections are used occasionally to supply a small amount of three-phase power to an otherwise single-phase load. In such a case, the connection in Figure 2–40 can be used, where transformer T_2 is much larger than transformer T_1.

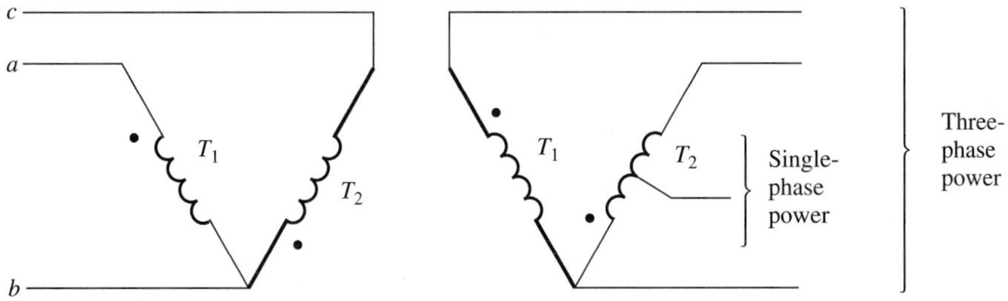

FIGURE 2–40
Using an open-Δ transformer connection to supply a small amount of three-phase power along with a lot of single-phase power. Transformer T_2 is much larger than transformer T_1.

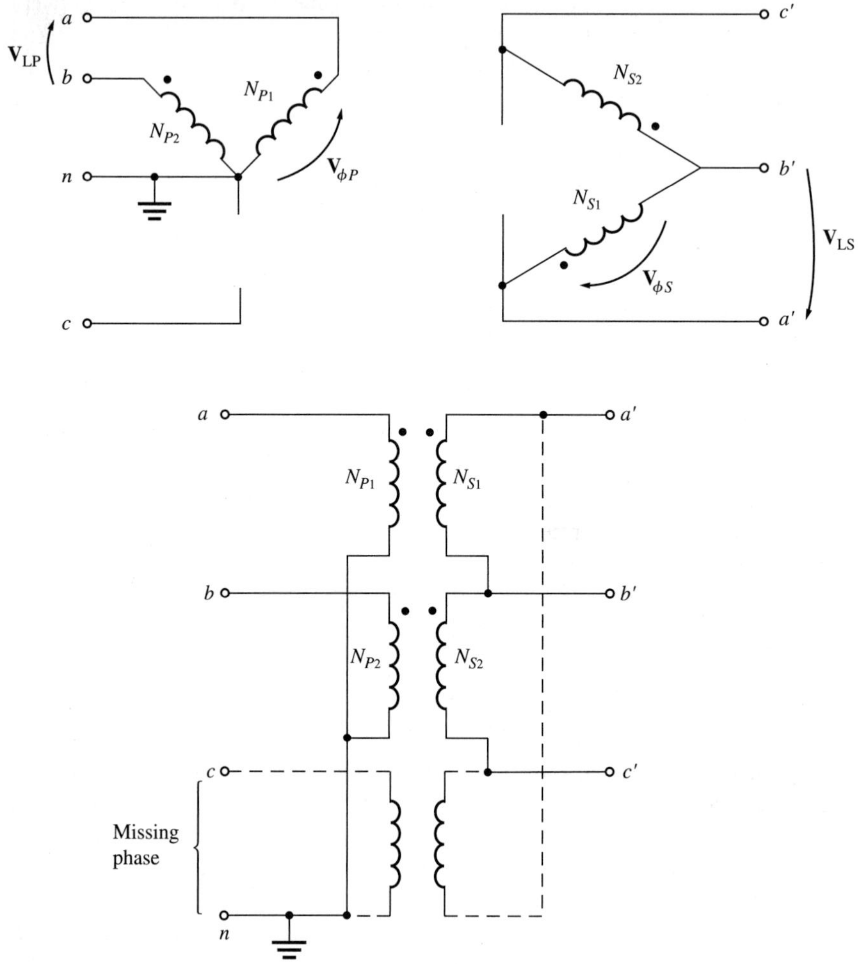

FIGURE 2–41
The open-Y–open-Δ transformer connection and wiring diagram. Note that this connection is identical to the Y–Δ connection in Figure 2–37b, except for the absence of the third transformer and the presence of the neutral lead.

The Open-Wye–Open-Delta Connection

The open-wye–open-delta connection is very similar to the open-delta connection except that the primary voltages are derived from two phases and the neutral. This type of connection is shown in Figure 2–41. It is used to serve small commercial customers needing three-phase service in rural areas where all three phases are not yet present on the power poles. With this connection, a customer can get three-phase service in a makeshift fashion until demand requires installation of the third phase on the power poles.

A major disadvantage of this connection is that a very large return current must flow in the neutral of the primary circuit.

The Scott-T Connection

The Scott-T connection is a way to derive two phases 90° apart from a three-phase power supply. In the early history of ac power transmission, two-phase and three-phase power systems were quite common. In those days, it was routinely necessary to interconnect two- and three-phase power systems, and the Scott-T transformer connection was developed for that purpose.

Today, two-phase power is primarily limited to certain control applications, but the Scott T is still used to produce the power needed to operate them.

The Scott T consists of two single-phase transformers with identical ratings. One has a tap on its primary winding at 86.6 percent of full-load voltage. They are connected as shown in Figure 2–42a. The 86.6 percent tap of transformer T_2 is connected to the center tap of transformer T_1. The voltages applied to the primary winding are shown in Figure 2–42b, and the resulting voltages applied to the primaries of the two transformers are shown in Figure 2–42c. Since these voltages are 90° apart, they result in a two-phase output.

It is also possible to convert two-phase power into three-phase power with this connection, but since there are very few two-phase generators in use, this is rarely done.

The Three-Phase T Connection

The Scott-T connection uses two transformers to convert *three-phase power* to *two-phase power* at a different voltage level. By a simple modification of that connection, the same two transformers can also convert *three-phase power* to *three-phase power* at a different voltage level. Such a connection is shown in Figure 2–43. Here both the primary and the secondary windings of transformer T_2 are tapped at the 86.6 percent point, and the taps are connected to the center taps of the corresponding windings on transformer T_1. In this connection T_1 is called the *main transformer* and T_2 is called the *teaser transformer*.

As in the Scott T, the three-phase input voltage produces two voltages 90° apart on the primary windings of the transformers. These primary voltages produce secondary voltages which are also 90° apart. Unlike the Scott T, though, the secondary voltages are recombined into a three-phase output.

One major advantage of the three-phase T connection over the other three-phase two-transformer connections (the open-delta and open-wye–open-delta) is that a neutral can be connected to both the primary side and the secondary side of the transformer bank. This connection is sometimes used in self-contained three-phase distribution transformers, since its construction costs are lower than those of a full three-phase transformer bank.

Since the bottom parts of the teaser transformer windings are not used on either the primary or the secondary sides, they could be left off with no change in performance. This is, in fact, typically done in distribution transformers.

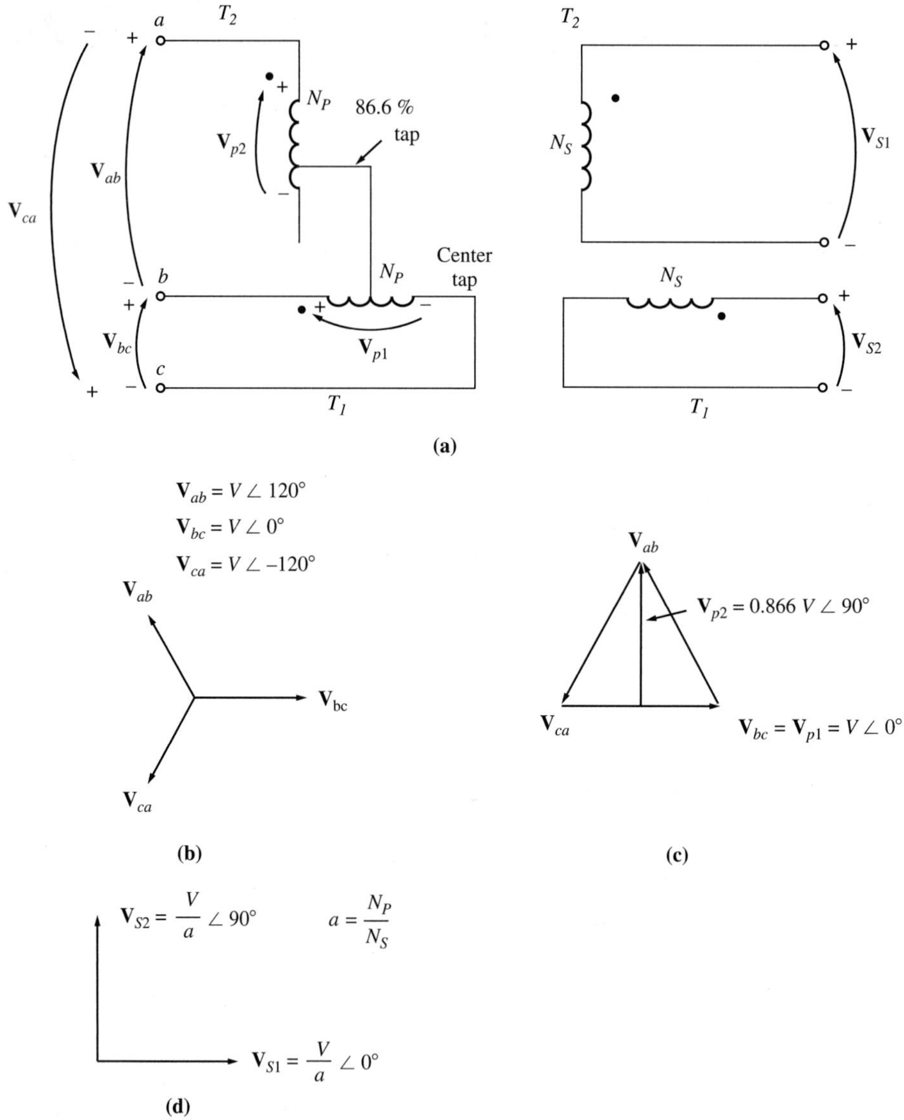

FIGURE 2–42
The Scott-T transformer connection. (a) Wiring diagram; (b) the three-phase input voltages; (c) the voltages on the transformer primary windings; (d) the two-phase secondary voltages.

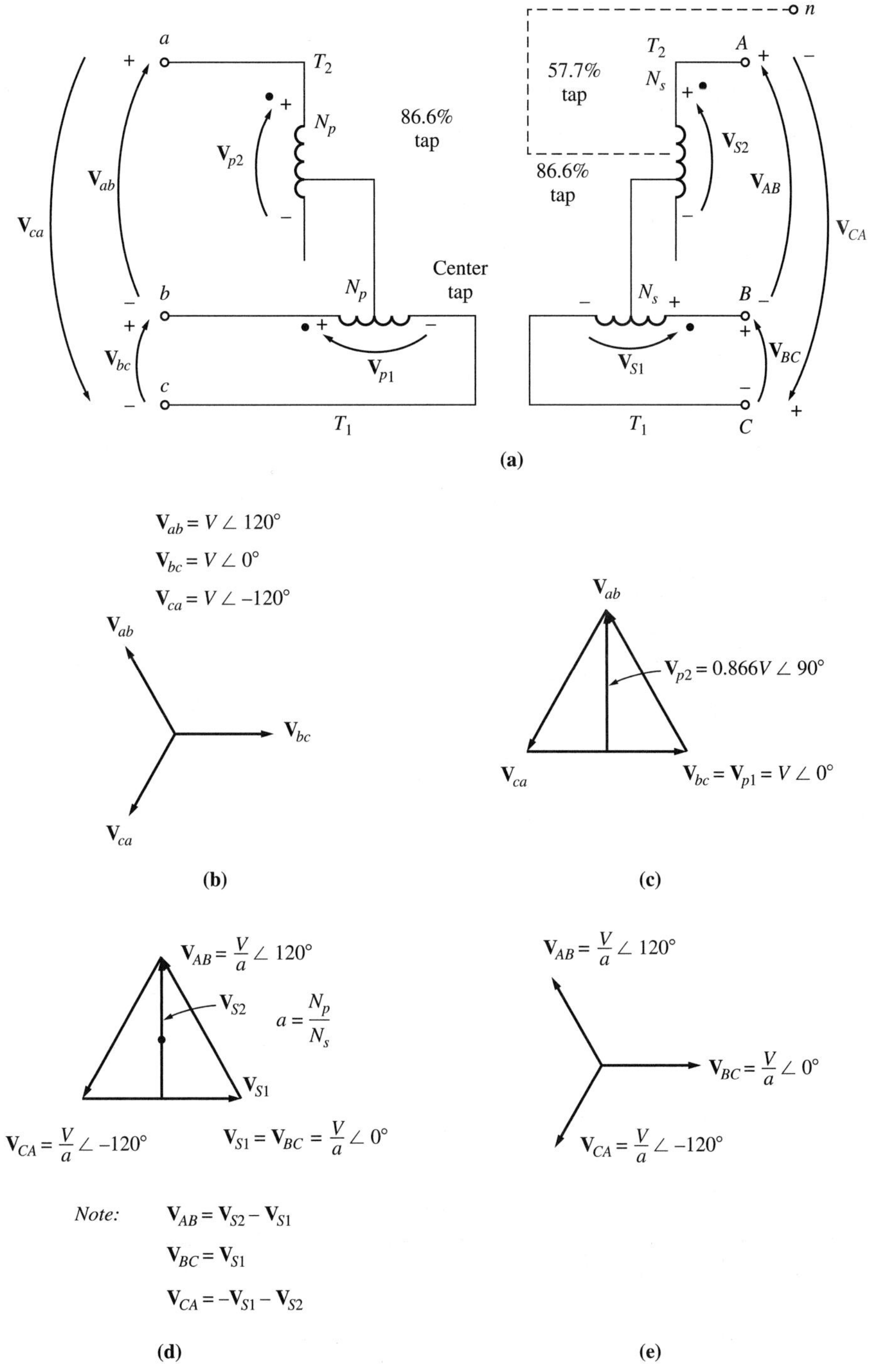

FIGURE 2–43
The three-phase T transformer connection. (a) Wiring diagram; (b) the three-phase input voltages; (c) the voltages on the transformer primary windings; (d) the voltages on the transformer secondary windings; (e) the resulting three-phase secondary voltages.

2.12 TRANSFORMER RATINGS AND RELATED PROBLEMS

Transformers have four major ratings:

1. Apparent power (kVA, or MVA)
2. Primary and secondary voltage (V)
3. Frequency (Hz)
4. Per-unit series resistance and reactance

These ratings can be found on the nameplates of most transformers. This section examines why these ratings are used to characterize a transformer. It also considers the related question of the current inrush that occurs when a transformer is first connected to the line.

The Voltage and Frequency Ratings of a Transformer

The voltage rating of a transformer serves two functions. One is to protect the winding insulation from breakdown due to an excessive voltage applied to it. This is not the most serious limitation in practical transformers. The second function is related to the magnetization curve and magnetization current of the transformer. Figure 2–11 shows a magnetization curve for a transformer. If a steady-state voltage

$$v(t) = V_M \sin \omega t \qquad \text{V}$$

is applied to a transformer's primary winding, the flux of the transformer is given by

$$\phi(t) = \frac{1}{N_P} \int v(t)\, dt$$

$$= \frac{1}{N_P} \int V_M \sin \omega t \, dt$$

$$\boxed{\phi(t) = -\frac{V_M}{\omega N_P} \cos \omega t} \tag{2–101}$$

If the applied voltage $v(t)$ is increased by 10 percent, the resulting maximum flux in the core also increases by 10 percent. Above a certain point on the magnetization curve, though, a 10 percent increase in flux requires an increase in magnetization current *much* larger than 10 percent. This concept is illustrated in Figure 2–44. As the voltage increases, the high-magnetization currents soon become unacceptable. The maximum applied voltage (and therefore the rated voltage) is set by the maximum acceptable magnetization current in the core.

Notice that voltage and frequency are related in a reciprocal fashion if the maximum flux is to be held constant:

$$\phi_{\max} = \frac{V_{\max}}{\omega N_P} \tag{2–102}$$

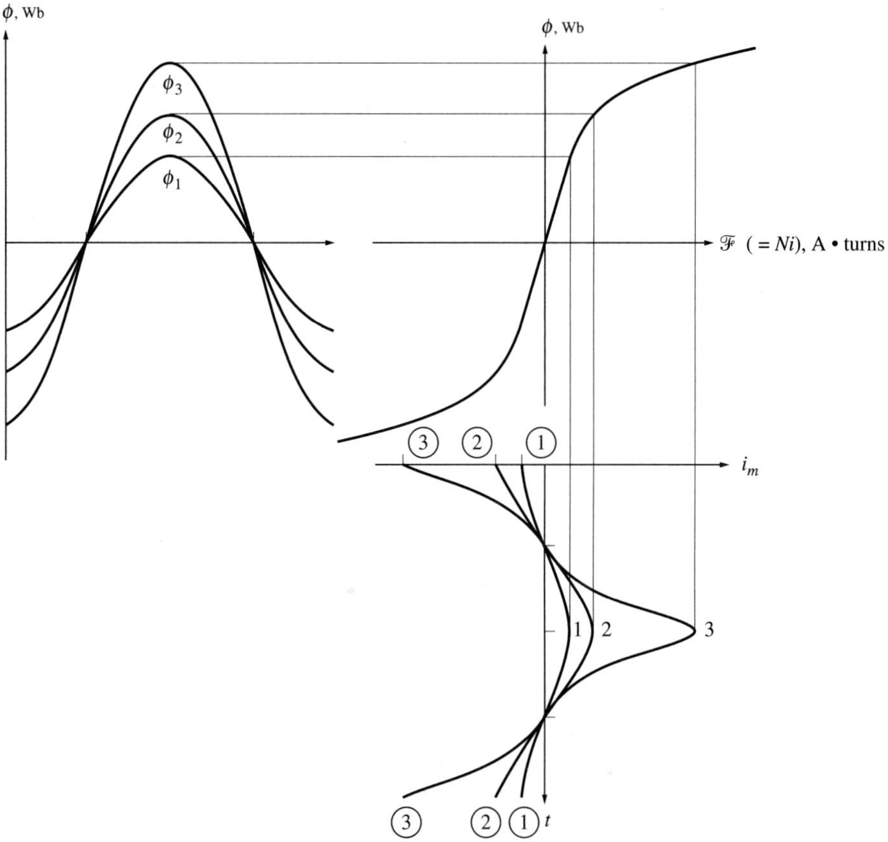

FIGURE 2–44
The effect of the peak flux in a transformer core upon the required magnetization current.

Thus, *if a 60-Hz transformer is to be operated on 50 Hz, its applied voltage must also be reduced by one-sixth* or the peak flux in the core will be too high. This reduction in applied voltage with frequency is called *derating*. Similarly, a 50-Hz transformer may be operated at a 20 percent higher voltage on 60 Hz if this action does not cause insulation problems.

Example 2–10. A 1-kVA, 230/115-V, 60-Hz single-phase transformer has 850 turns on the primary winding and 425 turns on the secondary winding. The magnetization curve for this transformer is shown in Figure 2–45.

(a) Calculate and plot the magnetization current of this transformer when it is run at 230 V on a 60-Hz power source. What is the rms value of the magnetization current?

(b) Calculate and plot the magnetization current of this transformer when it is run at 230 V on a 50-Hz power source. What is the rms value of the magnetization current? How does this current compare to the magnetization current at 60 Hz?

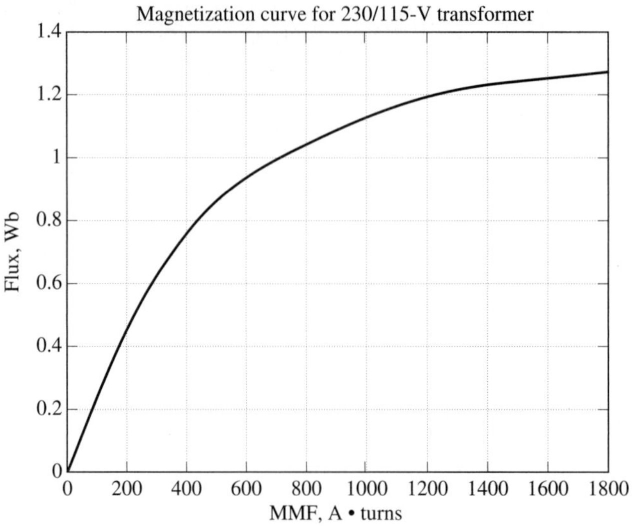

FIGURE 2–45
Magnetization curve for the 230/115-V transformer of Example 2–10.

Solution

The best way to solve this problem is to calculate the flux as a function of time for this core, and then use the magnetization curve to transform each flux value to a corresponding magnetomotive force. The magnetizing current can then be determined from the equation

$$i = \frac{\mathcal{F}}{N_P} \tag{2–103}$$

Assuming that the voltage applied to the core is $v(t) = V_M \sin \omega t$ volts, the flux in the core as a function of time is given by Equation (2–102):

$$\boxed{\phi(t) = -\frac{V_M}{\omega N_P} \cos \omega t} \tag{2–101}$$

The magnetization curve for this transformer is available electronically in a file called `mag_curve_1.dat`. This file can be used by MATLAB to translate these flux values into corresponding mmf values, and Equation (2–102) can be used to find the required magnetization current values. Finally, the rms value of the magnetization current can be calculated from the equation

$$I_{\text{rms}} = \sqrt{\frac{1}{T}\int_0^T i^2\,dt} \tag{2–104}$$

A MATLAB program to perform these calculations follows:

```
% M-file: mag_current.m
% M-file to calculate and plot the magnetization
% current of a 230/115 transformer operating at
% 230 volts and 50/60 Hz. This program also
% calculates the rms value of the mag. current.
```

```
% Load the magnetization curve. It is in two
% columns, with the first column being mmf and
% the second column being flux.
load mag_curve_1.dat;
mmf_data = mag_curve_1(:,1);
flux_data = mag_curve_1(:,2);

% Initialize values
VM = 325;                         % Maximum voltage (V)
NP = 850;                         % Primary turns

% Calculate angular velocity for 60 Hz
freq = 60;                        % Freq (Hz)
w = 2 * pi * freq;

% Calculate flux versus time
time = 0:1/3000:1/30;             % 0 to 1/30 sec
flux = -VM/(w*NP) * cos(w .* time);

% Calculate the mmf corresponding to a given flux
% using the flux's interpolation function.
mmf = interp1(flux_data,mmf_data,flux);

% Calculate the magnetization current
im = mmf / NP;

% Calculate the rms value of the current
irms = sqrt(sum(im.^2)/length(im));
disp(['The rms current at 60 Hz is ', num2str(irms)]);

% Plot the magnetization current.
figure(1)
subplot(2,1,1);
plot(time,im);
title ('\bfMagnetization Current at 60 Hz');
xlabel ('\bfTime (s)');
ylabel ('\bf\itI_{m} \rm(A)');
axis([0 0.04 -2 2]);
grid on;

% Calculate angular velocity for 50 Hz
freq = 50;                          % Freq (Hz)
w = 2 * pi * freq;

% Calculate flux versus time
time = 0:1/2500:1/25;               % 0 to 1/25 sec
flux = -VM/(w*NP) * cos(w .* time);

% Calculate the mmf corresponding to a given flux
% using the flux's interpolation function.
mmf = interp1(flux_data,mmf_data,flux);

% Calculate the magnetization current
im = mmf / NP;
```

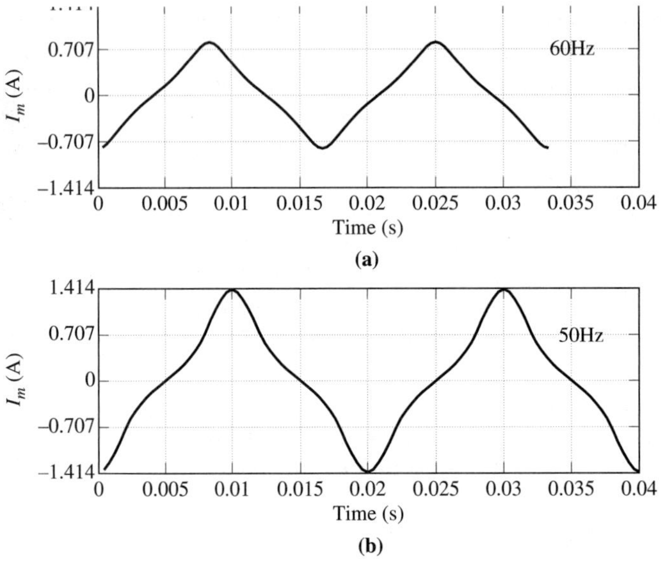

FIGURE 2–46
(a) Magnetization current for the transformer operating at 60 Hz. (b) Magnetization current for the transformer operating at 50 Hz.

```
% Calculate the rms value of the current
irms = sqrt(sum(im.^2)/length(im));
disp(['The rms current at 50 Hz is ', num2str(irms)]);

% Plot the magnetization current.
subplot(2,1,2);
plot(time,im);
title ('\bfMagnetization Current at 50 Hz');
xlabel ('\bfTime (s)');
ylabel ('\bf\itI_{m} \rm(A)');
axis([0 0.04 -2 2]);
grid on;
```

When this program executes, the results are

```
» mag_current
The rms current at 60 Hz is 0.4894
The rms current at 50 Hz is 0.79252
```

The resulting magnetization currents are shown in Figure 2–46. Note that the rms magnetization current increases by more than 60 percent when the frequency changes from 60 Hz to 50 Hz.

The Apparent Power Rating of a Transformer

The principal purpose of the apparent power rating is that, together with the voltage rating, it limits the current flow through the transformer windings. The current flow is important because it controls the i^2R losses in the transformer, which in turn control the heating of the transformer coils. It is the heating that is critical, since overheating the coils of a transformer *drastically* shortens the life of its insulation.

Transformers are rated in apparent power instead of real or reactive powers because the same amount of heating occurs for a given amount of current, regardless of its phase with respect to the terminal voltage. The magnitude of the current affects the heating, not the phase of the current.

The actual apparent power rating of a transformer may be more than a single value. In real transformers, there may be an apparent power rating for the transformer by itself, and another (higher) rating for the transformer with forced cooling. The key idea behind the power rating is that the hot-spot temperature in the transformer windings *must* be limited to protect the life of the transformer.

If a transformer's voltage is reduced for any reason (e.g., if it is operated at a lower frequency than normal), then the transformer's apparent power rating must be reduced by an equal amount. If this is not done, then the current in the transformer's windings will exceed the maximum permissible level and cause overheating.

The Problem of Current Inrush

A problem related to the voltage level in the transformer is the problem of current inrush at starting. Suppose that the voltage

$$v(t) = V_M \sin(\omega t + \theta) \qquad \text{V} \tag{2–105}$$

is applied at the moment the transformer is first connected to the power line. The maximum flux height reached on the first half-cycle of the applied voltage depends on the phase of the voltage at the time the voltage is applied. If the initial voltage is

$$v(t) = V_M \sin(\omega t + 90°) = V_M \cos \omega t \qquad \text{V} \tag{2–106}$$

and if the initial flux in the core is zero, then the maximum flux during the first half-cycle will just equal the maximum flux at steady state:

$$\phi_{\text{max}} = \frac{V_{\text{max}}}{\omega N_P} \tag{2–102}$$

This flux level is just the steady-state flux, so it causes no special problems. But if the applied voltage happens to be

$$v(t) = V_M \sin \omega t \qquad \text{V}$$

the maximum flux during the first half-cycle is given by

$$\begin{aligned}
\phi(t) &= \frac{1}{N_P}\int_0^{\pi/\omega} V_M \sin \omega t \, dt \\
&= -\frac{V_M}{\omega N_P} \cos \omega t \bigg|_0^{\pi/\omega} \\
&= -\frac{V_M}{\omega N_P}[(-1) - (1)]
\end{aligned}$$

$$\boxed{\phi_{\text{max}} = \frac{2V_{\text{max}}}{\omega N_P}} \tag{2–107}$$

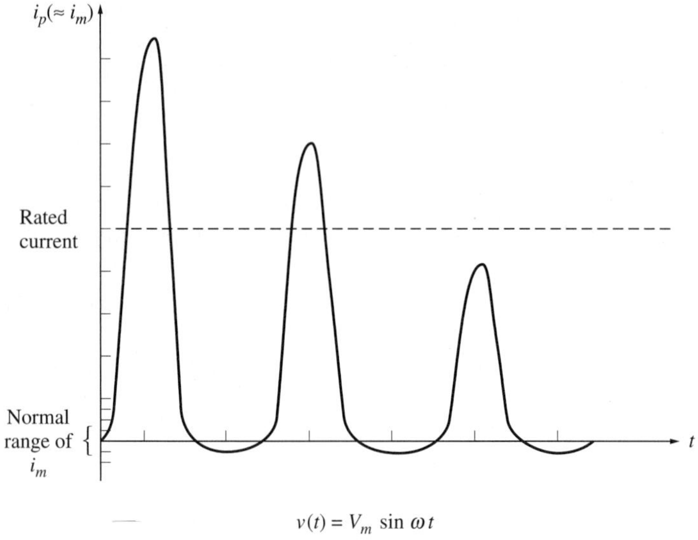

FIGURE 2–47
The current inrush due to a transformer's magnetization current on starting.

This maximum flux is twice as high as the normal steady-state flux. If the magnetization curve in Figure 2–11 is examined, it is easy to see that doubling the maximum flux in the core results in an *enormous* magnetization current. In fact, for part of the cycle, the transformer looks like a short circuit, and a very large current flows (see Figure 2–47).

For any other phase angle of the applied voltage between 90°, which is no problem, and 0°, which is the worst case, there is some excess current flow. The applied phase angle of the voltage is not normally controlled on starting, so there can be huge inrush currents during the first several cycles after the transformer is connected to the line. The transformer and the power system to which it is connected must be able to withstand these currents.

The Transformer Nameplate

A typical nameplate from a distribution transformer is shown in Figure 2–48. The information on such a nameplate includes rated voltage, rated kilovoltamperes, rated frequency, and the transformer per-unit series impedance. It also shows the voltage ratings for each tap on the transformer and the wiring schematic of the transformer.

Nameplates such as the one shown also typically include the transformer type designation and references to its operating instructions.

2.13 INSTRUMENT TRANSFORMERS

Two special-purpose transformers are used with power systems for taking measurements. One is the potential transformer, and the other is the current transformer.

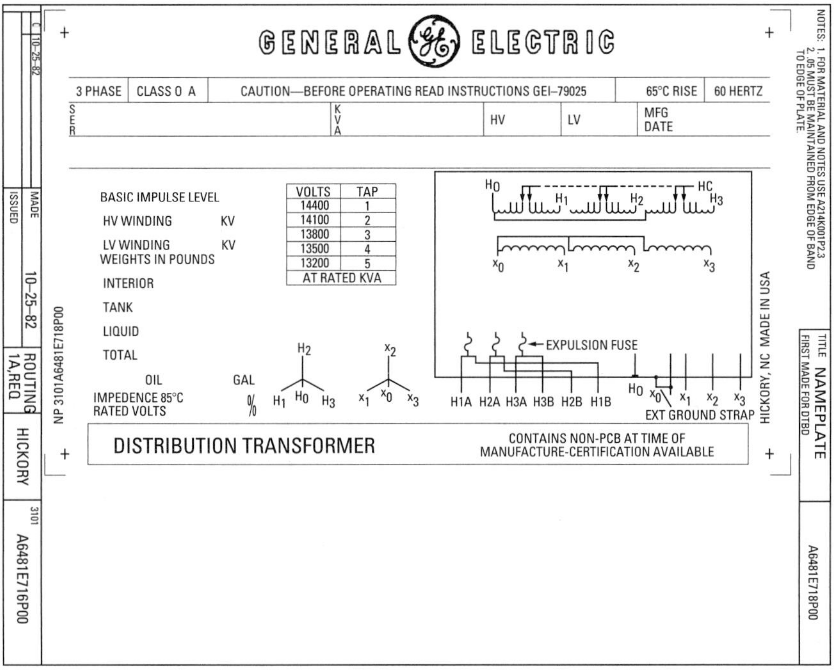

FIGURE 2–48
A sample distribution transformer nameplate. Note the ratings listed: voltage, frequency, apparent power, and tap settings. (*Courtesy of General Electric Company.*)

A *potential transformer* is a specially wound transformer with a high-voltage primary and a low-voltage secondary. It has a very low power rating, and its sole purpose is to provide a *sample* of the power system's voltage to the instruments monitoring it. Since the principal purpose of the transformer is voltage sampling, it must be very accurate so as not to distort the true voltage values too badly. Potential transformers of several *accuracy classes* may be purchased, depending on how accurate the readings must be for a given application.

Current transformers sample the current in a line and reduce it to a safe and measurable level. A diagram of a typical current transformer is shown in Figure 2–49. The current transformer consists of a secondary winding wrapped around a ferromagnetic ring, with the single primary line running through the center of the ring. The ferromagnetic ring holds and concentrates a small sample of the flux from the primary line. That flux then induces a voltage and current in the secondary winding.

A current transformer differs from the other transformers described in this chapter in that its windings are *loosely coupled.* Unlike all the other transformers, the mutual flux ϕ_M in the current transformer is smaller than the leakage flux ϕ_L. Because of the loose coupling, the voltage and current ratios of Equations (2–1) to (2–5) do not apply to a current transformer. Nevertheless, the secondary current

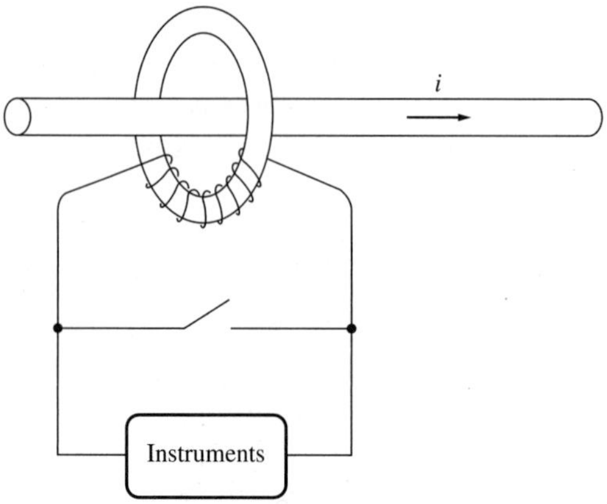

FIGURE 2–49
Sketch of a current transformer.

in a current transformer is directly proportional to the much larger primary current, and the device can provide an accurate sample of a line's current for measurement purposes.

Current transformer ratings are given as ratios of primary to secondary current. A typical current transformer ratio might be 600:5,800:5, or 1000:5. A 5-A rating is standard on the secondary of a current transformer.

It is important to keep a current transformer short-circuited at all times, since extremely high voltages can appear across its open secondary terminals. In fact, most relays and other devices using the current from a current transformer have a *shorting interlock* which must be shut before the relay can be removed for inspection or adjustment. Without this interlock, very dangerous high voltages will appear at the secondary terminals as the relay is removed from its socket.

2.14 SUMMARY

A transformer is a device for converting electric energy at one voltage level to electric energy at another voltage level through the action of a magnetic field. It plays an extremely important role in modern life by making possible the economical long-distance transmission of electric power.

When a voltage is applied to the primary of a transformer, a flux is produced in the core as given by Faraday's law. The changing flux in the core then induces a voltage in the secondary winding of the transformer. Because transformer cores have very high permeability, the net magnetomotive force required in the core to produce its flux is very small. Since the net magnetomotive force is very small, the primary circuit's magnetomotive force must be approximately equal and opposite to the secondary circuit's magnetomotive force. This fact yields the transformer current ratio.

A real transformer has leakage fluxes that pass through either the primary or the secondary winding, but not both. In addition there are hysteresis, eddy current, and copper losses. These effects are accounted for in the equivalent circuit of the transformer. Transformer imperfections are measured in a real transformer by its voltage regulation and its efficiency.

The per-unit system of measurement is a convenient way to study systems containing transformers, because in this system the different system voltage levels disappear. In addition, the per-unit impedances of a transformer expressed to its own ratings base fall within a relatively narrow range, providing a convenient check for reasonableness in problem solutions.

An autotransformer differs from a regular transformer in that the two windings of the autotransformer are connected. The voltage on one side of the transformer is the voltage across a single winding, while the voltage on the other side of the transformer is the sum of the voltages across *both* windings. Because only a portion of the power in an autotransformer actually passes through the windings, an autotransformer has a power rating advantage compared to a regular transformer of equal size. However, the connection destroys the electrical isolation between a transformer's primary and secondary sides.

The voltage levels of three-phase circuits can be transformed by a proper combination of two or three transformers. Potential transformers and current transformers can sample the voltages and currents present in a circuit. Both devices are very common in large power distribution systems.

QUESTIONS

2–1. Is the turns ratio of a transformer the same as the ratio of voltages across the transformer? Why or why not?

2–2. Why does the magnetization current impose an upper limit on the voltage applied to a transformer core?

2–3. What components compose the excitation current of a transformer? How are they modeled in the transformer's equivalent circuit?

2–4. What is the leakage flux in a transformer? Why is it modeled in a transformer equivalent circuit as an inductor?

2–5. List and describe the types of losses that occur in a transformer.

2–6. Why does the power factor of a load affect the voltage regulation of a transformer?

2–7. Why does the short-circuit test essentially show only i^2R losses and not excitation losses in a transformer?

2–8. Why does the open-circuit test essentially show only excitation losses and not i^2R losses?

2–9. How does the per-unit system of measurement eliminate the problem of different voltage levels in a power system?

2–10. Why can autotransformers handle more power than conventional transformers of the same size?

2–11. What are transformer taps? Why are they used?

2–12. What are the problems associated with the Y–Y three-phase transformer connection?

2–13. What is a TCUL transformer?

2–14. How can three-phase transformation be accomplished using only two transformers? What types of connections can be used? What are their advantages and disadvantages?

2–15. Explain why the open-Δ transformer connection is limited to supplying 57.7 percent of a normal Δ–Δ transformer bank's load.

2–16. Can a 60-Hz transformer be operated on a 50-Hz system? What actions are necessary to enable this operation?

2–17. What happens to a transformer when it is first connected to a power line? Can anything be done to mitigate this problem?

2–18. What is a potential transformer? How is it used?

2–19. What is a current transformer? How is it used?

2–20. A distribution transformer is rated at 18 kVA, 20,000/480 V, and 60 Hz. Can this transformer safely supply 15 kVA to a 415-V load at 50 Hz? Why or why not?

2–21. Why does one hear a hum when standing near a large power transformer?

PROBLEMS

2–1. A 100-kVA, 8000/277-V distribution transformer has the following resistances and reactances:

$$R_P = 5\ \Omega \qquad R_S = 0.005\ \Omega$$

$$X_P = 6\ \Omega \qquad X_S = 0.006\ \Omega$$

$$R_C = 50\ \text{k}\Omega \qquad X_M = 10\ \text{k}\Omega$$

The excitation branch impedances are given referred to the high-voltage side of the transformer.

(a) Find the equivalent circuit of this transformer referred to the low-voltage side.
(b) Find the per-unit equivalent circuit of this transformer.
(c) Assume that this transformer is supplying rated load at 277 V and 0.85 PF lagging. What is this transformer's input voltage? What is its voltage regulation?
(d) What are the copper losses and core losses in this transformer under the conditions of part *(c)*?
(e) What is the transformer's efficiency under the conditions of part *(c)*?

2–2. A single-phase power system is shown in Figure P2–1. The power source feeds a 100-kVA, 14/2.4-kV transformer through a feeder impedance of $38.2 + j140\ \Omega$. The transformer's equivalent series impedance referred to its low-voltage side is $0.10 + j0.40\ \Omega$. The load on the transformer is 90 kW at 0.80 PF lagging and 2300 V.

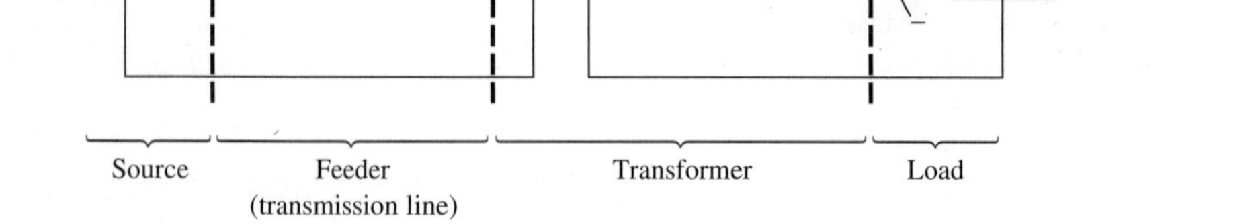

FIGURE P2–1
The circuit of Problem 2–2.

(a) What is the voltage at the power source of the system?
(b) What is the voltage regulation of the transformer?
(c) How efficient is the overall power system?

2–3. The secondary winding of an ideal transformer has a terminal voltage of $v_s(t) = 282.8 \sin 377t$ V. The turns ratio of the transformer is 100:200 ($a = 0.50$). If the secondary current of the transformer is $i_s(t) = 7.07 \sin (377t - 36.87°)$ A, what is the primary current of this transformer? What are its voltage regulation and efficiency?

2–4. The secondary winding of a real transformer has a terminal voltage of $v_s(t) = 282.8 \sin 377t$ V. The turns ratio of the transformer is 100:200 ($a = 0.50$). If the secondary current of the transformer is $i_s(t) = 7.07 \sin (377t - 36.87°)$ A, what is the primary current of this transformer? What are its voltage regulation and efficiency? The impedances of this transformer referred to the primary side are

$$R_{eq} = 0.20\ \Omega \qquad R_C = 300\ \Omega$$
$$X_{eq} = 0.80\ \Omega \qquad X_M = 100\ \Omega$$

2–5. When travelers from the USA and Canada visit Europe, they encounter a different power distribution system. Wall voltages in North America are 120 V rms at 60 Hz, while typical wall voltages in Europe are 230 V at 50 Hz. Many travelers carry small step-up/step-down transformers so that they can use their appliances in the countries that they are visiting. A typical transformer might be rated at 1 kVA and 115/230 V. It has 500 turns of wire on the 115-V side and 1000 turns of wire on the 230-V side. The magnetization curve for this transformer is shown in Figure P2–2, and can be found in file `p22.mag` at this book's website.

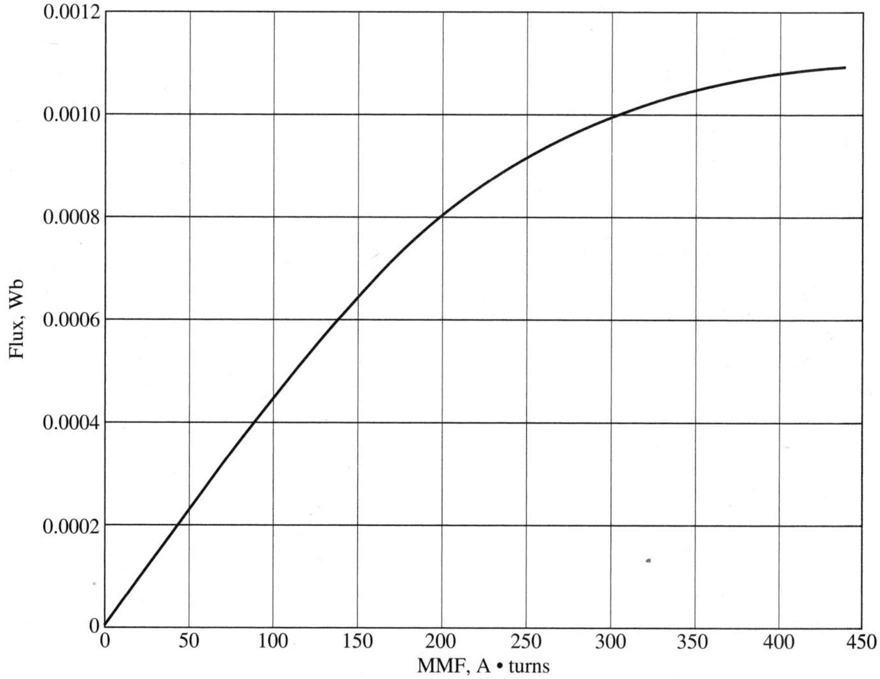

FIGURE P2–2
Magnetization curve for the transformer of Problem 2–5.

(*a*) Suppose that this transformer is connected to a 120-V, 60-Hz power source with no load connected to the 240-V side. Sketch the magnetization current that would flow in the transformer. (Use MATLAB to plot the current accurately, if it is available.) What is the rms amplitude of the magnetization current? What percentage of full-load current is the magnetization current?

(*b*) Now suppose that this transformer is connected to a 240-V, 50-Hz power source with no load connected to the 120-V side. Sketch the magnetization current that would flow in the transformer. (Use MATLAB to plot the current accurately, if it is available.) What is the rms amplitude of the magnetization current? What percentage of full-load current is the magnetization current?

(*c*) In which case is the magnetization current a higher percentage of full-load current? Why?

2–6. A 1000-VA, 230/115-V transformer has been tested to determine its equivalent circuit. The results of the tests are shown below.

Open-circuit test (on secondary side)	**Short-circuit test (on primary side)**
$V_{OC} = 115$ V	$V_{SC} = 17.1$ V
$I_{OC} = 0.11$ A	$I_{SC} = 8.7$ A
$P_{OC} = 3.9$ W	$P_{SC} = 38.1$ W

(*a*) Find the equivalent circuit of this transformer referred to the low-voltage side of the transformer.

(*b*) Find the transformer's voltage regulation at rated conditions and (1) 0.8 PF lagging, (2) 1.0 PF, (3) 0.8 PF leading.

(*c*) Determine the transformer's efficiency at rated conditions and 0.8 PF lagging.

2–7. A 30-kVA, 8000/230-V distribution transformer has an impedance referred to the primary of $20 + j100\ \Omega$. The components of the excitation branch referred to the primary side are $R_C = 100\ \text{k}\Omega$ and $X_M = 20\ \text{k}\Omega$.

(*a*) If the primary voltage is 7967 V and the load impedance is $Z_L = 2.0 + j0.7\ \Omega$, what is the secondary voltage of the transformer? What is the voltage regulation of the transformer?

(*b*) If the load is disconnected and a capacitor of $-j3.0\ \Omega$ is connected in its place, what is the secondary voltage of the transformer? What is its voltage regulation under these conditions?

2–8. A 150-MVA, 15/200-kV, single-phase power transformer has a per-unit resistance of 1.2 percent and a per-unit reactance of 5 percent (data taken from the transformer's nameplate). The magnetizing impedance is $j80$ per unit.

(*a*) Find the equivalent circuit referred to the low-voltage side of this transformer.

(*b*) Calculate the voltage regulation of this transformer for a full-load current at power factor of 0.8 lagging.

(*c*) Calculate the copper and core losses in the transformer at the conditions in (*b*).

(*d*) Assume that the primary voltage of this transformer is a constant 15 kV, and plot the secondary voltage as a function of load current for currents from no-load to full-load. Repeat this process for power factors of 0.8 lagging, 1.0, and 0.8 leading.

2–9. A 5000-kVA, 230/13.8-kV, single-phase power transformer has a per-unit resistance of 1 percent and a per-unit reactance of 5 percent (data taken from the transformer's nameplate). The open-circuit test performed on the low-voltage side of the transformer yielded the following data:

$$V_{OC} = 13.8 \text{ kV} \qquad I_{OC} = 21.1 \text{ A} \qquad P_{OC} = 90.8 \text{ kW}$$

(a) Find the equivalent circuit referred to the low-voltage side of this transformer.
(b) If the voltage on the secondary side is 13.8 kV and the power supplied is 4000 kW at 0.8 PF lagging, find the voltage regulation of the transformer. Find its efficiency.

2–10. A three-phase transformer bank is to handle 500 kVA and have a 34.5/11-kV voltage ratio. Find the rating of each individual transformer in the bank (high voltage, low voltage, turns ratio, and apparent power) if the transformer bank is connected to *(a)* Y–Y, *(b)* Y–Δ, *(c)* Δ–Y, *(d)* Δ–Δ, *(e)* open–Δ, *(f)* open-Y–open-Δ.

2–11. A 100-MVA, 230/115-kV, Δ–Y three-phase power transformer has a per-unit resistance of 0.015 pu and a per-unit reactance of 0.06 pu. The excitation branch elements are R_C = 100 pu and X_M = 20 pu.
(a) If this transformer supplies a load of 80 MVA at 0.8 PF lagging, draw the phasor diagram of one phase of the transformer.
(b) What is the voltage regulation of the transformer bank under these conditions?
(c) Sketch the equivalent circuit referred to the low-voltage side of one phase of this transformer. Calculate all the transformer impedances referred to the low-voltage side.
(d) Determine the losses in the transformer and the efficiency of the transformer under the conditions of part *(b)*.

2–12. Three 20-kVA, 24,000/277-V distribution transformers are connected in Δ–Y. The open-circuit test was performed on the low-voltage side of this transformer bank, and the following data were recorded:

$$V_{\text{line,OC}} = 480 \text{ V} \qquad I_{\text{line,OC}} = 4.10 \text{ A} \qquad P_{3\phi,\text{OC}} = 945 \text{ W}$$

The short-circuit test was performed on the high-voltage side of this transformer bank, and the following data were recorded:

$$V_{\text{line,SC}} = 1400 \text{ V} \qquad I_{\text{line,SC}} = 1.80 \text{ A} \qquad P_{3\phi,\text{SC}} = 912 \text{ W}$$

(a) Find the per-unit equivalent circuit of this transformer bank.
(b) Find the voltage regulation of this transformer bank at the rated load and 0.90 PF lagging.
(c) What is the transformer bank's efficiency under these conditions?

2–13. A 14,000/480-V, three-phase, Δ–Y-connected transformer bank consists of three identical 100-kVA, 8314/480-V transformers. It is supplied with power directly from a large constant-voltage bus. In the short-circuit test, the recorded values on the high-voltage side for one of these transformers are

$$V_{SC} = 510 \text{ V} \qquad I_{SC} = 12.6 \text{ A} \qquad P_{SC} = 3000 \text{ W}$$

(a) If this bank delivers a rated load at 0.8 PF lagging and rated voltage, what is the line-to-line voltage on the primary of the transformer bank?
(b) What is the voltage regulation under these conditions?
(c) Assume that the primary phase voltage of this transformer is a constant 8314 V, and plot the secondary voltage as a function of load current for currents from no-load to full-load. Repeat this process for power factors of 0.8 lagging, 1.0, and 0.8 leading.

(d) Plot the voltage regulation of this transformer as a function of load current for currents from no-load to full-load. Repeat this process for power factors of 0.8 lagging, 1.0, and 0.8 leading.
(e) Sketch the per-unit equivalent circuit of this transformer.

2–14. A 13.8-kV, single-phase generator supplies power to a load through a transmission line. The load's impedance is $Z_{load} = 500 \angle 36.87°\Omega$, and the transmission line's impedance is $Z_{line} = 60 \angle 60°\Omega$.

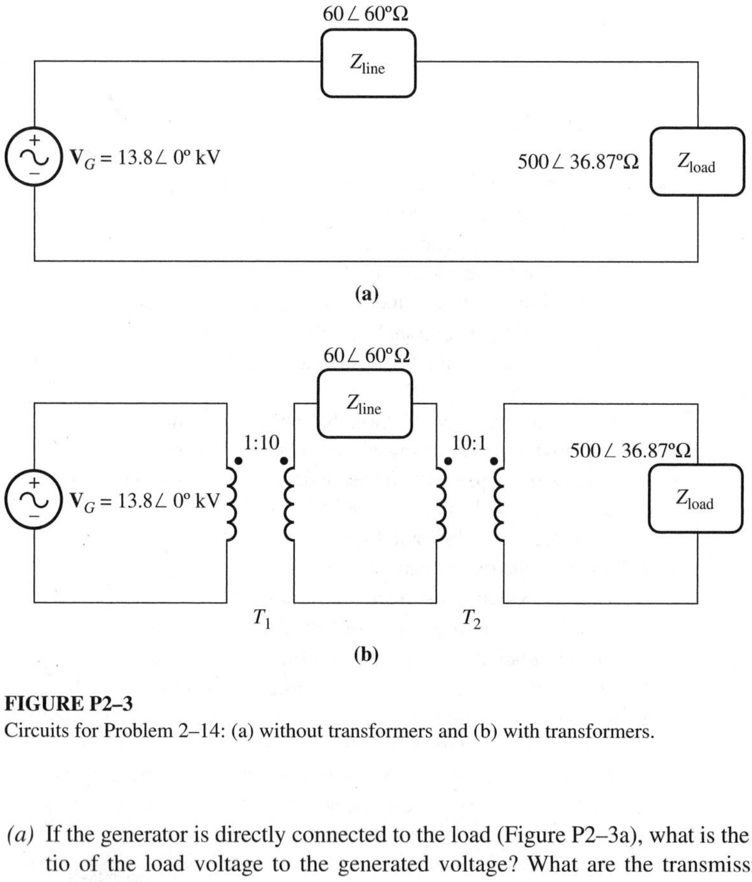

FIGURE P2–3
Circuits for Problem 2–14: (a) without transformers and (b) with transformers.

(a) If the generator is directly connected to the load (Figure P2–3a), what is the ratio of the load voltage to the generated voltage? What are the transmission losses of the system?
(b) What percentage of the power supplied by the source reaches the load (what is the efficiency of the transmission system)?
(c) If a 1:10 step-up transformer is placed at the output of the generator and a 10:1 transformer is placed at the load end of the transmission line, what is the new ratio of the load voltage to the generated voltage? What are the transmission losses of the system now? (*Note*: The transformers may be assumed to be ideal.)
(d) What percentage of the power supplied by the source reaches the load now?
(e) Compare the efficiencies of the transmission system with and without transformers.

2–15. An autotransformer is used to connect a 12.6-kV distribution line to a 13.8-kV distribution line. It must be capable of handling 2000 kVA. There are three phases, connected Y–Y with their neutrals solidly grounded.

(a) What must the N_C/N_{SE} turns ratio be to accomplish this connection?

(b) How much apparent power must the windings of each autotransformer handle?

(c) What is the power advantage of this autotransformer system?

(d) If one of the autotransformers were reconnected as an ordinary transformer, what would its ratings be?

2–16. Prove the following statement: If a transformer having a series impedance Z_{eq} is connected as an autotransformer, its per-unit series impedance Z'_{eq} as an autotransformer will be

$$Z'_{eq} = \frac{N_{SE}}{N_{SE} + N_C} Z_{eq}$$

Note that this expression is the reciprocal of the autotransformer power advantage.

2–17. A 10-kVA, 480/120-V conventional transformer is to be used to supply power from a 600-V source to a 120-V load. Consider the transformer to be ideal, and assume that all insulation can handle 600 V.

(a) Sketch the transformer connection that will do the required job.

(b) Find the kilovoltampere rating of the transformer in the configuration.

(c) Find the maximum primary and secondary currents under these conditions.

2–18. A 10-kVA, 480/120-V conventional transformer is to be used to supply power from a 600-V source to a 480-V load. Consider the transformer to be ideal, and assume that all insulation can handle 600 V.

(a) Sketch the transformer connection that will do the required job.

(b) Find the kilovoltampere rating of the transformer in the configuration.

(c) Find the maximum primary and secondary currents under these conditions.

(d) The transformer in Problem 2–18 is identical to the transformer in Problem 2–17, but there is a significant difference in the apparent power capability of the transformer in the two situations. Why? What does that say about the best circumstances in which to use an autotransformer?

2–19. Two phases of a 14.4-kV, three-phase distribution line serve a remote rural road (the neutral is also available). A farmer along the road has a 480 V feeder supplying 200 kW at 0.85 PF lagging of three-phase loads, plus 60 kW at 0.9 PF lagging of single-phase loads. The single-phase loads are distributed evenly among the three phases. Assuming that the open-Y–open-Δ connection is used to supply power to his farm, find the voltages and currents in each of the two transformers. Also find the real and reactive powers supplied by each transformer. Assume the transformers are ideal. What is the minimum required kVA rating of each transformer?

2–20. A 50-kVA, 20,000/480-V, 60-Hz, single-phase distribution transformer is tested with the following results:

Open-circuit test (measured from secondary side)	**Short-circuit test (measured from primary side)**
$V_{OC} = 480$ V	$V_{SC} = 1130$ V
$I_{OC} = 4.1$ A	$I_{SC} = 1.30$ A
$P_{OC} = 620$ W	$P_{SC} = 550$ W

(*a*) Find the per-unit equivalent circuit for this transformer at 60 Hz.
(*b*) What is the efficiency of the transformer at rated conditions and unity power factor? What is the voltage regulation at those conditions?
(*c*) What would the rating of this transformer be if it were operated on a 50-Hz power system?
(*d*) Sketch the equivalent circuit of this transformer referred to the primary side *if it is operating at 50 Hz.*
(*e*) What is the efficiency of the transformer at rated conditions on a 50-Hz power system, with unity power factor? What is the voltage regulation at those conditions?
(*f*) How does the efficiency of a transformer at rated conditions and 60 Hz compare to the same physical device running at 50 Hz?

2–21. Prove that the three-phase system of voltages on the secondary of the Y–Δ transformer shown in Figure 2–37b lags the three-phase system of voltages on the primary of the transformer by 30°.

2–22. Prove that the three-phase system of voltages on the secondary of the Δ–Y transformer shown in Figure 2–37c lags the three-phase system of voltages on the primary of the transformer by 30°.

2–23. A single-phase, 10-kVA, 480/120-V transformer is to be used as an autotransformer tying a 600-V distribution line to a 480-V load. When it is tested as a conventional transformer, the following values are measured on the primary (480-V) side of the transformer:

Open-circuit test (measured on secondary side)	**Short-circuit test (measured on primary side)**
$V_{OC} = 120$ V	$V_{SC} = 10.0$ V
$I_{OC} = 1.60$ A	$I_{SC} = 10.6$ A
$P_{OC} = 38$ W	$P_{SC} = 25$ W

(*a*) Find the per-unit equivalent circuit of this transformer when it is connected in the conventional manner. What is the efficiency of the transformer at rated conditions and unity power factor? What is the voltage regulation at those conditions?
(*b*) Sketch the transformer connections when it is used as a 600/480-V step-down autotransformer.
(*c*) What is the kilovoltampere rating of this transformer when it is used in the autotransformer connection?
(*d*) Answer the questions in (*a*) for the autotransformer connection.

2–24. Figure P2–4 shows a one-line diagram of a power system consisting of a three-phase, 480-V, 60-Hz generator supplying two loads through a transmission line with a pair of transformers at either end. (NOTE: One-line diagrams are described in Appendix A, the discussion of three-phase power circuits.)

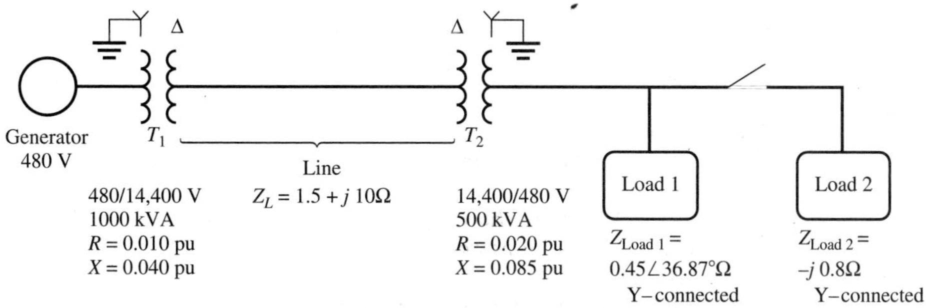

FIGURE P2–4
A one-line diagram of the power system of Problem 2–24. Note that some impedance values are given in the per-unit system, while others are given in ohms.

(a) Sketch the per-phase equivalent circuit of this power system.
(b) With the switch opened, find the real power P, reactive power Q, and apparent power S supplied by the generator. What is the power factor of the generator?
(c) With the switch closed, find the real power P, reactive power Q, and apparent power S supplied by the generator. What is the power factor of the generator?
(d) What are the transmission losses (transformer plus transmission line losses) in this system with the switch open? With the switch closed? What is the effect of adding Load 2 to the system?

REFERENCES

1. Beeman, Donald: *Industrial Power Systems Handbook,* McGraw-Hill, New York, 1955.
2. Del Toro, V.: *Electric Machines and Power Systems,* Prentice-Hall, Englewood Cliffs, N.J., 1985.
3. Feinberg, R.: *Modern Power Transformer Practice,* Wiley, New York, 1979.
4. Fitzgerald, A. E., C. Kingsley, Jr., and S. D. Umans: *Electric Machinery,* 6th ed., McGraw-Hill, New York, 2003.
5. McPherson, George: *An Introduction to Electrical Machines and Transformers,* Wiley, New York, 1981.
6. M.I.T. Staff: *Magnetic Circuits and Transformers,* Wiley, New York, 1943.
7. Slemon, G. R., and A. Straughen: *Electric Machines,* Addison-Wesley, Reading, Mass., 1980.
8. *Electrical Transmission and Distribution Reference Book,* Westinghouse Electric Corporation, East Pittsburgh, 1964.

CHAPTER

3

AC MACHINERY FUNDAMENTALS

LEARNING OBJECTIVES

- Learn how to generate an ac voltage in a loop rotating in a uniform magnetic field.
- Learn how to generate torque in a loop carrying a current in a uniforrn magnetic field.
- Learn how to create a rotating magnetic field from a three-phase stator.
- Understand how a rotating rotor with a magnetic field induces ac voltages in stator windings.
- Understand the relationship between electrical frequency, the number of poles, and the rotational speed of an electrical machine.
- Understand how torque is induced in an ac machine.
- Understand the effects of winding insulation on machine lifetimes.
- Understand the types of losses in a machine, and the power flow diagram.

Ac machines are generators that convert mechanical energy to ac electrical energy and motors that convert ac electrical energy to mechanical energy. The fundamental principles of ac machines are very simple, but unfortunately, they are somewhat obscured by the complicated construction of real machines. This chapter will first explain the principles of ac machine operation using simple examples, and then consider some of the complications that occur in real ac machines.

There are two major classes of ac machines—synchronous machines and induction machines. *Synchronous machines* are motors and generators whose magnetic field current is supplied by a separate dc power source, while *induction machines* are motors and generators whose field current is supplied by magnetic induction (transformer action) into their field windings. The field circuits of most synchronous and induction machines are located on their rotors. This chapter covers some of the fundamentals common to both types of three-phase ac machines. Synchronous machines will be covered in detail in Chapters 4 and 5, and induction machines will be covered in Chapter 6.

3.1 A SIMPLE LOOP IN A UNIFORM MAGNETIC FIELD

We will start our study of ac machines with a simple loop of wire rotating within a uniform magnetic field. A loop of wire in a uniform magnetic field is the simplest possible machine that produces a sinusoidal ac voltage. This case is not representative of real ac machines, since the flux in real ac machines is not constant in either magnitude or direction. However, the factors that control the voltage and torque on the loop will be the same as the factors that control the voltage and torque in real ac machines.

Figure 3–1 shows a simple machine consisting of a large stationary magnet producing an essentially constant and uniform magnetic field and a rotating loop of wire within that field. The rotating part of the machine is called the *rotor*, and the stationary part of the machine is called the *stator*. We will now determine the voltages present in the rotor as it rotates within the magnetic field.

The Voltage Induced in a Simple Rotating Loop

If the rotor of this machine is rotated, a voltage will be induced in the wire loop. To determine the magnitude and shape of the voltage, examine Figure 3–2. The loop of wire shown is rectangular, with sides *ab* and *cd* perpendicular to the plane of the page and with sides *bc* and *da* parallel to the plane of the page. The magnetic field is constant and uniform, pointing from left to right across the page.

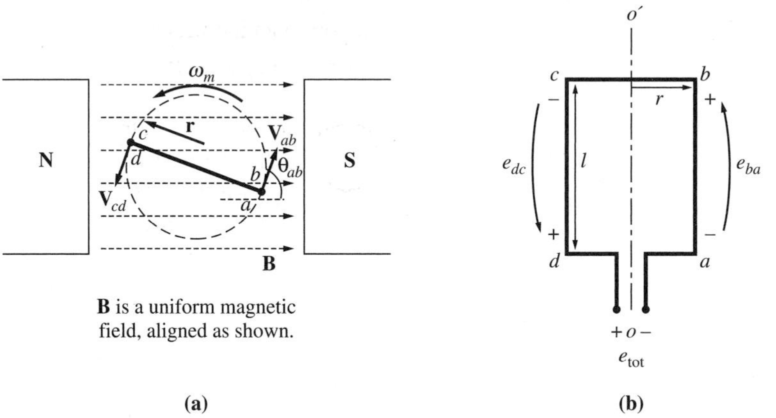

FIGURE 3–1
A simple rotating loop in a uniform magnetic field. (a) Front view; (b) view of coil.

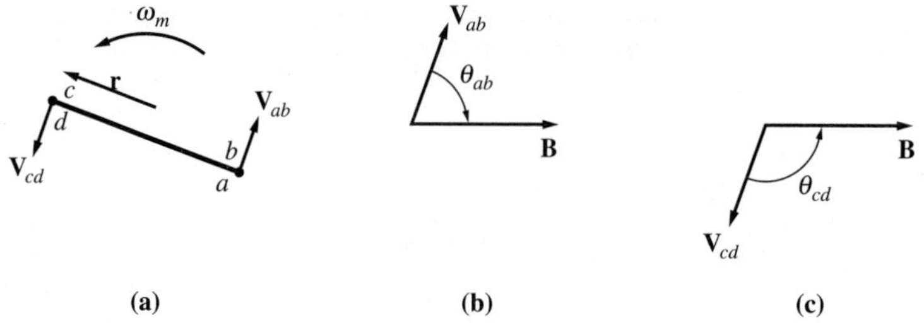

FIGURE 3–2
(a) Velocities and orientations of the sides of the loop with respect to the magnetic field. (b) The direction of motion with respect to the magnetic field for side *ab*. (c) The direction of motion with respect to the magnetic field for side *cd*.

To determine the total voltage e_{tot} on the loop, we will examine each segment of the loop separately and sum all the resulting voltages. The voltage on each segment is given by Equation (1–45):

$$e_{\text{ind}} = (\mathbf{v} \times \mathbf{B}) \bullet \mathbf{l} \tag{1–45}$$

1. *Segment ab.* In this segment, the velocity of the wire is tangential to the path of rotation, while the magnetic field **B** points to the right, as shown in Figure 3–2b. The quantity **v** × **B** points into the page, which is the same direction as segment *ab*. Therefore, the induced voltage on this segment of the wire is

$$\begin{aligned} e_{ba} &= (\mathbf{v} \times \mathbf{B}) \bullet \mathbf{l} \\ &= vBl \sin \theta_{ab} \qquad \text{into the page} \end{aligned} \tag{3–1}$$

2. *Segment bc.* In the first half of this segment, the quantity **v** × **B** points into the page, and in the second half of this segment, the quantity **v** × **B** points out of the page. Since the length **l** is in the plane of the page, **v** × **B** is perpendicular to **l** for both portions of the segment. Therefore the voltage in segment *bc* will be zero:

$$e_{cb} = 0 \tag{3–2}$$

3. *Segment cd.* In this segment, the velocity of the wire is tangential to the path of rotation, while the magnetic field **B** points to the right, as shown in Figure 3–2c. The quantity **v** × **B** points into the page, which is the same direction as segment *cd*. Therefore, the induced voltage on this segment of the wire is

$$\begin{aligned} e_{dc} &= (\mathbf{v} \times \mathbf{B}) \bullet \mathbf{l} \\ &= vBl \sin \theta_{cd} \qquad \text{out of the page} \end{aligned} \tag{3–3}$$

4. *Segment da.* Just as in segment *bc*, **v** × **B** is perpendicular to **l**. Therefore the voltage in this segment will also be zero:

$$e_{ad} = 0 \tag{3–4}$$

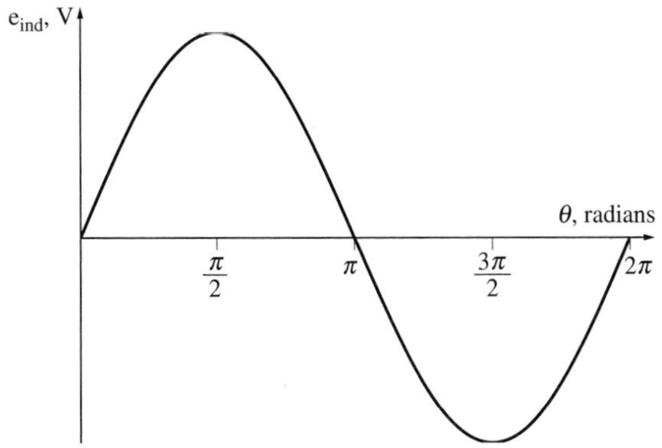

FIGURE 3–3
Plot of e_{ind} versus θ.

The total induced voltage on the loop e_{ind} is the sum of the voltages on each of its sides:

$$\begin{aligned} e_{ind} &= e_{ba} + e_{cb} + e_{dc} + e_{ad} \\ &= vBl \sin \theta_{ab} + vBl \sin \theta_{cd} \end{aligned} \tag{3–5}$$

Note that $\theta_{ab} = 180° - \theta_{cd}$, and recall the trigonometric identity $\sin \theta = \sin (180° - \theta)$. Therefore, the induced voltage becomes

$$e_{ind} = 2vBl \sin \theta \tag{3–6}$$

The resulting voltage e_{ind} is shown as a function of time in Figure 3–3.

There is an alternative way to express Equation (3–6), which clearly relates the behavior of the single loop to the behavior of larger, real ac machines. To derive this alternative expression, examine Figure 3–1 again. If the loop is rotating at a constant angular velocity ω, then angle θ of the loop will increase linearly with time. In other words,

$$\theta = \omega t$$

Also, the tangential velocity v of the edges of the loop can be expressed as

$$v = r\omega \tag{3–7}$$

where r is the radius from axis of rotation out to the edge of the loop and ω is the angular velocity of the loop. Substituting these expressions into Equation (3–6) gives

$$e_{ind} = 2r\omega Bl \sin \omega t \tag{3–8}$$

Notice also from Figure 3–1b that the area A of the loop is just equal to $2rl$. Therefore,

$$e_{ind} = AB\omega \sin \omega t \tag{3–9}$$

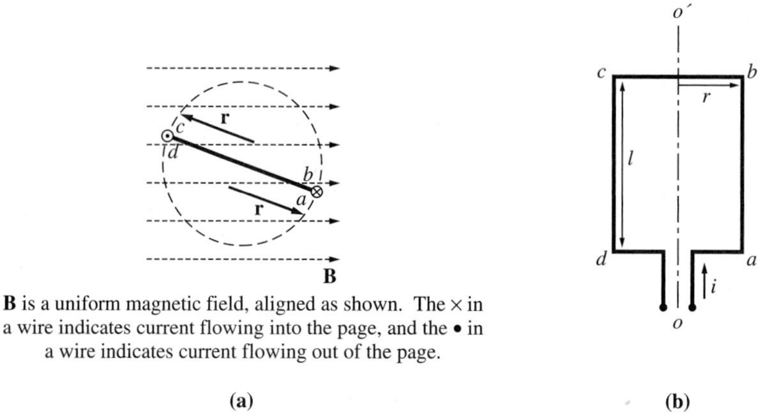

FIGURE 3–4
A current-carrying loop in a uniform magnetic field. (a) Front view; (b) view of coil.

Finally, note that the maximum flux through the loop occurs when the loop is perpendicular to the magnetic flux density lines. This flux is just the product of the loop's surface area and the flux density through the loop.

$$\phi_{\max} = AB \tag{3–10}$$

Therefore, the final form of the voltage equation is

$$\boxed{e_{\text{ind}} = \phi_{\max}\omega \sin \omega t} \tag{3–11}$$

Thus, *the voltage generated in the loop is a sinusoid whose magnitude is equal to the product of the flux inside the machine and the speed of rotation of the machine.* This is also true of real ac machines. In general, the voltage in any real machine will depend on three factors:

1. The flux in the machine
2. The speed of rotation
3. A constant representing the construction of the machine (the number of loops, etc.)

The Torque Induced in a Current-Carrying Loop

Now assume that the rotor loop is at some arbitrary angle θ with respect to the magnetic field, and that a current i is flowing in the loop, as shown in Figure 3–4. If a current flows in the loop, then a torque will be induced on the wire loop. To determine the magnitude and direction of the torque, examine Figure 3–5. The force on each segment of the loop will be given by Equation (1–43),

$$\mathbf{F} = i(\mathbf{l} \times \mathbf{B}) \tag{1–43}$$

where i = magnitude of current in the segment

$\mathbf{l}$ = length of the segment, with direction of $\mathbf{l}$ defined to be in the direction of current flow

$\mathbf{B}$ = magnetic flux density vector

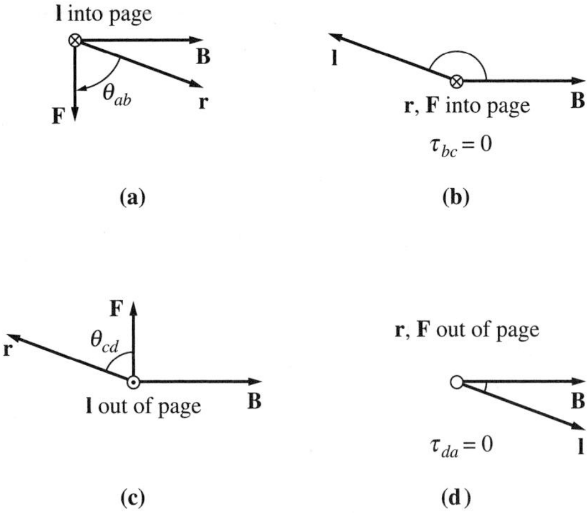

FIGURE 3–5
(a) Derivation of force and torque on segment *ab*. (b) Derivation of force and torque on segment *bc*. (c) Derivation of force and torque on segment *cd*. (d) Derivation of force and torque on segment *da*.

The torque on that segment will then be given by

$$\begin{aligned} \tau &= \text{(force applied) (perpendicular distance)} \\ &= (F)\,(r \sin \theta) \\ &= rF \sin \theta \end{aligned} \tag{1–6}$$

where θ is the angle between the vector **r** and the vector **F**. The direction of the torque is clockwise if it would tend to cause a clockwise rotation and counterclockwise if it would tend to cause a counterclockwise rotation.

1. *Segment ab.* In this segment, the direction of the current is into the page, while the magnetic field **B** points to the right, as shown in Figure 3–5a. The quantity **l** × **B** points down. Therefore, the induced force on this segment of the wire is

$$\begin{aligned} \mathbf{F} &= i(\mathbf{l} \times \mathbf{B}) \\ &= ilB \qquad \text{down} \end{aligned}$$

   The resulting torque is

$$\begin{aligned} \tau_{ab} &= (F)\,(r \sin \theta_{ab}) \\ &= rilB \sin \theta_{ab} \qquad \text{clockwise} \end{aligned} \tag{3–12}$$

2. *Segment bc.* In this segment, the direction of the current is in the plane of the page, while the magnetic field **B** points to the right, as shown in Figure 3–5b. The quantity **l** × **B** points into the page. Therefore, the induced force on this segment of the wire is

$$\begin{aligned} \mathbf{F} &= i(\mathbf{l} \times \mathbf{B}) \\ &= ilB \qquad \text{into the page} \end{aligned}$$

For this segment, the resulting torque is 0, since vectors **r** and **l** are parallel (both point into the page), and the angle θ_{bc} is 0.

$$\begin{aligned}\tau_{bc} &= (F)\,(r \sin \theta_{ab}) \\ &= 0\end{aligned} \tag{3–13}$$

3. *Segment cd.* In this segment, the direction of the current is out of the page, while the magnetic field **B** points to the right, as shown in Figure 3–5c. The quantity **l** × **B** points up. Therefore, the induced force on this segment of the wire is

$$\begin{aligned}\mathbf{F} &= i(\mathbf{l} \times \mathbf{B}) \\ &= ilB \qquad \text{up}\end{aligned}$$

The resulting torque is

$$\begin{aligned}\tau_{cd} &= (F)\,(r \sin \theta_{cd}) \\ &= rilB \sin \theta_{cd} \qquad \text{clockwise}\end{aligned} \tag{3–14}$$

4. *Segment da.* In this segment, the direction of the current is in the plane of the page, while the magnetic field **B** points to the right, as shown in Figure 3–5d. The quantity **l** × **B** points out of the page. Therefore, the induced force on this segment of the wire is

$$\begin{aligned}\mathbf{F} &= i(\mathbf{l} \times \mathbf{B}) \\ &= ilB \qquad \text{out of the page}\end{aligned}$$

For this segment, the resulting torque is 0, since vectors **r** and **l** are parallel (both point out of the page), and the angle θ_{da} is 0.

$$\begin{aligned}\tau_{da} &= (F)\,(r \sin \theta_{da}) \\ &= 0\end{aligned} \tag{3–15}$$

The total induced torque on the loop τ_{ind} is the sum of the torques on each of its sides:

$$\begin{aligned}\tau_{\text{ind}} &= \tau_{ab} + \tau_{bc} + \tau_{cd} + \tau_{da} \\ &= rilB \sin \theta_{ab} + rilB \sin \theta_{cd}\end{aligned} \tag{3–16}$$

Note that $\theta_{ab} = \theta_{cd}$, so the induced torque becomes

$$\tau_{\text{ind}} = 2rilB \sin \theta \tag{3–17}$$

The resulting torque τ_{ind} is shown as a function of angle in Figure 3–6. Note that the torque is maximum when the plane of the loop is parallel to the magnetic field, and the torque is zero when the plane of the loop is perpendicular to the magnetic field.

There is an alternative way to express Equation (3–17), which clearly relates the behavior of the single loop to the behavior of larger, real ac machines. To derive this alternative expression, examine Figure 3–7. If the current in the loop is as shown in the figure, that current will generate a magnetic flux density $\mathbf{B}_{\text{loop}}$ with the direction shown. The magnitude of $\mathbf{B}_{\text{loop}}$ will be

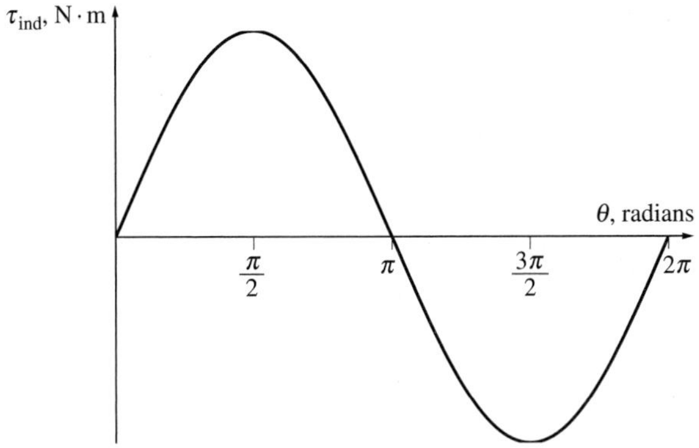

FIGURE 3–6
Plot of τ_{ind} versus θ.

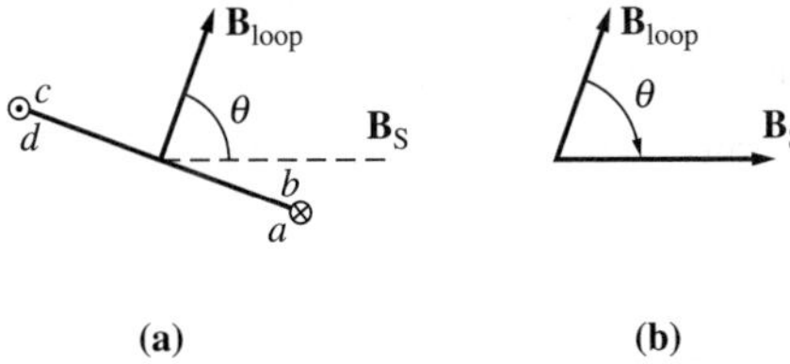

FIGURE 3–7
Derivation of the induced torque equation. (a) The current in the loop produces a magnetic flux density $\mathbf{B}_{loop}$ perpendicular to the plane of the loop; (b) geometric relationship between $\mathbf{B}_{loop}$ and $\mathbf{B}_S$.

$$B_{loop} = \frac{\mu i}{G}$$

where G is a factor that depends on the geometry of the loop.[1] Also, note that the area of the loop A is just equal to $2rl$. Substituting these two equations into Equation (3–17) yields the result

$$\tau_{ind} = \frac{AG}{\mu} B_{loop} B_S \sin\theta \tag{3–18}$$

$$= kB_{loop} B_S \sin\theta \tag{3–19}$$

where $k = AG/\mu$ is a factor depending on the construction of the machine, B_S is used for the stator magnetic field to distinguish it from the magnetic field generated by the rotor, and θ is the angle between $\mathbf{B}_{loop}$ and $\mathbf{B}_S$. The angle between $\mathbf{B}_{loop}$ and $\mathbf{B}_S$ can be seen by trigonometric identities to be the same as the angle θ in Equation (3–17).

Both the magnitude and the direction of the induced torque can be determined by expressing Equation (3–19) as a cross product:

$$\boxed{\tau_{ind} = k\mathbf{B}_{loop} \times \mathbf{B}_S} \tag{3–20}$$

[1]If the loop were a circle, then $G = 2r$, where r is the radius of the circle, so $B_{loop} = \mu i/2r$. For a rectangular loop, the value of G will vary depending on the exact length-to-width ratio of the loop.

Applying this equation to the loop in Figure 3–7 produces a torque vector into the page, indicating that the torque is clockwise, with the magnitude given by Equation (3–19).

Thus, *the torque induced in the loop is proportional to the strength of the loop's magnetic field, the strength of the external magnetic field, and the sine of the angle between them.* This is also true of real ac machines. In general, the torque in any real machine will depend on four factors:

1. The strength of the rotor magnetic field
2. The strength of the external magnetic field
3. The sine of the angle between them
4. A constant representing the construction of the machine (geometry, etc.)

3.2 THE ROTATING MAGNETIC FIELD

In Section 3.1, we showed that if two magnetic fields are present in a machine, then a torque will be created which will tend to line up the two magnetic fields. If one magnetic field is produced by the stator of an ac machine and the other one is produced by the rotor of the machine, then a torque will be induced in the rotor which will cause the rotor to turn and align itself with the stator magnetic field.

If there were some way to make the stator magnetic field rotate, then the induced torque in the rotor would cause it to constantly "chase" the stator magnetic field around in a circle. This, in a nutshell, is the basic principle of all ac motor operation.

How can the stator magnetic field be made to rotate? The fundamental principle of ac machine operation is that *if a three-phase set of currents, each of equal magnitude and differing in phase by 120°, flows in a three-phase winding, then it will produce a rotating magnetic field of constant magnitude.* The three-phase winding consists of three separate windings spaced 120 electrical degrees apart around the surface of the machine.

The rotating magnetic field concept is illustrated in the simplest case by an empty stator containing just three coils, each 120° apart (see Figure 3–8a). Since such a winding produces only one north and one south magnetic pole, it is a two-pole winding.

To understand the concept of the rotating magnetic field, we will apply a set of currents to the stator of Figure 3–8 and see what happens at specific instants of time. Assume that the currents in the three coils are given by the equations

$$i_{aa'}(t) = I_M \sin \omega t \qquad \text{A} \tag{3–21a}$$

$$i_{bb'}(t) = I_M \sin (\omega t - 120°) \qquad \text{A} \tag{3–21b}$$

$$i_{cc'}(t) = I_M \sin (\omega t - 240°) \qquad \text{A} \tag{3–21c}$$

The current in coil aa' flows into the a end of the coil and out the a' end of the coil. It produces the magnetic field intensity

$$\mathbf{H}_{aa'}(t) = H_M \sin \omega t \angle 0° \qquad \text{A} \bullet \text{turns / m} \tag{3–22a}$$

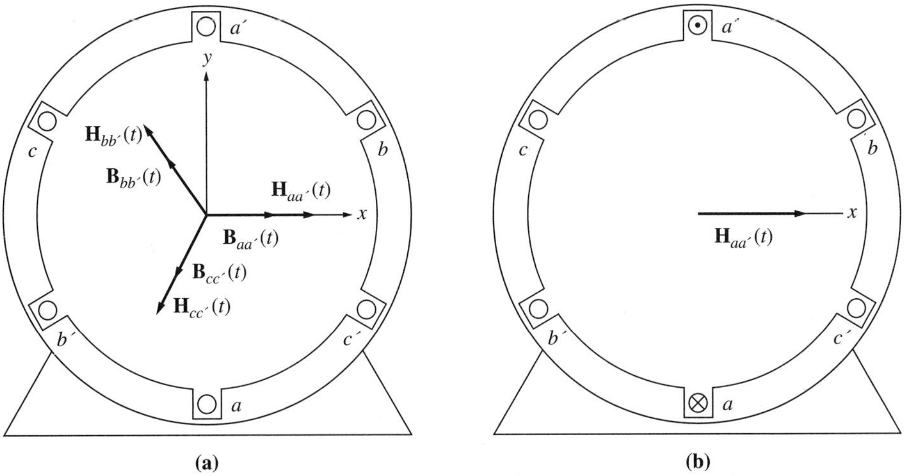

FIGURE 3–8
(a) A simple three-phase stator. Currents in this stator are assumed positive if they flow into the unprimed end and out the primed end of the coils. The magnetizing intensities produced by each coil are also shown. (b) The magnetizing intensity vector $H_{aa'}(t)$ produced by a current flowing in coil *aa′*.

where 0° is the *spatial* angle of the magnetic field intensity vector, as shown in Figure 3–8b. The direction of the magnetic field intensity vector $\mathbf{H}_{aa'}(t)$ is given by the right-hand rule: If the fingers of the right hand curl in the direction of the current flow in the coil, then the resulting magnetic field is in the direction that the thumb points. Notice that the magnitude of the magnetic field intensity vector $\mathbf{H}_{aa'}(t)$ varies sinusoidally in time, but the direction of $\mathbf{H}_{aa'}(t)$ is always constant. Similarly, the magnetic field intensity vectors $\mathbf{H}_{bb'}(t)$ and $\mathbf{H}_{cc'}(t)$ are

$$\mathbf{H}_{bb'}(t) = H_M \sin(\omega t - 120°) \angle 120° \quad \text{A} \bullet \text{turns / m} \quad (3\text{–}22\text{b})$$

$$\mathbf{H}_{cc'}(t) = H_M \sin(\omega t - 240°) \angle 240° \quad \text{A} \bullet \text{turns / m} \quad (3\text{–}22\text{c})$$

The flux densities resulting from these magnetic field intensities are given by Equation (1–21):

$$\mathbf{B} = \mu \mathbf{H} \quad (1\text{–}21)$$

They are

$$\mathbf{B}_{aa'}(t) = B_M \sin \omega t \angle 0° \quad \text{T} \quad (3\text{–}23\text{a})$$
$$\mathbf{B}_{bb'}(t) = B_M \sin(\omega t - 120°) \angle 120° \quad \text{T} \quad (3\text{–}23\text{b})$$
$$\mathbf{B}_{cc'}(t) = B_M \sin(\omega t - 240°) \angle 240° \quad \text{T} \quad (3\text{–}23\text{c})$$

where $B_M = \mu H_M$. The currents and their corresponding flux densities can be examined at specific times to determine the resulting net magnetic field in the stator.

For example, at time $\omega t = 0°$, the magnetic field from coil aa' will be

$$\mathbf{B}_{aa'} = 0 \tag{3–24a}$$

The magnetic field from coil bb' will be

$$\mathbf{B}_{bb'} = B_M \sin(-120°) \angle 120° \tag{3–24b}$$

and the magnetic field from coil cc' will be

$$\mathbf{B}_{cc'} = B_M \sin(-240°) \angle 240° \tag{3–24c}$$

The total magnetic field from all three coils added together will be

$$\begin{aligned}
\mathbf{B}_{\text{net}} &= \mathbf{B}_{aa'} + \mathbf{B}_{bb'} + \mathbf{B}_{cc'} \\
&= 0 + \left(-\frac{\sqrt{3}}{2}B_M\right)\angle 120° + \left(\frac{\sqrt{3}}{2}B_M\right)\angle 240° \\
&= \left(\frac{\sqrt{3}}{2}B_M\right)\left[-(\cos 120°\,\hat{\mathbf{x}} + \sin 120°\,\hat{\mathbf{y}}) + (\cos 240°\,\hat{\mathbf{x}} + \sin 240°\,\hat{\mathbf{y}})\right] \\
&= \left(\frac{\sqrt{3}}{2}B_M\right)\left(\frac{1}{2}\,\hat{\mathbf{x}} - \frac{\sqrt{3}}{2}\,\hat{\mathbf{y}} - \frac{1}{2}\,\hat{\mathbf{x}} - \frac{\sqrt{3}}{2}\,\hat{\mathbf{y}}\right) \\
&= \left(\frac{\sqrt{3}}{2}B_M\right)(-\sqrt{3}\,\hat{\mathbf{y}}) \\
&= -1.5B_M\,\hat{\mathbf{y}} \\
&= 1.5B_M \angle -90°
\end{aligned}$$

where $\hat{\mathbf{x}}$ is the unit vector in the x direction, and $\hat{\mathbf{y}}$ is the unit vector in the y direction in Figure 3–8. The resulting net magnetic field is shown in Figure 3–9a.

As another example, look at the magnetic field at time $\omega t = 90°$. At that time, the currents are

$$\begin{aligned}
i_{aa'} &= I_M \sin 90° \quad \text{A} \\
i_{bb'} &= I_M \sin(-30°) \quad \text{A} \\
i_{cc'} &= I_M \sin(-150°) \quad \text{A}
\end{aligned}$$

and the magnetic fields are

$$\begin{aligned}
\mathbf{B}_{aa'} &= B_M \angle 0° \\
\mathbf{B}_{bb'} &= -0.5\,B_M \angle 120° \\
\mathbf{B}_{cc'} &= -0.5\,B_M \angle 240°
\end{aligned}$$

The resulting net magnetic field is

$$\begin{aligned}
\mathbf{B}_{\text{net}} &= \mathbf{B}_{aa'} + \mathbf{B}_{bb'} + \mathbf{B}_{cc'} \\
&= B_M \angle 0° + \left(-\frac{1}{2}B_M\right)\angle 120° + \left(-\frac{1}{2}B_M\right)\angle 240° \\
&= B_M\left[\hat{\mathbf{x}} - \frac{1}{2}(\cos 120°\,\hat{\mathbf{x}} + \sin 120°\,\hat{\mathbf{y}}) - \frac{1}{2}(\cos 240°\,\hat{\mathbf{x}} + \sin 240°\,\hat{\mathbf{y}})\right]
\end{aligned}$$

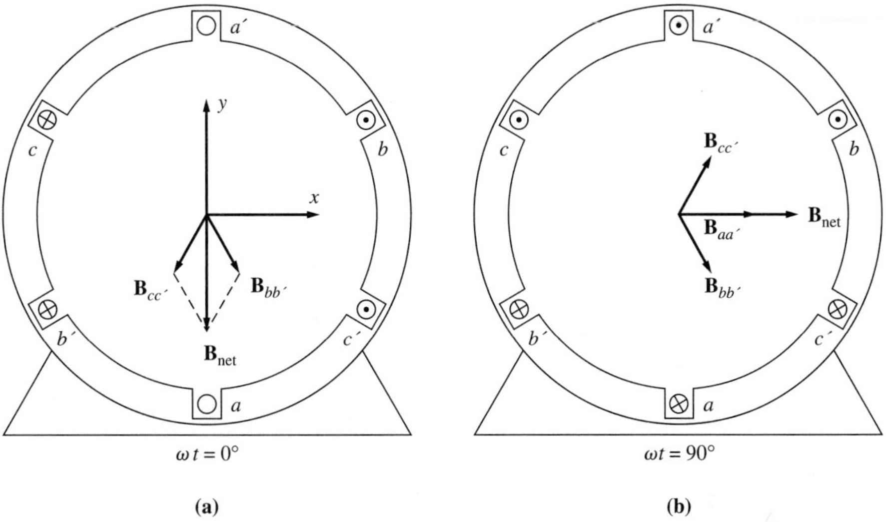

FIGURE 3–9
(a) The vector magnetic field in a stator at time $\omega t = 0°$. (b) The vector magnetic field in a stator at time $\omega t = 90°$.

$$= B_M\left(\hat{\mathbf{x}} - \frac{1}{2}\,\hat{\mathbf{x}} - \frac{\sqrt{3}}{2}\,\hat{\mathbf{y}} - \frac{1}{2}\,\hat{\mathbf{x}} - \frac{\sqrt{3}}{2}\,\hat{\mathbf{y}}\right)$$

$$= \left(\frac{\sqrt{3}}{2}B_M\right)(-\sqrt{3}\,\hat{\mathbf{y}})$$

$$= -\frac{3}{2}B_M\,\hat{\mathbf{y}}$$

$$= 1.5B_M \angle -90°$$

The resulting magnetic field is shown in Figure 3–9b. Notice that although the *direction* of the magnetic field has changed, the *magnitude* is constant. The magnetic field is maintaining a constant magnitude while rotating in a counterclockwise direction.

Proof of the Rotating Magnetic Field Concept

At any time t, the magnetic field will have the same magnitude $1.5B_M$, and it will continue to rotate at angular velocity ω. A proof of this statement for all time t is now given.

Refer again to the stator shown in Figure 3–8. In the coordinate system shown in the figure, the x direction is to the right and the y direction is upward. The vector $\hat{\mathbf{x}}$ is the unit vector in the horizontal direction, and the vector $\hat{\mathbf{y}}$ is the unit vector in the vertical direction. To find the total magnetic flux density in the stator, simply add vectorially the three component magnetic fields and determine their sum.

The net magnetic flux density in the stator is given by

$$\mathbf{B}_{\text{net}}(t) = \mathbf{B}_{aa'}(t) + \mathbf{B}_{bb'}(t) + \mathbf{B}_{cc'}(t)$$
$$= B_M \sin \omega t \angle 0° + B_M \sin (\omega t - 120°) \angle 120° + B_M \sin (\omega t - 240°) \angle 240° \text{ T}$$

Each of the three component magnetic fields can now be broken down into its x and y components.

$$\begin{aligned}\mathbf{B}_{\text{net}}(t) &= B_M \sin \omega t \, \hat{\mathbf{x}} \\ &\quad - [0.5B_M \sin (\omega t - 120°)]\hat{\mathbf{x}} + \left[\frac{\sqrt{3}}{2} B_M \sin (\omega t - 120°)\right]\hat{\mathbf{y}} \\ &\quad - [0.5B_M \sin (\omega t - 240°)]\hat{\mathbf{x}} - \left[\frac{\sqrt{3}}{2} B_M \sin (\omega t - 240°)\right]\hat{\mathbf{y}}\end{aligned}$$

Combining x and y components yields

$$\begin{aligned}\mathbf{B}_{\text{net}}(t) &= [B_M \sin \omega t - 0.5B_M \sin (\omega t - 120°) - 0.5B_M \sin (\omega t - 240°)]\,\hat{\mathbf{x}} \\ &\quad + \left[\frac{\sqrt{3}}{2} B_M \sin (\omega t - 120°) - \frac{\sqrt{3}}{2} B_M \sin (\omega t - 240°)\right]\hat{\mathbf{y}}\end{aligned}$$

By the angle-addition trigonometric identities,

$$\begin{aligned}\mathbf{B}_{\text{net}}(t) &= \left[B_M \sin \omega t + \frac{1}{4}B_M \sin \omega t + \frac{\sqrt{3}}{4}B_M \cos \omega t + \frac{1}{4}B_M \sin \omega t - \frac{\sqrt{3}}{4}B_M \cos \omega t\right]\hat{\mathbf{x}} \\ &\quad + \left[-\frac{\sqrt{3}}{4}B_M \sin \omega t - \frac{3}{4}B_M \cos \omega t + \frac{\sqrt{3}}{4}B_M \sin \omega t - \frac{3}{4}B_M \cos \omega t\right]\hat{\mathbf{y}}\end{aligned}$$

$$\boxed{\mathbf{B}_{\text{net}}(t) = (1.5B_M \sin \omega t)\hat{\mathbf{x}} - (1.5B_M \cos \omega t)\hat{\mathbf{y}}} \tag{3–25}$$

Equation (3–25) is the final expression for the net magnetic flux density. Notice that the magnitude of the field is a constant $1.5B_M$ and that the angle changes continually in a counterclockwise direction at angular velocity ω. Notice also that at $\omega t = 0°$, $\mathbf{B}_{\text{net}} = 1.5B_M \angle -90°$ and that at $\omega t = 90°$, $\mathbf{B}_{\text{net}} = 1.5B_M \angle 0°$. These results agree with the specific examples examined previously.

The Relationship between Electrical Frequency and the Speed of Magnetic Field Rotation

Figure 3–10 shows that the rotating magnetic field in this stator can be represented as a north pole (where the flux leaves the stator) and a south pole (where the flux enters the stator). These magnetic poles complete one mechanical rotation around the stator surface for each electrical cycle of the applied current. Therefore, the mechanical speed of rotation of the magnetic field in revolutions per second is equal to the electric frequency in hertz:

$$f_{se} = f_{sm} \qquad \text{two poles} \tag{3–26}$$

$$\omega_{se} = \omega_{sm} \qquad \text{two poles} \tag{3–27}$$

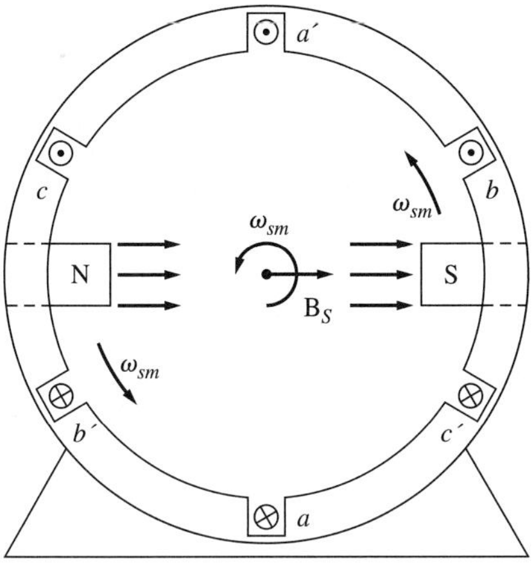

FIGURE 3–10
The rotating magnetic field in a stator represented as moving north and south stator poles.

Here f_{sm} and ω_{sm} are the mechanical speed of the stator magnetic fields in revolutions per second and radians per second, while f_{se} and ω_{se} are the electrical frequency of the stator currents in hertz and radians per second.

Notice that the windings on the two-pole stator in Figure 3–10 occur in the order (taken counterclockwise)

$$a\text{-}c'\text{-}b\text{-}a'\text{-}c\text{-}b'$$

What would happen in a stator if this pattern were repeated twice within it? Figure 3–11a shows such a stator. There, the pattern of windings (taken counterclockwise) is

$$a\text{-}c'\text{-}b\text{-}a'\text{-}c\text{-}b'\text{-}a\text{-}c'\text{-}b\text{-}a'\text{-}c\text{-}b'$$

which is just the pattern of the previous stator repeated twice. When a three-phase set of currents is applied to this stator, *two* north poles and *two* south poles are produced in the stator winding, as shown in Figure 3–11b. In this winding, a pole moves only halfway around the stator surface in one electrical cycle. Since one electrical cycle is 360 electrical degrees, and since the mechanical motion is 180 mechanical degrees, the relationship between the stator electrical angle θ_{se} and the mechanical angle θ_{sm} in this stator is

$$\theta_{se} = 2\theta_{sm} \tag{3–28}$$

Thus for the four-pole winding, the electrical frequency of the current is twice the mechanical frequency of rotation:

$$f_{se} = 2f_{sm} \quad \text{four poles} \tag{3–29}$$

$$\omega_{se} = 2\omega_{sm} \quad \text{four poles} \tag{3–30}$$

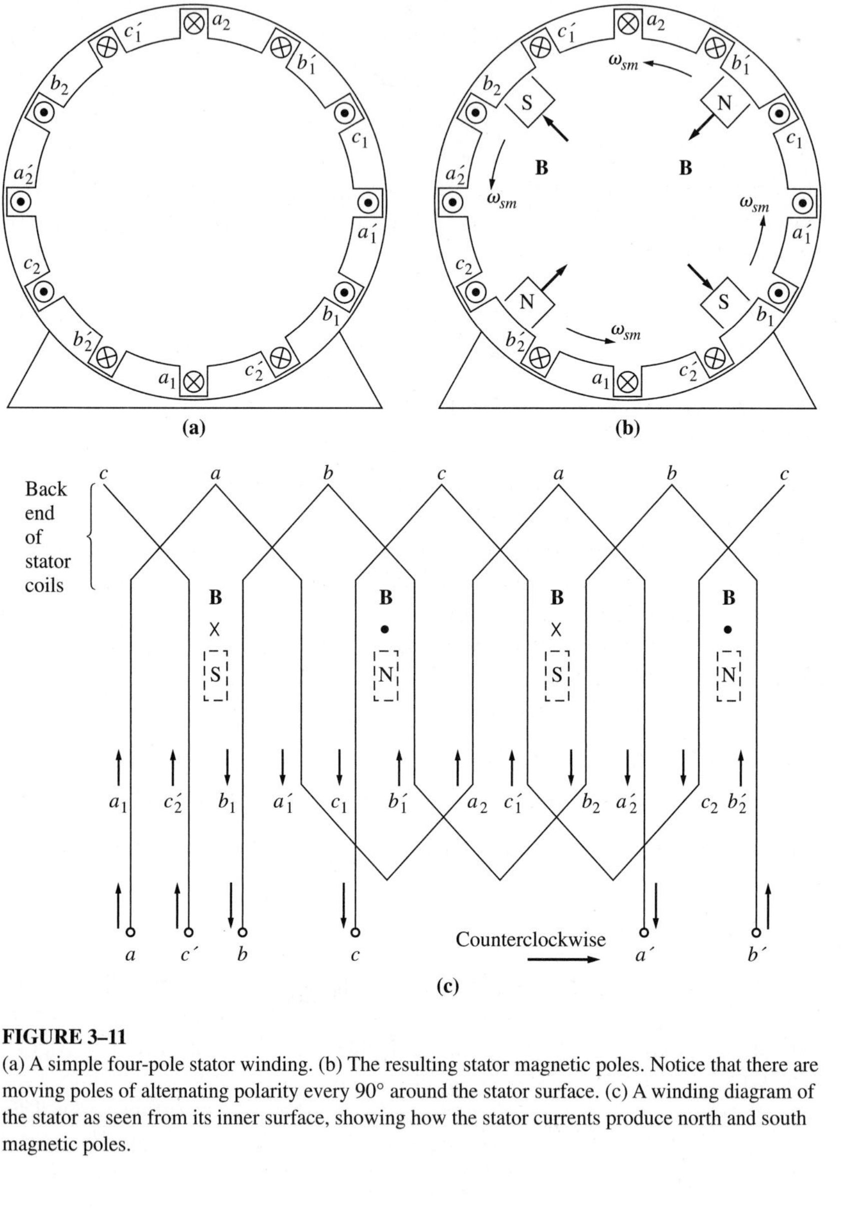

FIGURE 3–11
(a) A simple four-pole stator winding. (b) The resulting stator magnetic poles. Notice that there are moving poles of alternating polarity every 90° around the stator surface. (c) A winding diagram of the stator as seen from its inner surface, showing how the stator currents produce north and south magnetic poles.

In general, if the number of magnetic poles on an ac machine stator is P, then there are $P/2$ repetitions of the winding sequence a-c'-b-a'-c-b' around its inner surface, and the electrical and mechanical quantities on the stator are related by

$$\boxed{\theta_{se} = \frac{P}{2}\theta_{sm}} \tag{3–31}$$

$$\boxed{f_{se} = \frac{P}{2} f_{sm}} \tag{3–32}$$

$$\boxed{\omega_{se} = \frac{P}{2} \omega_{sm}} \tag{3–33}$$

Also, noting that $f_{sm} = n_{sm}/60$, it is possible to relate the electrical frequency of the stator in hertz to the resulting mechanical speed of the magnetic fields in revolutions per minute. This relationship is

$$\boxed{f_{se} = \frac{n_{sm}P}{120}} \tag{3–34}$$

Reversing the Direction of Magnetic Field Rotation

Another interesting fact can be observed about the resulting magnetic field. *If the current in any two of the three coils is swapped, the direction of the magnetic field's rotation will be reversed.* This means that it is possible to reverse the direction of rotation of an ac motor just by switching the connections on any two of the three coils. This result is verified below.

To prove that the direction of rotation is reversed, phases bb' and cc' in Figure 3–8 are switched and the resulting flux density $\mathbf{B}_{\text{net}}$ is calculated.

The net magnetic flux density in the stator is given by

$$\begin{aligned}\mathbf{B}_{\text{net}}(t) &= \mathbf{B}_{aa'}(t) + \mathbf{B}_{bb'}(t) + \mathbf{B}_{cc'}(t)\\ &= B_M \sin \omega t \angle 0° + B_M \sin (\omega t - 240°) \angle 120° + B_M \sin (\omega t - 120°) \angle 240° \text{ T}\end{aligned}$$

Each of the three component magnetic fields can now be broken down into its x and y components:

$$\begin{aligned}\mathbf{B}_{\text{net}}(t) = {}& B_M \sin \omega t \,\hat{\mathbf{x}}\\ &- [0.5B_M \sin (\omega t - 240°)]\hat{\mathbf{x}} + \left[\frac{\sqrt{3}}{2} B_M \sin (\omega t - 240°)\right]\hat{\mathbf{y}}\\ &- [0.5B_M \sin (\omega t - 120°)]\hat{\mathbf{x}} - \left[\frac{\sqrt{3}}{2} B_M \sin (\omega t - 120°)\right]\hat{\mathbf{y}}\end{aligned}$$

Combining x and y components yields

$$\begin{aligned}\mathbf{B}_{\text{net}}(t) = {}& [B_M \sin \omega t - 0.5B_M \sin (\omega t - 240°) - 0.5B_M \sin(\omega t - 120°]\hat{\mathbf{x}}\\ &+ \left[\frac{\sqrt{3}}{2} B_M \sin (\omega t - 240°) - \frac{\sqrt{3}}{2} B_M \sin (\omega t - 120°)\right]\hat{\mathbf{y}}\end{aligned}$$

By the angle-addition trigonometric identities,

$$\begin{aligned}\mathbf{B}_{\text{net}}(t) = {}& \left[B_M \sin \omega t + \frac{1}{4}B_M \sin \omega t - \frac{\sqrt{3}}{4}B_M \cos \omega t + \frac{1}{4}B_M \sin \omega t + \frac{\sqrt{3}}{4}B_M \cos \omega t\right]\hat{\mathbf{x}}\\ &+ \left[-\frac{\sqrt{3}}{4}B_M \sin \omega t + \frac{3}{4}B_M \cos \omega t + \frac{\sqrt{3}}{4}B_M \sin \omega t + \frac{3}{4}B_M \cos \omega t\right]\hat{\mathbf{y}}\end{aligned}$$

$$\boxed{\mathbf{B}_{\text{net}}(t) = (1.5B_M \sin \omega t)\hat{\mathbf{x}} + (1.5B_M \cos \omega t)\hat{\mathbf{y}}} \tag{3–35}$$

This time the magnetic field has the same magnitude but rotates in a clockwise direction. Therefore, *switching the currents in two stator phases reverses the direction of magnetic field rotation in an ac machine.*

Example 3–1. Create a MATLAB program that models the behavior of a rotating magnetic field in the three-phase stator shown in Figure 3–9.

Solution

The geometry of the loops in this stator is fixed as shown in Figure 3–9. The currents in the loops are

$$i_{aa'}(t) = I_M \sin \omega t \qquad \text{A} \tag{3–21a}$$

$$i_{bb'}(t) = I_M \sin (\omega t - 120°) \qquad \text{A} \tag{3–21b}$$

$$i_{cc'}(t) = I_M \sin (\omega t - 240°) \qquad \text{A} \tag{3–21c}$$

and the resulting magnetic flux densities are

$$\mathbf{B}_{aa'}(t) = B_M \sin \omega t \angle 0° \qquad \text{T} \tag{3–23a}$$

$$\mathbf{B}_{bb'}(t) = B_M \sin (\omega t - 120°) \angle 120° \qquad \text{T} \tag{3–23b}$$

$$\mathbf{B}_{cc'}(t) = B_M \sin (\omega t - 240°) \angle 240° \qquad \text{T} \tag{3–23c}$$

$$\phi = 2rlB = dlB$$

A simple MATLAB program that plots $\mathbf{B}_{aa'}$, $\mathbf{B}_{bb'}$, $\mathbf{B}_{cc'}$, and $\mathbf{B}_{\text{net}}$ as a function of time is shown below:

```
% M-file: mag_field.m
% M-file to calculate the net magnetic field produced
% by a three-phase stator.

% Set up the basic conditions
bmax = 1;                % Normalize bmax to 1
freq = 60;               % 60 Hz
w = 2*pi*freq;           % angular velocity (rad/s)

% First, generate the three component magnetic fields
t = 0:1/6000:1/60;
Baa = sin(w*t) .* (cos(0) + j*sin(0));
Bbb = sin(w*t-2*pi/3) .* (cos(2*pi/3) + j*sin(2*pi/3));
Bcc = sin(w*t+2*pi/3) .* (cos(-2*pi/3) + j*sin(-2*pi/3));

% Calculate Bnet
Bnet = Baa + Bbb + Bcc;

% Calculate a circle representing the expected maximum
% value of Bnet
circle = 1.5 * (cos(w*t) + j*sin(w*t));
```

```
% Plot the magnitude and direction of the resulting magnetic
% fields.  Note that Baa is black, Bbb is blue, Bcc is
% magenta, and Bnet is red.
for ii = 1:length(t)

   % Plot the reference circle
   plot(circle,'k');
   hold on;

   % Plot the four magnetic fields
   plot([0 real(Baa(ii))],[0 imag(Baa(ii))],'k','LineWidth',2);
   plot([0 real(Bbb(ii))],[0 imag(Bbb(ii))],'b','LineWidth',2);
   plot([0 real(Bcc(ii))],[0 imag(Bcc(ii))],'m','LineWidth',2);
   plot([0 real(Bnet(ii))],[0 imag(Bnet(ii))],'r','LineWidth',3);
   axis square;
   axis([-2 2 -2 2]);
   drawnow;
   hold off;

end
```

When this program is executed, it draws lines corresponding to the three component magnetic fields as well as a line corresponding to the net magnetic field. Execute this program and observe the behavior of $\mathbf{B}_{net}$.

3.3 MAGNETOMOTIVE FORCE AND FLUX DISTRIBUTION ON AC MACHINES

In Section 3.2, the flux produced inside an ac machine was treated as if it were in free space. The direction of the flux density produced by a coil of wire was assumed to be perpendicular to the plane of the coil, with the direction of the flux given by the right-hand rule.

The *flux in a real machine does not behave in the simple manner assumed above,* since there is a ferromagnetic rotor in the center of the machine, with a small air gap between the rotor and the stator. The rotor can be cylindrical, like the one shown in Figure 3–12a, or it can have pole faces projecting out from its surface, as shown in Figure 3–12b. If the rotor is cylindrical, the machine is said to have *nonsalient poles*; if the rotor has pole faces projecting out from it, the machine is said to have *salient poles.* Cylindrical rotor or nonsalient-pole machines are easier to understand and analyze than salient-pole machines, and this discussion will be restricted to machines with cylindrical rotors. Machines with salient poles are discussed briefly in Appendix C and more extensively in References 1 and 2.

Refer to the cylindrical-rotor machine in Figure 3–12a. The reluctance of the air gap in this machine is much higher than the reluctances of either the rotor or the stator, so *the flux density vector* **B** *takes the shortest possible path across the air gap* and jumps perpendicularly between the rotor and the stator.

To produce a sinusoidal voltage in a machine like this, *the magnitude of the flux density vector* **B** *must vary in a sinusoidal manner* along the surface of the air

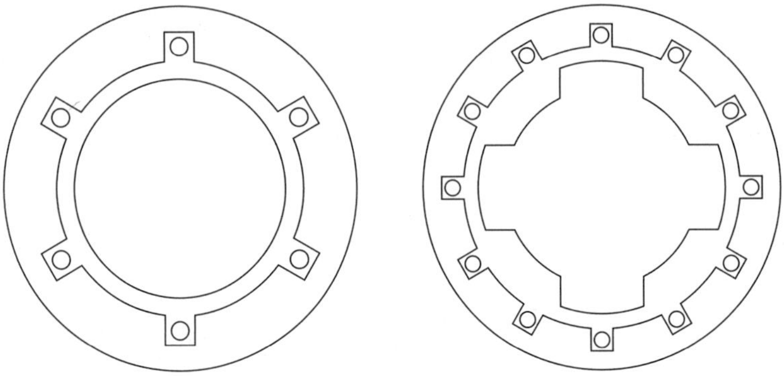

FIGURE 3–12
(a) An ac machine with a cylindrical or nonsalient-pole rotor. (b) An ac machine with a salient-pole rotor.

gap. The flux density will vary sinusoidally only if the magnetizing intensity **H** (and magnetomotive force $\mathscr{F}$) varies in a sinusoidal manner along the surface of the air gap (see Figure 3–13).

The most straightforward way to achieve a sinusoidal variation of magnetomotive force along the surface of the air gap is to distribute the turns of the winding that produces the magnetomotive force in closely spaced slots around the surface of the machine and to vary the number of conductors in each slot in a sinusoidal manner. Figure 3–14a shows such a winding, and Figure 3–14b shows the magnetomotive force resulting from the winding. The number of conductors in each slot is given by the equation

$$n_C = N_C \cos \alpha \tag{3–36}$$

where N_C is the number of conductors at an angle of 0°. As Figure 3–14b shows, this distribution of conductors produces a close approximation to a sinusoidal distribution of magnetomotive force. Furthermore, the more slots there are around the surface of the machine and the more closely spaced the slots are, the better this approximation becomes.

In practice, it is not possible to distribute windings exactly in accordance with Equation (3–36), since there are only a finite number of slots in a real machine and since only integral numbers of conductors can be included in each slot. The resulting magnetomotive force distribution is only approximately sinusoidal, and higher-order harmonic components will be present. Fractional-pitch windings are used to suppress these unwanted harmonic components, as explained in Appendix B.1.

Furthermore, it is often convenient for the machine designer to include equal numbers of conductors in each slot instead of varying the number in accordance with Equation (3–36). Windings of this type are described in Appendix B.2; they have stronger high-order harmonic components than windings designed in

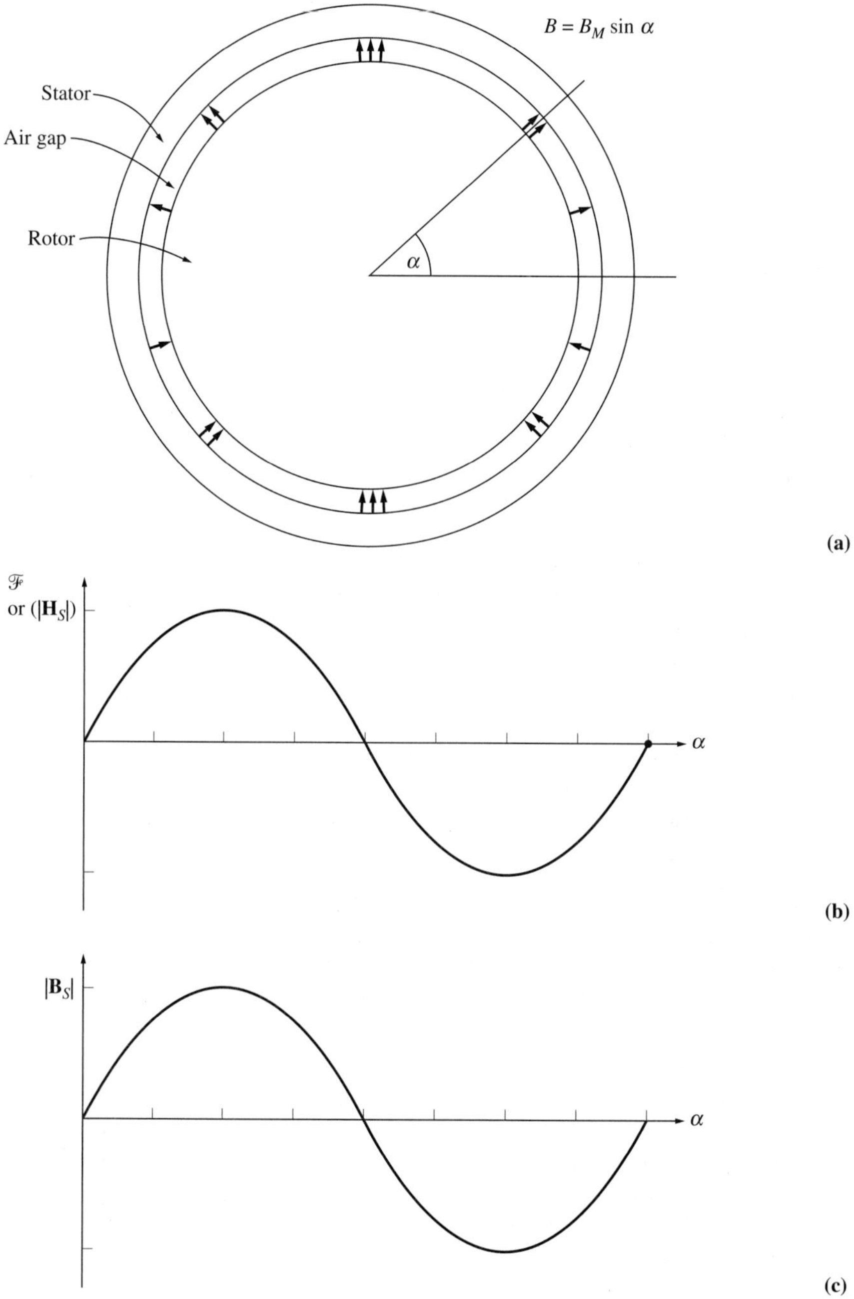

FIGURE 3–13
(a) A cylindrical rotor with sinusoidally varying air-gap flux density. (b) The magnetomotive force or magnetizing intensity as a function of angle α in the air gap. (c) The flux density as a function of angle α in the air gap.

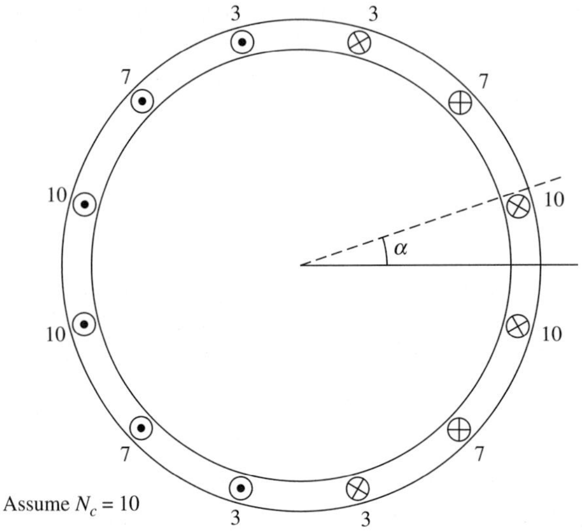

(a)

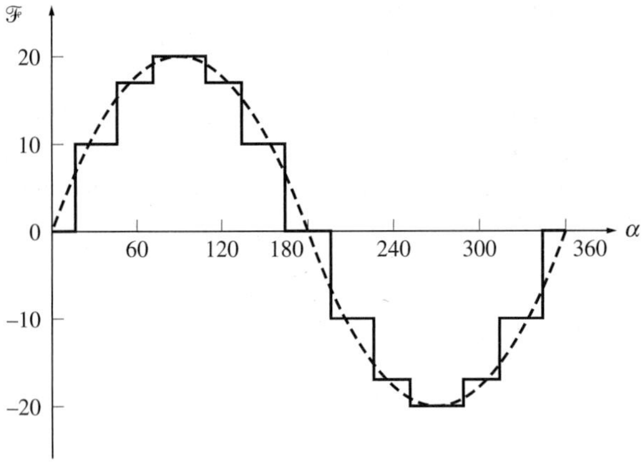

(b)

FIGURE 3–14
(a) An ac machine with a distributed stator winding designed to produce a sinusoidally varying air-gap flux density. The number of conductors in each slot is indicated on the diagram. (b) The magnetomotive force distribution resulting from the winding, compared to an ideal distribution.

accordance with Equation (3–36). The harmonic-suppression techniques of Appendix B.1 are especially important for such windings.

3.4 INDUCED VOLTAGE IN AC MACHINES

Just as a three-phase set of currents in a stator can produce a rotating magnetic field, a rotating magnetic field can produce a three-phase set of voltages in the coils of a stator. The equations governing the induced voltage in a three-phase stator will be developed in this section. To make the development easier, we will begin by looking at just one single-turn coil and then expand the results to a more general three-phase stator.

The Induced Voltage in a Coil on a Two-Pole Stator

Figure 3–15 shows a *rotating* rotor with a sinusoidally distributed magnetic field in the center of a *stationary* coil. Notice that this is the reverse of the situation studied in Section 3.1, which involved a stationary magnetic field and a rotating loop.

We will assume that the magnitude of the flux density vector **B** in the air gap between the rotor and the stator varies sinusoidally with mechanical angle, while the direction of **B** is always radially outward. This sort of flux distribution is the ideal to which machine designers aspire. (What happens when they don't achieve it is described in Appendix B.2.) If α is the angle measured from the direction of the peak rotor flux density, then the magnitude of the flux density vector **B** at a point around the *rotor* is given by

$$B = B_M \cos \alpha \tag{3–37a}$$

Note that at some locations around the air gap, the flux density vector will really point in toward the rotor; in those locations, the sign of Equation (3–37a) is negative. Since the rotor is itself rotating within the stator at an angular velocity ω_m, the magnitude of the flux density vector **B** at any angle α around the *stator* is given by

$$\boxed{B = B_M \cos(\omega t - \alpha)} \tag{3–37b}$$

The equation for the induced voltage in a wire is

$$e = (\mathbf{v} \times \mathbf{B}) \cdot \mathbf{l} \tag{1–45}$$

where $\mathbf{v}$ = velocity of the wire *relative to the magnetic field*

$\mathbf{B}$ = magnetic flux density vector

$\mathbf{l}$ = length of conductor in the magnetic field

However, this equation was derived for the case of a *moving wire* in a *stationary magnetic field*. In this case, the wire is stationary and the magnetic field is moving, so the equation does not directly apply. To use it, we must be in a frame of reference where the magnetic field appears to be stationary. If we "sit on the magnetic field" so that the field appears to be stationary, the sides of the coil will appear to go by at an apparent velocity $\mathbf{v}_{\text{rel}}$, and the equation can be applied. Figure 3–15b shows the vector magnetic field and velocities from the point of view of a stationary magnetic field and a moving wire.

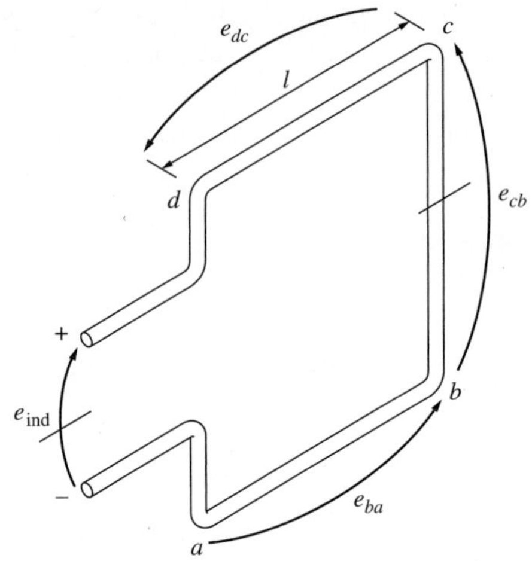

(a)

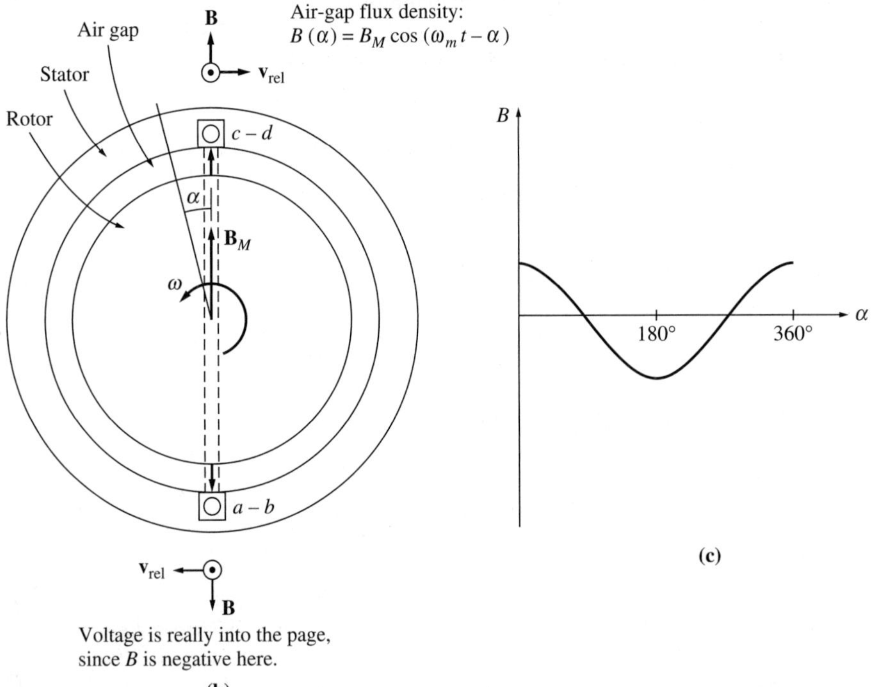

(b) **(c)**

FIGURE 3–15
(a) A rotating rotor magnetic field inside a stationary stator coil. Detail of coil. (b) The vector magnetic flux densities and velocities on the sides of the coil. The velocities shown are from a frame of reference in which the magnetic field is stationary. (c) The flux density distribution in the air gap.

The total voltage induced in the coil will be the sum of the voltages induced in each of its four sides. These voltages are determined below:

1. *Segment ab*. For segment *ab*, $\alpha = 180°$. Assuming that **B** is directed radially outward from the rotor, the angle between **v** and **B** in segment *ab* is 90°, while the quantity **v × B** is in the direction of **l**, so

$$\begin{aligned} e_{ba} &= (\mathbf{v} \times \mathbf{B}) \bullet \mathbf{l} \\ &= vBl \qquad \text{directed out of the page} \\ &= -v[B_M \cos (\omega_m t - 180°)]l \\ &= -vB_M l \cos (\omega_m t - 180°) \end{aligned} \tag{3–38}$$

where the minus sign comes from the fact that the voltage is built up with a polarity opposite to the assumed polarity.

2. Segment *bc*. The voltage on segment *bc* is zero, since the vector quantity **v × B** is perpendicular to **l**, so

$$e_{cb} = (\mathbf{v} \times \mathbf{B}) \bullet \mathbf{l} = 0 \tag{3–39}$$

3. Segment *cd*. For segment *cd*, the angle $\alpha = 0°$. Assuming that **B** is directed radially outward from the rotor, the angle between **v** and **B** in segment *cd* is 90°, while the quantity **v × B** is in the direction of **l**, so

$$\begin{aligned} e_{dc} &= (\mathbf{v} \times \mathbf{B}) \bullet \mathbf{l} \\ &= vBl \qquad \text{directed out of the page} \\ &= v(B_M \cos \omega_m t)l \\ &= vB_M l \cos \omega_m t \end{aligned} \tag{3–40}$$

4. Segment *da*. The voltage on segment *da* is zero, since the vector quantity **v × B** is perpendicular to **l**, so

$$e_{ad} = (\mathbf{v} \times \mathbf{B}) \bullet \mathbf{l} = 0 \tag{3–41}$$

Therefore, the total voltage on the coil will be

$$\begin{aligned} e_{\text{ind}} &= e_{ba} + e_{dc} \\ &= -vB_M l \cos(\omega_m t - 180°) + vB_M l \cos \omega_m t \end{aligned} \tag{3–42}$$

Since $\cos \theta = -\cos (\theta - 180°)$,

$$\begin{aligned} e_{\text{ind}} &= vB_M l \cos \omega_m t + vB_M l \cos \omega_m t \\ &= 2vB_M l \cos \omega_m t \end{aligned} \tag{3–43}$$

Since the velocity of the end conductors is given by $v = r\omega_m$, Equation (3–43) can be rewritten as

$$\begin{aligned} e_{\text{ind}} &= 2(r\omega_m)B_M l \cos \omega_m t \\ &= 2rlB_M\omega_m \cos \omega_m t \end{aligned}$$

Finally, the flux passing through the coil can be expressed as $\phi = 2rlB_m$ (see Problem 3–9), while $\omega_m = \omega_e = \omega$ for a two-pole stator, so the induced voltage can be expressed as

$$\boxed{e_{\text{ind}} = \phi\omega \cos \omega t} \tag{3–44}$$

Equation (3–44) describes the voltage induced in a single-turn coil. If the coil in the stator has N_C turns of wire, then the total induced voltage of the coil will be

$$\boxed{e_{\text{ind}} = N_C\phi\omega \cos \omega t} \tag{3–45}$$

Notice that the voltage produced in the stator of this simple ac machine winding is sinusoidal with an amplitude which depends on the flux ϕ in the machine, the angular velocity ω of the rotor, and a constant depending on the construction of the machine (N_C in this simple case). This is the same as the result that we obtained for the simple rotating loop in Section 3.1.

Note that Equation (3–45) contains the term cos ωt instead of the sin ωt found in some of the other equations in this chapter. The cosine term has no special significance compared to the sine—it resulted from our choice of reference direction for α in this derivation. If the reference direction for α had been rotated by 90° we would have had a sin ωt term.

The Induced Voltage in a Three-Phase Set of Coils

If *three coils*, each of N_C turns, are placed around the rotor magnetic field as shown in Figure 3–16, then the voltages induced in each of them will be the same in magnitude but will differ in phase by 120°. The resulting voltages in each of the three coils are

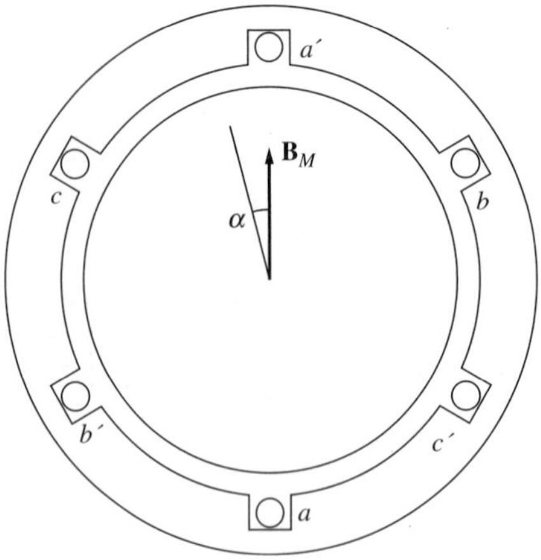

FIGURE 3–16
The production of three-phase voltages from three coils spaced 120° apart.

$$e_{aa'}(t) = N_C\,\phi\omega \sin \omega t \qquad \text{V} \tag{3–46a}$$

$$e_{bb'}(t) = N_C\,\phi\omega \sin (\omega t - 120°) \qquad \text{V} \tag{3–46b}$$

$$e_{cc'}(t) = N_C\,\phi\omega \sin (\omega t - 240°) \qquad \text{V} \tag{3–46c}$$

Therefore, a three-phase set of currents can generate a uniform rotating magnetic field in a machine stator, and a uniform rotating magnetic field can generate a three-phase set of voltages in such a stator.

The RMS Voltage in a Three-Phase Stator

The peak voltage in any phase of a three-phase stator of this sort is

$$E_{\text{max}} = N_C\,\phi\omega \tag{3–47}$$

Since $\omega = 2\pi f$, this equation can also be written as

$$E_{\text{max}} = 2\pi N_C\,\phi f \tag{3–48}$$

Therefore, the rms voltage of any phase of this three-phase stator is

$$E_A = \frac{2\pi}{\sqrt{2}} N_C \phi f \tag{3–49}$$

$$\boxed{E_A = \sqrt{2}\pi N_C \phi f} \tag{3–50}$$

The rms voltage at the *terminals* of the machine will depend on whether the stator is Y- or Δ-connected. If the machine is Y-connected, then the terminal voltage will be $\sqrt{3}$ times E_A; if the machine is Δ-connected, then the terminal voltage will just be equal to E_A.

Example 3–2. The following information is known about the simple two-pole generator in Figure 3–16. The peak flux density of the rotor magnetic field is 0.2 T, and the mechanical rate of rotation of the shaft is 3600 r/min. The stator diameter of the machine is 0.5 m, its coil length is 0.3 m, and there are 15 turns per coil. The machine is Y-connected.

(*a*) What are the three phase voltages of the generator as a function of time?
(*b*) What is the rms phase voltage of this generator?
(*c*) What is the rms terminal voltage of this generator?

Solution

The flux in this machine is given by

$$\phi = 2rlB = dlB$$

where d is the diameter and l is the length of the coil. Therefore, the flux in the machine is given by

$$\phi = (0.5 \text{ m})(0.3 \text{ m})(0.2 \text{ T}) = 0.03 \text{ Wb}$$

The speed of the rotor is given by

$$\omega = (3600 \text{ r/min})(2\pi \text{ rad})(1 \text{ min}/60 \text{ s}) = 377 \text{ rad/s}$$

(a) The magnitudes of the peak phase voltages are thus

$$E_{max} = N_C\phi\omega$$
$$= (15 \text{ turns})(0.03 \text{ Wb})(377 \text{ rad/s}) = 169.7 \text{ V}$$

and the three phase voltages are

$$e_{aa'}(t) = 169.7 \sin 377t \qquad V$$
$$e_{bb'}(t) = 169.7 \sin (377t - 120°) \qquad \text{V}$$
$$e_{cc'}(t) = 169.7 \sin (377t - 240°) \qquad \text{V}$$

(b) The rms phase voltage of this generator is

$$E_A = \frac{E_{max}}{\sqrt{2}} = \frac{169.7 \text{ V}}{\sqrt{2}} = 120 \text{ V}$$

(c) Since the generator is Y-connected,

$$V_T = \sqrt{3}E_A = \sqrt{3}(120 \text{ V}) = 208 \text{ V}$$

3.5 INDUCED TORQUE IN AN AC MACHINE

In ac machines under normal operating conditions, there are two magnetic fields present—a magnetic field from the rotor circuit and another magnetic field from the stator circuit. The interaction of these two magnetic fields produces the torque in the machine, just as two permanent magnets near each other will experience a torque which causes them to line up.

Figure 3–17 shows a simplified ac machine with a sinusoidal stator flux distribution that peaks in the upward direction and a single coil of wire mounted on the rotor. The stator flux distribution in this machine is

$$B_S(\alpha) = B_S \sin \alpha \tag{3–51}$$

where B_S is the magnitude of the peak flux density; $B_S(\alpha)$ is positive when the flux density vector points radially outward from the rotor surface to the stator surface. How much torque is produced in the rotor of this simplified ac machine? To find out, we will analyze the force and torque on each of the two conductors separately.

The induced force on conductor 1 is

$$\mathbf{F} = i(\mathbf{l} \times \mathbf{B}) \tag{1–43}$$
$$= ilB_S \sin \alpha \qquad \text{with direction as shown}$$

The torque on the conductor is

$$\tau_{\text{ind},1} = (\mathbf{r} \times \mathbf{F})$$
$$= rilB_S \sin \alpha \qquad \text{counterclockwise}$$

The induced force on conductor 2 is

$$\mathbf{F} = i(\mathbf{l} \times \mathbf{B}) \tag{1–43}$$
$$= ilB_S \sin \alpha \qquad \text{with direction as shown}$$

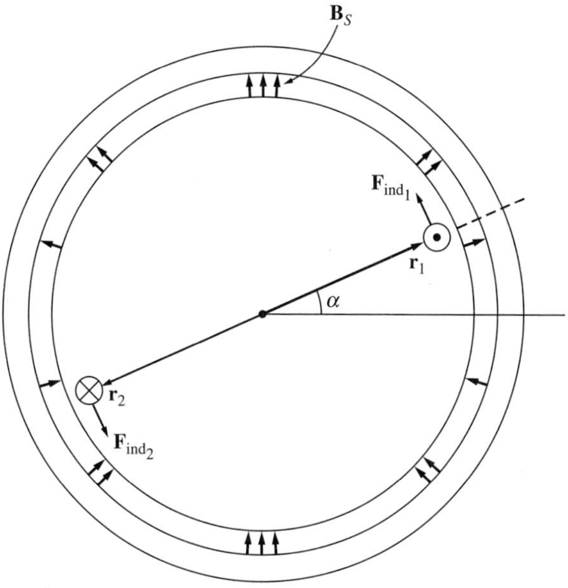

FIGURE 3–17
A simplified ac machine with a sinusoidal stator flux distribution and a single coil of wire mounted in the rotor.

The torque on the conductor is

$$\tau_{\text{ind},1} = (\mathbf{r} \times \mathbf{F})$$
$$= rilB_S \sin \alpha \qquad \text{counterclockwise}$$

Therefore, the torque on the rotor loop is

$$\boxed{\tau_{\text{ind}} = 2rilB_S \sin \alpha \qquad \text{counterclockwise}} \tag{3–52}$$

Equation (3–52) can be expressed in a more convenient form by examining Figure 3–18 and noting two facts:

1. The current i flowing in the rotor coil produces a magnetic field of its own. The direction of the peak of this magnetic field is given by the right-hand rule, and the magnitude of its magnetizing intensity $\mathbf{H}_R$ is directly proportional to the current flowing in the rotor:

$$H_R = Ci \tag{3–53}$$

where C is a constant of proportionality.

2. The angle between the peak of the stator flux density $\mathbf{B}_S$ and the peak of the rotor magnetizing intensity $\mathbf{H}_R$ is γ. Furthermore,

$$\gamma = 180° - \alpha \tag{3–54}$$

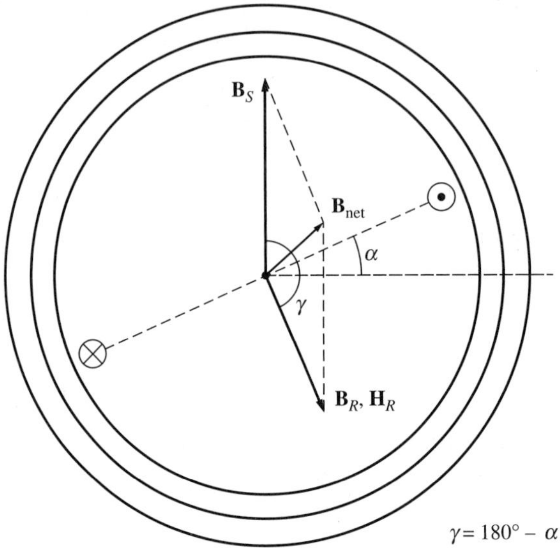

FIGURE 3–18
The components magnetic flux density inside the machine of Figure 3–17.

$$\sin \gamma = \sin (180° - \alpha) = \sin \alpha \tag{3–55}$$

By combining these two observations, the torque on the loop can be expressed as

$$\tau_{\text{ind}} = KH_R B_S \sin \alpha \qquad \text{counterclockwise} \tag{3–56}$$

where K is a constant dependent on the construction of the machine. Note that both the magnitude and the direction of the torque can be expressed by the equation

$$\boxed{\tau_{\text{ind}} = K\mathbf{H}_R \times \mathbf{B}_S} \tag{3–57}$$

Finally, since $B_R = \mu H_R$, this equation can be reexpressed as

$$\boxed{\tau_{\text{ind}} = k\mathbf{B}_R \times \mathbf{B}_S} \tag{3–58}$$

where $k = K/\mu$. Note that in general k will not be constant, since the magnetic permeability μ varies with the amount of magnetic saturation in the machine.

Equation (3–58) is just the same as Equation (3–20), which we derived for the case of a single loop in a uniform magnetic field. It can apply to any ac machine, not just to the simple one-loop rotor just described. Only the constant k will differ from machine to machine. This equation will be used only for a *qualitative* study of torque in ac machines, so the actual value of k is unimportant for our purposes.

The net magnetic field in this machine is the vector sum of the rotor and stator fields (assuming no saturation):

$$\mathbf{B}_{\text{net}} = \mathbf{B}_R + \mathbf{B}_S \tag{3–59}$$

This fact can be used to produce an equivalent (and sometimes more useful) expression for the induced torque in the machine. From Equation (3–58)

$$\tau_{\text{ind}} = k\mathbf{B}_R \times \mathbf{B}_S \tag{3–58}$$

But from Equation (3–59), $\mathbf{B}_S = \mathbf{B}_{\text{net}} - \mathbf{B}_R$, so

$$\begin{aligned} \tau_{\text{ind}} &= k\mathbf{B}_R \times (\mathbf{B}_{\text{net}} - \mathbf{B}_R) \\ &= k(\mathbf{B}_R \times \mathbf{B}_{\text{net}}) - k(\mathbf{B}_R \times \mathbf{B}_R) \end{aligned}$$

Since the cross product of any vector with itself is zero, this reduces to

$$\boxed{\tau_{\text{ind}} = k\mathbf{B}_R \times \mathbf{B}_{\text{net}}} \tag{3–60}$$

so the induced torque can also be expressed as a cross product of $\mathbf{B}_R$ and $\mathbf{B}_{\text{net}}$ with the same constant k as before. The magnitude of this expression is

$$\boxed{\tau_{\text{ind}} = kB_R B_{\text{net}} \sin \delta} \tag{3–61}$$

where δ is the angle between $\mathbf{B}_R$ and $\mathbf{B}_{\text{net}}$.

Equations (3–58) to (3–61) will be used to help develop a qualitative understanding of the torque in ac machines. For example, look at the simple synchronous machine in Figure 3–19. Its magnetic fields are rotating in a counterclockwise direction. What is the direction of the torque on the shaft of the machine's rotor? By applying the right-hand rule to Equation (3–58) or (3–60), the induced torque is found to be clockwise, or opposite the direction of rotation of the rotor. Therefore, this machine must be acting as a generator.

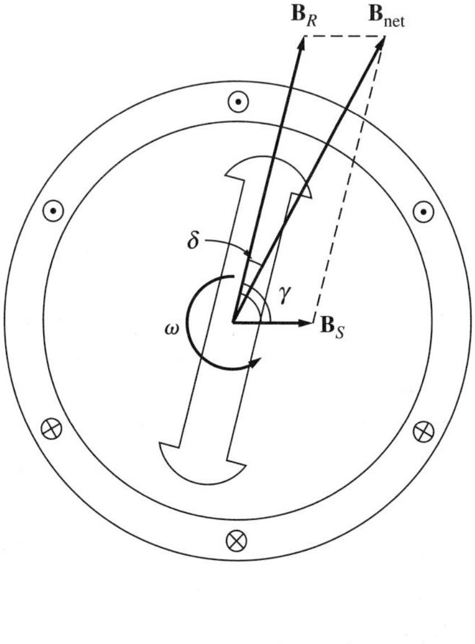

FIGURE 3–19
A simplified synchronous machine showing its rotor and stator magnetic fields.

3.6 WINDING INSULATION IN AN AC MACHINE

One of the most critical parts of an ac machine design is the insulation of its windings. If the insulation of a motor or generator breaks down, the machine shorts out. The repair of a machine with shorted insulation is quite expensive, if it is even possible. To prevent the winding insulation from breaking down as a result of overheating, it is necessary to limit the temperature of the windings. This can be partially done by providing a cooling air circulation over them, but ultimately the maximum winding temperature limits the maximum power that can be supplied continuously by the machine.

Insulation rarely fails from immediate breakdown at some critical temperature. Instead, the increase in temperature produces a gradual degradation of the insulation, making it subject to failure from another cause such as shock, vibration, or electrical stress. There was an old rule of thumb that said that the life expectancy of a motor with a given type of insulation is halved for each 10 percent rise in temperature above the rated temperature of the winding. This rule still applies to some extent today.

To standardize the temperature limits of machine insulation, the National Electrical Manufacturers Association (NEMA) in the United States has defined a series of *insulation system classes*. Each insulation system class specifies the maximum temperature rise permissible for that class of insulation. There are three common NEMA insulation classes for integral-horsepower ac motors: B, F, and H. Each class represents a higher permissible winding temperature than the one before it. For example, the armature winding temperature rise above ambient temperature in one type of continuously operating ac induction motor must be limited to 80°C for class B, 105°C for class F, and 125°C for class H insulation.

The effect of operating temperature on insulation life for a typical machine can be quite dramatic. A typical curve is shown in Figure 3–20. This curve shows the mean life of a machine in thousands of hours versus the temperature of the windings, for several different insulation classes.

The specific temperature specifications for each type of ac motor and generator are set out in great detail in NEMA Standard MG1-1993, *Motors and Generators*. Similar standards have been defined by the International Electrotechnical Commission (IEC) and by various national standards organizations in other countries.

3.7 AC MACHINE POWER FLOWS AND LOSSES

AC generators take in mechanical power and produce electric power, while ac motors take in electric power and produce mechanical power. In either case, not all the power input to the machine appears in useful form at the other end—there is *always* some loss associated with the process.

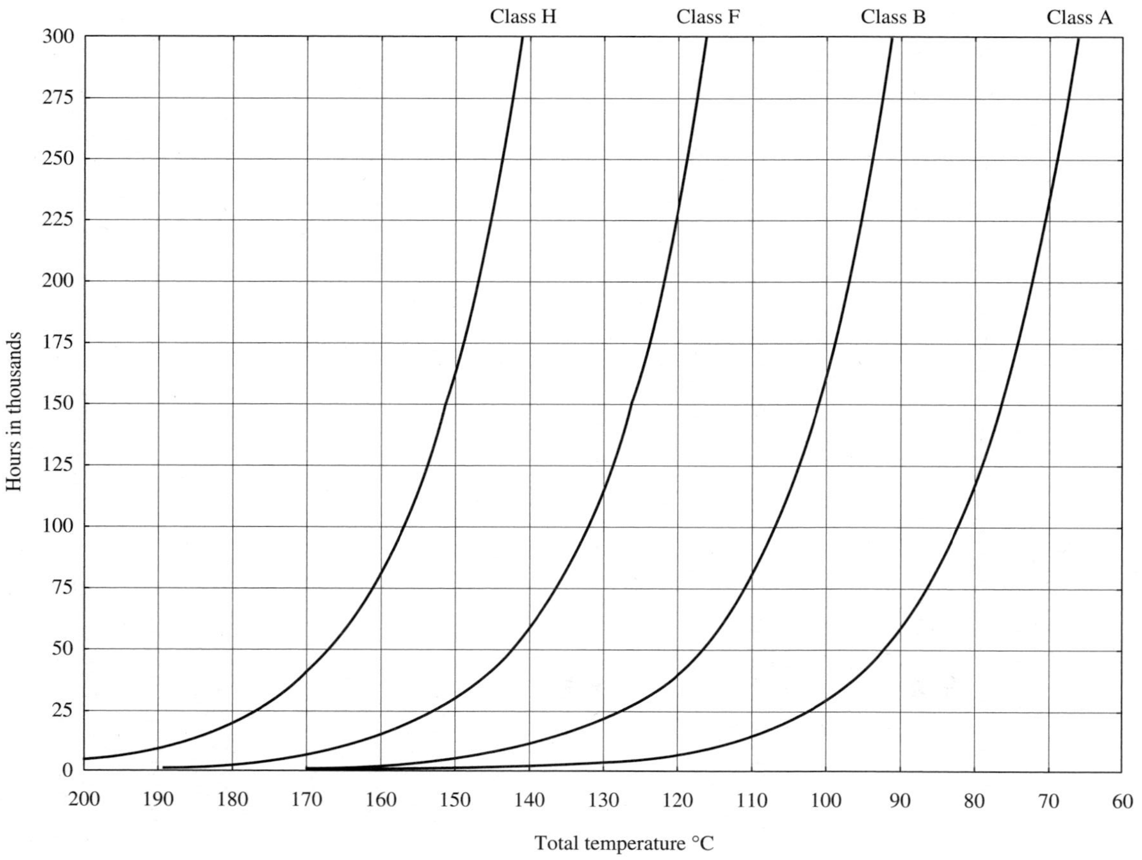

FIGURE 3–20
Plot of mean insulation life versus winding temperature for various insulation classes. *(Courtesy of Marathon Electric Company.)*

The efficiency of an ac machine is defined by the equation

$$\boxed{\eta = \frac{P_{\text{out}}}{P_{\text{in}}} \times 100\%} \tag{3–62}$$

The difference between the input power and the output power of a machine is the losses that occur inside it. Therefore,

$$\boxed{\eta = \frac{P_{\text{in}} - P_{\text{loss}}}{P_{\text{in}}} \times 100\%} \tag{3–63}$$

The Losses in AC Machines

The losses that occur in ac machines can be divided into four basic categories:

1. Electrical or copper losses (I^2R losses)
2. Core losses
3. Mechanical losses
4. Stray load losses

ELECTRICAL OR COPPER LOSSES. Copper losses are the resistive heating losses that occur in the stator (armature) and rotor (field) windings of the machine. The stator copper losses (SCL) in a three-phase ac machine are given by the equation

$$\boxed{P_{\text{SCL}} = 3I_A^2 R_A} \tag{3–64}$$

where I_A is the current flowing in each armature phase and R_A is the resistance of each armature phase.

The rotor copper losses (RCL) of a synchronous ac machine (induction machines will be considered separately in Chapter 7) are given by

$$\boxed{P_{\text{RCL}} = I_F^2 R_F} \tag{3–65}$$

where I_F is the current flowing in the field winding on the rotor and R_F is the resistance of the field winding. The resistance used in these calculations is usually the winding resistance at normal operating temperature.

CORE LOSSES. The core losses are the hysteresis losses and eddy current losses occurring in the metal of the motor. These losses were described in Chapter 1. These losses vary as the square of the flux density (B^2) and, for the stator, as the 1.5th power of the speed of rotation of the magnetic fields ($n^{1.5}$).

MECHANICAL LOSSES. The mechanical losses in an ac machine are the losses associated with mechanical effects. There are two basic types of mechanical losses: *friction* and *windage*. Friction losses are losses caused by the friction of the bearings in the machine, while windage losses are caused by the friction between the moving parts of the machine and the air inside the motor's casing. These losses vary as the cube of the speed of rotation of the machine.

The mechanical and core losses of a machine are often lumped together and called the *no-load rotational loss* of the machine. At no load, all the input power must be used to overcome these losses. Therefore, measuring the input power to the stator of an ac machine acting as a motor at no load will give an approximate value for these losses.

STRAY LOSSES (OR MISCELLANEOUS LOSSES). Stray losses are losses that cannot be placed in one of the previous categories. No matter how carefully losses are accounted for, some always escape inclusion in one of the above categories. All such losses are lumped into stray losses. For most machines, stray losses are taken by convention to be 1 percent of full load.

The Power-Flow Diagram

One of the most convenient techniques for accounting for power losses in a machine is the *power-flow diagram*. A power-flow diagram for an ac generator is shown in Figure 3–21a. In this figure, mechanical power is input into the machine, and then the stray losses, mechanical losses, and core losses are subtracted. After

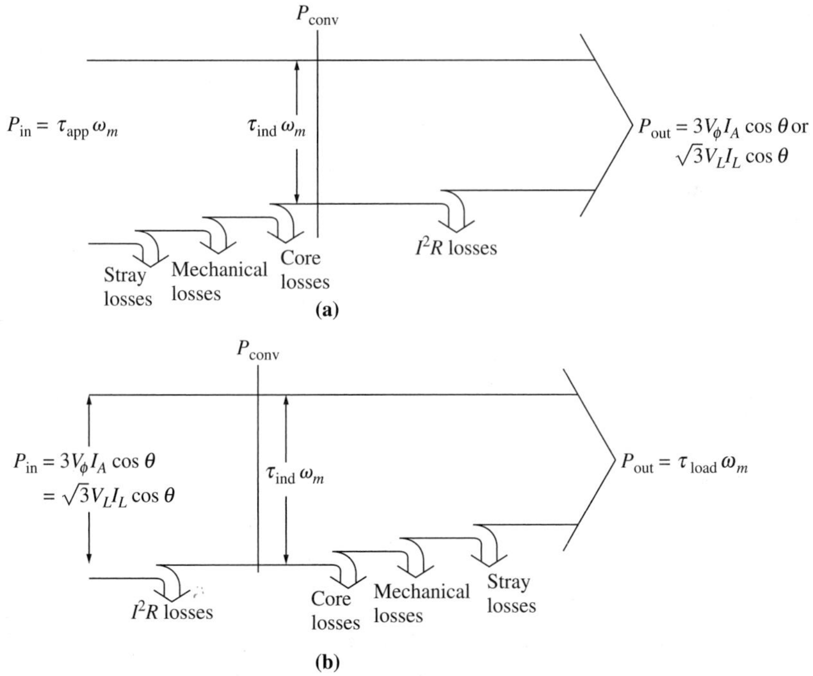

FIGURE 3–21
(a) The power-flow diagram of a three-phase ac generator. (b) The power-flow diagram of a three-phase ac motor.

they have been subtracted, the remaining power is ideally converted from mechanical to electrical form at the point labeled P_{conv}. The mechanical power that is converted is given by

$$\boxed{P_{conv} = \tau_{ind}\omega_m} \tag{3–66}$$

and the same amount of electrical power is produced. However, this is not the power that appears at the machine's terminals. Before the terminals are reached, the electrical I^2R losses must be subtracted.

In the case of ac motors, this power-flow diagram is simply reversed. The power-flow diagram for a motor is shown in Figure 3–21b.

Example problems involving the calculation of ac motor and generator efficiencies will be given in the next three chapters.

3.8 VOLTAGE REGULATION AND SPEED REGULATION

Generators are often compared to each other using a figure of merit called *voltage regulation*. Voltage regulation (VR) is a measure of the ability of a generator to keep a constant voltage at its terminals as load varies. It is defined by the equation

$$\boxed{VR = \frac{V_{nl} - V_{fl}}{V_{fl}} \times 100\%} \tag{3–67}$$

where V_{nl} is the no-load terminal voltage of the generator and V_{fl} is the full-load terminal voltage of the generator. It is a rough measure of the shape of the generator's voltage-current characteristic—a positive voltage regulation means a drooping characteristic, and a negative voltage regulation means a rising characteristic. A small VR is "better" in the sense that the voltage at the terminals of the generator is more constant with variations in load.

Similarly, motors are often compared to each other by using a figure of merit called *speed regulation*. Speed regulation (SR) is a measure of the ability of a motor to keep a constant shaft speed as load varies. It is defined by the equation

$$\boxed{SR = \frac{n_{nl} - n_{fl}}{n_{fl}} \times 100\%} \tag{3–68}$$

or

$$\boxed{SR = \frac{\omega_{nl} - \omega_{fl}}{\omega_{fl}} \times 100\%} \tag{3–69}$$

It is a rough measure of the shape of a motor's torque-speed characteristic—a positive speed regulation means that a motor's speed drops with increasing load, and a negative speed regulation means a motor's speed increases with increasing load. The magnitude of the speed regulation tells approximately how steep the slope of the torque-speed curve is.

3.9 SUMMARY

There are two major types of ac machines: synchronous machines and induction machines. The principal difference between the two types is that synchronous machines require a dc field current to be supplied to their rotors, while induction machines have the field current induced in their rotors by transformer action. They will be explored in detail in the next three chapters.

A three-phase system of currents supplied to a system of three coils spaced 120 electrical degrees apart on a stator will produce a uniform rotating magnetic field within the stator. The *direction of rotation* of the magnetic field can be *reversed* by simply swapping the connections to any two of the three phases. Conversely, a rotating magnetic field will produce a three-phase set of voltages within such a set of coils.

In stators of more than two poles, one complete mechanical rotation of the magnetic fields produces more than one complete electrical cycle. For such a stator, one mechanical rotation produces $P/2$ electrical cycles. Therefore, the electrical angle of the voltages and currents in such a machine is related to the mechanical angle of the magnetic fields by

$$\theta_{se} = \frac{P}{2}\theta_{sm}$$

The relationship between the electrical frequency of the stator and the mechanical rate of rotation of the magnetic fields is

$$f_{se} = \frac{n_{sm}P}{120}$$

The types of losses that occur in ac machines are electrical or copper losses (I^2R losses), core losses, mechanical losses, and stray losses. Each of these losses was described in this chapter, along with the definition of overall machine efficiency. Finally, voltage regulation was defined for generators as

$$\boxed{\text{VR} = \frac{V_{\text{nl}} - V_{\text{fl}}}{V_{\text{fl}}} \times 100\%}$$

and speed regulation was defined for motors as

$$\boxed{\text{SR} = \frac{n_{\text{nl}} - n_{\text{fl}}}{n_{\text{fl}}} \times 100\%}$$

QUESTIONS

3–1. What is the principal difference between a synchronous machine and an induction machine?

3–2. Why does switching the current flows in any two phases reverse the direction of rotation of a stator's magnetic field?

3–3. What is the relationship between electrical frequency and magnetic field speed for an ac machine?

3–4. What is the equation for the induced torque in an ac machine?

PROBLEMS

3–1. The simple loop rotating in a uniform magnetic field shown in Figure 3–1 has the following characteristics:

$$\mathbf{B} = 1.0 \text{ T to the right} \qquad r = 0.1 \text{ m}$$
$$l = 0.3 \text{ m} \qquad \omega_m = 377 \text{ rad/s}$$

(a) Calculate the voltage $e_{\text{tot}}(t)$ induced in this rotating loop.
(b) What is the frequency of the voltage produced in this loop?
(c) Suppose that a 10 Ω resistor is connected as a load across the terminals of the loop. Calculate the current that would flow through the resistor.
(d) Calculate the magnitude and direction of the induced torque on the loop for the conditions in *(c)*.
(e) Calculate the instantaneous and average electric power being generated by the loop for the conditions in *(c)*.
(f) Calculate the mechanical power being consumed by the loop for the conditions in *(c)*. How does this number compare to the amount of electric power being generated by the loop?

3–2. Develop a table showing the speed of magnetic field rotation in ac machines of 2, 4, 6, 8, 10, 12, and 14 poles operating at frequencies of 50, 60, and 400 Hz.

3–3. The first ac power system in the United States ran at a frequency of 133 Hz. If the ac power for this system were produced by a four-pole generator, how fast would the shaft of the generator have to rotate?

3–4. A three-phase, Y-connected, four-pole winding is installed in 24 slots on a stator. There are 40 turns of wire in each slot of the windings. All coils in each phase are connected in series. The flux per pole in the machine is 0.060 Wb, and the speed of rotation of the magnetic field is 1800 r/min.
(a) What is the frequency of the voltage produced in this winding?
(b) What are the resulting phase and terminal voltages of this stator?

3–5. A three-phase, Δ-connected, six-pole winding is installed in 36 slots on a stator. There are 150 turns of wire in each slot of the windings. All coils in each phase are connected in series. The flux per pole in the machine is 0.060 Wb, and the speed of rotation of the magnetic field is 1000 r/min.
(a) What is the frequency of the voltage produced in this winding?
(b) What are the resulting phase and terminal voltages of this stator?

3–6. A three-phase, Y-connected, 60 Hz, two-pole synchronous machine has a stator with 5000 turns of wire per phase. What rotor flux would be required to produce a terminal (line-to-line) voltage of 13.2 kV?

3–7. Modify the MATLAB in Example 3–1 by swapping the currents flowing in any two phases. What happens to the resulting net magnetic field?

3–8. If an ac machine has the rotor and stator magnetic fields shown in Figure P3–1, what is the direction of the induced torque in the machine? Is the machine acting as a motor or generator?

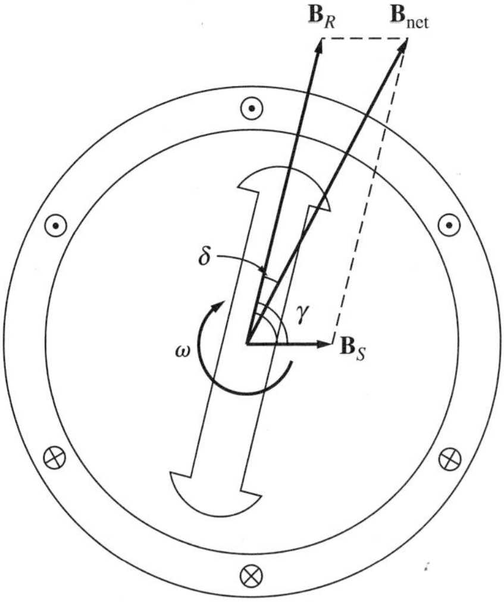

FIGURE P3–1
The ac machine of Problem 3–8.

3–9. The flux density distribution over the surface of a two-pole stator of radius r and length l is given by

$$B = B_M \cos (\omega_m t - \alpha) \qquad (3\text{–}37b)$$

Prove that the total flux under each pole face is

$$\phi = 2rlB_M$$

3–10. In the early days of ac motor development, machine designers had great difficulty controlling the core losses (hysteresis and eddy currents) in machines. They had not yet developed steels with low hysteresis, and were not making laminations as thin as the ones used today. To help control these losses, early ac motors in the United States were run from a 25 Hz ac power supply, while lighting systems were run from a separate 60 Hz ac power supply.

(a) Develop a table showing the speed of magnetic field rotation in ac machines of 2, 4, 6, 8, 10, 12, and 14 poles operating at 25 Hz. What was the fastest rotational speed available to these early motors?

(b) For a given motor operating at a constant flux density B, how would the core losses of the motor running at 25 Hz compare to the core losses of the motor running at 60 Hz?

(c) Why did the early engineers provide a separate 60-Hz power system for lighting?

3–11. In later years, motors improved and could be run directly from a 60 Hz power supply. As a result, 25 Hz power systems shrank and disappeared. However, there were many perfectly good working 25 Hz motors in factories around the country that owners were not ready to discard. To keep them running, some users created their own 25 Hz power in the plant using *motor-generator sets*. A motor-generator set consists of two machines connected on a common shaft, one acting as a motor and the other acting as a generator. If the two machines have different numbers of poles but exactly the same shaft speed, then the electrical frequency of the two machines will be different due to Equation (3–34). What combination of poles on the two machines could convert 60 Hz power to 25 Hz power?

$$f_{se} = \frac{n_{sm}P}{120} \tag{3–34}$$

REFERENCES

1. Del Toro, Vincent: *Electric Machines and Power Systems*, Prentice-Hall, Englewood Cliffs, N.J., 1985.
2. Fitzgerald, A. E., and Charles Kingsley: *Electric Machinery*, McGraw-Hill, New York, 1952.
3. Fitzgerald, A. E., Charles Kingsley, and S. D. Umans: *Electric Machinery*, 5th Ed., McGraw-Hill, New York, 1990.
4. International Electrotechnical Commission, *Rotating Electrical Machines Part 1: Rating and Performance*, IEC 33–1 (R1994), 1994.
5. Liwschitz-Garik, Michael, and Clyde Whipple: *Alternating-Current Machinery*, Van Nostrand, Princeton, N.J., 1961.
6. McPherson, George: *An Introduction to Electrical Machines and Transformers*, Wiley, New York, 1981.
7. National Electrical Manufacturers Association: *Motors and Generators,* Publication MG1-1993, Washington, 1993.
8. Werninck. E. H. (ed.): *Electric Motor Handbook*, McGraw-Hill Book Company, London, 1978.

CHAPTER

4

SYNCHRONOUS GENERATORS

LEARNING OBJECTIVES

- Understand the equivalent circuit of a synchronous generator.
- Be able to sketch phasor diagrams for a synchronous generator.
- Know the equations for power and torque in a synchronous generator.
- Know how to derive the characteristics of a synchronous machine from measurements (OCC and SCC).
- Understand how terminal voltage varies with load in a synchronous generator operating alone. Be able to calculate the terminal voltage at various load conditions.
- Understand the conditions required to parallel two or more synchronous generators.
- Understand the procedure for paralleling synchronous generators.
- Understand the operation of synchronous generators in parallel with a very large power system (or infinite bus).
- Understand the static stability limit of a synchronous generator, and why the transient stability limit is less than the static stability limit.
- Understand the transient currents that flow under fault (short-circuit) conditions.
- Understand synchronous generator ratings, and what condition limits each rating value.

Synchronous generators or *alternators* are synchronous machines used to convert mechanical power to ac electric power. This chapter explores the operation of synchronous generators, both when operating alone and when operating together with other generators.

4.1 SYNCHRONOUS GENERATOR CONSTRUCTION

In a synchronous generator, a rotor magnetic field is produced either by designing the rotor as a permanent magnet or by applying a dc current to a rotor winding to create an electromagnet. The rotor of the generator is then turned by a prime mover, producing a rotating magnetic field within the machine. This rotating magnetic field induces a three-phase set of voltages within the stator windings of the generator.

Two terms commonly used to describe the windings on a machine are *field windings* and *armature windings*. In general, the term *field windings* applies to the windings that produce the main magnetic field in a machine, and the term *armature windings* applies to the windings where the main voltage is induced. For synchronous machines, the field windings are on the rotor, so the terms *rotor windings* and *field windings* are used interchangeably. Similarly, the terms *stator windings* and *armature windings* are used interchangeably.

The rotor of a synchronous generator is essentially a large electromagnet. The magnetic poles on the rotor can be of either salient or nonsalient construction. The term *salient* means "protruding" or "sticking out," and a *salient pole* is a magnetic pole that sticks out radially from the shaft of the rotor. On the other hand, a *nonsalient pole* is a magnetic pole with windings embedded flush with the surface of the rotor. A nonsalient-pole rotor is shown in Figure 4–1. Note that the windings of the electromagnet are embedded in notches on the surface of the rotor. A salient-pole rotor is shown in Figure 4–2. Note that here the windings of the electromagnet are wrapped around the pole itself, instead of being embedded in notches on the surface of the rotor. Nonsalient-pole rotors are normally used for two- and four-pole rotors, while salient-pole rotors are normally used for rotors with four or more poles.

Because the rotor is subjected to changing magnetic fields, it is constructed of thin laminations to reduce eddy current losses.

A dc current must be supplied to the field circuit on the rotor if it is an electromagnet. Since the rotor is rotating, a special arrangement is required to get the

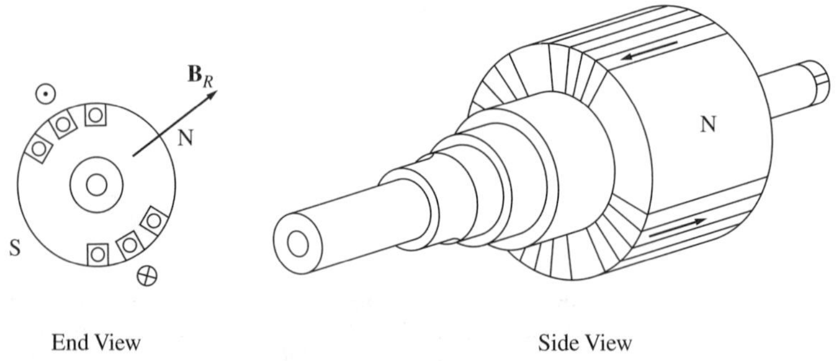

FIGURE 4–1
A nonsalient two-pole rotor for a synchronous machine.

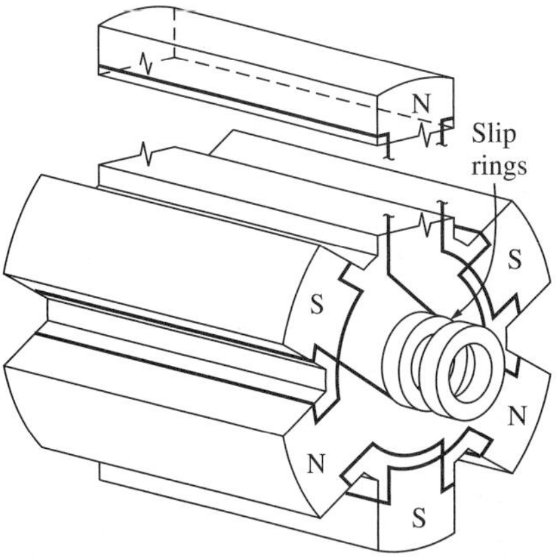

(a)

(b)

(c)

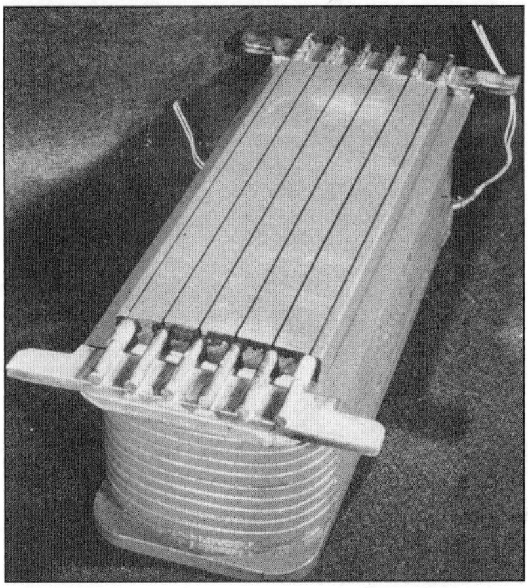

(d)

FIGURE 4–2
(a) A salient six-pole rotor for a synchronous machine. (b) Photograph of a salient eight-pole synchronous machine rotor showing the windings on the individual rotor poles. (*Courtesy of General Electric Company.*) (c) Photograph of a single salient pole from a rotor with the field windings not yet in place. (*Courtesy of General Electric Company.*) (d) A single salient pole shown after the field windings are installed but before it is mounted on the rotor. (*Courtesy of Westinghouse Electric Company.*)

dc power to its field windings. There are two common approaches to supplying this dc power:

1. Supply the dc power from an external dc source to the rotor by means of *slip rings* and *brushes*.
2. Supply the dc power from a special dc power source mounted directly on the shaft of the synchronous generator.

Slip rings are metal rings completely encircling the shaft of a machine but insulated from it. One end of the dc rotor winding is tied to each of the two slip rings on the shaft of the synchronous machine, and a stationary brush rides on each slip ring. A "brush" is a block of graphitelike carbon compound that conducts electricity freely but has very low friction, so that it doesn't wear down the slip ring. If the positive end of a dc voltage source is connected to one brush and the negative end is connected to the other, then the same dc voltage will be applied to the field winding at all times regardless of the angular position or speed of the rotor.

Slip rings and brushes create a few problems when they are used to supply dc power to the field windings of a synchronous machine. They increase the amount of maintenance required on the machine, since the brushes must be checked for wear regularly. In addition, brush voltage drop can be the cause of significant power losses on machines with larger field currents. Despite these problems, slip rings and brushes are used on all smaller synchronous machines, because no other method of supplying the dc field current is cost-effective.

On larger generators and motors, *brushless exciters* are used to supply the dc field current to the machine. A brushless exciter is a small ac generator with its field circuit mounted on the stator and its armature circuit mounted on the rotor shaft. The three-phase output of the exciter generator is rectified to direct current by a three-phase rectifier circuit also mounted on the shaft of the generator, and is then fed into the main dc field circuit. By controlling the small dc field current of the exciter generator (located on the stator), it is possible to adjust the field current on the main machine *without slip rings and brushes*. This arrangement is shown schematically in Figure 4–3, and a

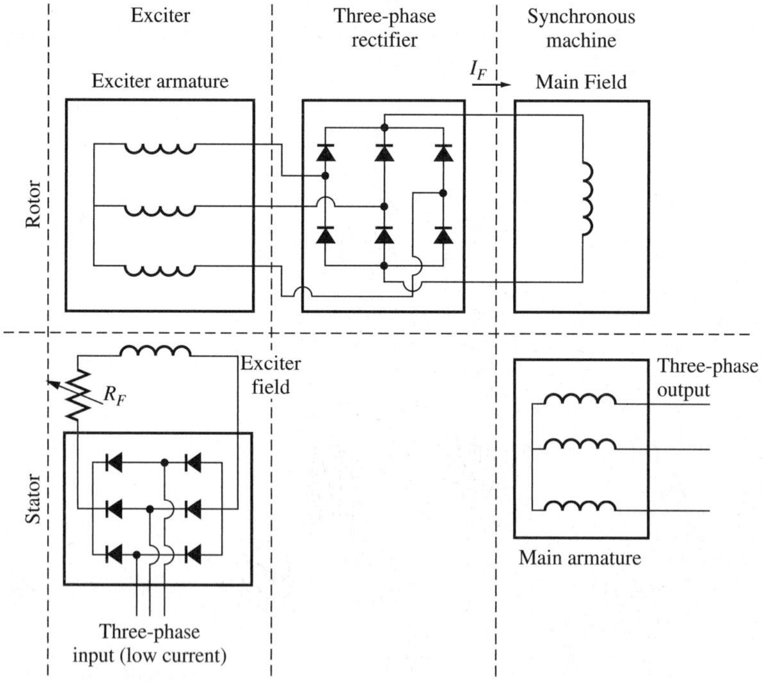

FIGURE 4–3
A brushless exciter circuit. A small three-phase current is rectified and used to supply the field circuit of the exciter, which is located on the stator. The output of the armature circuit of the exciter (on the rotor) is then rectified and used to supply the field current of the main machine.

synchronous machine rotor with a brushless exciter mounted on the same shaft is shown in Figure 4–4. Since no mechanical contacts ever occur between the rotor and the stator, a brushless exciter requires much less maintenance than slip rings and brushes.

To make the excitation of a generator *completely* independent of any external power sources, a small pilot exciter is often included in the system. A *pilot exciter* is a small ac generator with *permanent magnets* mounted on the rotor shaft and a three-phase winding on the stator. It produces the power for the field circuit of the exciter, which in turn controls the field circuit of the main machine. If a pilot exciter is included on the generator shaft, then *no external electric power* is required to run the generator (see Figure 4–5).

Many synchronous generators that include brushless exciters also have slip rings and brushes, so that an auxiliary source of dc field current is available in emergencies.

The stator of a synchronous generator has already been described in Chapter 3, and more details of stator construction are found in Appendix B. Synchronous generator stators are normally made of preformed stator coils in a double-layer winding. The winding itself is distributed and chorded in order to reduce the harmonic content of the output voltages and currents, as described in Appendix B.

A cutaway diagram of a complete large synchronous machine is shown in Figure 4–6. This drawing shows an eight-pole salient-pole rotor, a stator with distributed double-layer windings, and a brushless exciter.

FIGURE 4–4
Photograph of a synchronous machine rotor with a brushless exciter mounted on the same shaft. Notice the rectifying electronics visible next to the armature of the exciter. (Courtesy of Westinghouse Electric Company.)

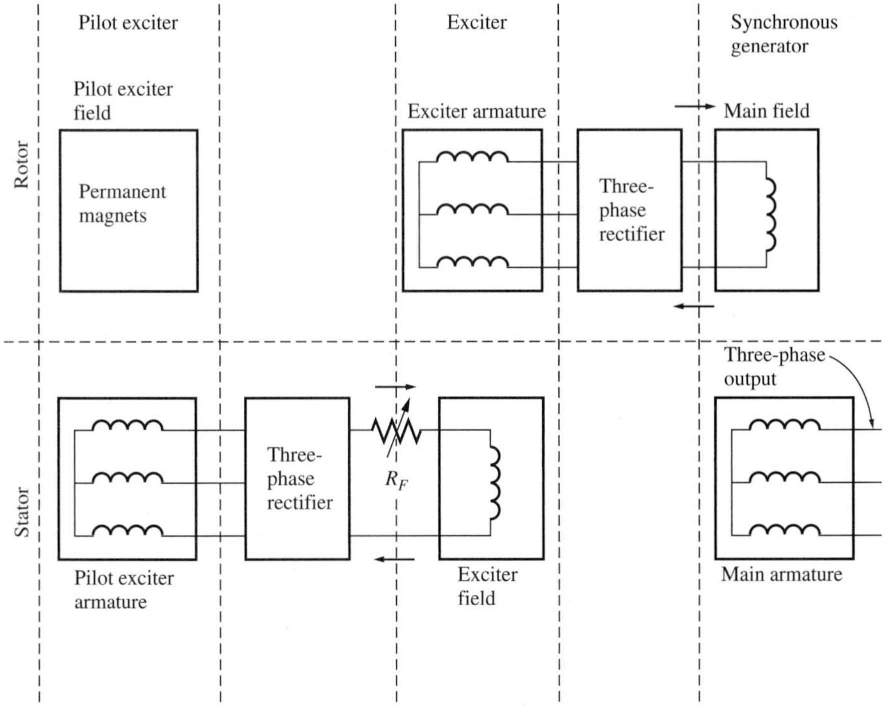

FIGURE 4–5
A brushless excitation scheme that includes a pilot exciter. The permanent magnets of the pilot exciter produce the field current of the exciter, which in turn produces the field current of the main machine.

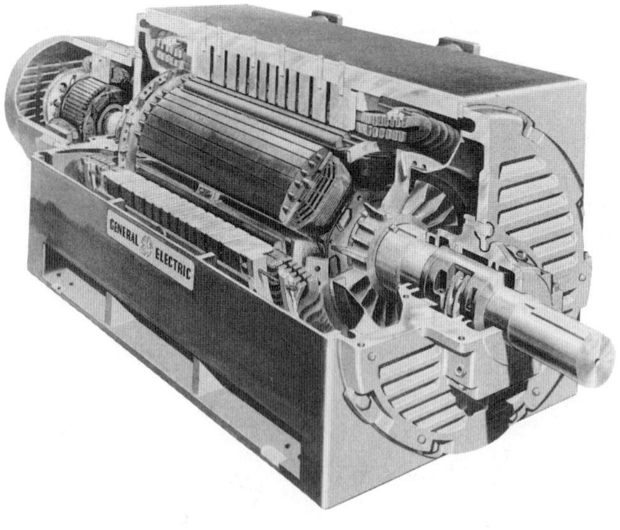

FIGURE 4–6
A cutaway diagram of a large synchronous machine. Note the salient-pole construction and the on-shaft exciter. (*Courtesy of General Electric Company.*)

4.2 THE SPEED OF ROTATION OF A SYNCHRONOUS GENERATOR

Synchronous generators are by definition *synchronous*, meaning that the electrical frequency produced is locked in or synchronized with the mechanical rate of rotation of the generator. A synchronous generator's rotor consists of an electromagnet to which direct current is supplied. The rotor's magnetic field points in whatever direction the rotor is turned. Now, the rate of rotation of the magnetic fields in the machine is related to the stator electrical frequency by Equation (3–34):

$$\boxed{f_{se} = \frac{n_m P}{120}} \tag{3–34}$$

where f_{se} = electrical frequency, in Hz

n_m = mechanical speed of magnetic field, in r/min (equals speed of rotor for synchronous machines)

P = number of poles

Since the rotor turns at the same speed as the magnetic field, *this equation relates the speed of rotor rotation to the resulting electrical frequency.* Electric power is generated at 50 or 60 Hz, so the generator must turn at a fixed speed depending on the number of poles on the machine. For example, to generate 60-Hz power in a two-pole machine, the rotor *must* turn at 3600 r/min. To generate 50-Hz power in a four-pole machine, the rotor *must* turn at 1500 r/min. The required rate of rotation for a given frequency can always be calculated from Equation (3–34).

4.3 THE INTERNAL GENERATED VOLTAGE OF A SYNCHRONOUS GENERATOR

In Chapter 3, the magnitude of the voltage induced in a given stator phase was found to be

$$E_A = \sqrt{2}\pi N_C \phi f \tag{3–50}$$

This voltage depends on the flux ϕ in the machine, the frequency or speed of rotation, and the machine's construction. In solving problems with synchronous machines, this equation is sometimes rewritten in a simpler form that emphasizes the quantities that are variable during machine operation. This simpler form is

$$\boxed{E_A = K\phi\omega} \tag{4–1}$$

where K is a constant representing the construction of the machine. If ω is expressed in *electrical* radians per second, then

$$K = \frac{N_c}{\sqrt{2}} \tag{4–2}$$

while if ω is expressed in *mechanical* radians per second, then

$$K = \frac{N_c P}{\sqrt{2}} \tag{4–3}$$

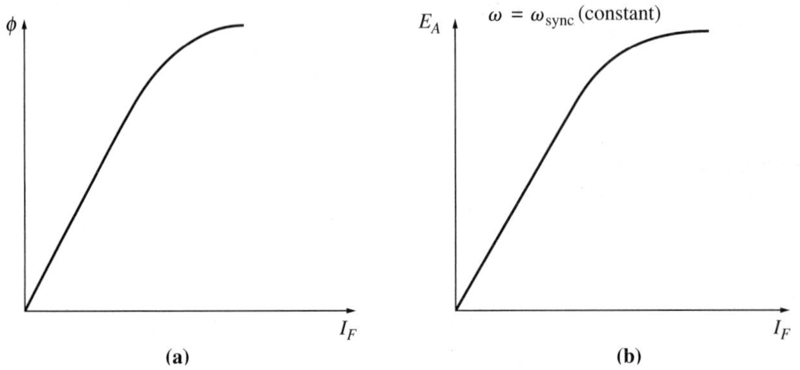

FIGURE 4–7
(a) Plot of flux versus field current for a synchronous generator. (b) The magnetization curve for the synchronous generator.

The internal generated voltage E_A is directly proportional to the flux and to the speed, but the flux itself depends on the current flowing in the rotor field circuit. The field circuit I_F is related to the flux ϕ in the manner shown in Figure 4–7a. Since E_A is directly proportional to the flux, the internal generated voltage E_A is related to the field current as shown in Figure 4–7b. This plot is called the *magnetization curve* or the *open-circuit characteristic* of the machine.

4.4 THE EQUIVALENT CIRCUIT OF A SYNCHRONOUS GENERATOR

The voltage $\mathbf{E}_A$ is the internal generated voltage produced in one phase of a synchronous generator. However, this voltage $\mathbf{E}_A$ is *not* usually the voltage that appears at the terminals of the generator. In fact, the only time the internal voltage $\mathbf{E}_A$ is the same as the output voltage $\mathbf{V}_\phi$ of a phase is when there is no armature current flowing in the machine. Why is the output voltage $\mathbf{V}_\phi$ from a phase not equal to $\mathbf{E}_A$, and what is the relationship between the two voltages? The answer to these questions yields the equivalent circuit model of a synchronous generator.

There are a number of factors that cause the difference between $\mathbf{E}_A$ and $\mathbf{V}_\phi$:

1. The distortion of the air-gap magnetic field by the current flowing in the stator, called *armature reaction.*
2. The self-inductance of the armature coils.
3. The resistance of the armature coils.
4. The effect of salient-pole rotor shapes.

We will explore the effects of the first three factors and derive a machine model from them. In this chapter, the effects of a salient-pole shape on the operation of a synchronous machine will be ignored; in other words, all the machines in this chapter are assumed to have nonsalient or cylindrical rotors. Making this assumption will cause the calculated answers to be slightly inaccurate if a machine does indeed have salient-pole rotors, but the errors are relatively minor. A discussion of the effects of rotor pole saliency is included in Appendix C.

The first effect mentioned, and normally the largest one, is armature reaction. When a synchronous generator's rotor is spun, a voltage $\mathbf{E}_A$ is induced in the generator's stator windings. If a load is attached to the terminals of the generator, a current flows. But a three-phase stator current flow will produce a magnetic field of its own in the machine. This *stator* magnetic field distorts the original rotor magnetic field, changing the resulting phase voltage. This effect is called *armature reaction* because the armature (stator) current affects the magnetic field which produced it in the first place.

To understand armature reaction, refer to Figure 4–8. Figure 4–8a shows a two-pole rotor spinning inside a three-phase stator. There is no load connected to the stator. The rotor magnetic field $\mathbf{B}_R$ produces an internal generated voltage $\mathbf{E}_A$ whose peak value coincides with the direction of $\mathbf{B}_R$. As was shown in the last chapter, the voltage will be positive out of the conductors at the top and negative into the conductors at the bottom of the figure. With no load on the generator, there is no armature current flow, and $\mathbf{E}_A$ will be equal to the phase voltage $\mathbf{V}_\phi$.

Now suppose that the generator is connected to a lagging load. Because the load is lagging, the peak current will occur at an angle *behind* the peak voltage. This effect is shown in Figure 4–8b.

The current flowing in the stator windings produces a magnetic field of its own. This stator magnetic field is called $\mathbf{B}_S$ and its direction is given by the right-hand rule to be as shown in Figure 4–8c. The stator magnetic field $\mathbf{B}_S$ produces a voltage of its own in the stator, and this voltage is called $\mathbf{E}_{\text{stat}}$ on the figure.

With two voltages present in the stator windings, the total voltage in a phase is just the *sum* of the internal generated voltage $\mathbf{E}_A$ and the armature reaction voltage $\mathbf{E}_{\text{stat}}$:

$$\mathbf{V}_\phi = \mathbf{E}_A + \mathbf{E}_{\text{stat}} \tag{4–4}$$

The net magnetic field $\mathbf{B}_{\text{net}}$ is just the sum of the rotor and stator magnetic fields:

$$\mathbf{B}_{\text{net}} = \mathbf{B}_R + \mathbf{B}_S \tag{4–5}$$

Since the angles of $\mathbf{E}_A$ and $\mathbf{B}_R$ are the same and the angles of $\mathbf{E}_{\text{stat}}$ and $\mathbf{B}_S$ are the same, the resulting magnetic field $\mathbf{B}_{\text{net}}$ will coincide with the net voltage $\mathbf{V}_\phi$. The resulting voltages and currents are shown in Figure 4–8d.

The angle between $\mathbf{B}_R$ and $\mathbf{B}_{\text{net}}$ is known as the *internal angle* or *torque angle* Δ (gr Δ) for the machine. This angle is proportional to the amount of power being supplied by the generator, as we shall see in Section 4.6.

How can the effects of armature reaction on the phase voltage be modeled? First, note that the voltage $\mathbf{E}_{\text{stat}}$ lies at an angle of 90° behind the plane of maximum current $\mathbf{I}_A$. Second, the voltage $\mathbf{E}_{\text{stat}}$ is directly proportional to the current $\mathbf{I}_A$.

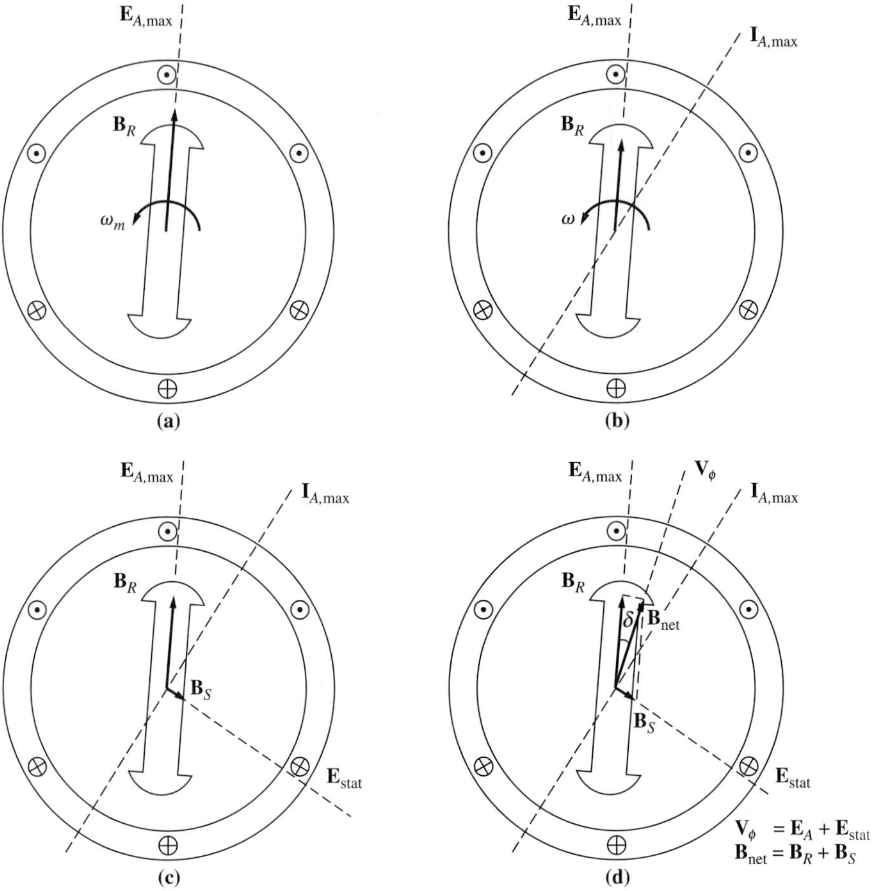

FIGURE 4–8
The development of a model for armature reaction: (a) A rotating magnetic field produces the internal generated voltage $\mathbf{E}_A$. (b) The resulting voltage produces a lagging *current flow* when connected to a lagging load. (c) The stator current produces its own magnetic field $\mathbf{B}_S$, which produces its own voltage $\mathbf{E}_{\text{stat}}$ in the stator windings of the machine. (d) The field $\mathbf{B}_S$ adds to $\mathbf{B}_R$, distorting it into $\mathbf{B}_{\text{net}}$. The voltage $\mathbf{E}_{\text{stat}}$ adds to $\mathbf{E}_A$, producing $\mathbf{V}_\phi$ at the output of the phase.

If X is a constant of proportionality, then *the armature reaction voltage can be expressed as*

$$\mathbf{E}_{\text{stat}} = -jX\mathbf{I}_A \tag{4–6}$$

The voltage on a phase is thus

$$\boxed{\mathbf{V}_\phi = \mathbf{E}_A - jX\mathbf{I}_A} \tag{4–7}$$

Look at the circuit shown in Figure 4–9. The Kirchhoff's voltage law equation for this circuit is

$$\mathbf{V}_\phi = \mathbf{E}_A - jX\mathbf{I}_A \tag{4–8}$$

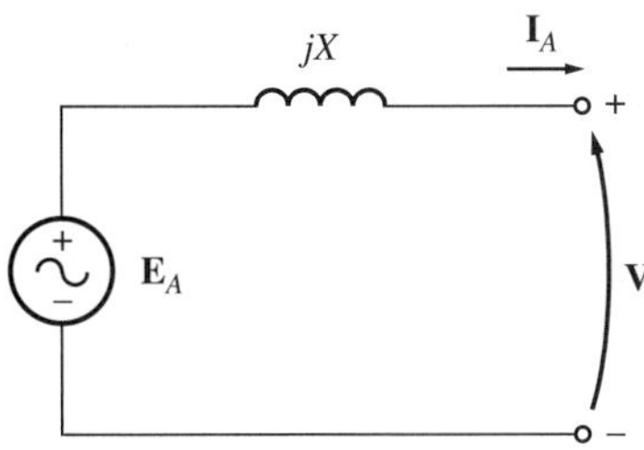

FIGURE 4–9
A simple circuit (see text).

This is exactly the same equation as the one describing the armature reaction voltage. Therefore, the armature reaction voltage can be modeled as an inductor in series with the internal generated voltage.

In addition to the effects of armature reaction, the stator coils have a self-inductance and a resistance. If the stator self-inductance is called L_A (and its corresponding reactance is called X_A) while the stator resistance is called R_A, then the total difference between $\mathbf{E}_A$ and $\mathbf{V}_\phi$ is given by

$$\mathbf{V}_\phi = \mathbf{E}_A - jX\mathbf{I}_A - jX_A\mathbf{I}_A - R_A\mathbf{I}_A \tag{4–9}$$

The armature reaction effects and the self-inductance in the machine are both represented by reactances, and it is customary to combine them into a single reactance, called the *synchronous reactance* of the machine:

$$X_S = X + X_A \tag{4–10}$$

Therefore, the final equation describing $\mathbf{V}_\phi$ is

$$\boxed{\mathbf{V}_\phi = \mathbf{E}_A - jX_S\mathbf{I}_A - R_A\mathbf{I}_A} \tag{4–11}$$

It is now possible to sketch the equivalent circuit of a three-phase synchronous generator. The full equivalent circuit of such a generator is shown in Figure 4–10. This figure shows a dc power source supplying the rotor field circuit, which is modeled by the coil's inductance and resistance in series. In series with R_F is an adjustable resistor R_{adj} which controls the flow of field current. The rest of the equivalent circuit consists of the models for each phase. Each phase has an internal generated voltage with a series inductance X_S (consisting of the sum of the armature reactance and the coil's self-inductance) and a series resistance R_A. The voltages and currents of the three phases are 120° apart in angle, but otherwise the three phases are identical.

These three phases can be either Y- or Δ-connected as shown in Figure 4–11. If they are Y-connected, then the terminal voltage V_T (which is the same as the time-to-line voltage V_L is related to the phase voltage by

$$V_T = V_L = \sqrt{3}V_\phi \tag{4–12}$$

If they are Δ-connected, then

$$V_T = V_\phi \tag{4–13}$$

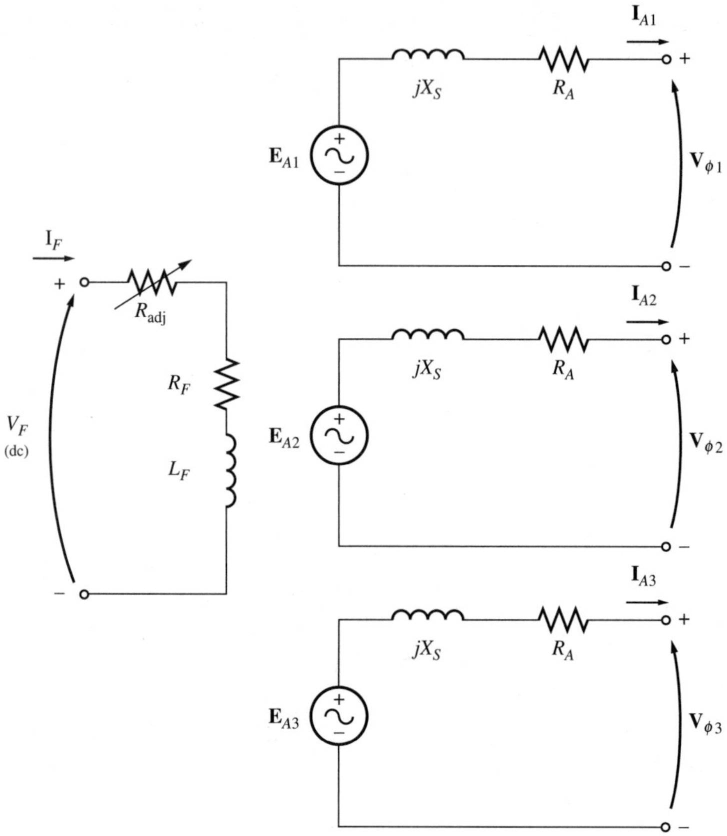

FIGURE 4–10
The full equivalent circuit of a three-phase synchronous generator.

The fact that the three phases of a synchronous generator are identical in all respects except for phase angle normally leads to the use of a *per-phase equivalent circuit*. The per-phase equivalent circuit of this machine is shown in Figure 4–12. One important fact must be kept in mind when the per-phase equivalent circuit is used: The three phases have the same voltages and currents *only* when the loads attached to them are *balanced*. If the generator's loads are not balanced, more sophisticated techniques of analysis are required. These techniques are beyond the scope of this book.

4.5 THE PHASOR DIAGRAM OF A SYNCHRONOUS GENERATOR

Because the voltages in a synchronous generator are ac voltages, they are usually expressed as phasors. Since phasors have both a magnitude and an angle, the

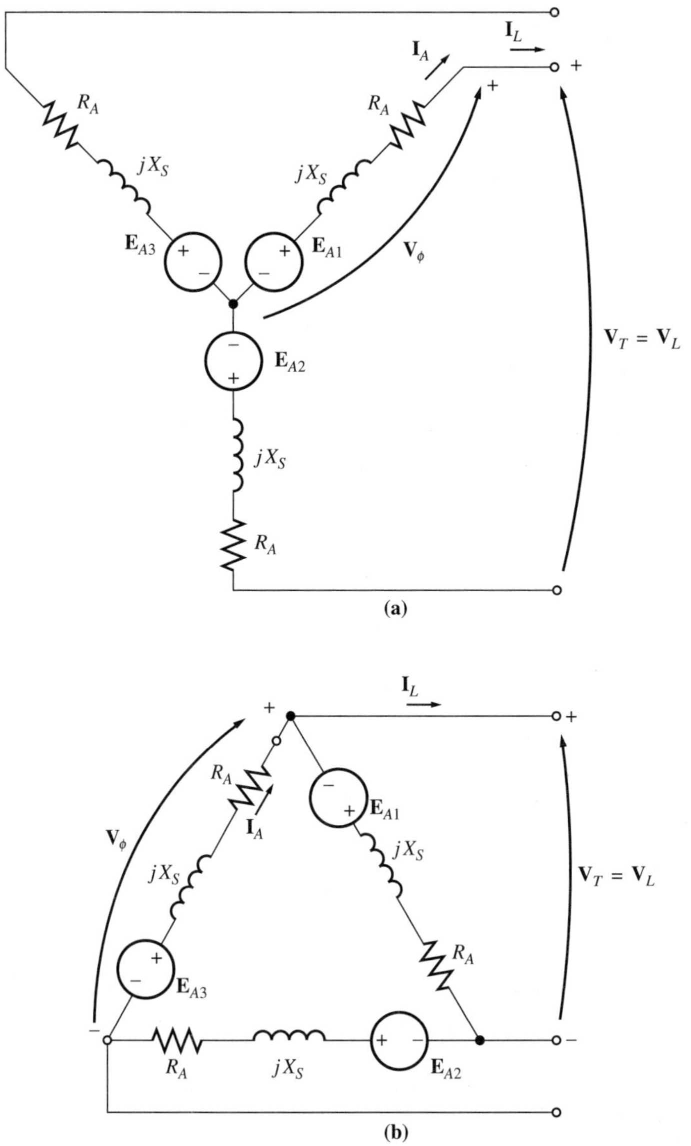

FIGURE 4–11
The generator equivalent circuit connected in (a) Y and (b) Δ.

relationship between them must be expressed by a two-dimensional plot. When the voltages within a phase ($\mathbf{E}_A$, $\mathbf{V}_\phi$, $jX_S\mathbf{I}_A$, and $R_A\mathbf{I}_A$) and the current $\mathbf{I}_A$ in the phase are plotted in such a fashion as to show the relationships among them, the resulting plot is called a *phasor diagram*.

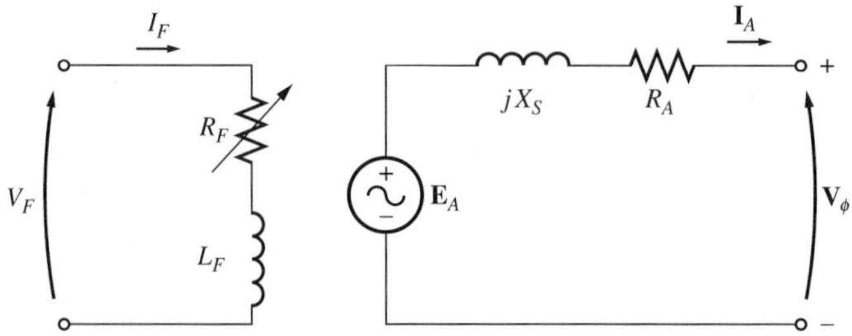

FIGURE 4–12
The per-phase equivalent circuit of a synchronous generator. The internal field circuit resistance and the external variable resistance have been combined into a single resistor R_F.

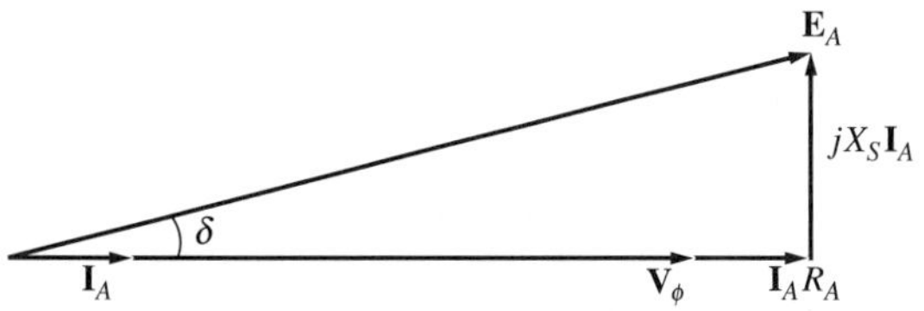

FIGURE 4–13
The phasor diagram of a synchronous generator at unity power factor.

For example, Figure 4–13 shows these relationships when the generator is supplying a load at unity power factor (a purely resistive load). From Equation (4–11), the total voltage $\mathbf{E}_A$ differs from the terminal voltage of the phase $\mathbf{V}_\phi$ by the resistive and inductive voltage drops. All voltages and currents are referenced to $\mathbf{V}_\phi$, which is arbitrarily assumed to be at an angle of 0°.

This phasor diagram can be compared to the phasor diagrams of generators operating at lagging and leading power factors. These phasor diagrams are shown in Figure 4–14. Notice that, *for a given phase voltage and armature current,* a larger internal generated voltage $\mathbf{E}_A$ is needed for lagging loads than for leading loads. Therefore, a larger field current is needed with lagging loads to get the same terminal voltage, because

$$E_A = K\phi\omega \tag{4–1}$$

and ω must be constant to keep a constant frequency.

Alternatively, *for a given field current and magnitude of load current, the terminal voltage is lower for lagging loads and higher for leading loads.*

In real synchronous machines, the synchronous reactance is normally much larger than the winding resistance R_A, so R_A is often neglected in the *qualitative* study of voltage variations. For accurate numerical results, R_A must of course be considered.

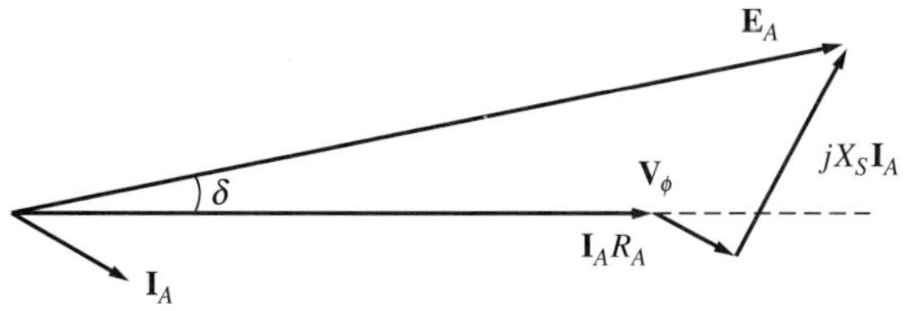

(a)

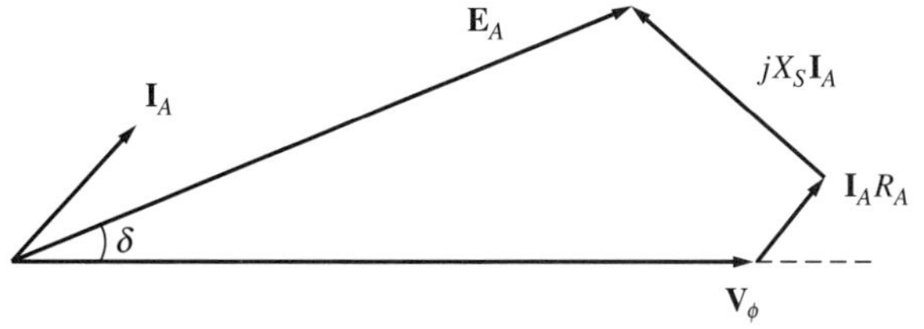

(b)

FIGURE 4–14
The phasor diagram of a synchronous generator at (a) lagging and (b) leading power factor.

4.6 POWER AND TORQUE IN SYNCHRONOUS GENERATORS

A synchronous generator is a synchronous machine used as a generator. It converts mechanical power to three-phase electrical power. The source of mechanical power, the *prime mover*, may be a diesel engine, a steam turbine, a water turbine, or any similar device. Whatever the source, it must have the basic property that its speed is almost constant regardless of the power demand. If that were not so, then the resulting power system's frequency would wander.

Not all the mechanical power going into a synchronous generator becomes electrical power out of the machine. The difference between input power and output power represents the losses of the machine. A power-flow diagram for a synchronous generator is shown in Figure 4–15. The input mechanical power is the shaft power in the generator $P_{\text{in}} = \tau_{\text{app}}\omega_{\text{m}}$, while the power converted from mechanical to electrical form internally is given by

$$P_{\text{conv}} = \tau_{\text{ind}}\omega_m \tag{4–14}$$

$$= 3E_A I_A \cos \gamma \tag{4–15}$$

where γ is the angle between $\mathbf{E}_A$ and $\mathbf{I}_A$. The difference between the input power to the generator and the power converted in the generator represents the mechanical, core, and stray losses of the machine.

The real electrical output power of the synchronous generator can be expressed in line quantities as

$$P_{\text{out}} = \sqrt{3}V_L I_L \cos \theta \tag{4–16}$$

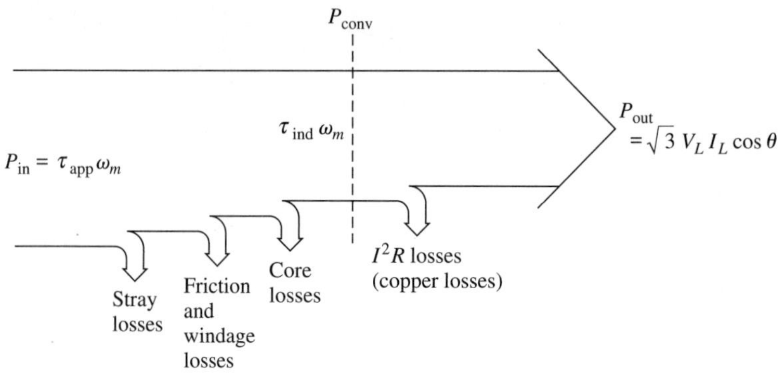

FIGURE 4–15
The power-flow diagram of a synchronous generator.

and in phase quantities as

$$P_{out} = 3V_\phi I_A \cos\theta \tag{4–17}$$

The reactive power output can be expressed in line quantities as

$$Q_{out} = \sqrt{3}V_L I_L \sin\theta \tag{4–18}$$

or in phase quantities as

$$Q_{out} = 3V_\phi I_A \sin\theta \tag{4–19}$$

If the armature resistance R_A is ignored (since $X_S >> R_A$), then a very useful equation can be derived to approximate the output power of the generator. To derive this equation, examine the phasor diagram in Figure 4–16. Figure 4–16 shows a simplified phasor diagram of a generator with the stator resistance ignored. Notice that the vertical segment *bc* can be expressed as either $E_A \sin\delta$ or $X_S I_A \cos\theta$. Therefore,

$$I_A \cos\theta = \frac{E_A \sin\delta}{X_S}$$

and substituting this expression into Equation (4–17) gives

$$\boxed{P_{conv} = \frac{3V_\phi E_A}{X_S}\sin\delta} \tag{4–20}$$

Since the resistances are assumed to be zero in Equation (4–20), there are no electrical losses in this generator, and this equation is both P_{conv} and P_{out}.

Equation (4–20) shows that the power produced by a synchronous generator depends on the angle δ between $\mathbf{V}_\phi$ and $\mathbf{E}_A$. The angle δ is known as the *internal angle* or *torque angle* of the machine. Notice also that the maximum power that the generator can supply occurs when $\delta = 90°$. At $\delta = 90°$, $\sin\delta = 1$, and

$$P_{max} = \frac{3V_\phi E_A}{X_S} \tag{4–21}$$

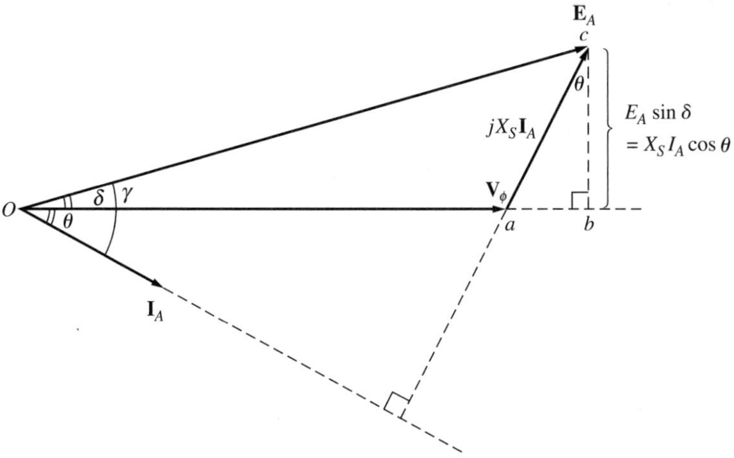

FIGURE 4–16
Simplified phasor diagram with armature resistance ignored.

The maximum power indicated by this equation is called the *static stability limit* of the generator. Normally, real generators never even come close to that limit. Full-load torque angles of 20 to 30 degrees are more typical of real machines.

Now take another look at Equations (4–17), (4–19), and (4–20). If $\mathbf{V}_\phi$ is assumed constant, then the *real power output is directly proportional* to the quantities $I_A \cos \theta$ and $E_A \sin \delta$, and the reactive power output is directly proportional to the quantity $I_A \sin \theta$. These facts are useful in plotting phasor diagrams of synchronous generators as loads change.

From Chapter 3, the induced torque in this generator can be expressed as

$$\tau_{\text{ind}} = k\mathbf{B}_R \times \mathbf{B}_S \tag{3–58}$$

or as

$$\tau_{\text{ind}} = k\mathbf{B}_R \times \mathbf{B}_{\text{net}} \tag{3–60}$$

The magnitude of Equation (3–60) can be expressed as

$$\tau_{\text{ind}} = kB_R B_{\text{net}} \sin \delta \tag{3–61}$$

where δ is the angle between the rotor and net magnetic fields (the so-called *torque angle*). Since $\mathbf{B}_R$ produces the voltage $\mathbf{E}_A$ and $\mathbf{B}_{\text{net}}$ produces the voltage $\mathbf{V}_\phi$, the angle δ between $\mathbf{E}_A$ *and* $\mathbf{V}_\phi$ is the same as the angle δ between $\mathbf{B}_R$ and $\mathbf{B}_{\text{net}}$.

An alternative expression for the induced torque in a synchronous generator can be derived from Equation (4–20). Because $P_{\text{conv}} = \tau_{\text{ind}}\omega_m$, the induced torque can be expressed as

$$\boxed{\tau_{\text{ind}} = \frac{3V_\phi E_A}{\omega_m X_S} \sin \delta} \tag{4–22}$$

This expression describes the induced torque in terms of electrical quantities, whereas Equation (3–60) gives the same information in terms of magnetic quantities.

Note that both the power converted from mechanical form to electrical form P_{conv} in a synchronous generator and the torque induced τ_{ind} in the rotor of the generator are dependent on the torque angle δ.

$$\boxed{P_{\text{conv}} = \frac{3V_\phi E_A}{X_S} \sin \delta} \tag{4–20}$$

$$\boxed{\tau_{\text{ind}} = \frac{3V_\phi E_A}{\omega_m X_S} \sin \delta} \tag{4–22}$$

Both of these quantities reach their maximum values when the torque angle δ reaches 90°. The generator is not capable of exceeding those limits even instantaneously. Real generators typically have full-load torque angles of 20–30°, so the absolute maximum instantaneous power and torque that they can supply is at least twice their full-load values. This reserve of power and torque is essential for the stability of power systems containing these generators, as we will see in Section 4.10.

4.7 MEASURING SYNCHRONOUS GENERATOR MODEL PARAMETERS

The equivalent circuit of a synchronous generator that has been derived contains three quantities that must be determined in order to completely describe the behavior of a real synchronous generator:

1. The relationship between field current and flux (and therefore between the field current and E_A)
2. The synchronous reactance
3. The armature resistance

This section describes a simple technique for determining these quantities in a synchronous generator.

The first step in the process is to perform the *open-circuit test* on the generator. To perform this test, the generator is turned at the rated speed, the terminals are disconnected from all loads, and the field current is set to zero. Then the field current is gradually increased in steps, and the terminal voltage is measured at each step along the way. With the terminals open, $I_A = 0$, so E_A is equal to V_ϕ. It is thus possible to construct a plot of E_A (or V_T) versus I_F from this information. This plot is the so-called *open-circuit characteristic* (OCC) of a generator. With this characteristic, it is possible to find the internal generated voltage of the generator for any given field current. A typical open-circuit characteristic is shown in Figure 4–17a. Notice that at first the curve is almost perfectly linear, until some saturation is observed at high field currents. The unsaturated iron in the frame of the synchronous machine has a reluctance several thousand times lower than the air-gap reluctance,

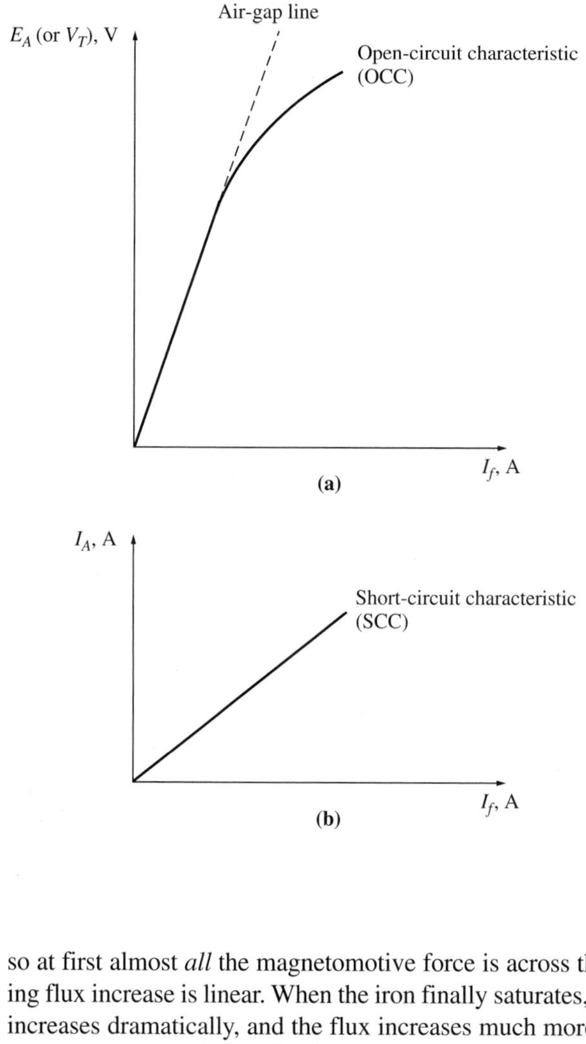

FIGURE 4–17
(a) The open-circuit characteristic (OCC) of a synchronous generator. (b) The short-circuit characteristic (SCC) of a synchronous generator.

so at first almost *all* the magnetomotive force is across the air gap, and the resulting flux increase is linear. When the iron finally saturates, the reluctance of the iron increases dramatically, and the flux increases much more slowly with an increase in magnetomotive force. The linear portion of an OCC is called the *air-gap line* of the characteristic.

The second step in the process is to conduct the *short-circuit test.* To perform the short-circuit test, adjust the field current to zero again and short-circuit the terminals of the generator through a set of ammeters. Then the armature current I_A or the line current I_L is measured as the field current is increased. Such a plot is called a *short-circuit characteristic* (SCC) and is shown in Figure 4–17b. It is essentially a straight line. To understand why this characteristic is a straight line, look at the equivalent circuit in Figure 4–12 when the terminals of the machine are short-circuited. Such a circuit is shown in Figure 4–18a. Notice that when the terminals are short-circuited, the armature current $\mathbf{I}_A$ is given by

$$\mathbf{I}_A = \frac{\mathbf{E}_A}{R_A + jX_S} \tag{4–23}$$

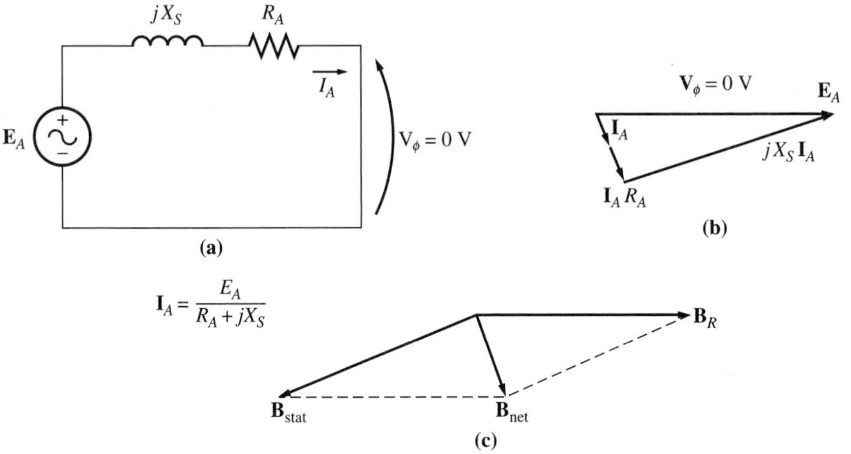

FIGURE 4–18
(a) The equivalent circuit of a synchronous generator during the short-circuit test. (b) The resulting phasor diagram. (c) The magnetic fields during the short-circuit test.

and its magnitude is just given by

$$I_A = \frac{E_A}{\sqrt{R_A^2 + X_S^2}} \tag{4–24}$$

The resulting phasor diagram is shown in Figure 4–18b, and the corresponding magnetic fields are shown in Figure 4–18c. Since $\mathbf{B}_S$ almost cancels $\mathbf{B}_R$, the net magnetic field $\mathbf{B}_{net}$ is *very* small (corresponding to internal resistive and inductive drops only). Since the net magnetic field in the machine is so small, the machine is unsaturated and the SCC is linear.

To understand what information these two characteristics yield, notice that, with $\mathbf{V}_\phi$ equal to zero in Figure 4–18, the *internal machine impedance* is given by

$$Z_S = \sqrt{R_A^2 + X_S^2} = \frac{E_A}{I_A} \tag{4–25}$$

Since $X_S >> R_A$, this equation reduces to

$$X_S \approx \frac{E_A}{I_A} = \frac{V_{\phi,oc}}{I_A} \tag{4–26}$$

If E_A and I_A are known for a given situation, then the synchronous reactance X_S can be found.

Therefore, an *approximate* method for determining the synchronous reactance X_S at a given field current is

1. Get the internal generated voltage E_A from the OCC at that field current.
2. Get the short-circuit current flow $I_{A,SC}$ at that field current from the SCC.
3. Find X_S by applying Equation (4–26).

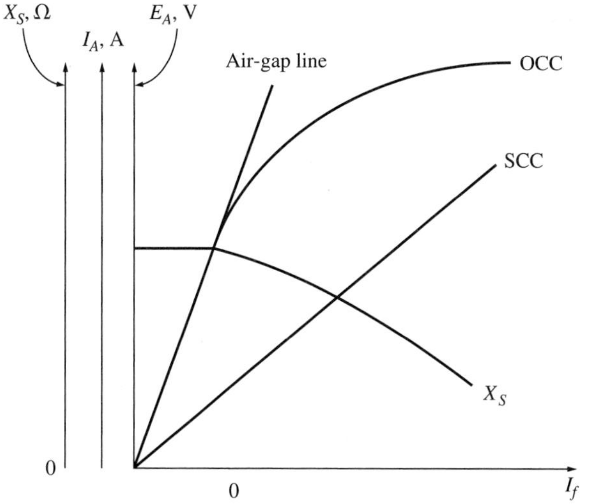

FIGURE 4–19
A sketch of the approximate synchronous reactance of a synchronous generator as a function of the field current in the machine. The constant value of reactance found at low values of field current is the *unsaturated* synchronous reactance of the machine.

There is a problem with this approach, however. The internal generated voltage E_A comes from the OCC, where the machine is partially *saturated* for large field currents, while I_A is taken from the SCC, where the machine is *unsaturated* at all field currents. Therefore, at higher field currents, the E_A taken from the OCC at a given field current is *not* the same as the E_A at the same field current under short-circuit conditions, and this difference makes the resulting value of X_S only approximate.

However, the answer given by this approach *is* accurate up to the point of saturation, so the *unsaturated synchronous reactance* $X_{S,u}$ of the machine can be found simply by applying Equation (4–26) at any field current in the linear portion (on the air-gap line) of the OCC curve.

The approximate value of synchronous reactance varies with the degree of saturation of the OCC, so the value of the synchronous reactance to be used in a given problem should be one calculated at the approximate load on the machine. A plot of approximate synchronous reactance as a function of field current is shown in Figure 4–19.

To get a more accurate estimation of the saturated synchronous reactance, refer to Section 5–3 of Reference 2.

If it is important to know a winding's resistance as well as its synchronous reactance, the resistance can be approximated by applying a dc voltage to the windings while the machine is stationary and measuring the resulting current flow. The use of dc voltage means that the reactance of the windings will be zero during the measurement process.

This technique is not perfectly accurate, since the ac resistance will be slightly larger than the dc resistance (as a result of the skin effect at higher frequencies). The measured value of the resistance can even be plugged into Equation (4–26) to improve the estimate of X_S, if desired. (Such an improvement is not much help in the approximate approach—saturation causes a much larger error in the X_S calculation than ignoring R_A does.)

The Short-Circuit Ratio

Another parameter used to describe synchronous generators is the short-circuit ratio. The *short-circuit ratio* of a generator is defined as the ratio of the *field current required for the rated voltage at open circuit* to the *field current required for the rated armature current at short circuit*. It can be shown that this quantity is just the reciprocal of the per-unit value of the approximate saturated synchronous reactance calculated by Equation (4–26).

Although the short-circuit ratio adds no new information about the generator that is not already known from the saturated synchronous reactance, it is important to know what it is, since the term is occasionally encountered in industry.

Example 4–1. A 200-kVA, 480-V, 50-Hz, Y-connected synchronous generator with a rated field current of 5 A was tested, and the following data were taken:

1. $V_{T,OC}$ at the rated I_F was measured to be 540 V.
2. $I_{L,SC}$ at the rated I_F was found to be 300 A.
3. When a dc voltage of 10 V was applied to two of the terminals, a current of 25 A was measured.

Find the values of the armature resistance and the approximate synchronous reactance in ohms that would be used in the generator model at the rated conditions.

Solution

The generator described above is Y-connected, so the direct current in the resistance test flows through two windings. Therefore, the resistance is given by

$$2R_A = \frac{V_{\text{DC}}}{I_{\text{DC}}}$$

$$R_A = \frac{V_{\text{DC}}}{2I_{\text{DC}}} = \frac{10 \text{ V}}{(2)(25 \text{ A})} = 0.2 \ \Omega$$

The internal generated voltage at the rated field current is equal to

$$E_A = V_{\phi,\text{OC}} = \frac{V_T}{\sqrt{3}}$$

$$= \frac{540 \text{ V}}{\sqrt{3}} = 311.8 \text{ V}$$

The short-circuit current I_A is just equal to the line current, since the generator is Y-connected:

$$I_{A,SC} = I_{L,SC} = 300 \text{ A}$$

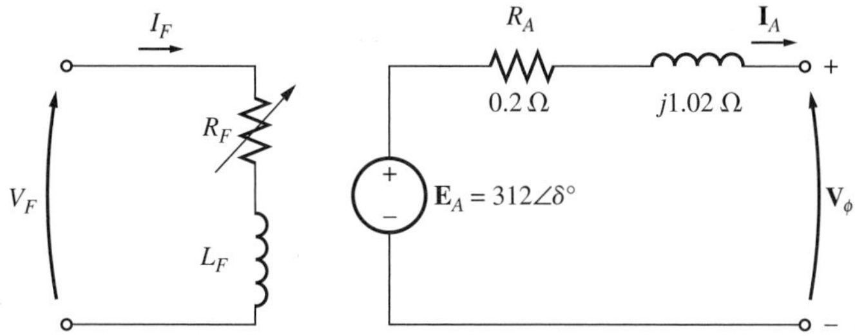

FIGURE 4–20
The per-phase equivalent circuit of the generator in Example 4–1.

Therefore, the synchronous reactance at the rated field current can be calculated from Equation (4–25):

$$\sqrt{R_A^2 + X_S^2} = \frac{E_A}{I_A} \tag{4–25}$$

$$\sqrt{(0.2\ \Omega)^2 + X_S^2} = \frac{311.8\ \text{V}}{300\ \text{A}}$$

$$\sqrt{(0.2\ \Omega)^2 + X_S^2} = 1.039\ \Omega$$

$$0.04 + X_S^2 = 1.08$$

$$X_S^2 = 1.04$$

$$X_S = 1.02\ \Omega$$

How much effect did the inclusion of R_A have on the estimate of X_S? Not much. If X_S is evaluated by Equation (4–26), the result is

$$X_S = \frac{E_A}{I_A} = \frac{311.8\ \text{V}}{300\ \text{A}} = 1.04\ \Omega$$

Since the error in X_S due to ignoring R_A is much less than the error due to saturation effects, approximate calculations are normally done with Equation (4–26).

The resulting per-phase equivalent circuit is shown in Figure 4–20.

4.8 THE SYNCHRONOUS GENERATOR OPERATING ALONE

The behavior of a synchronous generator under load varies greatly depending on the power factor of the load and on whether the generator is operating alone or in parallel with other synchronous generators. In this section, we will study the behavior of synchronous generators operating alone. We will study the behavior of synchronous generators operating in parallel in Section 4.9.

Throughout this section, concepts will be illustrated with simplified phasor diagrams ignoring the effect of R_A. In some of the numerical examples the resistance R_A will be included.

Unless otherwise stated in this section, the speed of the generators will be assumed constant, and all terminal characteristics are drawn assuming constant

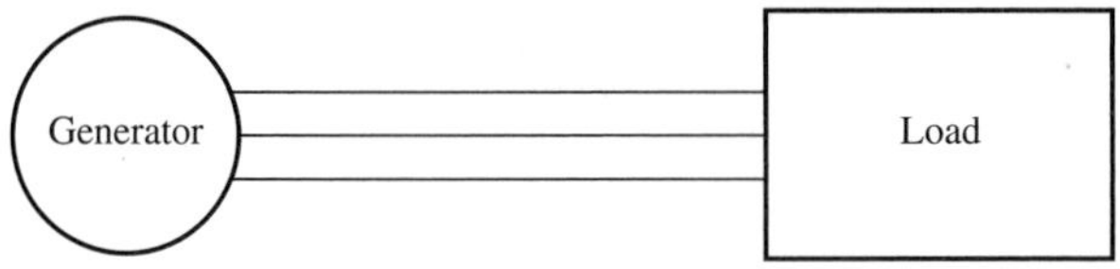

FIGURE 4–21
A single generator supplying a load.

speed. Also, the rotor flux in the generators is assumed constant unless their field current is explicitly changed.

The Effect of Load Changes on a Synchronous Generator Operating Alone

To understand the operating characteristics of a synchronous generator operating alone, examine a generator supplying a load. A diagram of a single generator supplying a load is shown in Figure 4–21. What happens when we increase the load on this generator?

An increase in the load is an increase in the real and/or reactive power drawn from the generator. Such a load increase increases the load current drawn from the generator. Because the field resistor has not been changed, the field current is constant, and therefore the flux ϕ is constant. Since the prime mover also keeps a constant speed ω, the *magnitude of the internal generated voltage* $E_A = K\phi\omega$ *is constant.*

If E_A is constant, just what does vary with a changing load? The way to find out is to construct phasor diagrams showing an increase in the load, keeping the constraints on the generator in mind.

First, examine a generator operating at a lagging power factor. If more load is added at the *same power factor*, then $|\mathbf{I}_A|$ increases but remains at the same angle θ with respect to $\mathbf{V}_\phi$ as before. Therefore, the armature reaction voltage $jX_S\mathbf{I}_A$ is larger than before but at the same angle. Now since

$$\mathbf{E}_A = \mathbf{V}_\phi + jX_S\mathbf{I}_A$$

$jX_S\mathbf{I}_A$ must stretch between $\mathbf{V}_\phi$ at an angle of 0° and $\mathbf{E}_A$, which is constrained to be of the same magnitude as before the load increase. If these constraints are plotted on a phasor diagram, there is one and only one point at which the armature reaction voltage can be parallel to its original position while increasing in size. The resulting plot is shown in Figure 4–22a.

If the constraints are observed, then it is seen that as the load increases, the voltage $\mathbf{V}_\phi$ decreases rather sharply.

Now suppose the generator is loaded with unity-power-factor loads. What happens if new loads are added at the same power factor? With the same constraints as before, it can be seen that this time $\mathbf{V}_\phi$ decreases only slightly (see Figure 4–22b).

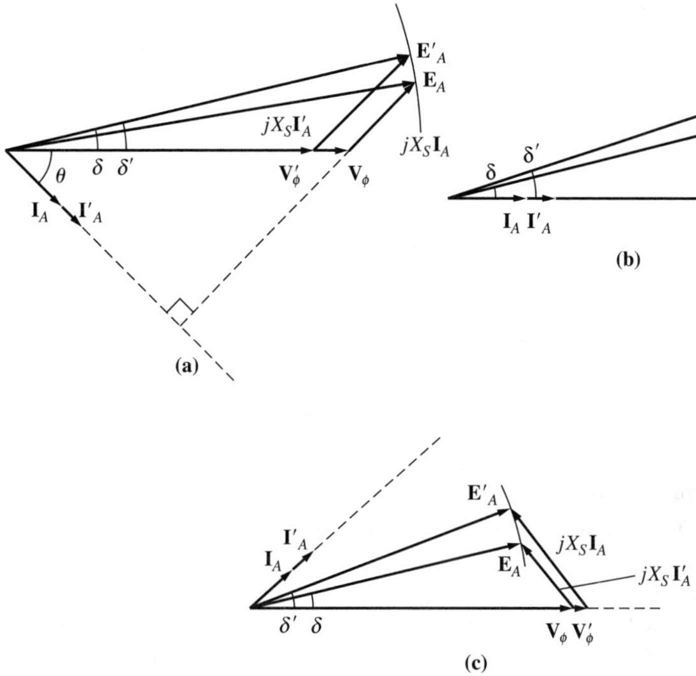

FIGURE 4–22
The effect of an increase in generator loads at constant power factor upon its terminal voltage. (a) Lagging power factor; (b) unity power factor; (c) leading power factor.

Finally, let the generator be loaded with leading-power-factor loads. If new loads are added at the same power factor this time, the armature reaction voltage lies outside its previous value, and $\mathbf{V}_\phi$ actually *rises* (see Figure 4–22c). In this last case, an increase in the load in the generator produced an increase in the terminal voltage. Such a result is not something one would expect on the basis of intuition alone.

General conclusions from this discussion of synchronous generator behavior are

1. If lagging loads ($+Q$ or inductive reactive power loads) are added to a generator, $\mathbf{V}_\phi$ and the terminal voltage V_T decrease significantly.
2. If unity-power-factor loads (no reactive power) are added to a generator, there is a slight decrease in $\mathbf{V}_\phi$ and the terminal voltage.
3. If leading loads ($-Q$ or capacitive reactive power loads) are added to a generator, $\mathbf{V}_\phi$ and the terminal voltage will rise.

A convenient way to compare the voltage behavior of two generators is by their *voltage regulation*. The voltage regulation (VR) of a generator is defined by the equation

$$\boxed{\mathrm{VR} = \frac{V_{\mathrm{nl}} - V_{\mathrm{fl}}}{V_{\mathrm{fl}}} \times 100\%} \tag{3–67}$$

where V_{nl} is the no-load voltage of the generator and V_{fl} is the full-load voltage of the generator. A synchronous generator operating at a lagging power factor has a fairly large positive voltage regulation, a synchronous generator operating at a unity power factor has a small positive voltage regulation, and a synchronous generator operating at a leading power factor often has a negative voltage regulation.

Normally, it is desirable to keep the voltage supplied to a load constant, even though the load itself varies. How can terminal voltage variations be corrected for? The obvious approach is to vary the magnitude of $\mathbf{E}_A$ to compensate for changes in the load. Recall that $E_A = K\phi\omega$. Since the frequency should not be changed in a normal system, E_A must be controlled by varying the flux in the machine.

For example, suppose that a lagging load is added to a generator. Then the terminal voltage will fall, as was previously shown. To restore it to its previous level, decrease the field resistor R_F. If R_F decreases, the field current will increase. An increase in I_F increases the flux, which in turn increases E_A, and an increase in E_A increases the phase and terminal voltage. This idea can be summarized as follows:

1. Decreasing the field resistance in the generator increases its field current.
2. An increase in the field current increases the flux in the machine.
3. An increase in the flux increases the internal generated voltage $E_A = K\phi\omega$.
4. An increase in E_A increases V_ϕ and the terminal voltage of the generator.

The process can be reversed to decrease the terminal voltage. It is possible to regulate the terminal voltage of a generator throughout a series of load changes simply by adjusting the field current.

Example Problems

The following three problems illustrate simple calculations involving voltages, currents, and power flows in synchronous generators. The first problem is an example that includes the armature resistance in its calculations, while the next two ignore R_A. Part of the first example problem addresses the question: *How must a generator's field current be adjusted to keep V_T constant as the load changes?* On the other hand, part of the second example problem asks the question: *If the load changes and the field is left alone, what happens to the terminal voltage?* You should compare the calculated behavior of the generators in these two problems to see if it agrees with the qualitative arguments of this section. Finally, the third example illustrates the use of a MATLAB program to derive the terminal characteristics of synchronous generator.

Example 4–2. A 480-V, 60-Hz, Δ-connected, four-pole synchronous generator has the OCC shown in Figure 4–23a. This generator has a synchronous reactance of 0.1 Ω and

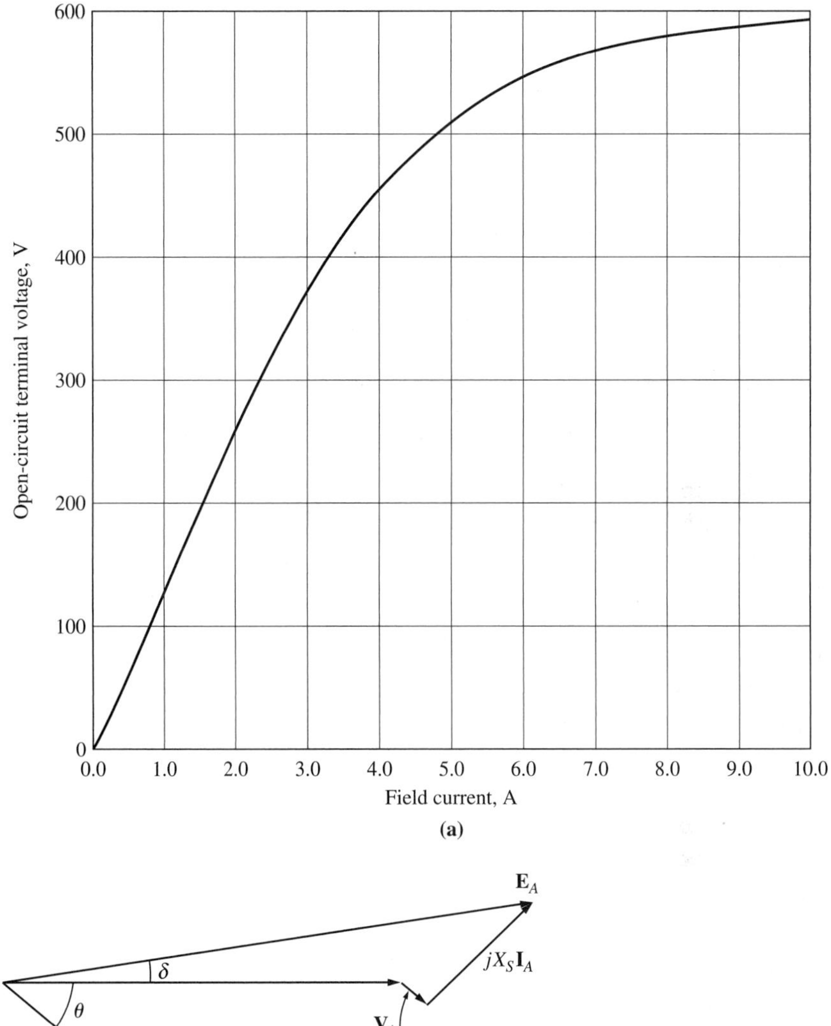

(a)

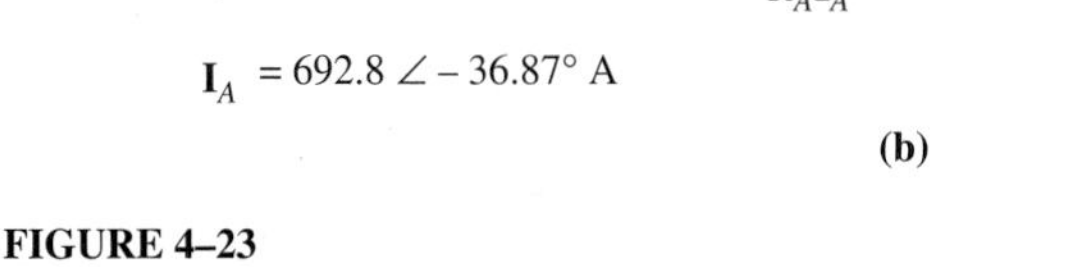

(b)

FIGURE 4–23
(a) Open-circuit characteristic of the generator in Example 4–2. (b) Phasor diagram of the generator in Example 4–2.

an armature resistance of 0.015 Ω. At full load, the machine supplies 1200 A at 0.8 PF lagging. Under full-load conditions, the friction and windage losses are 40 kW, and the core losses are 30 kW. Ignore any field circuit losses.

(*a*) What is the speed of rotation of this generator?
(*b*) How much field current must be supplied to the generator to make the terminal voltage 480 V at no load?
(*c*) If the generator is now connected to a load and the load draws 1200 A at 0.8 PF lagging, how much field current will be required to keep the terminal voltage equal to 480 V?
(*d*) How much power is the generator now supplying? How much power is supplied to the generator by the prime mover? What is this machine's overall efficiency?
(*e*) If the generator's load were suddenly disconnected from the line, what would happen to its terminal voltage?
(*f*) Finally, suppose that the generator is connected to a load drawing 1200 A at 0.8 PF *leading*. How much field current would be required to keep V_T at 480 V?

Solution

This synchronous generator is Δ-connected, so its phase voltage is equal to its line voltage $V_\phi = V_T$, while its phase current is related to its line current by the equation $I_L = \sqrt{3} I_\phi$.

(*a*) The relationship between the electrical frequency produced by a synchronous generator and the mechanical rate of shaft rotation is given by Equation (3–34):

$$f_{se} = \frac{n_m P}{120} \tag{3–34}$$

Therefore,

$$n_m = \frac{120 f_{se}}{P}$$

$$= \frac{120(60\text{ Hz})}{4\text{ poles}} = 1800\text{ r/min}$$

(*b*) In this machine, $V_T = V_\phi$. Since the generator is at no load, $\mathbf{I}_A = 0$ and $\mathbf{E}_A = \mathbf{V}_\phi$. Therefore, $V_T = V_\phi = E_A = 480$ V, and from the open-circuit characteristic, $I_F = 4.5$ A.
(*c*) If the generator is supplying 1200 A, then the armature current in the machine is

$$I_A = \frac{1200\text{ A}}{\sqrt{3}} = 692.8\text{ A}$$

The phasor diagram for this generator is shown in Figure 4–23b. If the terminal voltage is adjusted to be 480 V, the size of the internal generated voltage $\mathbf{E}_A$ is given by

$$\begin{aligned}
\mathbf{E}_A &= \mathbf{V}_\phi + R_A\mathbf{I}_A + jX_S\mathbf{I}_A \\
&= 480\angle 0^\circ\text{ V} + (0.015\ \Omega)(692.8\angle -36.87^\circ\text{ A}) + (j0.1\ \Omega)(692.8\angle -36.87^\circ\text{ A}) \\
&= 480\angle 0^\circ\text{ V} + 10.39\angle -36.87^\circ\text{ V} + 69.28\angle 53.13^\circ\text{ V} \\
&= 529.9 + j49.2\text{ V} = 532\angle 5.3^\circ\text{ V}
\end{aligned}$$

To keep the terminal voltage at 480 V, $\mathbf{E}_A$ must be adjusted to 532 V. From Figure 4–23, the required field current is 5.7 A.
(*d*) The power that the generator is now supplying can be found from Equation (4–16):

$$P_{\text{out}} = \sqrt{3} V_L I_L \cos\theta \tag{4–16}$$

$$= \sqrt{3}(480 \text{ V})(1200 \text{ A}) \cos 36.87°$$
$$= 798 \text{ kW}$$

To determine the power input to the generator, use the power-flow diagram (Figure 4–15). From the power-flow diagram, the mechanical input power is given by

$$P_{\text{in}} = P_{\text{out}} + P_{\text{elec loss}} + P_{\text{core loss}} + P_{\text{mech loss}} + P_{\text{stray loss}}$$

The stray losses were not specified here, so they will be ignored. In this generator, the electrical losses are

$$P_{\text{elec loss}} = 3I_A^2 R_A$$
$$= 3(692.8 \text{ A})^2(0.015 \ \Omega) = 21.6 \text{ kW}$$

The core losses are 30 kW, and the friction and windage losses are 40 kW, so the total input power to the generator is

$$P_{\text{in}} = 798 \text{ kW} + 21.6 \text{ kW} + 30 \text{ kW} + 40 \text{ kW} = 889.6 \text{ kW}$$

Therefore, the machine's overall efficiency is

$$\eta = \frac{P_{\text{out}}}{P_{\text{in}}} \times 100\% = \frac{798 \text{ kW}}{889.6 \text{ kW}} \times 100\% = 89.75\%$$

(e) If the generator's load were suddenly disconnected from the line, the current $\mathbf{I}_A$ would drop to zero, making $\mathbf{E}_A = \mathbf{V}_\phi$. Since the field current has not changed, $|\mathbf{E}_A|$ has not changed and $\mathbf{V}_\phi$ and V_T must rise to equal $\mathbf{E}_A$. Therefore, if the load were suddenly dropped, the terminal voltage of the generator would rise to 532 V.

(f) If the generator were loaded down with 1200 A at 0.8 PF leading while the terminal voltage was 480 V, then the internal generated voltage would have to be

$$\mathbf{E}_A = \mathbf{V}_\phi + R_A\mathbf{I}_A + jX_S\mathbf{I}_A$$
$$= 480 \angle 0° \text{ V} + (0.015 \ \Omega)(692.8 \angle 36.87° \text{ A}) + (j0.1 \ \Omega)(692.8 \angle 36.87° \text{ A})$$
$$= 480 \angle 0° \text{ V} + 10.39 \angle 36.87° \text{ V} + 69.28 \angle 126.87° \text{ V}$$
$$= 446.7 + j61.7 \text{ V} = 451 \angle 7.1° \text{ V}$$

Therefore, the internal generated voltage E_A must be adjusted to provide 451 V if V_T is to remain 480 V. Using the open-circuit characteristic, the field current would have to be adjusted to 4.1 A.

Which type of load (leading or lagging) needed a larger field current to maintain the rated voltage? Which type of load (leading or lagging) placed more thermal stress on the generator? Why?

Example 4–3. A 480-V, 50-Hz, Y-connected, six-pole synchronous generator has a per-phase synchronous reactance of 1.0 Ω. Its full-load armature current is 60 A at 0.8 PF lagging. This generator has friction and windage losses of 1.5 kW and core losses of 1.0 kW at 60 Hz at full load. Since the armature resistance is being ignored, assume that the I^2R losses are negligible. The field current has been adjusted so that the terminal voltage is 480 V at no load.

(a) What is the speed of rotation of this generator?
(b) What is the terminal voltage of this generator if the following are true?

1. It is loaded with the rated current at 0.8 PF lagging.
2. It is loaded with the rated current at 1.0 PF.
3. It is loaded with the rated current at 0.8 PF leading.

(c) What is the efficiency of this generator (ignoring the unknown electrical losses) when it is operating at the rated current and 0.8 PF lagging?

(d) How much shaft torque must be applied by the prime mover at full load? How large is the induced countertorque?

(e) What is the voltage regulation of this generator at 0.8 PF lagging? At 1.0 PF? At 0.8 PF leading?

Solution

This generator is Y-connected, so its phase voltage is given by $V_\phi = V_T/\sqrt{3}$. That means that when V_T is adjusted to 480 V, $V_\phi = 277$ V. The field current has been adjusted so that $V_{T,\text{nl}} = 480$ V, so $V_\phi = 277$ V. At *no load,* the armature current is zero, so the armature reaction voltage and the $I_A R_A$ drops are zero. Since $\mathbf{I}_A = 0$, the internal generated voltage $E_A = V_\phi = 277$ V. The internal generated voltage $E_A(= K\phi\omega)$ varies only when the field current changes. Since the problem states that the field current is adjusted initially and then left alone, the magnitude of the internal generated voltage is $E_A = 277$ V and will not change in this example.

(a) The speed of rotation of a synchronous generator in revolutions per minute is given by Equation (3–34):

$$f_{se} = \frac{n_m P}{120} \tag{3–34}$$

Therefore,

$$n_m = \frac{120 f_{se}}{P}$$

$$= \frac{120(50\text{ Hz})}{6\text{ poles}} = 1000\text{ r/min}$$

Alternatively, the speed expressed in radians per second is

$$\omega_m = (1000\text{ r/min})\left(\frac{1\text{ min}}{60\text{ s}}\right)\left(\frac{2\pi\text{ rad}}{1\text{r}}\right)$$

$$= 104.7\text{ rad/s}$$

(b) 1. If the generator is loaded down with rated current at 0.8 PF lagging, the resulting phasor diagram looks like the one shown in Figure 4–24a. In this phasor diagram, we know that $\mathbf{V}_\phi$ is at an angle of 0°, that the magnitude of $\mathbf{E}_A$ is 277 V, and that the quantity $jX_S\mathbf{I}_A$ is

$$jX_S\mathbf{I}_A = j(1.0\ \Omega)(60\angle -36.87°\text{ A}) = 60\angle 53.13°\text{ V}$$

The two quantities not known on the voltage diagram are the magnitude of $\mathbf{V}_\phi$ and the angle δ of $\mathbf{E}_A$. To find these values, the easiest approach is to construct a right triangle on the phasor diagram, as shown in the figure. From Figure 4–24a, the right triangle gives

$$E_A^2 = (V_\phi + X_S I_A \sin\theta)^2 + (X_S I_A \cos\theta)^2$$

Therefore, the phase voltage at the rated load and 0.8 PF lagging is

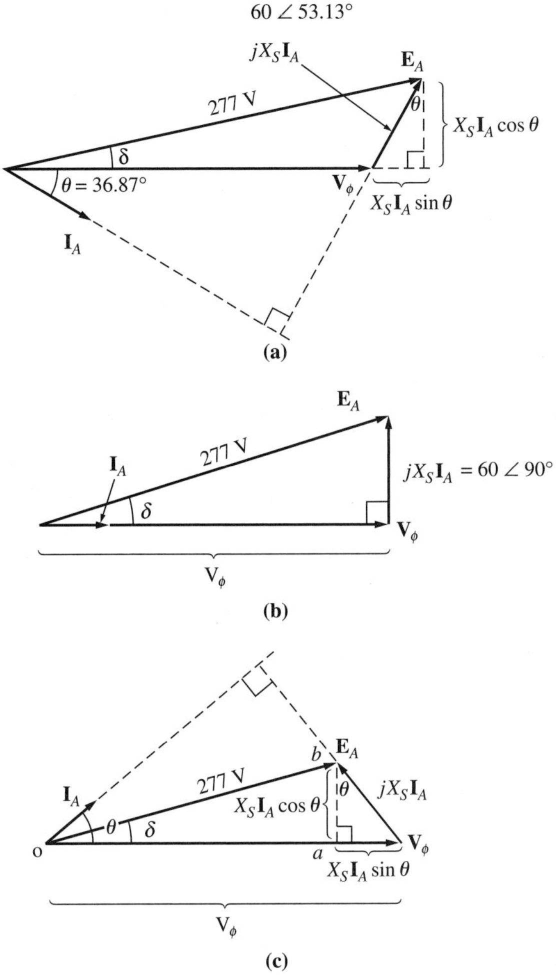

FIGURE 4–24
Generator phasor diagrams for Example 4–3. (a) Lagging power factor; (b) unity power factor; (c) leading power factor.

$$(277\text{ V})^2 = [V_\phi + (1.0\ \Omega)(60\text{ A})\sin 36.87°]^2 + [(1.0\ \Omega)(60\text{ A})\cos 36.87°]^2$$
$$76{,}729 = (V_\phi + 36)^2 + 2304$$
$$74{,}425 = (V_\phi + 36)^2$$
$$272.8 = V_\phi + 36$$
$$V_\phi = 236.8\text{ V}$$

Since the generator is Y-connected, $V_T = \sqrt{3}V_\phi = 410$ V.

2. If the generator is loaded with the rated current at unity power factor, then the phasor diagram will look like Figure 4–24b. To find $\mathbf{V}_\phi$ here the right triangle is

$$E_A^2 = V_\phi^2 + (X_S I_A)^2$$
$$(277\text{ V})^2 = V_\phi^2 + [(1.0\ \Omega)(60\text{ A})]^2$$
$$76{,}729 = V_\phi^2 + 3600$$
$$V_\phi^2 = 73{,}129$$
$$V_\phi = 270.4\text{ V}$$

Therefore, $V_T = \sqrt{3}V_\phi = 468.4$ V.

3. When the generator is loaded with the rated current at 0.8 PF leading, the resulting phasor diagram is the one shown in Figure 4–24c. To find $\mathbf{V}_\phi$ in this situation, we construct the triangle *OAB* shown in the figure. The resulting equation is

$$E_A^2 = (V_\phi - X_S I_A \sin\theta)^2 + (X_S I_A \cos\theta)$$

Therefore, the phase voltage at the rated load and 0.8 PF leading is

$$(277\text{ V})^2 = [V_\phi - (1.0\ \Omega)(60\text{ A})\sin 36.87°]^2 + [(1.0\ \Omega)(60\text{ A})\cos 36.87°]^2$$
$$76{,}729 = (V_\phi - 36)^2 + 2304$$
$$74{,}425 = (V_\phi - 36)^2$$
$$272.8 = V_\phi - 36$$
$$V_\phi = 308.8\text{ V}$$

Since the generator is Y-connected, $V_T = \sqrt{3}V_\phi = 535$ V.

(c) The output power of this generator at 60 A and 0.8 PF lagging is

$$P_{\text{out}} = 3V_\phi I_A \cos\theta$$
$$= 3(236.8\text{ V})(60\text{ A})(0.8) = 34.1\text{ kW}$$

The mechanical input power is given by

$$P_{\text{in}} = P_{\text{out}} + P_{\text{elec loss}} + P_{\text{core loss}} + P_{\text{mech loss}}$$
$$= 34.1\text{ kW} + 0 + 1.0\text{ kW} + 1.5\text{ kW} = 36.6\text{ kW}$$

The efficiency of the generator is thus

$$\eta = \frac{P_{\text{out}}}{P_{\text{in}}} \times 100\% = \frac{34.1\text{ kW}}{36.6\text{ kW}} \times 100\% = 93.2\%$$

(d) The input torque to this generator is given by the equation

$$P_{\text{in}} = \tau_{\text{app}}\omega_m$$

so
$$\tau_{\text{app}} = \frac{P_{\text{in}}}{\omega_m} = \frac{36.6\text{ kW}}{125.7\text{ rad/s}} = 291.2\text{ N}\cdot\text{m}$$

The induced countertorque is given by

$$P_{\text{conv}} = \tau_{\text{ind}}\omega_m$$

so
$$\tau_{\text{ind}} = \frac{P_{\text{conv}}}{\omega_V} = \frac{34.1\text{ kW}}{125.7\text{ rad/s}} = 271.3\text{ N}\cdot\text{m}$$

(e) The voltage regulation of a generator is defined as

$$\text{VR} = \frac{V_{\text{nl}} - V_{\text{fl}}}{V_{\text{fl}}} \times 100\% \tag{3–67}$$

By this definition, the voltage regulation for the lagging, unity, and leading power-factor cases are

1. Lagging case: $\text{VR} = \dfrac{480\text{ V} - 410\text{ V}}{410\text{ V}} \times 100\% = 17.1\%$

2. Unity case: $\text{VR} = \dfrac{480\text{ V} - 468\text{ V}}{468\text{ V}} \times 100\% = 2.6\%$

3. Leading case: $\text{VR} = \dfrac{480\text{ V} - 535\text{ V}}{535\text{ V}} \times 100\% = -10.3\%$

In Example 4–3, lagging loads resulted in a drop in terminal voltage, unity-power-factor loads caused little effect on V_T, and leading loads resulted in an increase in terminal voltage.

Example 4–4. Assume that the generator of Example 4–3 is operating at no load with a terminal voltage of 480 V. Plot the terminal characteristic (terminal voltage versus line current) of this generator as its armature current varies from no-load to full load at a power factor of (*a*) 0.8 lagging and (*b*) 0.8 leading. Assume that the field current remains constant at all times.

Solution

The terminal characteristic of a generator is a plot of its terminal voltage versus line current. Since this generator is Y-connected, its phase voltage is given by $V_\phi = V_T/\sqrt{3}$. If V_T is adjusted to 480 V at no-load conditions, then $V_\phi = E_A = 277$ V. Because the field current remains constant, E_A will remain 277 V at all times. The output current I_L from this generator will be the same as its armature current I_A because it is Y-connected.

(*a*) If the generator is loaded with a 0.8 PF lagging current, the resulting phasor diagram looks like the one shown in Figure 4–24a. In this phasor diagram, we know that $\mathbf{V}_\phi$ is at an angle of 0°, that the magnitude of $\mathbf{E}_A$ is 277 V, and that the quantity $jX_S\mathbf{I}_A$ stretches between $\mathbf{V}_\phi$ and $\mathbf{E}_A$ as shown. The two quantities not known on the phasor diagram are the magnitude of $\mathbf{V}_\phi$ and the angle δ of $\mathbf{E}_A$. To find V_ϕ, the easiest approach is to construct a right triangle on the phasor diagram, as shown in the figure. From Figure 4–24a, the right triangle gives

$$E_A^2 = (V_\phi + X_S I_A \sin\theta)^2 + (X_S I_A \cos\theta)^2$$

This equation can be used to solve for V_ϕ as a function of the current I_A:

$$V_\phi = \sqrt{E_A^2 - (X_S I_A \cos\theta)^2} - X_S I_A \sin\theta$$

A simple MATLAB M-file can be used to calculate V_ϕ (and hence V_T) as a function of current. Such an M-file is shown below:

```
% M-file: term_char_a.m
% M-file to plot the terminal characteristics of the
%   generator of Example 4-4 with an 0.8 PF lagging load.

% First, initialize the current amplitudes (21 values
% in the range 0-60 A)
i_a = (0:1:20) * 3;
```

```
% Now initialize all other values
v_phase = zeros(1,21);
e_a = 277.0;
x_s = 1.0;
theta = 36.87 * (pi/180);   % Converted to radians

% Now calculate v_phase for each current level
for ii = 1:21
 v_phase(ii) = sqrt(e_a^2 - (x_s * i_a(ii) * cos(theta))^2) ...
                           - (x_s * i_a(ii) * sin(theta));
end

% Calculate terminal voltage from the phase voltage
v_t = v_phase * sqrt(3);

% Plot the terminal characteristic, remembering the
% the line current is the same as i_a
plot(i_a,v_t,'Color','k','Linewidth',2.0);
xlabel('Line Current (A)','Fontweight','Bold');
ylabel('Terminal Voltage (V)','Fontweight','Bold');
title ('Terminal Characteristic for 0.8 PF lagging load', ...
    'Fontweight','Bold');
grid on;
axis([0 60 400 550]);
```

The plot resulting when this M-file is executed is shown in Figure 4–25a.

(b) If the generator is loaded with a 0.8 PF leading current, the resulting phasor diagram looks like the one shown in Figure 4–24c. To find V_ϕ, the easiest approach is to construct a right triangle on the phasor diagram, as shown in the figure. From Figure 4–24c, the right triangle gives

$$E_A^2 = (V_\phi - X_S I_A \sin\theta)^2 + (X_S I_A \cos\theta)^2$$

This equation can be used to solve for V_ϕ as a function of the current I_A:

$$V_\phi = \sqrt{E_A^2 - (X_S I_A \cos\theta)^2} + X_S I_A \sin\theta$$

This equation can be used to calculate and plot the terminal characteristic in a manner similar to that in part *a* above. The resulting terminal characteristic is shown in Figure 4–25b.

4.9 PARALLEL OPERATION OF AC GENERATORS

In today's world, an isolated synchronous generator supplying its own load independently of other generators is very rare. Such a situation is found in only a few out-of-the-way applications such as emergency generators. For all usual generator applications, there is more than one generator operating in parallel to supply the power demanded by the loads. An extreme example of this situation is the U.S. power grid, in which literally thousands of generators share the load on the system.

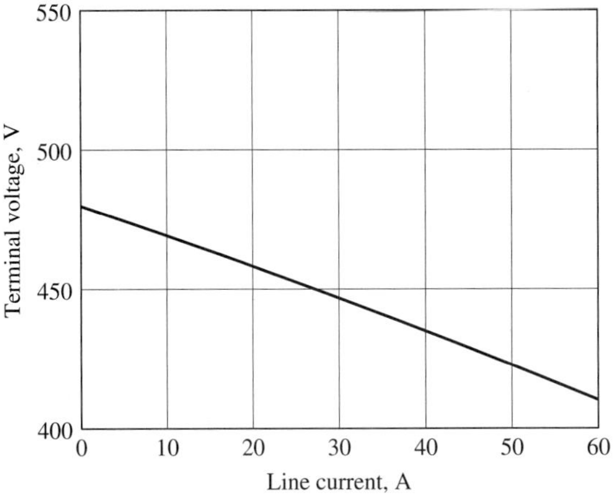

(a)

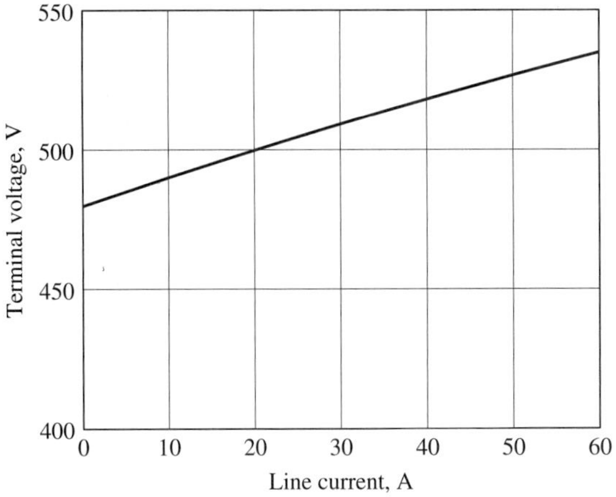

(b)

FIGURE 4–25
(a) Terminal characteristic for the generator of Example 4–4 when loaded with a 0.8 PF lagging load. (b) Terminal characteristic for the generator when loaded with a 0.8 PF leading load.

Why are synchronous generators operated in parallel? There are several major advantages to such operation:

1. Several generators can supply a bigger load than one machine by itself.
2. Having many generators increases the reliability of the power system, since the failure of any one of them does not cause a total power loss to the load.
3. Having many generators operating in parallel allows one or more of them to be removed for shutdown and preventive maintenance.

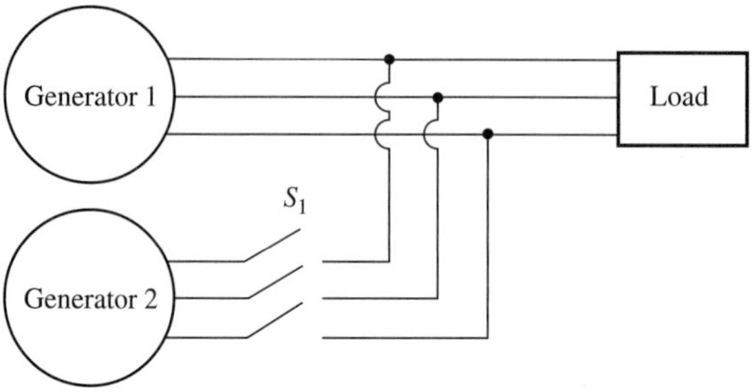

FIGURE 4–26
A generator being paralleled with a running power system.

4. If only one generator is used and it is not operating at near full load, then it will be relatively inefficient. With several smaller machines in parallel, it is possible to operate only a fraction of them. The ones that do operate are operating near full load and thus more efficiently.

This section explores the requirements for paralleling ac generators, and then looks at the behavior of synchronous generators operated in parallel.

The Conditions Required for Paralleling

Figure 4–26 shows a synchronous generator G_1 supplying power to a load, with another generator G_2 about to be paralleled with G_1 by closing the switch S_1. What conditions must be met before the switch can be closed and the two generators connected?

If the switch is closed arbitrarily at some moment, the generators are liable to be severely damaged, and the load may lose power. If the voltages are not exactly the same in each conductor being tied together, there will be a *very* large current flow when the switch is closed. To avoid this problem, each of the three phases must have *exactly the same voltage magnitude and phase angle* as the conductor to which it is connected. In other words, the voltage in phase *a* must be *exactly* the same as the voltage in phase *a′*, and so forth for phases *b-b′* and *c-c′*. To achieve this match, the following *paralleling conditions* must be met:

1. The rms *line voltages* of the two generators must be equal.
2. The two generators must have the same *phase sequence*.
3. The phase angles of the two *a* phases must be equal.
4. The frequency of the new generator, called the *oncoming generator*, must be slightly higher than the frequency of the running system.

These paralleling conditions require some explanation. Condition 1 is obvious—in order for two sets of voltages to be identical, they must of course have the same rms magnitude of voltage. The voltage in phases *a* and *a′* will be

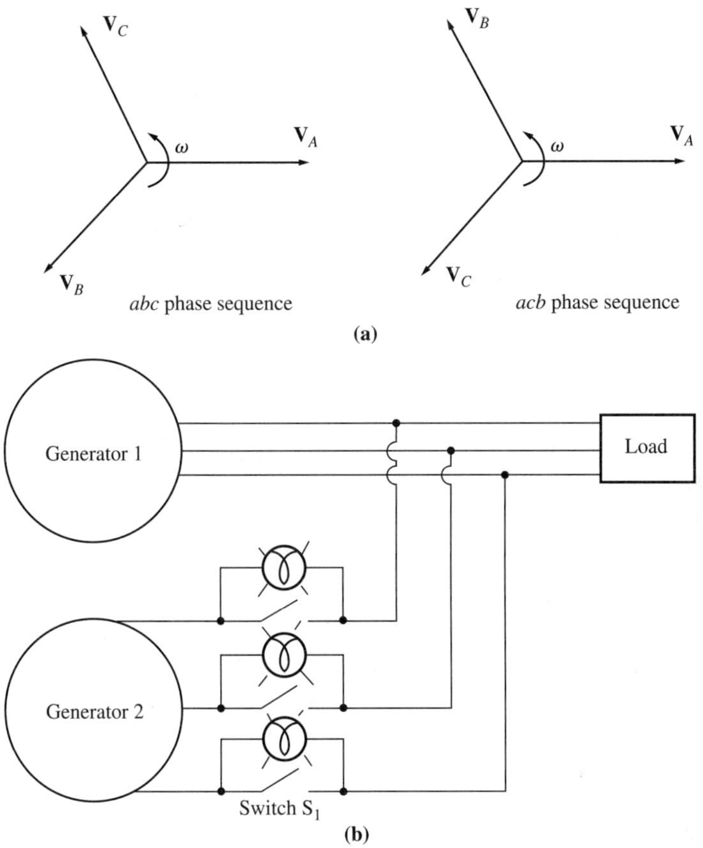

FIGURE 4–27
(a) The two possible phase sequences of a three-phase system. (b) The three-light-bulb method for checking phase sequence.

completely identical at all times if both their magnitudes and their angles are the same, which explains condition 3.

Condition 2 ensures that the sequence in which the phase voltages peak in the two generators is the same. If the phase sequence is different (as shown in Figure 4–27a), then even though one pair of voltages (the *a* phases) are in phase, the other two pairs of voltages are 120° out of phase. If the generators were connected in this manner, there would be no problem with phase *a*, but huge currents would flow in phases *b* and *c*, damaging both machines. To correct a phase sequence problem, simply swap the connections on any two of the three phases on one of the machines.

If the frequencies of the generators are not very nearly equal when they are connected together, large power transients will occur until the generators stabilize at a common frequency. The frequencies of the two machines must be very nearly equal, but they cannot be exactly equal. They must differ by a small amount so that the phase angles of the oncoming machine will change slowly with respect to the phase angles of the running system. In that way, the angles between the voltages can be observed and switch S_1 can be closed when the systems are exactly in phase.

The General Procedure for Paralleling Generators

Suppose that generator G_2 is to be connected to the running system shown in Figure 4–27. The following steps should be taken to accomplish the paralleling.

First, using voltmeters, the field current of the oncoming generator should be adjusted until its terminal voltage is equal to the line voltage of the running system.

Second, the phase sequence of the oncoming generator must be compared to the phase sequence of the running system. The phase sequence can be checked in a number of different ways. One way is to alternately connect a small induction motor to the terminals of each of the two generators. If the motor rotates in the same direction each time, then the phase sequence is the same for both generators. If the motor rotates in opposite directions, then the phase sequences differ, and two of the conductors on the incoming generator must be reversed.

Another way to check the phase sequence is the *three-light-bulb method.* In this approach, three light bulbs are stretched across the open terminals of the switch connecting the generator to the system as shown in Figure 4–27b. As the phase changes between the two systems, the light bulbs first get bright (large phase difference) and then get dim (small phase difference). *If all three bulbs get bright and dark together, then the systems have the same phase sequence.* If the bulbs brighten in succession, then the systems have the opposite phase sequence, and one of the sequences must be reversed.

Next, the frequency of the oncoming generator is adjusted to be slightly higher than the frequency of the running system. This is done first by watching a frequency meter until the frequencies are close and then by observing changes in phase between the systems. The oncoming generator is adjusted to a slightly higher frequency so that when it is connected, it will come on the line supplying power as a generator, instead of consuming it as a motor would (this point will be explained later).

Once the frequencies are very nearly equal, the voltages in the two systems will change phase with respect to each other very slowly. The phase changes are observed, and when the phase angles are equal, the switch connecting the two systems together is shut.

How can one tell when the two systems are finally in phase? A simple way is to watch the three light bulbs described above in connection with the discussion of phase sequence. When the three light bulbs all go out, the voltage difference across them is zero and the systems are in phase. This simple scheme works, but it is not very accurate. A better approach is to employ a synchroscope. A *synchroscope* is a meter that measures the difference in phase angle between the *a* phases of the two systems. The face of a synchroscope is shown in Figure 4–28. The dial shows the phase difference between the two *a* phases, with 0 (meaning in phase)

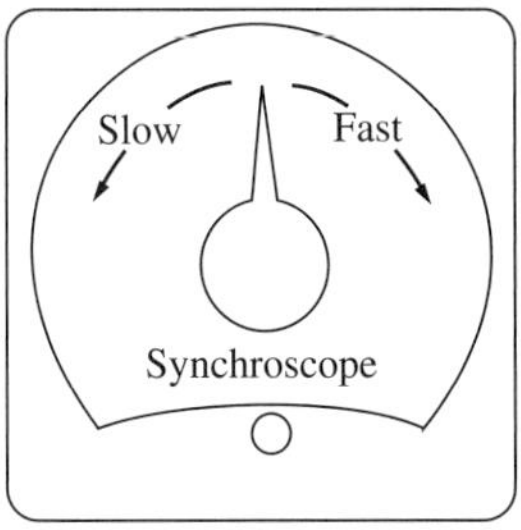

FIGURE 4–28
A synchroscope.

at the top and 180° at the bottom. Since the frequencies of the two systems are slightly different, the phase angle on the meter changes slowly. If the oncoming generator or system is faster than the running system (the desired situation), then the phase angle advances and the synchroscope needle rotates clockwise. If the oncoming machine is slower, the needle rotates counterclockwise. When the synchroscope needle is in the vertical position, the voltages are in phase, and the switch can be shut to connect the systems.

Notice, though, that a *synchroscope checks the relationships on only one phase*. It gives no information about phase sequence.

In large generators belonging to power systems, this whole process of paralleling a new generator to the line is automated, and a computer does this job. For smaller generators, though, the operator manually goes through the paralleling steps just described.

Frequency–Power and Voltage–Reactive Power Characteristics of a Synchronous Generator

All generators are driven by a *prime mover*, which is the generator's source of mechanical power. The most common type of prime mover is a steam turbine, but other types include diesel engines, gas turbines, water turbines, and even wind turbines.

Regardless of the original power source, all prime movers tend to behave in a similar fashion—as the power drawn from them increases, the speed at which they turn decreases. The decrease in speed is in general nonlinear, but some form of governor mechanism is usually included to make the decrease in speed linear with an increase in power demand.

Whatever governor mechanism is present on a prime mover, it will always be adjusted to provide a slight drooping characteristic with increasing load. The speed droop (SD) of a prime mover is defined by the equation

$$\boxed{SD = \frac{n_{nl} - n_{fl}}{n_{fl}} \times 100\%} \tag{4–27}$$

where n_{nl} is the no-load prime-mover speed and n_{fl} is the full-load prime-mover speed. Most generator prime movers have a speed droop of 2 to 4 percent, as defined in Equation (4–27). In addition, most governors have some type of set point

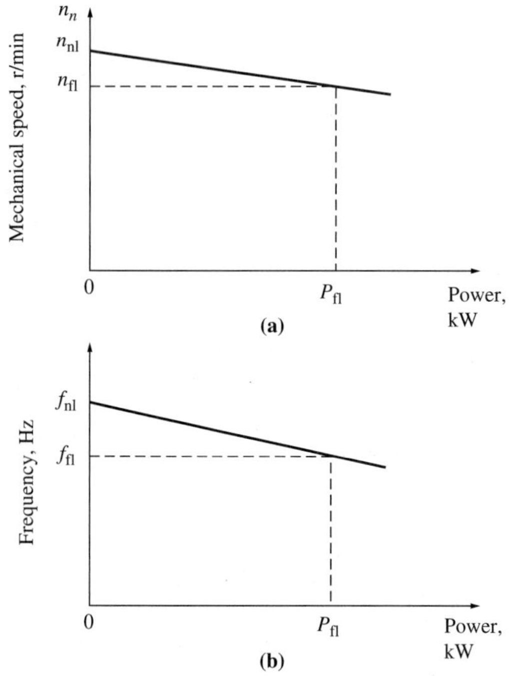

FIGURE 4–29
(a) The speed-versus-power curve for a typical prime mover. (b) The resulting frequency-versus-power curve for the generator.

adjustment to allow the no-load speed of the turbine to be varied. A typical speed-versus-power plot is shown in Figure 4–29.

Since the shaft speed is related to the resulting electrical frequency by Equation (3–34),

$$f_{se} = \frac{n_m P}{120} \tag{3–34}$$

the power output of a synchronous generator is related to its frequency. An example plot of frequency versus power is shown in Figure 4–29b. Frequency–power characteristics of this sort play an essential role in the parallel operation of synchronous generators.

The relationship between frequency and power can be described quantitatively by the equation

$$\boxed{P = s_P(f_{nl} - f_{sys})} \tag{4–28}$$

where P = power output of the generator
f_{nl} = no-load frequency of the generator
f_{sys} = operating frequency of system
s_P = slope of curve, in kW/Hz or MW/Hz

A similar relationship can be derived for the reactive power Q and terminal voltage V_T. As previously seen, when a lagging load is added to a synchronous

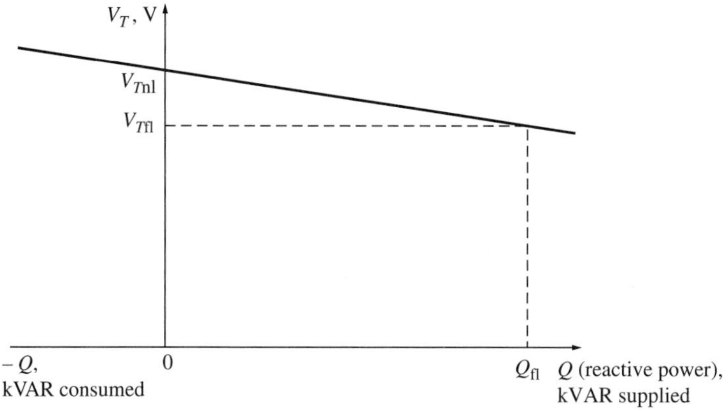

FIGURE 4–30
The curve of terminal voltage (V_T) versus reactive power (Q) for a synchronous generator.

generator, its terminal voltage drops. Likewise, when a leading load is added to a synchronous generator, its terminal voltage increases. It is possible to make a plot of terminal voltage versus reactive power, and such a plot has a drooping characteristic like the one shown in Figure 4–30. This characteristic is not intrinsically linear, but many generator voltage regulators include a feature to make it so. The characteristic curve can be moved up and down by changing the no-load terminal voltage set point on the voltage regulator. As with the frequency–power characteristic, this curve plays an important role in the parallel operation of synchronous generators.

The relationship between the terminal voltage and reactive power can be expressed by an equation similar to the frequency–power relationship [Equation (4–28)] if the voltage regulator produces an output that is linear with changes in reactive power.

It is important to realize that when a single generator is operating alone, the real power P and reactive power Q supplied by the generator will be the amount demanded by the load attached to the generator—the P and Q supplied cannot be controlled by the generator's controls. Therefore, for any given real power, the governor set points control the generator's operating frequency f_e and for any given reactive power, the field current controls the generator's terminal voltage V_T.

Example 4–5. Figure 4–31 shows a generator supplying a load. A second load is to be connected in parallel with the first one. The generator has a no-load frequency of 61.0 Hz and a slope s_P of 1 MW/Hz. Load 1 consumes a real power of 1000 kW at 0.8 PF lagging, while load 2 consumes a real power of 800 kW at 0.707 PF lagging.

(a) Before the switch is closed, what is the operating frequency of the system?
(b) After load 2 is connected, what is the operating frequency of the system?
(c) After load 2 is connected, what action could an operator take to restore the system frequency to 60 Hz?

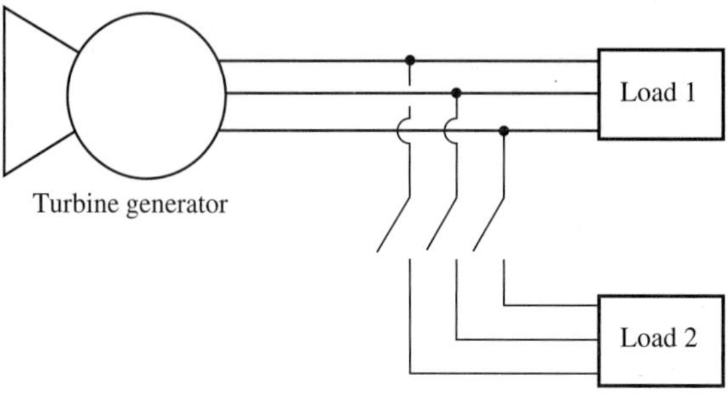

FIGURE 4–31
The power system in Example 4–5.

Solution

This problem states that the slope of the generator's characteristic is 1 MW/Hz and that its no-load frequency is 61 Hz. Therefore, the power produced by the generator is given by

$$P = s_P(f_{\text{nl}} - f_{\text{sys}}) \tag{4–28}$$

so

$$f_{\text{sys}} = f_{\text{nl}} - \frac{P}{s_P}$$

(a) The initial system frequency is given by

$$\begin{aligned} f_{\text{sys}} &= f_{\text{nl}} - \frac{P}{s_P} \\ &= 61\text{ Hz} - \frac{1000\text{ kW}}{1\text{ MW/Hz}} = 61\text{ Hz} - 1\text{ Hz} = 60\text{ Hz} \end{aligned}$$

(b) After load 2 is connected,

$$\begin{aligned} f_{\text{sys}} &= f_{\text{nl}} - \frac{P}{s_P} \\ &= 61\text{ Hz} - \frac{1800\text{ kW}}{1\text{ MW/Hz}} = 61\text{ Hz} - 1.8\text{ Hz} = 59.2\text{ Hz} \end{aligned}$$

(c) After the load is connected, the system frequency falls to 59.2 Hz. To restore the system to its proper operating frequency, the operator should increase the governor no-load set points by 0.8 Hz, to 61.8 Hz. This action will restore the system frequency to 60 Hz.

To summarize, when a generator is operating by itself supplying the system loads, then

1. The real and reactive power supplied by the generator will be the amount demanded by the attached load.
2. The governor set points of the generator will control the operating frequency of the power system.

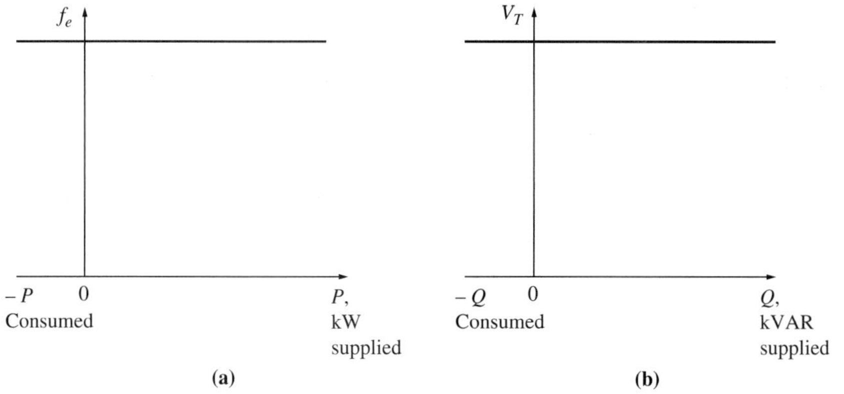

FIGURE 4–32
Curves for an infinite bus: (a) frequency versus power and (b) terminal voltage versus reactive power.

3. The field current (or the field regulator set points) controls the terminal voltage of the power system.

This is the situation found in isolated generators in remote field environments.

Operation of Generators in Parallel with Large Power Systems

When a synchronous generator is connected to a power system, the power system is often so large that *nothing* the operator of the generator does will have much of an effect on the power system. An example of this situation is the connection of a single generator to the U.S. power grid. The U.S. power grid is so large that no reasonable action on the part of the one generator can cause an observable change in overall grid frequency.

This idea is idealized in the concept of an infinite bus. An *infinite bus* is a power system so large that its voltage and frequency do not vary regardless of how much real and reactive power is drawn from or supplied to it. The power–frequency characteristic of such a system is shown in Figure 4–32a, and the reactive power–voltage characteristic is shown in Figure 4–32b.

To understand the behavior of a generator connected to such a large system, examine a system consisting of a generator and an infinite bus in parallel supplying a load. Assume that the generator's prime mover has a governor mechanism, but that the field is controlled manually by a resistor. It is easier to explain generator operation without considering an automatic field current regulator, so this discussion will ignore the slight differences caused by the field regulator when one is present. Such a system is shown in Figure 4–33a.

When a generator is connected in parallel with another generator or a large system, *the frequency and terminal voltage of all the machines must be the same,*

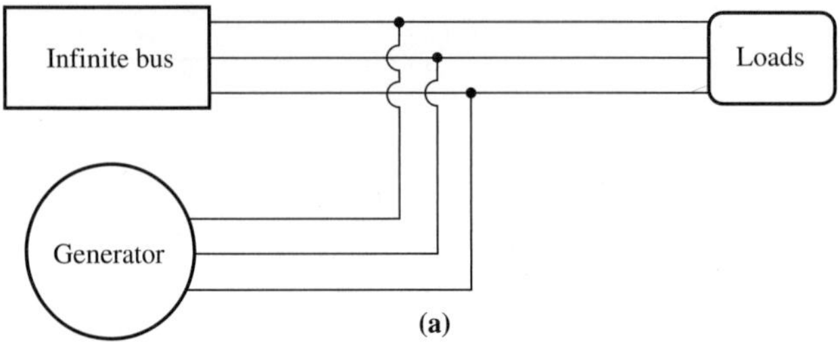

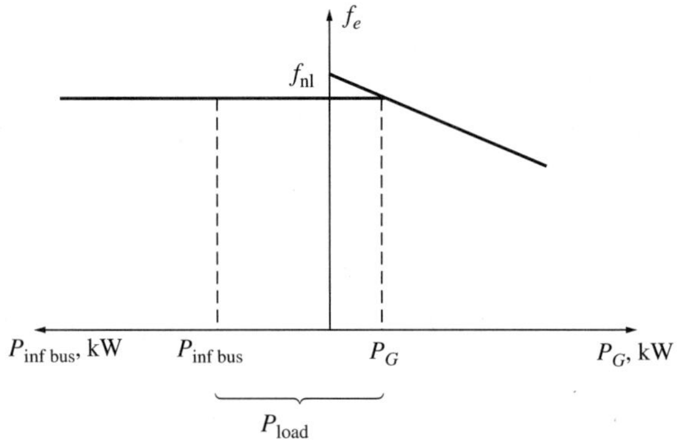

FIGURE 4–33
(a) A synchronous generator operating in parallel with an infinite bus. (b) The frequency-versus-power diagram (or *house diagram*) for a synchronous generator in parallel with an infinite bus.

since their output conductors are tied together. Therefore, their real power–frequency and reactive power–voltage characteristics can be plotted back to back, with a common vertical axis. Such a sketch, sometimes informally called a *house diagram,* is shown in Figure 4–33b.

Assume that the generator has just been paralleled with the infinite bus according to the procedure described previously. Then the generator will be essentially "floating" on the line, supplying a small amount of real power and little or no reactive power. This situation is shown in Figure 4–34.

Suppose the generator had been paralleled to the line but, instead of being at a slightly higher frequency than the running system, it was at a slightly lower frequency. In this case, when paralleling is completed, the resulting situation is shown in Figure 4–35. Notice that here the no-load frequency of the generator is less than the system's operating frequency. At this frequency, the power supplied by the generator is actually negative. In other words, when the generator's no-load frequency is less than the system's operating frequency, the generator actually consumes electric power and runs as a motor. It is to ensure that a generator comes on line supplying power instead of consuming it that the oncoming machine's frequency is adjusted higher than the running system's frequency. *Many real generators have a*

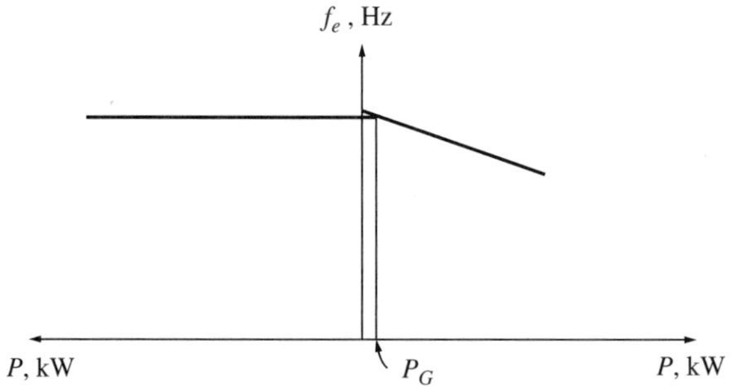

FIGURE 4–34
The frequency-versus-power diagram at the moment just after paralleling.

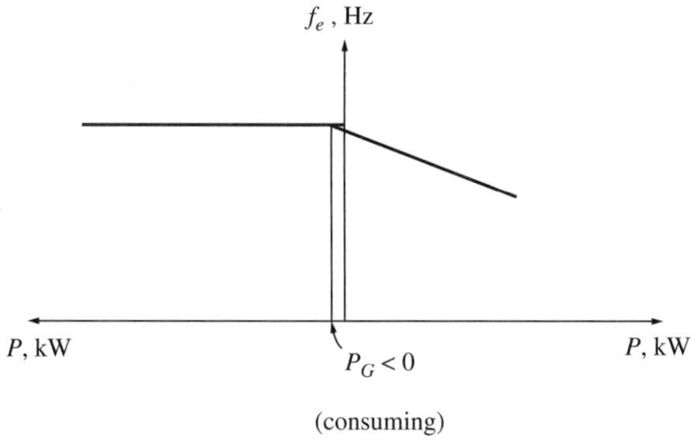

FIGURE 4–35
The frequency-versus-power diagram if the no-load frequency of the generator were slightly *less* than system frequency before paralleling.

reverse-power trip connected to them, so it is imperative that they be paralleled with their frequency higher than that of the running system. If such a generator ever starts to consume power, it will be automatically disconnected from the line.

Once the generator has been connected, what happens when its governor set points are increased? The effect of this increase is to shift the no-load frequency of the generator upward. Since the frequency of the system is unchanged (the frequency of an infinite bus cannot change), the power supplied by the generator increases. This is shown by the house diagram in Figure 4–36a and by the phasor diagram in Figure 4–36b. Notice in the phasor diagram that $E_A \sin \delta$ (which is proportional to the power supplied as long as V_T is constant) has increased, while the magnitude of E_A ($= K\phi\omega$) remains constant, since both the field current I_F and the speed of rotation ω are unchanged. As the governor set points are further increased, the no-load frequency increases and the power supplied by the generator increases. As the power output increases, E_A remains at constant magnitude while $E_A \sin \delta$ is further increased.

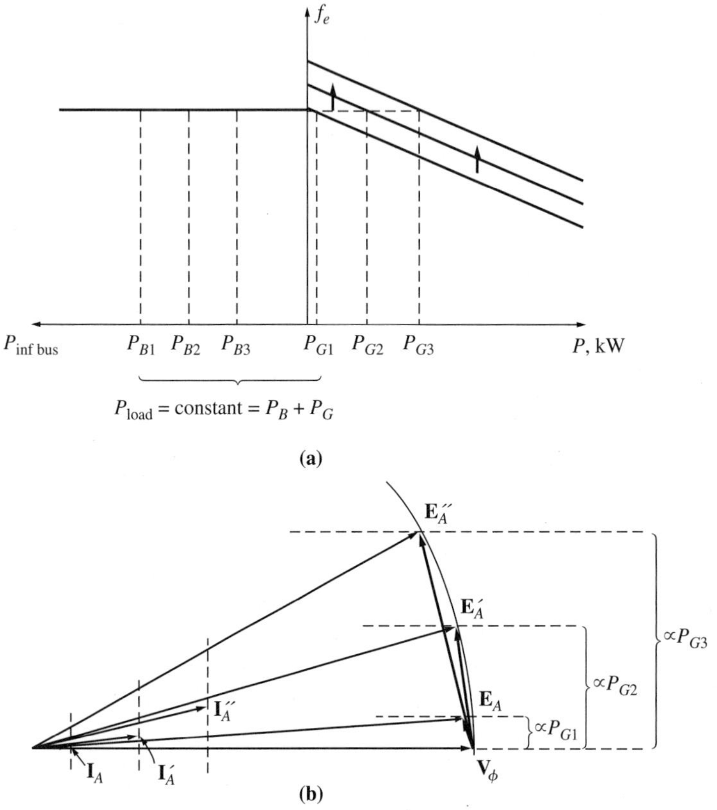

FIGURE 4–36
The effect of increasing the governor's set points on (a) the house diagram; (b) the phasor diagram.

What happens in this system if the power output of the generator is increased until it exceeds the power consumed by the load? If this occurs, the extra power generated flows back into the infinite bus. The infinite bus, by definition, can supply or consume any amount of power without a change in frequency, so the extra power is consumed.

After the real power of the generator has been adjusted to the desired value, the phasor diagram of the generator looks like Figure 4–36b. Notice that at this time the generator is actually operating at a slightly leading power factor, supplying negative reactive power. Alternatively, the generator can be said to be consuming reactive power. How can the generator be adjusted so that it will supply some reactive power Q to the system? This can be done by adjusting the field current of the machine. To understand why this is true, it is necessary to consider the constraints on the generator's operation under these circumstances.

The first constraint on the generator is that *the power must remain constant* when I_F is changed. The power into a generator (ignoring losses) is given by the equation $P_{\text{in}} = \tau_{\text{ind}}\omega_m$. Now, the prime mover of a synchronous generator has a

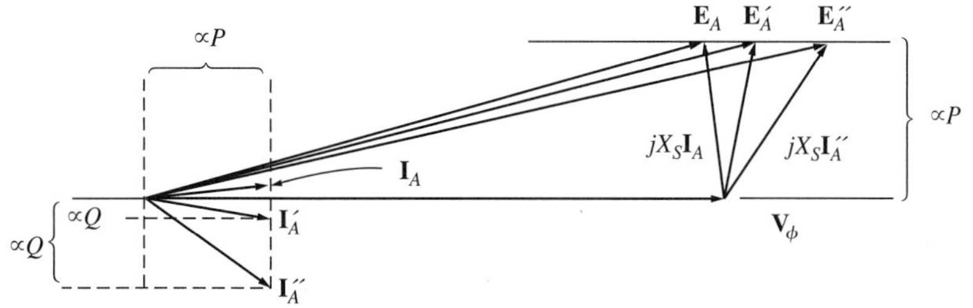

FIGURE 4–37
The effect of increasing the generator's field current on the phasor diagram of the machine.

fixed torque–speed characteristic for any given governor setting. This curve changes only when the governor set points are changed. Since the generator is tied to an infinite bus, its speed *cannot* change. If the generator's speed does not change and the governor set points have not been changed, the power supplied by the generator must remain constant.

If the power supplied is constant as the field current is changed, then the distances proportional to the power in the phasor diagram ($I_A \cos \theta$ and $E_A \sin \delta$) cannot change. When the field current is increased, the flux ϕ increases, and therefore E_A ($= K\phi\uparrow\omega$) increases. If E_A increases, but $E_A \sin \delta$ must remain constant, then the phasor $\mathbf{E}_A$ must "slide" along the line of constant power, as shown in Figure 4–37. Since $\mathbf{V}_\phi$ is constant, the angle of $jX_S\mathbf{I}_A$ changes as shown, and therefore the angle and magnitude of $\mathbf{I}_A$ change. Notice that as a result the distance proportional to Q ($I_A \sin \theta$) increases. In other words, *increasing the field current in a synchronous generator operating in parallel with an infinite bus increases the reactive power output of the generator.*

To summarize, when a generator is operating in parallel with an infinite bus:

1. The frequency and terminal voltage of the generator are controlled by the system to which it is connected.
2. The governor set points of the generator control the real power supplied by the generator to the system.
3. The field current in the generator controls the reactive power supplied by the generator to the system.

This situation is much the way real generators operate when connected to a very large power system.

Operation of Generators in Parallel with Other Generators of the Same Size

When a single generator operated alone, the real and reactive powers (P and Q) supplied by the generator were fixed, constrained to be equal to the power demanded by the load, and the frequency and terminal voltage were varied by the governor set points and the field current. When a generator operated in parallel with an infinite bus, the frequency and terminal voltage were constrained to be constant by the infinite bus, and the real and reactive powers were varied by the governor set points and the field current. What happens when a synchronous generator is connected in parallel not with an infinite bus, but rather with another generator of the same size? What will be the effect of changing governor set points and field currents?

If a generator is connected in parallel with another one of the same size, the resulting system is as shown in Figure 4–38a. In this system, the basic constraint is that *the sum of the real and reactive powers supplied by the two generators must equal the P and Q demanded by the load.* The system frequency is not constrained to be constant, and neither is the power of a given generator constrained to be constant. The power–frequency diagram for such a system immediately after G_2 has been paralleled to the line is shown in Figure 4–38b. Here, the total power P_{tot} (which is equal to P_{load}) is given by

$$P_{\text{tot}} = P_{\text{load}} = P_{G1} + P_{G2} \tag{4–29a}$$

and the total reactive power is given by

$$Q_{\text{tot}} = Q_{\text{load}} = Q_{G1} + Q_{G2} \tag{4–29b}$$

What happens if the governor set points of G_2 are increased? When the governor set points of G_2 are increased, the power–frequency curve of G_2 shifts upward, as shown in Figure 4–38c. Remember, the total power supplied to the load must not change. At the original frequency f_1, the power supplied by G_1 and G_2 will now be larger than the load demand, so the system cannot continue to operate at the same frequency as before. In fact, there is only one frequency at which the sum of the powers out of the two generators is equal to P_{load}. That frequency f_2 is higher than the original system operating frequency. At that frequency, G_2 supplies more power than before, and G_1 supplies less power than before.

Therefore, when two generators are operating together, an increase in governor set points on one of them

1. *Increases the system frequency.*
2. *Increases the power supplied by that generator, while reducing the power supplied by the other one.*

What happens if the field current of G_2 is increased? The resulting behavior is analogous to the real-power situation and is shown in Figure 4–38d. When two generators are operating together and the field current of G_2 is increased,

1. *The system terminal voltage is increased.*
2. *The reactive power Q supplied by that generator is increased, while the reactive power supplied by the other generator is decreased.*

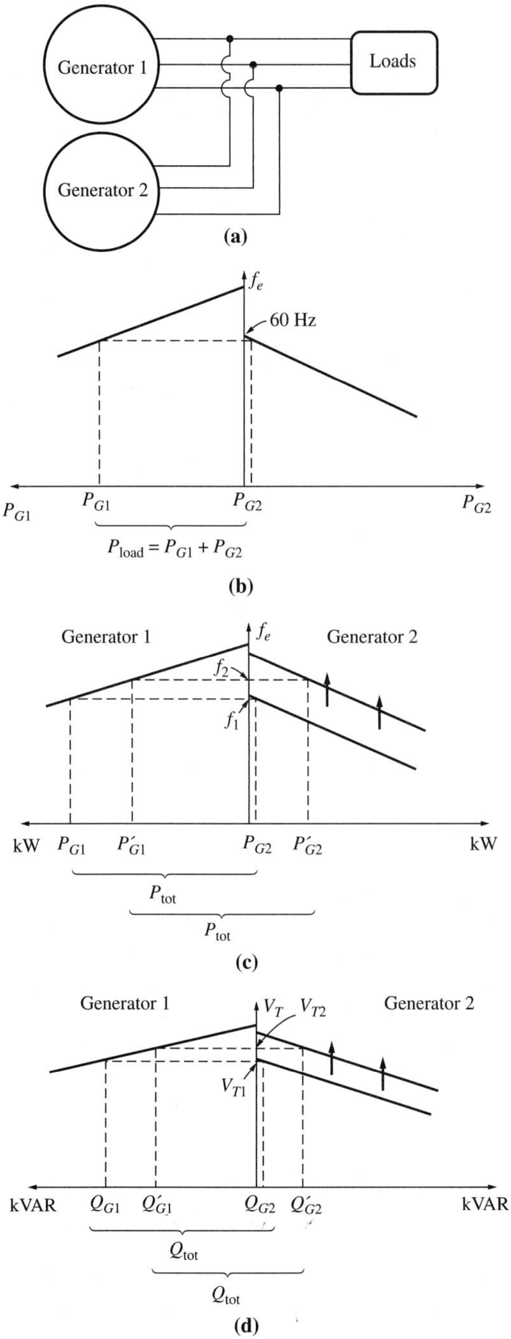

FIGURE 4–38
(a) A generator connected in parallel with another machine of the same size. (b) The corresponding house diagram at the moment generator 2 is paralleled with the system. (c) The effect of increasing generator 2's governor set points on the operation of the system. (d) The effect of increasing generator 2's field current on the operation of the system.

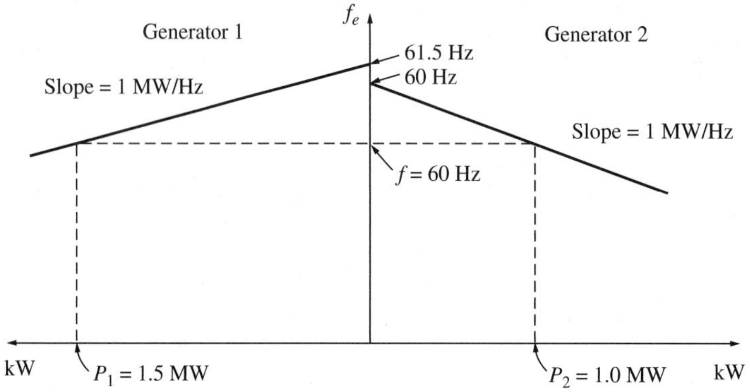

FIGURE 4–39
The house diagram for the system in Example 4–6.

If the slopes and no-load frequencies of the generator's speed droop (frequency–power) curves are known, then the powers supplied by each generator and the resulting system frequency can be determined quantitatively. Example 4–6 shows how this can be done.

Example 4–6. Figure 4–38a shows two generators supplying a load. Generator 1 has a no-load frequency of 61.5 Hz and a slope s_{P1} of 1 MW/Hz. Generator 2 has a no-load frequency of 61.0 Hz and a slope s_{P2} of 1 MW/Hz. The two generators are supplying a real load totaling 2.5 MW at 0.8 PF lagging. The resulting system power-frequency or house diagram is shown in Figure 4–39.

(a) At what frequency is this system operating, and how much power is supplied by each of the two generators?

(b) Suppose an additional 1-MW load were attached to this power system. What would the new system frequency be, and how much power would G_1 and G_2 supply now?

(c) With the system in the configuration described in part *b*, what will the system frequency and generator powers be if the governor set points on G_2 are increased by 0.5 Hz?

Solution

The power produced by a synchronous generator with a given slope and no-load frequency is given by Equation (4–28):

$$P_1 = s_{P1}(f_{\text{nl},1} - f_{\text{sys}})$$
$$P_2 = s_{P2}(f_{\text{nl},2} - f_{\text{sys}})$$

Since the total power supplied by the generators must equal the power consumed by the loads,

$$P_{\text{load}} = P_1 + P_2$$

These equations can be used to answer all the questions asked.

(a) In the first case, both generators have a slope of 1 MW/Hz, and G_1 has a no-load frequency of 61.5 Hz, while G_2 has a no-load frequency of 61.0 Hz. The total load is 2.5 MW. Therefore, the system frequency can be found as follows:

$$\begin{aligned} P_{\text{load}} &= P_1 + P_2 \\ &= s_{P1}(f_{\text{nl},1} - f_{\text{sys}}) + s_{P2}(f_{\text{nl},2} - f_{\text{sys}}) \\ 2.5 \text{ MW} &= (1 \text{ MW/Hz})(61.5 \text{ Hz} - f_{\text{sys}}) + (1 \text{ MW/Hz})(61 \text{ Hz} - f_{\text{sys}}) \\ &= 61.5 \text{ MW} - (1 \text{ MW/Hz})f_{\text{sys}} + 61 \text{ MW} - (1 \text{ MW/Hz})f_{\text{sys}} \\ &= 122.5 \text{ MW} - (2 \text{ MW/Hz})f_{\text{sys}} \end{aligned}$$

therefore $$f_{\text{sys}} = \frac{122.5 \text{ MW} - 2.5 \text{ MW}}{(2\text{MW/Hz})} = 60.0 \text{ Hz}$$

The resulting powers supplied by the two generators are

$$\begin{aligned} P_1 &= s_{P1}(f_{\text{nl},1} - f_{\text{sys}}) \\ &= (1 \text{ MW/Hz})(61.5 \text{ Hz} - 60.0 \text{ Hz}) = 1.5 \text{ MW} \\ P_2 &= s_{P2}(f_{\text{nl},2} - f_{\text{sys}}) \\ &= (1 \text{ MW/Hz})(61.0 \text{ Hz} - 60.0 \text{ Hz}) = 1 \text{ MW} \end{aligned}$$

(b) When the load is increased by 1 MW, the total load becomes 3.5 MW. The new system frequency is now given by

$$\begin{aligned} P_{\text{load}} &= s_{P1}(f_{\text{nl},1} - f_{\text{sys}}) + s_{P2}(f_{\text{nl},2} - f_{\text{sys}}) \\ 3.5 \text{ MW} &= (1 \text{ MW/Hz})(61.5 \text{ Hz} - f_{\text{sys}}) + (1 \text{ MW/Hz})(61 \text{ Hz} - f_{\text{sys}}) \\ &= 61.5 \text{ MW} - (1 \text{ MW/Hz})f_{\text{sys}} + 61 \text{ MW} - (1 \text{ MW/Hz})f_{\text{sys}} \\ &= 122.5 \text{ MW} - (2 \text{ MW/Hz})f_{\text{sys}} \end{aligned}$$

therefore $$f_{\text{sys}} = \frac{122.5 \text{ MW} - 3.5 \text{ MW}}{(2\text{MW/Hz})} = 59.5 \text{ Hz}$$

The resulting powers are

$$\begin{aligned} P_1 &= s_{P1}(f_{\text{nl},1} - f_{\text{sys}}) \\ &= (1 \text{ MW/Hz})(61.5 \text{ Hz} - 59.5 \text{ Hz}) = 2.0 \text{ MW} \\ P_2 &= s_{P2}(f_{\text{nl},2} - f_{\text{sys}}) \\ &= (1 \text{ MW/Hz})(61.0 \text{ Hz} - 59.5 \text{ Hz}) = 1.5 \text{ MW} \end{aligned}$$

(c) If the no-load governor set points of G_2 are increased by 0.5 Hz, the new system frequency becomes

$$\begin{aligned} P_{\text{load}} &= s_{P1}(f_{\text{nl},1} - f_{\text{sys}}) + s_{P2}(f_{\text{nl},2} - f_{\text{sys}}) \\ 3.5 \text{ MW} &= (1 \text{ MW/Hz})(61.5 \text{ Hz} - f_{\text{sys}}) + (1 \text{ MW/Hz})(61.5 \text{ Hz} - f_{\text{sys}}) \\ &= 123 \text{ MW} - (2 \text{ MW/Hz})f_{\text{sys}} \end{aligned}$$

$$f_{\text{sys}} = \frac{123 \text{ MW} - 3.5 \text{ MW}}{(2\text{MW/Hz})} = 59.75 \text{ Hz}$$

The resulting powers are

$$\begin{aligned} P_1 = P_2 &= s_{P1}(f_{\text{nl},1} - f_{\text{sys}}) \\ &= (1 \text{ MW/Hz})(61.5 \text{ Hz} - 59.75 \text{ Hz}) = 1.75 \text{ MW} \end{aligned}$$

Notice that the system frequency rose, the power supplied by G_2 rose, and the power supplied by G_1 fell.

When two generators of similar size are operating in parallel, a change in the governor set points of one of them changes both the system frequency and the power sharing between them. It would normally be desired to adjust only one of these quantities at a time. How can the power sharing of the power system be adjusted independently of the system frequency, and vice versa?

The answer is very simple. An increase in governor set points on one generator increases that machine's power and increases system frequency. A decrease in governor set points on the other generator decreases that machine's power and decreases the system frequency. Therefore, to adjust power sharing without changing the system frequency, *increase the governor set points of one generator and simultaneously decrease the governor set points of the other generator* (see Figure 4–40a). Similarly, *to adjust the system frequency without changing the power sharing, simultaneously increase or decrease both governor set points* (see Figure 4–40b).

Reactive power and terminal voltage adjustments work in an analogous fashion. To shift the reactive power sharing without changing V_T, *simultaneously increase the field current on one generator and decrease the field current on the other* (see Figure 4–40c). To change the terminal voltage without affecting the reactive power sharing, *simultaneously increase or decrease both field currents* (see Figure 4–40d).

To summarize, in the case of two generators operating together:

1. The system is constrained in that the total power supplied by the two generators together must equal the amount consumed by the load. Neither f_{sys} nor V_T is constrained to be constant.
2. To adjust the real power sharing between generators without changing f_{sys}, simultaneously increase the governor set points on one generator while decreasing the governor set points on the other. The machine whose governor set point was increased will assume more of the load.
3. To adjust f_{sys} without changing the real power sharing, simultaneously increase or decrease both generators' governor set points.
4. To adjust the reactive power sharing between generators without changing V_T, simultaneously increase the field current on one generator while decreasing the field current on the other. The machine whose field current was increased will assume more of the reactive load.
5. To adjust V_T without changing the reactive power sharing, simultaneously increase or decrease both generators' field currents.

It is very important that any synchronous generator intended to operate in parallel with other machines have a *drooping* frequency–power characteristic. If two generators have flat or nearly flat characteristics, then the power sharing between

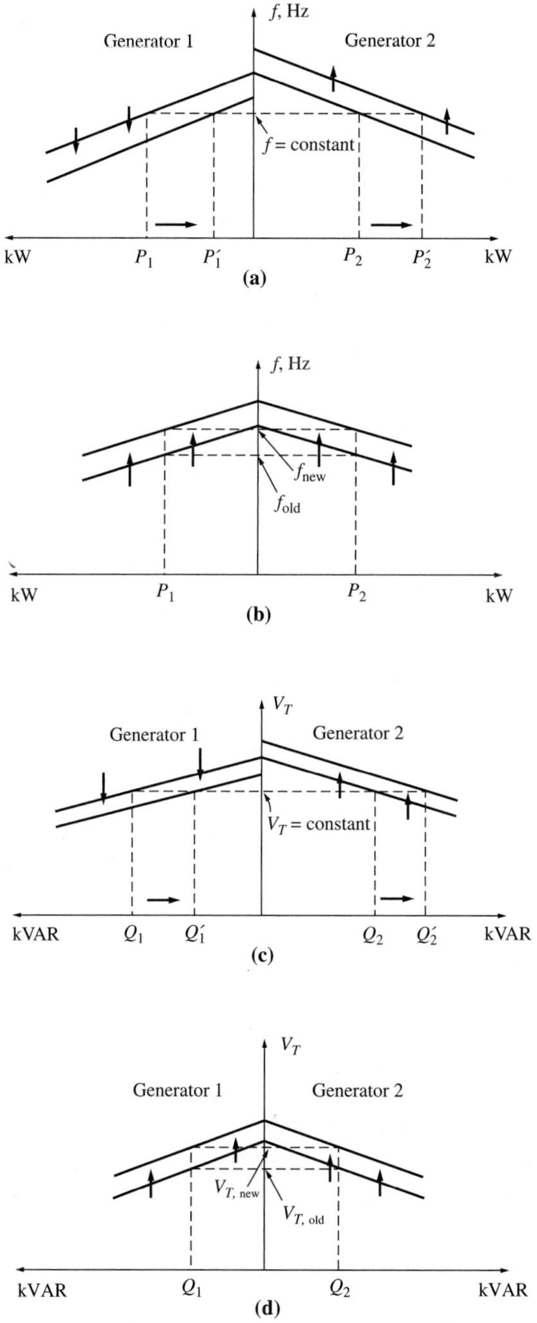

FIGURE 4–40
(a) Shifting power sharing without affecting system frequency. (b) Shifting system frequency without affecting power sharing. (c) Shifting reactive power sharing without affecting terminal voltage. (d) Shifting terminal voltage without affecting reactive power sharing.

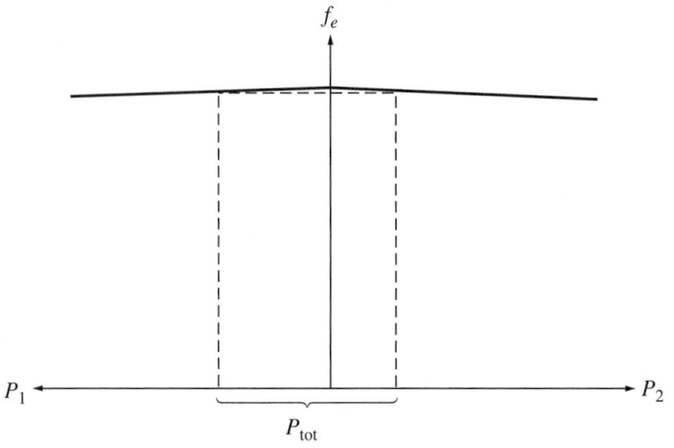

FIGURE 4–41
Two synchronous generators with flat frequency–power characteristics. A very tiny change in the no-load frequency of either of these machines could cause huge shifts in the power sharing.

them can vary widely with only the tiniest changes in no-load speed. This problem is illustrated by Figure 4–41. Notice that even very tiny changes in f_{nl} in one of the generators would cause wild shifts in power sharing. To ensure good control of power sharing between generators, they should have speed droops in the range of 2 to 5 percent.

4.10 SYNCHRONOUS GENERATOR TRANSIENTS

When the shaft torque applied to a generator or the output load on a generator changes suddenly, there is always a transient lasting for a finite period of time before the generator returns to steady state. For example, when a synchronous generator is paralleled with a running power system, it is initially turning faster and has a higher frequency than the power system does. Once it is paralleled, there is a transient period before the generator steadies down on the line and runs at line frequency while supplying a small amount of power to the load.

To illustrate this situation, refer to Figure 4–42. Figure 4–42a shows the magnetic fields and the phasor diagram of the generator at the moment just before it is paralleled with the power system. Here, the oncoming generator is supplying no load, its stator current is zero, $\mathbf{E}_A = \mathbf{V}_\phi$, and $\mathbf{B}_R = \mathbf{B}_{\text{net}}$.

At exactly time $t = 0$, the switch connecting the generator to the power system is shut, causing a stator current to flow. Since the generator's rotor is still turning faster than the system speed, it continues to move out ahead of the system's voltage $\mathbf{V}_\phi$. The induced torque on the shaft of the generator is given by

$$\tau_{\text{ind}} = k\mathbf{B}_R \times \mathbf{B}_{\text{net}} \tag{3–60}$$

The direction of this torque is opposite to the direction of motion, and it increases as the phase angle between $\mathbf{B}_R$ and $\mathbf{B}_{\text{net}}$ (or $\mathbf{E}_A$ and $\mathbf{V}_\phi$) increases. This torque *opposite*

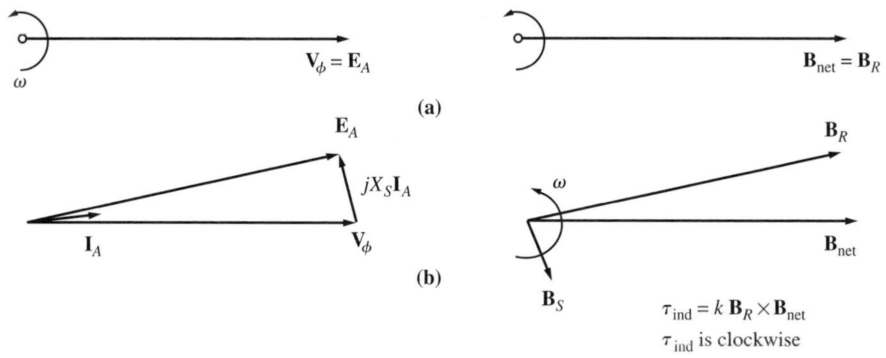

FIGURE 4–42
(a) The phasor diagram and magnetic fields of a generator at the moment of paralleling with a large power system. (b) The phasor diagram and house diagram shortly after (a). Here, the rotor has moved on ahead of the net magnetic fields, producing a clockwise torque. This torque is slowing the rotor down to the synchronous speed of the power system.

the direction of motion slows down the generator until it finally turns at synchronous speed with the rest of the power system.

Similarly, if the generator were turning at a speed *lower* than synchronous speed when it was paralleled with the power system, then the rotor would fall behind the net magnetic fields, and an induced torque *in the direction of motion* would be induced on the shaft of the machine. This torque would speed up the rotor until it again began turning at synchronous speed.

Transient Stability of Synchronous Generators

We learned earlier that the static stability limit of a synchronous generator is the maximum power that the generator can supply under any circumstances. The maximum power that the generator can supply is given by Equation (4–21):

$$P_{max} = \frac{3V_\phi E_A}{X_S} \tag{4–21}$$

and the corresponding maximum torque is

$$\tau_{max} = \frac{3V_\phi E_A}{\omega_m X_S} \tag{4–30}$$

In theory, a generator should be able to supply up to this amount of power and torque before becoming unstable. In practice, however, the maximum load that can be supplied by the generator is limited to a much lower level by its *dynamic stability limit*.

To understand the reason for this limitation, consider the generator in Figure 4–42 again. If the torque applied by the prime mover (τ_{app}) is suddenly increased, the shaft of the generator will begin to speed up, and the torque angle δ will increase as described. As the angle δ increases, the induced torque τ_{ind} of the generator will

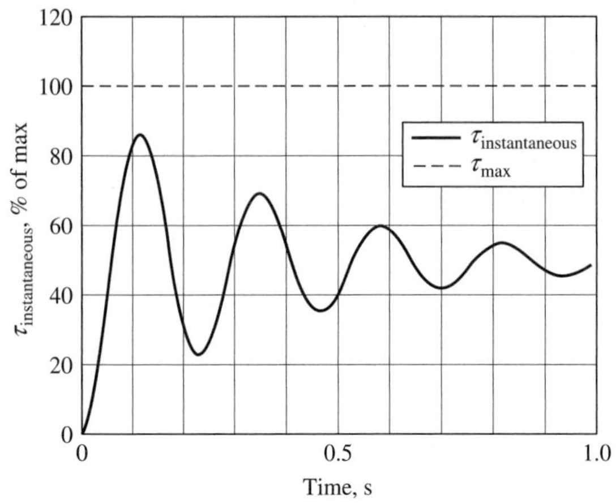

FIGURE 4–43
The dynamic response when an applied torque equal to 50% of τ_{max} is suddenly added to a synchronous generator.

increase until an angle δ is reached at which τ_{ind} is equal and opposite to τ_{app}. This is the steady-state operating point of the generator with the new load. However, the rotor of the generator has a great deal of inertia, so its torque angle δ actually *overshoots* the steady-state position, and gradually settles out in a damped oscillation, as shown in Figure 4–43. The exact shape of this damped oscillation can be determined by solving a nonlinear differential equation, which is beyond the scope of this book. For more information, see Reference 4, p. 345.

The important point about Figure 4–43 is that *if at any point in the transient response the instantaneous torque exceeds* τ_{max}, *the synchronous generator will be unstable*. The size of the oscillations depends on how suddenly the additional torque is applied to the synchronous generator. If it is added very gradually, the machine should be able to almost reach the static stability limit. On the other hand, if the load is added sharply, the machine will be stable only up to a much lower limit, which is very complicated to calculate. For very abrupt changes in torque or load, the dynamic stability limit may be less than half of the static stability limit.

Short-Circuit Transients in Synchronous Generators

By far the severest transient condition that can occur in a synchronous generator is the situation where the three terminals of the generator are suddenly shorted out. Such a short on a power system is called a *fault*. There are several components of current present in a shorted synchronous generator, which will be described below. The same effects occur in less severe transients like load changes, but they are much more obvious in the extreme case of a short circuit.

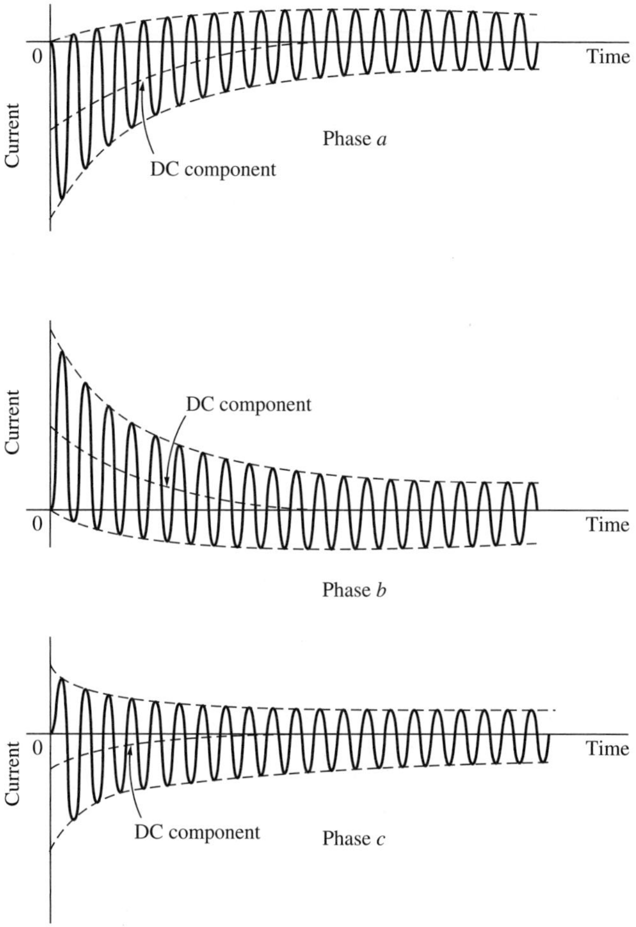

FIGURE 4–44
The total fault currents as a function of time during a three-phase fault at the terminals of a synchronous generator.

When a fault occurs on a synchronous generator, the resulting current flow in the phases of the generator can appear as shown in Figure 4–44. The current in each phase shown in Figure 4–42 can be represented as a dc transient component added on top of a symmetrical ac component. The symmetrical ac component by itself is shown in Figure 4–45.

Before the fault, only ac voltages and currents were present within the generator, while after the fault, both ac and dc currents are present. Where did the dc currents come from? Remember that the synchronous generator is basically inductive—it is modeled by an internal generated voltage in series with the synchronous reactance. Also, recall that *a current cannot change instantaneously in an inductor*. When the fault occurs, the ac component of current jumps to a very

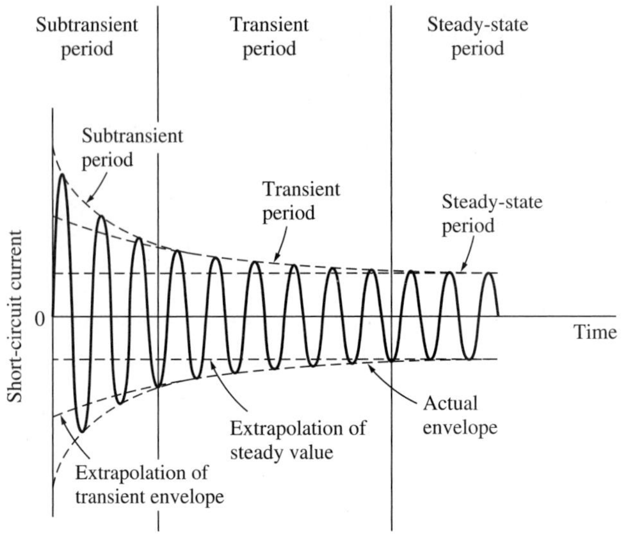

FIGURE 4–45
The symmetric ac component of the fault current.

large value, but the total current cannot change at that instant. The dc component of current is just large enough that the *sum* of the ac and dc components just after the fault equals the ac current flowing just before the fault. Since the instantaneous values of current at the moment of the fault are different in each phase, the magnitude of the dc component of current will be different in each phase.

These dc components of current decay fairly quickly, but they initially average about 50 or 60 percent of the ac current flow the instant after the fault occurs. The total initial current is therefore typically 1.5 or 1.6 times the ac component taken alone.

The ac symmetrical component of current is shown in Figure 4–45. It can be divided into roughly three periods. During the first cycle or so after the fault occurs, the ac current is very large and falls very rapidly. This period of time is called the *subtransient period*. After it is over, the current continues to fall at a slower rate, until at last it reaches a steady state. The period of time during which it falls at a slower rate is called the *transient period*, and the time after it reaches steady state is known as the *steady-state period*.

If the rms magnitude of the ac component of current is plotted as a function of time on a semilogarithmic scale, it is possible to observe the three periods of fault current. Such a plot is shown in Figure 4–46. It is possible to determine the time constants of the decays in each period from such a plot.

The ac rms current flowing in the generator during the subtransient period is called the *subtransient current* and is denoted by the symbol I''. This current is caused by the damper windings on synchronous generators (see Chapter 5 for a discussion of damper windings). The time constant of the subtransient current is

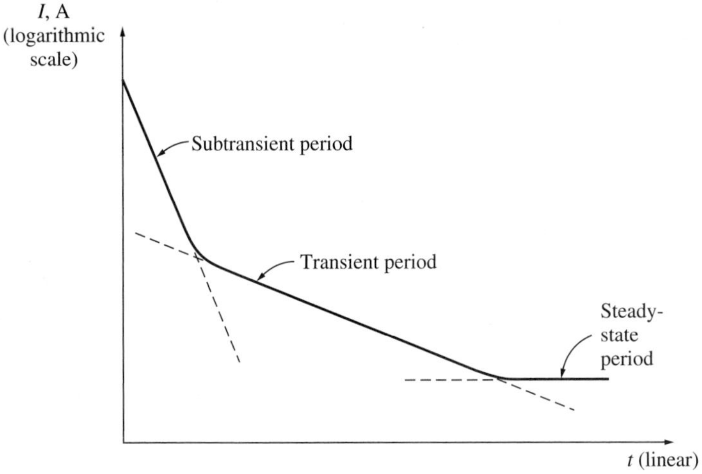

FIGURE 4–46
A semilogarithmic plot of the magnitude of the ac component of fault current as a function of time. The subtransient and transient time constants of the generator can be determined from such a plot.

given the symbol T'', and it can be determined from the slope of the subtransient current in the plot in Figure 4–46. This current can often be 10 times the size of the steady-state fault current.

The rms current flowing in the generator during the transient period is called the *transient current* and is denoted by the symbol I'. It is caused by a dc component of current induced in *the field circuit* at the time of the short. This field current increases the internal generated voltage and causes an increased fault current. Since the time constant of the dc field circuit is much longer than the time constant of the damper windings, the transient period lasts much longer than the subtransient period. This time constant is given the symbol T'. The average rms current during the transient period is often as much as 5 times the steady-state fault current.

After the transient period, the fault current reaches a steady-state condition. The steady-state current during a fault is denoted by the symbol I_{ss}. It is given approximately by the fundamental frequency component of the internal generated voltage E_A within the machine divided by its synchronous reactance:

$$I_{ss} = \frac{E_A}{X_S} \qquad \text{steady state} \tag{4–31}$$

The rms magnitude of the ac fault current in a synchronous generator varies continuously as a function of time. If I'' is the subtransient component of current at the instant of the fault, I' is the transient component of current at the instant of the fault, and I_{ss} is the steady-state fault current, then the rms magnitude of the current at any time after a fault occurs at the terminals of the generator is

$$I(t) = (I'' - I')e^{-t/T''} + (I' - I_{ss})e^{-t/T'} + I_{ss} \tag{4–32}$$

It is customary to define subtransient and transient reactances for a synchronous machine as a convenient way to describe the subtransient and transient components of fault current. The *subtransient reactance* of a synchronous generator is defined as the ratio of the fundamental component of the internal generated voltage to the subtransient component of current at the beginning of the fault. It is given by

$$X'' = \frac{E_A}{I''} \qquad \text{subtransient} \tag{4–33}$$

Similarly, the *transient reactance* of a synchronous generator is defined as the ratio of the fundamental component of E_A to the transient component of current I' at the beginning of the fault. This value of current is found by extrapolating the subtransient region in Figure 4–46 back to time zero:

$$X' = \frac{E_A}{I'} \qquad \text{transient} \tag{4–34}$$

For the purposes of sizing protective equipment, the subtransient current is often assumed to be E_A/X'', and the transient current is assumed to be E_A/X', since these are the maximum values that the respective currents take on.

Note that the preceding discussion of faults assumes that all three phases were shorted out simultaneously. If the fault does not involve all three phases equally, then more complex methods of analysis are required to understand it. These methods (known as symmetrical components) are beyond the scope of this book.

Example 4–7. A 100-MVA, 13.5–kV, Y-connected, three-phase, 60-Hz synchronous generator is operating at the rated voltage and no load when a three-phase fault develops at its terminals. Its reactances per unit to the machine's own base are

$$X_S = 1.0 \qquad X' = 0.25 \qquad X'' = 0.12$$

and its time constants are

$$T' = 1.10\,\text{s} \qquad T'' = 0.04\,\text{s}$$

The initial dc component in this machine averages 50 percent of the initial ac component.

(*a*) What is the ac component of current in this generator the instant after the fault occurs?

(*b*) What is the total current (ac plus dc) flowing in the generator right after the fault occurs?

(*c*) What will the ac component of the current be after two cycles? After 5 s?

Solution

The base current of this generator is given by the equation

$$I_{L,\text{base}} = \frac{S_\text{base}}{\sqrt{3}\,V_{L,\text{base}}} \tag{2–95}$$

$$= \frac{100\text{ MVA}}{\sqrt{3}(13.8\text{ kV})} = 4184\text{ A}$$

The subtransient, transient, and steady-state currents, per unit and in amperes, are

$$I'' = \frac{E_A}{X''} = \frac{1.0}{0.12} = 8.333$$

$$= (8.333)(4184 \text{ A}) = 34{,}900 \text{ A}$$

$$I' = \frac{E_A}{X'} = \frac{1.0}{0.25} = 4.00$$

$$= (4.00)(4184 \text{ A}) = 16{,}700 \text{ A}$$

$$I_{ss} = \frac{E_A}{X'} = \frac{1.0}{1.0} = 1.00$$

$$= (1.00)(4184 \text{ A}) = 4184 \text{ A}$$

(*a*) The initial ac component of current is $I'' = 34{,}900$ A.
(*b*) The total current (ac plus dc) at the beginning of the fault is

$$I_{\text{tot}} = 1.5I'' = 52{,}350 \text{ A}$$

(*c*) The ac component of current as a function of time is given by Equation (4–32):

$$I(t) = (I'' - I')e^{-t/T''} + (I' - I_{ss})e^{-t/T'} + I_{ss} \tag{4–32}$$

$$= 18{,}200e^{-t/0.04\text{ s}} + 12{,}516e^{-t/1.1\text{ s}} + 4184 \text{ A}$$

At two cycles, $t = 1/30$ s, the total current is

$$I\left(\frac{1}{30}\right) = 7910 \text{ A} + 12{,}142 \text{ A} + 4184 \text{ A} = 24{,}236 \text{ A}$$

After two cycles, the transient component of current is clearly the largest one and this time is in the transient period of the short circuit. At 5 s, the current is down to

$$I(5) = 0 \text{ A} + 133 \text{ A} + 4184 \text{ A} = 4317 \text{ A}$$

This is part of the steady-state period of the short circuit.

4.11 SYNCHRONOUS GENERATOR RATINGS

There are certain basic limits to the speed and power that may be obtained from a synchronous generator. These limits are expressed as *ratings* on the machine. The purpose of the ratings is to protect the generator from damage due to improper operation. To this end, each machine has a number of ratings listed on a nameplate attached to it.

Typical ratings on a synchronous machine are *voltage, frequency, speed, apparent power (kilovoltamperes), power factor, field current,* and *service factor.* These ratings, and the interrelationships among them, will be discussed in the following sections.

The Voltage, Speed, and Frequency Ratings

The rated frequency of a synchronous generator depends on the power system to which it is connected. The commonly used power system frequencies today are

50 Hz (in Europe, Asia, etc.), 60 Hz (in the Americas), and 400 Hz (in special-purpose and control applications). Once the operating frequency is known, there is only one possible rotational speed for a given number of poles. The fixed relationship between frequency and speed is given by Equation (3–34):

$$f_{se} = \frac{n_m P}{120} \tag{3–34}$$

as previously described.

Perhaps the most obvious rating is the voltage at which a generator is designed to operate. A generator's voltage depends on the flux, the speed of rotation, and the mechanical construction of the machine. For a given mechanical frame size and speed, the higher the desired voltage, the higher the machine's required flux. However, flux cannot be increased forever, since there is always a maximum allowable field current.

Another consideration in setting the maximum allowable voltage is the breakdown value of the winding insulation—normal operating voltages must not approach breakdown too closely.

Is it possible to operate a generator rated for one frequency at a different frequency? For example, is it possible to operate a 60-Hz generator at 50 Hz? The answer is a *qualified* yes, as long as certain conditions are met. Basically, the problem is that there is a maximum flux achievable in any given machine, and since $E_A = K\phi\omega$, the maximum allowable E_A changes when the speed is changed. Specifically, if a 60-Hz generator is to be operated at 50 Hz, then the operating voltage must be *derated* to 50/60, or 83.3 percent, of its original value. Just the opposite effect happens when a 50-Hz generator is operated at 60 Hz.

Apparent Power and Power-Factor Ratings

There are two factors that determine the power limits of electric machines. One is the mechanical torque on the shaft of the machine, and the other is the heating of the machine's windings. In all practical synchronous motors and generators, the shaft is strong enough mechanically to handle a much larger steady-state power than the machine is rated for, so the practical steady-state limits are set by heating in the machine's windings.

There are two windings in a synchronous generator, and each one must be protected from overheating. These two windings are the armature winding and the field winding. The maximum acceptable armature current sets the apparent power rating for a generator, since the apparent power S is given by

$$S = 3V_\phi I_A \tag{4–35}$$

If the rated voltage is known, then the maximum acceptable armature current determines the rated kilovoltamperes of the generator:

$$S_{\text{rated}} = 3V_{\phi,\text{rated}} I_{A,\text{max}} \tag{4–36}$$

or

$$S_{\text{rated}} = \sqrt{3} V_{L,\text{rated}} I_{L,\text{max}} \tag{4–37}$$

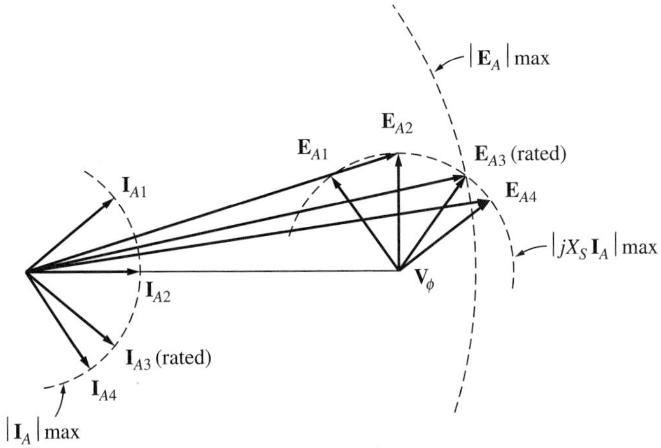

FIGURE 4–47
How the rotor field current limit sets the rated power factor of a generator.

It is important to realize that, for heating the armature windings, *the power factor of the armature current is irrelevant*. The heating effect of the stator copper losses is given by

$$P_{\text{SCL}} = 3I_A^2R_A \tag{4–38}$$

and is independent of the angle of the current with respect to $\mathbf{V}_\phi$. Because the current angle is irrelevant to the armature heating, these machines are rated in kilovoltamperes instead of kilowatts.

The other winding of concern is the field winding. The field copper losses are given by

$$P_{\text{RCL}} = I_F^2R_F \tag{4–39}$$

so the maximum allowable heating sets a maximum field current for the machine. Since $E_A = K\phi\omega$ this sets the maximum acceptable size for E_A.

The effect of having a maximum I_F and a maximum E_A translates directly into a restriction on the lowest acceptable power factor of the generator when it is operating at the rated kilovoltamperes. Figure 4–47 shows the phasor diagram of a synchronous generator with the rated voltage and armature current. The current can assume many different angles, as shown. The internal generated voltage $\mathbf{E}_A$ is the sum of $\mathbf{V}_\phi$ and $jX_S\mathbf{I}_A$. Notice that for some possible current angles the required E_A exceeds $E_{A,\text{max}}$. If the generator were operated at the rated armature current and these power factors, the field winding would burn up.

The angle of $\mathbf{I}_A$ that requires the maximum possible $\mathbf{E}_A$ while $\mathbf{V}_\phi$ remains at the rated value gives the rated power factor of the generator. It is possible to operate the generator at a lower (more lagging) power factor than the rated value, but only by cutting back on the kilovoltamperes supplied by the generator.

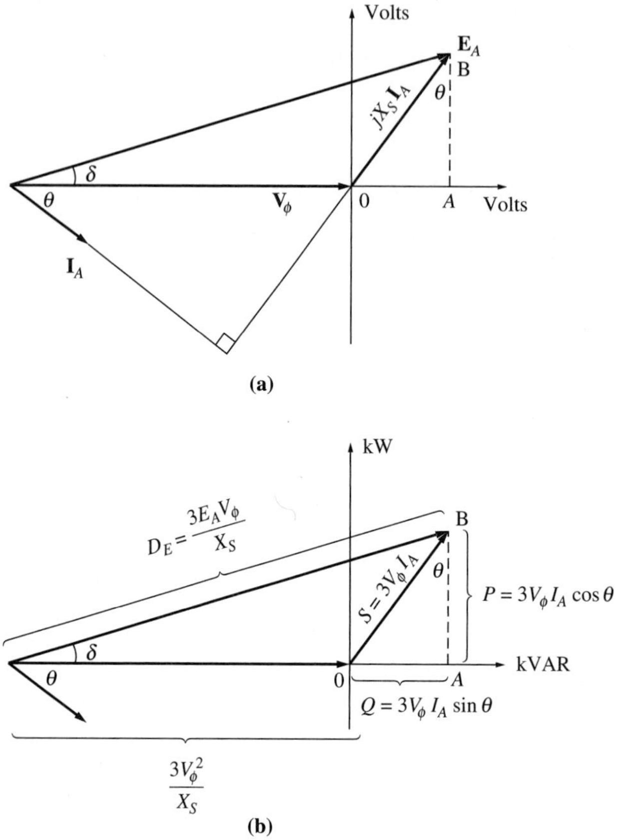

FIGURE 4–48
Derivation of a synchronous generator capability curve. (a) The generator phasor diagram; (b) the corresponding power units.

Synchronous Generator Capability Curves

The stator and rotor heat limits, together with any external limits on a synchronous generator, can be expressed in graphical form by a generator *capability diagram*. A capability diagram is a plot of complex power $S = P + jQ$. It is derived from the phasor diagram of the generator, assuming that $\mathbf{V}_\phi$ is constant at the machine's rated voltage.

Figure 4–48a shows the phasor diagram of a synchronous generator operating at a lagging power factor and its rated voltage. An orthogonal set of axes is drawn on the diagram with its origin at the tip of $\mathbf{V}_\phi$ and with units of volts. On this diagram, vertical segment AB has a length $X_S I_A \cos\theta$, and horizontal segment OA has a length $X_S I_A \sin\theta$.

The real power output of the generator is given by

$$P = 3V_\phi I_A \cos\theta \tag{4–17}$$

the reactive power output is given by

$$Q = 3V_\phi I_A \sin\theta \tag{4–19}$$

and the apparent power output is given by

$$S = 3V_\phi I_A \tag{4–35}$$

so the vertical and horizontal axes of this figure can be recalibrated in terms of real and reactive power (Figure 4–48b). The conversion factor needed to change the scale of the axes from volts to voltamperes (power units) is $3V_\phi/X_S$:

$$P = 3V_\phi I_A \cos\theta = \frac{3V_\phi}{X_S}(X_S I_A \cos\theta) \tag{4–40}$$

and

$$Q = 3V_\phi I_A \sin\theta = \frac{3V_\phi}{X_S}(X_S I_A \sin\theta) \tag{4–41}$$

On the voltage axes, the origin of the phasor diagram is at $-V_\phi$ on the horizontal axis, so the origin on the power diagram is at

$$\begin{aligned} Q &= \frac{3V_\phi}{X_S}(-V_\phi) \\ &= -\frac{3V_\phi^2}{X_S} \end{aligned} \tag{4–42}$$

The field current is proportional to the machine's flux, and the flux is proportional to $E_A = K\phi\omega$. The length corresponding to E_A on the power diagram is

$$D_E = -\frac{3E_A V_\phi}{X_S} \tag{4–43}$$

The armature current I_A is proportional to $X_S I_A$, and the length corresponding to $X_S I_A$ on the power diagram is $3V_\phi I_A$.

The final synchronous generator capability curve is shown in Figure 4–49. It is a plot of P versus Q, with real power P on the horizontal axis and reactive power Q on the vertical axis. Lines of constant armature current I_A appear as lines of constant $S = 3V_\phi I_A$, which are concentric circles around the origin. Lines of constant field current correspond to lines of constant E_A, which are shown as circles of magnitude $3E_A V_\phi/X_S$ centered on the point

$$Q = -\frac{3V_\phi^2}{X_S} \tag{4–42}$$

The armature current limit appears as the circle corresponding to the rated I_A or rated kilovoltamperes, and the field current limit appears as a circle corresponding to the rated I_F or E_A. *Any point that lies within both circles is a safe operating point for the generator.*

It is also possible to show other constraints on the diagram, such as the maximum prime-mover power and the static stability limit. A capability curve that also reflects the maximum prime-mover power is shown in Figure 4–50.

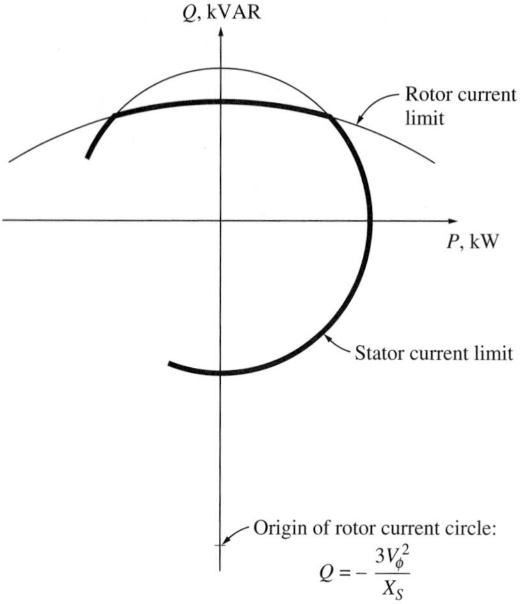

FIGURE 4–49
The resulting generator capability curve.

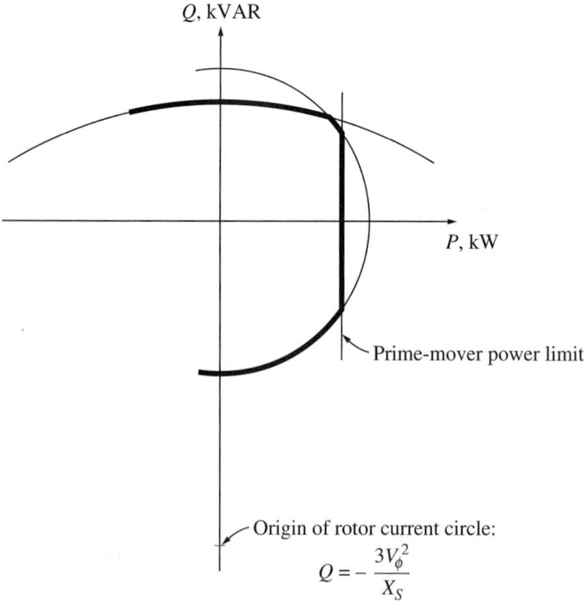

FIGURE 4–50
A capability diagram showing the prime-mover power limit.

Example 4–8. A 480-V, 50-Hz, Y-connected, six-pole synchronous generator is rated at 50 kVA at 0.8 PF lagging. It has a synchronous reactance of 1.0 Ω per phase. Assume that this generator is connected to a steam turbine capable of supplying up to 45 kW. The friction and windage losses are 1.5 kW, and the core losses are 1.0 kW.

(*a*) Sketch the capability curve for this generator, including the prime-mover power limit.

(*b*) Can this generator supply a line current of 56 A at 0.7 PF lagging? Why or why not?

(*c*) What is the maximum amount of reactive power this generator can produce?

(*d*) If the generator supplies 30 kW of real power, what is the maximum amount of reactive power that can be simultaneously supplied?

Solution

The maximum current in this generator can be found from Equation (4–36):

$$S_{\text{rated}} = 3V_{\phi,\text{rated}} I_{A,\max} \tag{4–36}$$

The voltage V_ϕ of this machine is

$$V_\phi = \frac{V_T}{\sqrt{3}} = \frac{480 \text{ V}}{\sqrt{3}} = 277 \text{ V}$$

so the maximum armature current is

$$I_{A,\max} = \frac{S_{\text{rated}}}{3V_\phi} = \frac{50 \text{ kVA}}{3(277 \text{ V})} = 60 \text{ A}$$

With this information, it is now possible to answer the questions.

(*a*) The maximum permissible apparent power is 50 kVA, which specifies the maximum safe armature current. The center of the E_A circles is at

$$Q = -\frac{3V_\phi^2}{X_S} \tag{4–42}$$

$$= -\frac{3(277 \text{ V})^2}{1.0 \ \Omega} = -230 \text{ kVAR}$$

The maximum size of E_A is given by

$$\mathbf{E}_A = \mathbf{V}_\phi + jX_S\mathbf{I}_A$$
$$= 277 \angle 0° \text{ V} + (j1.0 \ \Omega)(60 \angle -36.87° \text{ A})$$
$$= 313 + j48 \text{ V} = 317 \angle 8.7° \text{ V}$$

Therefore, the magnitude of the distance proportional to E_A is

$$D_E = \frac{3E_A V_\phi}{X_S} \tag{4–43}$$

$$= \frac{3(317 \text{ V})(277 \text{ V})}{1.0 \ \Omega} = 263 \text{ kVAR}$$

The maximum output power available with a prime-mover power of 45 kW is approximately

$$P_{\max,\text{out}} = P_{\max,\text{in}} - P_{\text{mech loss}} - P_{\text{core loss}}$$
$$= 45 \text{ kW} - 1.5 \text{ kW} - 1.0 \text{ kW} = 42.5 \text{ kW}$$

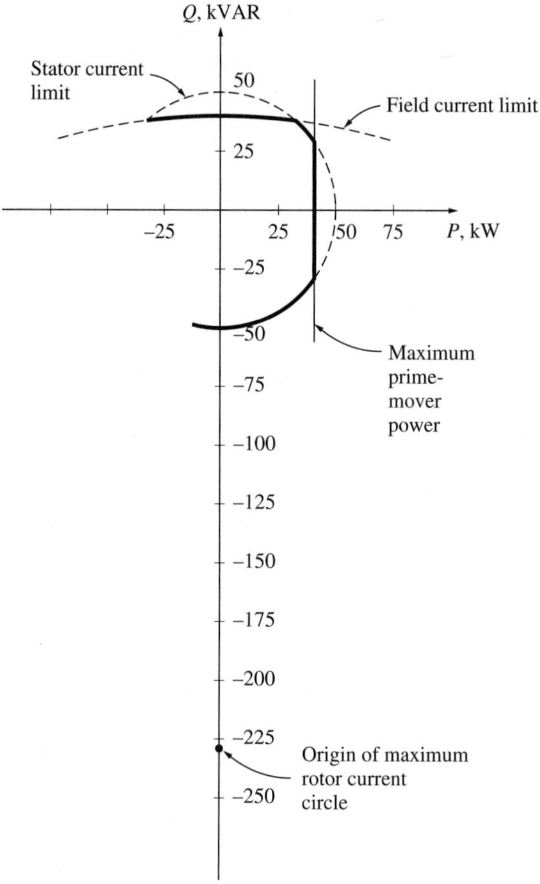

FIGURE 4–51
The capability diagram for the generator in Example 4–8.

(This value is approximate because the I^2R loss and the stray load loss were not considered.) The resulting capability diagram is shown in Figure 4–51.

(b) A current of 56 A at 0.7 PF lagging produces a real power of

$$P = 3V_\phi I_A \cos\theta$$
$$= 3(277\text{ V})(56\text{ A})(0.7) = 32.6\text{ kW}$$

and a reactive power of

$$Q = 3V_\phi I_A \sin\theta$$
$$= 3(277\text{ V})(56\text{ A})(0.714) = 33.2\text{ kVAR}$$

Plotting this point on the capability diagram shows that it is safely within the maximum I_A curve but outside the maximum I_F curve. Therefore, this point is *not* a safe operating condition.

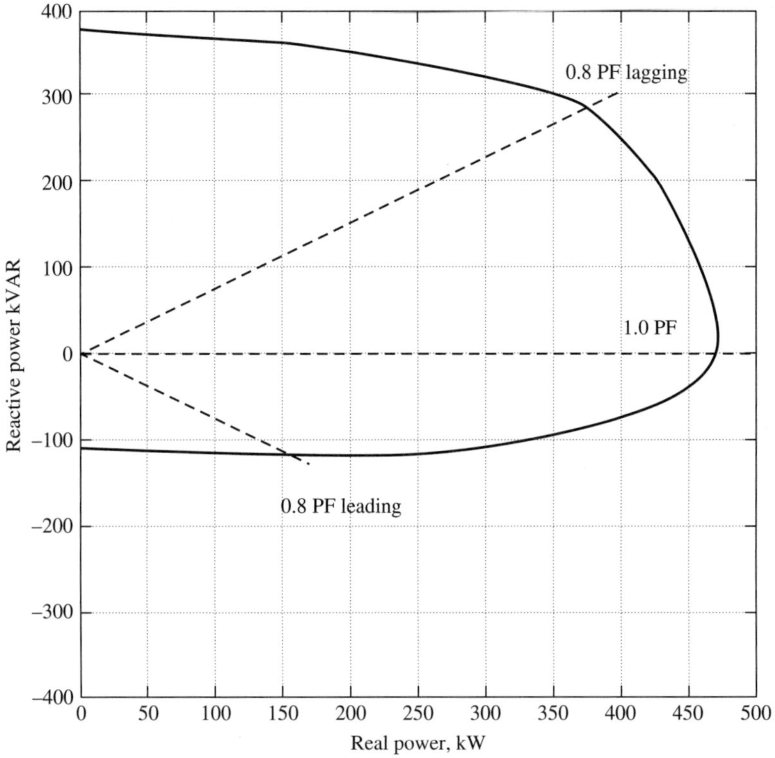

FIGURE 4–52
Capability curve for a real synchronous generator rated at 470 kVA. (*Courtesy of Marathon Electric Company.*)

(*c*) When the real power supplied by the generator is zero, the reactive power that the generator can supply will be maximum. This point is right at the peak of the capability curve. The Q that the generator can supply there is

$$Q = 263 \text{ kVAR} - 230 \text{ kVAR} = 33 \text{ kVAR}$$

(*d*) If the generator is supplying 30 kW of real power, the maximum reactive power that the generator can supply is 31.5 kVAR. This value can be found by entering the capability diagram at 30 kW and going up the constant-kilowatt line until a limit is reached. The limiting factor in this case is the field current—the armature will be safe up to 39.8 kVAR.

Figure 4–52 shows a typical capability for a real synchronous generator. Note that the capability boundaries are not a perfect circle for a real generator. This is true because real synchronous generators with salient poles have additional effects that we have not modeled. These effects are described in Appendix C.

Short-Time Operation and Service Factor

The most important limit in the steady-state operation of a synchronous generator is the heating of its armature and field windings. However, the heating limit usually occurs at a point much less than the maximum power that the generator is magnetically and mechanically able to supply. In fact, a typical synchronous generator is often able to supply up to 300 percent of its rated power for a while (until its windings burn up). This ability to supply power above the rated amount is used to supply momentary power surges during motor starting and similar load transients.

It is also possible to use a generator at powers exceeding the rated values for longer periods of time, as long as the windings do not have time to heat up too much before the excess load is removed. For example, a generator that could supply 1 MW indefinitely might be able to supply 1.5 MW for a couple of minutes without serious harm, and for progressively longer periods at lower power levels. However, the load must finally be removed, or the windings will overheat. The higher the power over the rated value, the shorter the time a machine can tolerate it.

Figure 4–53 illustrates this effect. This figure shows the time in seconds required for an overload to cause thermal damage to a typical electrical machine, whose windings were at normal operating temperature before the overload occurred. In this particular machine, a 20 percent overload can be tolerated for 1000 seconds, a 100 percent overload can be tolerated for about 30 seconds, and a 200 percent overload can be tolerated for about 10 seconds before damage occurs.

The maximum temperature rise that a machine can stand depends on the *insulation class* of its windings. There are four standard insulation classes: A, B, F, and H. While there is some variation in acceptable temperature depending on a machine's particular construction and the method of temperature measurement, these classes generally correspond to temperature rises of 60, 80, 105, and 125°C, respectively, above ambient temperature. The higher the insulation class of a given machine, the greater the power that can be drawn out of it without overheating its windings.

Overheating of windings is a *very serious* problem in a motor or generator. It was an old rule of thumb that for each 10°C temperature rise above the rated windings temperature, the average lifetime of a machine is cut in half (see Figure 3–20). Modern insulating materials are less susceptible to breakdown than that, but temperature rises still drastically shorten their lives. For this reason, a synchronous machine should not be overloaded unless absolutely necessary.

A question related to the overheating problem is: Just how well is the power requirement of a machine known? Before installation, there are often only approximate estimates of load. Because of this, general-purpose machines usually have a *service factor*. The service factor is defined as the ratio of the actual maximum power of the machine to its nameplate rating. A generator with a service factor of 1.15 can actually be operated at 115 percent of the rated load indefinitely without harm. The service factor on a machine provides a margin of error in case the loads were improperly estimated.

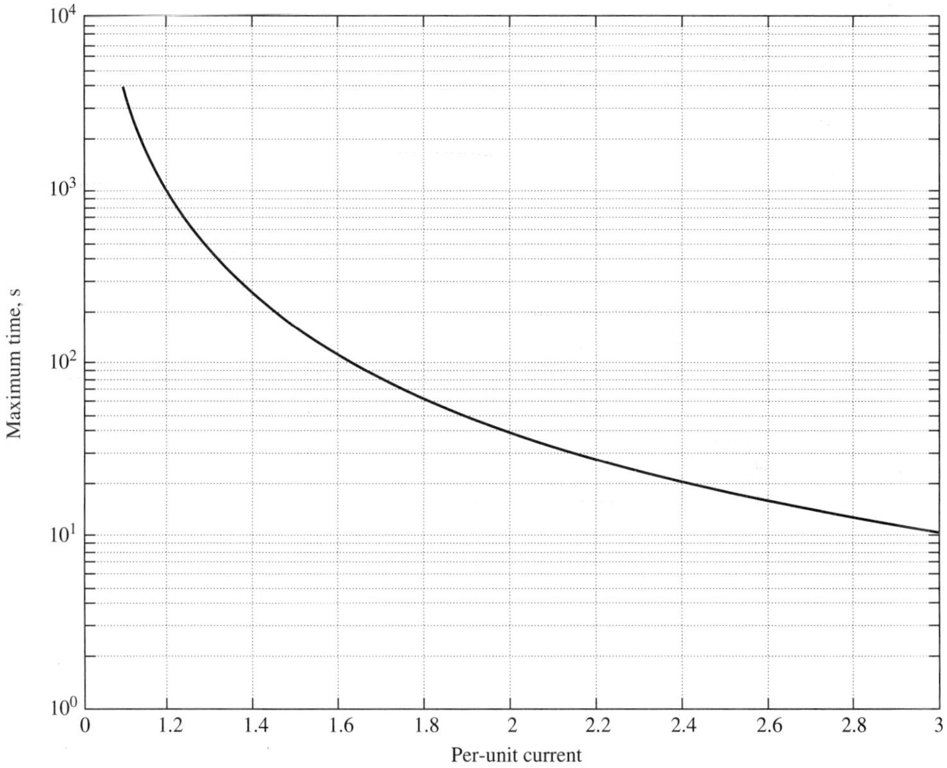

FIGURE 4–53
Thermal damage curve for a typical synchronous machine, assuming that the windings were already at operational temperature when the overload is applied. (*Courtesy of Marathon Electric Company.*)

4.12 SUMMARY

A synchronous generator is a device for converting mechanical power from a prime mover to ac electric power at a specific voltage and frequency. The term *synchronous* refers to the fact that this machine's electrical frequency is locked in or synchronized with its mechanical rate of shaft rotation. The synchronous generator is used to produce the vast majority of electric power used throughout the world.

The internal generated voltage of this machine depends on the rate of shaft rotation and on the magnitude of the field flux. The phase voltage of the machine differs from the internal generated voltage by the effects of armature reaction in the generator and also by the internal resistance and reactance of the armature windings. The terminal voltage of the generator will either equal the phase voltage or be related to it by $\sqrt{3}$, depending on whether the machine is Δ- or Y-connected.

The way in which a synchronous generator operates in a real power system depends on the constraints on it. When a generator operates alone, the real and reactive powers that must be supplied are determined by the load attached to it, and the governor set points and field current control the frequency and terminal voltage, respectively. When the generator is connected to an infinite bus, its frequency and voltage are fixed, so the governor set points and field current control the real and reactive power flow from the generator. In real systems containing generators of approximately equal size, the governor set points affect both frequency and power flow, and the field current affects both terminal voltage and reactive power flow.

A synchronous generator's ability to produce electric power is primarily limited by heating within the machine. When the generator's windings overheat, the life of the machine can be severely shortened. Since there are two different windings (armature and field), there are two separate constraints on the generator. The maximum allowable heating in the armature windings sets the maximum kilovoltamperes allowable from the machine, and the maximum allowable heating in the field windings sets the maximum size of E_A. The maximum size of E_A and the maximum size of I_A together set the rated power factor of the generator.

QUESTIONS

4–1. Why is the frequency of a synchronous generator locked into its rate of shaft rotation?

4–2. Why does an alternator's voltage drop sharply when it is loaded down with a lagging load?

4–3. Why does an alternator's voltage rise when it is loaded down with a leading load?

4–4. Sketch the phasor diagrams and magnetic field relationships for a synchronous generator operating at (*a*) unity power factor, (*b*) lagging power factor, (*c*) leading power factor.

4–5. Explain just how the synchronous impedance and armature resistance can be determined in a synchronous generator.

4–6. Why must a 60-Hz generator be derated if it is to be operated at 50 Hz? How much derating must be done?

4–7. Would you expect a 400-Hz generator to be larger or smaller than a 60-Hz generator of the same power and voltage rating? Why?

4–8. What conditions are necessary for paralleling two synchronous generators?

4–9. Why must the oncoming generator on a power system be paralleled at a higher frequency than that of the running system?

4–10. What is an infinite bus? What constraints does it impose on a generator paralleled with it?

4–11. How can the real power sharing between two generators be controlled without affecting the system's frequency? How can the reactive power sharing between two generators be controlled without affecting the system's terminal voltage?

4–12. How can the system frequency of a large power system be adjusted without affecting the power sharing among the system's generators?

4–13. How can the concepts of Section 4.9 be expanded to calculate the system frequency and power sharing among three or more generators operating in parallel?

4–14. Why is overheating such a serious matter for a generator?

4–15. Explain in detail the concept behind capability curves.

4–16. What are short-time ratings? Why are they important in regular generator operation?

PROBLEMS

4–1. At a location in Europe, it is necessary to supply 1000 kW of 60-Hz power. The only power sources available operate at 50 Hz. It is decided to generate the power by means of a motor-generator set consisting of a synchronous motor driving a synchronous generator. How many poles should each of the two machines have in order to convert 50-Hz power to 60-Hz power?

4–2. A 13.8-kV, 50-MVA, 0.9-power-factor-lagging, 60-Hz, four-pole Y-connected synchronous generator has a synchronous reactance of 2.5 Ω and an armature resistance of 0.2 Ω. At 60 Hz, its friction and windage losses are 1 MW, and its core losses are 1.5 MW. The field circuit has a dc voltage of 120 V, and the maximum I_F is 10 A. The current of the field circuit is adjustable over the range from 0 to 10 A. The OCC of this generator is shown in Figure P4–1.

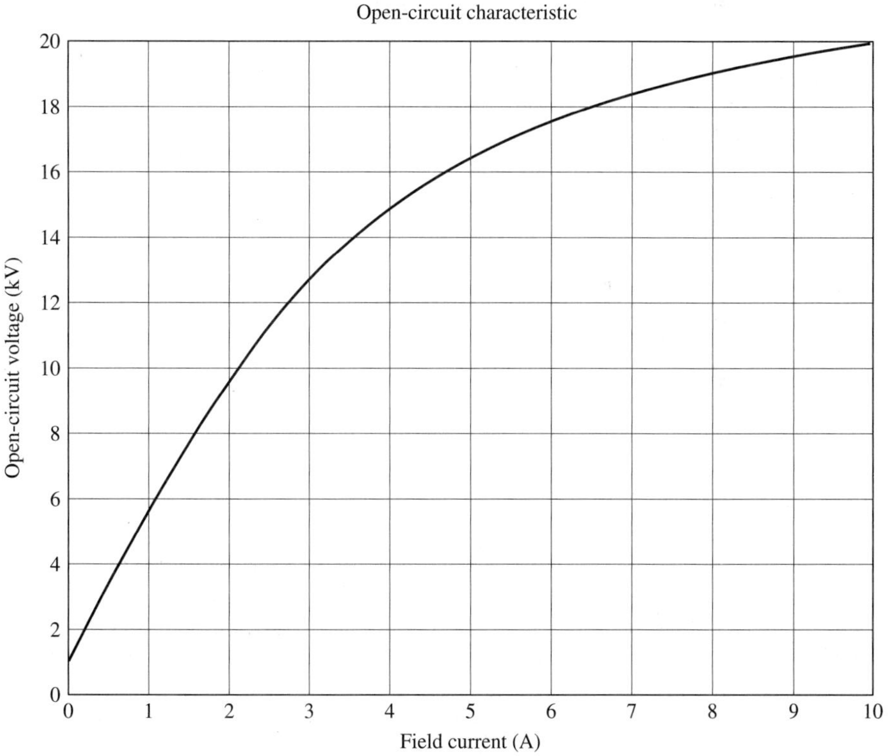

FIGURE P4–1
Open-circuit characteristic curve for the generator in Problem 4–2.

(*a*) How much field current is required to make the terminal voltage V_T (or line voltage V_L) equal to 13.8 kV when the generator is running at no load?

(*b*) What is the internal generated voltage E_A of this machine at rated conditions?

(c) What is the phase voltage V_ϕ of this generator at rated conditions?
(d) How much field current is required to make the terminal voltage V_T equal to 13.8 kV when the generator is running at rated conditions?
(e) Suppose that this generator is running at rated conditions, and then the load is removed without changing the field current. What would the terminal voltage of the generator be?
(f) How much steady-state power and torque must the generator's prime mover be capable of supplying to handle the rated conditions?
(g) Construct a capability curve for this generator.

4–3. Assume that the field current of the generator in Problem 4–2 has been adjusted to a value of 5 A.
(a) What will the terminal voltage of this generator be if it is connected to a Δ-connected load with an impedance of $24 \angle 25° \ \Omega$?
(b) Sketch the phasor diagram of this generator.
(c) What is the efficiency of the generator at these conditions?
(d) Now assume that another identical Δ-connected load is to be paralleled with the first one. What happens to the phasor diagram for the generator?
(e) What is the new terminal voltage after the load has been added?
(f) What must be done to restore the terminal voltage to its original value?

4–4. Assume that the field current of the generator in Problem 4–2 is adjusted to achieve rated voltage (13.8 kV) at full-load conditions in each of the following questions.
(a) What is the efficiency of the generator at rated load?
(b) What is the voltage regulation of the generator if it is loaded to rated kilo-voltamperes with 0.9-PF-lagging loads?
(c) What is the voltage regulation of the generator if it is loaded to rated kilo-voltamperes with 0.9-PF-leading loads?
(d) What is the voltage regulation of the generator if it is loaded to rated kilo-voltamperes with unity-power-factor loads?
(e) Use MATLAB to plot the terminal voltage of the generator as a function of load for all three power factors.

4–5. Assume that the field current of the generator in Problem 4–2 has been adjusted so that it supplies rated voltage when loaded with rated current at unity power factor.
(a) What is the torque angle δ of the generator when supplying rated current at unity power factor?
(b) What is the maximum power that this generator can deliver to a unity power factor load when the field current is adjusted to the current value?
(c) When this generator is running at full load with unity power factor, how close is it to the static stability limit of the machine?

4–6. The internal generated voltage E_A of a Y-connected, three-phase synchronous generator is 14.4 kV, and the terminal voltage V_T is 12.8 kV. The synchronous reactance of this machine is 4 Ω, and the armature resistance can be ignored.
(a) If the torque angle of the generator $\delta = 18°$, how much power is being supplied by this generator at the current time?
(b) What is the power factor of the generator at this time?
(c) Sketch the phasor diagram under these circumstances.
(d) Ignoring losses in this generator, what torque must be applied to its shaft by the prime mover at these conditions?

4–7. A 100-MVA, 14.4–kV, 0.8-PF-lagging, 50-Hz, two-pole, Y-connected synchronous generator has a per-unit synchronous reactance of 1.1 and a per-unit armature resistance of 0.011.

(a) What are its synchronous reactance and armature resistance in ohms?

(b) What is the magnitude of the internal generated voltage E_A at the rated conditions? What is its torque angle δ at these conditions?

(c) Ignoring losses in this generator, what torque must be applied to its shaft by the prime mover at full load?

4–8. A 200-MVA, 12-kV, 0.85-PF-lagging, 50-Hz, 20-pole, Y-connected water turbine generator has a per-unit synchronous reactance of 0.9 and a per-unit armature resistance of 0.1. This generator is operating in parallel with a large power system (infinite bus).

(a) What is the speed of rotation of this generator's shaft?

(b) What is the magnitude of the internal generated voltage E_A at rated conditions?

(c) What is the torque angle of the generator at rated conditions?

(d) What are the values of the generator's synchronous reactance and armature resistance in ohms?

(e) If the field current is held constant, what is the maximum power possible out of this generator? How much reserve power or torque does this generator have at full load?

(f) At the absolute maximum power possible, how much reactive power will this generator be supplying or consuming? Sketch the corresponding phasor diagram. (Assume I_F is still unchanged.)

4–9. A 480-V, 250-kVA, 0.8-PF-lagging, two-pole, three-phase, 60-Hz synchronous generator's prime mover has a no-load speed of 3650 r/min and a full-load speed of 3570 r/min. It is operating in parallel with a 480-V, 250-kVA, 0.85-PF-lagging, four-pole, 60-Hz synchronous generator whose prime mover has a no-load speed of 1800 r/min and a full-load speed of 1780 r/min. The loads supplied by the two generators consist of 300 kW at 0.8 PF lagging.

(a) Calculate the speed droops of generator 1 and generator 2.

(b) Find the operating frequency of the power system.

(c) Find the power being supplied by each of the generators in this system.

(d) What must the generator's operators do to adjust the operating frequency to 60 Hz?

(e) If the current line voltage is 460 V, what must the generator's operators do to correct for the low terminal voltage?

4–10. Three physically identical synchronous generators are operating in parallel. They are all rated for a full load of 100 MW at 0.8 PF lagging. The no-load frequency of generator A is 61 Hz, and its speed droop is 3 percent. The no-load frequency of generator B is 61.5 Hz, and its speed droop is 3.4 percent. The no-load frequency of generator C is 60.5 Hz, and its speed droop is 2.6 percent.

(a) If a total load consisting of 230 MW is being supplied by this power system, what will the system frequency be, and how will the power be shared among the three generators?

(b) Create a plot showing the power supplied by each generator as a function of the total power supplied to all loads (you may use MATLAB to create this plot). At what load does one of the generators exceed its ratings? Which generator exceeds its ratings first?

(*c*) Is this power sharing in (*a*) acceptable? Why or why not?

(*d*) What actions could an operator take to improve the real power sharing among these generators?

4–11. A paper mill has installed three steam generators (boilers) to provide process steam and also to use some its waste products as an energy source. Since there is extra capacity, the mill has installed three 10-MW turbine generators to take advantage of the situation. Each generator is a 4160-V, 12.5 MVA, 60 Hz, 0.8-PF-lagging, two-pole, Y-connected synchronous generator with a synchronous reactance of 1.10 Ω and an armature resistance of 0.03 Ω. Generators 1 and 2 have a characteristic power–frequency slope s_P of 5 MW/Hz, and generators 3 has a slope of 6 MW/Hz.

(*a*) If the no-load frequency of each of the three generators is adjusted to 61 Hz, how much power will the three machines be supplying when the actual system frequency is 60 Hz?

(*b*) What is the maximum power the three generators can supply in this condition without the ratings of one of them being exceeded? At what frequency does this limit occur? How much power does each generator supply at that point?

(*c*) What would have to be done to get all three generators to supply their rated real and reactive powers at an overall operating frequency of 60 Hz?

(*d*) What would the internal generated voltages of the three generators be under this condition?

4–12. Suppose that you were an engineer planning a new electric co-generation facility for a plant with excess process steam. You have a choice of either two 10-MW turbine-generators or a single 20-MW turbine-generator. What would be the advantages and disadvantages of each choice?

4–13. A 25-MVA, 12.2-kV, 0.9-PF-lagging, three-phase, two-pole, Y-connected, 60-Hz synchronous generator was tested by the open-circuit test, and its air-gap voltage was extrapolated with the following results:

Open-circuit test					
Field current, A	320	365	380	475	570
Line voltage, kV	13.0	13.8	14.1	15.2	16.0
Extrapolated air-gap voltage, kV	15.4	17.5	18.3	22.8	27.4

The short-circuit test was then performed with the following results:

Short-circuit test					
Field current, A	320	365	380	475	570
Armature current, A	1040	1190	1240	1550	1885

The armature resistance is 0.6 Ω per phase.

(*a*) Find the unsaturated synchronous reactance of this generator in ohms per phase and ohms per unit.

(*b*) Find the approximate saturated synchronous reactance X_S at a field current of 380 A. Express the answer both in ohms per phase and per unit.

(*c*) Find the approximate saturated synchronous reactance at a field current of 475 A. Express the answer both in ohms per phase and in per unit.
(*d*) Find the short-circuit ratio for this generator.
(*e*) What is the internal generated voltage of this generator at rated conditions?
(*f*) What field current is required to achieve rated voltage at rated load?

4–14. During a short-circuit test, a Y-connected synchronous generator produces 100 A of short-circuit armature current per phase at a field current of 2.5 A. At the same field current, the open-circuit line voltage is measured to be 440 V.
(*a*) Calculate the saturated synchronous reactance under these conditions.
(*b*) If the armature resistance is 0.3 Ω per phase, and the generator supplies 60 A to a purely resistive Y-connected load of 3 Ω per phase at this field current setting, determine the voltage regulation under these load conditions.

4–15. A three-phase, Y-connected synchronous generator is rated 120 MVA, 13.8 kV, 0.8-PF-lagging, and 60 Hz. Its synchronous reactance is 1.2 Ω per phase, and its armature resistance is 0.1 Ω per phase.
(*a*) What is its voltage regulation?
(*b*) What would the voltage and apparent power rating of this generator be if it were operated at 50 Hz with the same armature and field losses as it had at 60 Hz?
(*c*) What would the voltage regulation of the generator be at 50 Hz?

Problems 4–16 to 4–26 refer to a six-pole, Y-connected synchronous generator rated at 1 MVA, 3.2 kV, 0.9 PF lagging, and 60 Hz. Its armature resistance R_A is 0.7 Ω. The core losses of this generator at rated conditions are 8 kW, and the friction and windage losses are 10 kW. The open-circuit and short-circuit characteristics are shown in Figure P4–2.

4–16. (*a*) What is the saturated synchronous reactance of this generator at the rated conditions?
(*b*) What is the unsaturated synchronous reactance of this generator?
(*c*) Plot the saturated synchronous reactance of this generator as a function of load.

4–17. (*a*) What are the rated current and internal generated voltage of this generator?
(*b*) What field current does this generator require to operate at the rated voltage, current, and power factor?

4–18. What is the voltage regulation of this generator at the rated current and power factor?

4–19. If this generator is operating at the rated conditions and the load is suddenly removed, what will the terminal voltage be?

4–20. What are the electrical losses in this generator at rated conditions?

4–21. If this machine is operating at rated conditions, what input torque must be applied to the shaft of this generator? Express your answer both in newton-meters and in pound-feet.

4–22. What is the torque angle δ of this generator at rated conditions?

4–23. Assume that the generator field current is adjusted to supply 3200 V under rated conditions. What is the static stability limit of this generator? (*Note:* You may ignore R_A to make this calculation easier.) How close is the full-load condition of this generator to the static stability limit?

4–24. Assume that the generator field current is adjusted to supply 3200 V under rated conditions. Plot the power supplied by the generator as a function of the torque angle δ.

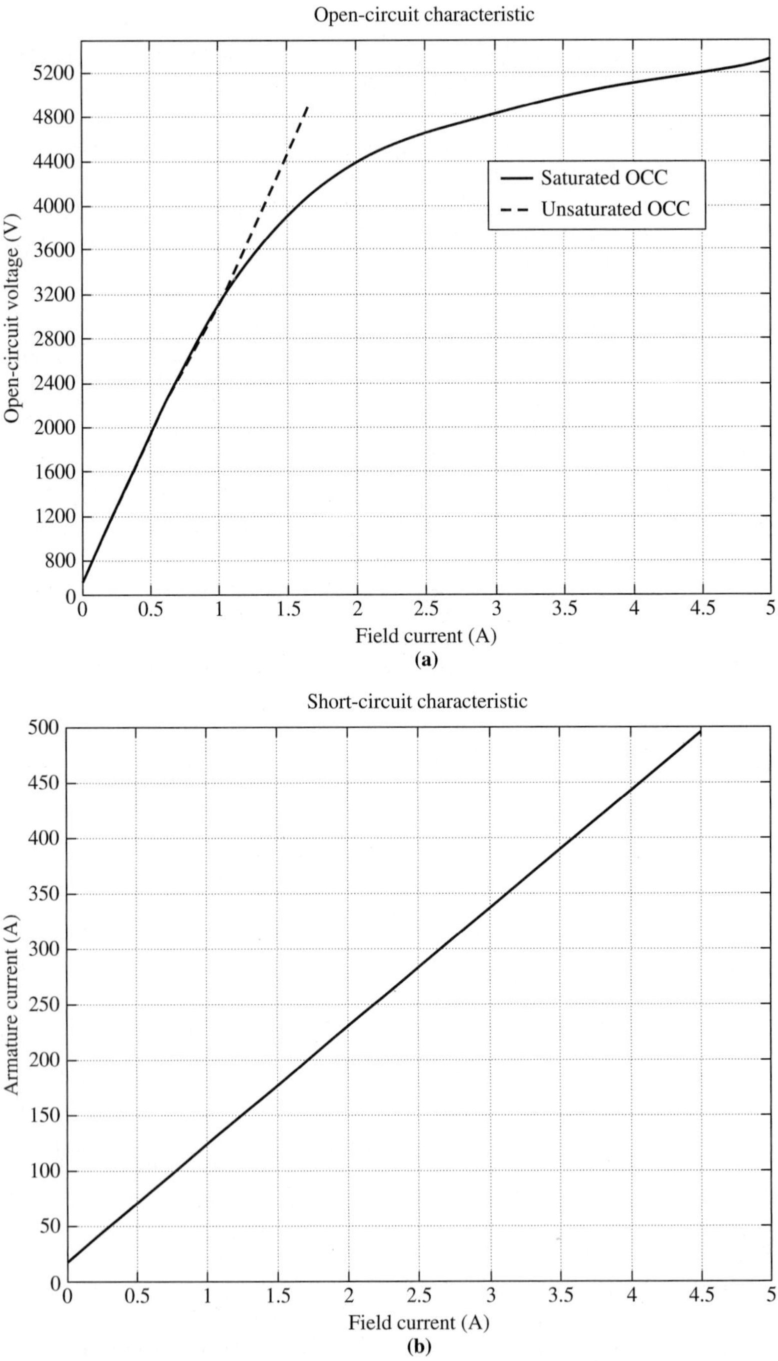

FIGURE P4–2
(a) Open-circuit characteristic curve for the generator in Problems 4–16 to 4–26. *(b)* Short-circuit characteristic curve for the generator in Problems 4–16 to 4–26.

4–25. Assume that the generator's field current is adjusted so that the generator supplies rated voltage at the rated load current and power factor. If the field current and the magnitude of the load current are held constant, how will the terminal voltage change as the load power factor varies from 0.9 PF lagging to 0.9 PF leading? Make a plot of the terminal voltage versus the load power factor.

4–26. Assume that the generator is connected to a 3200-V infinite bus, and that its field current has been adjusted so that it is supplying rated power and power factor to the bus. You may ignore the armature resistance R_A when answering the following questions.

(a) What will happen to the real and reactive power supplied by this generator if the field flux (and therefore E_A) is reduced by 5 percent?

(b) Plot the real power supplied by this generator as a function of the flux ϕ as the flux is varied from 80 percent to 100 percent of the flux at rated conditions.

(c) Plot the reactive power supplied by this generator as a function of the flux ϕ as the flux is varied from 80 percent to 100 percent of the flux at rated conditions.

(d) Plot the line current supplied by this generator as a function of the flux ϕ as the flux is varied from 80 percent to 100 percent of the flux at rated conditions.

4–27. Two identical 2.5-MVA, 1200-V 0.8-PF-lagging, 60-Hz, three-phase synchronous generators are connected in parallel to supply a load. The prime movers of the two generators happen to have different speed droop characteristics. When the field currents of the two generators are equal, one delivers 1200 A at 0.9 PF lagging, while the other delivers 900 A at 0.75 PF lagging.

(a) What are the real power and the reactive power supplied by each generator to the load?

(b) What is the overall power factor of the load?

(c) In what direction must the field current on each generator be adjusted in order for them to operate at the same power factor?

4–28. A generating station for a power system consists of four 300-MVA, 15-kV, 0.85-PF-lagging synchronous generators with identical speed droop characteristics operating in parallel. The governors on the generators' prime movers are adjusted to produce a 3-Hz drop from no load to full load. Three of these generators are each supplying a steady 200 MW at a frequency of 60 Hz, while the fourth generator (called the *swing generator*) handles all incremental load changes on the system while maintaining the system's frequency at 60 Hz.

(a) At a given instant, the total system loads are 650 MW at a frequency of 60 Hz. What are the no-load frequencies of each of the system's generators?

(b) If the system load rises to 725 MW and the generator's governor set points do not change, what will the new system frequency be?

(c) To what frequency must the no-load frequency of the swing generator be adjusted in order to restore the system frequency to 60 Hz?

(d) If the system is operating at the conditions described in part *(c)*, what would happen if the swing generator were tripped off the line (disconnected from the power line)?

4–29. A 100-MVA, 14.4-kV, 0.8-PF-lagging, Y-connected synchronous generator has a negligible armature resistance and a synchronous reactance of 1.0 per unit. The generator is connected in parallel with a 60-Hz, 14.4-kV infinite bus that is capable of supplying or consuming any amount of real or reactive power with no change in frequency or terminal voltage.

(a) What is the synchronous reactance of the generator in ohms?

(b) What is the internal generated voltage $\mathbf{E}_A$ of this generator under rated conditions?

(c) What is the armature current $\mathbf{I}_A$ in this machine at rated conditions?
(d) Suppose that the generator is initially operating at rated conditions. If the internal generated voltage $\mathbf{E}_A$ is decreased by 5 percent, what will the new armature current $\mathbf{I}_A$ be?
(e) Repeat part *(d)* for 10, 15, 20, and 25 percent reductions in $\mathbf{E}_A$.
(f) Plot the magnitude of the armature current I_A as a function of E_A. (You may wish to use MATLAB to create this plot.)

REFERENCES

1. Chaston, A. N.: *Electric Machinery,* Reston Publishing, Reston, Va., 1986.
2. Del Toro, V.: *Electric Machines and Power Systems*, Prentice-Hall, Englewood Cliffs, N.J., 1985.
3. Fitzgerald, A. E., and C. Kingsley, Jr.: *Electric Machinery*, McGraw-Hill Book Company, New York, 1952.
4. Fitzgerald, A. E., C. Kingsley, Jr., and S. D. Umans: *Electric Machinery*, 5th ed., McGraw-Hill Book Company, New York, 1990.
5. Kosow, Irving L.: *Electric Machinery and Transformers*, Prentice-Hall, Englewood Cliffs, N.J., 1972.
6. Liwschitz-Garik, Michael, and Clyde Whipple: *Alternating-Current Machinery*, Van Nostrand, Princeton, N.J., 1961.
7. McPherson, George: *An Introduction to Electrical Machines and Transformers*, Wiley, New York, 1981.
8. Slemon, G. R., and A. Straughen: *Electric Machines*, Addison-Wesley, Reading, Mass., 1980.
9. Werninck, E. H. (ed.): *Electric Motor Handbook*, McGraw-Hill Book Company, London, 1978.

CHAPTER 5

SYNCHRONOUS MOTORS

LEARNING OBJECTIVES

- Understand the equivalent circuit of a synchronous motor.
- Be able to sketch phasor diagrams for a synchronous motor.
- Know the equations for power and torque in a synchronous motor.
- Understand how and why power factor varies as synchronous motor load increases.
- Understand how and why power factor varies as synchronous motor field current varies—the "V" curve.
- Understand how synchronous motors can be started.
- Be able to tell whether a synchronous machine is acting as a motor or a generator and whether it is supplying or consuming reactive power by examining its phasor diagram.
- Understand synchronous motor ratings.

Synchronous motors are synchronous machines used to convert electrical power to mechanical power. This chapter explores the basic operation of synchronous motors and relates their behavior to that of synchronous generators.

5.1 BASIC PRINCIPLES OF MOTOR OPERATION

To understand the basic concept of a synchronous motor, look at Figure 5–1, which shows a two-pole synchronous motor. The field current I_F of the motor produces a steady-state magnetic field $\mathbf{B}_R$. A three-phase set of voltages is applied to the stator of the machine, which produces a three-phase current flow in the windings.

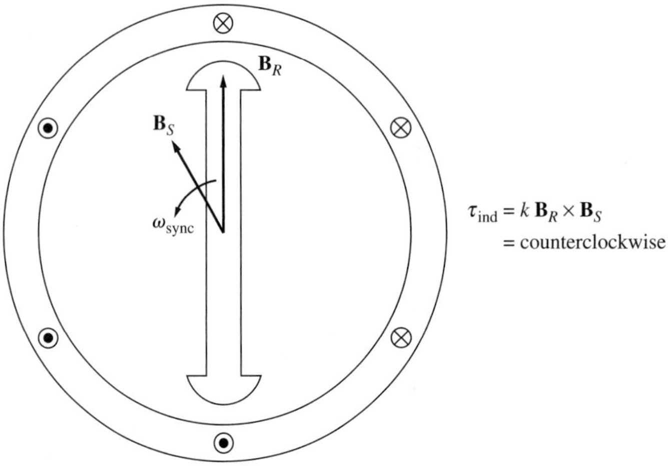

FIGURE 5–1
A two-pole synchronous motor.

As was shown in Chapter 3, a three-phase set of currents in an armature winding produces a uniform rotating magnetic field $\mathbf{B}_S$. Therefore, there are two magnetic fields present in the machine, and the *rotor field will tend to line up with the stator field*, just as two bar magnets will tend to line up if placed near each other. Since the stator magnetic field is rotating, the rotor magnetic field (and the rotor itself) will constantly try to catch up. The larger the angle between the two magnetic fields (up to a certain maximum), the greater the torque on the rotor of the machine. The basic principle of synchronous motor operation is that the rotor "chases" the rotating stator magnetic field around in a circle, never quite catching up with it.

Since a synchronous motor is the same physical machine as a synchronous generator, all of the basic speed, power, and torque equations of Chapters 3 and 4 apply to synchronous motors also.

The Equivalent Circuit of a Synchronous Motor

A synchronous motor is the same in all respects as a synchronous generator, except that the direction of power flow is reversed. Since the direction of power flow in the machine is reversed, the direction of current flow in the stator of the motor may be expected to reverse also. Therefore, the equivalent circuit of a synchronous motor is exactly the same as the equivalent circuit of a synchronous generator, *except* that the reference direction of $\mathbf{I}_A$ is *reversed*. The resulting full equivalent circuit is shown in Figure 5–2a, and the per-phase equivalent circuit is shown in Figure 5–2b. As before, the three phases of the equivalent circuit may be either Y- or Δ-connected.

Because of the change in direction of $\mathbf{I}_A$, the Kirchhoff's voltage law equation for the equivalent circuit changes, too. Writing a Kirchhoff's voltage law equation for the new equivalent circuit yields

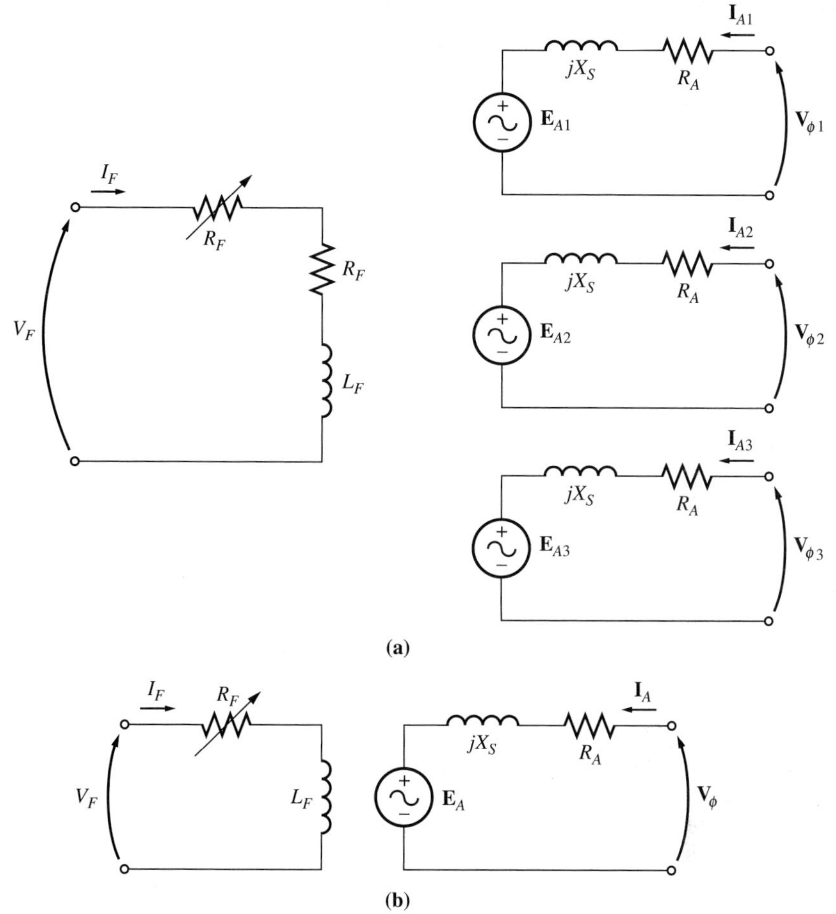

FIGURE 5–2
(a) The full equivalent circuit of a three-phase synchronous motor. (b) The per-phase equivalent circuit.

$$\boxed{\mathbf{V}_\phi = \mathbf{E}_A + jX_S\mathbf{I}_A + R_A\mathbf{I}_A} \tag{5–1}$$

or

$$\boxed{\mathbf{E}_A = \mathbf{V}_\phi - jX_S\mathbf{I}_A - R_A\mathbf{I}_A} \tag{5–2}$$

This is exactly the same as the equation for a generator, except that the sign on the current term has been reversed.

The Synchronous Motor from a Magnetic Field Perspective

To begin to understand synchronous motor operation, take another look at a synchronous generator connected to an infinite bus. The generator has a prime mover turning its shaft, causing it to rotate. The direction of the applied torque τ_{app} from the prime mover is in the direction of motion, because the prime mover makes the generator rotate in the first place.

The phasor diagram of the generator operating with a large field current is shown in Figure 5–3a, and the corresponding magnetic field diagram is shown in Figure 5–3b. As described before, $\mathbf{B}_R$ corresponds to (produces) $\mathbf{E}_A$, $\mathbf{B}_{\text{net}}$ corresponds to (produces) $\mathbf{V}_\phi$, and $\mathbf{B}_S$ corresponds to $\mathbf{E}_{\text{stat}}$ $(= -jX_S\mathbf{I}_A)$. The rotation of both the phasor diagram and magnetic field diagram is counterclockwise in the figure, following the standard mathematical convention of increasing angle.

The induced torque in the generator can be found from the magnetic field diagram. From Equations (3–60) and (3–61) the induced torque is given by

$$\tau_{\text{ind}} = k\mathbf{B}_R \times \mathbf{B}_{\text{net}} \tag{3–60}$$

or

$$\tau_{\text{ind}} = kB_R B_{\text{net}} \sin \delta \tag{3–61}$$

Notice that from the magnetic field diagram *the induced torque in this machine is clockwise*, opposing the direction of rotation. In other words, the induced torque in the generator is a countertorque, opposing the rotation caused by the external applied torque τ_{app}.

Suppose that, instead of turning the shaft in the direction of motion, the prime mover suddenly loses power and starts to drag on the machine's shaft. What happens to the machine now? The rotor slows down because of the drag on its shaft and falls behind the net magnetic field in the machine (see Figure 5–4a). As the rotor, and therefore $\mathbf{B}_R$, slows down and falls behind $\mathbf{B}_{\text{net}}$ the operation of the machine suddenly changes. By Equation (3–60), when $\mathbf{B}_R$ is behind $\mathbf{B}_{\text{net}}$, the induced torque's direction reverses and becomes counterclockwise. In other words, the machine's torque is now in the direction of motion, and the machine is acting as a motor. The increasing torque angle δ results in a larger and larger torque in the direction of rotation, until eventually the motor's induced torque equals the load

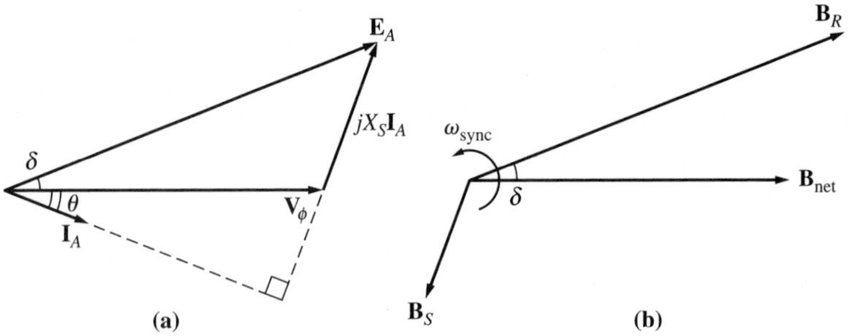

FIGURE 5–3
(a) Phasor diagram of a synchronous generator operating at a lagging power factor. (b) The corresponding magnetic field diagram.

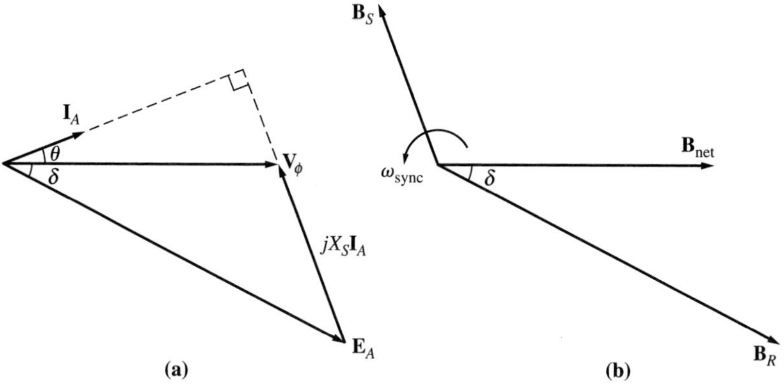

FIGURE 5–4
(a) Phasor diagram of a synchronous motor. (b) The corresponding magnetic field diagram.

torque on its shaft. At that point, the machine will be operating at steady state and synchronous speed again, but now as a motor.

The phasor diagram corresponding to generator operation is shown in Figure 5–3a, and the phasor diagram corresponding to motor operation is shown in Figure 5–4a. The reason that the quantity $jX_S\mathbf{I}_A$ points from $\mathbf{V}_\phi$, to $\mathbf{E}_A$ in the generator and from $\mathbf{E}_A$ to $\mathbf{V}_\phi$ in the motor is that the reference direction of $\mathbf{I}_A$ was reversed in the definition of the motor equivalent circuit. The basic difference between motor and generator operation in synchronous machines can be seen either in the magnetic field diagram or in the phasor diagram. *In a generator*, $\mathbf{E}_A$ lies ahead of $\mathbf{V}_\phi$, and $\mathbf{B}_R$ lies ahead of $\mathbf{B}_{\text{net}}$. *In a motor*, $\mathbf{E}_A$ lies behind $\mathbf{V}_\phi$, and $\mathbf{B}_R$ lies behind $\mathbf{B}_{\text{net}}$. In a motor the induced torque is in the direction of motion, and in a generator the induced torque is a countertorque opposing the direction of motion.

5.2 STEADY-STATE SYNCHRONOUS MOTOR OPERATION

This section explores the behavior of synchronous motors under varying conditions of load and field current as well as the question of power-factor correction with synchronous motors. The following discussions will generally ignore the armature resistance of the motors for simplicity. However, R_A will be considered in some of the worked numerical calculations.

The Synchronous Motor Torque-Speed Characteristic Curve

Synchronous motors supply power to loads that are basically constant-speed devices. They are usually connected to power systems *very* much larger than the individual motors, so the power systems appear as infinite buses to the motors. This means that

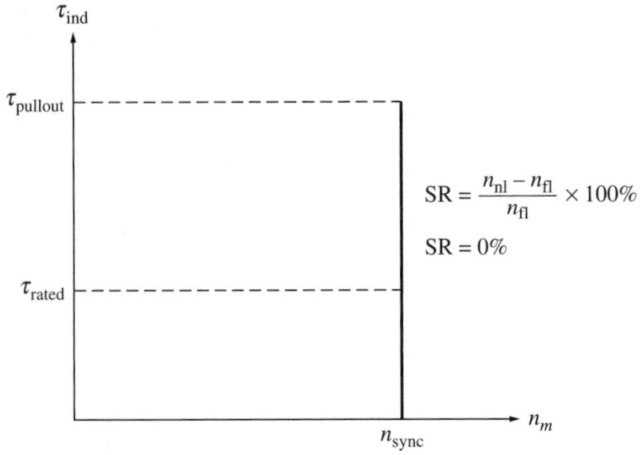

FIGURE 5–5
The torque–speed characteristic of a synchronous motor. Since the speed of the motor is constant, its speed regulation is zero.

the terminal voltage and the system frequency will be constant regardless of the amount of power drawn by the motor. The speed of rotation of the motor is locked to the rate of rotation of the magnetic fields, and the rate of rotation of the applied mechanical fields is locked to the applied electrical frequency, so *the speed of the synchronous motor will be constant regardless of the load.* This fixed rate of rotation is given

$$n_m = \frac{120 f_{se}}{P} \tag{5–3}$$

where n_m is the mechanical rate of rotation, f_{se} is the stator electrical frequency, and P is the number of poles in the motor.

The resulting torque–speed characteristic curve is shown in Figure 5–5. The steady-state speed of the motor is constant from no load all the way up to the maximum torque that the motor can supply (called the *pullout torque*), so the speed regulation of this motor [Equation (3–68)] is 0%. The torque equation is

$$\tau_{\text{ind}} = kB_R B_{\text{net}} \sin \delta \tag{3–61}$$

or

$$\boxed{\tau_{\text{ind}} = \frac{3V_\phi E_A \sin \delta}{\omega_m X_S}} \tag{4–22}$$

The maximum or pullout torque occurs when $\delta = 90°$. Normal full-load torques are much less than that, however. In fact, the pullout torque may typically be three times the full-load torque of the machine.

When the torque on the shaft of a synchronous motor exceeds the pullout torque, the rotor can no longer remain locked to the stator and net magnetic fields. Instead, the rotor starts to slip behind them. As the rotor slows down, the stator magnetic field "laps" it repeatedly, and the direction of the induced torque in the rotor reverses with each pass. The resulting huge torque surges, first one way and then the other way, cause the whole motor to vibrate severely. The loss of synchronization after the pullout torque is exceeded is known as *slipping poles.*

The maximum or pullout torque of the motor is given by

$$\tau_{\max} = kB_R B_{\text{net}} \tag{5–4a}$$

or

$$\boxed{\tau_{\max} = \frac{3V_\phi E_A}{\omega_m X_S}} \tag{5–4b}$$

These equations indicate that the larger the field current (and hence E_A), *the greater the maximum torque of the motor.* There is therefore a stability advantage in operating the motor with a large field current or a large E_A.

The Effect of Load Changes on a Synchronous Motor

If a load is attached to the shaft of a synchronous motor, the motor will develop enough torque to keep the motor and its load turning at a synchronous speed. What happens when the load is changed on a synchronous motor?

To find out, examine a synchronous motor operating initially with a leading power factor, as shown in Figure 5–6. If the load on the shaft of the motor is increased, the rotor will initially slow down. As it does, the torque angle δ becomes larger, and the induced torque increases. The increase in induced torque eventually speeds the rotor back up, and the motor again turns at synchronous speed but with a larger torque angle δ.

What does the phasor diagram look like during this process? To find out, examine the constraints on the machine during a load change. Figure 5–6a shows the motor's phasor diagram before the loads are increased. The internal generated voltage E_A is equal to $K\phi\omega$ and so depends on *only* the field current in the machine and the speed of the machine. The speed is constrained to be constant by the input power supply, and since no one has touched the field circuit, the field current is constant as well. Therefore, $|\mathbf{E}_A|$ *must be constant as the load changes.* The distances proportional to power ($E_A \sin \delta$ and $I_A \cos \theta$) will increase, but the magnitude of $\mathbf{E}_A$ must remain constant. As the load increases, $\mathbf{E}_A$ swings down in the manner shown in Figure 5–6b. As $\mathbf{E}_A$ swings down further and further, the quantity $jX_S\mathbf{I}_A$ has to increase to reach from the tip of $\mathbf{E}_A$ to $\mathbf{V}_\phi$, and therefore the armature current $\mathbf{I}_A$ also increases. Notice that the power-factor angle θ changes too, becoming less and less leading and then more and more lagging.

Example 5–1. A 208-V, 45-hp, 0.8-PF-leading, Δ-connected, 60-Hz synchronous machine has a synchronous reactance of 2.5 Ω and a negligible armature resistance. Its friction and windage losses are 1.5 kW, and its core losses are 1.0 kW. Initially, the shaft is supplying a 15-hp load, and the motor's power factor is 0.80 leading.

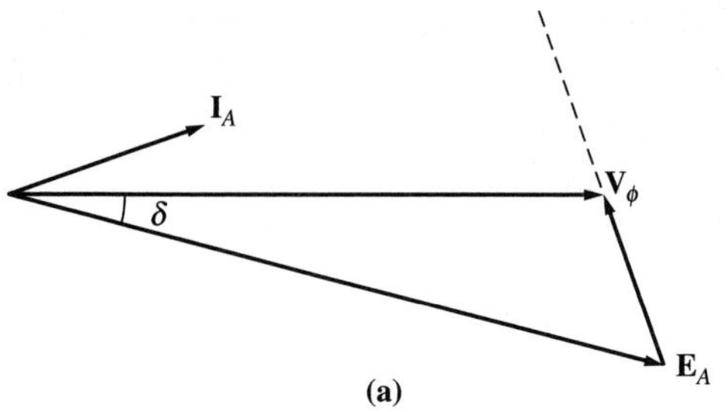

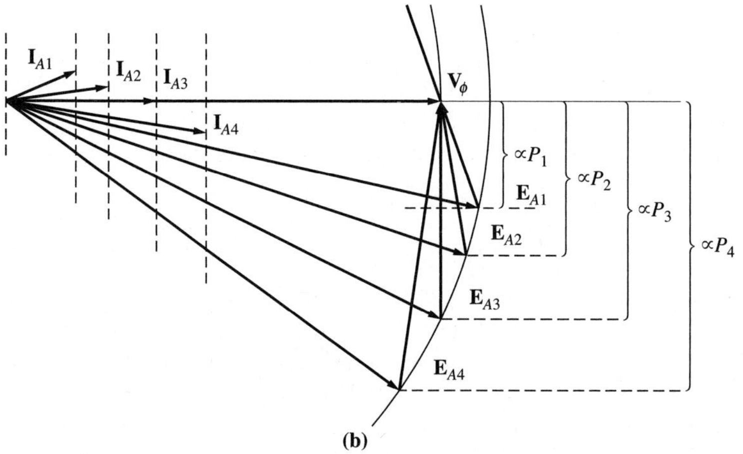

FIGURE 5–6
(a) Phasor diagram of a motor operating at a leading power factor. (b) The effect of an increase in load on the operation of a synchronous motor.

(a) Sketch the phasor diagram of this motor, and find the values of $\mathbf{I}_A$, I_L, and $\mathbf{E}_A$.
(b) Assume that the shaft load is now increased to 30 hp. Sketch the behavior of the phasor diagram in response to this change.
(c) Find $\mathbf{I}_A$, I_L, and $\mathbf{E}_A$ after the load change. What is the new motor power factor?

Solution

(a) Initially, the motor's output power is 15 hp. This corresponds to an output of

$$P_{\text{out}} = (15 \text{ hp})(0.746 \text{ KW/hp}) = 11.19 \text{ kW}$$

Therefore, the electric power supplied to the machine is

$$\begin{aligned} P_{\text{in}} &= P_{\text{out}} + P_{\text{mech loss}} + P_{\text{core loss}} + P_{\text{elec loss}} \\ &= 11.19 \text{ kW} + 1.5 \text{ kW} + 1.0 \text{ kW} + 0 \text{ kW} = 13.69 \text{ kW} \end{aligned}$$

Since the motor's power factor is 0.80 leading, the resulting line current flow is

$$\begin{aligned} I_L &= \frac{P_{\text{in}}}{\sqrt{3}\, V_T \cos \theta} \\ &= \frac{13.69 \text{ kW}}{\sqrt{3}(208 \text{ V})(0.80)} = 47.5 \text{ A} \end{aligned}$$

and the armature current is $I_L/\sqrt{3}$, with 0.8 leading power factor, which gives the result

$$\mathbf{I}_A = 27.4 \angle 36.87° \text{ A}$$

To find $\mathbf{E}_A$, apply Kirchhoff's voltage law [Equation (5–2)]:

$$\begin{aligned}\mathbf{E}_A &= \mathbf{V}_\phi - jX_S\mathbf{I}_A \\ &= 208 \angle 0° \text{ V} - (j2.5\ \Omega)(27.4 \angle 36.87° \text{ A}) \\ &= 208 \angle 0° \text{ V} - 68.5 \angle 126.87° \text{ V} \\ &= 249.1 - j54.8 \text{ V} = 255 \angle -12.4° \text{ V}\end{aligned}$$

The resulting phasor diagram is shown in Figure 5–7a.

(b) As the power on the shaft is increased to 30 hp, the shaft slows momentarily, and the internal generated voltage $\mathbf{E}_A$ swings out to a larger angle δ while maintaining a constant magnitude. The resulting phasor diagram is shown in Figure 5–7b.

(c) After the load changes, the electric input power of the machine becomes

$$\begin{aligned}P_{\text{in}} &= P_{\text{out}} + P_{\text{mech loss}} + P_{\text{core loss}} + P_{\text{elec loss}} \\ &= (30 \text{ hp})(0.746 \text{ kW/hp}) + 1.5 \text{ kW} + 1.0 \text{ kW} + 0 \text{ kW} \\ &= 24.88 \text{ kW}\end{aligned}$$

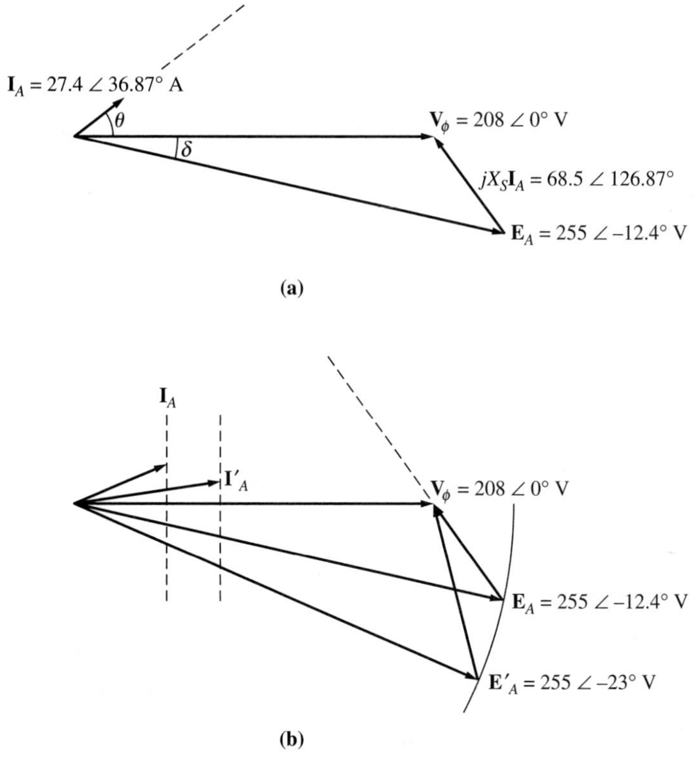

FIGURE 5–7
(a) The motor phasor diagram for Example 5–1a. (b) The motor phasor diagram for Example 5–1b.

From the equation for power in terms of torque angle [Equation (4–20)], it is possible to find the magnitude of the angle δ (remember that the magnitude of $\mathbf{E}_A$ is constant):

$$P = \frac{3V_\phi E_A \sin\delta}{X_S} \tag{4–20}$$

so

$$\delta = \sin^{-1}\frac{X_S P}{3V_\phi E_A}$$

$$= \sin^{-1}\frac{(2.5\ \Omega)(24.88\ \text{kW})}{3(208\ \text{V})(255\ \text{V})}$$

$$= \sin^{-1} 0.391 = 23°$$

The internal generated voltage thus becomes $\mathbf{E}_A = 355\ \angle -23°$ V. Therefore, $\mathbf{I}_A$ will be given by

$$\mathbf{I}_A = \frac{\mathbf{V}_\phi - \mathbf{E}_A}{jX_S}$$

$$= \frac{208\ \angle 0°\ \text{V} - 255\ \angle -23°\ \text{V}}{j2.5\ \Omega}$$

$$= \frac{103.1\ \angle 105°\ \text{V}}{j2.5\ \Omega} = 41.2\ \angle 15°\ \text{A}$$

and I_L will become

$$I_L = \sqrt{3} I_A = 71.4\ \text{A}$$

The final power factor will be cos (−15°) or 0.966 leading.

The Effect of Field Current Changes on a Synchronous Motor

We have seen how a change in shaft load on a synchronous motor affects the motor. There is one other quantity on a synchronous motor that can be readily adjusted—its field current. What effect does a change in field current have on a synchronous motor?

To find out, look at Figure 5–8. Figure 5–8a shows a synchronous motor initially operating at a lagging power factor. Now, increase its field current and see what happens to the motor. Note that *an increase in field current increases the magnitude of* $\mathbf{E}_A$ *but does not affect the real power supplied by the motor*. The power supplied by the motor changes only when the shaft load torque changes. Since a change in I_F does not affect the shaft speed n_m, and since the load attached to the shaft is unchanged, the real power supplied is unchanged. Of course, V_T is also constant, since it is kept constant by the power source supplying the motor. The distances proportional to power on the phasor diagram ($E_A \sin\delta$ and $I_A \cos\theta$) must therefore be constant. When the field current is increased, $\mathbf{E}_A$ must increase, but it can only do so by sliding out along the line of constant power. This effect is shown in Figure 5–8b.

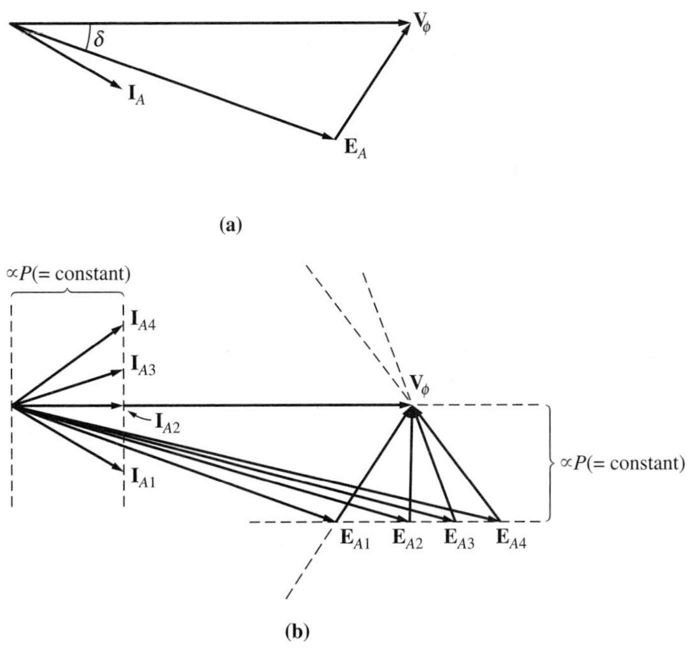

FIGURE 5–8
(a) A synchronous motor operating at a lagging power factor. (b) The effect of an increase in field current on the operation of this motor.

Notice that as the value of $\mathbf{E}_A$ increases, the magnitude of the armature current $\mathbf{I}_A$ first decreases and then increases again. At low $\mathbf{E}_A$, the armature current is lagging, and the motor is an inductive load. It is acting like an inductor-resistor combination, consuming reactive power Q. As the field current is increased, the armature current eventually lines up with $\mathbf{V}_\phi$, and the motor looks purely resistive. As the field current is increased further, the armature current becomes leading, and the motor becomes a capacitive load. It is now acting like a capacitor-resistor combination, consuming negative reactive power $-Q$ or, alternatively, supplying reactive power Q to the system.

A plot of I_A versus I_F for a synchronous motor is shown in Figure 5–9. Such a plot is called a *synchronous motor V curve*, for the obvious reason that it is shaped like the letter V. There are several V curves drawn, corresponding to different real power levels. For each curve, the minimum armature current occurs at unity power factor, when only real power is being supplied to the motor. At any other point on the curve, some reactive power is being supplied to or by the motor as well. For field currents *less* than the value giving minimum I_A, the armature current is lagging, consuming Q. For field currents *greater* than the value giving the minimum I_A, the armature current is leading, supplying Q to the power system as a capacitor would. Therefore, by controlling the field current of a synchronous

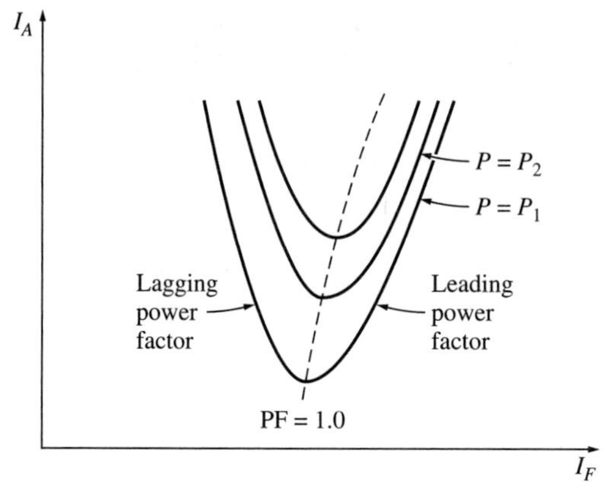

FIGURE 5–9
Synchronous motor V curves.

motor, the *reactive power* supplied to or consumed by the power system can be controlled.

When the projection of the phasor $\mathbf{E}_A$ onto $\mathbf{V}_\phi$ ($E_A \cos \delta$) is *shorter* than $\mathbf{V}_\phi$ itself, a synchronous motor has a lagging current and consumes Q. Since the field current is small in this situation, the motor is said to be *underexcited*. On the other hand, when the projection of $\mathbf{E}_A$ onto $\mathbf{V}_\phi$ is *longer* than $\mathbf{V}_\phi$ itself, a synchronous motor has a leading current and supplies Q to the power system. Since the field current is large in this situation, the motor is said to be *overexcited*. Phasor diagrams illustrating these concepts are shown in Figure 5–10.

Example 5–2. The 208-V, 45-hp, 0.8-PF-leading, Δ-connected, 60-Hz synchronous motor of the previous example is supplying a 15-hp load with an initial power factor of 0.85 PF lagging. The field current I_F at these conditions is 4.0 A.

(a) Sketch the initial phasor diagram of this motor, and find the values $\mathbf{I}_A$ and $\mathbf{E}_A$.
(b) If the motor's flux is increased by 25 percent, sketch the new phasor diagram of the motor. What are $\mathbf{E}_A$, $\mathbf{I}_A$, and the power factor of the motor now?

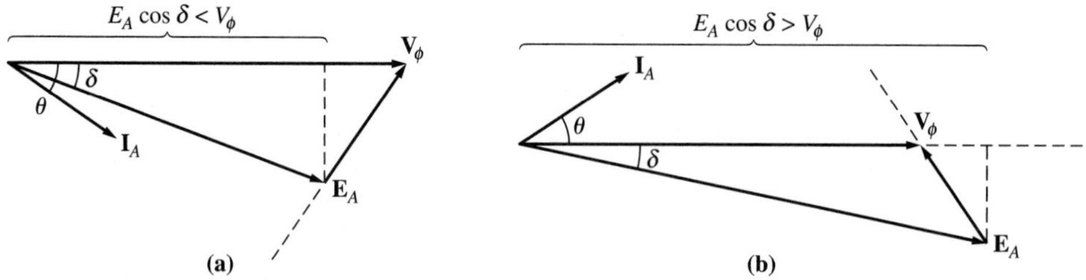

FIGURE 5–10
(a) The phasor diagram of an *underexcited* synchronous motor. (b) The phasor diagram of an *overexcited* synchronous motor.

(*c*) Assume that the flux in the motor varies linearly with the field current I_F. Make a plot of I_A versus I_F for the synchronous motor with a 15-hp load.

Solution

(*a*) From the previous example, the electric input power with all the losses included is $P_{in} = 13.69$ kW. Since the motor's power factor is 0.85 lagging, the resulting armature current flow is

$$I_A = \frac{P_{in}}{3V_\phi \cos\theta}$$

$$= \frac{13.69 \text{ kW}}{3(208 \text{ V})(0.85)} = 25.8 \text{ A}$$

The angle θ is $\cos^{-1} 0.85 = 31.8°$, so the phasor current I_A is equal to

$$\mathbf{I}_A = 25.8 \angle -31.8° \text{ A}$$

To find $\mathbf{E}_A$, apply Kirchhoff's voltage law [Equation (5–2)]:

$$\mathbf{E}_A = \mathbf{V}_\phi - jX_S\mathbf{I}_A$$

$$= 208 \angle 0° \text{ V} - (j2.5\ \Omega)(25.8 \angle -31.8° \text{ A})$$

$$= 208 \angle 0° \text{ V} - 64.5 \angle 58.2° \text{ V}$$

$$= 182 \angle -17.5° \text{ V}$$

The resulting phasor diagram is shown in Figure 5–11, together with the results for part *b*.

(*b*) If the flux ϕ is increased by 25 percent, then $E_A = K\phi\omega$ will increase by 25 percent too:

$$E_{A2} = 1.25\ E_{A1} = 1.25(182 \text{ V}) = 227.5 \text{ V}$$

However, the power supplied to the load must remain constant. Since the distance $E_A \sin\delta$ is proportional to the power, that distance on the phasor diagram must be constant from the original flux level to the new flux level. Therefore,

$$E_{A1} \sin\delta_1 = E_{A2} \sin\delta_2$$

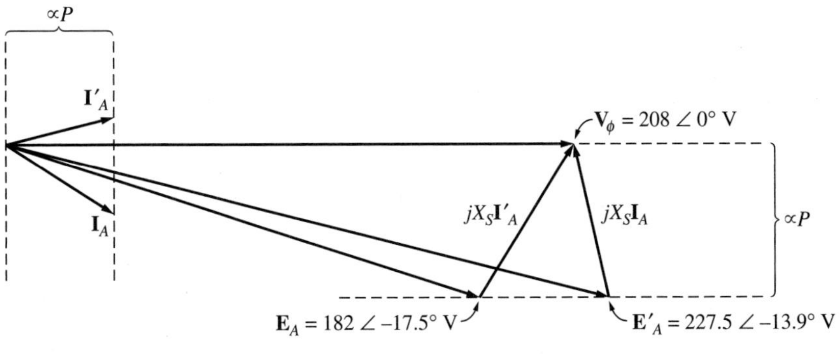

FIGURE 5–11
The phasor diagram of the motor in Example 5–2.

$$\delta_2 = \sin^{-1}\left(\frac{E_{A1}}{E_{A2}} \sin \delta_1\right)$$

$$= \sin^{-1}\left[\frac{182 \text{ V}}{227.5 \text{ V}} \sin(-17.5°)\right] = -13.9°$$

The armature current can now be found from Kirchhoff's voltage law:

$$\mathbf{I}_{A2} = \frac{\mathbf{V}_\phi - \mathbf{E}_{A2}}{jX_S}$$

$$\mathbf{I}_A = \frac{208 \angle 0° \text{ V} - 227.5 \angle -13.9° \text{ V}}{j2.5 \ \Omega}$$

$$= \frac{56.2 \angle 103.2° \text{ V}}{j2.5 \ \Omega} = 22.5 \angle 13.2° \text{ A}$$

Finally, the motor's power factor is now

$$\text{PF} = \cos(13.2°) = 0.974 \qquad \text{leading}$$

The resulting phasor diagram is also shown in Figure 5–11.

(*c*) Because the flux is assumed to vary linearly with field current, E_A will also vary linearly with field current. We know that E_A is 182 V for a field current of 4.0 A, so E_A for any given field current can be found from the ratio

$$\frac{E_{A2}}{182 \text{ V}} = \frac{I_{F2}}{4.0 \text{ A}}$$

or

$$E_{A2} = 45.5 \, I_{F2} \tag{5–5}$$

The torque angle δ for any given field current can be found from the fact that the power supplied to the load must remain constant:

$$E_{A1} \sin \delta_1 = E_{A2} \sin \delta_2$$

so

$$\delta_2 = \sin^{-1}\left(\frac{E_{A1}}{E_{A2}} \sin \delta_1\right) \tag{5–6}$$

These two pieces of information give us the phasor voltage $\mathbf{E}_A$. Once $\mathbf{E}_A$ is available, the new armature current can be calculated from Kirchhoff's voltage law:

$$\mathbf{I}_{A2} = \frac{\mathbf{V}_\phi - \mathbf{E}_{A2}}{jX_S} \tag{5–7}$$

A MATLAB M-file to calculate and plot I_A versus I_F using Equations (5–5) through (5–7) is shown below:

```
% M-file: v_curve.m
% M-file create a plot of armature current versus field
% current for the synchronous motor of Example 5-2.

% First, initialize the field current values (21 values
% in the range 3.8-5.8 A)
i_f = (38:1:58) / 10;

% Now initialize all other values
i_a = zeros(1,21);              % Pre-allocate i_a array
x_s = 2.5;                      % Synchronous reactance
v_phase = 208;                  % Phase voltage at 0 degrees
```

```
delta1 = -17.5 * pi/180;          % delta 1 in radians
e_a1 = 182 * (cos(delta1) + j * sin(delta1));

% Calculate the armature current for each value
for ii = 1:21
    % Calculate magnitude of e_a2
    e_a2 = 45.5 * i_f(ii);

    % Calculate delta2
    delta2 = asin ( abs(e_a1) / abs(e_a2) * sin(delta1) );

    % Calculate the phasor e_a2
    e_a2 = e_a2 * (cos(delta2) + j * sin(delta2));

    % Calculate i_a
    i_a(ii) = ( v_phase - e_a2 ) / ( j * x_s);
end

% Plot the v-curve
plot(i_f,abs(i_a),'Color','k','Linewidth',2.0);
xlabel('Field Current (A)','Fontweight','Bold');
ylabel('Armature Current (A)','Fontweight','Bold');
title ('Synchronous Motor V-Curve','Fontweight','Bold');
grid on;
```

The plot produced by this M-file is shown in Figure 5–12. Note that for a field current of 4.0 A, the armature current is 25.8 A. This result agrees with part *a* of this example.

The Synchronous Motor and Power-Factor Correction

Figure 5–13 shows an infinite bus whose output is connected through a transmission line to an industrial plant at a distant point. The industrial plant shown consists

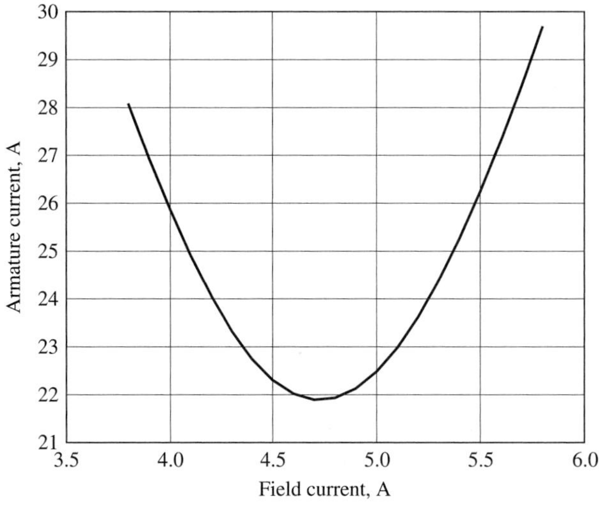

FIGURE 5–12
V curve for the synchronous motor of Example 5–2.

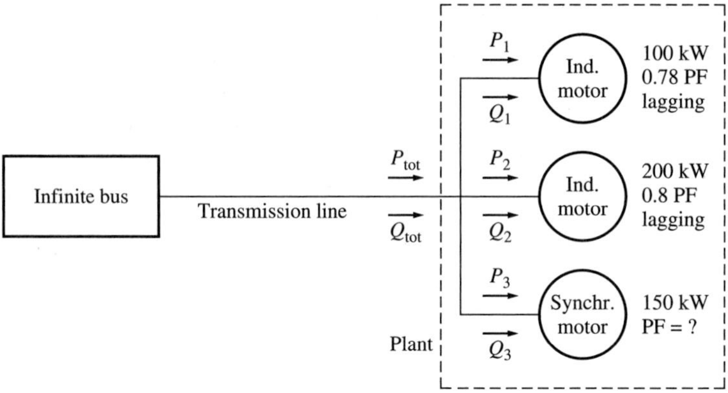

FIGURE 5–13
A simple power system consisting of an infinite bus supplying an industrial plant through a transmission line.

of three loads. Two of the loads are induction motors with lagging power factors, and the third load is a synchronous motor with a variable power factor.

What does the ability to set the power factor of one of the loads do for the power system? To find out, examine the following example problem. (*Note:* A review of the three-phase power equations and their uses is given in Appendix A. Some readers may wish to consult it when studying this problem.)

Example 5–3. The infinite bus in Figure 5–13 operates at 480 V. Load 1 is an induction motor consuming 100 kW at 0.78 PF lagging, and load 2 is an induction motor consuming 200 kW at 0.8 PF lagging. Load 3 is a synchronous motor whose real power consumption is 150 kW.

(a) If the synchronous motor is adjusted to operate at 0.85 PF lagging, what is the transmission line current in this system?
(b) If the synchronous motor is adjusted to operate at 0.85 PF leading, what is the transmission line current in this system?
(c) Assume that the transmission line losses are given by

$$P_{\text{LL}} = 3I_L^2R_L \qquad \text{line loss}$$

where LL stands for line losses. How do the transmission losses compare in the two cases?

Solution

(a) In the first case, the real power of load 1 is 100 kW, and the reactive power of load 1 is

$$\begin{aligned} Q_1 &= P_1 \tan \theta \\ &= (100 \text{ kW}) \tan (\cos^{-1} 0.78) = (100 \text{ kW}) \tan 38.7° \\ &= 80.2 \text{ kVAR} \end{aligned}$$

The real power of load 2 is 200 kW, and the reactive power of load 2 is

$$\begin{aligned} Q_2 &= P_2 \tan \theta \\ &= (200 \text{ kW}) \tan (\cos^{-1} 0.80) = (200 \text{ kW}) \tan 36.87° \\ &= 150 \text{ kVAR} \end{aligned}$$

The real power load 3 is 150 kW, and the reactive power of load 3 is

$$\begin{aligned} Q_3 &= P_3 \tan \theta \\ &= (150 \text{ kW}) \tan (\cos^{-1} 0.85) = (150 \text{ kW}) \tan 31.8° \\ &= 93 \text{ kVAR} \end{aligned}$$

Thus, the total real load is

$$\begin{aligned} P_{\text{tot}} &= P_1 + P_2 + P_3 \\ &= 100 \text{ kW} + 200 \text{ kW} + 150 \text{ kW} = 450 \text{ kW} \end{aligned}$$

and the total reactive load is

$$\begin{aligned} Q_{\text{tot}} &= Q_1 + Q_2 + Q_3 \\ &= 80.2 \text{ kVAR} + 150 \text{ kVAR} + 93 \text{ kVAR} = 323.2 \text{ kVAR} \end{aligned}$$

The equivalent system power factor is thus

$$\begin{aligned} \text{PF} &= \cos \theta = \cos \left(\tan^{-1} \frac{Q}{P}\right) = \cos \left(\tan^{-1} \frac{323.2 \text{ kVAR}}{450 \text{ kW}}\right) \\ &= \cos 35.7° = 0.812 \text{ lagging} \end{aligned}$$

Finally, the line current is given by

$$I_L = \frac{P_{\text{tot}}}{\sqrt{3} V_L \cos \theta} = \frac{450 \text{ kW}}{\sqrt{3}(480 \text{ V})(0.812)} = 667 \text{ A}$$

(b) The real and reactive powers of loads 1 and 2 are unchanged, as is the real power of load 3. The reactive power of load 3 is

$$\begin{aligned} Q_3 &= P_3 \tan \theta \\ &= (150 \text{ kW}) \tan (-\cos^{-1} 0.85) = (150 \text{ kW}) \tan (-31.8°) \\ &= -93 \text{ kVAR} \end{aligned}$$

Thus, the total real load is

$$\begin{aligned} P_{\text{tot}} &= P_1 + P_2 + P_3 \\ &= 100 \text{ kW} + 200 \text{ kW} + 150 \text{ kW} = 450 \text{ kW} \end{aligned}$$

and the total reactive load is

$$\begin{aligned} Q_{\text{tot}} &= Q_1 + Q_2 + Q_3 \\ &= 80.2 \text{ kVAR} + 150 \text{ kVAR} - 93 \text{ kVAR} = 137.2 \text{ kVAR} \end{aligned}$$

The equivalent system power factor is thus

$$\begin{aligned} \text{PF} &= \cos \theta = \cos \left(\tan^{-1} \frac{Q}{P}\right) = \cos \left(\tan^{-1} \frac{137.2 \text{ kVAR}}{450 \text{ kW}}\right) \\ &= \cos 16.96° = 0.957 \text{ lagging} \end{aligned}$$

Finally, the line current is given by

$$I_L = \frac{P_{\text{tot}}}{\sqrt{3}V_L \cos\theta} = \frac{450 \text{ kW}}{\sqrt{3}(480 \text{ V})(0.957)} = 566 \text{ A}$$

(c) The transmission losses in the first case are

$$P_{\text{LL}} = 3I_L^2R_L = 3(667 \text{ A})^2R_L = 1{,}344{,}700\ R_L$$

The transmission losses in the second case are

$$P_{\text{LL}} = 3I_L^2R_L = 3(566 \text{ A})^2R_L = 961{,}070\ R_L$$

Notice that in the second case the transmission power losses are 28 percent less than in the first case, while the power supplied to the loads is the same.

As seen in Example 5–3, the ability to adjust the power factor of one or more loads in a power system can significantly affect the operating efficiency of the power system. The lower the power factor of a system, the greater the losses in the power lines feeding it. Most loads on a typical power system are induction motors, so power systems are almost invariably lagging in power factor. Having one or more leading loads (overexcited synchronous motors) on the system can be useful for the following reasons:

1. A leading load can supply some reactive power Q for nearby lagging loads, instead of it coming from the generator. Since the reactive power does not have to travel over the long and fairly high-resistance transmission lines, the transmission line current is reduced and the power system losses are much lower. (This was shown by the previous example.)
2. Since the transmission lines carry less current, they can be smaller for a given rated power flow. A lower equipment current rating reduces the cost of a power system significantly.
3. In addition, requiring a synchronous motor to operate with a leading power factor means that the motor must be run *overexcited.* This mode of operation increases the motor's maximum torque and reduces the chance of accidentally exceeding the pullout torque.

The use of synchronous motors or other equipment to increase the overall power factor of a power system is called *power-factor correction.* Since a synchronous motor can provide power-factor correction and lower power system costs, many loads that can accept a constant-speed motor (even though they do not necessarily *need* one) are driven by synchronous motors. Even though a synchronous motor may cost more than an induction motor on an individual basis, the ability to operate a synchronous motor at leading power factors for power-factor correction saves money for industrial plants. This results in the purchase and use of synchronous motors.

Any synchronous motor that exists in a plant is run overexcited as a matter of course to achieve power-factor correction and to increase its pullout torque.

However, running a synchronous motor overexcited requires a high field current and flux, which causes significant rotor heating. An operator must be careful not to overheat the field windings by exceeding the rated field current.

The Synchronous Capacitor or Synchronous Condenser

A synchronous motor purchased to drive a load can be operated overexcited to supply reactive power Q for a power system. In fact, at some times in the past a synchronous motor was purchased and run without a load, *simply for power-factor correction*. The phasor diagram of a synchronous motor operating overexcited at no load is shown in Figure 5–14.

Since there is no power being drawn from the motor, the distances proportional to power ($E_A \sin \delta$ and $I_A \cos \theta$) are zero. Since the Kirchhoff's voltage law equation for a synchronous motor is

$$\mathbf{V}_\phi = \mathbf{E}_A + jX_S\mathbf{I}_A \tag{5–1}$$

the quantity $jX_S\mathbf{I}_A$ points to the left, and therefore the armature current $\mathbf{I}_A$ points straight up. If $\mathbf{V}_\phi$ and $\mathbf{I}_A$ are examined, the voltage–current relationship between them looks like that of a capacitor. An overexcited synchronous motor at no load looks just like a large capacitor to the power system.

Some synchronous motors used to be sold specifically for power-factor correction. These machines had shafts that did not even come through the frame of the motor—no load could be connected to them even if one wanted to do so. Such special-purpose synchronous motors were often called *synchronous condensers* or *synchronous capacitors*. (*Condenser* is an old name for capacitor.)

The V curve for a synchronous capacitor is shown in Figure 5–15a. Since the real power supplied to the machine is zero (except for losses), at unity power factor the current $I_A = 0$. As the field current is increased above that point, the line current (and the reactive power supplied by the motor) increases in a nearly linear fashion until saturation is reached. Figure 5–15b shows the effect of increasing the field current on the motor's phasor diagram.

Today, conventional static capacitors are more economical to buy and use than synchronous capacitors. However, some synchronous capacitors may still be in use in older industrial plants.

FIGURE 5–14
The phasor diagram of a *synchronous capacitor* or synchronous condenser.

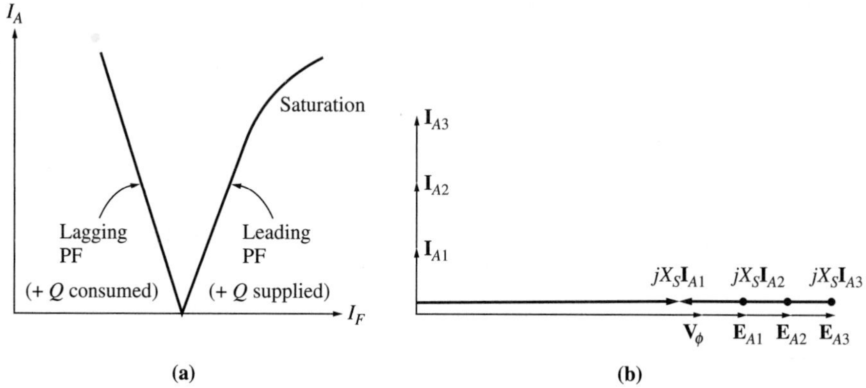

FIGURE 5–15
(a) The V curve of a synchronous capacitor. (b) The corresponding machine phasor diagram.

5.3 STARTING SYNCHRONOUS MOTORS

Section 5.2 explained the behavior of a synchronous motor under steady-state conditions. In that section, the motor was always assumed to be initially turning at *synchronous speed.* What has not yet been considered is the question: How did the motor get to synchronous speed in the first place?

To understand the nature of the starting problem, refer to Figure 5–16. This figure shows a 60-Hz synchronous motor at the moment power is applied to its stator windings. The rotor of the motor is stationary, and therefore the magnetic field $\mathbf{B}_R$ is stationary. The stator magnetic field $\mathbf{B}_S$ is starting to sweep around the motor at synchronous speed.

Figure 5–16a shows the machine at time $t = 0$ s, when $\mathbf{B}_R$ and $\mathbf{B}_S$ are exactly lined up. By the induced-torque equation

$$\tau_{\text{ind}} = k\mathbf{B}_R \times \mathbf{B}_S \tag{3–58}$$

the induced torque on the shaft of the rotor is zero. Figure 5–16b shows the situation at time $t = 1/240$ s. In such a short time, the rotor has barely moved, but the stator magnetic field now points to the left. By the induced-torque equation, the torque on the shaft of the rotor is now *counterclockwise.* Figure 5–16c shows the situation at time $t = 1/120$ s. At that point $\mathbf{B}_R$ and $\mathbf{B}_S$ point in opposite directions, and τ_{ind} again equals zero. At $t = 3/240$ s, the stator magnetic field now points to the right, and the resulting torque is *clockwise.*

Finally, at $t = 1/60$ s, the stator magnetic field is again lined up with the rotor magnetic field, and $\tau_{\text{ind}} = 0$. During one electrical cycle, the torque was first counterclockwise and then clockwise, and the average torque over the complete cycle was zero. What happens to the motor is that it vibrates heavily with each electrical cycle and finally overheats.

Such an approach to synchronous motor starting is hardly satisfactory—managers tend to frown on employees who burn up their expensive equipment. So just how *can* a synchronous motor be started?

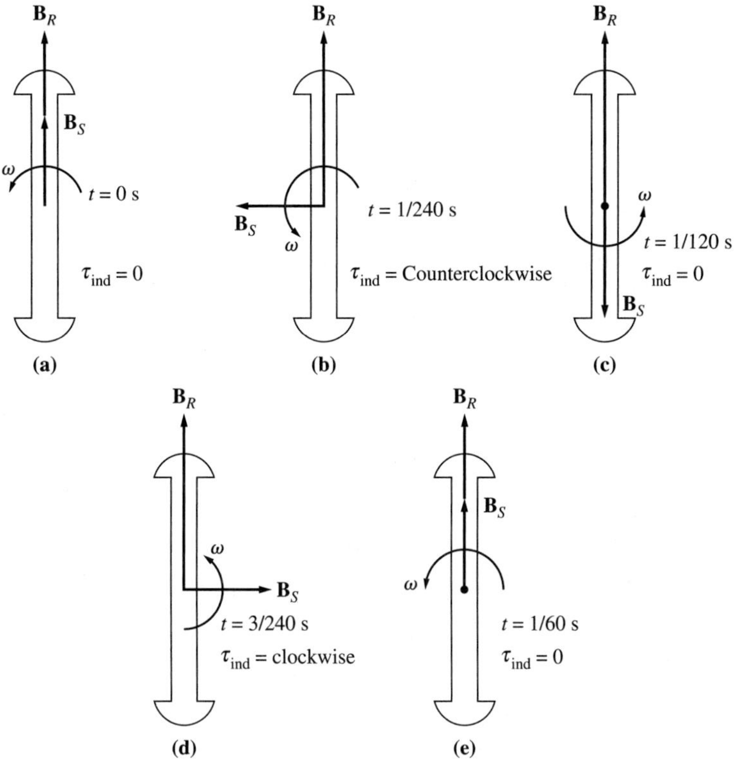

FIGURE 5–16
Starting problems in a synchronous motor—the torque alternates rapidly in magnitude and direction, so that the net starting torque is zero.

Three basic approaches can be used to safely start a synchronous motor:

1. *Reduce the speed of the stator magnetic field* to a low enough value that the rotor can accelerate and lock in with it during one half-cycle of the magnetic field's rotation. This can be done by reducing the frequency of the applied electric power.
2. *Use an external prime mover* to accelerate the synchronous motor up to synchronous speed, go through the paralleling procedure, and bring the machine on the line as a generator. Then, turning off or disconnecting the prime mover will make the synchronous machine a motor.
3. *Use damper windings or amortisseur windings*. The function of damper windings and their use in motor starting will be explained below.

Each of these approaches to synchronous motor starting will be described in turn.

Motor Starting by Reducing Electrical Frequency

If the stator magnetic fields in a synchronous motor rotate at a low enough speed, there will be no problem for the rotor to accelerate and to lock in with the stator magnetic field. The speed of the stator magnetic fields can then be increased to operating speed by gradually increasing f_{se} up to its normal 50- or 60-Hz value.

This approach to starting synchronous motors makes a lot of sense, but it does have one big problem: Where does the variable electrical frequency come from? Regular power systems are very carefully regulated at 50 or 60 Hz, so until recently any variable-frequency voltage source had to come from a dedicated generator. Such a situation was obviously impractical except for very unusual circumstances.

Today, things are different. Solid-state motor controllers can be used to convert a constant input frequency to any desired output frequency. With the development of modern solid-state variable-frequency drive packages, it is perfectly possible to continuously control the electrical frequency applied to the motor all the way from a fraction of a hertz up to and above full rated frequency. If such a variable-frequency drive unit is included in a motor-control circuit to achieve speed control, then starting the synchronous motor is very easy—simply adjust the frequency to a very low value for starting, and then raise it up to the desired operating frequency for normal running.

When a synchronous motor is operated at a speed lower than the rated speed, its internal generated voltage $E_A = K\phi\omega$ will be smaller than normal. If E_A is reduced in magnitude, then the terminal voltage applied to the motor must be reduced as well in order to keep the stator current at safe levels. The voltage in any variable-frequency drive or variable-frequency starter circuit must vary roughly linearly with the applied frequency.

To learn more about such solid-state motor-drive units, refer to Reference 9.

Motor Starting with an External Prime Mover

The second approach to starting a synchronous motor is to attach an external starting motor to it and bring the synchronous machine up to full speed with the external motor. Then the synchronous machine can be paralleled with its power system as a generator, and the starting motor can be detached from the shaft of the machine. Once the starting motor is turned off, the shaft of the machine slows down, the rotor magnetic field $\mathbf{B}_R$ falls behind $\mathbf{B}_{\text{net}}$, and the synchronous machine starts to act as a motor. Once paralleling is completed, the synchronous motor can be loaded down in an ordinary fashion.

This whole procedure is not as preposterous as it sounds, since many synchronous motors are parts of motor-generator sets, and the synchronous machine in the motor-generator set may be started with the other machine serving as the starting motor. Also, the starting motor only needs to overcome the inertia of the synchronous machine without a load—no load is attached until the motor is paralleled to the power system. Since only the motor's inertia must be overcome, the starting motor can have a *much* smaller rating than the synchronous motor it starts.

Since most large synchronous motors have brushless excitation systems mounted on their shafts, it is often possible to use these exciters as starting motors.

For many medium-size to large synchronous motors, an external starting motor or starting by using the exciter may be the only possible solution, because the power systems they are tied to may not be able to handle the starting currents needed to use the amortisseur winding approach described next.

Motor Starting by Using Amortisseur Windings

By far the most popular way to start a synchronous motor is to employ *amortisseur* or *damper* windings. Amortisseur windings are special bars laid into notches carved in the face of a synchronous motor's rotor and then shorted out on each end by a large *shorting ring*. A pole face with a set of amortisseur windings is shown in Figure 5–17, and amortisseur windings are visible in Figures 4–2 and 4–4.

To understand what a set of amortisseur windings does in a synchronous motor, examine the stylized salient two-pole rotor shown in Figure 5–18. This rotor shows an amortisseur winding with the shorting bars on the ends of the two rotor

FIGURE 5–17
A rotor field pole for a synchronous machine showing amortisseur windings in the pole face. (*Courtesy of General Electric Company.*)

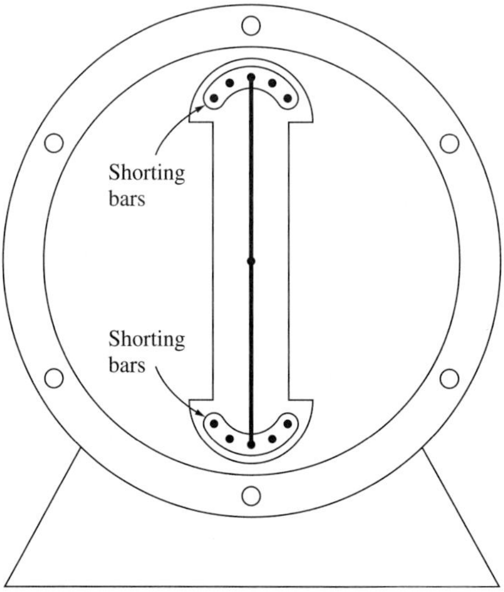

FIGURE 5–18
A simplified diagram of a salient two-pole machine showing amortisseur windings.

pole faces connected by wires. (This is not quite the way normal machines are constructed, but it will serve beautifully to illustrate the point of the windings.)

Assume initially that *the main rotor field winding is disconnected* and that a three-phase set of voltages is applied to the stator of this machine. When the power is first applied at time $t = 0$ s, assume that the magnetic field $\mathbf{B}_S$ is vertical, as shown in Figure 5–19a. As the magnetic field $\mathbf{B}_S$ sweeps along in a counterclockwise direction, it induces a voltage in the bars of the amortisseur winding given by Equation (1–45):

$$e_{\text{ind}} = (\mathbf{v} \times \mathbf{B}) \bullet \mathbf{l} \qquad (1\text{–}45)$$

where $\mathbf{v}$ = velocity of the bar *relative to the magnetic field*
$\mathbf{B}$ = magnetic flux density vector
$\mathbf{l}$ = length of conductor in the magnetic field

The bars at the top of the rotor are moving to the right *relative to the magnetic field*, so the resulting direction of the induced voltage is out of the page. Similarly, the induced voltage is into the page in the bottom bars. These voltages produce a current flow out of the top bars and into the bottom bars, resulting in a winding magnetic field $\mathbf{B}_W$ pointing to the right. By the induced-torque equation

$$\tau_{\text{ind}} = k\mathbf{B}_W \times \mathbf{B}_S$$

the resulting torque on the bars (and the rotor) is *counterclockwise.*

Figure 5–19b shows the situation at $t = 1/240$ s. Here, the stator magnetic field has rotated 90° while the rotor has barely moved (it simply cannot speed up in so short a time). At this point, the voltage induced in the amortisseur windings is zero, because $\mathbf{v}$ is parallel to $\mathbf{B}$. With no induced voltage, there is no current in the windings, and the induced torque is zero.

Figure 5–19c shows the situation at $t = 1/120$ s. Now the stator magnetic field has rotated 90°, and the rotor still has not moved yet. The induced voltage [given by Equation (1–45)] in the amortisseur windings is out of the page in the bottom bars and into the page in the top bars. The resulting current flow is out of the page in the bottom bars and into the page in the top bars, causing a magnetic field $\mathbf{B}_W$ to point to the left. The resulting induced torque, given by

$$\tau_{\text{ind}} = k\mathbf{B}_W \times \mathbf{B}_S$$

is counterclockwise.

Finally, Figure 5–19d shows the situation at time $t = 3/240$ s. Here, as at $t = 1/240$ s, the induced torque is zero.

Notice that sometimes the torque is counterclockwise and sometimes it is essentially zero, but it is *always unidirectional.* Since there is a net torque in a single direction, the motor's rotor speeds up. (This is entirely different from starting a synchronous motor with its normal field current, since in that case torque is first clockwise and then counterclockwise, averaging out to zero. In this case, torque is *always* in the same direction, so there is a nonzero average torque.)

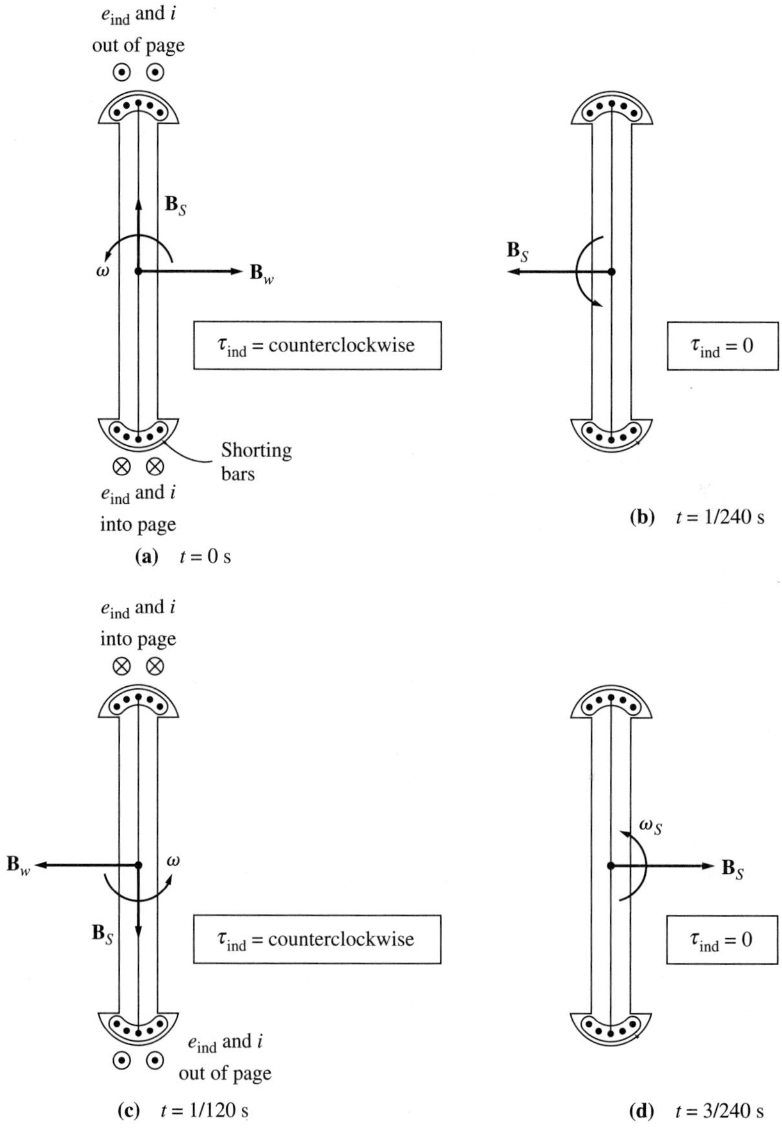

FIGURE 5–19
The development of a unidirectional torque with synchronous motor amortisseur windings.

Although the motor's rotor will speed up, it can never quite reach synchronous speed. This is easy to understand. Suppose that a rotor is turning at synchronous speed. Then the speed of the stator magnetic field $\mathbf{B}_S$ is the same as the rotor's speed, and there is *no relative motion* between $\mathbf{B}_S$ and the rotor. If there is no relative motion, the induced voltage in the windings will be zero, the resulting current flow will be zero, and the winding magnetic field will be zero. Therefore, there will be no torque on the rotor to keep it turning. Even though a rotor cannot speed up all the way to synchronous speed, it can get close. It gets close enough to n_{sync} that the regular field current can be turned on, and the rotor will pull into step with the stator magnetic fields.

In a real machine, the field windings are not open-circuited during the starting procedure. If the field windings were open-circuited, then very high voltages would be produced in them during starting. If the field winding is short-circuited during starting, no dangerous voltages are produced, and the induced field current actually contributes extra starting torque to the motor.

To summarize, if a machine has amortisseur windings, it can be started by the following procedure:

1. Disconnect the field windings from their dc power source and short them out.
2. Apply a three-phase voltage to the stator of the motor, and let the rotor accelerate up to near-synchronous speed. The motor should have no load on its shaft, so that its speed can approach n_{sync} as closely as possible.
3. Connect the dc field circuit to its power source. After this is done, the motor will lock into step at synchronous speed, and loads may then be added to its shaft.

The Effect of Amortisseur Windings on Motor Stability

If amortisseur windings are added to a synchronous machine for starting, we get a free bonus—an increase in machine stability. The stator magnetic field rotates at a constant speed n_{sync}, which varies only when the system frequency varies. If the rotor turns at n_{sync}, then the amortisseur windings have no induced voltage at all. If the rotor turns *slower* than n_{sync}, then there will be relative motion between the rotor and the stator magnetic field and a voltage will be induced in the windings. This voltage produces a current flow, and the current flow produces a magnetic field. The interaction of the two magnetic fields produces a torque that tends to speed the machine up again. On the other hand, if the rotor turns *faster* than the stator magnetic field, a torque will be produced that tries to slow the rotor down. Thus, *the torque produced by the amortisseur windings speeds up slow machines and slows down fast machines.*

These windings therefore tend to dampen out the load or other transients on the machine. It is for this reason that amortisseur windings are also called *damper windings*. Amortisseur windings are also used on synchronous generators, where they serve a similar stabilizing function when a generator is operating in parallel with other generators on an infinite bus. If a variation in shaft torque occurs on the generator, its rotor will momentarily speed up or slow down, and these changes will be opposed by the amortisseur windings. Amortisseur windings improve the overall stability of power systems by reducing the magnitude of power and torque transients.

Amortisseur windings are responsible for most of thc subtransient current in a faulted synchronous machine. A short circuit at the terminals of a generator is just another form of transient, and the amortisseur windings respond very quickly to it.

5.4 SYNCHRONOUS GENERATORS AND SYNCHRONOUS MOTORS

A synchronous generator is a synchronous machine that converts mechanical power to electric power, while a synchronous motor is a synchronous machine that converts electric power to mechanical power. In fact, they are both the same physical machine.

A synchronous machine can supply real power to or consume real power from a power system and can supply reactive power to or consume reactive power from a power system. All four combinations of real and reactive power flows are possible, and Figure 5–20 shows the phasor diagrams for these conditions.

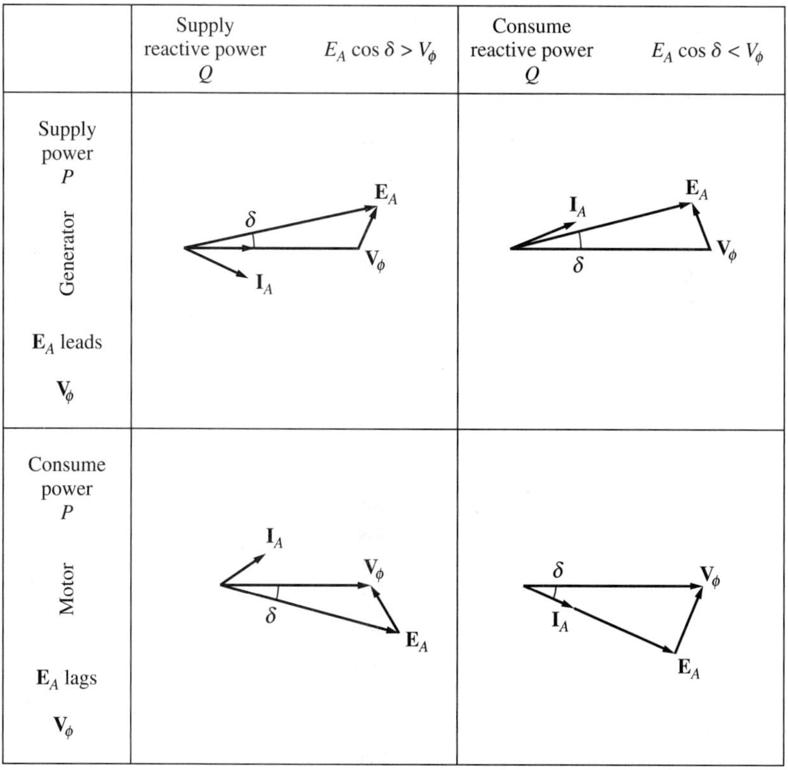

FIGURE 5–20
Phasor diagrams showing the generation and consumption of real power P and reactive power Q by synchronous generators and motors.

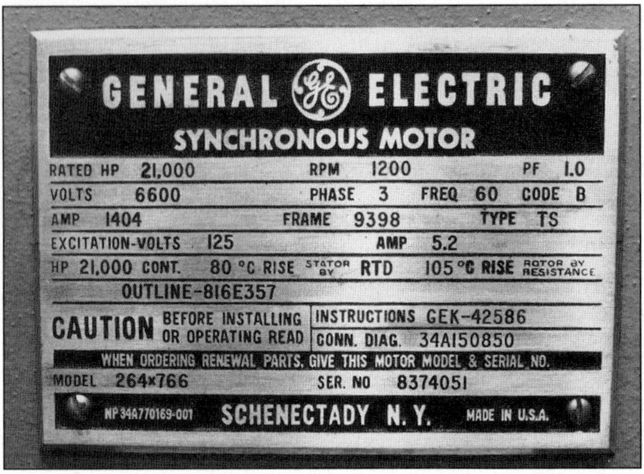

FIGURE 5–21
A typical nameplate for a large synchronous motor. (*Courtesy of General Electric Company.*)

Notice from the figure that

1. The distinguishing characteristic of a synchronous generator (supplying P) is that $\mathbf{E}_A$ *lies ahead of* $\mathbf{V}_\phi$ while for a motor $\mathbf{E}_A$ *lies behind* $\mathbf{V}_\phi$.
2. The distinguishing characteristic of a machine supplying reactive power Q is that $\mathbf{E}_A \cos \delta > \mathbf{V}_\phi$ regardless of whether the machine is acting as a generator or as a motor. A machine that is consuming reactive power Q has $\mathbf{E}_A \cos \delta < \mathbf{V}_\phi$.

5.5 SYNCHRONOUS MOTOR RATINGS

Since synchronous motors are the same physical machines as synchronous generators, the basic machine ratings are the same. The one major difference is that a large E_A gives a *leading* power factor instead of a lagging one, and therefore the effect of the maximum field current limit is expressed as a rating at a *leading* power factor. Also, since the output of a synchronous motor is mechanical power, a synchronous motor's power rating is usually given in output horsepower (in the USA) or output kilowatts (everywhere else in the world), instead of being specified by a voltampere rating and power factor the way generators are.

The nameplate of a large synchronous motor is shown in Figure 5–21. In addition to the information shown in the figure, a smaller synchronous motor would have a service factor on its nameplate.

In general, synchronous motors are more adaptable to low-speed, high-power applications than induction motors (see Chapter 6). They are therefore commonly used for low-speed, high-power loads.

5.6 SUMMARY

A synchronous motor is the same physical machine as a synchronous generator, except that the direction of real power flow is reversed. Since synchronous motors are usually connected to power systems containing generators much larger than the motors, the frequency and terminal voltage of a synchronous motor are fixed (i.e., the power system looks like an infinite bus to the motor).

The equivalent circuit of a synchronous motor is the same as the equivalent circuit of a synchronous generator, except that the assumed direction of the armature current is reversed.

The speed of a synchronous motor is constant from no load to the maximum possible load on the motor. The speed of rotation is

$$n_m = \frac{120 f_{se}}{P} \tag{5–3}$$

from no load all the way up to the maximum possible load. The maximum possible power a synchronous motor can produce is

$$P_{\max} = \frac{3V_\phi E_A}{X_S} \tag{4–21}$$

And the maximum possible torque is given by

$$\tau_{\max} = \frac{3V_\phi E_A}{\omega X_S} \tag{4–22}$$

If this value is exceeded, the rotor will not be able to stay locked in with the stator magnetic fields, and the motor will *slip poles.*

If we ignore the effect of electrical and mechanical losses, then power converted from electrical to mechanical form in the motor is given by

$$P_{\text{conv}} = \frac{3V_\phi E_A}{X_S} \sin \delta \tag{4–20}$$

If the input voltage $\mathbf{V}_\phi$ is constant, then the power converted (and thus the power supplied) is directly proportional to the quantity $E_A \sin \delta$. This relationship can be useful when plotting synchronous motor phasor diagrams. For example, if the field current is increased or decreased, the internal generated voltage of the motor will increase or decrease, but the quantity $E_A \sin \delta$ will remain constant. This constraint makes it easy to plot the changes in the motor's phasor diagram (see Figure 5–9), and to calculate synchronous motor V curves.

If the field current of a synchronous motor is varied while its shaft load remains constant, then the reactive power supplied or consumed by the motor will vary . If $E_A \cos \delta > V_\phi$, the motor will supply reactive power, while if $E_A \cos \delta < V_\phi$, the motor will consume reactive power. A synchronous motor is usually operated with $E_A \cos \delta > V_\phi$, so that the synchronous motor supplies reactive power to the power system and reduces the overall power factor of the loads.

A synchronous motor has no net starting torque and so cannot start by itself. There are three main ways to start a synchronous motor:

1. Reduce the stator frequency to a safe starting level.
2. Use an external prime mover.
3. Put amortisseur or damper windings on the motor to accelerate it to near-synchronous speed before a direct current is applied to the field windings.

If damper windings are present on a motor, they will also increase the stability of the motor during load transients.

QUESTIONS

5–1. What is the difference between a synchronous motor and a synchronous generator?

5–2. What is the speed regulation of a synchronous motor?

5–3. When would a synchronous motor be used even though its constant-speed characteristic was not needed?

5–4. Why can't a synchronous motor start by itself?

5–5. What techniques are available to start a synchronous motor?

5–6. What are amortisseur windings? Why is the torque produced by them unidirectional at starting, while the torque produced by the main field winding alternates direction?

5–7. What is a synchronous capacitor? Why would one be used?

5–8. Explain, using phasor diagrams, what happens to a synchronous motor as its field current is varied. Derive a synchronous motor V curve from the phasor diagram.

5–9. Is a synchronous motor's field circuit in more danger of overheating when it is operating at a leading or at a lagging power factor? Explain, using phasor diagrams.

5–10. A synchronous motor is operating at a fixed real load, and its field current is increased. If the armature current falls, was the motor initially operating at a lagging or a leading power factor?

5–11. Why must the voltage applied to a synchronous motor be derated for operation at frequencies lower than the rated value?

PROBLEMS

5–1. A 480-V, 60-Hz, 400-hp, 0.8-PF-leading, eight-pole, Δ-connected synchronous motor has a synchronous reactance of 0.6 Ω and negligible armature resistance. Ignore its friction, windage, and core losses for the purposes of this problem. Assume that $|\mathbf{E}_A|$ is directly proportional to the field current I_F (in other words, assume that the motor operates in the linear part of the magnetization curve), and that $|\mathbf{E}_A| = 480$ V when $I_F = 4$ A.

(a) What is the speed of this motor?

(b) If this motor is initially supplying 400 hp at 0.8 PF lagging, what are the magnitudes and angles of $\mathbf{E}_A$ and $\mathbf{I}_A$?

(c) How much torque is this motor producing? What is the torque angle δ? How near is this value to the maximum possible induced torque of the motor for this field current setting?

(d) If $|\mathbf{E}_A|$ is increased by 30 percent, what is the new magnitude of the armature current? What is the motor's new power factor?

(e) Calculate and plot the motor's V curve for this load condition.

5–2. Assume that the motor of Problem 5–1 is operating at rated conditions.

(a) What are the magnitudes and angles of $\mathbf{E}_A$ and $\mathbf{I}_A$, and I_F?

(b) Suppose the load is removed from the motor. What are the magnitudes and angles of $\mathbf{E}_A$ and $\mathbf{I}_A$ now?

5–3. A 230-V, 50-Hz, two-pole synchronous motor draws 40 A from the line at unity power factor and full load. Assuming that the motor is lossless, answer the following questions:

(*a*) What is the output torque of this motor? Express the answer both in newton-meters and in pound-feet.

(*b*) What must be done to change the power factor to 0.85 leading? Explain your answer, using phasor diagrams.

(*c*) What will the magnitude of the line current be if the power factor is adjusted to 0.85 leading?

5–4. A 2300-V, 1000-hp, 0.8-PF-leading, 60-Hz, two-pole, Y-connected synchronous motor has a synchronous reactance of 5.0 Ω and an armature resistance of 0.3 Ω. At 60 Hz, its friction and windage losses are 30 kW, and its core losses are 20 kW. The field circuit has a dc voltage of 200 V, and the maximum I_F is 10 A. The open-circuit characteristic of this motor is shown in Figure P5–1. Answer the following questions about the motor, assuming that it is being supplied by an infinite bus.

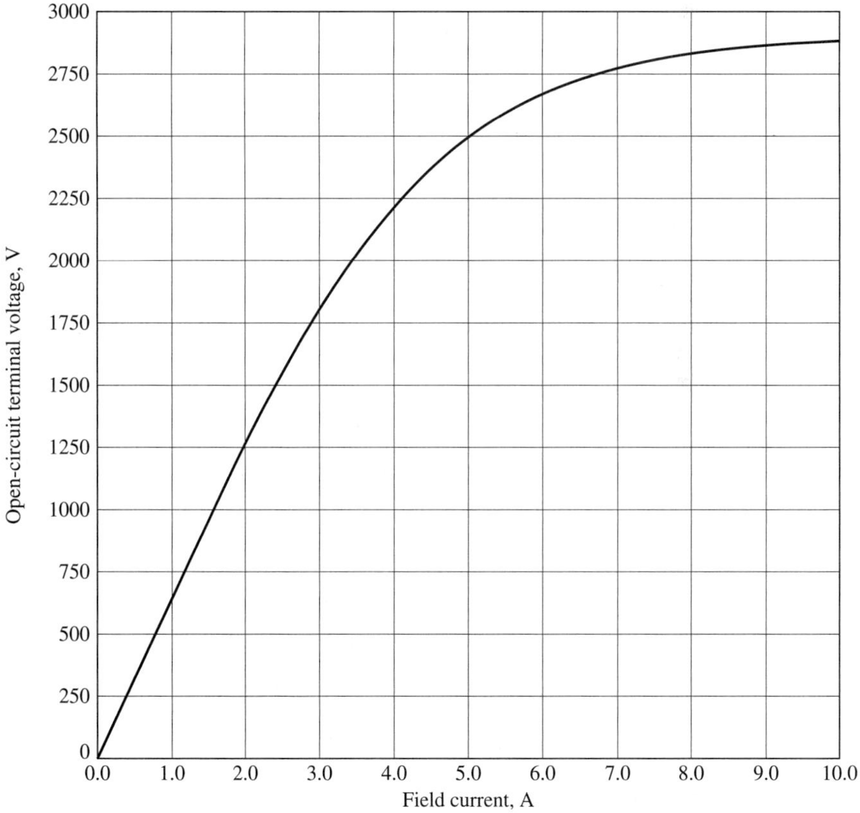

FIGURE P5–1
The open-circuit characteristic for the motor in Problems 5–4 and 5–5.

(a) How much field current would be required to make this machine operate at unity power factor when supplying full load?
(b) What is the motor's efficiency at full load and unity power factor?
(c) If the field current were increased by 5 percent, what would the new value of the armature current be? What would the new power factor be? How much reactive power is being consumed or supplied by the motor?
(d) What is the maximum torque this machine is theoretically capable of supplying at unity power factor? At 0.8 PF leading?

5–5. Plot the V curves (I_A versus I_F) for the synchronous motor of Problem 5–4 at no-load, half-load, and full-load conditions. (Note that an electronic version of the open-circuit characteristics in Figure P5–1 is available at the book's website. It may simplify the calculations required by this problem.)

5–6. If a 60-Hz synchronous motor is to be operated at 50 Hz, will its synchronous reactance be the same as at 60 Hz, or will it change? (*Hint:* Think about the derivation of X_s.)

5–7. A 208-V, Y-connected synchronous motor is drawing 50 A at unity power factor from a 208-V power system. The field current flowing under these conditions is 2.7 A. Its synchronous reactance is 1.6 Ω. Assume a linear open-circuit characteristic.
(a) Find $\mathbf{V}_\phi$ and $\mathbf{E}_A$ for these conditions.
(b) Find the torque angle δ.
(c) What is the static stability power limit under these conditions?
(d) How much field current would be required to make the motor operate at 0.80 PF leading?
(e) What is the new torque angle in part *(d)*?

5–8. A 4.12-kV, 60-Hz, 3000-hp, 0.8-PF-leading, Δ-connected, three-phase synchronous motor has a synchronous reactance of 1.1 per unit and an armature resistance of 0.1 per unit. If this motor is running at rated voltage with a line current of 300 A at 0.85 PF leading, what is the internal generated voltage per phase inside this motor? What is the torque angle δ?

5–9. Figure P5–2 shows a synchronous motor phasor diagram for a motor operating at a leading power factor with no R_A. For this motor, the torque angle is given by

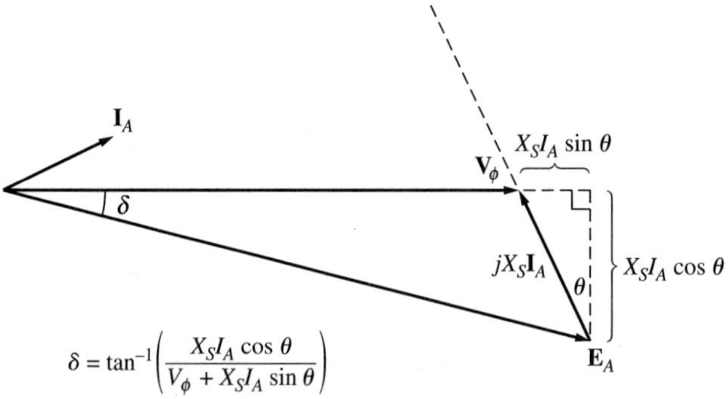

FIGURE P5–2
Phasor diagram of a motor at a leading power factor.

$$\tan \delta = \frac{X_S I_A \cos \theta}{V_\phi + X_S I_A \sin \theta}$$

$$\delta = \tan^{-1} \left(\frac{X_S I_A \cos \theta}{V_\phi + X_S I_A \sin \theta} \right)$$

Derive an equation for the torque angle of the synchronous motor *if the armature resistance is included.*

5–10. A synchronous machine has a synchronous reactance of 1.0 Ω per phase and an armature resistance of 0.1 Ω per phase. If $\mathbf{E}_A = 460 \angle -10° \text{ V}$ and $\mathbf{V}_\phi = 480 \angle 0° \text{ V}$, is this machine a motor or a generator? How much power P is this machine consuming from or supplying to the electrical system? How much reactive power Q is this machine consuming from or supplying to the electrical system?

5–11. A 500-kVA, 600-V, 0.8-PF-leading, Y-connected synchronous motor has a synchronous reactance of 1.0 per unit and an armature resistance of 0.1 per unit. At the current time, $\mathbf{E}_A = 1.00 \angle 12° \text{ pu}$ and $\mathbf{V}_\phi = 1 \angle 0° \text{ pu}$.

(a) Is this machine currently acting as a motor or a generator?
(b) How much power P is this machine consuming from or supplying to the electrical system?
(c) How much reactive power Q is this machine consuming from or supplying to the electrical system?
(d) Is this machine operating within its rated limits?

5–12. Figure P5–3 shows a small industrial plant supplied by an external 480-V, three-phase power supply. The plant includes three main loads as shown in the figure. Answer the following questions about the plant. The synchronous motor is rated at 100 hp, 460 V, and 0.8 PF leading. The synchronous reactance is 1.1 pu and armature resistance is 0.01 pu. The OCC for this motor is shown in Figure P5–4.

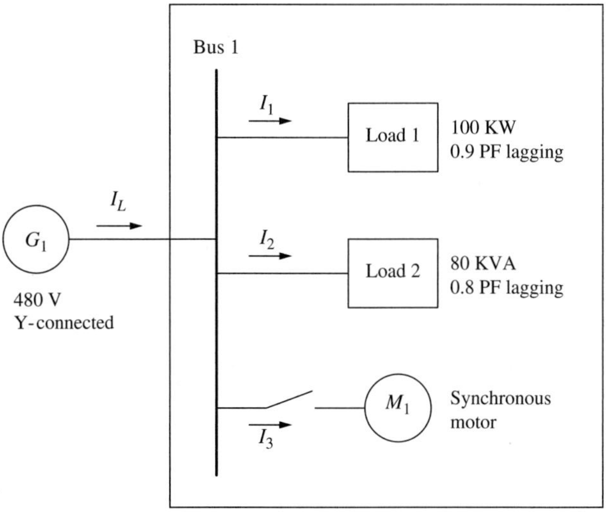

FIGURE P5–3
A small industrial facility.

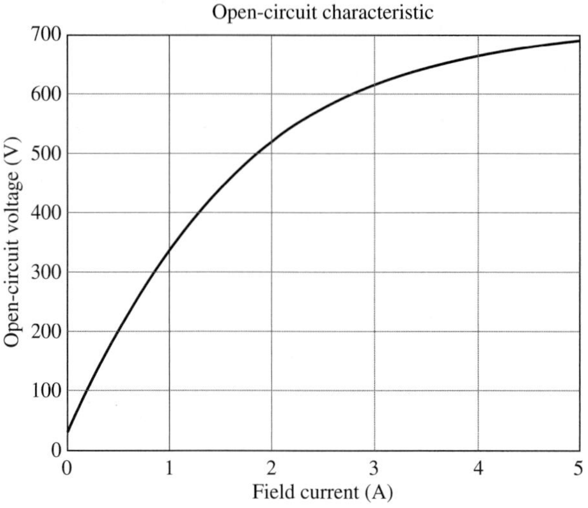

FIGURE P5–4
Open-circuit characteristic of synchronous motor.

(*a*) If the switch on the synchronous motor is open, how much real, reactive, and apparent power is being supplied to the plant? What is the current I_L in the transmission line?

The switch is now closed and the synchronous motor is supplying rated power with the field current adjusted to 1.5 A.

(*b*) What is the real and reactive power supplied to the motor?
(*c*) What is the torque angle of the motor?
(*d*) What is the power factor of the motor?
(*e*) How much real, reactive, and apparent power is being supplied to the plant now? What is the current I_L in the transmission line?

Now suppose that the field current is increased to 3.0 A.

(*f*) What is the real and reactive power supplied to the motor?
(*g*) What is the torque angle of the motor?
(*h*) What is the power factor of the motor?
(*i*) How much real, reactive, and apparent power is being supplied to the plant now? What is the current I_L in the transmission line?
(*j*) How does the line current when the field current is 1.5 A compare to the line current when the field current is 3.0 A?

5–13. A 480-V, 100-kW, 0.8-PF-leading, 50-Hz, four-pole, Y-connected synchronous motor has a synchronous reactance of 1.8 Ω and a negligible armature resistance. The rotational losses are also to be ignored. This motor is to be operated over a continuous range of speeds from 300 to 1500 r/min, where the speed changes are to be accomplished by controlling the system frequency with a solid-state drive.

(*a*) Over what range must the input frequency be varied to provide this speed control range?
(*b*) How large is E_A at the motor's rated conditions?
(*c*) What is the maximum power the motor can produce at rated speed with the E_A calculated in part (*b*)?

(*d*) What is the largest value that E_A could be at 300 r/min?
(*e*) Assuming that the applied voltage V_ϕ is derated by the same amount as E_A, what is the maximum power the motor could supply at 300 r/min?
(*f*) How does the power capability of a synchronous motor relate to its speed?

5–14. A 2300-V, 400-hp, 60-Hz, eight-pole, Y-connected synchronous motor has a rated power factor of 0.85 leading. At full load, the efficiency is 90 percent. The armature resistance is 0.8 Ω, and the synchronous reactance is 11 Ω. Find the following quantities for this machine when it is operating at full load:
(*a*) Output torque
(*b*) Input power
(*c*) n_m
(*d*) $\mathbf{E}_A$
(*e*) $|\mathbf{I}_A|$
(*f*) P_{conv}
(*g*) $P_{\text{mech}} + P_{\text{core}} + P_{\text{stray}}$

5–15. The Y-connected synchronous motor whose nameplate is shown in Figure 5–21 has a per-unit synchronous reactance of 0.70 and a per-unit resistance of 0.02.
(*a*) What is the rated input power of this motor?
(*b*) What is the magnitude of $\mathbf{E}_A$ at rated conditions?
(*c*) If the input power of this motor is 12 MW, what is the maximum reactive power the motor can simultaneously supply? Is it the armature current or the field current that limits the reactive power output?
(*d*) How much power does the field circuit consume at the rated conditions?
(*e*) What is the efficiency of this motor at full load?
(*f*) What is the output torque of the motor at the rated conditions? Express the answer both in newton-meters and in pound-feet.

5–16. A 480-V, 500-kVA, 0.8-PF-lagging, Y-connected synchronous generator has a synchronous reactance of 0.4 Ω and a negligible armature resistance. This generator is supplying power to a 480-V, 80-kW, 0.8-PF-leading, Y-connected synchronous motor with a synchronous reactance of 2.0 Ω and a negligible armature resistance. The synchronous generator is adjusted to have a terminal voltage of 480 V when the motor is drawing the rated power at unity power factor.
(*a*) Calculate the magnitudes and angles of $\mathbf{E}_A$ for both machines.
(*b*) If the flux of the motor is increased by 10 percent, what happens to the terminal voltage of the power system? What is its new value?
(*c*) What is the power factor of the motor after the increase in motor flux?

5–17. A 440-V, 60-Hz, three-phase, Y-connected synchronous motor has a synchronous reactance of 1.5 Ω per phase. The field current has been adjusted so that the torque angle δ is 25° when the power supplied by the generator is 90 kW.
(*a*) What is the magnitude of the internal generated voltage $\mathbf{E}_A$ in this machine?
(*b*) What are the magnitude and angle of the armature current in the machine? What is the motor's power factor?
(*c*) If the field current remains constant, what is the absolute maximum power this motor could supply?

5–18. A 460-V, 200-kVA, 0.85-PF-leading, 400-Hz, four-pole, Y-connected synchronous motor has negligible armature resistance and a synchronous reactance of 0.90 per unit. Ignore all losses.
(*a*) What is the speed of rotation of this motor?
(*b*) What is the output torque of this motor at the rated conditions?

(*c*) What is the internal generated voltage of this motor at the rated conditions?

(*d*) With the field current remaining at the value present in the motor in part (*c*), what is the maximum possible output power from the machine?

5–19. A 100-hp, 440-V, 0.8-PF-leading, Δ-connected synchronous motor has an armature resistance of 0.3 Ω and a synchronous reactance of 4.0 Ω. Its efficiency at full load is 96 percent.

(*a*) What is the input power to the motor at rated conditions?

(*b*) What is the line current of the motor at rated conditions? What is the phase current of the motor at rated conditions?

(*c*) What is the reactive power consumed by or supplied by the motor at rated conditions?

(*d*) What is the internal generated voltage $\mathbf{E}_A$ of this motor at rated conditions?

(*e*) What are the stator copper losses in the motor at rated conditions?

(*f*) What is P_{conv} at rated conditions?

(*g*) If E_A is decreased by 10 percent, how much reactive power will be consumed by or supplied by the motor?

5–20. Answer the following questions about the machine of Problem 5–19.

(*a*) If $\mathbf{E}_A = 430 \angle 15° \text{ V}$ and $\mathbf{V}_\phi = 440 \angle 0° \text{ V}$, is this machine consuming real power from or supplying real power to the power system? Is it consuming reactive power from or supplying reactive power to the power system?

(*b*) Calculate the real power P and reactive power Q supplied or consumed by the machine under the conditions in part (*a*). Is the machine operating within its ratings under these circumstances?

(*c*) If $\mathbf{E}_A = 470 \angle -20° \text{ V}$ and $\mathbf{V}_\phi = 440 \angle 0° \text{ V}$, is this machine consuming real power from or supplying real power to the power system? Is it consuming reactive power from or supplying reactive power to the power system?

(*d*) Calculate the real power P and reactive power Q supplied or consumed by the machine under the conditions in part (*c*). Is the machine operating within its ratings under these circumstances?

REFERENCES

1. Chaston, A. N. *Electric Machinery*. Reston, Va.: Reston Publishing, 1986.
2. Del Toro, V. *Electric Machines and Power Systems*. Englewood Cliffs, N.J.: Prentice-Hall, 1985.
3. Fitzgerald, A. E., and C. Kingsley, Jr. *Electric Machinery*. New York: McGraw-Hill, 1952.
4. Fitzgerald, A. E., C. Kingsley, Jr., and S. D. Umans. *Electric Machinery*, 6th ed. New York: McGraw-Hill, 2003.
5. Kosow, Irving L. *Control of Electric Motors*. Englewood Cliffs, N.J.: Prentice-Hall, 1972.
6. Liwschitz-Garik, Michael, and Clyde Whipple. *Alternating-Current Machinery*. Princeton, N.J.: Van Nostrand, 1961.
7. Nasar, Syed A. (ed.). *Handbook of Electric Machines*. New York: McGraw-Hill, 1987.
8. Slemon, G. R., and A. Straughen. *Electric Machines*. Reading, Mass.: Addison-Wesley, 1980.
9. Vithayathil, Joseph. *Power Electronics: Principles and Applications*. New York: McGraw-Hill, 1995.
10. Werninck, E. H. (ed.). *Electric Motor Handbook*. London: McGraw-Hill, 1978.

CHAPTER 6

INDUCTION MOTORS

LEARNING OBJECTIVES

- Understand the key differences between a synchronous motor and an induction motor.
- Understand the concept of rotor slip and its relationship to rotor frequency.
- Understand and know how to use the equivalent circuit of an induction motor.
- Understand power flows and the power flow diagram of an induction motor.
- Be able to use the equation for the torque–speed characteristic curve.
- Understand how the torque–speed characteristic curve varies with different rotor designs.
- Understand the techniques used for induction motor starting.
- Understand how the speed of induction motors can be controlled.
- Understand how to measure induction motor circuit model parameters.
- Understand the induction machine used as a generator.
- Understand induction motor ratings.

In Chapter 5, we saw how amortisseur windings on a synchronous motor could develop a starting torque without the necessity of supplying an external field current to them. In fact, amortisseur windings work so well that a motor could be built without the synchronous motor's main dc field circuit at all. A machine with only a continuous set of amortisseur windings is called an *induction machine*. Such machines are called induction machines because the rotor voltage (which produces the rotor current and the rotor magnetic field) is *induced* in the rotor windings rather than being physically connected by wires. The distinguishing feature of an induction motor is that *no dc field current is required* to run the machine.

Although it is possible to use an induction machine as either a motor or a generator, it has many disadvantages as a generator and so is only used as a generator in special applications. For this reason, induction machines are usually referred to as induction motors.

FIGURE 6–1
The stator of a typical induction motor, showing the stator windings. (*Courtesy of MagneTek, Inc.*)

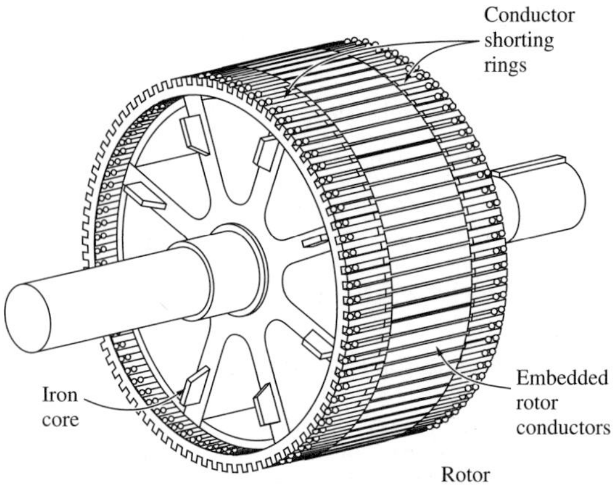

(a)

(b)

FIGURE 6–2
(a) Sketch of cage rotor. (b) A typical cage rotor. (*Courtesy of General Electric Company.*)

6.1 INDUCTION MOTOR CONSTRUCTION

An induction motor has the same physical stator as a synchronous machine, with a different rotor construction. A typical two-pole stator is shown in Figure 6–1. It looks (and is) the same as a synchronous machine stator. There are two different types of induction motor rotors which can be placed inside the stator. One is called a *cage rotor,* while the other is called a *wound rotor.*

Figures 6–2 and 6–3 show cage induction motor rotors. A cage induction motor rotor consists of a series of conducting bars laid into slots carved in the face

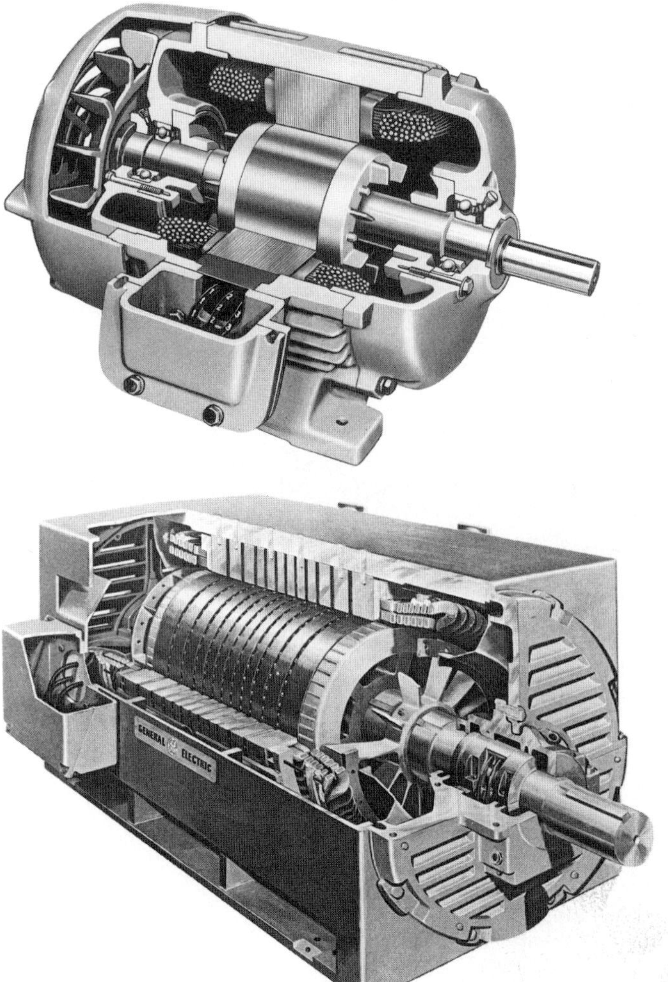

(a)

(b)

FIGURE 6–3
(a) Cutaway diagram of a typical small cage rotor induction motor. (*Courtesy of MagneTek, Inc.*)
(b) Cutaway diagram of a typical large cage rotor induction motor. (*Courtesy of General Electric Company.*)

of the rotor and shorted at either end by large *shorting rings*. This design is referred to as a cage rotor because the conductors, if examined by themselves, would look like one of the exercise wheels that squirrels or hamsters run on.

The other type of rotor is a wound rotor. A *wound rotor* has a complete set of three-phase windings that are similar to the windings on the stator. The three phases of the rotor windings are usually Y-connected, and the ends of the three rotor wires are tied to slip rings on the rotor's shaft. The rotor windings are shorted through brushes riding on the slip rings. Wound-rotor induction motors therefore have their rotor currents accessible at the stator brushes, where they can be examined and where extra resistance can be inserted into the rotor circuit. It is possible to take advantage of this feature to modify the torque–speed characteristic of the motor. Two wound rotors are shown in Figure 6–4, and a complete wound-rotor induction motor is shown in Figure 6–5.

Wound-rotor induction motors are more expensive than cage induction motors, and they require much more maintenance because of the wear associated with their brushes and slip rings. As a result, wound-rotor induction motors are rarely used.

(a)

(b)

FIGURE 6–4
Typical wound rotors for induction motors. Notice the slip rings and the bars connecting the rotor windings to the slip rings. (*Courtesy of General Electric Company.*)

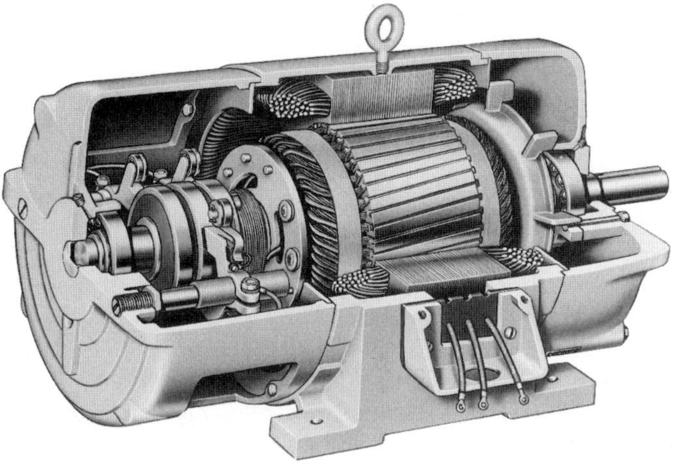

FIGURE 6–5
Cutaway diagram of a wound-rotor induction motor. Notice the brushes and slip rings. Also notice that the rotor windings are skewed to eliminate slot harmonics. (*Courtesy of MagneTek, Inc.*)

6.2 BASIC INDUCTION MOTOR CONCEPTS

Induction motor operation is basically the same as that of amortisseur windings on synchronous motors. That basic operation will now be reviewed, and some important induction motor terms will be defined.

The Development of Induced Torque in an Induction Motor

Figure 6–6 shows a cage rotor induction motor. A three-phase set of voltages has been applied to the stator, and a three-phase set of stator currents is flowing. These currents produce a magnetic field $\mathbf{B}_S$, which is rotating in a counterclockwise direction. The speed of the magnetic field's rotation is given by

$$\boxed{n_{\text{sync}} = \frac{120 f_{se}}{P}} \tag{6–1}$$

where f_{se} is the system frequency applied to the stator in hertz and P is the number of poles in the machine. This rotating magnetic field $\mathbf{B}_S$ passes over the rotor bars and induces a voltage in them.

The voltage induced in a given rotor bar is given by the equation

$$e_{\text{ind}} = (\mathbf{v} \times \mathbf{B}) \cdot \mathbf{l} \tag{1–45}$$

where $\mathbf{v}$ = velocity of the bar *relative to the magnetic field*

$\mathbf{B}$ = magnetic flux density vector

$\mathbf{l}$ = length of conductor in the magnetic field

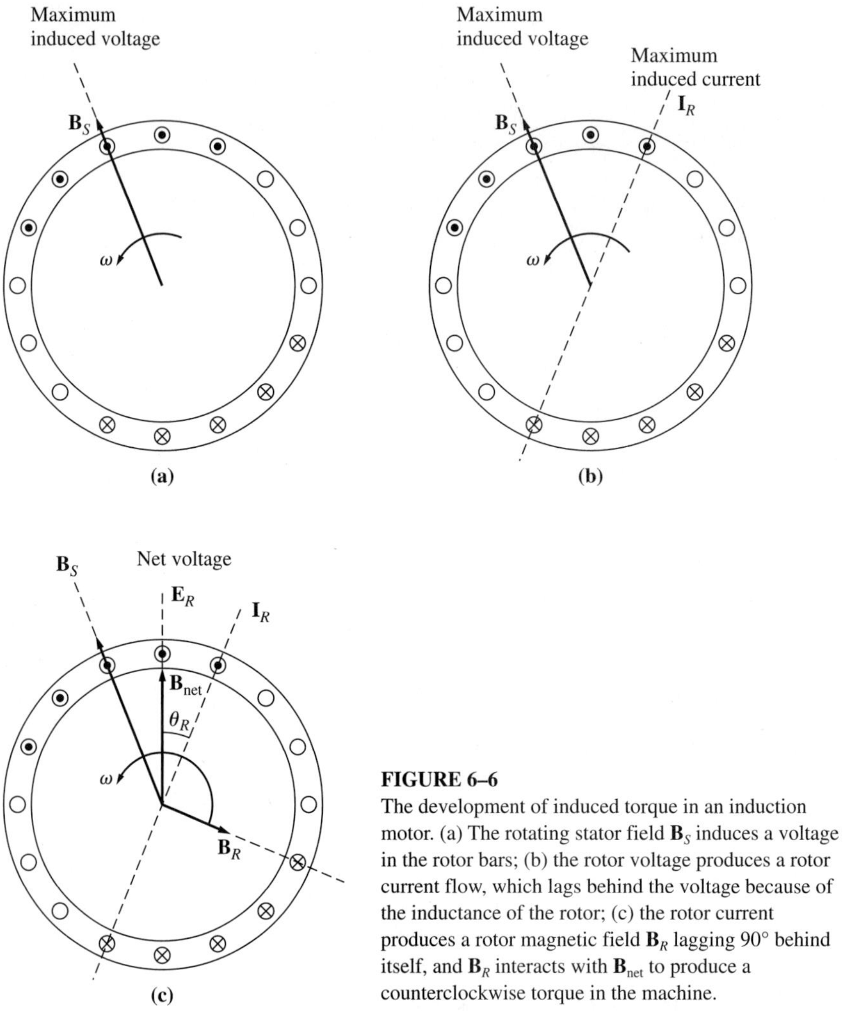

FIGURE 6–6
The development of induced torque in an induction motor. (a) The rotating stator field $\mathbf{B}_S$ induces a voltage in the rotor bars; (b) the rotor voltage produces a rotor current flow, which lags behind the voltage because of the inductance of the rotor; (c) the rotor current produces a rotor magnetic field $\mathbf{B}_R$ lagging 90° behind itself, and $\mathbf{B}_R$ interacts with $\mathbf{B}_{net}$ to produce a counterclockwise torque in the machine.

It is the *relative* motion of the rotor compared to the stator magnetic field that produces induced voltage in a rotor bar. The velocity of the upper rotor bars relative to the magnetic field is to the right, so the induced voltage in the upper bars is out of the page, while the induced voltage in the lower bars is into the page. This results in a current flow out of the upper bars and into the lower bars. However, since the rotor assembly is inductive, the peak rotor current lags behind the peak rotor voltage (see Figure 6–6b). The rotor current flow produces a rotor magnetic field $\mathbf{B}_R$.

Finally, since the induced torque in the machine is given by

$$\tau_{\text{ind}} = k\mathbf{B}_R \times \mathbf{B}_S \tag{3–58}$$

the resulting torque is counterclockwise. Since the rotor induced torque is counterclockwise, the rotor accelerates in that direction.

There is a finite upper limit to the motor's speed, however. If the induction motor's rotor were turning at *synchronous speed*, then the rotor bars would be stationary *relative to the magnetic field* and there would be no induced voltage. If e_{ind} were equal to 0, then there would be no rotor current and no rotor magnetic field. With no rotor magnetic field, the induced torque would be zero, and the rotor would slow down as a result of friction losses. An induction motor can thus speed up to near-synchronous speed, but it can never exactly reach synchronous speed.

Note that in normal operation *both the rotor and stator magnetic fields* $\mathbf{B}_R$ *and* $\mathbf{B}_S$ *rotate together at synchronous speed* n_{sync}, *while the rotor itself turns at a slower speed.*

The Concept of Rotor Slip

The voltage induced in a rotor bar of an induction motor depends on the speed of the rotor *relative to the magnetic fields*. Since the behavior of an induction motor depends on the rotor's voltage and current, it is often more logical to talk about this relative speed. Two terms are commonly used to define the relative motion of the rotor and the magnetic fields. One is *slip speed*, defined as the difference between synchronous speed and rotor speed:

$$\boxed{n_{\text{slip}} = n_{\text{sync}} - n_m} \tag{6–2}$$

where n_{slip} = slip speed of the machine
n_{sync} = speed of the magnetic fields
n_m = mechanical shaft speed of motor

The other term used to describe the relative motion is *slip*, which is the relative speed expressed on a per-unit or a percentage basis. That is, slip is defined as

$$s = \frac{n_{\text{slip}}}{n_{\text{sync}}}(\times\ 100\%) \tag{6–3}$$

$$\boxed{s = \frac{n_{\text{sync}} - n_m}{n_{\text{sync}}}(\times\ 100\%)} \tag{6–4}$$

This equation can also be expressed in terms of angular velocity ω (radians per second) as

$$\boxed{s = \frac{\omega_{\text{sync}} - \omega_m}{\omega_{\text{sync}}}(\times\ 100\%)} \tag{6–5}$$

Notice that if the rotor turns at synchronous speed, $s = 0$, while if the rotor is stationary, $s = 1$. All normal motor speeds fall somewhere between those two limits.

It is possible to express the mechanical speed of the rotor shaft in terms of synchronous speed and slip. Solving Equations (6–4) and (6–5) for mechanical speed yields

$$\boxed{n_m = (1 - s)n_{\text{sync}}} \tag{6–6}$$

or

$$\boxed{\omega_m = (1 - s)\omega_{\text{sync}}} \tag{6–7}$$

These equations are useful in the derivation of induction motor torque and power relationships.

The Electrical Frequency on the Rotor

An induction motor works by inducing voltages and currents in the rotor of the machine, and for that reason it has sometimes been called a *rotating transformer*. Like a transformer, the primary (stator) induces a voltage in the secondary (rotor), but *unlike* a transformer, the secondary frequency is not necessarily the same as the primary frequency.

If the rotor of a motor is locked so that it cannot move, then the rotor will have the same frequency as the stator. On the other hand, if the rotor turns at synchronous speed, the frequency on the rotor will be zero. What will the rotor frequency be for any arbitrary rate of rotor rotation?

At $n_m = 0$ r/min, the rotor frequency $f_{re} = f_{se}$, and the slip $s = 1$. At $n_m = n_{\text{sync}}$, the rotor frequency $f_{re} = 0$ Hz, and the slip $s = 0$. For any speed in between, the rotor frequency is directly proportional to the *difference* between the speed of the magnetic field n_{sync} and the speed of the rotor n_m. Since the slip of the rotor is defined as

$$s = \frac{n_{\text{sync}} - n_m}{n_{\text{sync}}} \tag{6–4}$$

the rotor frequency can be expressed as

$$\boxed{f_{re} = sf_{se}} \tag{6–8}$$

Several alternative forms of this expression exist that are sometimes useful. One of the more common expressions is derived by substituting Equation (6–4) for the slip into Equation (6–8) and then substituting for n_{sync} in the denominator of the expression:

$$f_{re} = \frac{n_{\text{sync}} - n_m}{n_{\text{sync}}} f_{se}$$

But $n_{\text{sync}} = 120 f_{se}/P$ [from Equation (6–1)], so

$$f_{re} = (n_{\text{sync}} - n_m)\frac{P}{120 f_{se}} f_{se}$$

Therefore,

$$\boxed{f_{re} = \frac{P}{120}(n_{\text{sync}} - n_m)} \tag{6–9}$$

Example 6–1. A 208-V, 10-hp, four-pole, 60-Hz, Y-connected induction motor has a full-load slip of 5 percent.

(*a*) What is the synchronous speed of this motor?

(*b*) What is the rotor speed of this motor at the rated load?
(*c*) What is the rotor frequency of this motor at the rated load?
(*d*) What is the shaft torque of this motor at the rated load?

Solution

(*a*) The synchronous speed of this motor is

$$n_{\text{sync}} = \frac{120 f_{se}}{P} \quad (6\text{–}1)$$
$$= \frac{120(60\text{ Hz})}{4\text{ poles}} = 1800\text{ r/min}$$

(*b*) The rotor speed of the motor is given by

$$n_m = (1 - s)n_{\text{sync}} \tag{6–6}$$
$$= (1 - 0.05)(1800\text{ r/min}) = 1710\text{ r/min}$$

(*c*) The rotor frequency of this motor is given by

$$f_{re} = sf_{se} = (0.05)(60\text{ Hz}) = 3\text{ Hz} \tag{6–8}$$

Alternatively, the frequency can be found from Equation (6–9):

$$f_{re} = \frac{P}{120}(n_{\text{sync}} - n_m) \tag{6–9}$$
$$= \frac{4}{120}(1800\text{ r/min} - 1710\text{ r/min}) = 3\text{ Hz}$$

(*d*) The shaft load torque is given by

$$\tau_{\text{load}} = \frac{P_{\text{out}}}{\omega_m}$$
$$= \frac{(10\text{ hp})(746\text{ W/hp})}{(1710\text{ r/min})(2\pi\text{ rad/r})(1\text{ min/60 s})} = 41.7\text{ N} \bullet \text{m}$$

The shaft load torque in English units is given by Equation (1–17):

$$\tau_{\text{load}} = \frac{5252P}{n}$$

where τ is in pound-feet, P is in horsepower, and n_m is in revolutions per minute. Therefore,

$$\tau_{\text{load}} = \frac{5252(10\text{ hp})}{1710\text{ r/min}} = 30.7\text{ lb} \bullet \text{ft}$$

6.3 THE EQUIVALENT CIRCUIT OF AN INDUCTION MOTOR

An induction motor relies for its operation on the induction of voltages and currents in its rotor circuit from the stator circuit (transformer action). Because the induction of voltages and currents in the rotor circuit of an induction motor is essentially a

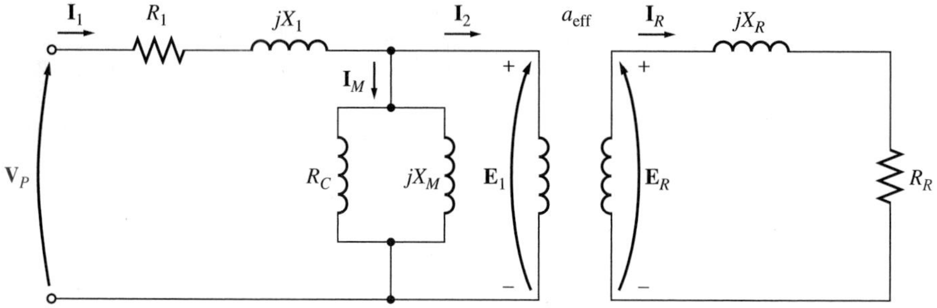

FIGURE 6–7
The transformer model of an induction motor, with rotor and stator connected by an ideal transformer of turns ratio a_{eff}.

transformer operation, the equivalent circuit of an induction motor will turn out to be very similar to the equivalent circuit of a transformer. An induction motor is called a *singly excited* machine (as opposed to a *doubly excited* synchronous machine), since power is supplied to only the stator circuit. Because an induction motor does not have an independent field circuit, its model will not contain an internal voltage source such as the internal generated voltage $\mathbf{E}_A$ in a synchronous machine.

It is possible to derive the equivalent circuit of an induction motor from a knowledge of transformers and from what we already know about the variation of rotor frequency with speed in induction motors. The induction motor model will be developed by starting with the transformer model in Chapter 2 and then deciding how to take the variable rotor frequency and other similar induction motor effects into account.

The Transformer Model of an Induction Motor

A transformer per-phase equivalent circuit, representing the operation of an induction motor, is shown in Figure 6–7. As in any transformer, there is a certain resistance and self-inductance in the primary (stator) windings, which must be represented in the equivalent circuit of the machine. The stator resistance will be called R_1, and the stator leakage reactance will be called X_1. These two components appear right at the input to the machine model.

Also, like any transformer with an iron core, the flux in the machine is related to the integral of the applied voltage $\mathbf{E}_1$. The curve of magnetomotive force versus flux (magnetization curve) for this machine is compared to a similar curve for a power transformer in Figure 6–8. Notice that the slope of the induction motor's magnetomotive force-flux curve is much shallower than the curve of a good transformer. This is because there must be an air gap in an induction motor, which greatly increases the reluctance of the flux path and therefore reduces the coupling between primary and secondary windings. The higher reluctance caused by the air gap means that a higher magnetizing current is required to obtain a given flux level. Therefore, the magnetizing reactance X_M in the equivalent circuit will have a much smaller value (or the susceptance B_M will have a much larger value) than it would in an ordinary transformer.

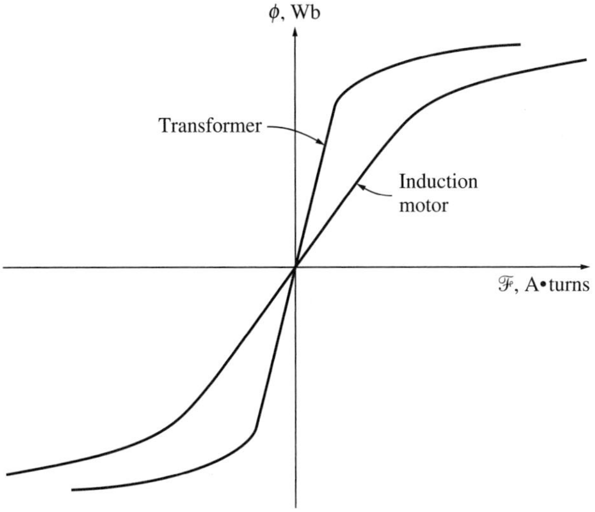

FIGURE 6–8
The magnetization curve of an induction motor compared to that of a transformer.

The primary internal stator voltage $\mathbf{E}_1$ is coupled to the secondary $\mathbf{E}_R$ by an ideal transformer with an effective turns ratio a_{eff}. The effective turns ratio a_{eff} is fairly easy to determine for a wound-rotor motor—it is basically the ratio of the conductors per phase on the stator to the conductors per phase on the rotor, modified by any pitch and distribution factor differences. It is rather difficult to see a_{eff} clearly in the case of a cage rotor motor because there are no distinct windings on the cage rotor. In either case, there *is* an effective turns ratio for the motor.

The voltage $\mathbf{E}_R$ produced in the rotor in turn produces a current flow in the shorted rotor (or secondary) circuit of the machine.

The primary impedances and the magnetization current of the induction motor are very similar to the corresponding components in a transformer equivalent circuit. An induction motor equivalent circuit differs from a transformer equivalent circuit primarily in the effects of varying rotor frequency on the rotor voltage $\mathbf{E}_R$ and the rotor impedances R_R and jX_R.

The Rotor Circuit Model

In an induction motor, when the voltage is applied to the stator windings, a voltage is induced in the rotor windings of the machine. In general, *the greater the relative motion between the rotor and the stator magnetic fields, the greater the resulting rotor voltage and rotor frequency*. The largest relative motion occurs when the rotor is stationary, called the *locked-rotor* or *blocked-rotor* condition, so the largest voltage and rotor frequency are induced in the rotor at that condition. The smallest voltage (0 V) and frequency (0 Hz) occur when the rotor moves at the same speed as the stator magnetic field, resulting in no relative motion. The magnitude and frequency of the voltage induced in the rotor at any speed between these extremes is *directly proportional to the slip of the rotor*. Therefore, if the magnitude of the induced rotor voltage at locked-rotor conditions is called E_{R0}, the magnitude of the induced voltage at any slip will be given by the equation

$$E_R = sE_{R0} \tag{6–10}$$

and the frequency of the induced voltage at any slip will be given by the equation

$$f_{re} = sf_{se} \tag{6–8}$$

This voltage is induced in a rotor containing both resistance and reactance. The rotor resistance R_R is a constant (except for the skin effect), independent of slip, while the rotor reactance is affected in a more complicated way by slip.

The reactance of an induction motor rotor depends on the inductance of the rotor and the frequency of the voltage and current in the rotor. With a rotor inductance of L_R, the rotor reactance is given by

$$X_R = \omega_{re} L_R = 2\pi f_{re} L_R$$

By Equation (6–8), $f_{re} = sf_{se}$, so

$$\begin{aligned} X_R &= 2\pi s f_{se} L_R \\ &= s(2\pi f_{se} L_R) \\ &= sX_{R0} \end{aligned} \tag{6–11}$$

where X_{R0} is the blocked-rotor rotor reactance.

The resulting rotor equivalent circuit is shown in Figure 6–9. The rotor current flow can be found as

$$\mathbf{I}_R = \frac{\mathbf{E}_R}{R_R + jX_R}$$

$$\boxed{\mathbf{I}_R = \frac{\mathbf{E}_R}{R_R + jsX_{R0}}} \tag{6–12}$$

or

$$\boxed{\mathbf{I}_R = \frac{\mathbf{E}_{R0}}{R_R/s + jX_{R0}}} \tag{6–13}$$

Notice from Equation (6–13) that it is possible to treat all of the rotor effects due to varying rotor speed as being caused by a *varying impedance* supplied with power from a constant-voltage source $\mathbf{E}_{R0}$. The equivalent rotor impedance from this point of view is

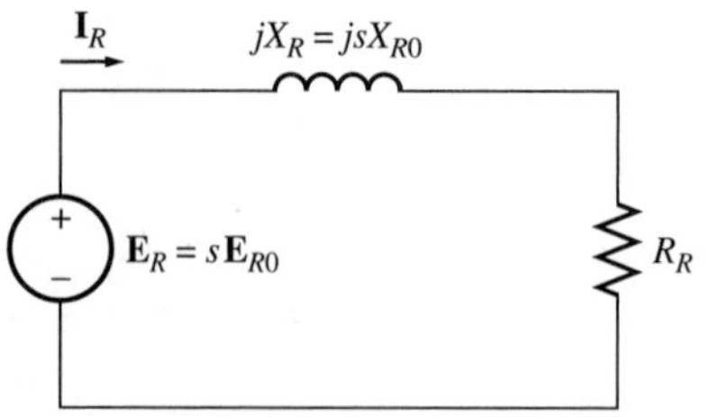

FIGURE 6–9
The rotor circuit model of an induction motor.

$$Z_{R,\text{eq}} = R_R/s + jX_{R0} \tag{6–14}$$

and the rotor equivalent circuit using this convention is shown in Figure 6–10. In the equivalent circuit in Figure 6–10, the rotor voltage is a constant $\mathbf{E}_{R0}$ V and the rotor impedance $Z_{R,\text{eq}}$ contains all the effects of varying rotor slip. A plot of the current flow in the rotor as developed in Equations (6–12) and (6–13) is shown in Figure 6–11.

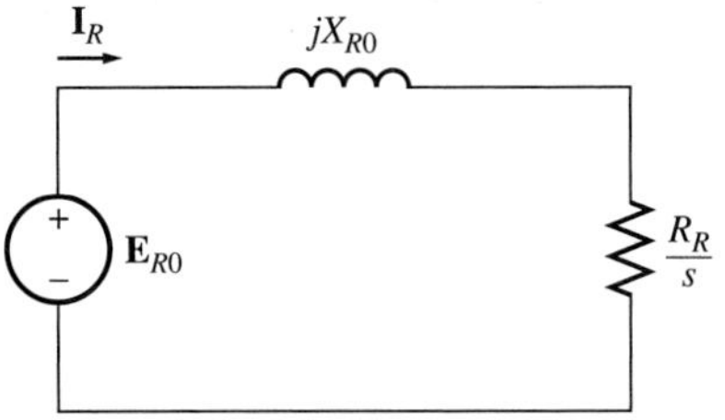

FIGURE 6–10
The rotor circuit model with all the frequency (slip) effects concentrated in resistor R_R.

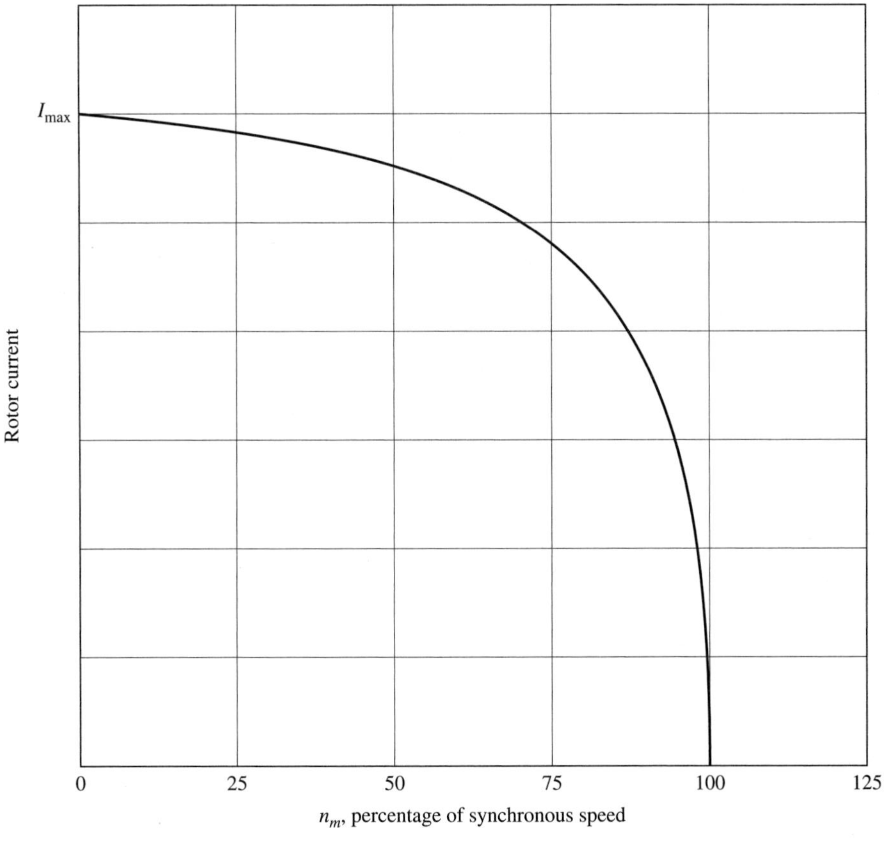

FIGURE 6–11
Rotor current as a function of rotor speed.

Notice that at very low slips the resistive term $R_R/s >> X_{R0}$, so the rotor resistance predominates and the rotor current varies *linearly* with slip. At high slips, X_{R0} is much larger than R_R/s, and the rotor current *approaches a steady-state value* as the slip becomes very large.

The Final Equivalent Circuit

To produce the final per-phase equivalent circuit for an induction motor, it is necessary to refer the rotor part of the model over to the stator side. The rotor circuit model that will be referred to the stator side is the model shown in Figure 6–10, which has all the speed variation effects concentrated in the impedance term.

In an ordinary transformer, the voltages, currents, and impedances on the secondary side of the device can be referred to the primary side by means of the turns ratio of the transformer:

$$\mathbf{V}_P = \mathbf{V}'_S = a\mathbf{V}_S \tag{6–15}$$

$$\mathbf{I}_P = \mathbf{I}'_S = \frac{\mathbf{I}_S}{a} \tag{6–16}$$

and

$$Z'_S = a^2 Z_S \tag{6–17}$$

where the prime refers to the referred values of voltage, current, and impedance.

Exactly the same sort of transformation can be done for the induction motor's rotor circuit. If the effective turns ratio of an induction motor is a_{eff}, then the transformed rotor voltage becomes

$$\mathbf{E}_1 = \mathbf{E}'_R = a_{\text{eff}} \mathbf{E}_{R0} \tag{6–18}$$

the rotor current becomes

$$\mathbf{I}_2 = \frac{\mathbf{I}_R}{a_{\text{eff}}} \tag{6–19}$$

and the rotor impedance becomes

$$Z_2 = a_{\text{eff}}^2 \left(\frac{R_R}{s} + jX_{R0} \right) \tag{6–20}$$

If we now make the following definitions:

$$R_2 = a_{\text{eff}}^2 R_R \tag{6–21}$$

$$X_2 = a_{\text{eff}}^2 X_{R0} \tag{6–22}$$

then the final per-phase equivalent circuit of the induction motor is as shown in Figure 6–12.

The rotor resistance R_R and the locked-rotor rotor reactance X_{R0} are very difficult or impossible to determine directly on cage rotors, and the effective turns ratio a_{eff} is also difficult to obtain for cage rotors. Fortunately, though, it is possible

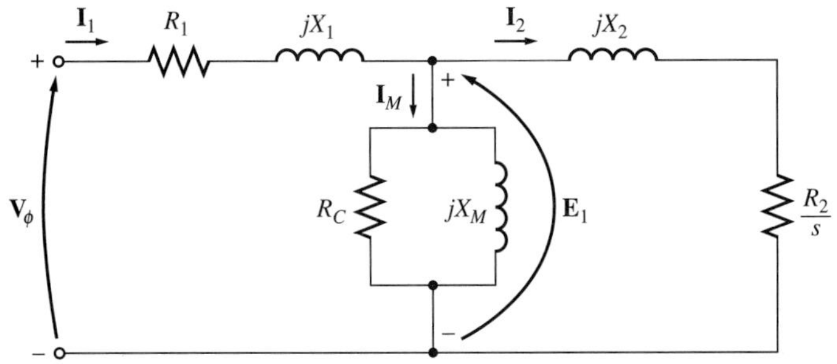

FIGURE 6–12
The per-phase equivalent circuit of an induction motor.

to make measurements that will directly give the *referred resistance and reactance* R_2 and X_2, even though R_R, X_{R0} and a_{eff} are not known separately. The measurement of induction motor parameters will be taken up in Section 6.7.

6.4 POWER AND TORQUE IN INDUCTION MOTORS

Because induction motors are singly excited machines, their power and torque relationships are considerably different from the relationships in the synchronous machines previously studied. This section reviews the power and torque relationships in induction motors.

Losses and the Power-Flow Diagram

An induction motor can be basically described as a rotating transformer. Its input is a three-phase system of voltages and currents. For an ordinary transformer, the output is electric power from the secondary windings. The secondary windings in an induction motor (the rotor) are shorted out, so no electrical output exists from normal induction motors. Instead, the output is mechanical. The relationship between the input electric power and the output mechanical power of this motor is shown in the power-flow diagram in Figure 6–13.

The input power to an induction motor P_{in} is in the form of three-phase electric voltages and currents. The first losses encountered in the machine are I^2R losses in the stator windings (the *stator copper loss* P_{SCL}). Then some amount of power is lost as hysteresis and eddy currents in the stator (P_{core}). The power remaining at this point is transferred to the rotor of the machine across the air gap between the stator and rotor. This power is called the *air-gap power* P_{AG} of the machine. After the power is transferred to the rotor, some of it is lost as I^2R losses (the *rotor copper loss* P_{RCL}), and the rest is converted from electrical to mechanical form (P_{conv}). Finally, friction and windage losses $P_{\text{F\&W}}$ and stray losses P_{misc} are subtracted. The remaining power is the output of the motor P_{out}.

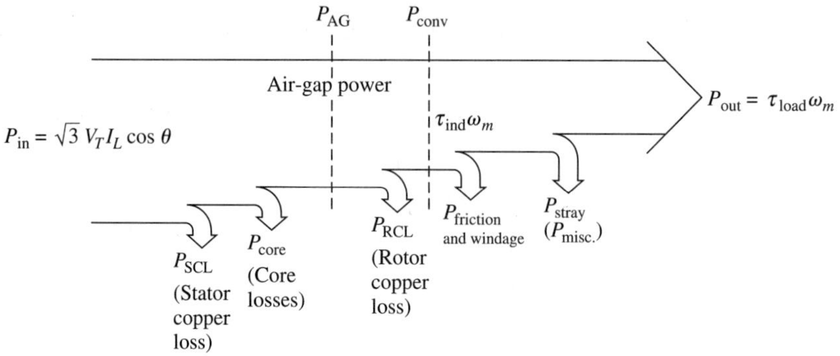

FIGURE 6–13
The power-flow diagram of an induction motor.

The *core losses* do not always appear in the power-flow diagram at the point shown in Figure 6–13. Because of the nature of core losses, where they are accounted for in the machine is somewhat arbitrary. The core losses of an induction motor come partially from the stator circuit and partially from the rotor circuit. Since an induction motor normally operates at a speed near synchronous speed, the relative motion of the magnetic fields over the rotor surface is quite slow, and the rotor core losses are very tiny compared to the stator core losses. Since the largest fraction of the core losses comes from the stator circuit, all the core losses are lumped together at that point on the diagram. These losses are represented in the induction motor equivalent circuit by the resistor R_C (or the conductance G_C). If core losses are just given by a number (X watts) instead of as a circuit element, they are often lumped together with the mechanical losses and subtracted at the point on the diagram where the mechanical losses are located.

The *higher* the speed of an induction motor, the *higher* its friction, windage, and stray losses. On the other hand, the *higher* the speed of the motor (up to n_{sync}), the *lower* its core losses. Therefore, these three categories of losses are sometimes lumped together and called *rotational losses*. The total rotational losses of a motor are often considered to be constant with changing speed, since the component losses change in opposite directions with a change in speed.

Example 6–2. A 480-V, 60-Hz, 50-hp, three-phase induction motor is drawing 60 A at 0.85 PF lagging. The stator copper losses are 2 kW, and the rotor copper losses are 700 W. The friction and windage losses are 600 W, the core losses are 1800 W, and the stray losses are negligible. Find the following quantities:

(a) The air-gap power P_{AG}
(b) The power converted P_{conv}
(c) The output power P_{out}
(d) The efficiency of the motor

Solution

To answer these questions, refer to the power-flow diagram for an induction motor (Figure 6–13).

(a) The air-gap power is just the input power minus the stator I^2R losses and core losses. The input power is given by

$$P_{\text{in}} = \sqrt{3}V_T I_L \cos\theta$$
$$= \sqrt{3}(480\text{ V})(60\text{ A})(0.85) = 42.4\text{ kW}$$

From the power-flow diagram, the air-gap power is given by

$$P_{\text{AG}} = P_{\text{in}} - P_{\text{SCL}} - P_{\text{core}}$$
$$= 42.4\text{ kW} - 2\text{ kW} - 1.8\text{ kW} = 38.6\text{ kW}$$

(b) From the power-flow diagram, the power converted from electrical to mechanical form is

$$P_{\text{conv}} = P_{\text{AG}} - P_{\text{RCL}}$$
$$= 38.6\text{ kW} - 700\text{ W} = 37.9\text{ kW}$$

(c) From the power-flow diagram, the output power is given by

$$P_{\text{out}} = P_{\text{conv}} - P_{\text{F\&W}} - P_{\text{misc}}$$
$$= 37.9\text{ kW} - 600\text{ W} - 0\text{ W} = 37.3\text{ kW}$$

or, in horsepower,

$$P_{\text{out}} = (37.3\text{ kW})\frac{1\text{ hp}}{0.746\text{ kW}} = 50\text{ hp}$$

(d) Therefore, the induction motor's efficiency is

$$\eta = \frac{P_{\text{out}}}{P_{\text{in}}} \times 100\%$$
$$= \frac{37.3\text{ kW}}{42.4\text{ kW}} \times 100\% = 88\%$$

Power and Torque in an Induction Motor

Figure 6–12 shows the per-phase equivalent circuit of an induction motor. If the equivalent circuit is examined closely, it can be used to derive the power and torque equations governing the operation of the motor.

The input current to a phase of the motor can be found by dividing the input voltage by the total equivalent impedance:

$$\mathbf{I}_1 = \frac{\mathbf{V}_\phi}{Z_{\text{eq}}} \tag{6–23}$$

where

$$Z_{\text{eq}} = R_1 + jX_1 + \frac{1}{G_C - jB_M + \dfrac{1}{V_2/s + jX_2}} \tag{6–24}$$

Therefore, the stator copper losses, the core losses, and the rotor copper losses can be found. The stator copper losses in the three phases are given by

$$\boxed{P_{\text{SCL}} = 3I_1^2 R_1} \tag{6–25}$$

The core losses are given by

$$\boxed{P_{\text{core}} = 3E_1^2 G_C} \tag{6–26}$$

so the air-gap power can be found as

$$\boxed{P_{\text{AG}} = P_{\text{in}} - P_{\text{SCL}} - P_{\text{core}}} \tag{6–27}$$

Look closely at the equivalent circuit of the rotor. The *only* element in the equivalent circuit where the air-gap power can be consumed is in the resistor R_2/s. Therefore, the *air-gap power* can also be given by

$$\boxed{P_{\text{AG}} = 3I_2^2 \frac{R_2}{s}} \tag{6–28}$$

The actual resistive losses in the rotor circuit are given by the equation

$$P_{\text{RCL}} = 3I_R^2 R_R \tag{6–29}$$

Since power is unchanged when referred across an ideal transformer, the rotor copper losses can also be expressed as

$$\boxed{P_{\text{RCL}} = 3I_2^2 R_2} \tag{6–30}$$

After stator copper losses, core losses, and rotor copper losses are subtracted from the input power to the motor, the remaining power is converted from electrical to mechanical form. This converted power, which is sometimes called *developed mechanical power*, is given by

$$\begin{aligned} P_{\text{conv}} &= P_{\text{AG}} - P_{\text{RCL}} \\ &= 3I_2^2 \frac{R_2}{s} - 3I_2^2 R_2 \\ &= 3I_2^2 R_2\left(\frac{1}{s} - 1\right) \end{aligned}$$

$$\boxed{P_{\text{conv}} = 3I_2^2 R_2\left(\frac{1 - s}{s}\right)} \tag{6–31}$$

Notice from Equations (6–28) and (6–30) that the rotor copper losses are equal to the air-gap power times the slip:

$$P_{\text{RCL}} = sP_{\text{AG}} \tag{6–32}$$

Therefore, the lower the slip of the motor, the lower the rotor losses in the machine. Note also that if the rotor is not turning, the slip $s = 1$ and the *air-gap power is entirely consumed in the rotor*. This is logical, since if the rotor is not turning, the output power P_{out} $(= \tau_{\text{load}}\,\omega_m)$ must be zero. Since $P_{\text{conv}} = P_{\text{AG}} - P_{\text{RCL}}$, this also gives another relationship between the air-gap power and the power converted from electrical to mechanical form:

$$P_{\text{conv}} = P_{\text{AG}} - P_{\text{RCL}}$$
$$= P_{\text{AG}} - sP_{\text{AG}}$$

$$\boxed{P_{\text{conv}} = (1 - s)P_{\text{AG}}} \tag{6–33}$$

Finally, if the friction and windage losses and the stray losses are known, the output power can be found as

$$\boxed{P_{\text{out}} = P_{\text{conv}} - P_{\text{F\&W}} - P_{\text{misc}}} \tag{6–34}$$

The *induced torque* τ_{ind} in a machine was defined as the torque generated by the internal electric-to-mechanical power conversion. This torque differs from the torque actually available at the terminals of the motor by an amount equal to the friction and windage torques in the machine. The induced torque is given by the equation

$$\tau_{\text{ind}} = \frac{P_{\text{conv}}}{\omega_m} \tag{6–35}$$

This torque is also called the *developed torque* of the machine.

The induced torque of an induction motor can be expressed in a different form as well. Equation (6–7) expresses actual speed in terms of synchronous speed and slip, while Equation (6–33) expresses P_{conv} in terms of P_{AG} and slip. Substituting these two equations into Equation (6–35) yields

$$\tau_{\text{ind}} = \frac{(1 - s)P_{\text{AG}}}{(1 - s)\omega_{\text{sync}}}$$

$$\boxed{\tau_{\text{ind}} = \frac{P_{\text{AG}}}{\omega_{\text{sync}}}} \tag{6–36}$$

The last equation is especially useful because it expresses induced torque directly in terms of air-gap power and *synchronous speed*, which does not vary. A knowledge of P_{AG} thus directly yields τ_{ind}.

Separating the Rotor Copper Losses and the Power Converted in an Induction Motor's Equivalent Circuit

Part of the power coming across the air gap in an induction motor is consumed in the rotor copper losses, and part of it is converted to mechanical power to drive the motor's shaft. It is possible to separate the two uses of the air-gap power and to indicate them separately on the motor equivalent circuit.

Equation (6–28) gives an expression for the total air-gap power in an induction motor, while Equation (6–30) gives the actual rotor losses in the motor. The air-gap power is the power which would be consumed in a resistor of value

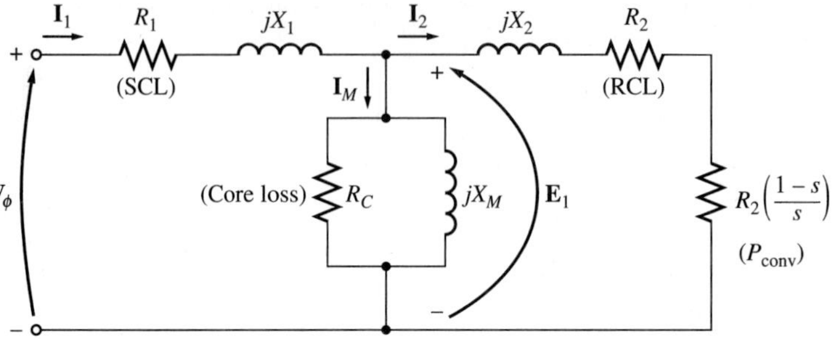

FIGURE 6–14
The per-phase equivalent circuit with rotor losses and P_{core} separated.

R_2/s, while the rotor copper losses are the power which would be consumed in a resistor of value R_2. The difference between them is P_{conv}, which must therefore be the power consumed in a resistor of value

$$R_{\text{conv}} = \frac{R_2}{s} - R_2 = R_2\left(\frac{1}{s} - 1\right)$$

$$\boxed{R_{\text{conv}} = R_2\left(\frac{1-s}{s}\right)} \tag{6–37}$$

Per-phase equivalent circuit with the rotor copper losses and the power converted to mechanical form separated into distinct elements is shown in Figure 6–14.

Example 6–3. A 460-V, 25-hp, 60-Hz, four-pole, Y-connected induction motor has the following impedances in ohms per phase referred to the stator circuit:

$$R_1 = 0.641\ \Omega \qquad R_2 = 0.332\ \Omega$$
$$X_1 = 1.106\ \Omega \qquad X_2 = 0.464\ \Omega \qquad X_M = 26.3\ \Omega$$

The total rotational losses are 1100 W and are assumed to be constant. The core loss is lumped in with the rotational losses. For a rotor slip of 2.2 percent at the rated voltage and rated frequency, find the motor's

(*a*) Speed
(*b*) Stator current
(*c*) Power factor
(*d*) P_{conv} and P_{out}
(*e*) τ_{ind} and τ_{load}
(*f*) Efficiency

Solution

The per-phase equivalent circuit of this motor is shown in Figure 6–12, and the power-flow diagram is shown in Figure 6–13. Since the core losses are lumped together with the friction and windage losses and the stray losses, they will be treated like the mechanical losses and be subtracted after P_{conv} in the power-flow diagram.

(a) The synchronous speed is

$$n_{sync} = \frac{120 f_{se}}{P} = \frac{120(60 \text{ Hz})}{4 \text{ poles}} = 1800 \text{ r/min}$$

or $$\omega_{sync} = (1800 \text{ r/min})\left(\frac{2\pi \text{ rad}}{1 \text{ r}}\right)\left(\frac{1 \text{ min}}{60 \text{ s}}\right) = 188.5 \text{ rad/s}$$

The rotor's mechanical shaft speed is

$$\begin{aligned} n_m &= (1 - s)n_{sync} \\ &= (1 - 0.022)(1800 \text{ r/min}) = 1760 \text{ r/min} \end{aligned}$$

or $$\begin{aligned} \omega_m &= (1 - s)\omega_{sync} \\ &= (1 - 0.022)(188.5 \text{ rad/s}) = 184.4 \text{ rad/s} \end{aligned}$$

(b) To find the stator current, get the equivalent impedance of the circuit. The first step is to combine the referred rotor impedance in parallel with the magnetization branch, and then to add the stator impedance to that combination in series. The referred rotor impedance is

$$\begin{aligned} Z_2 &= \frac{R_2}{s} + jX_2 \\ &= \frac{0.332}{0.022} + j0.464 \\ &= 15.09 + j0.464 \ \Omega = 15.10\angle 1.76^\circ \ \Omega \end{aligned}$$

The combined magnetization plus rotor impedance is given by

$$\begin{aligned} Z_f &= \frac{1}{1/jX_M + 1/Z_2} \\ &= \frac{1}{-j0.038 + 0.0662\angle -1.76^\circ} \\ &= \frac{1}{0.0773\angle -31.1^\circ} = 12.94\angle 31.1^\circ \ \Omega \end{aligned}$$

Therefore, the total impedance is

$$\begin{aligned} Z_{tot} &= Z_{stat} + Z_f \\ &= 0.641 + j1.106 + 12.94\angle 31.1^\circ \ \Omega \\ &= 11.72 + j7.79 = 14.07\angle 33.6^\circ \ \Omega \end{aligned}$$

The resulting stator current is

$$\begin{aligned} \mathbf{I}_1 &= \frac{\mathbf{V}_\phi}{Z_{tot}} \\ &= \frac{266\angle 0^\circ \text{ V}}{14.07\angle 33.6^\circ \ \Omega} = 18.88\angle -33.6^\circ \text{ A} \end{aligned}$$

(c) The power motor power factor is

$$\text{PF} = \cos 33.6^\circ = 0.833 \qquad \text{lagging}$$

(d) The input power to this motor is

$$\begin{aligned} P_{\text{in}} &= \sqrt{3} V_T I_L \cos \theta \\ &= \sqrt{3}(460 \text{ V})(18.88 \text{ A})(0.833) = 12{,}530 \text{ W} \end{aligned}$$

The stator copper losses in this machine are

$$\begin{aligned} P_{\text{SCL}} &= 3 I_1^2 R_1 \qquad (6\text{–}25) \\ &= 3(18.88 \text{ A})^2(0.641 \ \Omega) = 685 \text{ W} \end{aligned}$$

The air-gap power is given by

$$P_{\text{AG}} = P_{\text{in}} - P_{\text{SCL}} = 12{,}530 \text{ W} - 685 \text{ W} = 11{,}845 \text{ W}$$

Therefore, the power converted is

$$P_{\text{conv}} = (1 - s)P_{\text{AG}} = (1 - 0.022)(11{,}845 \text{ W}) = 11{,}585 \text{ W}$$

The power P_{out} is given by

$$\begin{aligned} P_{\text{out}} &= P_{\text{conv}} - P_{\text{rot}} = 11{,}585 \text{ W} - 1100 \text{ W} = 10{,}485 \text{ W} \\ &= 10{,}485 \text{ W}\left(\frac{1 \text{ hp}}{746 \text{ W}}\right) = 14.1 \text{ hp} \end{aligned}$$

(e) The induced torque is given by

$$\begin{aligned} \tau_{\text{ind}} &= \frac{P_{\text{AG}}}{\omega_{\text{sync}}} \\ &= \frac{11{,}845 \text{ W}}{188.5 \text{ rad/s}} = 62.8 \text{ N} \bullet \text{m} \end{aligned}$$

and the output torque is given by

$$\begin{aligned} \tau_{\text{load}} &= \frac{P_{\text{out}}}{\omega_m} \\ &= \frac{10{,}485 \text{ W}}{184.4 \text{ rad/s}} = 56.9 \text{ N} \bullet \text{m} \end{aligned}$$

(In English units, these torques are 46.3 and 41.9 lb-ft, respectively.)

(f) The motor's efficiency at this operating condition is

$$\begin{aligned} \eta &= \frac{P_{\text{out}}}{P_{\text{in}}} \times 100\% \\ &= \frac{10{,}485 \text{ W}}{12{,}530 \text{ W}} \times 100\% = 83.7\% \end{aligned}$$

6.5 INDUCTION MOTOR TORQUE–SPEED CHARACTERISTICS

How does the torque of an induction motor change as the load changes? How much torque can an induction motor supply at starting conditions? How much does the speed of an induction motor drop as its shaft load increases? To find out

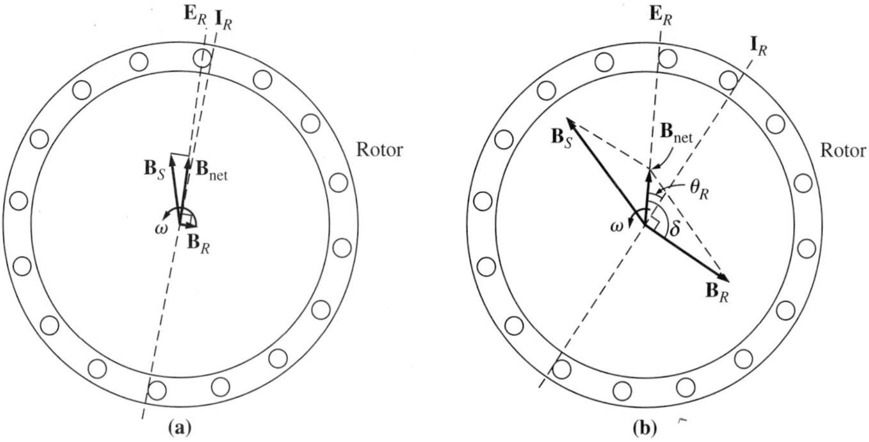

FIGURE 6–15
(a) The magnetic fields in an induction motor under light loads. (b) The magnetic fields in an induction motor under heavy loads.

the answers to these and similar questions, it is necessary to clearly understand the relationships among the motor's torque, speed, and power.

In the following material, the torque–speed relationship will be examined first from the physical viewpoint of the motor's magnetic field behavior. Then, a general equation for torque as a function of slip will be derived from the induction motor equivalent circuit (Figure 6–12).

Induced Torque from a Physical Standpoint

Figure 6–15a shows a cage rotor induction motor that is initially operating at no load and therefore very nearly at synchronous speed. The net magnetic field $\mathbf{B}_{\text{net}}$ in this machine is produced by the magnetization current $\mathbf{I}_M$ flowing in the motor's equivalent circuit (see Figure 6–12). The magnitude of the magnetization current and hence of $\mathbf{B}_{\text{net}}$ is directly proportional to the voltage $\mathbf{E}_1$. If $\mathbf{E}_1$ is constant, then the net magnetic field in the motor is constant. In an actual machine, $\mathbf{E}_1$ varies as the load changes, because the stator impedances R_1 and X_1 cause varying voltage drops with varying load. However, these drops in the stator windings are relatively small, so $\mathbf{E}_1$ (and hence $\mathbf{I}_M$ and $\mathbf{B}_{\text{net}}$) is approximately constant with changes in load.

Figure 6–15a shows the induction motor at no load. At no load, the rotor slip is very small, and so the relative motion between the rotor and the magnetic fields is very small and the rotor frequency is also very small. Since the relative motion is small, the voltage $\mathbf{E}_R$ induced in the bars of the rotor is very small, and the resulting current flow $\mathbf{I}_R$ is small. Also, because the rotor frequency is so very small, the reactance of the rotor is nearly zero, and the maximum rotor current $\mathbf{I}_R$ is almost in phase with the rotor voltage $\mathbf{E}_R$. The rotor current thus produces a small magnetic field $\mathbf{B}_R$ at an angle just slightly greater than 90° behind the net magnetic field $\mathbf{B}_{\text{net}}$. Notice that the stator current must be quite large even at no load, since it must supply most of $\mathbf{B}_{\text{net}}$. (This is why induction motors have large no-load currents compared to other types of machines. The no-load current of an induction motor is usually 30–60 percent of the full-load current.)

The induced torque, which keeps the rotor turning, is given by the equation

$$\tau_{\text{ind}} = k\mathbf{B}_R \times \mathbf{B}_{\text{net}} \tag{3–60}$$

Its magnitude is given by

$$\tau_{\text{ind}} = k\mathbf{B}_R\mathbf{B}_{\text{net}} \sin \delta \tag{3–61}$$

Since the rotor magnetic field is very small, the induced torque is also quite small—just large enough to overcome the motor's rotational losses.

Now suppose the induction motor is loaded down (Figure 6–15b). As the motor's load increases, its slip increases, and the rotor speed falls. Since the rotor speed is slower, there is now *more relative motion* between the rotor and the stator magnetic fields in the machine. Greater relative motion produces a stronger rotor voltage $\mathbf{E}_R$ which in turn produces a larger rotor current $\mathbf{I}_R$. With a larger rotor current, the rotor magnetic field $\mathbf{B}_R$ also increases. However, the angle of the rotor current and $\mathbf{B}_R$ changes as well. Since the rotor slip is larger, the rotor frequency rises ($f_{re} = sf_{se}$), and the rotor's reactance increases ($\omega_{re}L_R$). Therefore, the rotor current now lags further behind the rotor voltage, and the rotor magnetic field shifts with the current. Figure 6–15b shows the induction motor operating at a fairly high load. Notice that the rotor current has increased and that the angle δ has increased. The increase in B_R tends to increase the torque, while the increase in angle δ tends to decrease the torque (τ_{ind} is proportional to $\sin \delta$, and $\delta > 90°$). Since the first effect is larger than the second one, the overall induced torque increases to supply the motor's increased load.

When does an induction motor reach pullout torque? This happens when the point is reached where, as the load on the shaft is increased, the $\sin \delta$ term decreases more than the B_R term increases. At that point, a further increase in load decreases τ_{ind}, and the motor stops.

It is possible to use a knowledge of the machine's magnetic fields to approximately derive the output torque-versus-speed characteristic of an induction motor. Remember that the magnitude of the induced torque in the machine is given by

$$\tau_{\text{ind}} = kB_RB_{\text{net}} \sin \delta \tag{3–61}$$

Each term in this expression can be considered separately to derive the overall machine behavior. The individual terms are

1. B_R. The rotor magnetic field is directly proportional to the current flowing in the rotor, as long as the rotor is unsaturated. The current flow in the rotor increases with increasing slip (decreasing speed) according to Equation (6–13). This current flow was plotted in Figure 6–11 and is shown again in Figure 6–16a.
2. B_{net}. The net magnetic field in the motor is proportional to E_1 and therefore is approximately constant (E_1 actually decreases with increasing current flow, but this

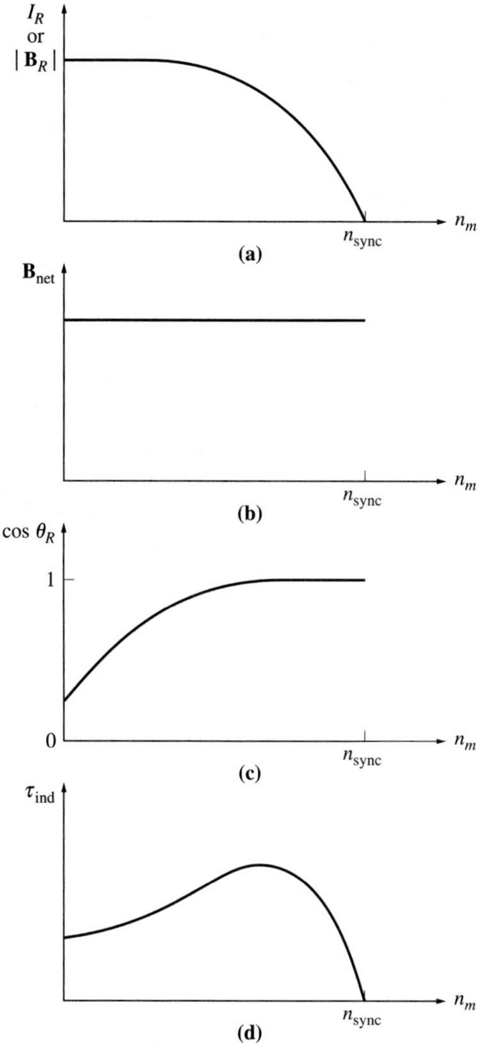

FIGURE 6–16
Graphical development of an induction motor torque–speed characteristic. (a) Plot of rotor current (and thus $|\mathbf{B}_R|$) versus speed for an induction motor; (b) plot of net magnetic field versus speed for the motor; (c) plot of rotor power factor versus speed for the motor; (d) the resulting torque–speed characteristic.

effect is small compared to the other two, and it will be ignored in this graphical development). The curve for B_{net} versus speed is shown in Figure 6–16b.

3. $\sin\delta$. The angle δ between the net and rotor magnetic fields can be expressed in a very useful way. Look at Figure 6–15b. In this figure, it is clear that *the angle δ is just equal to the power-factor angle of the rotor plus* 90°:

$$\delta = \theta_R + 90° \tag{6–38}$$

Therefore, $\sin \delta = \sin(\theta_R + 90°) = \cos \theta_R$. This term is the power factor of the rotor. The rotor power-factor angle can be calculated from the equation

$$\theta_R = \tan^{-1} \frac{X_R}{R_R} = \tan^{-1} \frac{sX_{R0}}{R_R} \tag{6–39}$$

The resulting rotor power factor is given by

$$PF_R = \cos \theta_R$$

$$\boxed{PF_R = \cos\left(\tan^{-1} \frac{sX_{R0}}{R_R}\right)} \tag{6–40}$$

A plot of rotor power factor versus speed is shown in Figure 6–16c.

Since the induced torque is proportional to the product of these three terms, the torque–speed characteristic of an induction motor can be constructed from the graphical multiplication of the previous three plots (Figure 6–16a to c). The torque–speed characteristic of an induction motor derived in this fashion is shown in Figure 6–16d.

This characteristic curve can be divided roughly into three regions. The first region is the *low-slip region* of the curve. In the low-slip region, the motor slip increases approximately linearly with increased load, and the rotor mechanical speed decreases approximately linearly with load. In this region of operation, the rotor reactance is negligible, so the rotor power factor is approximately unity, while the rotor current increases linearly with slip. *The entire normal steady-state operating range of an induction motor is included in this linear low-slip region.* Thus in normal operation, an induction motor has a linear speed droop.

The second region on the induction motor's curve can be called the *moderate-slip region.* In the moderate-slip region, the rotor frequency is higher than before, and the rotor reactance is on the same order of magnitude as the rotor resistance. In this region, the rotor current no longer increases as rapidly as before, and the power factor starts to drop. The peak torque (the *pullout torque*) of the motor occurs at the point where, for an incremental increase in load, the increase in the rotor current is exactly balanced by the decrease in the rotor power factor.

The third region on the induction motor's curve is called the *high-slip region.* In the high-slip region, the induced torque actually decreases with increased load, since the increase in rotor current is completely overshadowed by the decrease in rotor power factor.

For a typical induction motor, the pullout torque on the curve will be 200 to 250 percent of the rated full-load torque of the machine, and the *starting torque* (the torque at zero speed) will be 150 percent or so of the full-load torque. Unlike a synchronous motor, the induction motor can start with a full load attached to its shaft.

The Derivation of the Induction Motor Induced-Torque Equation

It is possible to use the equivalent circuit of an induction motor and the power-flow diagram for the motor to derive a general expression for induced torque as a function of speed. The induced torque in an induction motor is given by Equation (6–35) or (6–36):

$$\tau_{\text{ind}} = \frac{P_{\text{conv}}}{\omega_m} \tag{6–35}$$

$$\tau_{\text{ind}} = \frac{P_{\text{AG}}}{\omega_{\text{sync}}} \tag{6–36}$$

The latter equation is especially useful, since the synchronous speed is a constant for a given frequency and number of poles. Since ω_{sync} is constant, a knowledge of the air-gap power gives the induced torque of the motor.

The air-gap power is the power crossing the gap from the stator circuit to the rotor circuit. It is equal to the power absorbed in the resistance R_2/s. How can this power be found?

Refer to the equivalent circuit given in Figure 6–17. In this figure, the air-gap power supplied to one phase of the motor can be seen to be

$$P_{\text{AG},1\phi} = I_2^2 \frac{R_2}{s}$$

Therefore, the total air-gap power is

$$P_{\text{AG}} = 3I_2^2 \frac{R_2}{s}$$

If I_2 can be determined, then the air-gap power and the induced torque will be known.

Although there are several ways to solve the circuit in Figure 6–17 for the current I_2, perhaps the easiest one is to determine the Thevenin equivalent of the portion of the circuit to the left of the *X*'s in the figure. Thevenin's theorem states that any linear circuit that can be separated by two terminals from the rest of the system can be replaced by a single voltage source in series with an equivalent impedance. If this were done to the induction motor equivalent circuit, the resulting circuit would be a simple series combination of elements as shown in Figure 6–18c.

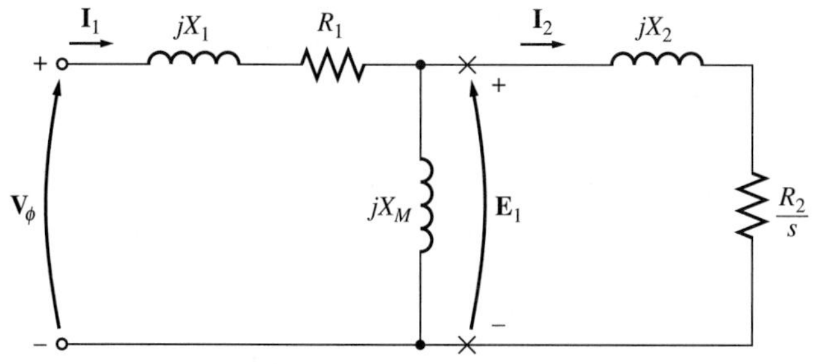

FIGURE 6–17
Per-phase equivalent circuit of an induction motor.

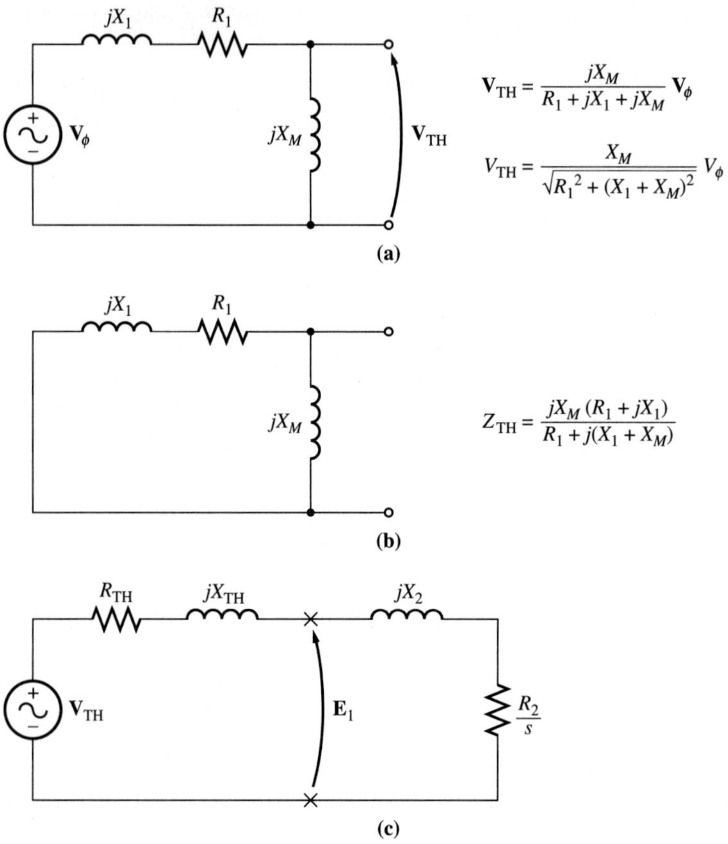

FIGURE 6–18
(a) The Thevenin equivalent voltage of an induction motor input circuit. (b) The Thevenin equivalent impedance of the input circuit. (c) The resulting simplified equivalent circuit of an induction motor.

To calculate the Thevenin equivalent of the input side of the induction motor equivalent circuit, first open-circuit the terminals at the *X*'s and find the resulting open-circuit voltage present there. Then, to find the Thevenin impedance, kill (short-circuit) the phase voltage and find the Z_{eq} seen "looking" into the terminals.

Figure 6–18a shows the open terminals used to find the Thevenin voltage. By the voltage divider rule,

$$\mathbf{V}_{TH} = \mathbf{V}_\phi \frac{Z_M}{Z_M + Z_1}$$

$$= \mathbf{V}_\phi \frac{jX_M}{R_1 + jX_1 + jX_M}$$

The magnitude of the Thevenin voltage $\mathbf{V}_{TH}$ is

$$\boxed{V_{TH} = V_\phi \frac{X_M}{\sqrt{R_1^2 + (X_1 + X_M)^2}}} \tag{6–41a}$$

Since the magnetization reactance $X_M >> X_1$ and $X_M >> R_1$, the magnitude of the Thevenin voltage is approximately

$$\boxed{V_{TH} \approx V_\phi \frac{X_M}{X_1 + X_M}} \tag{6–41b}$$

to quite good accuracy.

Figure 6–18b shows the input circuit with the input voltage source killed. The two impedances are in parallel, and the Thevenin impedance is given by

$$Z_{TH} = \frac{Z_1 Z_M}{Z_1 + Z_M} \tag{6–42}$$

This impedance reduces to

$$Z_{TH} = R_{TH} + jX_{TH} = \frac{jX_M(R_1 + jX_1)}{R_1 + j(X_1 + X_M)} \tag{6–43}$$

Because $X_M >> X_1$ and $X_M + X_1 >> R_1$, the Thevenin resistance and reactance are approximately given by

$$\boxed{R_{TH} \approx R_1 \left(\frac{X_M}{X_1 + X_M}\right)^2} \tag{6–44}$$

$$\boxed{X_{TH} \approx X_1} \tag{6–45}$$

The resulting equivalent circuit is shown in Figure 6–18c. From this circuit, the current $\mathbf{I}_2$ is given by

$$\mathbf{I}_2 = \frac{\mathbf{V}_{TH}}{Z_{TH} + Z_2} \tag{6–46}$$

$$= \frac{\mathbf{V}_{TH}}{R_{TH} + R_2/s + jX_{TH} + jX_2} \tag{6–47}$$

The magnitude of this current is

$$I_2 = \frac{V_{TH}}{\sqrt{(R_{TH} + R_2/s)^2 + (X_{TH} + X_2)^2}} \tag{6–48}$$

The air-gap power is therefore given by

$$P_{AG} = 3I_2^2 \frac{R_2}{s}$$

$$= \frac{3V_{TH}^2 R_2/s}{(R_{TH} + R_2/s)^2 + (X_{TH} + X_2)^2} \tag{6–49}$$

and the rotor-induced torque is given by

$$\tau_{ind} = \frac{P_{AG}}{\omega_{sync}}$$

$$\boxed{\tau_{ind} = \frac{3V_{TH}^2 R_2/s}{\omega_{sync}[(R_{TH} + R_2/s)^2 + (X_{TH} + X_2)^2]}} \tag{6–50}$$

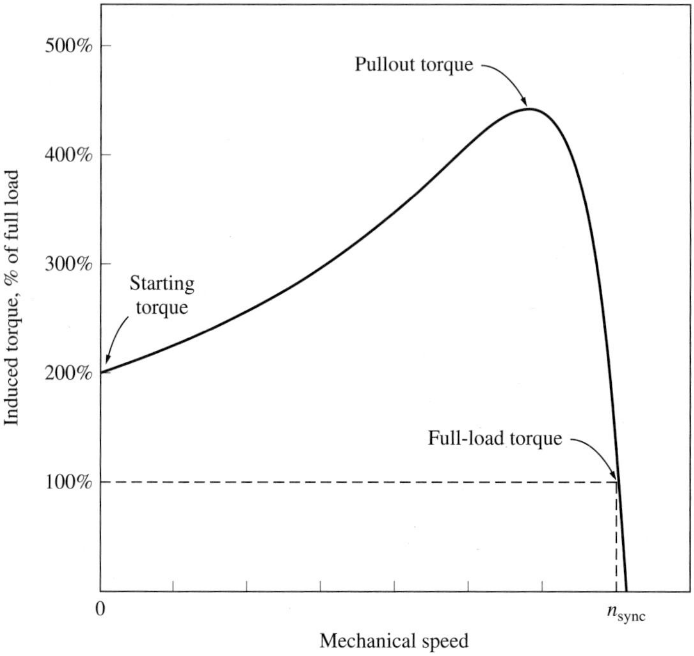

FIGURE 6–19
A typical induction motor torque–speed characteristic curve.

A plot of induction motor torque as a function of speed (and slip) is shown in Figure 6–19, and a plot showing speeds both above and below the normal motor range is shown in Figure 6–20.

Comments on the Induction Motor Torque–Speed Curve

The induction motor torque–speed characteristic curve plotted in Figures 6–19 and 6–20 provides several important pieces of information about the operation of induction motors. This information is summarized as follows:

1. The induced torque of the motor is zero at synchronous speed. This fact has been discussed previously.
2. The torque–speed curve is nearly linear between no load and full load. In this range, the rotor resistance is much larger than the rotor reactance, so the rotor current, the rotor magnetic field, and the induced torque increase linearly with increasing slip.
3. There is a maximum possible torque that cannot be exceeded. This torque, called the *pullout torque* or *breakdown torque*, is 2 to 3 times the rated full-load torque of the motor. The next section of this chapter contains a method for calculating pullout torque.

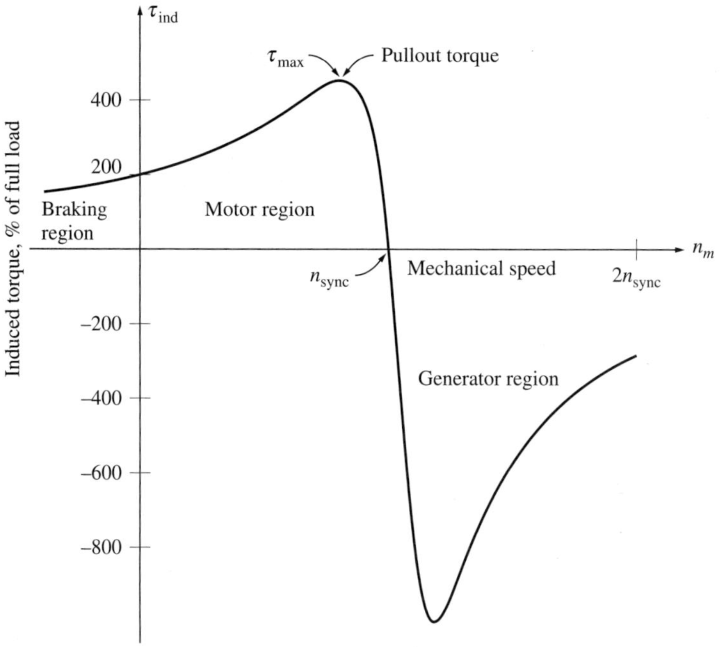

FIGURE 6–20
Induction motor torque–speed characteristic curve, showing the extended operating ranges (braking region and generator region).

4. The starting torque on the motor is slightly larger than its full-load torque, so this motor will start carrying any load that it can supply at full power.
5. Notice that the torque on the motor for a given slip varies as the square of the applied voltage. This fact is useful in one form of induction motor speed control that will be described later.
6. If the rotor of the induction motor is driven faster than synchronous speed, then the direction of the induced torque in the machine reverses and the machine becomes a *generator*, converting mechanical power to electric power. The use of induction machines as generators will be described later.
7. If the motor is turning backward relative to the direction of the magnetic fields, the induced torque in the machine will stop the machine very rapidly and will try to rotate it in the other direction. Since reversing the direction of magnetic field rotation is simply a matter of switching any two stator phases, this fact can be used as a way to very rapidly stop an induction motor. The act of switching two phases in order to stop the motor very rapidly is called *plugging*.

The power converted to mechanical form in an induction motor is equal to

$$P_{conv} = \tau_{ind}\,\omega_m$$

and is shown plotted in Figure 6–21. Notice that the peak power supplied by the induction motor occurs at a different speed than the maximum torque; and, of course, no power is converted to mechanical form when the rotor is at zero speed.

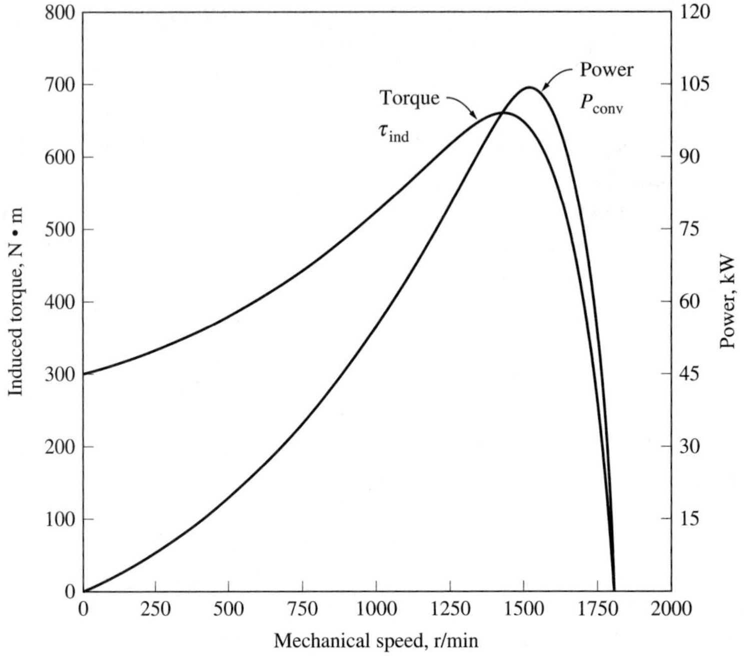

FIGURE 6–21
Induced torque and power converted versus motor speed in revolutions per minute for an example four-pole induction motor.

Maximum (Pullout) Torque in an Induction Motor

Since the induced torque is equal to P_{AG}/ω_{sync}, the maximum possible torque occurs when the air-gap power is maximum. Since the air-gap power is equal to the power consumed in the resistor R_2/s, the *maximum induced torque will occur when the power consumed by that resistor is maximum.*

When is the power supplied to R_2/s at its maximum? Refer to the simplified equivalent circuit in Figure 6–18c. In a situation where the angle of the load impedance is fixed, the maximum power transfer theorem states that maximum power transfer to the load resistor R_2/s will occur when the *magnitude* of that impedance is equal to the *magnitude* of the source impedance. The equivalent source impedance in the circuit is

$$Z_{source} = R_{TH} + jX_{TH} + jX_2 \tag{6–51}$$

so the maximum power transfer occurs when

$$\frac{R_2}{s} = \sqrt{R_{TH}^2 + (X_{TH} + X_2)^2} \tag{6–52}$$

Solving Equation (6–52) for slip, we see that *the slip at pullout torque is given by*

$$s_{\max} = \frac{R_2}{\sqrt{R_{TH}^2 + (X_{TH} + X_2)^2}} \tag{6–53}$$

Notice that the referred rotor resistance R_2 appears only in the numerator, so the slip of the rotor at maximum torque is directly proportional to the rotor resistance.

The value of the maximum torque can be found by inserting the expression for the slip at maximum torque into the torque equation [Equation (6–50)]. The resulting equation for the maximum or pullout torque is

$$\tau_{\max} = \frac{3V_{TH}^2}{2\omega_{sync}[R_{TH} + \sqrt{R_{TH}^2 + (X_{TH} + X_2)^2}]} \tag{6–54}$$

This torque is proportional to the square of the supply voltage and is also inversely related to the size of the stator impedances and the rotor reactance. The smaller a machine's reactances, the larger the maximum torque it is capable of achieving. Note that *slip* at which the maximum torque occurs is directly proportional to rotor resistance [Equation (6–53)], but the *value* of the maximum torque is independent of the value of rotor resistance [Equation (6–54)].

The torque–speed characteristic for a wound-rotor induction motor is shown in Figure 6–22. Recall that it is possible to insert resistance into the rotor circuit of a wound rotor because the rotor circuit is brought out to the stator through slip rings. Notice on the figure that as the rotor resistance is increased, the pullout speed of the motor decreases, but the maximum torque remains constant.

It is possible to take advantage of this characteristic of wound-rotor induction motors to start very heavy loads. If a resistance is inserted into the rotor circuit, the maximum torque can be adjusted to occur at starting conditions. Therefore, the maximum possible torque would be available to start heavy loads. On the other hand, once the load is turning, the extra resistance can be removed from the circuit, and the maximum torque will move up to near-synchronous speed for regular operation.

Example 6–4. A two-pole, 50-Hz induction motor supplies 15 kW to a load at a speed of 2950 r/min.

(a) What is the motor's slip?
(b) What is the induced torque in the motor in N • m under these conditions?
(c) What will the operating speed of the motor be if its torque is doubled?
(d) How much power will be supplied by the motor when the torque is doubled?

Solution

(a) The synchronous speed of this motor is

$$n_{sync} = \frac{120f_{se}}{P} = \frac{120(50\text{ Hz})}{2\text{ poles}} = 3000\text{ r/min}$$

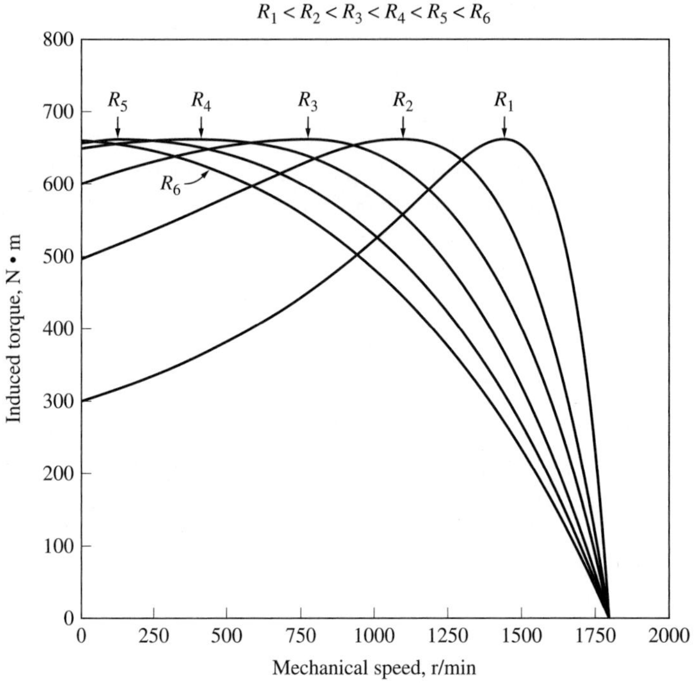

FIGURE 6–22
The effect of varying rotor resistance on the torque–speed characteristic of a wound-rotor induction motor.

Therefore, the motor's slip is

$$s = \frac{n_{\text{sync}} - n_{\text{m}}}{n_{\text{sync}}} (\times\ 100\%) \qquad (6\text{–}4)$$

$$= \frac{3000\ \text{r/min} - 2950\ \text{r/min}}{3000\ \text{r/min}} (\times\ 100\%)$$

$$= 0.0167 \text{ or } 1.67\%$$

(*b*) The induced torque in the motor must be assumed equal to the load torque, and P_{conv} must be assumed equal to P_{load}, since no value was given for mechanical losses. The torque is thus

$$\tau_{\text{ind}} = \frac{P_{\text{conv}}}{\omega_{\text{m}}}$$

$$= \frac{15\ \text{kW}}{(2950\ \text{r/min})(2\pi\ \text{rad/r})(1\ \text{min}/60\ \text{s})}$$

$$= 48.6\ \text{N} \bullet \text{m}$$

(*c*) In the low-slip region, the torque–speed curve is linear, and the induced torque is directly proportional to slip. Therefore, if the torque doubles, then the new slip will be 3.33 percent. The operating speed of the motor is thus

$$n_m = (1 - s)n_{\text{sync}} = (1 - 0.0333)(3000\ \text{r/min}) = 2900\ \text{r/min}$$

(d) The power supplied by the motor is given by

$$P_{\text{conv}} = \tau_{\text{ind}}\omega_m$$

$$= (97.2 \text{ N} \bullet \text{m})(2900 \text{ r/min})(2\pi \text{ rad/r})(1 \text{ min/60 s})$$

$$= 29.5 \text{ kW}$$

Example 6–5. A 460-V, 25-hp, 60-Hz, four-pole, Y-connected wound-rotor induction motor has the following impedances in ohms per phase referred to the stator circuit:

$$R_1 = 0.641\ \Omega \qquad R_2 = 0.332\ \Omega$$
$$X_1 = 1.106\ \Omega \qquad X_2 = 0.464\ \Omega \qquad X_M = 26.3\ \Omega$$

(a) What is the maximum torque of this motor? At what speed and slip does it occur?

(b) What is the starting torque of this motor?

(c) When the rotor resistance is doubled, what is the speed at which the maximum torque now occurs? What is the new starting torque of the motor?

(d) Calculate and plot the torque–speed characteristics of this motor both with the original rotor resistance and with the rotor resistance doubled.

Solution

The Thevenin voltage of this motor is

$$V_{\text{TH}} = V_\phi \frac{X_M}{\sqrt{R_1^2 + (X_1 + X_M)^2}} \tag{6–41a}$$

$$= \frac{(266 \text{ V})(26.3\ \Omega)}{\sqrt{(0.641\ \Omega)^2 + (1.106\ \Omega + 26.3\ \Omega)^2}} = 255.2 \text{ V}$$

The Thevenin resistance is

$$R_{\text{TH}} \approx R_1\left(\frac{X_M}{X_1 + X_M}\right)^2 \tag{6–44}$$

$$\approx (0.641\ \Omega)\left(\frac{26.3\ \Omega}{1.106\ \Omega + 26.3\ \Omega}\right)^2 = 0.590\ \Omega$$

The Thevenin reactance is

$$X_{\text{TH}} \approx X_1 = 1.106\ \Omega$$

(a) The slip at which maximum torque occurs is given by Equation (6–53):

$$s_{\max} = \frac{R_2}{\sqrt{R_{\text{TH}}^2 + (X_{\text{TH}} + X_2)^2}} \tag{6–53}$$

$$= \frac{0.332\ \Omega}{\sqrt{(0.590\ \Omega)^2 + (1.106\ \Omega + 0.464\ \Omega)^2}} = 0.198$$

This corresponds to a mechanical speed of

$$n_m = (1 - s)n_{\text{sync}} = (1 - 0.198)(1800 \text{ r/min}) = 1444 \text{ r/min}$$

The torque at this speed is

$$\tau_{max} = \frac{3V_{TH}^2}{2\omega_{sync}[R_{TH} + \sqrt{R_{TH}^2 + (X_{TH} + X_2)^2}]} \tag{6–54}$$

$$= \frac{3(255.2 \text{ V})^2}{2(188.5 \text{ rad/s})[0.590\ \Omega + \sqrt{(0.590\ \Omega)^2 + (1.106\ \Omega + 0.464\ \Omega)^2}]}$$

$$= 229 \text{ N} \bullet \text{m}$$

(b) The starting torque of this motor is found by setting $s = 1$ in Equation (6–50):

$$\tau_{start} = \frac{3V_{TH}^2 R_2}{\omega_{sync}[(R_{TH} + R_2)^2 + (X_{TH} + X_2)^2]}$$

$$= \frac{3(255.2 \text{ V})^2(0.332\ \Omega)}{(188.5 \text{ rad/s})[(0.590\ \Omega + 0.332\ \Omega)^2 + (1.106\ \Omega + 0.464\ \Omega)^2]}$$

$$= 104 \text{ N} \bullet \text{m}$$

(c) If the rotor resistance is doubled, then the slip at maximum torque doubles, too. Therefore,

$$s_{max} = 0.396$$

and the speed at maximum torque is

$$n_m = (1 - s)n_{sync} = (1 - 0.396)(1800 \text{ r/min}) = 1087 \text{ r/min}$$

The maximum torque is still

$$\tau_{max} = 229 \text{ N} \bullet \text{m}$$

The starting torque is now

$$\tau_{start} = \frac{3(255.2 \text{ V})^2(0.664\ \Omega)}{(188.5 \text{ rad/s})[(0.590\ \Omega + 0.664\ \Omega)^2 + (1.106\ \Omega + 0.464\ \Omega)^2]}$$

$$= 170 \text{ N} \bullet \text{m}$$

(d) We will create a MATLAB M-file to calculate and plot the torque–speed characteristic of the motor both with the original rotor resistance and with the doubled rotor resistance. The M-file will calculate the Thevenin impedance using the exact equations for V_{TH} and Z_{TH} [Equations (6–41a) and (6–43)] instead of the approximate equations, because the computer can easily perform the exact calculations. It will then calculate the induced torque using Equation (6–50) and plot the results. The resulting M-file follows:

```
% M-file: torque_speed_curve.m
% M-file create a plot of the torque-speed curve of the
%   induction motor of Example 6-5.

% First, initialize the values needed in this program.
r1 = 0.641;                  % Stator resistance
x1 = 1.106;                  % Stator reactance
r2 = 0.332;                  % Rotor resistance
x2 = 0.464;                  % Rotor reactance
xm = 26.3;                   % Magnetization branch reactance
v_phase = 460 / sqrt(3);     % Phase voltage
n_sync = 1800;               % Synchronous speed (r/min)
```

```
w_sync = 188.5;               % Synchronous speed (rad/s)

% Calculate the Thevenin voltage and impedance from Equations
% 6-41a and 6-43.
v_th = v_phase * ( xm / sqrt(r1^2 + (x1 + xm)^2) );
z_th = ((j*xm) * (r1 + j*x1)) / (r1 + j*(x1 + xm));
r_th = real(z_th);
x_th = imag(z_th);

% Now calculate the torque-speed characteristic for many
% slips between 0 and 1.  Note that the first slip value
% is set to 0.001 instead of exactly 0 to avoid divide-
% by-zero problems.
s = (0:1:50) / 50;                   % Slip
s(1) = 0.001;
nm = (1 - s) * n_sync;               % Mechanical speed

% Calculate torque for original rotor resistance
for ii = 1:51
  t_ind1(ii) = (3 * v_th^2 * r2 / s(ii)) / ...
          (w_sync * ((r_th + r2/s(ii))^2 + (x_th + x2)^2) );
end

% Calculate torque for doubled rotor resistance
for ii = 1:51
  t_ind2(ii) = (3 * v_th^2 * (2*r2) / s(ii)) / ...
          (w_sync * ((r_th + (2*r2)/s(ii))^2 + (x_th + x2)^2) );
end

% Plot the torque-speed curve
plot(nm,t_ind1,'Color','b','LineWidth',2.0);
hold on;
plot(nm,t_ind2,'Color','k','LineWidth',2.0,'LineStyle','-.');
xlabel('\bf\itn_{m}');
ylabel('\bf\tau_{ind}');
title ('\bfInduction motor torque-speed characteristic');
legend ('Original R_{2}','Doubled R_{2}');
grid on;
hold off;
```

The resulting torque–speed characteristics are shown in Figure 6–23. Note that the peak torque and starting torque values on the curves match the calculations of parts (*a*) through (*c*). Also, note that the starting torque of the motor rose as R_2 increased.

6.6 VARIATIONS IN INDUCTION MOTOR TORQUE–SPEED CHARACTERISTICS

Section 6.5 contained the derivation of the torque–speed characteristic for an induction motor. In fact, several characteristic curves were shown, depending on the rotor resistance. Example 6–5 illustrated an induction motor designer's dilemma—if a rotor is designed with high resistance, then the motor's starting torque is quite

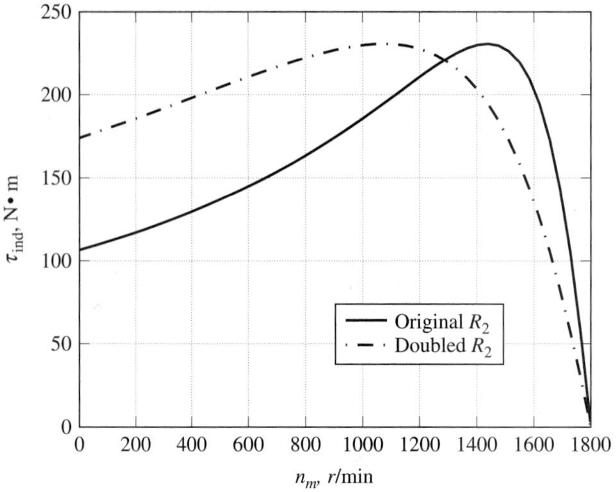

FIGURE 6–23
Torque–speed characteristics for the motor of Example 6–5.

high, but the slip is also quite high at normal operating conditions. Recall that $P_{conv} = (1 - s)P_{AG}$, so *the higher the slip, the smaller the fraction of air-gap power actually converted to mechanical form*, and thus the lower the motor's efficiency. A motor with high rotor resistance has a good starting torque but poor efficiency at normal operating conditions. On the other hand, a motor with low rotor resistance has a low starting torque and high starting current, but its efficiency at normal operating conditions is quite high. An induction motor designer is forced to compromise between the conflicting requirements of high starting torque and good efficiency.

One possible solution to this difficulty was suggested in passing in Section 6.5: use a wound-rotor induction motor and insert extra resistance into the rotor during starting. The extra resistance could be completely removed for better efficiency during normal operation. Unfortunately, wound-rotor motors are more expensive, need more maintenance, and require a more complex automatic control circuit than cage rotor motors. Also, it is sometimes important to completely seal a motor when it is placed in a hazardous or explosive environment, and this is easier to do with a completely self-contained rotor. It would be nice to figure out some way to add extra rotor resistance at starting and to remove it during normal running without slip rings and *without operator or control circuit intervention.*

Figure 6–24 illustrates the desired motor characteristic. This figure shows two wound-rotor motor characteristics, one with high resistance and one with low resistance. At high slips, the desired motor should behave like the high-resistance wound-rotor motor curve; at low slips, it should behave like the low-resistance wound-rotor motor curve.

Fortunately, it is possible to accomplish just this effect by properly taking advantage of *leakage reactance* in induction motor rotor design.

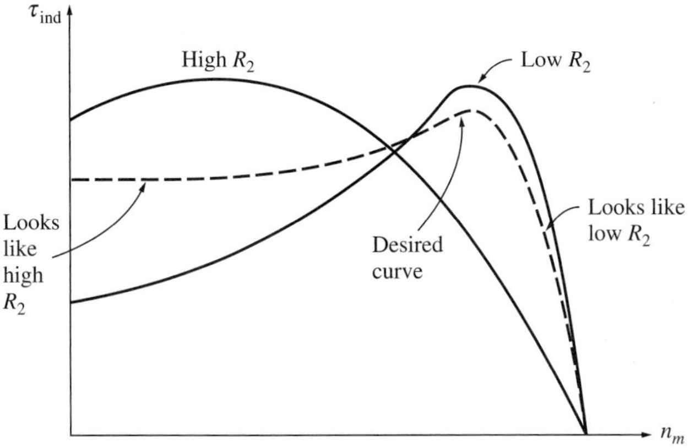

FIGURE 6–24
A torque–speed characteristic curve combining high-resistance effects at low speeds (high slip) with low-resistance effects at high speed (low slip).

Control of Motor Characteristics by Cage Rotor Design

The reactance X_2 in an induction motor equivalent circuit represents the referred form of the rotor's leakage reactance. Recall that leakage reactance is the reactance due to the rotor flux lines that do not also couple with the stator windings. In general, the farther away from the stator a rotor bar or part of a bar is, the greater its leakage reactance, since a smaller percentage of the bar's flux will reach the stator. Therefore, if the bars of a cage rotor are placed near the surface of the rotor, they will have only a small leakage flux and the reactance X_2 will be small in the equivalent circuit. On the other hand, if the rotor bars are placed deeper into the rotor surface, there will be more leakage and the rotor reactance X_2 will be larger.

For example, Figure 6–25a is a photograph of a rotor lamination showing the cross section of the bars in the rotor. The rotor bars in the figure are quite large and are placed near the surface of the rotor. Such a design will have a low resistance (due to its large cross section) and a low leakage reactance and X_2 (due to the bar's location near the stator). Because of the low rotor resistance, the pullout torque will be quite near synchronous speed [see Equation (6–53)], and the motor will be quite efficient. Remember that

$$P_{conv} = (1 - s)P_{AG} \tag{6–33}$$

so very little of the air-gap power is lost in the rotor resistance. However, since R_2 is small, the motor's starting torque will be small, and its starting current will be high. This type of design is called the National Electrical Manufacturers Association (NEMA) design class A. It is more or less a typical induction motor, and its characteristics are basically the same as those of a wound-rotor motor with no extra resistance inserted. Its torque–speed characteristic is shown in Figure 6–26.

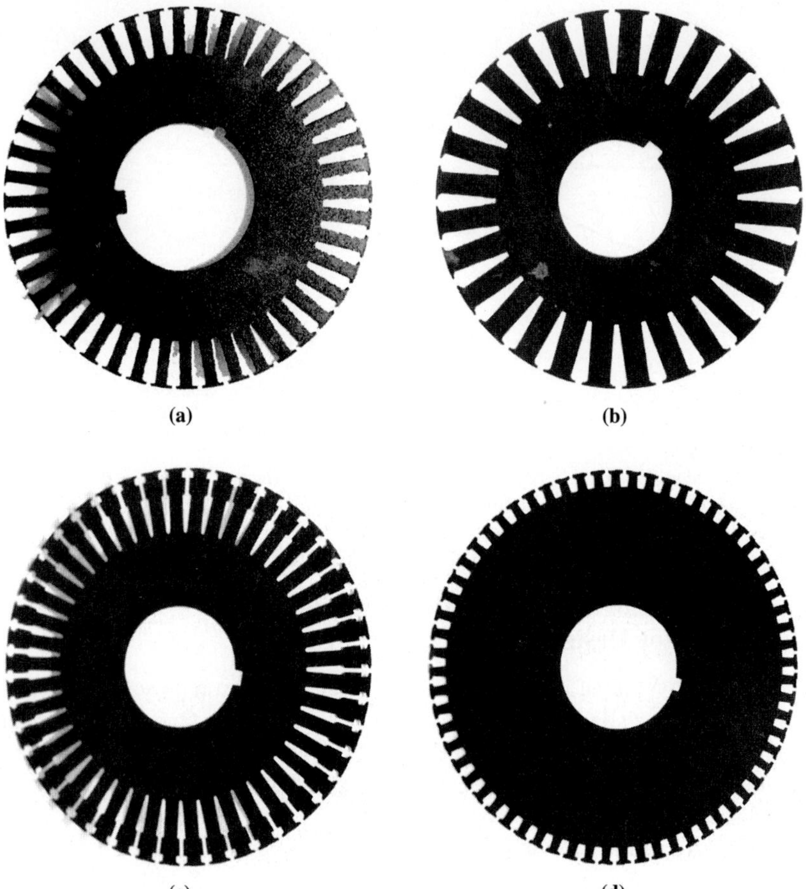

FIGURE 6–25
Laminations from typical cage induction motor rotors, showing the cross section of the rotor bars: (a) NEMA design class A—large bars near the surface; (b) NEMA design class B—large, deep rotor bars; (c) NEMA design class C—double-cage rotor design; (d) NEMA design class D—small bars near the surface. (*Courtesy of MagneTek, Inc.*)

Figure 6–25d, however, shows the cross section of an induction motor rotor with *small* bars placed near the surface of the rotor. Since the cross-sectional area of the bars is small, the rotor resistance is relatively high. Since the bars are located near the stator, the rotor leakage reactance is still small. This motor is very much like a wound-rotor induction motor with extra resistance inserted into the rotor. Because of the large rotor resistance, this motor has a pullout torque occurring at a high slip, and its starting torque is quite high. A cage motor with this type of rotor construction is called NEMA design class D. Its torque–speed characteristic is also shown in Figure 6–26.

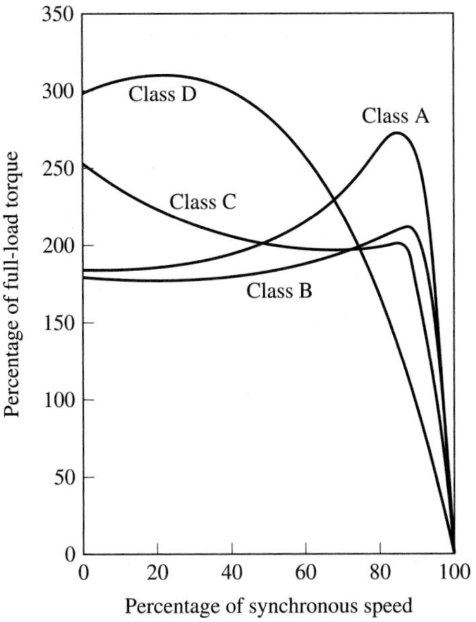

FIGURE 6–26
Typical torque–speed curves for different rotor designs.

Deep-Bar and Double-Cage Rotor Designs

Both of the previous rotor designs are essentially similar to a wound-rotor motor with a set rotor resistance. How can a *variable* rotor resistance be produced to combine the high starting torque and low starting current of a class D design with the low normal operating slip and high efficiency of a class A design?

It is possible to produce a variable rotor resistance by the use of deep rotor bars or double-cage rotors. The basic concept is illustrated with a deep-bar rotor in Figure 6–27. Figure 6–27a shows a current flowing through the upper part of a deep rotor bar. Since current flowing in that area is tightly coupled to the stator, the leakage inductance is small for this region. Figure 6–27b shows current flowing deeper in the bar. Here, the leakage inductance is higher. Since all parts of the rotor bar are in parallel electrically, the bar essentially represents a series of parallel electric circuits, the upper ones having a smaller inductance and the lower ones having a larger inductance (Figure 6–27c).

At low slip, the rotor's frequency is very small, and the reactances of all the parallel paths through the bar are small compared to their resistances. The impedances of all parts of the bar are approximately equal, so current flows through all parts of the bar equally. The resulting large cross-sectional area makes the rotor resistance quite small, resulting in good efficiency at low slips. At high slip (starting conditions), the reactances are large compared to the resistances in the rotor bars, so all the current is forced to flow in the low-reactance part of the bar near the stator. Since the *effective* cross section is lower, the rotor resistance is higher than before. With a high rotor resistance at starting conditions, the starting torque is relatively higher and the starting current is relatively lower than in

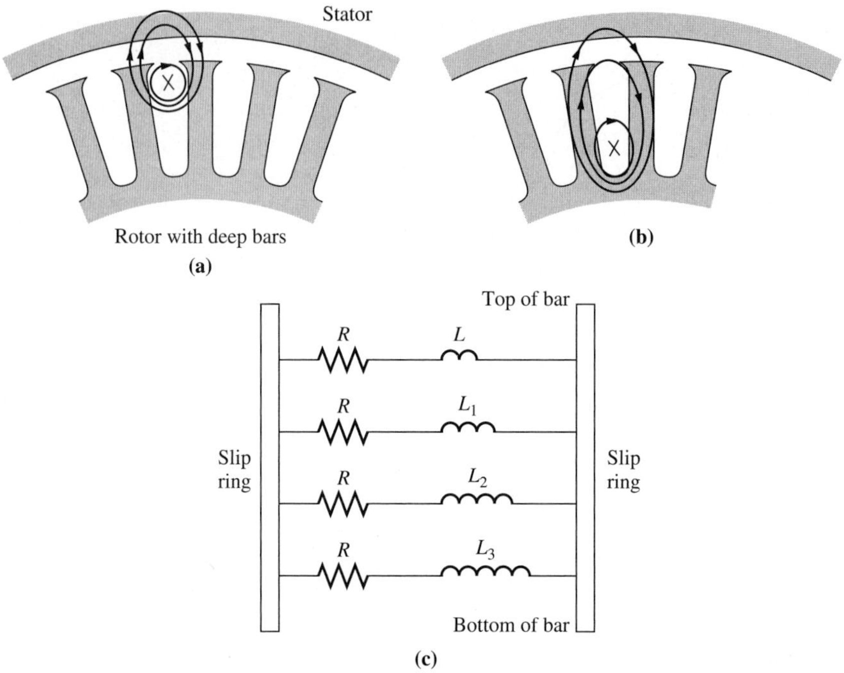

FIGURE 6–27
Flux linkage in a deep-bar rotor. (a) For a current flowing in the top of the bar, the flux is tightly linked to the stator, and leakage inductance is small; (b) for a current flowing in the bottom of the bar, the flux is loosely linked to the stator, and leakage inductance is large; (c) resulting equivalent circuit of the rotor bar as a function of depth in the rotor.

a class A design. A typical torque–speed characteristic for this construction is the design class B curve in Figure 6–26.

A cross-sectional view of a double-cage rotor is shown in Figure 6–25c. It consists of a large, low-resistance set of bars buried deeply in the rotor and a small, high-resistance set of bars set at the rotor surface. It is similar to the deep-bar rotor, except that the difference between low-slip and high-slip operation is even more exaggerated. At starting conditions, only the small bar is effective, and the rotor resistance is *quite* high. This high resistance results in a large starting torque. However, at normal operating speeds, both bars are effective, and the resistance is almost as low as in a deep-bar rotor. Double-cage rotors of this sort are used to produce NEMA class B and class C characteristics. Possible torque–speed characteristics for a rotor of this design are designated design class B and design class C in Figure 6–26.

Double-cage rotors have the disadvantage that they are more expensive than the other types of cage rotors, but they are cheaper than wound-rotor designs. They allow some of the best features possible with wound-rotor motors (high starting torque with a low starting current and good efficiency at normal operating conditions) at a lower cost and without the need of maintaining slip rings and brushes.

Induction Motor Design Classes

It is possible to produce a large variety of torque–speed curves by varying the rotor characteristics of induction motors. To help industry select appropriate motors for varying applications in the integral-horsepower range, NEMA in the United States and the International Electrotechnical Commission (IEC) in Europe have defined a series of standard designs with different torque–speed curves. These standard designs are referred to as *design classes*, and an individual motor may be referred to as a design class X motor. It is these NEMA and IEC design classes that were referred to earlier. Figure 6–26 shows typical torque–speed curves for the four standard NEMA design classes. The characteristic features of each standard design class are given below.

DESIGN CLASS A. Design class A motors are the standard motor design, with a normal starting torque, a normal starting current, and low slip. The full-load slip of design A motors must be less than 5 percent and must be less than that of a design B motor of equivalent rating. The pullout torque is 200 to 300 percent of the full-load torque and occurs at a low slip (less than 20 percent). The starting torque of this design is at least the rated torque for larger motors and is 200 percent or more of the rated torque for smaller motors. The principal problem with this design class is its extremely high inrush current on starting. Current flows at starting are typically 500 to 800 percent of the rated current. In sizes above about 7.5 hp, some form of reduced-voltage starting must be used with these motors to prevent voltage dip problems on starting in the power system they are connected to. In the past, design class A motors were the standard design for most applications below 7.5 hp and above about 200 hp, but they have largely been replaced by design class B motors in recent years. Typical applications for these motors are driving fans, blowers, pumps, lathes, and other machine tools.

DESIGN CLASS B. Design class B motors have a normal starting torque, a lower starting current, and low slip. This motor produces about the same starting torque as the class A motor with about 25 percent less current. The pullout torque is greater than or equal to 200 percent of the rated load torque, but less than that of the class A design because of the increased rotor reactance. Rotor slip is still relatively low (less than 5 percent) at full load. Applications are similar to those for design A, but design B is preferred because of its lower starting-current requirements. Design class B motors have largely replaced design class A motors in new installations.

DESIGN CLASS C. Design class C motors have a high starting torque with low starting currents and low slip (less than 5 percent) at full load. The pullout torque is slightly lower than that for class A motors, while the starting torque is up to 250 percent of the full-load torque. These motors are built from double-cage rotors, so they are more expensive than motors in the previous classes. They are used for high-starting-torque loads, such as loaded pumps, compressors, and conveyors.

DESIGN CLASS D. Design class D motors have a very high starting torque (275 percent or more of the rated torque) and a low starting current, but they also have a high slip at full load. They are essentially ordinary class A induction motors, but with the rotor bars made smaller and with a higher-resistance material. The high rotor resistance shifts the peak torque to a very low speed. It is even possible for the highest torque to occur at zero speed (100 percent slip). Full-load slip for these motors is quite high because of the high rotor resistance. It is typically 7 to 11 percent, but may go as high as 17 percent or more. These motors are used in applications requiring the acceleration of extremely high-inertia-type loads, especially large flywheels used in punch presses or shears. In such applications, these motors gradually accelerate a large flywheel up to full speed, which then drives the punch. After a punching operation, the motor then reaccelerates the flywheel over a fairly long time for the next operation.

In addition to these four design classes, NEMA used to recognize design classes E and F, which were called *soft-start* induction motors (see Figure 6–28). These designs were distinguished by having very low starting currents and were used for low-starting-torque loads in situations where starting currents were a problem. These designs are now obsolete.

Example 6–6. A 460-V, 30-hp, 60-Hz, four-pole, Y-connected induction motor has two possible rotor designs, a single-cage rotor and a double-cage rotor. (The stator is identical for either rotor design.) The motor with the single-cage rotor may be modeled by the following impedances in ohms per phase referred to the stator circuit:

$$R_1 = 0.641\ \Omega \qquad R_2 = 0.300\ \Omega$$
$$X_1 = 0.750\ \Omega \qquad X_2 = 0.500\ \Omega \qquad X_M = 26.3\ \Omega$$

The motor with the double-cage rotor may be modeled as a tightly coupled, high-resistance outer cage in parallel with a loosely coupled, low-resistance inner cage (similar to the structure of Figure 6–25c). The stator and magnetization resistance and reactances will be identical with those in the single-cage design.

The resistance and reactance of the rotor outer cage are:

$$R_{2o} = 3.200\ \Omega \qquad X_{2o} = 0.500\ \Omega$$

Note that the resistance is high because the outer bar has a small cross section, while the reactance is the same as the reactance of the single-cage rotor, since the outer cage is very close to the stator, and the leakage reactance is small.

The resistance and reactance of the inner cage are

$$R_{2i} = 0.400\ \Omega \qquad X_{2i} = 3.300\ \Omega$$

Here the resistance is low because the bars have a large cross-sectional area, but the leakage reactance is quite high.

Calculate the torque–speed characteristics associated with the two rotor designs. How do they compare?

Solution

The torque–speed characteristic of the motor with the single-cage rotor can be calculated in exactly the same manner as Example 6–5. The torque–speed characteristic of the motor with the double-cage rotor can also be calculated in the same fashion, *except* that at each

FIGURE 6–28
Rotor cross section, showing the construction of the former design class F induction motor. Since the rotor bars are deeply buried, they have a very high leakage reactance. The high leakage reactance reduces the starting torque and current of this motor, so it is called a *soft-start design*. (*Courtesy of MagneTek, Inc.*)

slip the rotor resistance and reactance will be the parallel combination of the impedances of the inner and outer cages. At low slips, the rotor reactance will be relatively unimportant, and the large inner cage will play a major part in the machine's operation. At high slips, the high reactance of the inner cage almost removes it from the circuit.

A MATLAB M-file to calculate and plot the two torque–speed characteristics follows:

```
% M-file: torque_speed_2.m
% M-file create and plot of the torque-speed curve of an
%   induction motor with a double-cage rotor design.

% First, initialize the values needed in this program.
r1 = 0.641;                % Stator resistance
x1 = 0.750;                % Stator reactance
r2 = 0.300;                % Rotor resistance for single-
                           %   cage motor
```

```
r2i = 0.400;                 % Rotor resistance for inner
                             %   cage of double-cage motor
r2o = 3.200;                 % Rotor resistance for outer
                             %   cage of double-cage motor
x2 = 0.500;                  % Rotor reactance for single-
                             %   cage motor
x2i = 3.300;                 % Rotor reactance for inner
                             %   cage of double-cage motor
x2o = 0.500;                 % Rotor reactance for outer
                             %   cage of double-cage motor
xm = 26.3;                   % Magnetization branch reactance
v_phase = 460 / sqrt(3);     % Phase voltage
n_sync = 1800;               % Synchronous speed (r/min)
w_sync = 188.5;              % Synchronous speed (rad/s)

% Calculate the Thevenin voltage and impedance from Equations
% 6-41a and 6-43.
v_th = v_phase * ( xm / sqrt(r1^2 + (x1 + xm)^2) );
z_th = ((j*xm) * (r1 + j*x1)) / (r1 + j*(x1 + xm));
r_th = real(z_th);
x_th = imag(z_th);

% Now calculate the motor speed for many slips between
% 0 and 1. Note that the first slip value is set to
% 0.001 instead of exactly 0 to avoid divide-by-zero
% problems.
s = (0:1:50) / 50;               % Slip
s(1) = 0.001;                    % Avoid division-by-zero
nm = (1 - s) * n_sync;           % Mechanical speed

% Calculate torque for the single-cage rotor.
for ii = 1:51
  t_ind1(ii) = (3 * v_th^2 * r2 / s(ii)) / ...
       (w_sync * ((r_th + r2/s(ii))^2 + (x_th + x2)^2) );
end

% Calculate resistance and reactance of the double-cage
% rotor at this slip, and then use those values to
% calculate the induced torque.
for ii = 1:51
  y_r = 1/(r2i + j*s(ii)*x2i) + 1/(r2o + j*s(ii)*x2o);
  z_r = 1/y_r;               % Effective rotor impedance
  r2eff = real(z_r);         % Effective rotor resistance
  x2eff = imag(z_r);         % Effective rotor reactance

  % Calculate induced torque for double-cage rotor.
  t_ind2(ii) = (3 * v_th^2 * r2eff / s(ii)) / ...
       (w_sync * ((r_th + r2eff/s(ii))^2 + (x_th + x2eff)^2) );
end

% Plot the torque-speed curves
plot(nm,t_ind1,'b-','LineWidth',2.0);
hold on;
plot(nm,t_ind2,'k-.','LineWidth',2.0);
xlabel('\bf\itn_{m}');
```

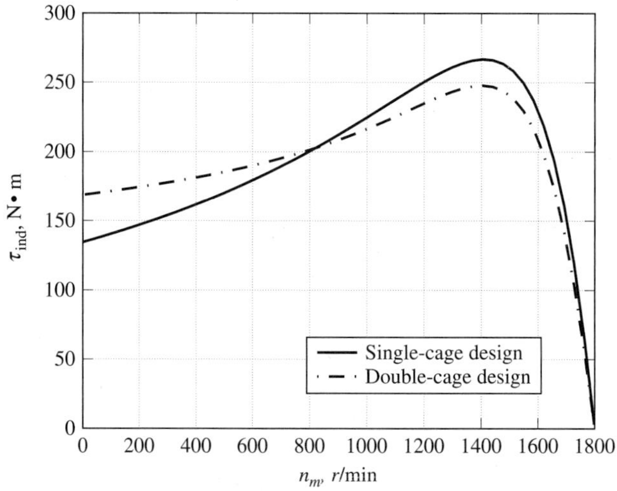

FIGURE 6–29
Comparison of torque–speed characteristics for the single- and double-cage rotors of Example 6–6.

```
ylabel('\bf\tau_{ind}');
title ('\bfInduction motor torque-speed characteristics');
legend ('Single-cage design','Double-cage design');
grid on;
hold off;
```

The resulting torque–speed characteristics are shown in Figure 6–29. Note that the double-cage design has a slightly higher slip in the normal operating range, a smaller maximum torque and a higher starting torque compared to the corresponding single-cage rotor design. This behavior matches our theoretical discussions in this section.

6.7 TRENDS IN INDUCTION MOTOR DESIGN

The fundamental ideas behind the induction motor were developed during the late 1880s by Nicola Tesla, who received a patent on his ideas in 1888. At that time, he presented a paper before the American Institute of Electrical Engineers [AIEE, predecessor of today's Institute of Electrical and Electronics Engineers (IEEE)] in which he described the basic principles of the wound-rotor induction motor, along with ideas for two other important ac motors—the synchronous motor and the reluctance motor.

Although the basic idea of the induction motor was described in 1888, the motor itself did not spring forth in full-fledged form. There was an initial period of rapid development, followed by a series of slow, evolutionary improvements which have continued to this day.

The induction motor assumed recognizable modern form between 1888 and 1895. During that period, two- and three-phase power sources were developed to produce the rotating magnetic fields within the motor, distributed stator windings

FIGURE 6–30
The evolution of the induction motor. The motors shown in this figure are all rated at 220 V and 15 hp. There has been a dramatic decrease in motor size and material requirements in induction motors since the first practical ones were produced in the 1890s. (*Courtesy of General Electric Company.*)

were developed, and the cage rotor was introduced. By 1896, fully functional and recognizable three-phase induction motors were commercially available.

Between then and the early 1970s, there was continual improvement in the quality of the steels, the casting techniques, the insulation, and the construction features used in induction motors. These trends resulted in a smaller motor for a given power output, yielding considerable savings in construction costs. In fact, a modern 100-hp motor is the same physical size as a 7.5-hp motor of 1897. This progression is vividly illustrated by the 15-hp induction motors shown in Figure 6–30. (See also Figure 6–31.)

However, these improvements in induction motor design did *not* necessarily lead to improvements in motor operating efficiency. The major design effort was directed toward reducing the initial materials cost of the machines, not toward increasing their efficiency. The design effort was oriented in that direction because electricity was so inexpensive, making the up-front cost of a motor the principal criterion used by purchasers in its selection.

Since the price of oil began its spectacular climb in 1973, the lifetime operating cost of machines has become more and more important, and the initial installation cost has become relatively less important. As a result of these trends, new emphasis has been placed on motor efficiency both by designers and by end users of the machines.

New lines of high-efficiency induction motors are now being produced by all major manufacturers, and they are forming an ever-increasing share of the induction

FIGURE 6–31
Typical early large induction motors. The motors shown were rated at 2000 hp. (*Courtesy of General Electric Company.*)

motor market. Several techniques are used to improve the efficiency of these motors compared to the traditional standard-efficiency designs. Among these techniques are

1. More copper is used in the stator windings to reduce copper losses.
2. The rotor and stator core length is increased to reduce the magnetic flux density in the air gap of the machine. This reduces the magnetic saturation of the machine, decreasing core losses.
3. More steel is used in the stator of the machine, allowing a greater amount of heat transfer out of the motor and reducing its operating temperature. The rotor's fan is then redesigned to reduce windage losses.
4. The steel used in the stator is a special high-grade electrical steel with low hysteresis losses.
5. The steel is made of an especially thin gauge (i.e., the laminations are very close together), and the steel has a very high internal resistivity. Both effects tend to reduce the eddy current losses in the motor.
6. The rotor is carefully machined to produce a uniform air gap, reducing the stray load losses in the motor.

In addition to the general techniques described above, each manufacturer has his own unique approaches to improving motor efficiency. A typical high-efficiency induction motor is shown in Figure 6–32.

To aid in the comparison of motor efficiencies, NEMA has adopted a standard technique for measuring motor efficiency based on Method B of the IEEE Standard 112, *Test Procedure for Polyphase Induction Motors and Generators*. NEMA has also introduced a rating called *NEMA nominal efficiency*, which appears on the nameplates of design class A, B, and C motors. The nominal efficiency identifies the average efficiency of a large number of motors of a given model, and it also guarantees a certain minimum efficiency for that type of motor. The standard NEMA nominal efficiencies are shown in Figure 6–33.

FIGURE 6–32
A General Electric Energy Saver motor, typical of modern high-efficiency induction motors. (*Courtesy of General Electric Company.*)

Nominal efficiency, %	Guaranteed minimum efficiency, %	Nominal efficiency, %	Guaranteed minimum efficiency, %
95.0	94.1	80.0	77.0
94.5	93.6	78.5	75.5
94.1	93.0	77.0	74.0
93.6	92.4	75.5	72.0
93.0	91.7	74.0	70.0
92.4	91.0	72.0	68.0
91.7	90.2	70.0	66.0
91.0	89.5	68.0	64.0
90.2	88.5	66.0	62.0
89.5	87.5	64.0	59.5
88.5	86.5	62.0	57.5
87.5	85.5	59.5	55.0
86.5	84.0	57.5	52.5
85.5	82.5	55.0	50.5
84.0	81.5	52.5	48.0
82.5	80.0	50.5	46.0
81.5	78.5		

FIGURE 6–33
Table of NEMA nominal efficiency standards. The nominal efficiency represents the mean efficiency of a large number of sample motors, and the guaranteed minimum efficiency represents the lowest permissible efficiency for any given motor of the class. (*Reproduced by permission from Motors and Generators, NEMA Publication MG-I, copyright 1987 by NEMA.*)

Other standards organizations have also established efficiency standards for induction motors, the most important of which are the British (BS-269), IEC (IEC 34-2), and Japanese (JEC-37) standards. However, the techniques prescribed for measuring induction motor efficiency are different in each standard and yield *different results for the same physical machine*. If two motors are each rated at 82.5 percent efficiency, but they are measured according to different standards, then they may not be equally efficient. When two motors are compared, it is important to compare efficiencies measured under the same standard.

6.8 STARTING INDUCTION MOTORS

Induction motors do not present the types of starting problems that synchronous motors do. In many cases, induction motors can be started by simply connecting them to the power line. However, there are sometimes good reasons for not doing this. For example, the starting current required may cause such a dip in the power system voltage that *across-the-line starting* is not acceptable.

For wound-rotor induction motors, starting can be achieved at relatively low currents by inserting extra resistance in the rotor circuit during starting. This extra resistance not only increases the starting torque but also reduces the starting current.

For cage induction motors, the starting current can vary widely depending primarily on the motor's rated power and on the effective rotor resistance at starting conditions. To estimate the rotor current at starting conditions, all cage motors now have a starting *code letter* (not to be confused with their *design class* letter) on their nameplates. The code letter sets limits on the amount of current the motor can draw at starting conditions.

These limits are expressed in terms of the starting apparent power of the motor as a function of its horsepower rating. Figure 6–34 is a table containing the starting kilovoltamperes per horsepower for each code letter.

To determine the starting current for an induction motor, read the rated voltage, horsepower, and code letter from its nameplate. Then the starting apparent power for the motor will be

$$S_{\text{start}} = (\text{rated horsepower})(\text{code letter factor}) \tag{6–55}$$

and the starting current can be found from the equation

$$I_L = \frac{S_{\text{start}}}{\sqrt{3}V_T} \tag{6–56}$$

Example 6–7. What is the starting current of a 15-hp, 208-V, code-letter-F, three-phase induction motor?

Solution

According to Figure 6–34, the maximum kilovoltamperes per horsepower is 5.6. Therefore, the maximum starting kilovoltamperes of this motor is

$$S_{\text{start}} = (15 \text{ hp})(5.6) = 84 \text{ kVA}$$

Nominal code letter	Locked rotor, kVA/hp	Nominal code letter	Locked rotor, kVA/hp
A	0–3.15	L	9.00–10.00
B	3.15–3.55	M	10.00–11.00
C	3.55–4.00	N	11.20–12.50
D	4.00–4.50	P	12.50–14.00
E	4.50–5.00	R	14.00–16.00
F	5.00–5.60	S	16.00–18.00
G	5.60–6.30	T	18.00–20.00
H	6.30–7.10	U	20.00–22.40
J	7.10–8.00	V	22.40 and up
K	8.00–9.00		

FIGURE 6–34
Table of NEMA code letters, indicating the starting kilovoltamperes per horsepower of rating for a motor. Each code letter extends up to, but does not include, the lower bound of the next higher class. (*Reproduced by permission from Motors and Generators, NEMA Publication MG-1, copyright 1987 by NEMA.*)

The starting current is thus

$$I_L = \frac{S_{\text{start}}}{\sqrt{3}V_T} \tag{6–56}$$

$$= \frac{84 \text{ kVA}}{\sqrt{3}(208 \text{ V})} = 233 \text{ A}$$

If necessary, the starting current of an induction motor may be reduced by a starting circuit. However, if this is done, it will also reduce the starting torque of the motor.

One way to reduce the starting current is to change a normally Δ-connected motor into a Y-connected motor during the starting process. If the stator winding from the motor is switched from a Δ-connection to a Y-connection, then the phase voltage across the winding will decrease from V_L to $V_L / \sqrt{3}$, reducing the maximum starting current by the same ratio. When the motor accelerates to close to full speed, the stator windings can be opened and reconnected in a Δ configuration (See Figure 6–35).

Another way to reduce the starting current is to insert extra inductors or resistors into the power line during starting. While formerly common, this approach is rare today. An alternative approach is to reduce the motor's terminal voltage during starting by using autotransformers to step it down. Figure 6–36 shows a typical reduced-voltage starting circuit using autotransformers. During starting, contacts 1 and 3 are shut, supplying a lower voltage to the motor. Once the motor is nearly up to speed, those contacts are opened and contacts 2 are shut. These contacts put full line voltage across the motor.

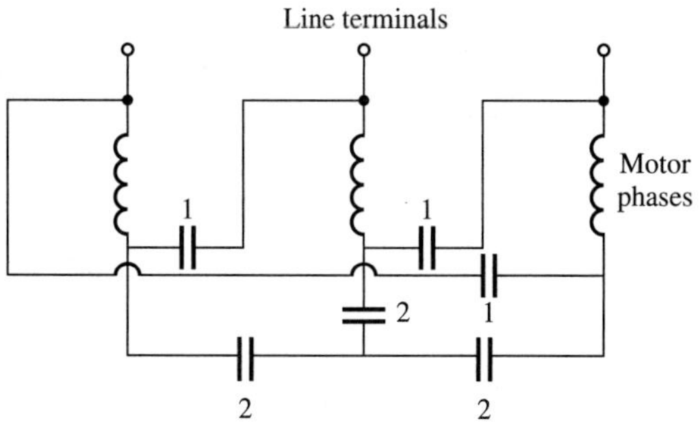

Starting sequence:

(*a*) Close 1
(*b*) Open 1 when motor is turning
(*c*) Close 2

FIGURE 6–35
A Y-Δ induction motor starter.

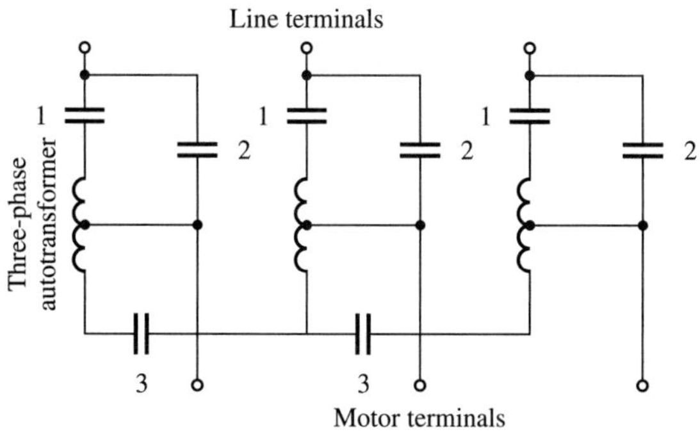

Starting sequence:

(*a*) Close 1 and 3
(*b*) Open 1 and 3
(*c*) Close 2

FIGURE 6–36
An autotransformer starter for an induction motor.

It is important to realize that while the starting current is reduced in direct proportion to the decrease in terminal voltage, the starting torque decreases as the *square* of the applied voltage. Therefore, only a certain amount of current reduction can be done if the motor is to start with a shaft load attached.

Induction Motor Starting Circuits

A typical full-voltage or across-the-line magnetic induction motor starter circuit is shown in Figure 6–37, and the meanings of the symbols used in the figure are explained in Figure 6–38. This operation of this circuit is very simple. When the start button is pressed, the relay (or *contactor*) coil M is energized, causing the

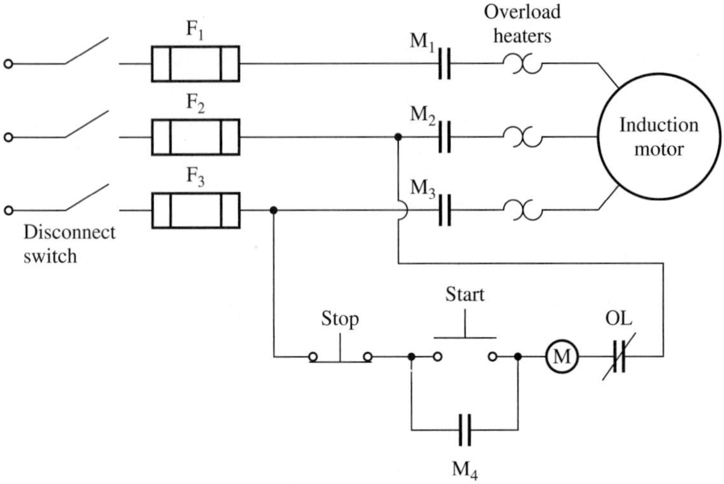

FIGURE 6–37
A typical across-the-line starter for an induction motor.

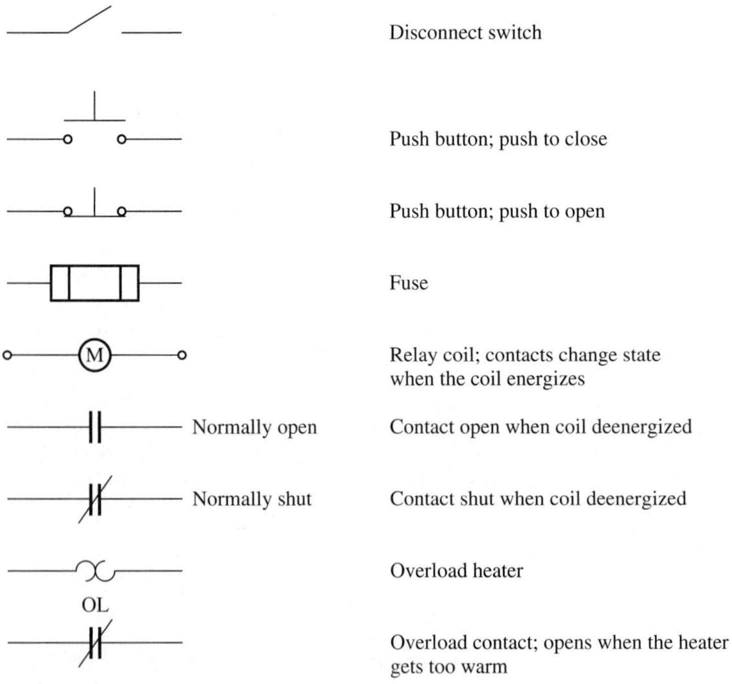

FIGURE 6–38
Typical components found in induction motor control circuits.

normally open contacts M_1, M_2, and M_3 to shut. When these contacts shut, power is applied to the induction motor, and the motor starts. Contact M_4 also shuts, which shorts out the starting switch, allowing the operator to release it without removing power from the M relay. When the stop button is pressed, the M relay is deenergized, and the M contacts open, stopping the motor.

A magnetic motor starter circuit of this sort has several built-in protective features:

1. Short-circuit protection
2. Overload protection
3. Undervoltage protection

Short-circuit protection for the motor is provided by fuses F_1, F_2, and F_3. If a sudden short circuit develops within the motor and causes a current flow many times larger than the rated current, these fuses will blow, disconnecting the motor from the power supply and preventing it from burning up. However, these fuses must *not* burn up during normal motor starting, so they are designed to require currents many times greater than the full-load current before they open the circuit. This means that short circuits through a high resistance and/or excessive motor loads will not be cleared by the fuses.

Overload protection for the motor is provided by the devices labeled OL in the figure. These overload protection devices consist of two parts, an overload heater element and overload contacts. Under normal conditions, the overload contacts are shut. However, when the temperature of the heater elements rises far enough, the OL contacts open, deenergizing the M relay, which in turn opens the normally open M contacts and removes power from the motor.

When an induction motor is overloaded, it is eventually damaged by the excessive heating caused by its high currents. However, this damage takes time, and an induction motor will not normally be hurt by brief periods of high currents (such as starting currents). Only if the high current is sustained will damage occur. The overload heater elements also depend on heat for their operation, so they will not be affected by brief periods of high current during starting, and yet they will operate during long periods of high current, removing power from the motor before it can be damaged.

Undervoltage protection is provided by the controller as well. Notice from the figure that the control power for the M relay comes from directly across the lines to the motor. If the voltage applied to the motor falls too much, the voltage applied to the M relay will also fall and the relay will deenergize. The M contacts then open, removing power from the motor terminals.

An induction motor starting circuit with resistors to reduce the starting current flow is shown in Figure 6–39. This circuit is similar to the previous one, except that there are additional components present to control removal of the starting resistor. Relays 1TD, 2TD, and 3TD in Figure 6–39 are so-called time-delay relays, meaning that when they are energized there is a set time delay before their contacts shut.

When the start button is pushed in this circuit, the M relay energizes and power is applied to the motor as before. Since the 1TD, 2TD, and 3TD contacts are all open, the full starting resistor is in series with the motor, reducing the starting current.

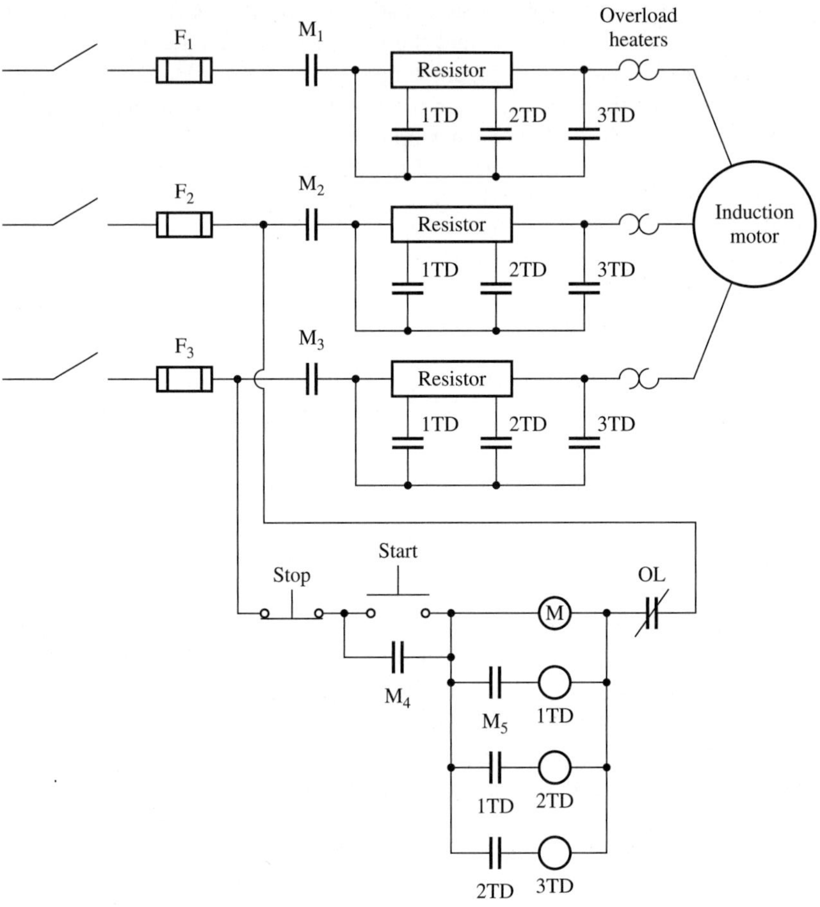

FIGURE 6–39
A three-step resistive starter for an induction motor.

When the M contacts close, notice that the 1TD relay is energized. However, there is a finite delay before the 1TD contacts close. During that time, the motor partially speeds up, and the starting current drops off some. After that time, the 1TD contacts close, cutting out part of the starting resistance and simultaneously energizing the 2TD relay. After another delay, the 2TD contacts shut, cutting out the second part of the resistor and energizing the 3TD relay. Finally, the 3TD contacts close, and the entire starting resistor is out of the circuit.

By a judicious choice of resistor values and time delays, this starting circuit can be used to prevent the motor starting current from becoming dangerously large, while still allowing enough current flow to ensure prompt acceleration to normal operating speeds.

6.9 SPEED CONTROL OF INDUCTION MOTORS

Until the advent of modern solid-state drives, induction motors in general were not good machines for applications requiring considerable speed control. The normal operating range of a typical induction motor (design classes A, B, and C) is confined to less than 5 percent slip, and the speed variation over that range is more or less directly proportional to the load on the shaft of the motor. Even if the slip could be made larger, the efficiency of the motor would become very poor, since the rotor copper losses are directly proportional to the slip on the motor (remember that $P_{\text{RCL}} = sP_{\text{AG}}$).

There are really only two techniques by which the speed of an induction motor can be controlled. One is to vary the synchronous speed, which is the speed of the stator and rotor magnetic fields, since the rotor speed always remains near n_{sync}. The other technique is to vary the slip of the motor for a given load. Each of these approaches will be taken up in more detail.

The synchronous speed of an induction motor is given by

$$n_{\text{sync}} = \frac{120 f_{se}}{P} \tag{6–1}$$

so the only ways in which the synchronous speed of the machine can be varied are (1) by changing the electrical frequency and (2) by changing the number of poles on the machine. Slip control may be accomplished by varying either the rotor resistance or the terminal voltage of the motor.

Induction Motor Speed Control by Pole Changing

There are two major approaches to changing the number of poles in an induction motor:

1. The method of consequent poles
2. Multiple stator windings

The *method of consequent poles* is quite an old method for speed control, having been originally developed in 1897. It relies on the fact that the number of poles in the stator windings of an induction motor can easily be changed by a factor of 2:1 with only simple changes in coil connections. Figure 6–40 shows a simple two-pole induction motor stator suitable for pole changing. Notice that the individual coils are of very short pitch (60 to 90°). Figure 6–41 shows phase *a* of these windings separately for more clarity of detail.

Figure 6–41a shows the current flow in phase *a* of the stator windings at an instant of time during normal operation. Note that the magnetic field leaves the stator in the upper phase group (a north pole) and enters the stator in the lower phase group (a south pole). This winding is thus producing two stator magnetic poles.

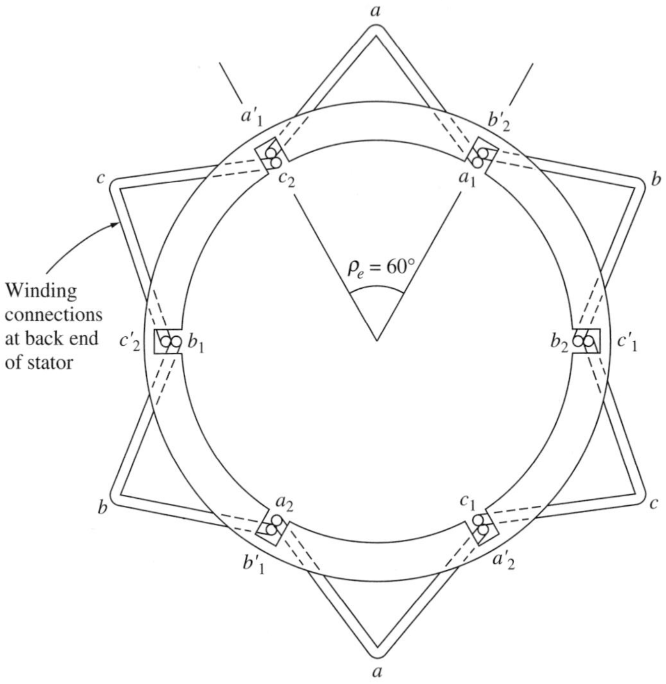

FIGURE 6–40
A two-pole stator winding for pole changing. Notice the very small rotor pitch of these windings.

Now suppose that the direction of current flow in the *lower* phase group on the stator is reversed (Figure 6–41b). Then the magnetic field will leave the stator in *both* the upper phase group *and* the lower phase group—each one will be a north magnetic pole. The magnetic flux in this machine must return to the stator *between* the two phase groups, producing a pair of *consequent* south magnetic poles. Notice that now the stator has four magnetic poles—twice as many as before.

The rotor in such a motor is of the cage design, since a cage rotor always has as many poles induced in it as there are in the stator and can thus adapt when the number of stator poles changes.

When the motor is reconnected from two-pole to four-pole operation, the resulting maximum torque of the induction motor can be the same as before (constant-torque connection), half of its previous value (square-law-torque connection, used for fans, etc.), or twice its previous value (constant-output-power connection), depending on how the stator windings are rearranged. Figure 6–42 shows the possible stator connections and their effect on the torque–speed curve.

The major disadvantage of the consequent-pole method of changing speed is that the speeds *must* be in a ratio of 2:1. The traditional approach to overcoming this limitation was to employ *multiple stator windings* with different numbers of poles and to energize only one set at a time. For example, a motor might be wound with a four-pole and a six-pole set of stator windings, and its synchronous

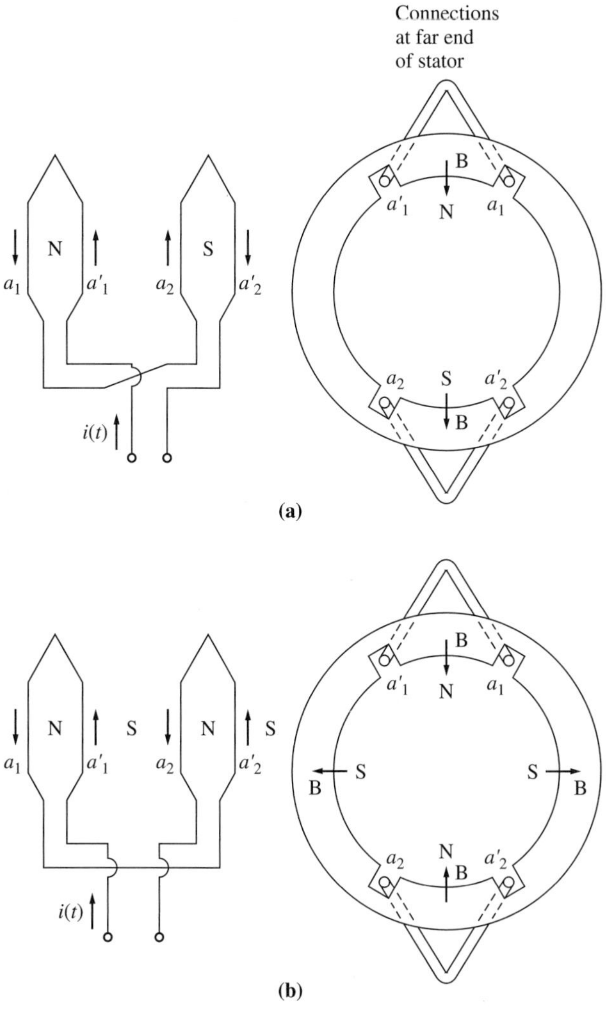

FIGURE 6–41
A close-up view of one phase of a pole-changing winding. (a) In the two-pole configuration, one coil is a north pole and the other one is a south pole. (b) When the connection on one of the two coils is reversed, they are both north poles, and the magnetic flux returns to the stator at points halfway between the two coils. The south poles are called *consequent poles*, and the winding is now a four-pole winding.

speed on a 60-Hz system could be switched from 1800 to 1200 r/min simply by supplying power to the other set of windings. Unfortunately, multiple stator windings increase the expense of the motor and are therefore used only when absolutely necessary.

By combining the method of consequent poles with multiple stator windings, it is possible to build a four-speed induction motor. For example, with separate

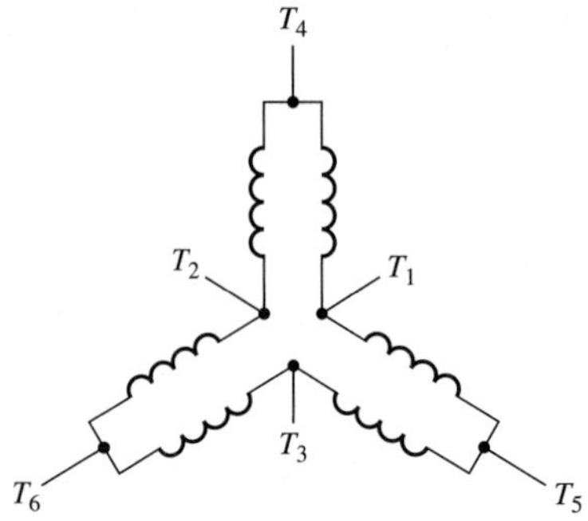

Speed	Lines L_1	Lines L_2	Lines L_3	
Low	T_1	T_2	T_3	T_4, T_5, T_6 open
High	T_4	T_5	T_6	T_1 - T_2 - T_3 together

(a)

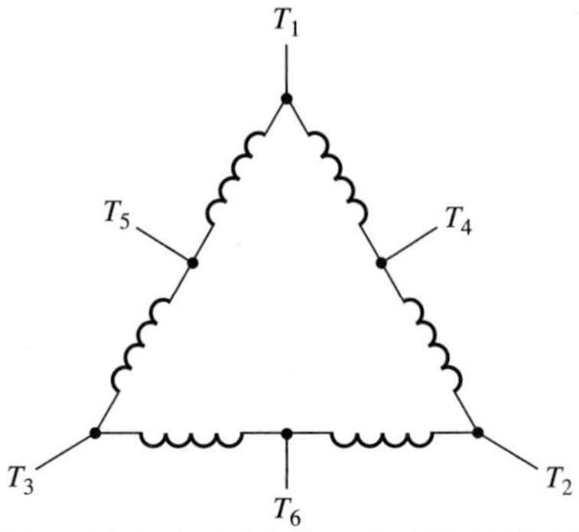

Speed	Lines L_1	Lines L_2	Lines L_3	
Low	T_4	T_5	T_6	T_1 - T_2 - T_3 together
High	T_1	T_2	T_3	T_4, T_5, T_6 open

(b)

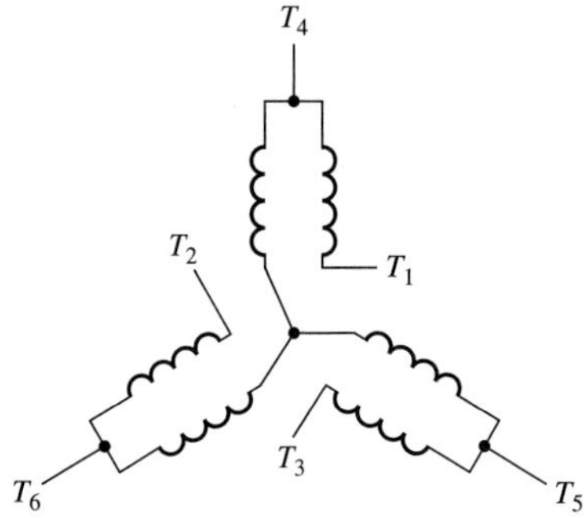

Speed	Lines L_1	Lines L_2	Lines L_3	
Low	T_1	T_2	T_3	T_4, T_5, T_6 open
High	T_4	T_5	T_6	T_1 - T_2 - T_3 together

(c)

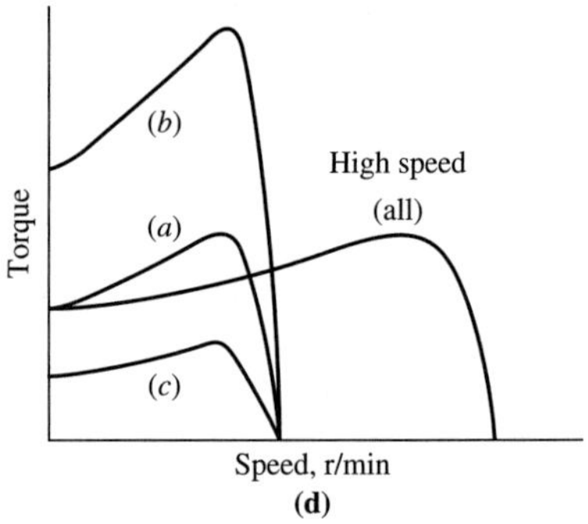

(d)

FIGURE 6–42
Possible connections of the stator coils in a pole-changing motor, together with the resulting torque–speed characteristics: (a) *Constant-torque connection*—the torque capabilities of the motor remain approximately constant in both high-speed and low-speed connections. (b) *Constant-horsepower connection*—the power capabilities of the motor remain approximately constant in both high-speed and low-speed connections. (c) *Fan torque connection*—the torque capabilities of the motor change with speed in the same manner as fan-type loads.

four- and six-pole windings, it is possible to produce a 60-Hz motor capable of running at 600, 900, 1200, and 1800 r/min.

Speed Control by Changing the Line Frequency

If the electrical frequency applied to the stator of an induction motor is changed, the rate of rotation of its magnetic fields n_{sync} will change in direct proportion to the change in electrical frequency, and the no-load point on the torque–speed characteristic curve will change with it (see Figure 6–43). The synchronous speed of the motor at rated conditions is known as the *base speed*. By using variable frequency control, it is possible to adjust the speed of the motor either above or below base speed. A properly designed variable-frequency induction motor drive can be *very* flexible. It can control the speed of an induction motor over a range from as little as 5 percent of base speed up to about twice base speed. However, it is important to maintain certain voltage and torque limits on the motor as the frequency is varied, to ensure safe operation.

When running at speeds below the base speed of the motor, it is necessary to reduce the terminal voltage applied to the stator for proper operation. The terminal voltage applied to the stator should be decreased linearly with decreasing stator frequency. This process is called *derating*. If it is not done, the steel in the core of the induction motor will saturate and excessive magnetization currents will flow in the machine.

To understand the necessity for derating, recall that an induction motor is basically a rotating transformer. As with any transformer, the flux in the core of an induction motor can be found from Faraday's law:

$$v(t) = -N\frac{d\phi}{dt} \tag{1–36}$$

If a voltage $v(t) = V_M \sin \omega t$ is applied to the core, the resulting flux ϕ is

$$\phi(t) = \frac{1}{N_P}\int v(t)\, dt$$

$$= \frac{1}{N_P}\int V_M \sin \omega t\, dt$$

$$\boxed{\phi(t) = -\frac{V_M}{\omega N_P}\cos \omega t} \tag{6–57}$$

Note that the electrical frequency appears in the *denominator* of this expression. Therefore, if the electrical frequency applied to the stator *decreases* by 10 percent while the magnitude of the voltage applied to the stator remains constant, the flux in the core of the motor will *increase* by about 10 percent and the magnetization current of the motor will increase. In the unsaturated region of the motor's magnetization curve, the increase in magnetization current will also be about 10 percent. However, in the saturated region of the motor's magnetization curve, a 10 percent increase in flux requires a much larger increase in magnetization current. Induction motors are normally designed to operate near the saturation point on their magnetization curves, so the increase in flux due to a decrease in frequency

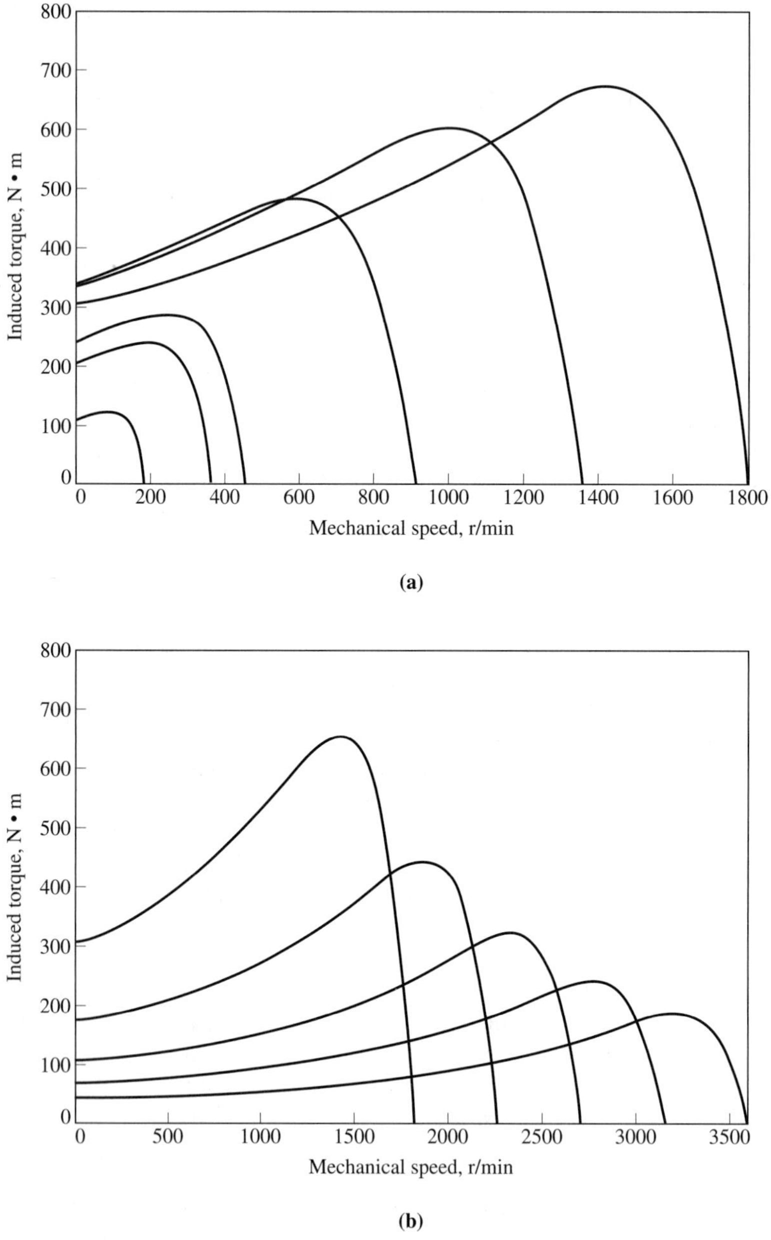

FIGURE 6–43
Variable-frequency speed control in an induction motor: (a) The family of torque–speed characteristic curves for speeds below base speed, assuming that the line voltage is derated linearly with frequency. (b) The family of torque–speed characteristic curves for speeds above base speed, assuming that the line voltage is held constant.

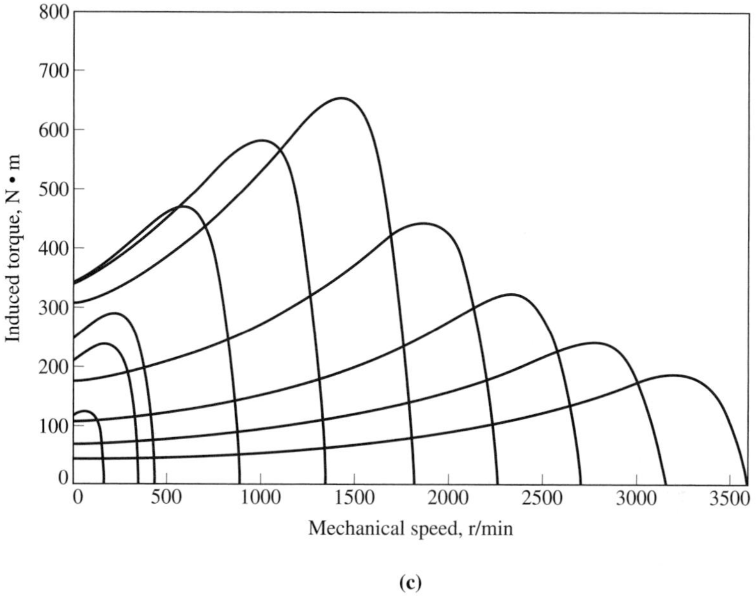

FIGURE 6–43 (concluded)
(c) The torque–speed characteristic curves for all frequencies.

will cause excessive magnetization currents to flow in the motor. (This same problem was observed in transformers; see Section 2.12.)

To avoid excessive magnetization currents, it is customary to decrease the applied stator voltage in direct proportion to the decrease in frequency whenever the frequency falls below the rated frequency of the motor. Since the applied voltage v appears in the numerator of Equation (6–57) and the frequency ω appears in the denominator of Equation (6–57), the two effects counteract each other, and the magnetization current is unaffected.

When the voltage applied to an induction motor is varied linearly with frequency below the base speed, the flux in the motor will remain approximately constant. Therefore, the maximum torque which the motor can supply remains fairly high. However, the maximum power rating of the motor must be decreased linearly with decreases in frequency to protect the stator circuit from overheating. The power supplied to a three-phase induction motor is given by

$$P = \sqrt{3}V_L I_L \cos\theta$$

If the voltage V_L is decreased, then the maximum power P must also be decreased, or else the current flowing in the motor will become excessive, and the motor will overheat.

Figure 6–43a shows a family of induction motor torque–speed characteristic curves for speeds below base speed, assuming that the magnitude of the stator voltage varies linearly with frequency.

When the electrical frequency applied to the motor exceeds the rated frequency of the motor, the stator voltage is held constant at the rated value. Although saturation considerations would permit the voltage to be raised above the rated value under these circumstances, it is limited to the rated voltage to protect the winding insulation of the motor. The higher the electrical frequency above base speed, the larger the denominator of Equation (6–57) becomes. Since the numerator term is held constant above rated frequency, the resulting flux in the machine decreases and the maximum torque decreases with it. Figure 6–43b shows a family of induction motor torque–speed characteristic curves for speeds above base speed, assuming that the stator voltage is held constant.

If the stator voltage is varied linearly with frequency below base speed and is held constant at rated value above base speed, then the resulting family of torque–speed characteristics is as shown in Figure 6–43c. The rated speed for the motor shown in Figure 6–43 is 1800 r/min.

In the past, the principal disadvantage of electrical frequency control as a method of speed changing was that a dedicated generator or mechanical frequency changer was required to make it operate. This problem has disappeared with the development of modern solid-state variable-frequency motor drives. In fact, changing the line frequency with solid-state motor drives has become the method of choice for induction motor speed control. Note that this method can be used with *any* induction motor, unlike the pole-changing technique, which requires a motor with special stator windings.

A typical solid-state variable-frequency induction motor drive will be described in Section 6.10.

Speed Control by Changing the Line Voltage

The torque developed by an induction motor is proportional to the square of the applied voltage. If a load has a torque–speed characteristic such as the one shown in Figure 6–44, then the speed of the motor may be controlled over a limited range by varying the line voltage. This method of speed control is sometimes used on small motors driving fans.

Speed Control by Changing the Rotor Resistance

In wound-rotor induction motors, it is possible to change the shape of the torque-speed curve by inserting extra resistances into the rotor circuit of the machine. The resulting torque–speed characteristic curves are shown in Figure 6–45. If the torque–speed curve of the load is as shown in the figure, then changing the rotor resistance will change the operating speed of the motor. However, inserting extra resistances into the rotor circuit of an induction motor seriously reduces the efficiency of the machine.

This method of speed control is mostly of historical interest, since very few wound-rotor induction motors are built anymore. When it is used, it is normally used only for short periods because of the efficiency problem mentioned in the previous paragraph.

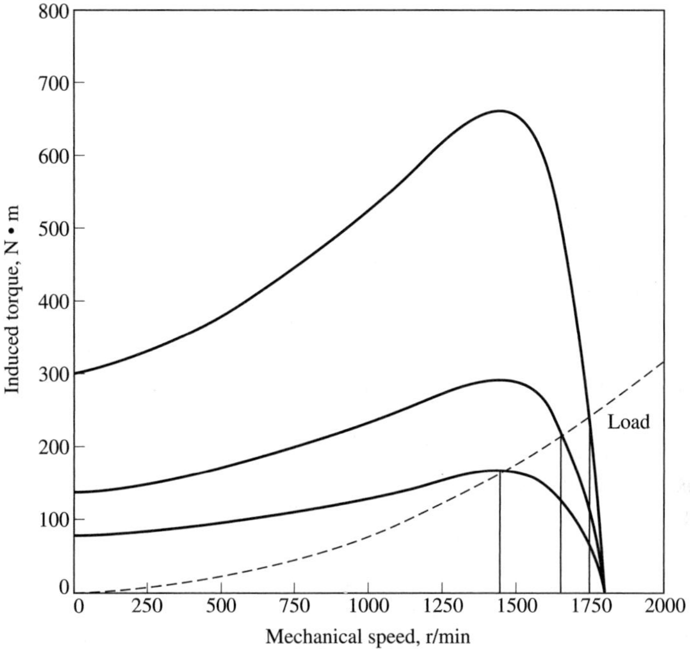

FIGURE 6–44
Variable-line-voltage speed control in an induction motor.

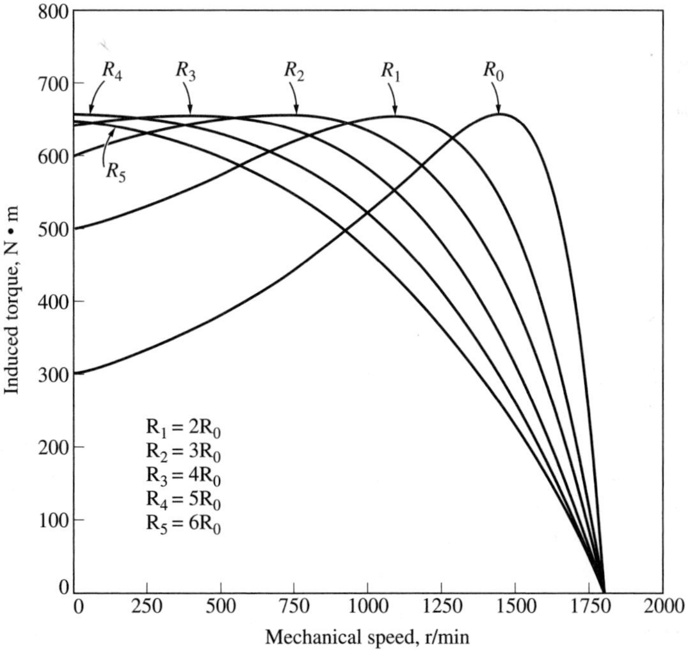

FIGURE 6–45
Speed control by varying the rotor resistance of a wound-rotor induction motor.

6.10 SOLID-STATE INDUCTION MOTOR DRIVES

As mentioned in the previous section, the method of choice today for induction motor speed control is the solid-state variable-frequency induction motor drive. A typical drive of this sort is shown in Figure 6–46. The drive is very flexible: its input power can be either single-phase or three-phase, either 50 or 60 Hz, and anywhere from 208 to 230 V. The output from this drive is a three-phase set of voltages whose frequency can be varied from 0 up to 120 Hz and whose voltage can be varied from 0 V up to the rated voltage of the motor.

The output voltage and frequency control are achieved by using the pulse-width modulation (PWM) techniques.[1] Both output frequency and output voltage can be controlled independently by pulse-width modulation. Figure 6–47 illustrates the manner in which the PWM drive can control the output frequency while maintaining a constant rms voltage level, while Figure 6–48 illustrates the manner in which the PWM drive can control the rms voltage level while maintaining a constant frequency.

As we described in Section 6.9, it is often desirable to vary the output frequency and output rms voltage together in a linear fashion. Figure 6–49 shows typical output voltage waveforms from one phase of the drive for the situation in which frequency and voltage are varied simultaneously in a linear fashion.[2]

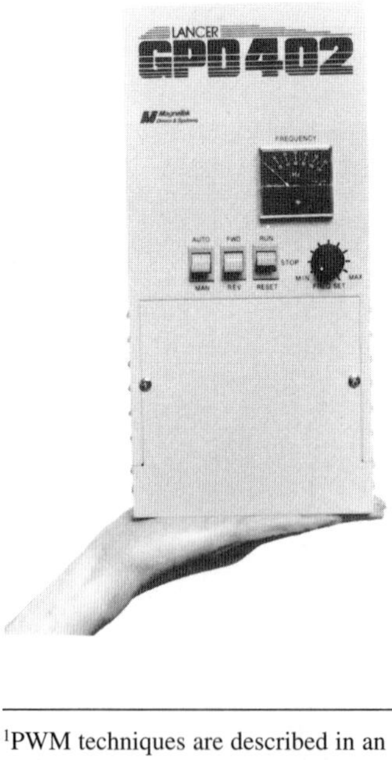

FIGURE 6–46
A typical solid-state variable-frequency induction motor drive. (*Courtesy of MagneTek, Inc.*)

---

[1]PWM techniques are described in an online supplement to this book, "Introduction to Power Electronics," which is available at the book's website.

[2]The output waveforms in Figure 6–48 are actually simplified waveforms. The real induction motor drive has a much higher carrier frequency than that shown in the figure.

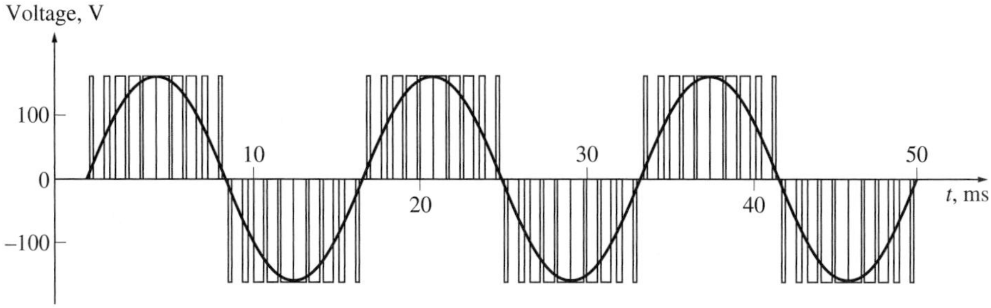

(a)

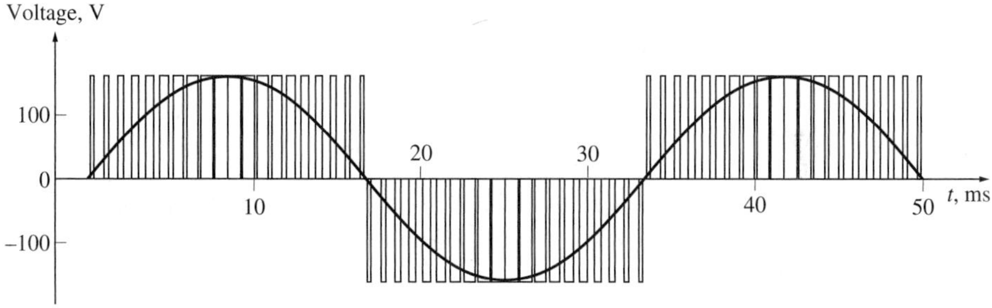

(b)

FIGURE 6–47
Variable-frequency control with a PWM waveform: (a) 60-Hz, 120-V PWM waveform; (b) 30-Hz, 120-V PWM waveform.

Figure 6–49a shows the output voltage adjusted for a frequency of 60 Hz and an rms voltage of 120 V. Figure 6–49b shows the output adjusted for a frequency of 30 Hz and an rms voltage of 60 V, and Figure 6–49c shows the output adjusted for a frequency of 20 Hz and an rms voltage of 40 V. Notice that the peak voltage out of the drive remains the same in all three cases; the rms voltage level is controlled by the fraction of time the voltage is switched on, and the frequency is controlled by the rate at which the polarity of the pulses switches from positive to negative and back again.

The typical induction motor drive shown in Figure 6–46 has many built-in features which contribute to its adjustability and ease of use. Here is a summary of some of these features.

Frequency (Speed) Adjustment

The output frequency of the drive can be controlled manually from a control mounted on the drive cabinet, or it can be controlled remotely by an external voltage or current signal. The ability to adjust the frequency of the drive in response to some external signal is very important, since it permits an external computer or

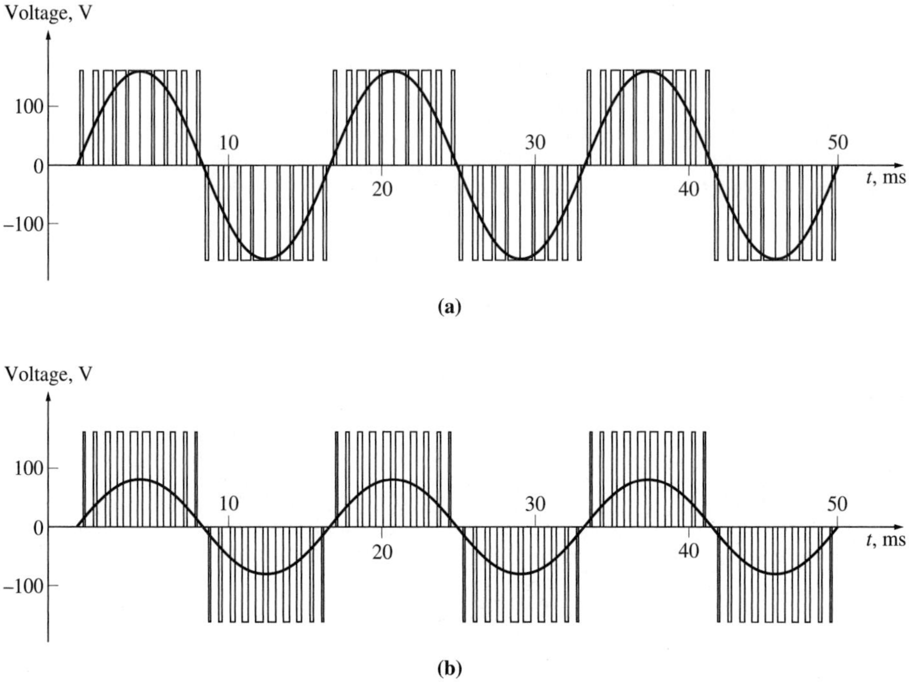

FIGURE 6–48
Variable voltage control with a PWM waveform: (a) 60-Hz, 120-V PWM waveform; (b) 60-Hz, 60-V PWM waveform.

process controller to control the speed of the motor in accordance with the overall needs of the plant in which it is installed.

A Choice of Voltage and Frequency Patterns

The types of mechanical loads which might be attached to an induction motor vary greatly. Some loads such as fans require very little torque when starting (or running at low speeds) and have torques which increase as the square of the speed. Other loads might be harder to start, requiring more than the rated full-load torque of the motor just to get the load moving. This drive provides a variety of voltage-versus-frequency patterns which can be selected to match the torque from the induction motor to the torque required by its load. Three of these patterns are shown in Figures 6–50 through 6–52.

Figure 6–50a shows the standard or general-purpose voltage-versus-frequency pattern, described in the previous section. This pattern changes the output voltage linearly with changes in output frequency for speeds below base speed and holds the output voltage constant for speeds above base speed. (The small

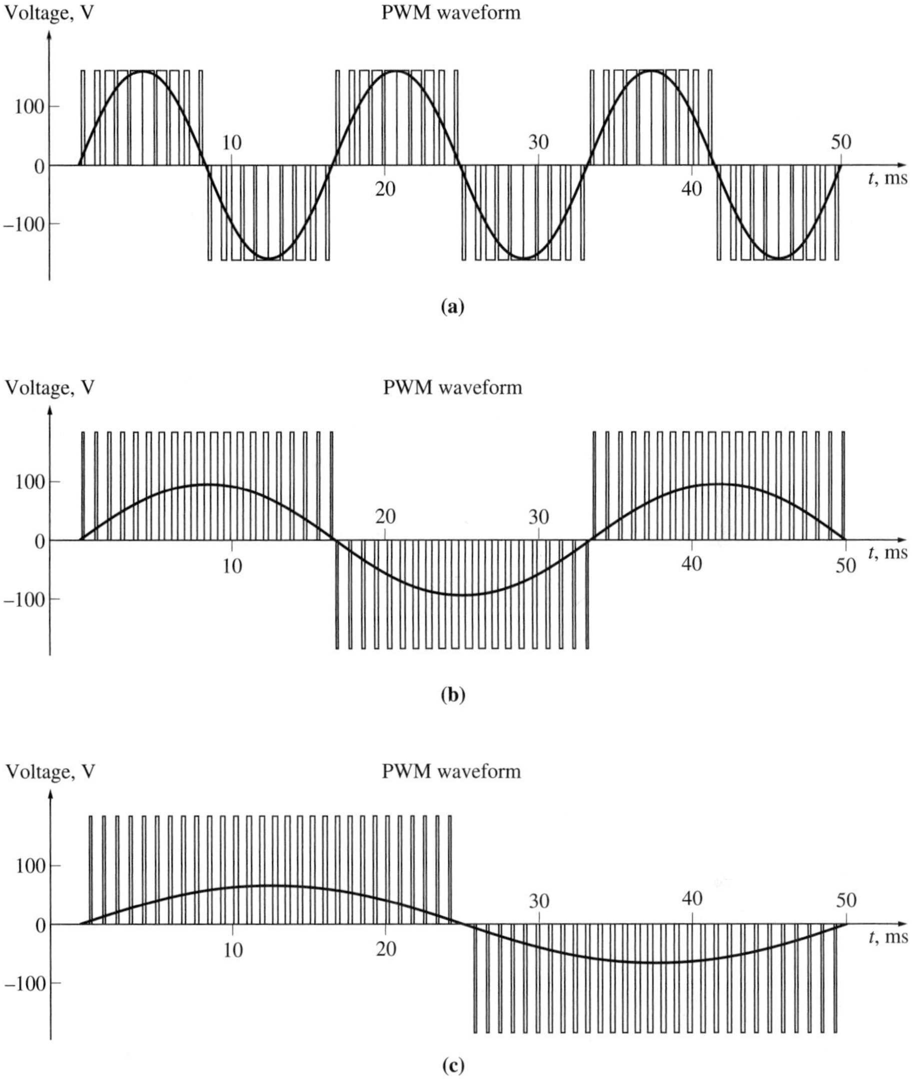

FIGURE 6–49
Simultaneous voltage and frequency control with a PWM waveform: (a) 60-Hz, 120-V PWM waveform; (b) 30-Hz, 60-V PWM waveform; (c) 20-Hz, 40-V PWM waveform.

constant-voltage region at very low frequencies is necessary to ensure that there will be some starting torque at the very lowest speeds.) Figure 6–50b shows the resulting induction motor torque–speed characteristics for several operating frequencies below base speed.

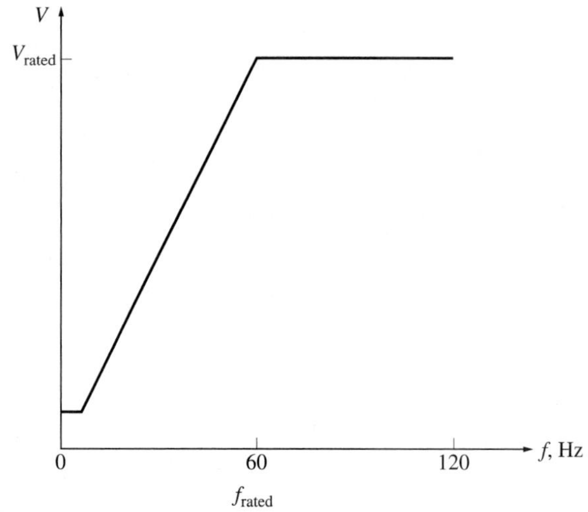

(a)

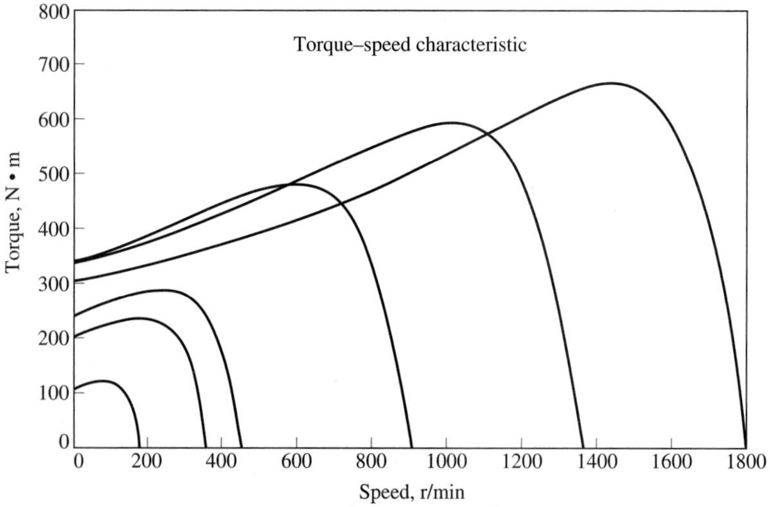

(b)

FIGURE 6–50
(a) Possible voltage-versus-frequency patterns for the solid-state variable-frequency induction motor drive: *general-purpose pattern.* This pattern consists of a linear voltage–frequency curve below rated frequency and a constant voltage above rated frequency. (b) The resulting torque–speed characteristic curves for speeds below rated frequency (speeds above rated frequency look like Figure 6–42b).

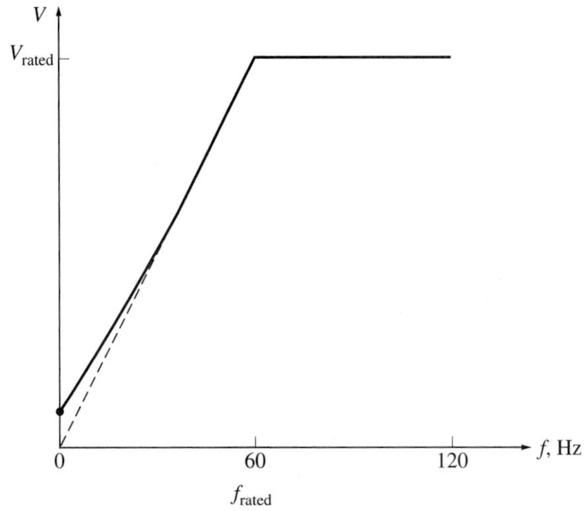

(a)

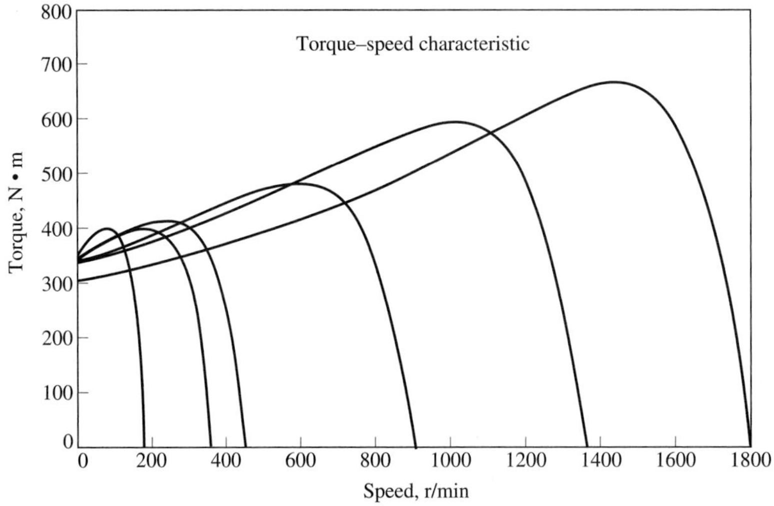

(b)

FIGURE 6–51
(a) Possible voltage-versus-frequency patterns for the solid-state variable-frequency induction motor drive: *high-starting-torque pattern.* This is a modified voltage–frequency pattern suitable for loads requiring high starting torques. It is the same as the linear voltage–frequency pattern except at low speeds. The voltage is disproportionately high at very low speeds, which produces extra torque at the cost of a higher magnetization current. (b) The resulting torque–speed characteristic curves for speeds below rated frequency (speeds above rated frequency look like Figure 6–42b).

Figure 6–51a shows the voltage-versus-frequency pattern used for loads with high starting torques. This pattern also changes the output voltage linearly with changes in output frequency for speeds below base speed, but it has a shallower slope at frequencies below 30 Hz. For any given frequency below 30 Hz, the output voltage will be *higher* than it was with the previous pattern. This higher voltage will produce a higher torque, but at the cost of increased magnetic saturation and higher magnetization currents. The increased saturation and higher currents are often acceptable for the short periods required to start heavy loads. Figure 6–51b shows the induction motor torque–speed characteristics for several operating frequencies below base speed. Notice the increased torque available at low frequencies compared to Figure 6–50b.

Figure 6–52a shows the voltage-versus-frequency pattern used for loads with low starting torques (called *soft-start loads*). This pattern changes the output voltage parabolically with changes in output frequency for speeds below base speed. For any given frequency below 60 Hz, the output voltage will be lower than it was with the standard pattern. This lower voltage will produce a lower torque, providing a slow, smooth start for low-torque loads. Figure 6–52b shows the induction motor torque–speed characteristics for several operating frequencies below base speed. Notice the decreased torque available at low frequencies compared to Figure 6–50.

Independently Adjustable Acceleration and Deceleration Ramps

When the desired operating speed of the motor is changed, the drive controlling it will change frequency to bring the motor to the new operating speed. If the speed change is sudden (e.g., an instantaneous jump from 900 to 1200 r/min), the drive does not try to make the motor instantaneously jump from the old desired speed to the new desired speed. Instead, the rate of motor acceleration or deceleration is limited to a safe level by special circuits built into the electronics of the drive. These rates can be adjusted independently for accelerations and decelerations.

Motor Protection

The induction motor drive has built into it a variety of features designed to protect the motor attached to the drive. The drive can detect excessive steady-state currents (an overload condition), excessive instantaneous currents, overvoltage conditions, or undervoltage conditions. In any of these cases, it will shut down the motor.

Induction motor drives like the one described above are now so flexible and reliable that induction motors with these drives are displacing dc motors in many applications which require a wide range of speed variation.

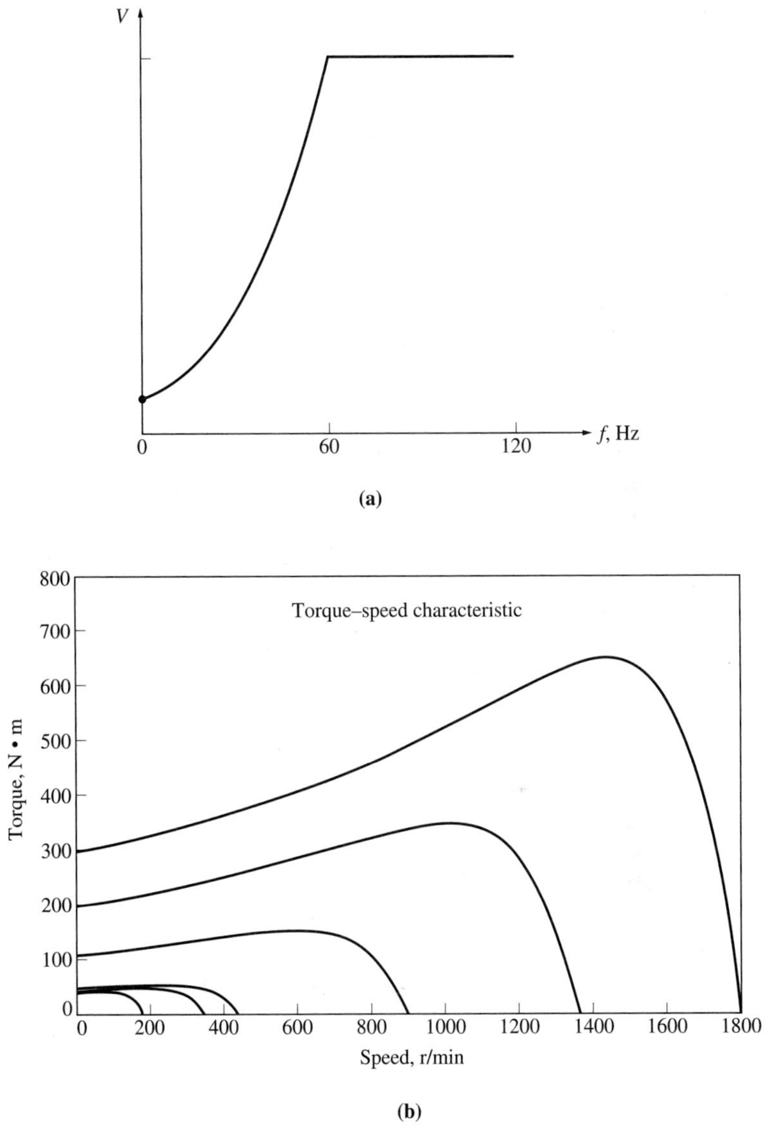

FIGURE 6–52
(a) Possible voltage-versus-frequency patterns for the solid-state variable-frequency induction motor drive: *fan torque pattern.* This is a voltage–frequency pattern suitable for use with motors driving fans and centrifugal pumps, which have a very low starting torque. (b) The resulting torque–speed characteristic curves for speeds below rated frequency (speeds above rated frequency look like Figure 6–42b).

6.11 DETERMINING CIRCUIT MODEL PARAMETERS

The equivalent circuit of an induction motor is a very useful tool for determining the motor's response to changes in load. However, if a model is to be used for a real machine, it is necessary to determine what the element values are that go into the model. How can R_1, R_2, X_1, X_2, and X_M be determined for a real motor?

These pieces of information may be found by performing a series of tests on the induction motor that are analogous to the short-circuit and open-circuit tests in a transformer. The tests must be performed under precisely controlled conditions, since the resistances vary with temperature and the rotor resistance also varies with rotor frequency. The exact details of how each induction motor test must be performed in order to achieve accurate results are described in IEEE Standard 112. Although the details of the tests are very complicated, the concepts behind them are relatively straightforward and will be explained here.

The No-Load Test

The no-load test of an induction motor measures the rotational losses of the motor and provides information about its magnetization current. The test circuit for this test is shown in Figure 6–53a. Wattmeters, a voltmeter, and three ammeters are connected to an induction motor, which is allowed to spin freely. The only load on the motor is the friction and windage losses, so all P_{conv} in this motor is consumed by mechanical losses, and the slip of the motor is very small (possibly as small as 0.001 or less). The equivalent circuit of this motor is shown in Figure 6–53b. With its very small slip, the resistance corresponding to its power converted, $R_2(1 - s)/s$, is much much larger than the resistance corresponding to the rotor copper losses R_2 and much larger than the rotor reactance X_2. In this case, the equivalent circuit reduces approximately to the last circuit in Figure 6–53b. There, the output resistor is in parallel with the magnetization reactance X_M and the core losses R_C.

In this motor at no-load conditions, the input power measured by the meters must equal the losses in the motor. The rotor copper losses are negligible because the current I_2 is *extremely* small [because of the large load resistance $R_2(1 - s)/s$], so they may be neglected. The stator copper losses are given by

$$\boxed{P_{\text{SCL}} = 3I_1^2 R_1} \tag{6–25}$$

so the input power must equal

$$\begin{aligned} P_{\text{in}} &= P_{\text{SCL}} + P_{\text{core}} + P_{\text{F\&W}} + P_{\text{misc}} \\ &= 3I_1^2 R_1 + P_{\text{rot}} \end{aligned} \tag{6–58}$$

where P_{rot} is the rotational losses of the motor:

$$P_{\text{rot}} = P_{\text{core}} + P_{\text{F\&W}} + P_{\text{misc}} \tag{6–59}$$

Thus, given the input power to the motor, the rotational losses of the machine may be determined.

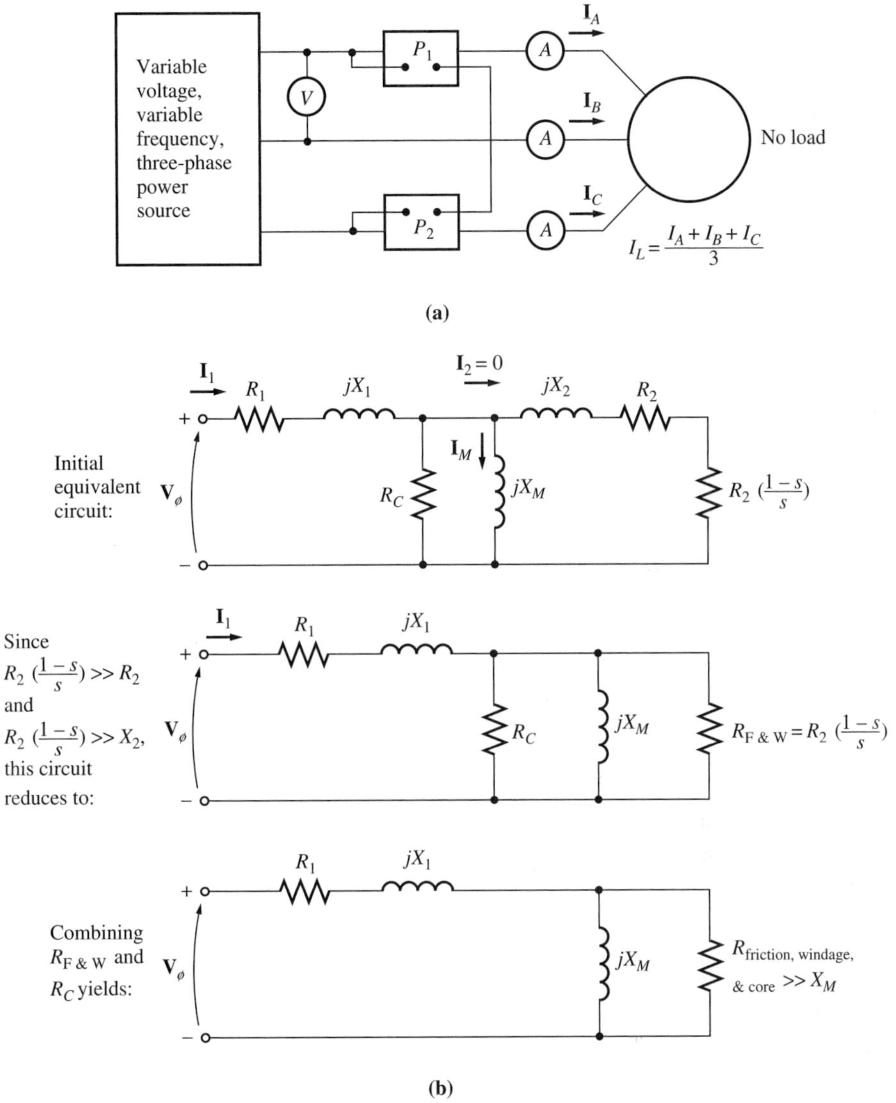

FIGURE 6–53
The no-load test of an induction motor: (a) test circuit; (b) the resulting motor equivalent circuit. Note that at no load the motor's impedance is essentially the series combination of R_1, jX_1, and jX_M.

The equivalent circuit that describes the motor operating in this condition contains resistors R_C and $R_2(1 - s)/s$ in parallel with the magnetizing reactance X_M. The current needed to establish a magnetic field is quite large in an induction motor, because of the high reluctance of its air gap, so the reactance X_M will be much smaller than the resistances in parallel with it and the overall input power factor will be very small. With the large lagging current, most of the voltage drop will be across the inductive components in the circuit. The equivalent input impedance is thus approximately

$$\boxed{|Z_{eq}| = \frac{V_\phi}{I_{1,nl}} \approx X_1 + X_M} \tag{6–60}$$

and if X_1 can be found in some other fashion, the magnetizing impedance X_M will be known for the motor.

The DC Test for Stator Resistance

The rotor resistance R_2 plays an extremely critical role in the operation of an induction motor. Among other things, R_2 determines the shape of the torque–speed curve, determining the speed at which the pullout torque occurs. A standard motor test called the *locked-rotor test* can be used to determine the total motor circuit resistance (this test is taken up in the next section). However, this test finds only the *total* resistance. To find the rotor resistance R_2 accurately, it is necessary to know R_1 so that it can be subtracted from the total.

There is a test for R_1 independent of R_2, X_1 and X_2. This test is called the *dc test*. Basically, a dc voltage is applied to the stator windings of an induction motor. Because the current is dc, there is no induced voltage in the rotor circuit and no resulting rotor current flow. Also, the reactance of the motor is zero at direct current. Therefore, the only quantity limiting current flow in the motor is the stator resistance, and that resistance can be determined.

The basic circuit for the dc test is shown in Figure 6–54. This figure shows a dc power supply connected to two of the three terminals of a Y-connected induction motor. To perform the test, the current in the stator windings is adjusted to the rated value, and the voltage between the terminals is measured. The current in the stator windings is adjusted to the rated value in an attempt to heat the windings to the same temperature they would have during normal operation (remember, winding resistance is a function of temperature).

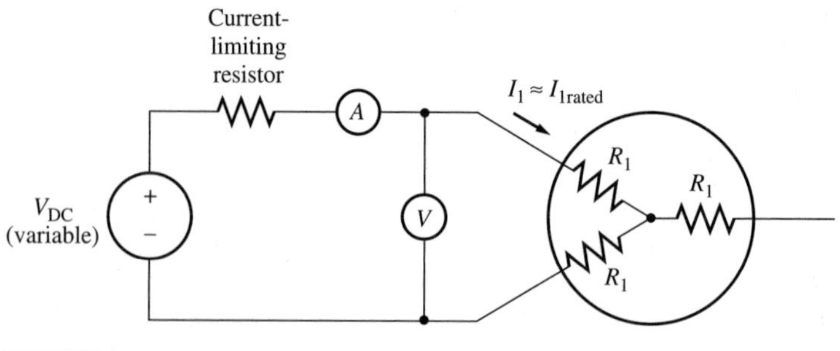

FIGURE 6–54
Test circuit for a dc resistance test.

The current in Figure 6–54 flows through two of the windings, so the total resistance in the current path is $2R_1$. Therefore,

$$2R_1 = \frac{V_{DC}}{I_{DC}}$$

or

$$\boxed{R_1 = \frac{V_{DC}}{2I_{DC}}} \tag{6–61}$$

With this value of R_1 the stator copper losses at no load may be determined, and the rotational losses may be found as the difference between the input power at no load and the stator copper losses.

The value of R_1 calculated in this fashion is not completely accurate, since it neglects the skin effect that occurs when an ac voltage is applied to the windings. More details concerning corrections for temperature and skin effect can be found in IEEE Standard 112.

The Locked-Rotor Test

The third test that can be performed on an induction motor to determine its circuit parameters is called the *locked-rotor test*, or sometimes the *blocked-rotor test*. This test corresponds to the short-circuit test on a transformer. In this test, the rotor is locked or blocked so that it *cannot* move, a voltage is applied to the motor, and the resulting voltage, current, and power are measured.

Figure 6–55a shows the connections for the locked-rotor test. To perform the locked-rotor test, an ac voltage is applied to the stator, and the current flow is adjusted to be approximately full-load value. When the current is full-load value, the voltage, current, and power flowing into the motor are measured. The equivalent circuit for this test is shown in Figure 6–55b. Notice that since the rotor is not moving, the slip $s = 1$, and so the rotor resistance R_2/s is just equal to R_2 (quite a small value). Since R_2 and X_2 are so small, almost all the input current will flow through them, instead of through the much larger magnetizing reactance X_M. Therefore, the circuit under these conditions looks like a series combination of X_1, R_1, X_2, and R_2.

There is one problem with this test, however. In normal operation, the stator frequency is the line frequency of the power system (50 or 60 Hz). At starting conditions, the rotor is also at line frequency. However, at normal operating conditions, the slip of most motors is only 2 to 4 percent, and the resulting rotor frequency is in the range of 1 to 3 Hz. This creates a problem in that *the line frequency does not represent the normal operating conditions of the rotor*. Since effective rotor resistance is a strong function of frequency for design class B and C motors, the incorrect rotor frequency can lead to misleading results in this test. A typical compromise is to use a frequency 25 percent or less of the rated frequency. While this approach is acceptable for essentially constant resistance rotors (design classes A and D), it leaves a lot to be desired when one is trying to find the normal rotor resistance of a variable-resistance rotor. Because of these and similar problems, a great deal of care must be exercised in taking measurements for these tests.

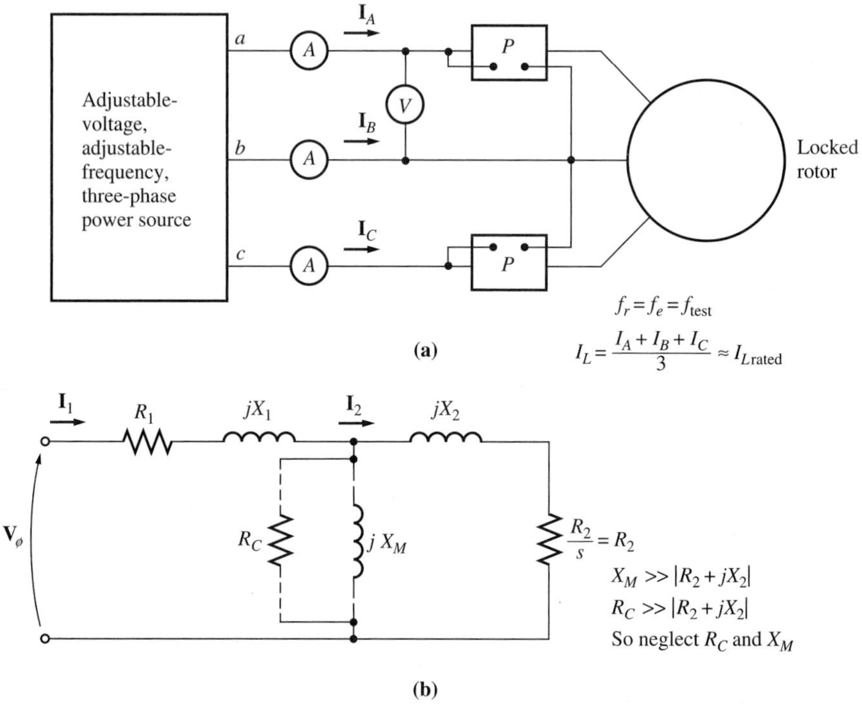

FIGURE 6–55
The locked-rotor test for an induction motor: (a) test circuit; (b) motor equivalent circuit.

After a test voltage and frequency have been set up, the current flow in the motor is quickly adjusted to about the rated value, and the input power, voltage, and current are measured before the rotor can heat up too much. The input power to the motor is given by

$$P = \sqrt{3}V_T I_L \cos\theta$$

so the locked-rotor power factor can be found as

$$\boxed{\text{PF} = \cos\theta = \frac{P_{\text{in}}}{\sqrt{3}V_T I_L}} \tag{6–62}$$

and the impedance angle θ is just equal to $\cos^{-1}$ PF.

The magnitude of the total impedance in the motor circuit at this time is

$$\boxed{|Z_{\text{LR}}| = \frac{V_\phi}{I_1} = \frac{V_T}{\sqrt{3}I_L}} \tag{6–63}$$

and the angle of the total impedance is θ. Therefore,

$$\begin{aligned} Z_{\text{LR}} &= R_{\text{LR}} + jX'_{\text{LR}} \\ &= |Z_{\text{LR}}|\cos\theta + j|Z_{\text{LR}}|\sin\theta \end{aligned} \tag{6–64}$$

	X_1 and X_2 as functions of X_{LR}	
Rotor Design	X_1	X_2
Wound rotor	$0.5\ X_{LR}$	$0.5\ X_{LR}$
Design A	$0.5\ X_{LR}$	$0.5\ X_{LR}$
Design B	$0.4\ X_{LR}$	$0.6\ X_{LR}$
Design C	$0.3\ X_{LR}$	$0.7\ X_{LR}$
Design D	$0.5\ X_{LR}$	$0.5\ X_{LR}$

FIGURE 6–56
Rules of thumb for dividing rotor and stator circuit reactance.

The locked-rotor resistance R_{LR} is equal to

$$\boxed{R_{LR} = R_1 + R_2} \tag{6–65}$$

while the locked-rotor reactance X'_{LR} is equal to

$$X'_{LR} = X'_1 + X'_2 \tag{6–66}$$

where X'_1 and X'_2 are the stator and rotor reactances *at the test frequency*, respectively.

The rotor resistance R_2 can now be found as

$$R_2 = R_{LR} - R_1 \tag{6–67}$$

where R_1 was determined in the dc test. The total rotor reactance referred to the stator can also be found. Since the reactance is directly proportional to the frequency, the total equivalent reactance at the normal operating frequency can be found as

$$\boxed{X_{LR} = \frac{f_{\text{rated}}}{f_{\text{test}}} X'_{LR} = X_1 + X_2} \tag{6–68}$$

Unfortunately, there is no simple way to separate the contributions of the stator and rotor reactances from each other. Over the years, experience has shown that motors of certain design types have certain proportions between the rotor and stator reactances. Figure 6–56 summarizes this experience. In normal practice, it really does not matter just how X_{LR} is broken down, since the reactance appears as the sum $X_1 + X_2$ in all the torque equations.

Example 6–8. The following test data were taken on a 7.5-hp, four-pole, 208-V, 60-Hz, design A, Y-connected induction motor having a rated current of 28 A.

DC test:

$V_{DC} = 13.6$ V $\qquad I_{DC} = 28.0$ A

No-load test:

$V_T = 208$ V $\qquad f = 60$ Hz

$I_A = 8.12$ A $\qquad P_{in} = 420$ W

$$I_B = 8.20 \text{ A}$$
$$I_C = 8.18 \text{ A}$$

Locked-rotor test:

$$V_T = 25 \text{ V} \qquad f = 15 \text{ Hz}$$
$$I_A = 28.1 \text{ A} \qquad P_{\text{in}} = 920 \text{ W}$$
$$I_B = 28.0 \text{ A}$$
$$I_C = 27.6 \text{ A}$$

(a) Sketch the per-phase equivalent circuit for this motor.

(b) Find the slip at the pullout torque, and find the value of the pullout torque itself.

Solution

(a) From the dc test,

$$R_1 = \frac{V_{\text{DC}}}{2I_{\text{DC}}} = \frac{13.6 \text{ V}}{2(28.0 \text{ A})} = 0.243 \ \Omega$$

From the no-load test,

$$I_{\text{L,av}} = \frac{8.12 \text{ A} + 8.20 \text{ A} + 8.18 \text{ A}}{3} = 8.17 \text{ A}$$
$$V_{\phi,\text{nl}} = \frac{208 \text{ V}}{\sqrt{3}} = 120 \text{ V}$$

Therefore,

$$|Z_{\text{nl}}| = \frac{120 \text{ V}}{8.17 \text{ A}} = 14.7 \ \Omega = X_1 + X_M$$

When X_1 is known, X_M can be found. The stator copper losses are

$$P_{\text{SCL}} = 3I_1^2 R_1 = 3(8.17 \text{ A})^2(0.243 \ \Omega) = 48.7 \text{ W}$$

Therefore, the no-load rotational losses are

$$P_{\text{rot}} = P_{\text{in,nl}} - P_{\text{SCL,nl}}$$
$$= 420 \text{ W} - 48.7 \text{ W} = 371.3 \text{ W}$$

From the locked-rotor test,

$$I_{\text{L,av}} = \frac{28.1 \text{ A} + 28.0 \text{ A} + 27.6 \text{ A}}{3} = 27.9 \text{ A}$$

The locked-rotor impedance is

$$|Z_{\text{LR}}| = \frac{V_\phi}{I_A} = \frac{V_T}{\sqrt{3}I_A} = \frac{25 \text{ V}}{\sqrt{3}(27.9 \text{ A})} = 0.517 \ \Omega$$

and the impedance angle θ is

$$\theta = \cos^{-1} \frac{P_{\text{in}}}{\sqrt{3}V_T I_L}$$
$$= \cos^{-1} \frac{920 \text{ W}}{\sqrt{3}(25 \text{ V})(27.9 \text{ A})}$$

$$= \cos^{-1} 0.762 = 40.4°$$

Therefore, $R_{LR} = 0.517 \cos 40.4° = 0.394\ \Omega = R_1 + R_2$. Since $R_1 = 0.243\ \Omega$, R_2 must be $0.151\ \Omega$. The reactance at 15 Hz is

$$X'_{LR} = 0.517 \sin 40.4° = 0.335\ \Omega$$

The equivalent reactance at 60 Hz is

$$X_{LR} = \frac{f_{rated}}{f_{test}} X'_{LR} = \left(\frac{60\text{ Hz}}{15\text{ Hz}}\right) 0.335\ \Omega = 1.34\ \Omega$$

For design class A induction motors, this reactance is assumed to be divided equally between the rotor and stator, so

$$X_1 = X_2 = 0.67\ \Omega$$

$$X_M = |Z_{nl}| - X_1 = 14.7\ \Omega - 0.67\ \Omega = 14.03\ \Omega$$

The final per-phase equivalent circuit is shown in Figure 6–57.

(b) For this equivalent circuit, the Thevenin equivalents are found from Equations (6–41b), (6–44), and (6–45) to be

$$V_{TH} = 114.6\text{ V} \qquad R_{TH} = 0.221\ \Omega \qquad X_{TH} = 0.67\ \Omega$$

Therefore, the slip at the pullout torque is given by

$$s_{max} = \frac{R_2}{\sqrt{R_{TH}^2 + (X_{TH} + X_2)^2}} \tag{6–53}$$

$$= \frac{0.151\ \Omega}{\sqrt{(0.243\ \Omega)^2 + (0.67\ \Omega + 0.67\ \Omega)^2}} = 0.111 = 11.1\%$$

The maximum torque of this motor is given by

$$\tau_{max} = \frac{3V_{TH}^2}{2\omega_{sync}[R_{TH} + \sqrt{R_{TH}^2 + (X_{TH} + X^2)}]} \tag{6–54}$$

$$= \frac{3(114.6\text{ V})^2}{2(188.5\text{ rad/s})[0.221\ \Omega + \sqrt{(0.221\ \Omega)^2 + (0.67\ \Omega + 0.67\ \Omega)^2}]}$$

$$= 66.2\text{ N} \bullet \text{m}$$

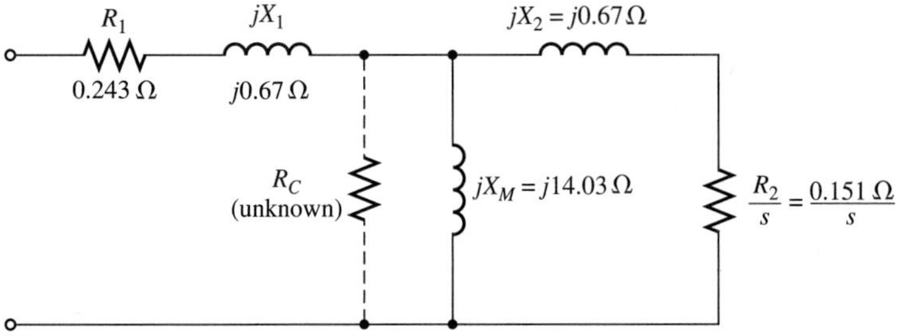

FIGURE 6–57
Motor per-phase equivalent circuit for Example 6–8.

6.12 THE INDUCTION GENERATOR

The torque–speed characteristic curve in Figure 6–20 shows that if an induction motor is driven at a speed *greater* than n_{sync} by an external prime mover, the direction of its inducted torque will reverse and it will act as a generator. As the torque applied to its shaft by the prime mover increases, the amount of power produced by the induction generator increases. As Figure 6–58 shows, there is a maximum possible induced torque in the generator mode of operation. This torque is known as the *pushover torque* of the generator. If a prime mover applies a torque greater than the pushover torque to the shaft of an induction generator, the generator will overspeed.

As a generator, an induction machine has severe limitations. Because it lacks a separate field circuit, an induction generator *cannot* produce reactive power. In fact, it consumes reactive power, and an external source of reactive power must be connected to it at all times to maintain its stator magnetic field. This external source of reactive power must also control the terminal voltage of the generator—with no field current, an induction generator cannot control its own output voltage. Normally, the generator's voltage is maintained by the external power system to which it is connected.

The one great advantage of an induction generator is its simplicity. An induction generator does not need a separate field circuit and does not have to be driven continuously at a fixed speed. As long as the machine's speed is some value greater than n_{sync} for the power system to which it is connected, it will function as a generator. The greater the torque applied to its shaft (up to a certain point), the

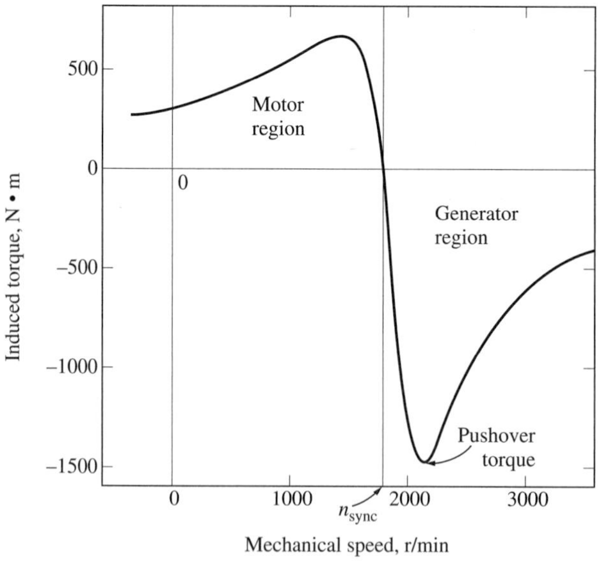

FIGURE 6–58
The torque–speed characteristic of an induction machine, showing the generator region of operation. Note the pushover torque.

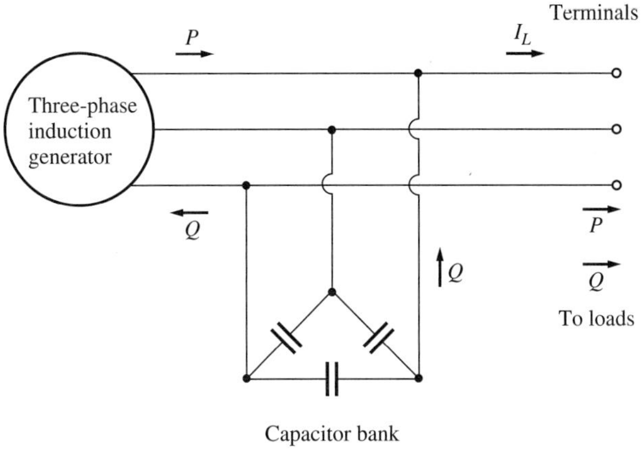

FIGURE 6–59
An induction generator operating alone with a capacitor bank to supply reactive power.

greater its resulting output power. The fact that no fancy regulation is required makes this generator a good choice for windmills, heat recovery systems, and similar supplementary power sources attached to an existing power system. In such applications, power-factor correction can be provided by capacitors, and the generator's terminal voltage can be controlled by the external power system.

The Induction Generator Operating Alone

It is also possible for an induction machine to function as an isolated generator, independent of any power system, as long as capacitors are available to supply the reactive power required by the generator and by any attached loads. Such an isolated induction generator is shown in Figure 6–59.

The magnetizing current I_M required by an induction machine as a function of terminal voltage can be found by running the machine as a motor at no load and measuring its armature current as a function of terminal voltage. Such a magnetization curve is shown in Figure 6–60a. To achieve a given voltage level in an induction generator, external capacitors must supply the magnetization current corresponding to that level.

Since the reactive current that a capacitor can produce is *directly proportional* to the voltage applied to it, the locus of all possible combinations of voltage and current through a capacitor is a straight line. Such a plot of voltage versus current for a given frequency is shown in Figure 6–60b. *If a three-phase set of capacitors is connected across the terminals of an induction generator, the no-load voltage of the induction generator will be the intersection of the generator's magnetization curve and the capacitor's load line.* The no-load terminal voltage of an induction generator for three different sets of capacitance is shown in Figure 6–60c.

How does the voltage build up in an induction generator when it is first started? When an induction generator first starts to turn, the residual magnetism in

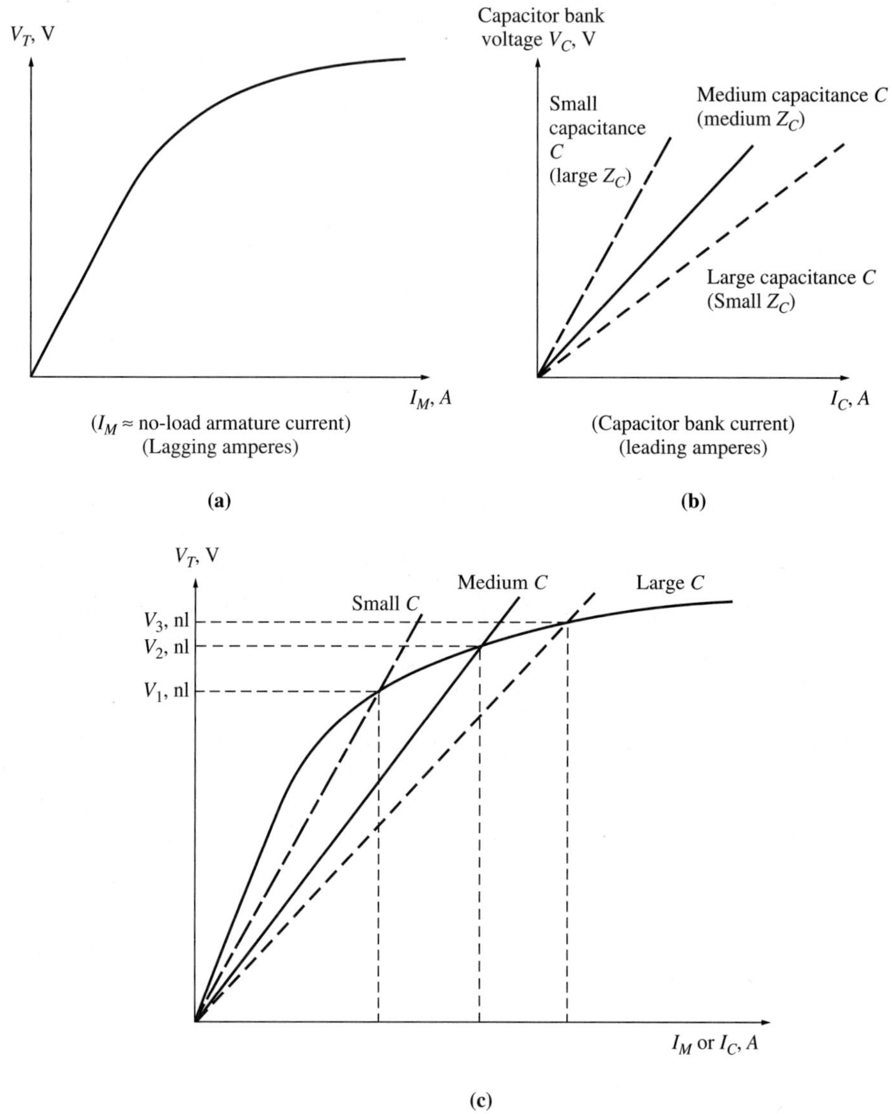

FIGURE 6–60
(a) The magnetization curve of an induction machine. It is a plot of the terminal voltage of the machine as a function of its magnetization current (which *lags* the phase voltage by approximately 90°). (b) Plot of the voltage–current characteristic of a capacitor bank. Note that the larger the capacitance, the greater its current for a given voltage. This current *leads* the phase voltage by approximately 90°. (c) The no-load terminal voltage for an isolated induction generator can be found by plotting the generator terminal characteristic and the capacitor voltage–current characteristic on a single set of axes. The intersection of the two curves is the point at which the reactive power demanded by the generator is exactly supplied by the capacitors, and this point gives the *no-load terminal voltage* of the generator.

its field circuit produces a small voltage. That small voltage produces a capacitive current flow, which increases the voltage, further increasing the capacitive current, and so forth until the voltage is fully built up. If no residual flux is present in the induction generator's rotor, then its voltage will not build up, and it must be magnetized by momentarily running it as a motor.

The most serious problem with an induction generator is that its voltage varies wildly with changes in load, especially reactive load. Typical terminal characteristics of an induction generator operating alone with a constant parallel capacitance are shown in Figure 6–61. Notice that, in the case of inductive loading, the voltage collapses *very* rapidly. This happens because the fixed capacitors must supply all the reactive power needed by both the load and the generator, and any reactive power diverted to the load moves the generator back along its magnetization curve, causing a major drop in generator voltage. It is therefore very difficult to start an induction motor on a power system supplied by an induction generator—special techniques must be employed to increase the effective capacitance during starting and then decrease it during normal operation.

Because of the nature of the induction machine's torque–speed characteristic, an induction generator's frequency varies with changing loads: but since the torque–speed characteristic is very steep in the normal operating range, the total frequency variation is usually limited to less than 5 percent. This amount of variation may be quite acceptable in many isolated or emergency generator applications.

Induction Generator Applications

Induction generators have been used since early in the twentieth century, but by the1960s and 1970s they had largely disappeared from use. However, the induction generator has made a comeback since the oil price shocks of 1973. With energy costs so high, energy recovery became an important part of the economics of most

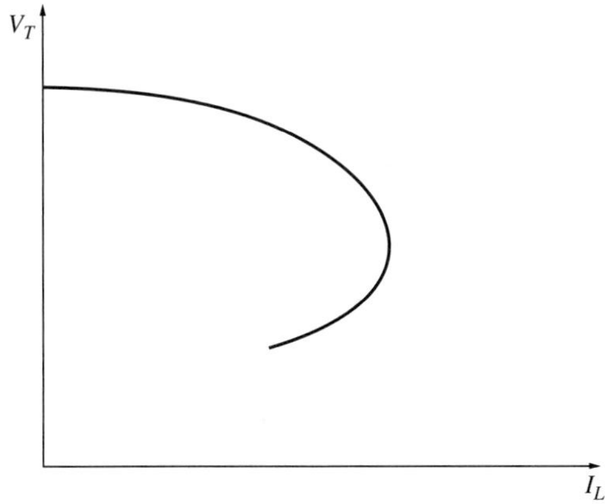

FIGURE 6–61
The terminal voltage–current characteristic of an induction generator for a load with a constant lagging power factor.

industrial processes. The induction generator is ideal for such applications because it requires very little in the way of control systems or maintenance.

Because of their simplicity and small size per kilowatt of output power, induction generators are also favored very strongly for small windmills. Many commercial windmills are designed to operate in parallel with large power systems, supplying a fraction of the customer's total power needs. In such operation, the power system can be relied on for voltage and frequency control, and static capacitors can be used for power-factor correction.

It is interesting that wound-rotor induction machines have been making a bit of a comeback as induction generators connected to windmills. As mentioned previously, wound-rotor machines are more expensive than cage rotor machines, and they require more maintenance because of the slip rings and brushes included in their design. However, wound-rotor machines allow rotor resistance control, as discussed in Section 6.9. Inserting or removing rotor resistance changes the shape of the torque–speed characteristic, and therefore the operating speed of the machine (see Figure 6–45).

This characteristic of wound-rotor machines can be very important for induction generators connected to windmills. Wind is a very fickle and uncertain power source: sometimes it blows strongly, sometimes it blows lightly, and sometimes it doesn't blow at all. To use an ordinary cage-rotor induction machine as a generator, the wind must be turning the machine's shaft at a speed between n_{sync} and the pushover speed (as shown in Figure 6–58). This is a relatively narrow range of speeds, which limits the wind conditions under which a wind generator can be used.

Wound-rotor machines are better here because it is possible to insert a rotor resistance and thus change the shape of the torque–speed characteristic. Figure 6–62

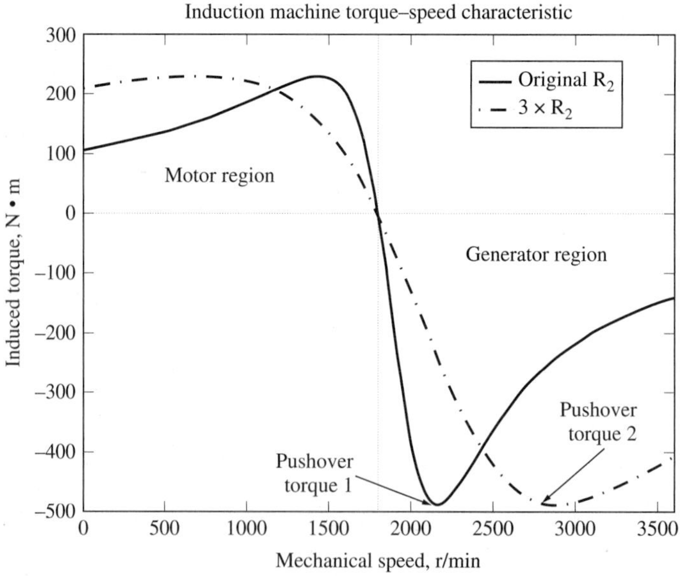

FIGURE 6–62
Torque–speed characteristic of a wound-rotor induction generator with the original rotor resistance and with three times the original rotor resistance. Note that the range of speeds at which the machine can operate as a generator is greatly increased by adding rotor resistance.

shows an example of a wound-rotor induction machine with both the original rotor resistance R_2 and trebled rotor resistance $3R_2$. Note that the pushover torque is the same for both cases, but the range of speeds between n_{sync} and the pushover speed is much greater for the generator with inserted rotor resistance. This allows the generator to produce useful power over a wider range of wind conditions.

In practice, solid-state controllers are used instead of actual resistors to adjust the effective rotor resistance of modern wound-rotor induction generators. However, the effect on the torque–speed characteristic is the same.

6.13 INDUCTION MOTOR RATINGS

A nameplate for a typical high-efficiency small-to-medium-sized induction motor is shown in Figure 6–63. The most important ratings present on the nameplate are

1. Output power (This will be horsepower in the USA and kilowatts everywhere else in the world.)

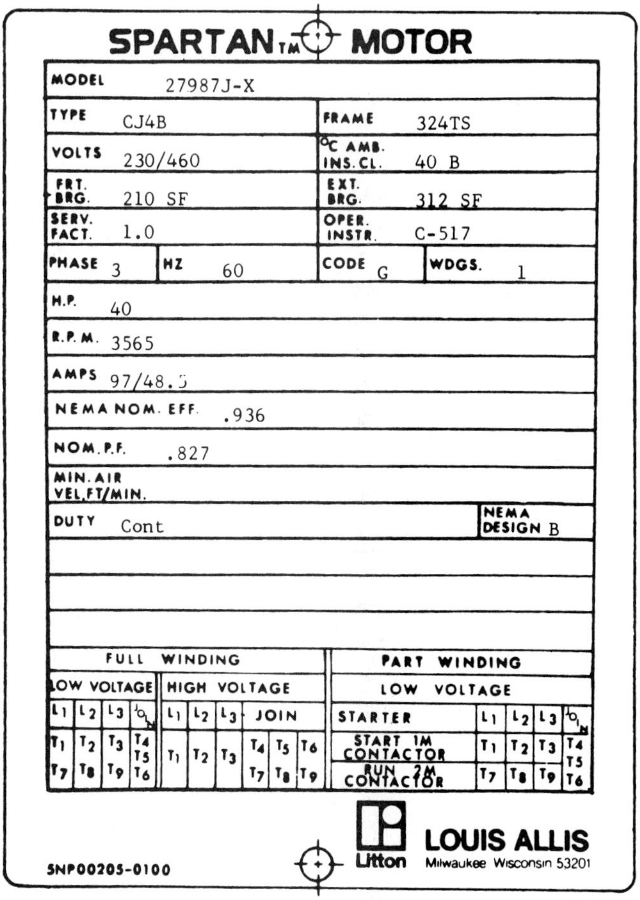

FIGURE 6–63
The nameplate of a typical high-efficiency induction motor. (*Courtesy of MagneTek, Inc.*)

2. Voltage
3. Current
4. Power factor
5. Speed
6. Nominal efficiency
7. NEMA design class
8. Starting code

A nameplate for a typical standard-efficiency induction motor would be similar, except that it might not show a nominal efficiency.

The voltage limit on the motor is based on the maximum acceptable magnetization current flow, since the higher the voltage gets, the more saturated the motor's iron becomes and the higher its magnetization current becomes. Just as in the case of transformers and synchronous machines, a 60-Hz induction motor may be used on a 50-Hz power system, but only if the voltage rating is decreased by an amount proportional to the decrease in frequency. This derating is necessary because the flux in the core of the motor is proportional to the integral of the applied voltage. To keep the maximum flux in the core constant while the period of integration is increasing, the average voltage level must decrease.

The current limit on an induction motor is based on the maximum acceptable heating in the motor's windings, and the power limit is set by the combination of the voltage and current ratings with the machine's power factor and efficiency.

NEMA design classes, starting code letters, and nominal efficiencies were discussed in previous sections of this chapter.

6.14 SUMMARY

The induction motor is the most popular type of ac motor because of its simplicity and ease of operation. An induction motor does not have a separate field circuit; instead, it depends on transformer action to induce voltages and currents in its field circuit. In fact, an induction motor is basically a rotating transformer. Its equivalent circuit is similar to that of a transformer, except for the effects of varying speed.

There are two types of induction motor rotors, cage rotors and wound rotors. Cage rotors consist of a series of parallel bars all around the rotor, shorted together at each end. Wound rotors are complete three-phase rotor windings, with the phases brought out of the rotor through slip rings and brushes. Wound rotors are more expensive and require more maintenance than cage rotors, so they are very rarely used (except sometimes for induction generators).

An induction motor normally operates at a speed near synchronous speed, but it can never operate at exactly n_{sync}. There must always be some relative motion in order to induce a voltage in the induction motor's field circuit. The rotor voltage induced by the relative motion between the rotor and the stator magnetic field produces a rotor current, and that rotor current interacts with the stator magnetic field to produce the induced torque in the motor.

In an induction motor, the slip or speed at which the maximum torque occurs can be controlled by varying the rotor resistance. The *value* of that maximum torque is independent of the rotor resistance. A high rotor resistance lowers the speed at which maximum torque occurs and thus increases the starting torque of the motor. However, it pays for this starting torque by having very poor speed regulation in its normal operating range. A low rotor resistance, on the other hand, reduces the motor's starting torque while improving its speed regulation. Any normal induction motor design must be a compromise between these two conflicting requirements.

One way to achieve such a compromise is to employ deep-bar or double-cage rotors. These rotors have a high effective resistance at starting and a low effective resistance under normal running conditions, thus yielding both a high starting torque and good speed regulation in the same motor. The same effect can be achieved with a wound-rotor induction motor if the rotor field resistance is varied.

Induction motors are classified by their torque–speed characteristics into a series of NEMA design classes. Design class A motors are standard induction motors, with normal starting torque, relatively high starting current, low slip, and high pullout torque. These motors can cause problems when started across the line due to the high starting currents. Design class B motors use a deep bar design to produce normal starting torque, lower starting current, a bit greater slip, and a bit lower pullout torque when compared to design class A motors. Since they require about 25 percent less starting current, they work better in many applications where the power system cannot supply high surge currents. Design class C motors use a deep-bar or double-cage design to produce a high starting torque with low starting current, at the expense of greater slip and lower pullout torque. These motors can be used in applications where high starting torque is needed without drawing excessive line currents. Design class D motors use high-resistance bars to produce very high starting torque with low starting currents, at the expense of very high slip. Pullout torque is quite high for this design, but it can occur at extremely high slips.

Speed control of induction motors can be accomplished by changing the number of poles on the machine, by changing the applied electrical frequency, by changing the applied terminal voltage, or by changing the rotor resistance in the case of a wound-rotor induction motor. All of these techniques are regularly used (except for changing rotor resistance), but by far the most common technique today is to change the applied electrical frequency using a solid-state controller.

An induction motor has a starting current that is many times the rated current of the motor, and this can cause problems for the power systems that the motors are connected to. The starting current of a given induction motor is specified by a NEMA code letter, which is printed on the motor's nameplate. When this starting current is too high to be handled by the power system, motor starter circuits are used to reduce the starting current to a safe level. Starter circuits can change motor connections from Δ to Y during starting, can insert extra resistors during starting, or can reduce the applied voltage (and frequency) during starting.

The induction machine can also be used as a generator as long as there is some source of reactive power (capacitors or a synchronous machine) available in the power system. An induction generator operating alone has serious voltage regulation problems, but when it operates in parallel with a large power system, the power system can control the machine's voltage. Induction generators are usually rather small machines and are used principally with alternative energy sources, such as windmills, or with energy recovery systems. Almost all the really large generators in use are synchronous generators.

QUESTIONS

6–1. What are slip and slip speed in an induction motor?

6–2. How does an induction motor develop torque?

6–3. Why is it impossible for an induction motor to operate at synchronous speed?

6–4. Sketch and explain the shape of a typical induction motor torque–speed characteristic curve.

6–5. What equivalent circuit element has the most direct control over the speed at which the pullout torque occurs?

6–6. What is a deep-bar cage rotor? Why is it used? What NEMA design class(es) can be built with it?

6–7. What is a double-cage cage rotor? Why is it used? What NEMA design class(es) can be built with it?

6–8. Describe the characteristics and uses of wound-rotor induction motors and of each NEMA design class of cage motors.

6–9. Why is the efficiency of an induction motor (wound-rotor or cage) so poor at high slips?

6–10. Name and describe four means of controlling the speed of induction motors.

6–11. Why is it necessary to reduce the voltage applied to an induction motor as electrical frequency is reduced?

6–12. Why is terminal voltage speed control limited in operating range?

6–13. What are starting code factors? What do they say about the starting current of an induction motor?

6–14. How does a resistive starter circuit for an induction motor work?

6–15. What information is learned in a locked-rotor test?

6–16. What information is learned in a no-load test?

6–17. What actions are taken to improve the efficiency of modern high-efficiency induction motors?

6–18. What controls the terminal voltage of an induction generator operating alone?

6–19. For what applications are induction generators typically used?

6–20. How can a wound-rotor induction motor be used as a frequency changer?

6–21. How do different voltage–frequency patterns affect the torque–speed characteristics of an induction motor?

6–22. Describe the major features of the solid-state induction motor drive featured in Section 6.10.

6–23. Two 480-V, 100-hp induction motors are manufactured. One is designed for 50-Hz operation, and one is designed for 60-Hz operation, but they are otherwise similar. Which of these machines is larger?

6–24. An induction motor is running at the rated conditions. If the shaft load is now increased, how do the following quantities change?

(a) Mechanical speed
(b) Slip
(c) Rotor induced voltage
(d) Rotor current
(e) Rotor frequency
(f) P_{RCL}
(g) Synchronous speed

PROBLEMS

6–1. A 220-V, three-phase, six-pole, 50-Hz induction motor is running at a slip of 3.5 percent. Find:

(a) The speed of the magnetic fields in revolutions per minute
(b) The speed of the rotor in revolutions per minute
(c) The slip speed of the rotor
(d) The rotor frequency in hertz

6–2. Answer the questions in Problem 6–1 for a 480-V, three-phase, two-pole, 60-Hz induction motor running at a slip of 0.025.

6–3. A three-phase, 60-Hz induction motor runs at 715 r/min at no load and at 670 r/min at full load.

(a) How many poles does this motor have?
(b) What is the slip at rated load?
(c) What is the speed at one-quarter of the rated load?
(d) What is the rotor's electrical frequency at one-quarter of the rated load?

6–4. A 50-kW, 460-V, 50-Hz, two-pole induction motor has a slip of 5 percent when operating at full-load conditions. At full-load conditions, the friction and windage losses are 700 W, and the core losses are 600 W. Find the following values for full-load conditions:

(a) The shaft speed n_m
(b) The output power in watts
(c) The load torque τ_{load} in newton-meters
(d) The induced torque τ_{ind} in newton-meters
(e) The rotor frequency in hertz

6–5. A 208-V, four-pole, 60-Hz, Y-connected wound-rotor induction motor is rated at 30 hp. Its equivalent circuit components are

$$R_1 = 0.100\ \Omega \qquad R_2 = 0.070\ \Omega \qquad X_M = 10.0\ \Omega$$
$$X_1 = 0.210\ \Omega \qquad X_2 = 0.210\ \Omega$$
$$P_{mech} = 500\ \text{W} \qquad P_{misc} \approx 0 \qquad P_{core} = 400\ \text{W}$$

For a slip of 0.05, find

(a) The line current I_L
(b) The stator copper losses P_{SCL}
(c) The air-gap power P_{AG}
(d) The power converted from electrical to mechanical form P_{conv}
(e) The induced torque τ_{ind}
(f) The load torque τ_{load}

(g) The overall machine efficiency η

(h) The motor speed in revolutions per minute and radians per second

6–6. For the motor in Problem 6–5, what is the slip at the pullout torque? What is the pullout torque of this motor?

6–7. (a) Calculate and plot the torque–speed characteristic of the motor in Problem 6–5.

(b) Calculate and plot the output-power-versus-speed curve of the motor in Problem 6–5.

6–8. For the motor of Problem 6–5, how much additional resistance (referred to the stator circuit) would it be necessary to add to the rotor circuit to make the maximum torque occur at starting conditions (when the shaft is not moving)? Plot the torque–speed characteristic of this motor with the additional resistance inserted.

6–9. If the motor in Problem 6–5 is to be operated on a 50-Hz power system, what must be done to its supply voltage? Why? What will the equivalent circuit component values be at 50 Hz? Answer the questions in Problem 6–5 for operation at 50 Hz with a slip of 0.05 and the proper voltage for this machine.

6–10. A three-phase, 60-Hz, two-pole induction motor runs at a no-load speed of 3580 r/min and a full-load speed of 3440 r/min. Calculate the slip and the electrical frequency of the rotor at no-load and full-load conditions. What is the speed regulation of this motor [Equation (3-68)]?

6–11. The input power to the rotor circuit of a six-pole, 60-Hz induction motor running at 1100 r/min is 5 kW. What is the rotor copper loss in this motor?

6–12. The power crossing the air gap of a 60-Hz, four-pole induction motor is 25 kW, and the power converted from electrical to mechanical form in the motor is 23.2 kW.

(a) What is the slip of the motor at this time?

(b) What is the induced torque in this motor?

(c) Assuming that the mechanical losses are 300 W at this slip, what is the load torque of this motor?

6–13. Figure 6–18a shows a simple circuit consisting of a voltage source, a resistor, and two reactances. Find the Thevenin equivalent voltage and impedance of this circuit at the terminals. Then derive the expressions for the magnitude of $\mathbf{V}_{TH}$ and for R_{TH} given in Equations (6–41*b*) and (6–44).

6–14. Figure P6–1 shows a simple circuit consisting of a voltage source, two resistors, and two reactances in parallel with each other. If the resistor R_L is allowed to vary, but all the other components are constant, at what value of R_L will the maximum possible power be supplied to it? *Prove* your answer. (*Hint:* Derive an expression for load power in terms of V, R_S, X_S, R_L, and X_L and take the partial derivative of that expression with respect to R_L.) Use this result to derive the expression for the pullout torque [Equation (6–54)].

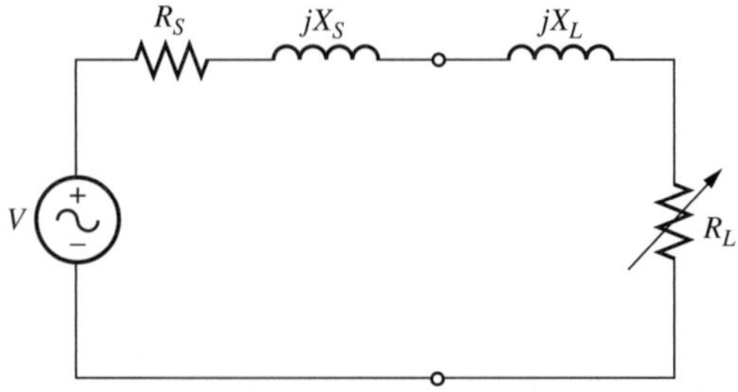

FIGURE P6–1
Circuit for Problem 6–14.

6–15. A 460-V, 60-Hz, four-pole, Y-connected induction motor is rated at 25 hp. The equivalent circuit parameters are

$R_1 = 0.15\ \Omega$	$R_2 = 0.154\ \Omega$	$X_M = 20\ \Omega$
$X_1 = 0.852\ \Omega$	$X_2 = 1.066\ \Omega$	
$P_{F\&W} = 400$ W	$P_{misc} = 150$ W	$P_{core} = 400$ W

For a slip of 0.02, find
(a) The line current I_L
(b) The stator power factor
(c) The rotor power factor
(d) The rotor frequency
(e) The stator copper losses P_{SCL}
(f) The air-gap power P_{AG}
(g) The power converted from electrical to mechanical form P_{conv}
(h) The induced torque τ_{ind}
(i) The load torque τ_{load}
(j) The overall machine efficiency η
(k) The motor speed in revolutions per minute and radians per second
(l) What is the starting code letter for this motor?

6–16. For the motor in Problem 6–15, what is the pullout torque? What is the slip at the pullout torque? What is the rotor speed at the pullout torque?

6–17. If the motor in Problem 6–15 is to be driven from a 460-V, 50-Hz power supply, what will the pullout torque be? What will the slip be at pullout?

6–18. Plot the following quantities for the motor in Problem 6–15 as slip varies from 0 percent to 10 percent: *(a)* τ_{ind} *(b)* P_{conv} *(c)* P_{out} *(d)* efficiency η. At what slip does P_{out} equal the rated power of the machine?

6–19. A dc test is performed on a 460-V, Δ-connected, 100-hp induction motor. If $V_{DC} =$ 21 V and $I_{DC} = 72$ A, what is the stator resistance R_1? *Why is this so?*

6–20. A 208-V, six-pole, Y-connected, 25-hp design class B induction motor is tested in the laboratory, with the following results:

No load:	208 V, 24.0 A, 1400 W, 60 Hz
Locked rotor:	24.6 V, 64.5 A, 2200 W, 15 Hz
dc test:	13.5 V, 64 A

Find the equivalent circuit of this motor, and plot its torque–speed characteristic curve.

6–21. A 460-V, 10-hp, four-pole, Y-connected, insulation class F, service factor 1.15 induction motor has the following parameters

$R_1 = 0.54\ \Omega$	$R_2 = 0.488\ \Omega$	$X_M = 51.12\ \Omega$
$X_1 = 2.093\ \Omega$	$X_2 = 3.209\ \Omega$	
$P_{F\&W} = 150$ W	$P_{misc} = 50$ W	$P_{core} = 150$ kW

For a slip of 0.02, find
(a) The line current I_L
(b) The stator power factor

(c) The rotor power factor
(d) The rotor frequency
(e) The stator copper losses P_{SCL}
(f) The air-gap power P_{AG}
(g) The power converted from electrical to mechanical form P_{conv}
(h) The induced torque τ_{ind}
(i) The load torque τ_{load}
(j) The overall machine efficiency η
(k) The motor speed in revolutions per minute and radians per second
(l) Sketch the power flow diagram for this motor.
(m) What is the starting code letter for this motor?
(n) What is the maximum acceptable temperature rise in this motor, given its insulation class?
(o) What does the service factor of this motor mean?

6–22. Plot the torque–speed characteristic of the motor in Problem 6–21. What is the starting torque of this motor?

6–23. A 460-V, four-pole, 75-hp, 60-Hz, Y-connected, three-phase induction motor develops its full-load induced torque at 1.2 percent slip when operating at 60 Hz and 460 V. The per-phase circuit model impedances of the motor are

$$R_1 = 0.058\ \Omega \qquad X_M = 18\ \Omega$$
$$X_1 = 0.32\ \Omega \qquad X_2 = 0.386\ \Omega$$

Mechanical, core, and stray losses may be neglected in this problem.
(a) Find the value of the rotor resistance R_2.
(b) Find τ_{max}, s_{max}, and the rotor speed at maximum torque for this motor.
(c) Find the starting torque of this motor.
(d) What code letter factor should be assigned to this motor?

6–24. Answer the following questions about the motor in Problem 6–21.
(a) If this motor is started from a 460-V infinite bus, how much current will flow in the motor at starting?
(b) If transmission line with an impedance of $0.50 + j0.35\ \Omega$ per phase is used to connect the induction motor to the infinite bus, what will the starting current of the motor be? What will the motor's terminal voltage be on starting?
(c) If an ideal 1.4:1 step-down autotransformer is connected between the transmission line and the motor, what will the current be in the transmission line during starting? What will the voltage be at the motor end of the transmission line during starting?

6–25. In this chapter, we learned that a step-down autotransformer could be used to reduce the starting current drawn by an induction motor. While this technique works, an autotransformer is relatively expensive. A much less expensive way to reduce the starting current is to use a device called *Y–Δ starter* (described earlier in this chapter). If an induction motor is normally Δ-connected, it is possible to reduce its phase voltage V_ϕ (and hence its starting current) by simply reconnecting the stator windings in Y during starting, and then restoring the connections to Δ when the motor comes up to speed. Answer the following questions about this type of starter.
(a) How would the phase voltage at starting compare with the phase voltage under normal running conditions?
(b) How would the starting current of the Y-connected motor compare to the starting current if the motor remained in a Δ-connection during starting?

6–26. A 460-V, 50-hp, six-pole, Δ-connected, 60-Hz, three-phase induction motor has a full-load slip of 4 percent, an efficiency of 91 percent, and a power factor of 0.87 lagging. At startup, the motor develops 1.75 times the full-load torque but draws 7 times the rated current at the rated voltage. This motor is to be started with an autotransformer-reduced voltage starter.

(*a*) What should the output voltage of the starter circuit be to reduce the starting torque until it equals the rated torque of the motor?

(*b*) What will the motor starting current and the current drawn from the supply be at this voltage?

6–27. A wound-rotor induction motor is operating at rated voltage and frequency with its slip rings shorted and with a load of about 25 percent of the rated value for the machine. If the rotor resistance of this machine is doubled by inserting external resistors into the rotor circuit, explain what happens to the following:

(*a*) Slip s

(*b*) Motor speed n_m

(*c*) The induced voltage in the rotor

(*d*) The rotor current

(*e*) τ_{ind}

(*f*) P_{out}

(*g*) P_{RCL}

(*h*) Overall efficiency η

6–28. A 460-V, 75-hp, four-pole, Y-connected induction motor has the following parameters:

$R_1 = 0.058\ \Omega$	$R_2 = 0.037\ \Omega$	$X_M = 9.24\ \Omega$
$X_1 = 0.320\ \Omega$	$X_2 = 0.386\ \Omega$	
$P_{\text{F\&W}} = 650\ \text{W}$	$P_{\text{misc}} = 150\ \text{W}$	$P_{\text{core}} = 600\ \text{kW}$

For a slip of 0.01, find

(*a*) The line current I_L

(*b*) The stator power factor

(*c*) The rotor power factor

(*d*) The rotor frequency

(*e*) The stator copper losses P_{SCL}

(*f*) The air-gap power P_{AG}

(*g*) The power converted from electrical to mechanical form P_{conv}

(*h*) The induced torque τ_{ind}

(*i*) The load torque τ_{load}

(*j*) The overall machine efficiency η

(*k*) The motor speed in revolutions per minute and radians per second

(*l*) Sketch the power flow diagram for this motor.

(*m*) What is the starting code letter for this motor?

6–29. Plot the torque–speed characteristic of the motor in Problem 6–28. What is the starting torque of this motor?

6–30. Answer the following questions about a 460-V, Δ-connected, two-pole, 100-hp, 60-Hz, starting code letter F induction motor:

(*a*) What is the maximum starting current that this machine's controller must be designed to handle?

(*b*) If the controller is designed to switch the stator windings from a Δ-connection to a Y-connection during starting, what is the maximum starting current that the controller must be designed to handle?

(*c*) If a 1.25:1 step-down autotransformer starter is used during starting, what is the maximum starting current that it must be designed to handle?

6–31. When it is necessary to stop an induction motor very rapidly, many induction motor controllers reverse the direction of rotation of the magnetic fields by switching any two stator leads. When the direction of rotation of the magnetic fields is reversed, the motor develops an induced torque opposite to the current direction of rotation, so it quickly stops and tries to start turning in the opposite direction. If power is removed from the stator circuit at the moment when the rotor speed goes through zero, then the motor has been stopped very rapidly. This technique for rapidly stopping an induction motor is called *plugging*. The motor of Problem 6–23 is running at rated conditions and is to be stopped by plugging.

(*a*) What is the slip s before plugging?

(*b*) What is the frequency of the rotor before plugging?

(*c*) What is the induced torque τ_{ind} before plugging?

(*d*) What is the slip s immediately after switching the stator leads?

(*e*) What is the frequency of the rotor immediately after switching the stator leads?

(*f*) What is the induced torque τ_{ind} immediately after switching the stator leads?

6–32. A 460-V, 10-hp, two-pole, Y-connected induction motor has the following parameters:

$$R_1 = 0.54\ \Omega \qquad X_1 = 2.093\ \Omega \qquad X_M = 51.12\ \Omega$$

$$P_{\text{F\&W}} = 150\text{ W} \qquad P_{\text{misc}} = 50\text{ W} \qquad P_{\text{core}} = 150\text{ kW}$$

The rotor is a dual-cage design, with a tightly coupled, high-resistance outer bar and a loosely coupled, low-resistance inner bar (see Figure 6–25c). The parameters of the outer bar are

$$R_2 = 3.20\ \Omega \qquad X_2 = 2.00\ \Omega$$

The resistance is high due to the lower cross-sectional area, and the reactance is relatively low due to the tight coupling between the rotor and stator. The parameters of the inner bar are

$$R_2 = 0.382\ \Omega \qquad X_2 = 5.10\ \Omega$$

The resistance is low due to the high cross-sectional area, but the reactance is relatively high due to the quite loose coupling between the rotor and stator.

Calculate the torque–speed characteristic for this induction motor, and compare it to the torque–speed characteristic for the single-cage design in Problem 6–21. How do the curves differ? Explain the differences.

REFERENCES

1. Alger, Phillip. *Induction Machines*, 2nd ed., Gordon and Breach, New York, 1970.
2. Del Toro, V.: *Electric Machines and Power Systems*. Prentice Hall, Englewood Cliffs, N.J., 1985.
3. Fitzgerald, A. E., and C. Kingsley, Jr.: *Electric Machinery*. McGraw-Hill, New York, 1952.
4. Fitzgerald, A. E., C. Kingsley, Jr., and S. D. Umans. *Electric Machinery*, 6th ed., McGraw-Hill, New York, 2003.
5. Institute of Electrical and Electronics Engineers. *Standard Test Procedure for Polyphase Induction Motors and Generators*, IEEE Standard 112-1996, IEEE, New York, 1996.

6. Kosow, Irving L.: *Control of Electric Motors*. Prentice Hall, Englewood Cliffs, N.J., 1972.
7. McPherson, George. *An Introduction to Electrical Machines and Transformers*. Wiley, New York, 1981.
8. National Electrical Manufacturers Association: *Motors and Generators*, Publication No. MG1-2006, NEMA, Washington, 2006.
9. Slemon, G. R., and A. Straughen. *Electric Machines*, Addison-Wesley, Reading, Mass., 1980.
10. Vithayathil, Joseph: *Power Electronics: Principles and Applications*, McGraw-Hill, New York, 1995.
11. Werninck, E. H. (ed.): *Electric Motor Handbook*, McGraw-Hill, London, 1978.

CHAPTER 7

DC MACHINERY FUNDAMENTALS

LEARNING OBJECTIVES

- Understand how voltage is induced in a rotating loop.
- Understand how curved pole faces contribute to a constant flux, and thus more constant output voltages.
- Understand and be able to use the equation for induced voltage and torque in a dc machine.
- Understand commutation.
- Understand problems with commutation, including armature reaction and $L\frac{di}{dt}$ effects.
- Understand the power flow diagram for dc machines.

DC machines are generators that convert mechanical energy to dc electric energy and motors that convert dc electric energy to mechanical energy. Most dc machines are like ac machines in that they have ac voltages and currents within them—dc machines have a dc output only because a mechanism exists that converts the internal ac voltages to dc voltages at their terminals. Since this mechanism is called a commutator, dc machinery is also known as *commutating machinery.*

The fundamental principles involved in the operation of dc machines are very simple. Unfortunately, they are usually somewhat obscured by the complicated construction of real machines. This chapter will first explain the principles of dc machine operation by using simple examples and then consider some of the complications that occur in real dc machines.

7.1 A SIMPLE ROTATING LOOP BETWEEN CURVED POLE FACES

The linear machine studied in Section 1.8 served as an introduction to basic machine behavior. Its response to loading and to changing magnetic fields closely resembles the behavior of the real dc generators and motors that we will study in Chapter 8. However, real generators and motors do not move in a straight line—they *rotate*. The next step toward understanding real dc machines is to study the simplest possible example of a rotating machine.

The simplest possible rotating dc machine is shown in Figure 7–1. It consists of a single loop of wire rotating about a fixed axis. The rotating part of this machine is called the *rotor*, and the stationary part is called the *stator*. The magnetic field for the machine is supplied by the magnetic north and south poles shown on the stator in Figure 7–1.

Notice that the loop of rotor wire lies in a slot carved in a ferromagnetic core. The iron rotor, together with the curved shape of the pole faces, provides a constant-width air gap between the rotor and stator. Remember from Chapter 1 that the reluctance of air is much higher than the reluctance of the iron in the machine. To minimize the reluctance of the flux path through the machine, the magnetic flux must take the shortest possible path through the air between the pole face and the rotor surface.

Since the magnetic flux must take the shortest path through the air, it is *perpendicular* to the rotor surface everywhere under the pole faces. Also, since the air gap is of uniform width, the reluctance is the same everywhere under the pole faces. The uniform reluctance means that the magnetic flux density is constant everywhere under the pole faces.

The Voltage Induced in a Rotating Loop

If the rotor of this machine is rotated, a voltage will be induced in the wire loop. To determine the magnitude and shape of the voltage, examine Figure 7–2. The loop of wire shown is rectangular, with sides *ab* and *cd* perpendicular to the plane of the page and with sides *bc* and *da* parallel to the plane of the page. The magnetic field is constant and perpendicular to the surface of the rotor everywhere under the pole faces and rapidly falls to zero beyond the edges of the poles.

To determine the total voltage e_{tot} on the loop, examine each segment of the loop separately and sum all the resulting voltages. The voltage on each segment is given by Equation (1–45):

$$e_{\text{ind}} = (\mathbf{v} \times \mathbf{B}) \cdot \mathbf{l} \tag{1–45}$$

1. *Segment ab*. In this segment, the velocity of the wire is tangential to the path of rotation. The magnetic field **B** points out perpendicular to the rotor surface everywhere under the pole face and is zero beyond the edges of the pole face. Under the pole face, velocity **v** is perpendicular to **B**, and the quantity **v** × **B** points into the page. Therefore, the induced voltage on the segment is

$$\begin{aligned} e_{ba} &= (\mathbf{v} \times \mathbf{B}) \cdot \mathbf{l} \\ &= \begin{cases} vBl & \text{positive into page} \quad \text{under the pole face} \\ 0 & \qquad\qquad\qquad\quad\;\; \text{beyond the pole edges} \end{cases} \end{aligned} \tag{7–1}$$

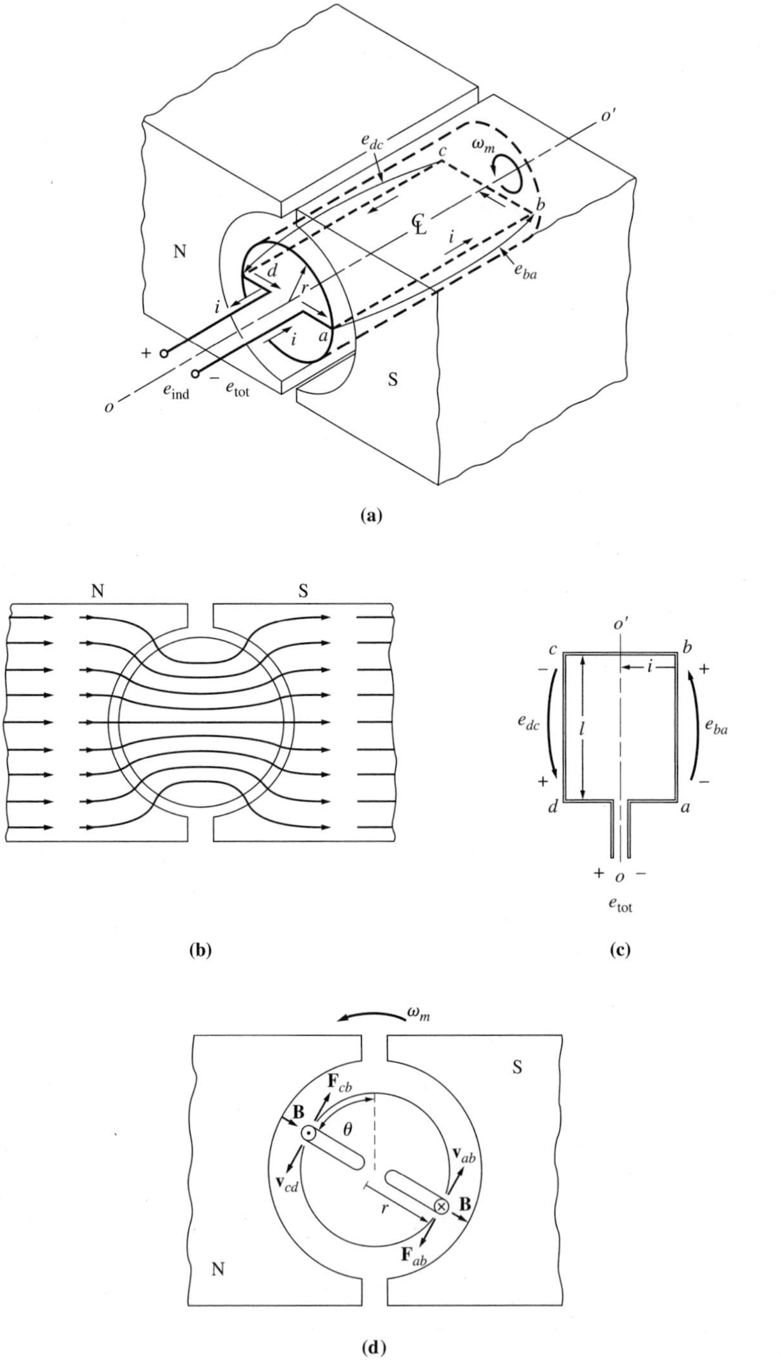

FIGURE 7–1
A simple rotating loop between curved pole faces. (a) Perspective view; (b) view of field lines; (c) top view; (d) front view.

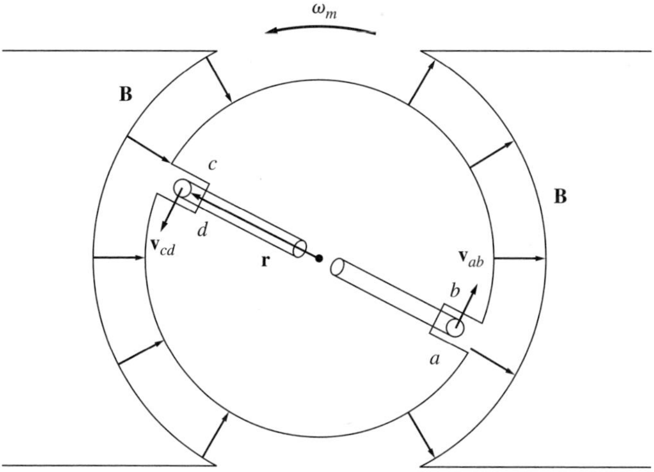

FIGURE 7–2
Derivation of an equation for the voltages induced in the loop.

2. *Segment bc.* In this segment, the quantity $\mathbf{v} \times \mathbf{B}$ is either into or out of the page, while length $\mathbf{l}$ is in the plane of the page, so $\mathbf{v} \times \mathbf{B}$ is perpendicular to $\mathbf{l}$. Therefore the voltage in segment *bc* will be zero:

$$e_{cb} = 0 \tag{7–2}$$

3. *Segment cd.* In this segment, the velocity of the wire is tangential to the path of rotation. The magnetic field $\mathbf{B}$ points *in* perpendicular to the rotor surface everywhere under the pole face and is zero beyond the edges of the pole face. Under the pole face, velocity $\mathbf{v}$ is perpendicular to $\mathbf{B}$, and the quantity $\mathbf{v} \times \mathbf{B}$ points out of the page. Therefore, the induced voltage on the segment is

$$\begin{aligned} e_{dc} &= (\mathbf{v} \times \mathbf{B}) \cdot \mathbf{l} \\ &= \begin{cases} vBl & \text{positive out of page} \quad \text{under the pole face} \\ 0 & \text{beyond the pole edges} \end{cases} \end{aligned} \tag{7–3}$$

4. *Segment da.* Just as in segment *bc*, $\mathbf{v} \times \mathbf{B}$ is perpendicular to $\mathbf{l}$. Therefore the voltage in this segment will be zero, too:

$$e_{ad} = 0 \tag{7–4}$$

The total induced voltage on the loop e_{ind} is given by

$$e_{\text{ind}} = e_{ba} + e_{cb} + e_{dc} + e_{ad}$$

$$\boxed{e_{\text{ind}} = \begin{cases} 2vBl & \text{under the pole faces} \\ 0 & \text{beyond the pole edges} \end{cases}} \tag{7–5}$$

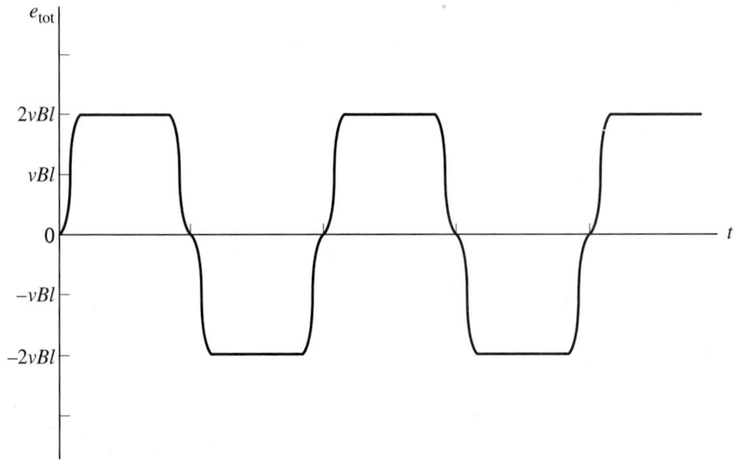

FIGURE 7–3
The output voltage of the loop.

When the loop rotates through 180°, segment *ab* is under the north pole face instead of the south pole face. At that time, the direction of the voltage on the segment reverses, but its magnitude remains constant. The resulting voltage e_{tot} is shown as a function of time in Figure 7–3.

There is an alternative way to express Equation (7–5) which clearly relates the behavior of the single loop to the behavior of larger, real dc machines. To

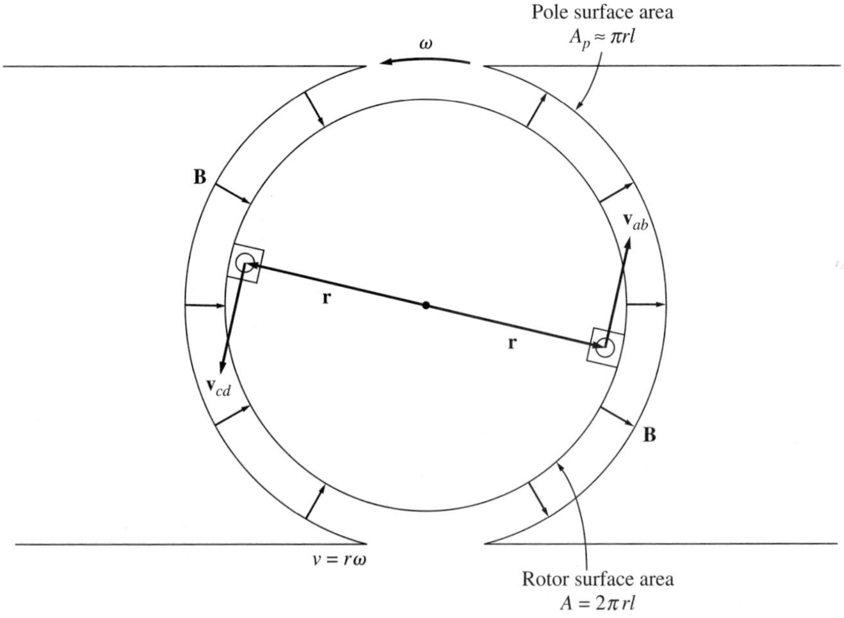

FIGURE 7–4
Derivation of an alternative form of the induced voltage equation.

derive this alternative expression, examine Figure 7–4. Notice that the tangential velocity v of the edges of the loop can be expressed as

$$v = r\omega_m$$

where r is the radius from the axis of rotation out to the edge of the loop and ω_m is the angular velocity of the loop. Substituting this expression into Equation (7–5) gives

$$e_{\text{ind}} = \begin{cases} 2r\omega_m Bl & \text{under the pole faces} \\ 0 & \text{beyond the pole edges} \end{cases}$$

$$e_{\text{ind}} = \begin{cases} 2rlB\omega_m & \text{under the pole faces} \\ 0 & \text{beyond the pole edges} \end{cases}$$

Notice also from Figure 7–4 that the rotor surface is a cylinder, so the area of the rotor surface A is just equal to $2\pi rl$. Since there are two poles, the area of the rotor *under each pole* (ignoring the small gaps between poles) is $A_P = \pi rl$. Therefore,

$$e_{\text{ind}} = \begin{cases} \dfrac{2}{\pi} A_P B\omega_m & \text{under the pole faces} \\ 0 & \text{beyond the pole edges} \end{cases}$$

Since the flux density B is constant everywhere in the air gap under the pole faces, the total flux under each pole is just the area of the pole times its flux density:

$$\phi = A_P B$$

Therefore, the final form of the voltage equation is

$$\boxed{e_{\text{ind}} = \begin{cases} \dfrac{2}{\pi} \phi\omega_m & \text{under the pole faces} \\ 0 & \text{beyond the pole edges} \end{cases}} \tag{7–6}$$

Thus, *the voltage generated in the machine is equal to the product of the flux inside the machine and the speed of rotation of the machine*, multiplied by a constant representing the mechanical construction of the machine. In general, the voltage in any real machine will depend on the same three factors:

1. The flux in the machine
2. The speed of rotation
3. A constant representing the construction of the machine

Getting DC Voltage Out of the Rotating Loop

Figure 7–3 is a plot of the voltage e_{tot} generated by the rotating loop. As shown, the voltage out of the loop is alternately a constant positive value and a constant negative value. How can this machine be made to produce a dc voltage instead of the ac voltage it now has?

One way to do this is shown in Figure 7–5a. Here two semicircular conducting segments are added to the end of the loop, and two fixed contacts are set up at

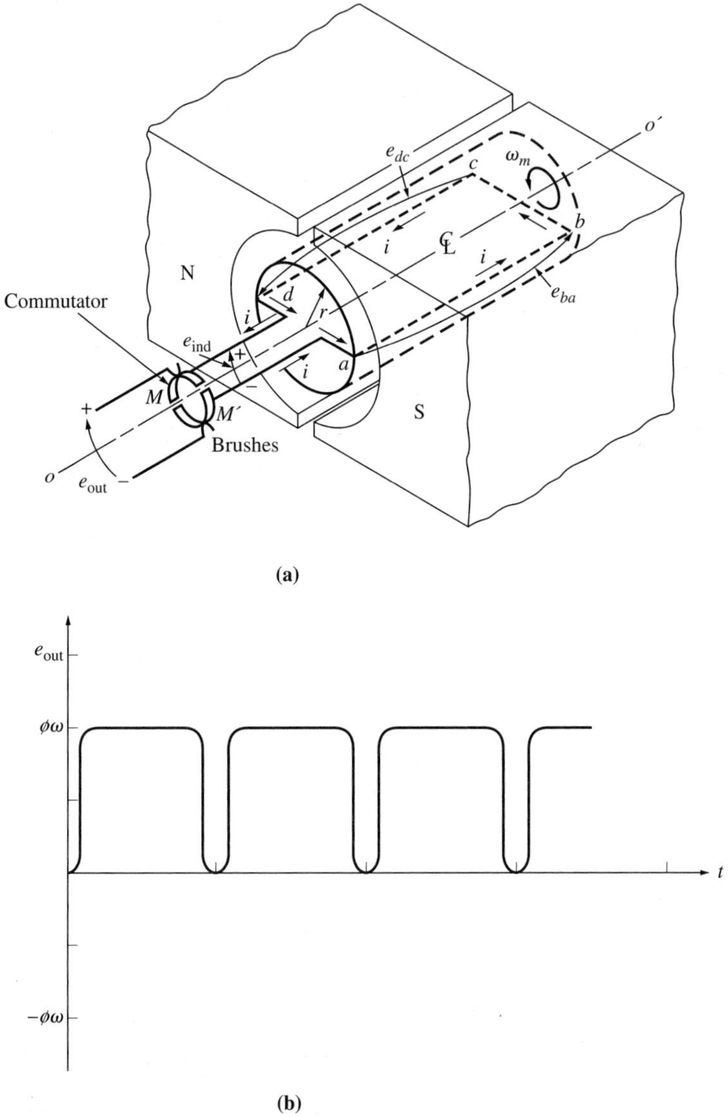

FIGURE 7–5
Producing a dc output from the machine with a commutator and brushes. (a) Perspective view; (b) the resulting output voltage.

an angle such that at the instant when the voltage in the loop is zero, the contacts short-circuit the two segments. In this fashion, *every time the voltage of the loop switches direction, the contacts also switch connections, and the output of the contacts is always built up in the same way* (Figure 7–5b). This connection-switching process is known as *commutation*. The rotating semicircular segments are called *commutator segments*, and the fixed contacts are called *brushes*.

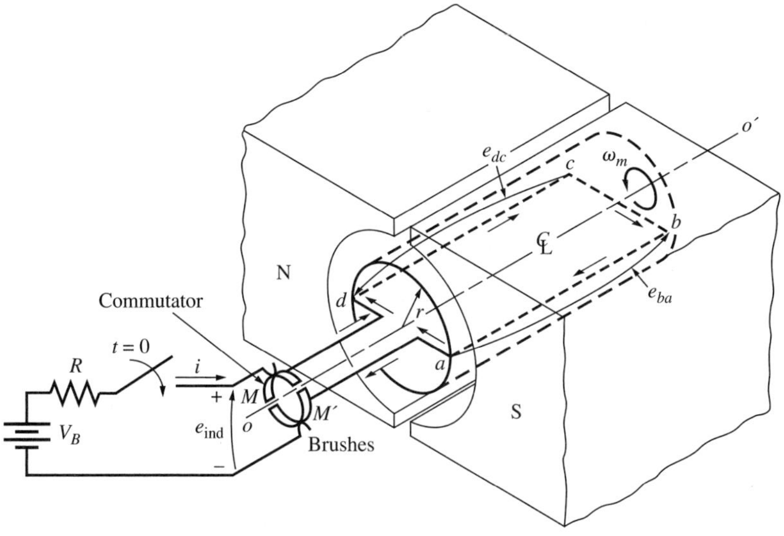

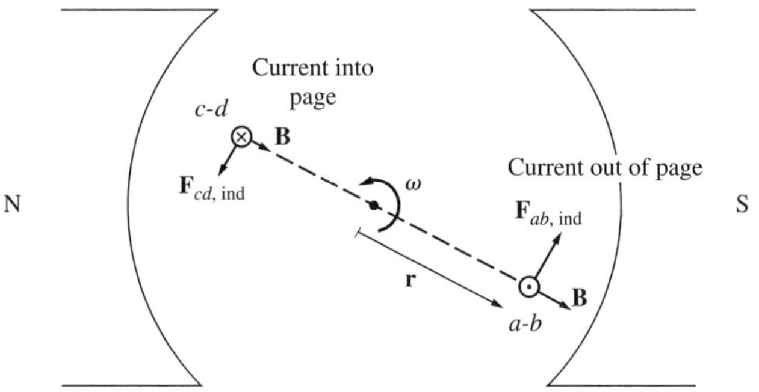

FIGURE 7–6
Derivation of an equation for the induced torque in the loop. Note that the iron core is not shown in part **b** for clarity.

The Induced Torque in the Rotating Loop

Suppose a battery is now connected to the machine in Figure 7–5. The resulting configuration is shown in Figure 7–6. How much torque will be produced in the loop when the switch is closed and a current is allowed to flow into it? To determine the torque, look at the close-up of the loop shown in Figure 7–6b.

The approach to take in determining the torque on the loop is to look at one segment of the loop at a time and then sum the effects of all the individual segments. The force on a segment of the loop is given by Equation (1–43):

$$\mathbf{F} = i(\mathbf{l} \times \mathbf{B}) \tag{1–43}$$

and the torque on the segment is given by

$$\tau = rF \sin \theta \tag{1–6}$$

where θ is the angle between **r** and **F**. The torque is essentially zero whenever the loop is beyond the pole edges.

While the loop is under the pole faces, the torque is

1. *Segment ab*. In segment *ab*, the current from the battery is directed out of the page. The magnetic field under the pole face is pointing radially out of the rotor, so the force on the wire is given by

$$\begin{aligned} \mathbf{F}_{ab} &= i(\mathbf{l} \times \mathbf{B}) \\ &= ilB \qquad \text{tangent to direction of motion} \end{aligned} \tag{7–7}$$

   The torque on the rotor caused by this force is

$$\begin{aligned} \tau_{\text{ab}} &= rF \sin \theta \\ &= r(ilB) \sin 90° \\ &= rilB \qquad \text{CCW} \end{aligned} \tag{7–8}$$

2. *Segment bc*. In segment *bc*, the current from the battery is flowing from the upper left to the lower right in the picture. The force induced on the wire is given by

$$\begin{aligned} \mathbf{F}_{bc} &= i(\mathbf{l} \times \mathbf{B}) \\ &= 0 \qquad \text{since } \mathbf{l} \text{ is parallel to } \mathbf{B} \end{aligned} \tag{7–9}$$

   Therefore,

$$\tau_{\text{bc}} = 0 \tag{7–10}$$

3. *Segment cd*. In segment *cd*, the current from the battery is directed into the page. The magnetic field under the pole face is pointing radially into the rotor, so the force on the wire is given by

$$\begin{aligned} \mathbf{F}_{cd} &= i(\mathbf{l} \times \mathbf{B}) \\ &= ilB \qquad \text{tangent to direction of motion} \end{aligned} \tag{7–11}$$

   The torque on the rotor caused by this force is

$$\begin{aligned} \tau_{\text{cd}} &= rF \sin \theta \\ &= r(ilB) \sin 90° \\ &= rilB \qquad \text{CCW} \end{aligned} \tag{7–12}$$

4. *Segment da*. In segment *da,* the current from the battery is flowing from the upper left to the lower right in the picture. The force induced on the wire is given by

$$\begin{aligned}\mathbf{F}_{da} &= i(\mathbf{l} \times \mathbf{B}) \\ &= 0 \qquad \text{since } \mathbf{l} \text{ is parallel to } \mathbf{B}\end{aligned} \tag{7–13}$$

Therefore,

$$\tau_{\text{da}} = 0 \tag{7–14}$$

The resulting total induced torque on the loop is given by

$$\tau_{\text{ind}} = \tau_{ab} + \tau_{bc} + \tau_{cd} + \tau_{da}$$

$$\boxed{\tau_{\text{ind}} = \begin{cases} 2rilB & \text{under the pole faces} \\ 0 & \text{beyond the pole edges} \end{cases}} \tag{7–15}$$

By using the facts that $A_P \approx \pi rl$ and $\phi = A_P B$, the torque expression can be reduced to

$$\boxed{\tau_{\text{ind}} = \begin{cases} \dfrac{2}{\pi}\phi i & \text{under the pole faces} \\ 0 & \text{beyond the pole edges} \end{cases}} \tag{7–16}$$

Thus, *the torque produced in the machine is the product of the flux in the machine and the current in the machine*, times some quantity representing the mechanical construction of the machine (the percentage of the rotor covered by pole faces). In general, the torque in *any* real machine will depend on the same three factors:

1. The flux in the machine
2. The current in the machine
3. A constant representing the construction of the machine

Example 7–1. Figure 7–6 shows a simple rotating loop between curved pole faces connected to a battery and a resistor through a switch. The resistor shown models the total resistance of the battery and the wire in the machine. The physical dimensions and characteristics of this machine are

$$\begin{aligned} r &= 0.5 \text{ m} & l &= 1.0 \text{ m} \\ R &= 0.3\ \Omega & B &= 0.25 \text{ T} \\ V_B &= 120 \text{ V} \end{aligned}$$

(*a*) What happens when the switch is closed?
(*b*) What is the machine's maximum starting current? What is its steady-state angular velocity at no load?
(*c*) Suppose a load is attached to the loop, and the resulting load torque is 10 N • m. What would the new steady-state speed be? How much power is supplied to the shaft of the machine? How much power is being supplied by the battery? Is this machine a motor or a generator?

(*d*) Suppose the machine is again unloaded, and a torque of 7.5 N • m is applied to the shaft in the direction of rotation. What is the new steady-state speed? Is this machine now a motor or a generator?

(*e*) Suppose the machine is running unloaded. What would the final steady-state speed of the rotor be if the flux density were reduced to 0.20 T?

Solution

(*a*) When the switch in Figure 7–6 is closed, a current will flow in the loop. Since the loop is initially stationary, $e_{ind} = 0$. Therefore, the current will be given by

$$i = \frac{V_B - e_{\text{ind}}}{R} = \frac{V_B}{R}$$

This current flows through the rotor loop, producing a torque

$$\tau_{\text{ind}} = \frac{2}{\pi}\phi i \qquad \text{CCW}$$

This induced torque produces an angular acceleration in a counterclockwise direction, so the rotor of the machine begins to turn. But as the rotor begins to turn, an induced voltage is produced in the motor, given by

$$e_{\text{ind}} = \frac{2}{\pi}\phi\omega_m$$

so the current i falls. As the current falls, $\tau_{\text{ind}} = (2/\pi)\phi i\downarrow$ decreases, and the machine winds up in steady state with $\tau_{\text{ind}} = 0$, and the battery voltage $V_B = e_{\text{ind}}$.

This is the same sort of starting behavior seen earlier in the linear dc machine.

(*b*) At *starting conditions,* the machine's current is

$$i = \frac{V_B}{R} = \frac{120 \text{ V}}{0.3\ \Omega} = 400 \text{ A}$$

At *no-load steady-state conditions,* the induced torque τ_{ind} must be zero. But $\tau_{\text{ind}} = 0$ implies that current i must equal zero, since $\tau_{\text{ind}} = (2/\pi)\phi i$, and the flux is nonzero. The fact that $i = 0$ A means that the battery voltage $V_B = e_{\text{ind}}$. Therefore, the speed of the rotor is

$$V_B = e_{\text{ind}} = \frac{2}{\pi}\phi\omega_m$$

$$\omega = \frac{V_B}{(2/\pi)\phi} = \frac{V_B}{2rlB}$$

$$= \frac{120 \text{ V}}{2(0.5 \text{ m})(1.0 \text{ m})(0.25 \text{ T})} = 480 \text{ rad/s}$$

(*c*) If a load torque of 10 N • m is applied to the shaft of the machine, it will begin to slow down. But as ω decreases, $e_{\text{ind}} = (2/\pi)\phi\,\omega\downarrow$ decreases and the rotor current increases $[i = (V_B - e_{\text{ind}}\downarrow)/R]$. As the rotor current increases, $|\tau_{\text{ind}}|$ increases too, until $|\tau_{\text{ind}}| = |\tau_{\text{load}}|$ at a lower speed ω.

At steady state, $|\tau_{\text{load}}| = |\tau_{\text{ind}}| = (2/\pi)\phi i$. Therefore,

$$i = \frac{\tau_{\text{ind}}}{(2/\pi)\phi} = \frac{\tau_{\text{ind}}}{2rlB}$$

$$= \frac{10 \text{ N} \bullet \text{m}}{(2)(0.5 \text{ m})(1.0 \text{ m})(0.25 \text{ T})} = 40 \text{ A}$$

By Kirchhoff's voltage law, $e_{\text{ind}} = V_B - iR$, so

$$e_{\text{ind}} = 120 \text{ V} - (40 \text{ A})(0.3 \ \Omega) = 108 \text{ V}$$

Finally, the speed of the shaft is

$$\omega = \frac{e_{\text{ind}}}{(2/\pi)\phi} = \frac{e_{\text{ind}}}{2rlB}$$

$$= \frac{108 \text{ V}}{(2)(0.5 \text{ m})(1.0 \text{ m})(0.25 \text{ T})} = 432 \text{ rad/s}$$

The power supplied to the shaft is

$$P = \tau\omega_m$$

$$= (10 \text{ N} \bullet \text{m})(432 \text{ rad/s}) = 4320 \text{ W}$$

The power out of the battery is

$$P = V_B i = (120 \text{ V})(40 \text{ A}) = 4800 \text{ W}$$

This machine is operating as a *motor,* converting electric power to mechanical power.

(d) If a torque is applied in the direction of motion, the rotor accelerates. As the speed increases, the internal voltage e_{ind} increases and exceeds V_B, so the current flows out of the top of the bar and into the battery. This machine is now a *generator*. This current causes an induced torque opposite to the direction of motion. The induced torque opposes the external applied torque, and eventually $|\tau_{\text{load}}| = |\tau_{\text{ind}}|$ at a higher speed ω_m.

The current in the rotor will be

$$i = \frac{\tau_{\text{ind}}}{(2/\pi)\phi} = \frac{\tau_{\text{ind}}}{2rlB}$$

$$= \frac{7.5 \text{ N} \cdot \text{m}}{(2)(0.5 \text{ m})(1.0 \text{ m})(0.25 \text{ T})} = 30 \text{ A}$$

The induced voltage e_{ind} is

$$e_{\text{ind}} = V_B + iR$$

$$= 120 \text{ V} + (30 \text{ A})(0.3 \ \Omega)$$

$$= 129 \text{ V}$$

Finally, the speed of the shaft is

$$\omega = \frac{e_{\text{ind}}}{(2/\pi)\phi} = \frac{e_{\text{ind}}}{2rlB}$$

$$= \frac{129 \text{ V}}{(2)(0.5 \text{ m})(1.0 \text{ m})(0.25 \text{ T})} = 516 \text{ rad/s}$$

(e) Since the machine is initially unloaded at the original conditions, the speed $\omega_m = 480$ rad/s. If the flux decreases, there is a transient. However, after the transient is over, the machine must again have zero torque, since there is still no load on its shaft. If $\tau_{\text{ind}} = 0$, then the current in the rotor must be zero, and $V_B = e_{\text{ind}}$. The shaft speed is thus

$$\omega = \frac{e_{\text{ind}}}{(2/\pi)\phi} = \frac{e_{\text{ind}}}{2rlB}$$

$$= \frac{120 \text{ V}}{(2)(0.5 \text{ m})(1.0 \text{ m})(0.20 \text{ T})} = 600 \text{ rad/s}$$

Notice that when the flux in the machine is decreased, its speed increases. This is the same behavior seen in the linear machine and the same way that real dc motors behave.

7.2 COMMUTATION IN A SIMPLE FOUR-LOOP DC MACHINE

Commutation is the process of converting the ac voltages and currents in the rotor of a dc machine to dc voltages and currents at its terminals. It is the most critical part of the design and operation of any dc machine. A more detailed study is necessary to determine just how this conversion occurs and to discover the problems associated with it. In this section, the technique of commutation will be explained for a machine more complex than the single rotating loop in Section 7.1 but less complex than a real dc machine. Section 7.3 will continue this development and explain commutation in real dc machines.

A simple four-loop, two-pole dc machine is shown in Figure 7–7. This machine has four complete loops buried in slots carved in the laminated steel of its rotor. The pole faces of the machine are curved to provide a uniform air-gap width and to give a uniform flux density everywhere under the faces.

The four loops of this machine are laid into the slots in a special manner. The "unprimed" end of each loop is the outermost wire in each slot, while the "primed" end of each loop is the innermost wire in the slot directly opposite. The winding's connections to the machine's commutator are shown in Figure 7–7b. Notice that loop 1 stretches between commutator segments a and b, loop 2 stretches between segments b and c, and so forth around the rotor.

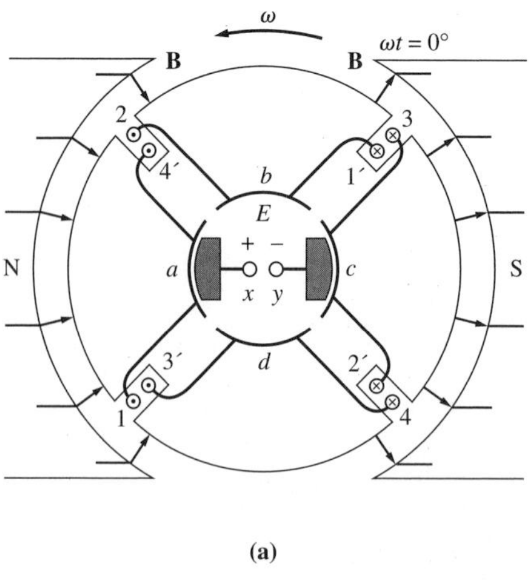

FIGURE 7–7
(a) A four-loop, two-pole dc machine shown at time $\omega t = 0°$. *(continues)*

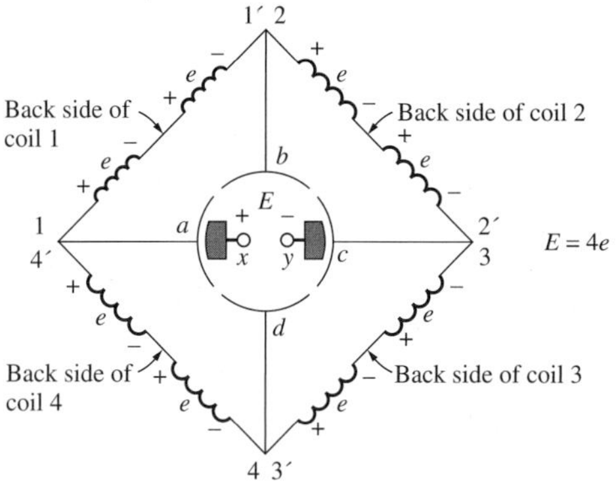

(b)

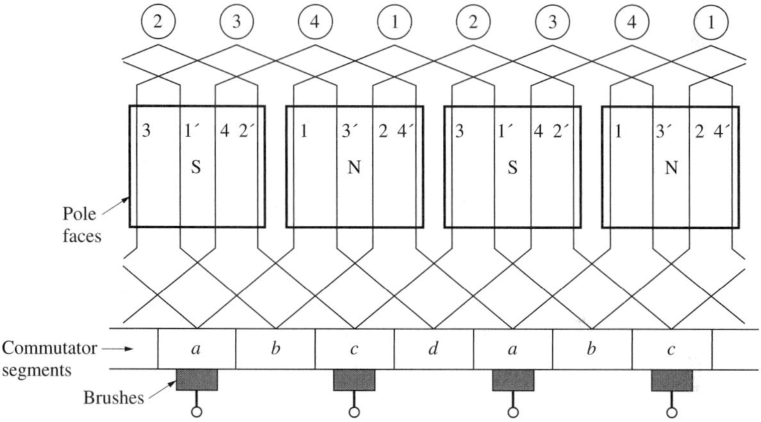

(c)

FIGURE 7–7 (concluded)
(b) The voltages on the rotor conductors at this time. (c) A winding diagram of this machine showing the interconnections of the rotor loops.

At the instant shown in Figure 7–7, the 1, 2, 3′, and 4′ ends of the loops are under the north pole face, while the 1′, 2′, 3, and 4 ends of the loops are under the south pole face. The voltage in each of the 1, 2, 3′, and 4′ ends of the loops is given by

$$e_{\text{ind}} = (\mathbf{v} \times \mathbf{B}) \bullet \mathbf{l} \tag{1–45}$$

$$e_{\text{ind}} = vBl \qquad \text{positive out of page} \tag{7–17}$$

The voltage in each of the 1′, 2′, 3, and 4 ends of the loops is given by

$$e_{\text{ind}} = (\mathbf{v} \times \mathbf{B}) \bullet \mathbf{l} \tag{1–45}$$

$$= vBl \qquad \text{positive into the page} \tag{7–18}$$

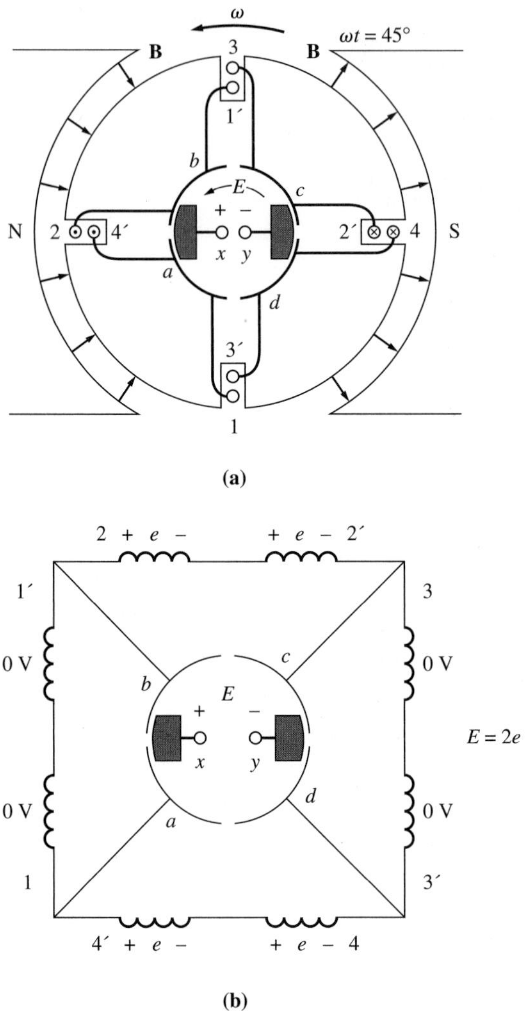

FIGURE 7–8
The same machine at time $\omega t = 45°$, showing the voltages on the conductors.

The overall result is shown in Figure 7–7b. In Figure 7–7b, each coil represents one side (or *conductor*) of a loop. If the induced voltage on any one side of a loop is called $e = vBl$, then the total voltage at the brushes of the machine is

$$\boxed{E = 4e \qquad \omega t = 0°} \tag{7–19}$$

Notice that there are two parallel paths for current through the machine. The existence of two or more parallel paths for rotor current is a common feature of all commutation schemes.

What happens to the voltage E of the terminals as the rotor continues to rotate? To find out, examine Figure 7–8. This figure shows the machine at time $\omega t = 45°$. At that time, loops 1 and 3 have rotated into the gap between the poles, so the voltage across each of them is zero. Notice that at this instant the brushes

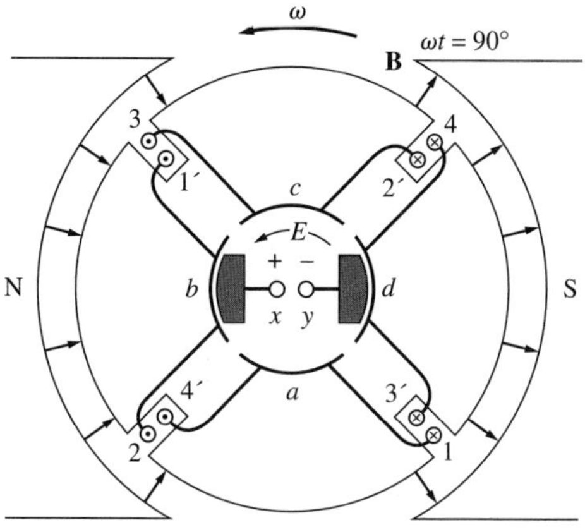

(a)

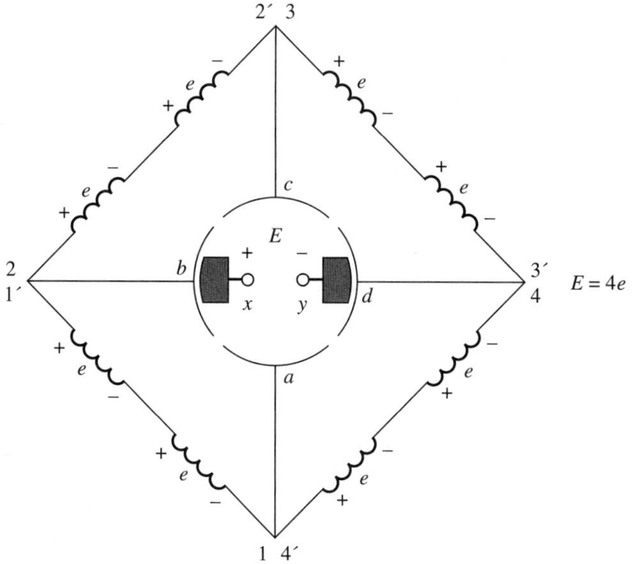

(b)

FIGURE 7–9
The same machine at time $\omega t = 90°$, showing the voltages on the conductors.

of the machine are shorting out commutator segments *ab* and *cd*. This happens just at the time when the loops between these segments have 0 V across them, so shorting out the segments creates no problem. At this time, only loops 2 and 4 are under the pole faces, so the terminal voltage *E* is given by

$$\boxed{E = 2e \qquad \omega t = 0°} \tag{7–20}$$

Now let the rotor continue to turn through another 45°. The resulting situation is shown in Figure 7–9. Here, the 1′, 2, 3, and 4′ ends of the loops are under

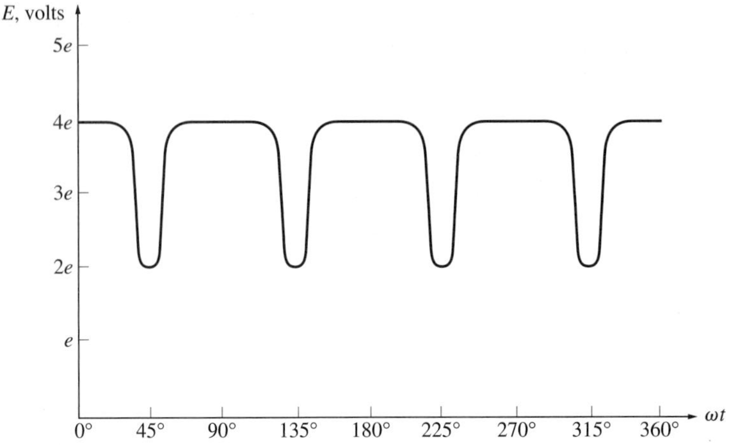

FIGURE 7–10
The resulting output voltage of the machine in Figure 7–7.

the north pole face, and the 1, 2′, 3′, and 4 ends of the loops are under the south pole face. The voltages are still built up out of the page for the ends under the north pole face and into the page for the ends under the south pole face. The resulting voltage diagram is shown in Figure 7–9b. There are now four voltage-carrying ends in each parallel path through the machine, so the terminal voltage E is given by

$$\boxed{E = 4e \qquad \omega t = 90^\circ} \tag{7–21}$$

Compare Figure 7–7 to Figure 7–9. Notice that *the voltages on loops 1 and 3 have reversed between the two pictures, but since their connections have also reversed, the total voltage is still being built up in the same direction as before.* This fact is at the heart of every commutation scheme. Whenever the voltage reverses in a loop, the connections of the loop are also switched, and the total voltage is still built up in the original direction.

The terminal voltage of this machine as a function of time is shown in Figure 7–10. It is a better approximation to a constant dc level than the single rotating loop in Section 7.1 produced. As the number of loops on the rotor increases, the approximation to a perfect dc voltage continues to get better and better.

In summary,

> *Commutation* is the process of switching the loop connections on the rotor of a dc machine just as the voltage in the loop switches polarity, in order to maintain an essentially constant dc output voltage.

As in the case of the simple rotating loop, the rotating segments to which the loops are attached are called *commutator segments*, and the stationary pieces that ride on top of the moving segments are called *brushes*. The commutator segments in real machines are typically made of copper bars. The brushes are made of a mixture containing graphite, so that they cause very little friction as they rub over the rotating commutator segments.

7.3 COMMUTATION AND ARMATURE CONSTRUCTION IN REAL DC MACHINES

In real dc machines, there are several ways in which the loops on the rotor (also called the *armature*) can be connected to its commutator segments. These different connections affect the number of parallel current paths within the rotor, the output voltage of the rotor, and the number and position of the brushes riding on the commutator segments. We will now examine the construction of the coils on a real dc rotor and then look at how they are connected to the commutator to produce a dc voltage.

The Rotor Coils

Regardless of the way in which the windings are connected to the commutator segments, most of the rotor windings themselves consist of diamond-shaped preformed coils which are inserted into the armature slots as a unit (see Figure 7–11). Each coil consists of a number of *turns* (loops) of wire, each turn taped and insulated from the other turns and from the rotor slot. Each side of a turn is called *a conductor*. The number of conductors on a machine's armature is given by

$$\boxed{Z = 2CN_C} \tag{7–22}$$

where Z = number of conductors on rotor

C = number of coils on rotor

N_C = number of turns per coil

Normally, a coil spans 180 electrical degrees. This means that when one side is under the center of a given magnetic pole, the other side is under the center of a pole of *opposite polarity*. The *physical* poles may not be located 180 mechanical degrees apart, but the magnetic field has completely reversed its polarity in traveling from under one pole to the next. The relationship between the electrical angle and mechanical angle in a given machine is given by

$$\boxed{\theta_e = \frac{P}{2}\theta_m} \tag{7–23}$$

where θ_e = electrical angle, in degrees

θ_m = mechanical angle, in degrees

P = number of magnetic poles on the machine

If a coil spans 180 electrical degrees, the voltages in the conductors on either side of the coil will be exactly the same in magnitude and opposite in direction at all times. Such a coil is called a *full-pitch coil.*

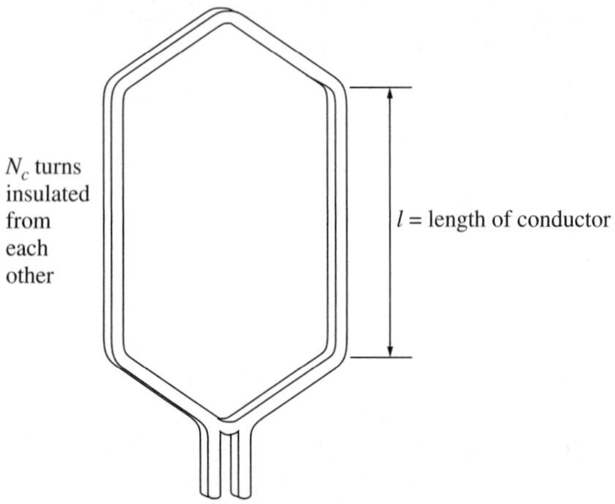

(a)

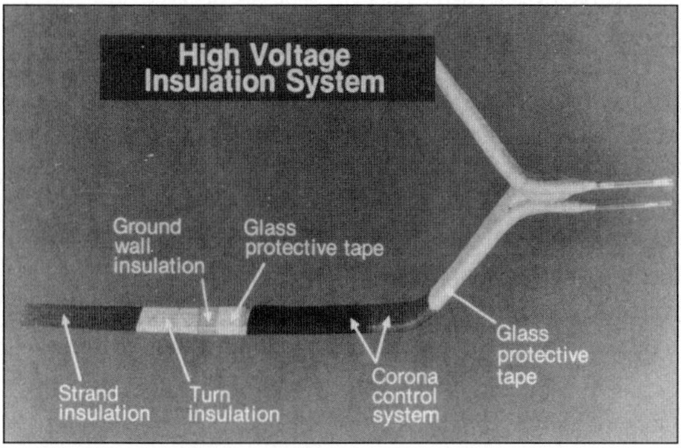

(b)

FIGURE 7–11
(a) The shape of a typical preformed rotor coil. (b) A typical coil insulation system showing the insulation between turns within a coil. (*Courtesy of General Electric Company.*)

Sometimes a coil is built that spans less than 180 electrical degrees. Such a coil is called a *fractional-pitch coil*, and a rotor winding wound with fractional-pitch coils is called a *chorded winding*. The amount of chording in a winding is described by a *pitch factor p*, which is defined by the equation

$$\boxed{p = \frac{\text{electrical angle of coil}}{180°} \times 100\%} \tag{7–24}$$

Sometimes a small amount of chording will be used in dc rotor windings to improve commutation.

Most rotor windings are *two-layer windings*, meaning that sides from two different coils are inserted into each slot. One side of each coil will be at the bottom of its slot, and the other side will be at the top of its slot. Such a construction requires the individual coils to be placed in the rotor slots by a very elaborate

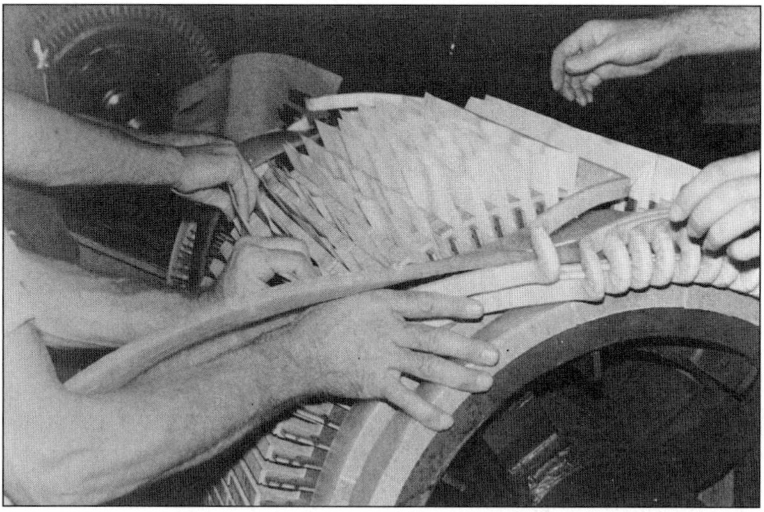

FIGURE 7–12
The installation of preformed rotor coils on a dc machine rotor. (*Courtesy of Westinghouse Electric Company.*)

procedure (see Figure 7–12). One side of each of the coils is placed in the bottom of its slot, and then after all the bottom sides are in place, the other side of each coil is placed in the top of its slot. In this fashion, all the windings are woven together, increasing the mechanical strength and uniformity of the final structure.

Connections to the Commutator Segments

Once the windings are installed in the rotor slots, they must be connected to the commutator segments. There are a number of ways in which these connections can be made, and the different winding arrangements which result have different advantages and disadvantages.

The distance (in number of segments) between the commutator segments to which the two ends of a coil are connected is called the *commutator pitch* y_c. If the end of a coil (or a set number of coils, for wave construction) is connected to a commutator segment ahead of the one its beginning is connected to, the winding is called a *progressive winding* (see Figure 7–13a). If the end of a coil is connected to a commutator segment behind the one its beginning is connected to, the winding is called a *retrogressive winding* (see Figure 7–13b). If everything else is identical, the direction of rotation of a progressive-wound rotor will be opposite to the direction of rotation of a retrogressive-wound rotor.

Rotor (armature) windings are further classified according to the *plex* of their windings. A *simplex* rotor winding is a single, complete, closed winding wound on a rotor. A *duplex* rotor winding is a rotor with *two complete and independent sets* of rotor windings. If a rotor has a duplex winding, then each of the windings will be associated with every other commutator segment: One winding

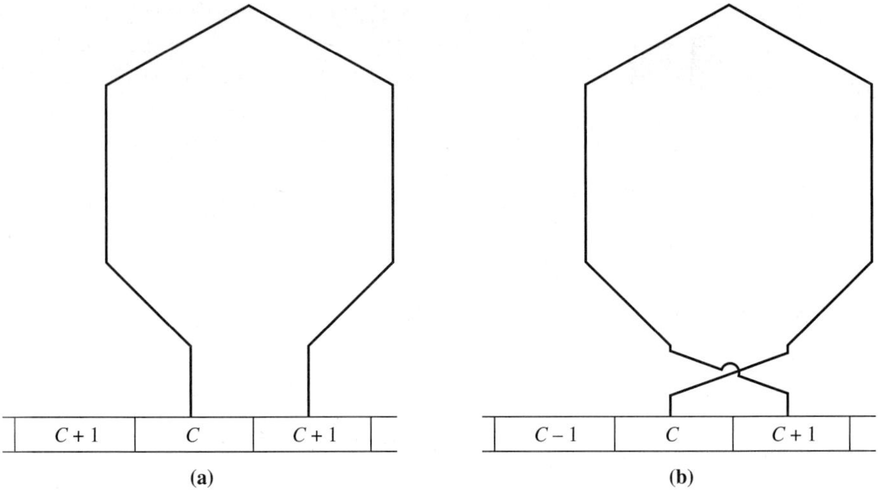

FIGURE 7–13
(a) A coil in a progressive rotor winding. (b) A coil in a retrogressive rotor winding.

will be connected to segments 1, 3, 5, etc., and the other winding will be connected to segments 2, 4, 6, etc. Similarly, a *triplex* winding will have three complete and independent sets of windings, each winding connected to every third commutator segment on the rotor. Collectively, all armatures with more than one set of windings are said to have *multiplex windings.*

Finally, armature windings are classified according to the sequence of their connections to the commutator segments. There are two basic sequences of armature winding connections—*lap windings* and *wave windings*. In addition, there is a third type of winding, called *a frog-leg winding*, which combines lap and wave windings on a single rotor. These windings will be examined individually below, and their advantages and disadvantages will be discussed.

The Lap Winding

The simplest type of winding construction used in modern dc machines is the simplex *series* or *lap winding*. A simplex lap winding is a rotor winding consisting of coils containing one or more turns of wire with the two ends of each coil coming out at *adjacent commutator segments* (Figure 7–13). If the end of the coil is connected to the segment after the segment that the beginning of the coil is connected to, the winding is a progressive lap winding and $y_c = 1$; if the end of the coil is connected to the segment before the segment that the beginning of the coil is connected to, the winding is a retrogressive lap winding and $y_c = -1$. A simple two-pole machine with lap windings is shown in Figure 7–14.

An interesting feature of simplex lap windings is that *there are as many parallel current paths through the machine as there are poles on the machine*. If C is the number of coils and commutator segments present in the rotor and P is the

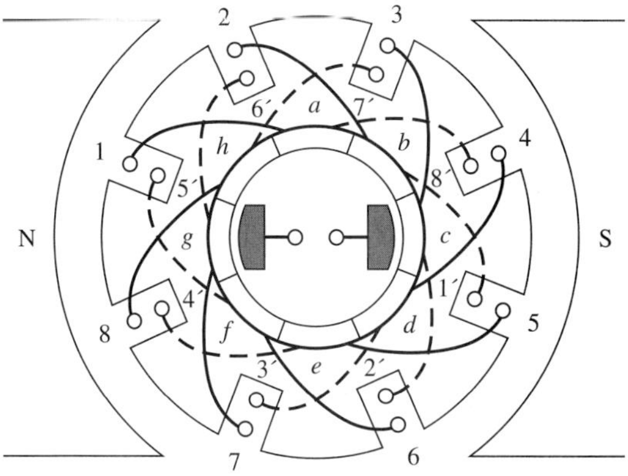

FIGURE 7–14
A simple two-pole lap-wound dc machine.

number of poles on the machine, then there will be C/P coils in each of the P parallel current paths through the machine. The fact that there are P current paths also requires that there be as many brushes on the machine as there are poles in order to tap all the current paths. This idea is illustrated by the simple four-pole motor in Figure 7–15. Notice that, for this motor, there are four current paths through the rotor, each having an equal voltage. The fact that there are many current paths in a multipole machine makes the lap winding an ideal choice for fairly low-voltage, high-current machines, since the high currents required can be split among the several different current paths. This current splitting permits the size of individual rotor conductors to remain reasonable even when the total current becomes extremely large.

The fact that there are many parallel paths through a multipole lap-wound machine can lead to a serious problem, however. To understand the nature of this problem, examine the six-pole machine in Figure 7–16. Because of long usage, there has been slight wear on the bearings of this machine, and the lower wires are closer to their pole faces than the upper wires are. As a result, there is a *larger* voltage in the current paths involving wires under the lower pole faces than in the paths involving wires under the upper pole faces. Since all the paths are connected in parallel, the result will be a circulating current flowing out some of the brushes in the machine and back into others, as shown in Figure 7–17. Needless to say, this is not good for the machine. Since the winding resistance of a rotor circuit is so small, a very tiny imbalance among the voltages in the parallel paths will cause large circulating currents through the brushes and potentially serious heating problems.

The problem of circulating currents within the parallel paths of a machine with four or more poles can never be entirely resolved, but it can be reduced somewhat by *equalizers* or *equalizing windings*. Equalizers are bars located on the rotor of a lap-wound dc machine that short together points at the same voltage

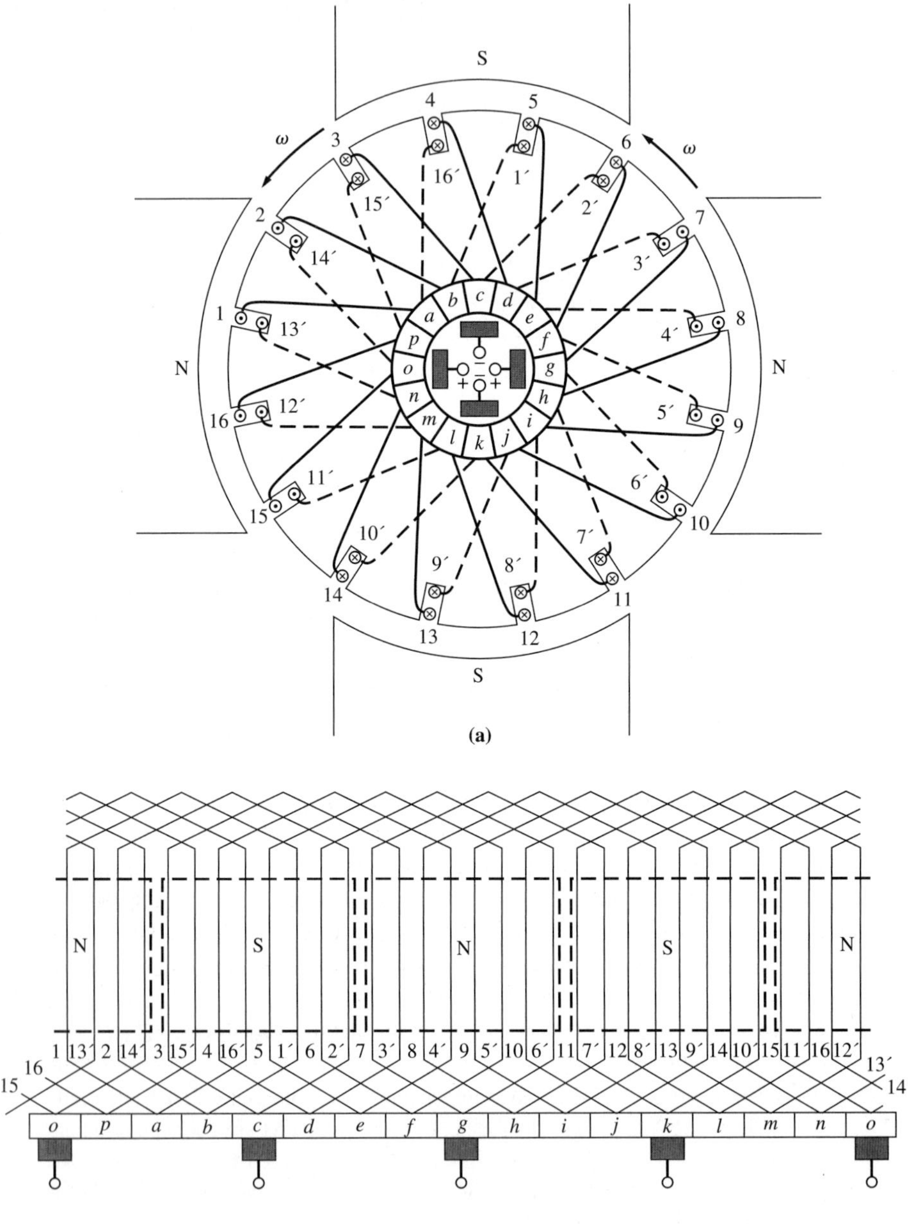

FIGURE 7–15
(a) A four-pole lap-wound dc motor. (b) The rotor winding diagram of this machine. Notice that each winding ends on the commutator segment just after the one it begins at. This is a progressive lap winding.

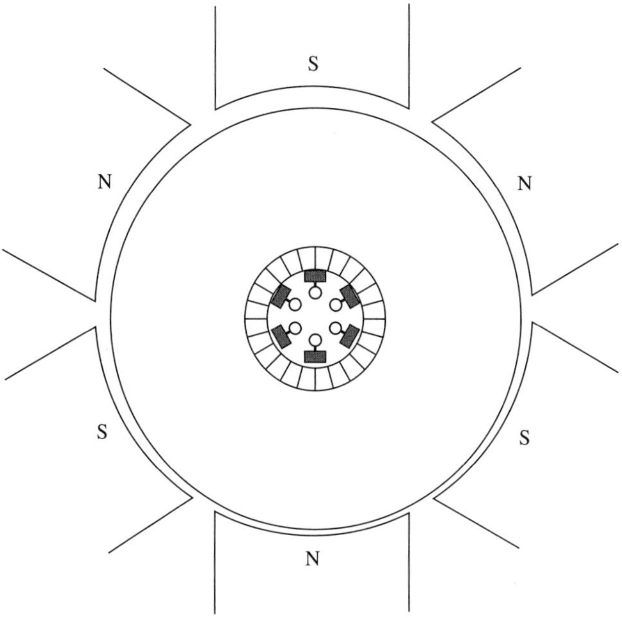

FIGURE 7–16
A six-pole dc motor showing the effects of bearing wear. Notice that the rotor is slightly closer to the lower poles than it is to the upper poles.

level in the different parallel paths. The effect of this shorting is to cause any circulating currents that occur to flow inside the small sections of windings thus shorted together and to prevent this circulating current from flowing through the brushes of the machine. These circulating currents even partially correct the flux imbalance that caused them to exist in the first place. An equalizer for the four-pole machine in Figure 7–15 is shown in Figure 7–18, and an equalizer for a large lap-wound dc machine is shown in Figure 7–19.

If a lap winding is duplex, then there are two completely independent windings wrapped on the rotor, and every other commutator segment is tied to one of the sets. Therefore, an individual coil ends on the second commutator segment down from where it started, and $y_c = \pm 2$ (depending on whether the winding is progressive or retrogressive). Since each set of windings has as many current paths as the machine has poles, there are *twice as many current paths* as the machine has poles in a duplex lap winding.

In general, for an m-plex lap winding, the commutator pitch y_c is

$$\boxed{y_c = \pm m} \qquad \text{lap winding} \tag{7–25}$$

and the number of current paths in a machine is

$$\boxed{a = mP} \qquad \text{lap winding} \tag{7–26}$$

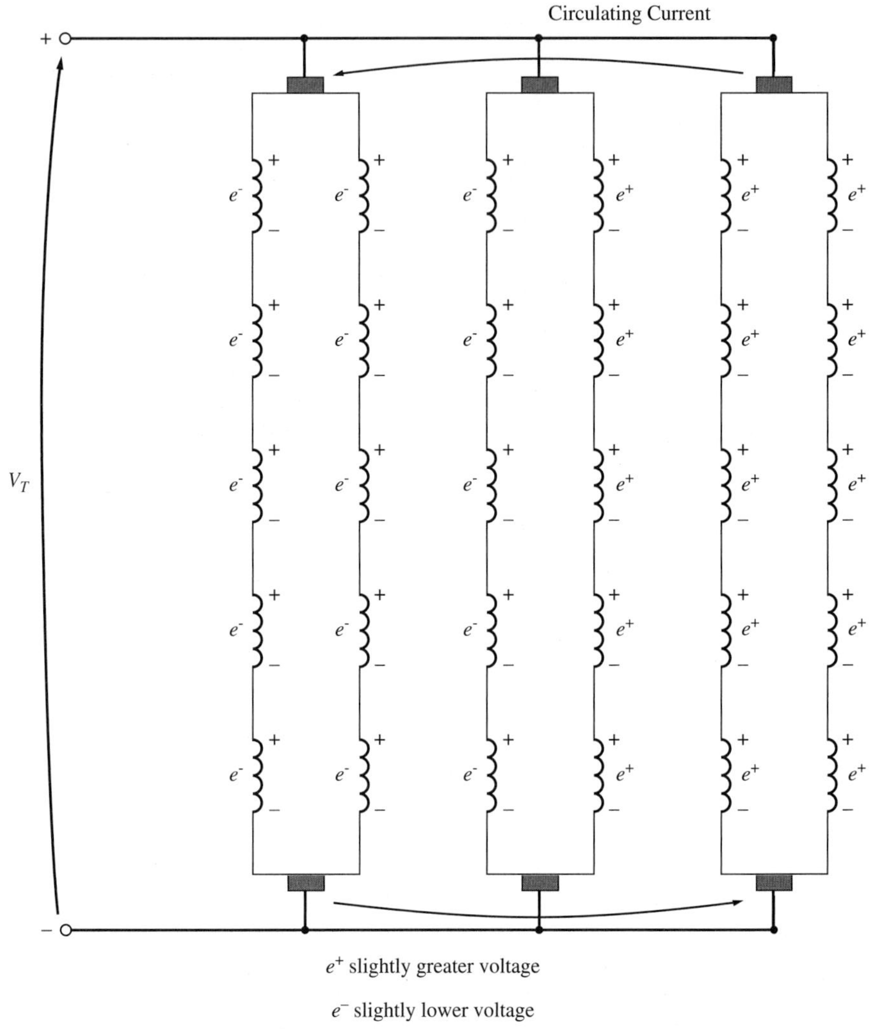

FIGURE 7–17
The voltages on the rotor conductors of the machine in Figure 7–16 are unequal, producing circulating currents flowing through its brushes.

where a = number of current paths in the rotor
m = plex of the windings (1, 2, 3, etc.)
P = number of poles on the machine

The Wave Winding

The *series* or *wave winding* is an alternative way to connect the rotor coils to the commutator segments. Figure 7–20 shows a simple four-pole machine with a

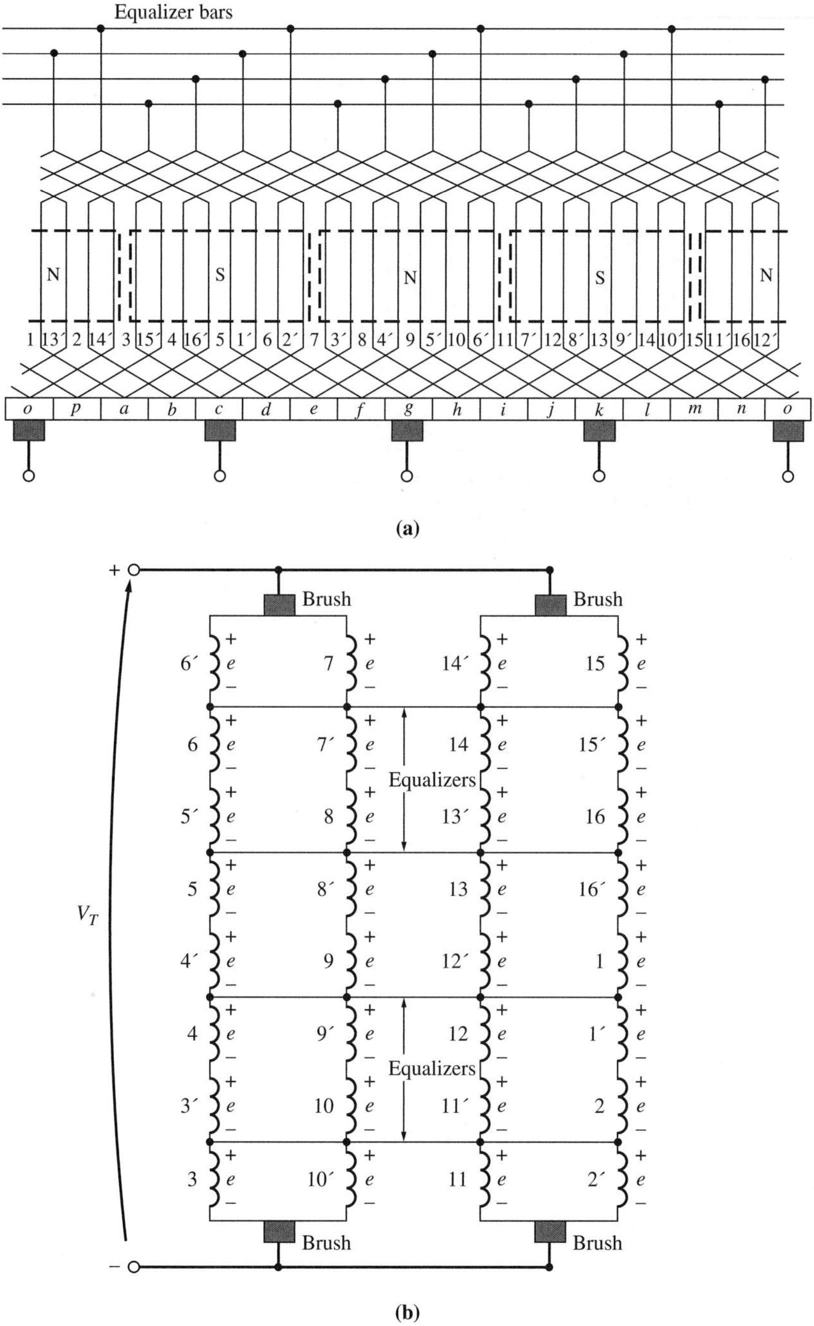

FIGURE 7–18
(a) An equalizer connection for the four-pole machine in Figure 7–15. (b) A voltage diagram for the machine shows the points shorted by the equalizers.

FIGURE 7–19
A closeup of the commutator of a large lap-wound dc machine. The equalizers are mounted in the small ring just in front of the commutator segments. (*Courtesy of General Electric Company.*)

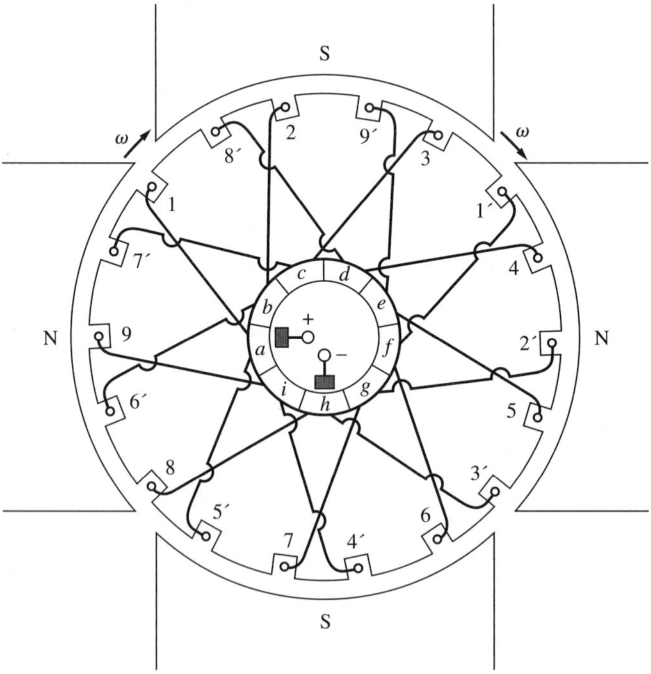

FIGURE 7–20
A simple four-pole wave-wound dc machine.

simplex wave winding. In this simplex wave winding, every *other* rotor coil connects back to a commutator segment adjacent to the beginning of the first coil. Therefore, *there are two coils in series* between the adjacent commutator segments. Furthermore, since each pair of coils between adjacent segments has a side under each pole face, all output voltages are the sum of the effects of every pole, and there can be no voltage imbalances.

The lead from the second coil may be connected to the segment either ahead of or behind the segment at which the first coil begins. If the second coil is connected to the segment ahead of the first coil, the winding is progressive; if it is connected to the segment behind the first coil, it is retrogressive.

In general, if there are P poles on the machine, then there are $P/2$ coils in series between adjacent commutator segments. If the $(P/2)$th coil is connected to the segment ahead of the first coil, the winding is progressive. If the $(P/2)$th coil is connected to the segment behind the first coil, the winding is retrogressive.

In a simplex wave winding, there are only two current paths. There are $C/2$ or one-half of the windings in each current path. The brushes in such a machine will be located a full pole pitch apart from each other.

What is the commutator pitch for a wave winding? Figure 7–20 shows a progressive nine-coil winding, and the end of a coil occurs five segments down from its starting point. In a retrogressive wave winding, the end of the coil occurs four segments down from its starting point. Therefore, the end of a coil in a four-pole wave winding must be connected just before or just after the point halfway around the circle from its starting point.

The general expression for commutator pitch in any simplex wave winding is

$$\boxed{y_c = \frac{2(C \pm 1)}{P}} \qquad \text{simplex wave} \qquad (7\text{–}27)$$

where C is the number of coils on the rotor and P is the number of poles on the machine. The plus sign is associated with progressive windings, and the minus sign is associated with retrogressive windings. A simplex wave winding is shown in Figure 7–21.

Since there are only two current paths through a simplex wave-wound rotor, only two brushes are needed to draw off the current. This is because the segments undergoing commutation connect the points with equal voltage under all the pole faces. More brushes can be added at points 180 electrical degrees apart if desired, since they are at the same potential and are connected together by the wires undergoing commutation in the machine. Extra brushes are usually added to a wave-wound machine, even though they are not necessary, because they reduce the amount of current that must be drawn through a given brush set.

Wave windings are well suited to building higher-voltage dc machines, since the number of coils in series between commutator segments permits a high voltage to be built up more easily than with lap windings.

A multiplex wave winding is a winding with multiple *independent* sets of wave windings on the rotor. These extra sets of windings have two current paths each, so the number of current paths on a multiplex wave winding is

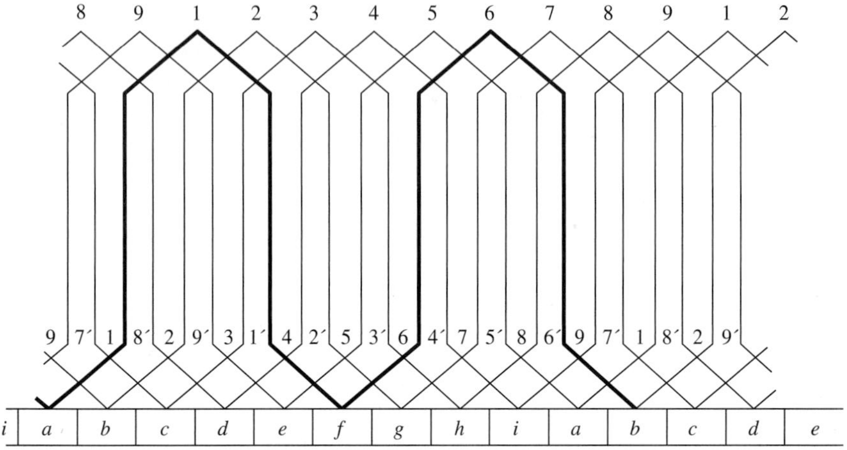

FIGURE 7–21
The rotor winding diagram for the machine in Figure 7–20. Notice that the end of every second coil in series connects to the segment after the beginning of the first coil. This is a progressive wave winding.

$$\boxed{a = 2m} \qquad \text{multiplex wave} \qquad (7\text{–}28)$$

The Frog-Leg Winding

The *frog-leg winding or self-equalizing winding* gets its name from the shape of its coils, as shown in Figure 7–22. It consists of a lap winding and a wave winding combined.

The equalizers in an ordinary lap winding are connected at points of equal voltage on the windings. Wave windings reach between points of essentially equal voltage under successive pole faces of the same polarity, which are the same locations that equalizers tie together. A frog-leg or self-equalizing winding combines a lap winding with a wave winding, so that the wave windings can function as equalizers for the lap winding.

The number of current paths present in a frog-leg winding is

$$\boxed{a = 2Pm_{\text{lap}}} \qquad \text{frog-leg winding} \qquad (7\text{–}29)$$

where P is the number of poles on the machine and m_{lap} is the plex of the lap winding.

Example 7–2. Describe the rotor winding arrangement of the four-loop machine in Section 7.2.

Solution
The machine described in Section 7.2 has four coils, each containing one turn, resulting in a total of eight conductors. It has a progressive lap winding.

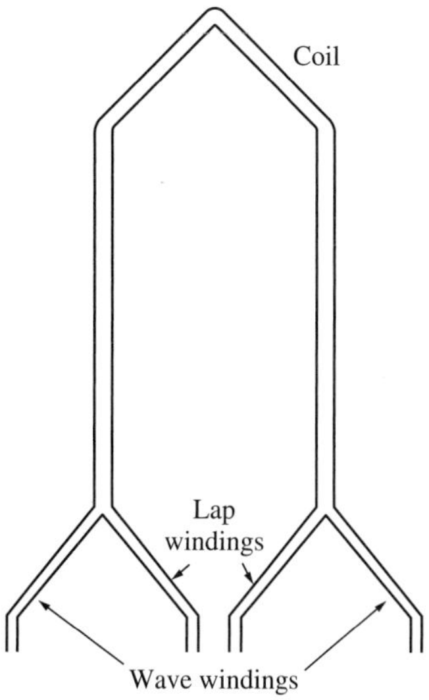

FIGURE 7–22
A frog-leg or self-equalizing winding coil.

7.4 PROBLEMS WITH COMMUTATION IN REAL MACHINES

The commutation process as described in Sections 7.2 and 7.3 is not as simple in practice as it seems in theory, because two major effects occur in the real world to disturb it:

1. Armature reaction
2. *L di/dt* voltages

This section explores the nature of these problems and the solutions employed to mitigate their effects.

Armature Reaction

If the magnetic field windings of a dc machine are connected to a power supply and the rotor of the machine is turned by an external source of mechanical power, then a voltage will be induced in the conductors of the rotor. This voltage will be rectified into a dc output by the action of the machine's commutator.

Now connect a load to the terminals of the machine, and a current will flow in its armature windings. This current flow will produce a magnetic field of its own, which will distort the original magnetic field from the machine's poles. This distortion of the flux in a machine as the load is increased is called *armature reaction*. It causes two serious problems in real dc machines.

The first problem caused by armature reaction is *neutral-plane shift*. The *magnetic neutral plane* is defined as the plane within the machine where the

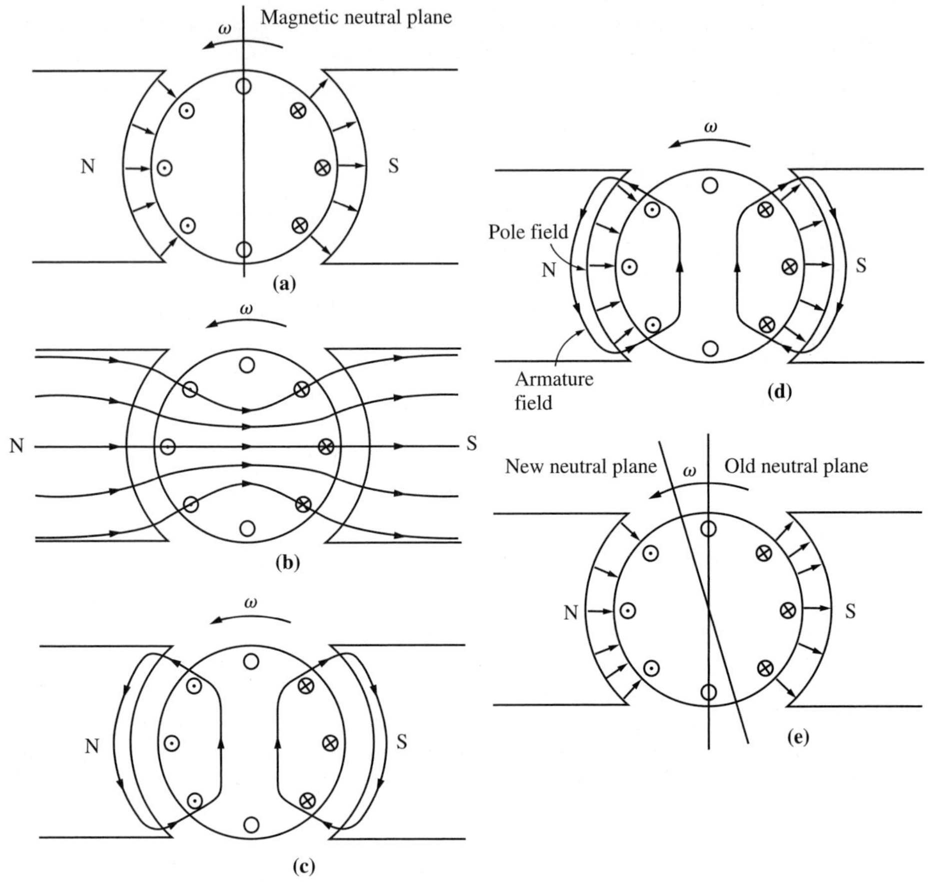

FIGURE 7–23
The development of armature reaction in a dc generator. (a) Initially the pole flux is uniformly distributed, and the magnetic neutral plane is vertical; (b) the effect of the air gap on the pole flux distribution; (c) the armature magnetic field resulting when a load is connected to the machine; (d) both rotor and pole fluxes are shown, indicating points where they add and subtract; (e) the resulting flux under the poles. The neutral plane has shifted in the direction of motion.

velocity of the rotor wires is exactly parallel to the magnetic flux lines, so that e_{ind} in the conductors in the plane is exactly zero.

To understand the problem of neutral-plane shift, examine Figure 7–23. Figure 7–23a shows a two-pole dc machine. Notice that the flux is distributed uniformly under the pole faces. The rotor windings shown have voltages built up out of the page for wires under the north pole face and into the page for wires under the south pole face. The neutral plane in this machine is exactly vertical.

Now suppose a load is connected to this machine so that it acts as a generator. Current will flow out of the positive terminal of the generator, so current will be flowing out of the page for wires under the north pole face and into the page for wires under the south pole face. This current flow produces a magnetic field from the rotor windings, as shown in Figure 7–23c. This rotor magnetic field affects the original magnetic field from the poles that produced the generator's voltage in the first place. In some places under the pole surfaces, it subtracts from the pole flux, and in other places it adds to the pole flux. The overall result is that the magnetic flux in the air gap of the machine is skewed as shown in Figure 7–23d and e. Notice that the place on the rotor where the induced voltage in a conductor would be zero (the neutral plane) has shifted.

For the generator shown in Figure 7–23, the magnetic neutral plane shifted in the direction of rotation. If this machine had been a motor, the current in its rotor would be reversed and the flux would bunch up in the opposite corners from the bunches shown in the figure. As a result, the magnetic neutral plane would shift the other way.

In general, the neutral plane shifts in the direction of motion for a generator and opposite to the direction of motion for a motor. Furthermore, the amount of the shift depends on the amount of rotor current and hence on the load of the machine.

So what's the big deal about neutral-plane shift? It's just this: The commutator must short out commutator segments just at the moment when the voltage across them is equal to zero. If the brushes are set to short out conductors in the vertical plane, then the voltage between segments is indeed zero *until the machine is loaded*. When the machine is loaded, the neutral plane shifts, and the brushes short out commutator segments with a finite voltage across them. The result is a current flow circulating between the shorted segments and large sparks at the brushes when the current path is interrupted as the brush leaves a segment. The end result is *arcing and sparking at the brushes*. This is a very serious problem, since it leads to drastically reduced brush life, pitting of the commutator segments, and greatly increased maintenance costs. Notice that this problem cannot be fixed even by placing the brushes over the full-load neutral plane, because then they would spark at no load.

In extreme cases, the neutral-plane shift can even lead to *flashover* in the commutator segments near the brushes. The air near the brushes in a machine is normally ionized as a result of the sparking on the brushes. Flashover occurs when the voltage of adjacent commutator segments gets large enough to sustain an arc in the ionized air above them. If flashover occurs, the resulting arc can even melt the commutator's surface.

The second major problem caused by armature reaction is called *flux weakening*. To understand flux weakening, refer to the magnetization curve shown in Figure 7–24. Most machines operate at flux densities near the saturation point. Therefore, at locations on the pole surfaces where the rotor magnetomotive force adds to the pole magnetomotive force, only a small increase in flux occurs. But at locations on the pole surfaces where the rotor magnetomotive force subtracts from the pole magnetomotive force, there is a larger decrease in flux. The net result is that *the total average flux under the entire pole face is decreased* (see Figure 7–25).

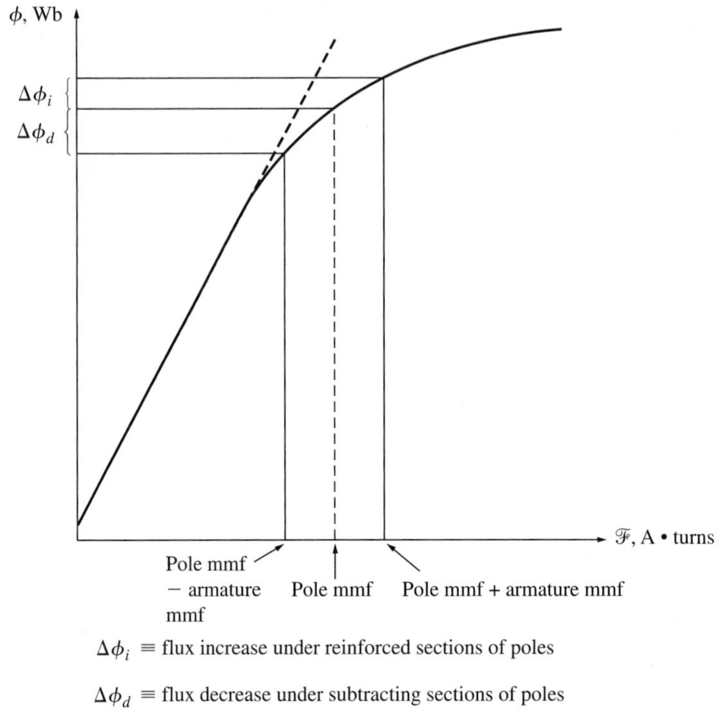

FIGURE 7–24
A typical magnetization curve shows the effects of pole saturation where armature and pole magnetomotive forces add.

Flux weakening causes problems in both generators and motors. In generators, the effect of flux weakening is simply to reduce the voltage supplied by the generator for any given load. In motors, the effect can be more serious. As the early examples in this chapter showed, when the flux in a motor is decreased, its speed increases. But increasing the speed of a motor can increase its load, resulting in more flux weakening. It is possible for some shunt dc motors to reach a runaway condition as a result of flux weakening, where the speed of the motor just keeps increasing until the machine is disconnected from the power line or until it destroys itself.

L di/dt Voltages

The second major problem is the *L di/dt* voltage that occurs in commutator segments being shorted out by the brushes, sometimes called *inductive kick*. To understand this problem, look at Figure 7–26. This figure represents a series of commutator segments and the conductors connected between them. Assuming that the current in the brush is 400 A, the current in each path is 200 A. Notice that when a commutator segment is shorted out, the current flow through that commutator

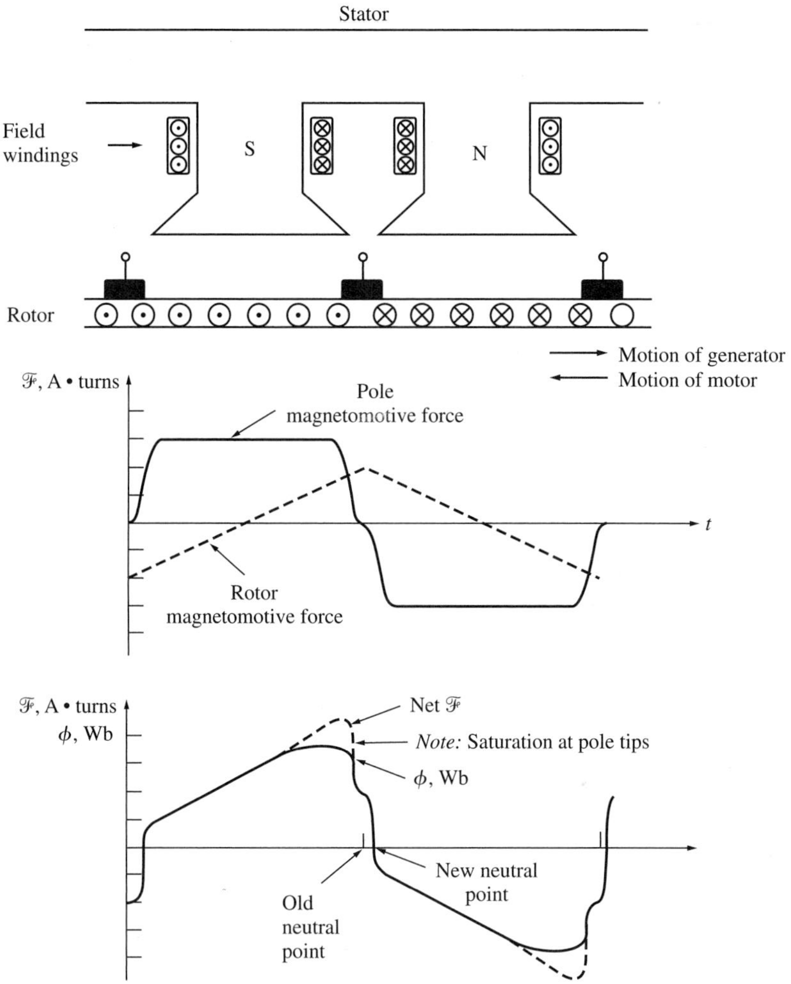

FIGURE 7–25
The flux and magnetomotive force under the pole faces in a dc machine. At those points where the magnetomotive forces subtract, the flux closely follows the net magnetomotive force in the iron; but at those points where the magnetomotive forces add, saturation limits the total flux present. Note also that the neutral point of the rotor has shifted.

segment must reverse. How fast must this reversal occur? Assuming that the machine is turning at 800 r/min and that there are 50 commutator segments (a reasonable number for a typical motor), each commutator segment moves under a brush and clears it again in $t = 0.0015$ s. Therefore, the rate of change in current with respect to time in the shorted loop must *average*

$$\frac{di}{dt} = \frac{400 \text{ A}}{0.0015 \text{ s}} = 266{,}667 \text{ A/s} \tag{7–30}$$

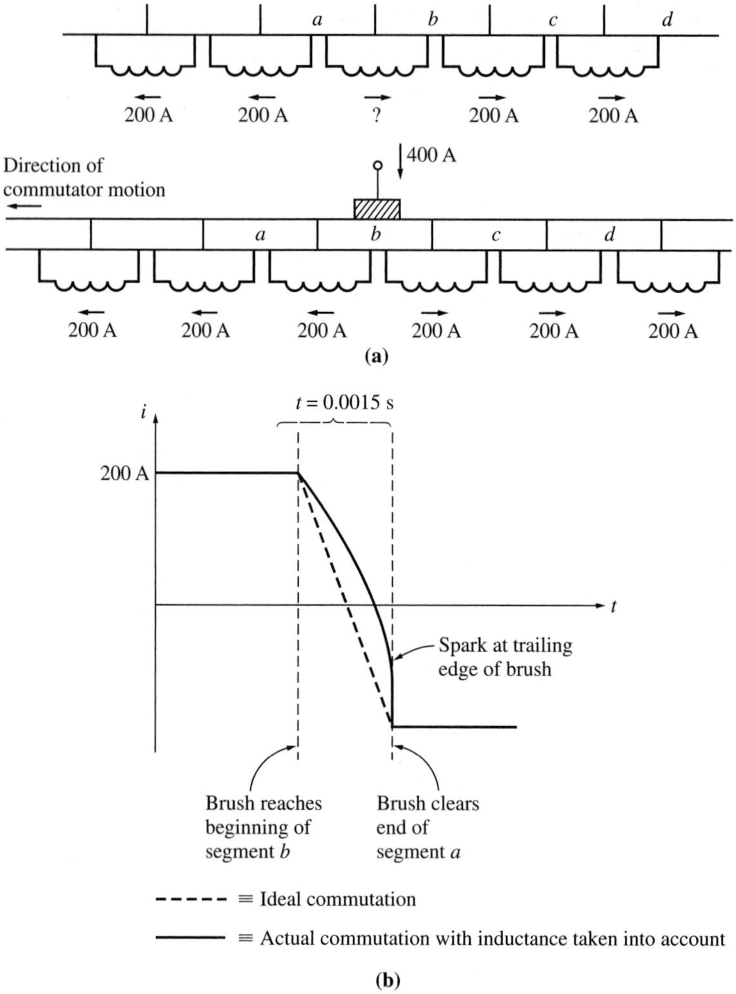

FIGURE 7–26
(a) The reversal of current flow in a coil undergoing commutation. Note that the current in the coil between segments a and b must reverse direction while the brush shorts together the two commutator segments. (b) The current reversal in the coil undergoing commutation as a function of time for both ideal commutation and real commutation, with the coil inductance taken into account.

With even a tiny inductance in the loop, a very significant inductive voltage kick $v = L\, di/dt$ will be induced in the shorted commutator segment. This high voltage naturally causes sparking at the brushes of the machine, resulting in the same arcing problems that the neutral-plane shift causes.

Solutions to the Problems with Commutation

Three approaches have been developed to partially or completely correct the problems of armature reaction and *L di/dt* voltages:

1. Brush shifting
2. Commutating poles or interpoles
3. Compensating windings

Each of these techniques is explained below, together with its advantages and disadvantages.

BRUSH SHIFTING. Historically, the first attempts to improve the process of commutation in real dc machines started with attempts to stop the sparking at the brushes caused by the neutral-plane shifts and *L di/dt* effects. The first approach taken by machine designers was simple: If the neutral plane of the machine shifts, why not shift the brushes with it in order to stop the sparking? It certainly seemed like a good idea, but there are several serious problems associated with it. For one thing, the neutral plane moves with every change in load, and the shift direction reverses when the machine goes from motor operation to generator operation. Therefore, someone had to adjust the brushes every time the load on the machine changed. In addition, shifting the brushes may have stopped the brush sparking, but it actually *aggravated* the flux-weakening effect of the armature reaction in the machine. This is true because of two effects:

1. The rotor magnetomotive force now has a vector component that opposes the magnetomotive force from the poles (see Figure 7–27).
2. The change in armature current distribution causes the flux to bunch up even more at the saturated parts of the pole faces.

Another slightly different approach sometimes taken was to fix the brushes in a compromise position (say, one that caused no sparking at two-thirds of full load). In this case, the motor sparked at no load and somewhat at full load, but if it spent most of its life operating at about two-thirds of full load, then sparking was minimized. Of course, such a machine could not be used as a generator at all—the sparking would have been horrible.

By about 1910, the brush-shifting approach to controlling sparking was already obsolete. Today, brush shifting is only used in very small machines that always run as motors. This is done because better solutions to the problem are simply not economical in such small motors.

COMMUTATING POLES OR INTERPOLES. Because of the disadvantages noted above and especially because of the requirement that a person must adjust the brush positions of machines as their loads change, another solution to the problem of brush sparking was developed. The basic idea behind this new approach is that

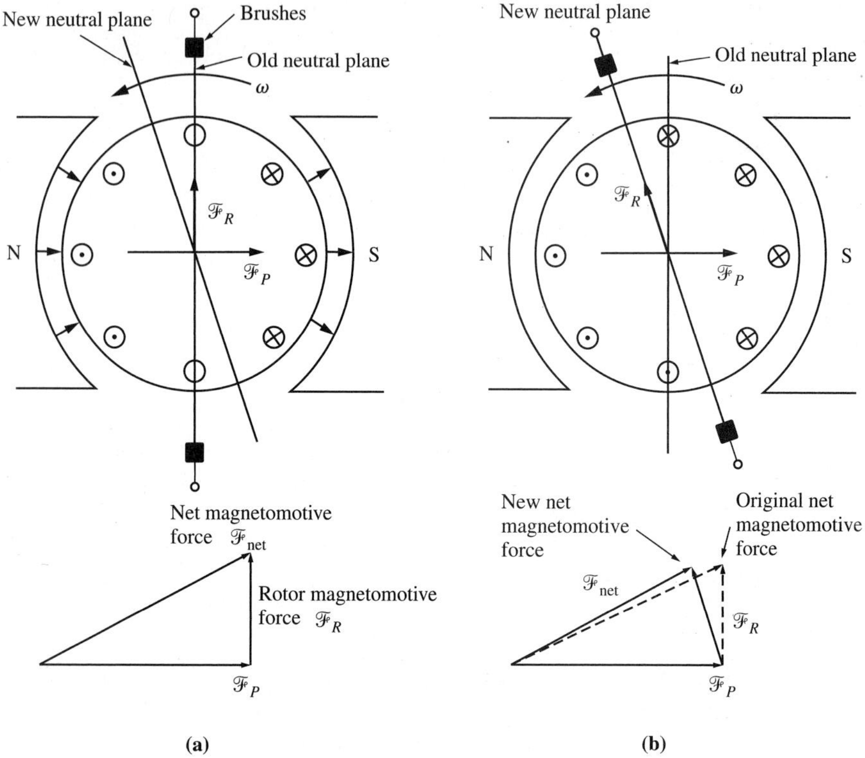

FIGURE 7–27
(a) The net magnetomotive force in a dc machine with its brushes in the vertical plane. (b) The net magnetomotive force in a dc machine with its brushes over the shifted neutral plane. Notice that now there is a component of armature magnetomotive force *directly opposing* the poles' magnetomotive force, and the net magnetomotive force in the machine is reduced.

if the voltage in the wires undergoing commutation can be made zero, then there will be no sparking at the brushes. To accomplish this, small poles, called *commutating poles* or *interpoles*, are placed midway between the main poles. These commutating poles are located *directly over* the conductors being commutated. By providing a flux from the commutating poles, the voltage in the coils undergoing commutation can be exactly canceled. If the cancellation is exact, then there will be no sparking at the brushes.

The commutating poles do not otherwise change the operation of the machine, because they are so small that they affect only the few conductors about to undergo commutation. Notice that the *armature reaction* under the main pole faces is unaffected, since the effects of the commutating poles do not extend that far. This means that the flux weakening in the machine is unaffected by commutating poles.

How is cancellation of the voltage in the commutator segments accomplished for all values of loads? This is done by simply connecting the interpole windings in *series* with the windings on the rotor, as shown in Figure 7–28. As the load increases and the rotor current increases, the magnitude of the neutral-plane shift and the size of the $L\,di/dt$ effects increase as well. Both these effects increase the voltage in the conductors undergoing commutation. However, the interpole flux also increases, producing a larger voltage in the conductors that opposes the voltage due to the neutral-plane shift. The net result is that their effects cancel over a broad range of loads. Note that interpoles work for both motor and generator operation, since when the machine changes from motor to generator, the current both in its rotor and in its interpoles reverses direction. Therefore, the voltage effects from them still cancel.

What polarity must the flux in the interpoles be? The interpoles must induce a voltage in the conductors undergoing commutation that is *opposite* to the voltage caused by neutral-plane shift and $L\,di/dt$ effects. In the case of a generator, the neutral plane shifts in the direction of rotation, meaning that the conductors undergoing commutation have the same polarity of voltage as the pole they just left (see Figure 7–29). To oppose this voltage, the interpoles must have the opposite flux, which is the flux of the upcoming pole. In a motor, however, the neutral plane shifts opposite to the direction of rotation, and the conductors undergoing commutation have the same flux as the pole they are approaching. In order to oppose this voltage, the interpoles must have the same polarity as the previous main pole. Therefore,

1. The interpoles must be of the same polarity as the next upcoming main pole in a generator.

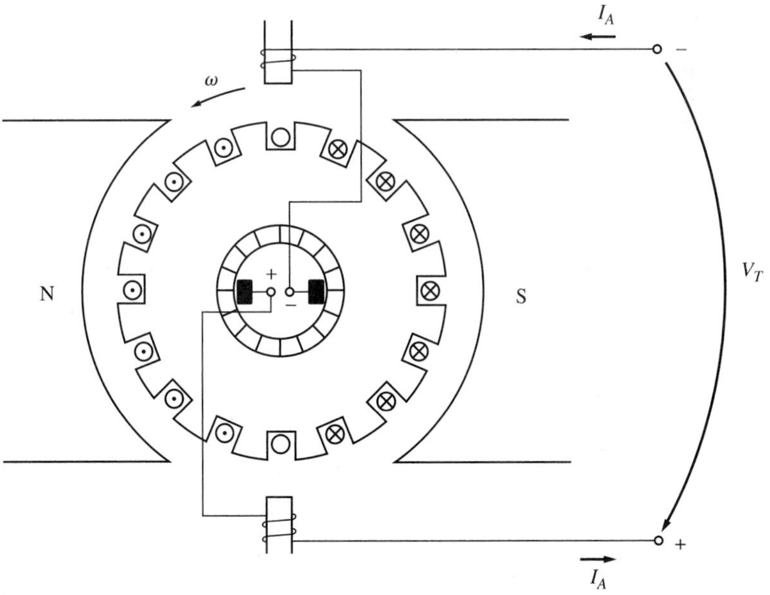

FIGURE 7–28
A dc machine with interpoles.

2. The interpoles must be of the same polarity as the previous main pole in a motor.

The use of commutating poles or interpoles is very common, because they correct the sparking problems of dc machines at a fairly low cost. They are almost always found in any dc machine of 1 hp or larger. It is important to realize, though, that they do *nothing* for the flux distribution under the pole faces, so the flux-weakening problem is still present. Most medium-sized general-purpose motors correct for sparking problems with interpoles and just live with the flux-weakening effects.

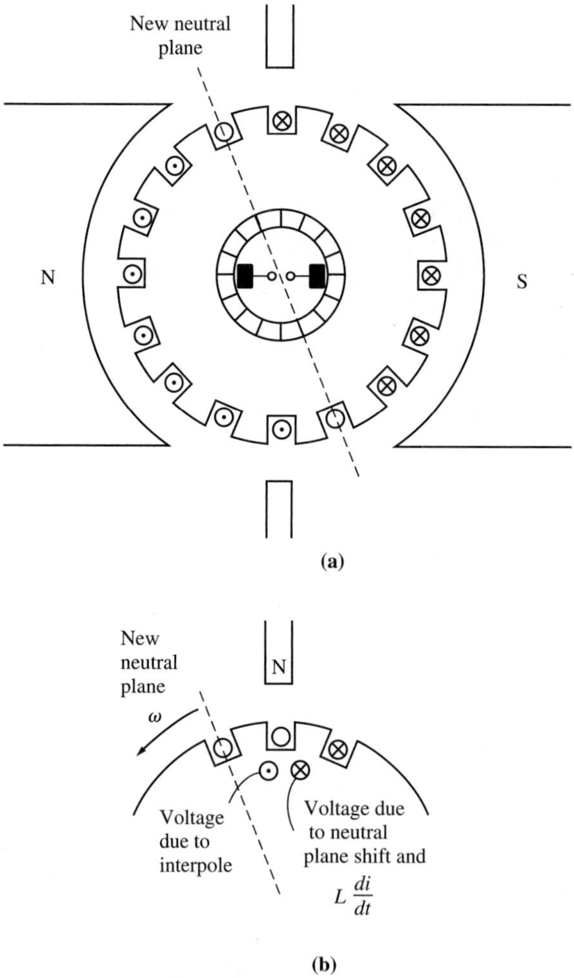

FIGURE 7–29
Determining the required polarity of an interpole. The flux from the interpole must produce a voltage that opposes the existing voltage in the conductor.

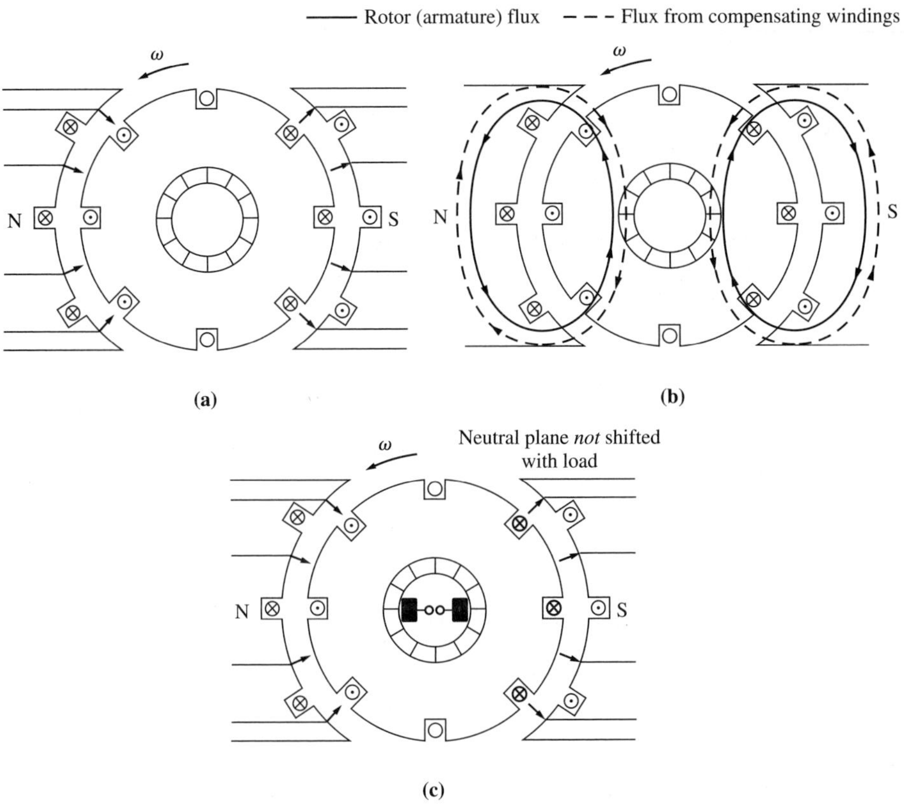

FIGURE 7–30
The effect of compensating windings in a dc machine. (a) The pole flux in the machine; (b) the fluxes from the armature and compensating windings. Notice that they are equal and opposite; (c) the net flux in the machine, which is just the original pole flux.

COMPENSATING WINDINGS. For very heavy, severe duty cycle motors, the flux-weakening problem can be very serious. To completely cancel armature reaction and thus eliminate both neutral-plane shift and flux weakening, a different technique was developed. This third technique involves placing *compensating windings* in slots carved in the faces of the poles parallel to the rotor conductors, to cancel the distorting effect of armature reaction. These windings are connected in series with the rotor windings, so that whenever the load changes in the rotor, the current in the compensating windings changes, too. Figure 7–30 shows the basic concept. In Figure 7–30a, the pole flux is shown by itself. In Figure 7–30b, the rotor flux and the compensating winding flux are shown. Figure 7–30c represents the sum of these three fluxes, which is just equal to the original pole flux by itself.

Figure 7–31 shows a more careful development of the effect of compensating windings on a dc machine. Notice that the magnetomotive force due to the

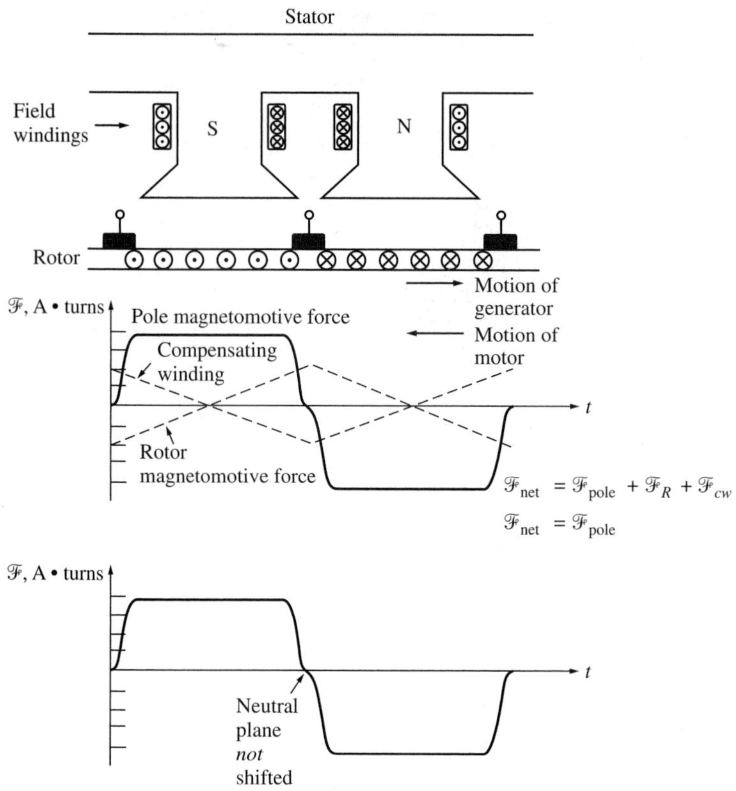

FIGURE 7–31
The flux and magnetomotive forces in a dc machine with compensating windings.

compensating windings is equal and opposite to the magnetomotive force due to the rotor at every point under the pole faces. The resulting net magnetomotive force is just the magnetomotive force due to the poles, so the flux in the machine is unchanged regardless of the load on the machine. The stator of a large dc machine with compensating windings is shown in Figure 7–32.

The major disadvantage of compensating windings is that they are expensive, since they must be machined into the faces of the poles. Any motor that uses them must also have interpoles, since compensating windings do not cancel $L\,di/dt$ effects. The interpoles do not have to be as strong, though, since they are canceling only $L\,di/dt$ voltages in the windings, and not the voltages due to neutral-plane shifting. Because of the expense of having both compensating windings and interpoles on such a machine, these windings are used only where the extremely severe nature of a motor's duty demands them.

FIGURE 7–32
The stator of a six-pole dc machine with interpoles and compensating windings. (*Courtesy of Westinghouse Electric Company.*)

7.5 THE INTERNAL GENERATED VOLTAGE AND INDUCED TORQUE EQUATIONS OF REAL DC MACHINES

How much voltage is produced by a real dc machine? The induced voltage in any given machine depends on three factors:

1. The flux ϕ in the machine
2. The speed ω_m of the machine's rotor
3. A constant depending on the construction of the machine

How can the voltage in the rotor windings of a real machine be determined? The voltage out of the armature of a real machine is equal to the number of conductors per current path times the voltage on each conductor. The voltage in *any single conductor under the pole faces* was previously shown to be

$$e_{\text{ind}} = e = vBl \tag{7–31}$$

The voltage out of the armature of a real machine is thus

$$E_A = \frac{ZvBl}{a} \tag{7–32}$$

where Z is the total number of conductors and a is the number of current paths. The velocity of each conductor in the rotor can be expressed as $v = r\omega_m$, where r is the radius of the rotor, so

$$E_A = \frac{Zr\omega_m Bl}{a} \tag{7–33}$$

This voltage can be reexpressed in a more convenient form by noting that the flux of a pole is equal to the flux density under the pole times the pole's area:

$$\phi = BA_P$$

The rotor of the machine is shaped like a cylinder, so its area is

$$A = 2\pi rl \tag{7–34}$$

If there are P poles on the machine, then the portion of the area associated with each pole is the total area A divided by the number of poles P:

$$A_P = \frac{A}{P} = \frac{2\pi rl}{P} \tag{7–35}$$

The total *flux per pole* in the machine is thus

$$\phi = BA_P = \frac{B(2\pi rl)}{P} = \frac{2\pi rlB}{P} \tag{7–36}$$

Therefore, the internal generated voltage in the machine can be expressed as

$$E_A = \frac{Zr\omega_m Bl}{a} \tag{7–33}$$

$$= \left(\frac{ZP}{2\pi a}\right)\left(\frac{2\pi rlB}{P}\right)\omega_m$$

$$\boxed{E_A = \frac{ZP}{2\pi a}\phi\omega_m} \tag{7–37}$$

Finally,

$$\boxed{E_A = K\phi\omega_m} \tag{7–38}$$

where

$$\boxed{K = \frac{ZP}{2\pi a}} \tag{7–39}$$

In modern industrial practice, it is common to express the speed of a machine in revolutions per minute instead of radians per second. The conversion from revolutions per minute to radians per second is

$$\omega_m = \frac{2\pi}{60}n_m \tag{7–40}$$

so the voltage equation with speed expressed in terms of revolutions per minute is

$$\boxed{E_A = K'\phi n_m} \tag{7–41}$$

where

$$\boxed{K' = \frac{ZP}{60a}} \tag{7–42}$$

How much torque is induced in the armature of a real dc machine? The torque in any dc machine depends on three factors:

1. The flux ϕ in the machine
2. The armature (or rotor) current I_A in the machine
3. A constant depending on the construction of the machine

How can the torque on the rotor of a real machine be determined? The torque on the armature of a real machine is equal to the number of conductors Z times the torque on each conductor. The torque in *any single conductor under the pole faces* was previously shown to be

$$\tau_{\text{cond}} = rI_{\text{cond}}lB \tag{7–43}$$

If there are a current paths in the machine, then the total armature current I_A is split among the a current paths, so the current in a single conductor is given by

$$I_{\text{cond}} = \frac{I_A}{a} \tag{7–44}$$

and the torque in a single conductor on the motor may be expressed as

$$\tau_{\text{cond}} = \frac{rI_AlB}{a} \tag{7–45}$$

Since there are Z conductors, the total induced torque in a dc machine rotor is

$$\tau_{\text{ind}} = \frac{ZrlBI_A}{a} \tag{7–46}$$

The flux per pole in this machine can be expressed as

$$\phi = BA_P = \frac{B(2\pi rl)}{P} = \frac{2\pi rlB}{P} \tag{7–47}$$

so the total induced torque can be reexpressed as

$$\tau_{\text{ind}} = \frac{ZP}{2\pi a}\phi I_A \tag{7–48}$$

Finally,

$$\boxed{\tau_{\text{ind}} = K\phi I_A} \tag{7–49}$$

where

$$\boxed{K = \frac{ZP}{2\pi a}} \tag{7–39}$$

Both the internal generated voltage and the induced torque equations just given are only approximations, because not all the conductors in the machine are under the pole faces at any given time and also because the surfaces of each pole do not cover an entire $1/P$ of the rotor's surface. To achieve greater accuracy, the number of conductors under the pole faces could be used instead of the total number of conductors on the rotor.

Example 7–3. A duplex lap-wound armature is used in a six-pole dc machine with six brush sets, each spanning two commutator segments. There are 72 coils on the armature, each containing 12 turns. The flux per pole in the machine is 0.039 Wb, and the machine spins at 400 r/min.

(a) How many current paths are there in this machine?
(b) What is its induced voltage E_A?

Solution

(a) The number of current paths in this machine is

$$a = mP = 2(6) = 12 \text{ current paths} \tag{7–26}$$

(b) The induced voltage in the machine is

$$E_A = K'\phi n_m \tag{7–41}$$

and

$$K' = \frac{ZP}{60a} \tag{7–42}$$

The number of conductors in this machine is

$$\begin{aligned} Z &= 2CN_C \\ &= 2(72)(12) = 1728 \text{ conductors} \end{aligned} \tag{7–22}$$

Therefore, the constant K' is

$$K' = \frac{ZP}{60a} = \frac{(1728)(6)}{(60)(12)} = 14.4$$

and the voltage E_A is

$$\begin{aligned} E_A &= K'\phi n_m \\ &= (14.4)(0.039 \text{ Wb})(400 \text{ r/min}) \\ &= 224.6 \text{ V} \end{aligned}$$

Example 7–4. A 12-pole dc generator has a simplex wave-wound armature containing 144 coils of 10 turns each. The resistance of each turn is 0.011 Ω. Its flux per pole is 0.05 Wb, and it is turning at a speed of 200 r/min.

(a) How many current paths are there in this machine?
(b) What is the induced armature voltage of this machine?
(c) What is the effective armature resistance of this machine?
(d) If a 1-kΩ resistor is connected to the terminals of this generator, what is the resulting induced countertorque on the shaft of the machine? (Ignore the internal armature resistance of the machine.)

Solution

(a) There are $a = 2m = 2$ current paths in this winding.

(b) There are $Z = 2CN_c = 2(144)(10) = 2880$ conductors on this generator's rotor. Therefore,

$$K' = \frac{ZP}{60a} = \frac{(2880)(12)}{(60)(2)} = 288$$

Therefore, the induced voltage is

$$\begin{aligned} E_A &= K'\phi n_m \\ &= (288)(0.05 \text{ Wb})(200 \text{ r/min}) \\ &= 2880 \text{ V} \end{aligned}$$

(c) There are two parallel paths through the rotor of this machine, each one consisting of $Z/2 = 1440$ conductors, or 720 turns. Therefore, the resistance in each current path is

$$\text{Resistance/path} = (720 \text{ turns})(0.011\ \Omega/\text{turn}) = 7.92\ \Omega$$

Since there are two parallel paths, the effective armature resistance is

$$R_A = \frac{7.92\ \Omega}{2} = 3.96\ \Omega$$

(d) If a 1000-Ω load is connected to the terminals of the generator, and if R_A is ignored, then a current of $I = 2880 \text{ V}/1000\ \Omega = 2.88$ A flows. The constant K is given by

$$K = \frac{ZP}{2\pi a} = \frac{(2880)(12)}{(2\pi)(2)} = 2750.2$$

Therefore, the countertorque on the shaft of the generator is

$$\begin{aligned} \tau_{\text{ind}} &= K\phi I_A = (2750.2)(0.05 \text{ Wb})(2.88 \text{ A}) \\ &= 396 \text{ N} \bullet \text{m} \end{aligned}$$

7.6 THE CONSTRUCTION OF DC MACHINES

A simplified sketch of a dc machine is shown in Figure 7–33, and a more detailed cutaway diagram of a dc machine is shown in Figure 7–34.

The physical structure of the machine consists of two parts: the *stator* or stationary part and the *rotor* or rotating part. The stationary part of the machine consists of the *frame*, which provides physical support, and the *pole pieces*, which project inward and provide a path for the magnetic flux in the machine. The ends of the pole pieces that are near the rotor spread out over the rotor surface to distribute its flux evenly over the rotor surface. These ends are called the *pole shoes*. The exposed surface of a pole shoe is called a *pole face*, and the distance between the pole face and the rotor is called the *air gap*.

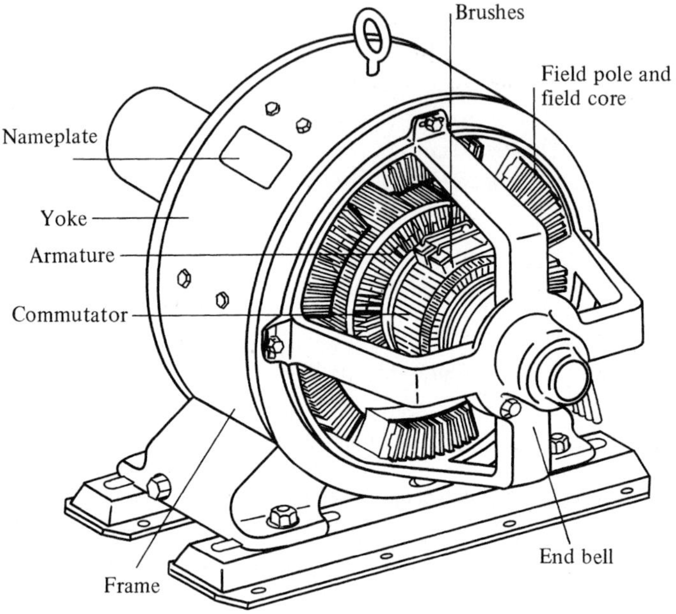

FIGURE 7–33
A simplified diagram of a dc machine.

FIGURE 7–34
(a) A cutaway view of a 4000-hp, 700-V, 18–pole dc machine showing compensating windings, interpoles, equalizer, and commutator. (*Courtesy of General Electric Company.*) (b) A cutaway view of a smaller four-pole dc motor including interpoles but without compensating windings. (*Courtesy of MagneTek Incorporated.*)

There are two principal windings on a dc machine: the armature windings and the field windings. The *armature windings* are defined as the windings in which a voltage is induced, and the *field windings* are defined as the windings that produce the main magnetic flux in the machine. In a normal dc machine, the armature windings are located on the rotor, and the field windings are located on the stator. Because the armature windings are located on the rotor, a dc machine's rotor itself is sometimes called an *armature*.

Some major features of typical dc motor construction are described below.

Pole and Frame Construction

The main poles of older dc machines were often made of a single cast piece of metal, with the field windings wrapped around it. They often had bolted-on laminated tips to reduce core losses in the pole faces. Since solid-state drive packages have become common, the main poles of newer machines are made entirely of laminated material (see Figure 7–35). This is true because there is a much higher ac content in the power supplied to dc motors driven by solid-state drive packages, resulting in much higher eddy current losses in the stators of the machines. The pole faces are typically either *chamfered* or *eccentric* in construction, meaning that the outer tips of a pole face are spaced slightly further from the rotor's surface than the center of the pole face is (see Figure 7–36). This action increases the reluctance at the tips of a pole face and therefore reduces the flux-bunching effect of armature reaction on the machine.

FIGURE 7–35
Main field pole assembly for a dc motor. Note the pole laminations and compensating windings. (*Courtesy of General Electric Company.*)

The poles on dc machines are called *salient poles*, because they stick out from the surface of the stator.

The interpoles in dc machines are located between the main poles. They are more and more commonly of laminated construction, because of the same loss problems that occur in the main poles.

Some manufacturers are even constructing the portion of the frame that serves as the magnetic flux's return path (the *yoke*) with laminations, to further reduce core losses in electronically driven motors.

Rotor or Armature Construction

The rotor or armature of a dc machine consists of a shaft machined from a steel bar with a core built up over it. The core is composed of many laminations stamped from a steel plate, with notches along its outer surface to hold the armature windings. The commutator is built onto the shaft of the rotor at one end of the core. The armature coils are laid into the slots on the core, as described in Section 7.4, and their ends are connected to the commutator segments. A large dc machine rotor is shown in Figure 7–37.

Commutator and Brushes

The commutator in a dc machine (Figure 7–38) is typically made of copper bars insulated by a mica-type material. The copper bars are made sufficiently thick to permit normal wear over the lifetime of the motor. The mica insulation between commutator segments is harder than the commutator material itself, so as a machine ages, it is often necessary to *undercut* the commutator insulation to ensure that it does not stick up above the level of the copper bars.

The brushes of the machine are made of carbon, graphite, metal graphite, or a mixture of carbon and graphite. They have a high conductivity to reduce electrical losses and a low coefficient of friction to reduce excessive wear. They are

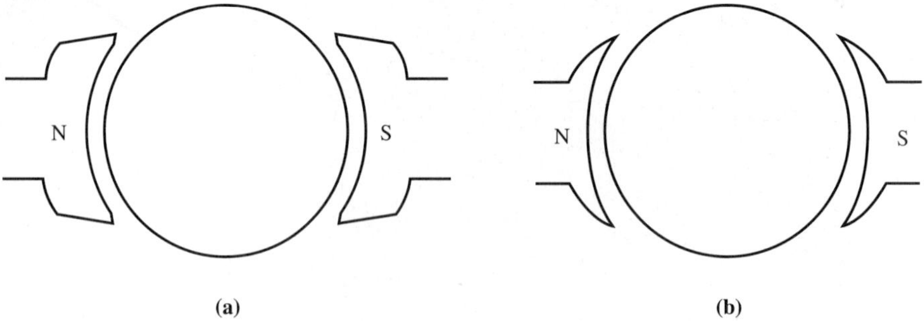

FIGURE 7–36
Poles with extra air-gap width at the tips to reduce armature reaction. (a) Chamfered poles; (b) eccentric or uniformly graded poles.

FIGURE 7–37
Photograph of a dc machine with the upper stator half removed shows the construction of its rotor. (*Courtesy of General Electric Company.*)

FIGURE 7–38
Close-up view of commutator and brushes in a large dc machine. (*Courtesy of General Electric Company.*)

deliberately made of much softer material than that of the commutator segments, so that the commutator surface will experience very little wear. The choice of brush hardness is a compromise: If the brushes are too soft, they will have to be replaced too often; but if they are too hard, the commutator surface will wear excessively over the life of the machine.

All the wear that occurs on the commutator surface is a direct result of the fact that the brushes must rub over them to convert the ac voltage in the rotor wires to dc voltage at the machine's terminals. If the pressure of the brushes is too great, both the brushes and commutator bars wear excessively. However, if the brush pressure is too small, the brushes tend to jump slightly and a great deal of sparking occurs at the brush-commutator segment interface. This sparking is equally bad for the brushes and the commutator surface. Therefore, the brush pressure on the commutator surface must be carefully adjusted for maximum life.

Another factor which affects the wear on the brushes and segments in a dc machine commutator is the amount of current flowing in the machine. The brushes normally ride over the commutator surface on a thin oxide layer, which lubricates the motion of the brush over the segments. However, if the current is very small, that layer breaks down, and the friction between the brushes and the commutator is greatly increased. This increased friction contributes to rapid wear. For maximum brush life, a machine should be at least partially loaded all the time.

Winding Insulation

Other than the commutator, the most critical part of a dc motor's design is the insulation of its windings. If the insulation of the motor windings breaks down, the motor shorts out. The repair of a machine with shorted insulation is quite expensive, if it is even possible. To prevent the insulation of the machine windings from breaking down as a result of overheating, it is necessary to limit the temperature of the windings. This can be partially done by providing a cooling air circulation over them, but ultimately the maximum winding temperature limits the maximum power that can be supplied continuously by the machine.

Insulation rarely fails from immediate breakdown at some critical temperature. Instead, the increase in temperature produces a gradual degradation of the insulation, making it subject to failure due to another cause such as shock, vibration, or electrical stress. There is an old rule of thumb which says that the life expectancy of a motor with a given insulation is halved for each 10 percent rise in winding temperature. This rule still applies to some extent today.

To standardize the temperature limits of machine insulation, the National Electrical Manufacturers Association (NEMA) in the United States has defined a series of *insulation system classes*. Each insulation system class specifies the maximum temperature rise permissible for each type of insulation. There are four standard NEMA insulation classes for integral-horsepower dc motors: A, B, F, and H. Each class represents a higher permissible winding temperature than the one before it. For example, if the armature winding temperature rise above ambient temperature in one type of continuously operating dc motor is measured by thermometer, it must be limited to 70°C for class A, 100°C for class B, 130°C for class F, and 155°C for class H insulation.

These temperature specifications are set out in great detail in NEMA Standard MG1-1993, *Motors and Generators*. Similar standards have been defined by the International Electrotechnical Commission (IEC) and by various national standards organizations in other countries.

7.7 POWER FLOW AND LOSSES IN DC MACHINES

DC generators take in mechanical power and produce electric power, while dc motors take in electric power and produce mechanical power. In either case, not all the power input to the machine appears in useful form at the other end—there is *always* some loss associated with the process.

The efficiency of a dc machine is defined by the equation

$$\boxed{\eta = \frac{P_{\text{out}}}{P_{\text{in}}} \times 100\%} \tag{7–50}$$

The difference between the input power and the output power of a machine is the losses that occur inside it. Therefore,

$$\boxed{\eta = \frac{P_{\text{out}} - P_{\text{loss}}}{P_{\text{in}}} \times 100\%} \tag{7–51}$$

The Losses in DC Machines

The losses that occur in dc machines can be divided into five basic categories:

1. Electrical or copper losses (I^2R losses)
2. Brush losses
3. Core losses
4. Mechanical losses
5. Stray load losses

ELECTRICAL OR COPPER LOSSES. Copper losses are the losses that occur in the armature and field windings of the machine. The copper losses for the armature and field windings are given by

$$\text{Armature loss:} \qquad P_A = I_A^2 R_A \tag{7–52}$$

$$\text{Field loss:} \qquad P_F = I_F^2 R_F \tag{7–53}$$

where P_A = armature loss
P_F = field circuit loss

I_A = armature current
I_F = field current
R_A = armature resistance
R_F = field resistance

The resistance used in these calculations is usually the winding resistance at normal operating temperature.

BRUSH LOSSES. The brush drop loss is the power lost across the contact potential at the brushes of the machine. It is given by the equation

$$\boxed{P_{BD} = V_{BD} I_A} \tag{7–54}$$

where P_{BD} = brush drop loss
V_{BD} = brush voltage drop
I_A = armature current

The reason that the brush losses are calculated in this manner is that the voltage drop across a set of brushes is approximately constant over a large range of armature currents. Unless otherwise specified, the brush voltage drop is usually assumed to be about 2 V.

CORE LOSSES. The core losses are the hysteresis losses and eddy current losses occurring in the metal of the motor. These losses are described in Chapter 1. These losses vary as the square of the flux density (B^2) and, for the rotor, as the 1.5th power of the speed of rotation ($n^{1.5}$).

MECHANICAL LOSSES. The mechanical losses in a dc machine are the losses associated with mechanical effects. There are two basic types of mechanical losses: *friction* and *windage*. Friction losses are losses caused by the friction of the bearings in the machine, while windage losses are caused by the friction between the moving parts of the machine and the air inside the motor's casing. These losses vary as the cube of the speed of rotation of the machine.

STRAY LOSSES (OR MISCELLANEOUS LOSSES). Stray losses are losses that cannot be placed in one of the previous categories. No matter how carefully losses are accounted for, some always escape inclusion in one of the above categories. All such losses are lumped into stray losses. For most machines, stray losses are taken by convention to be 1 percent of full load.

The Power-Flow Diagram

One of the most convenient techniques for accounting for power losses in a machine is the *power-flow diagram*. A power-flow diagram for a dc generator is shown in Figure 7–39a. In this figure, mechanical power is input into the machine,

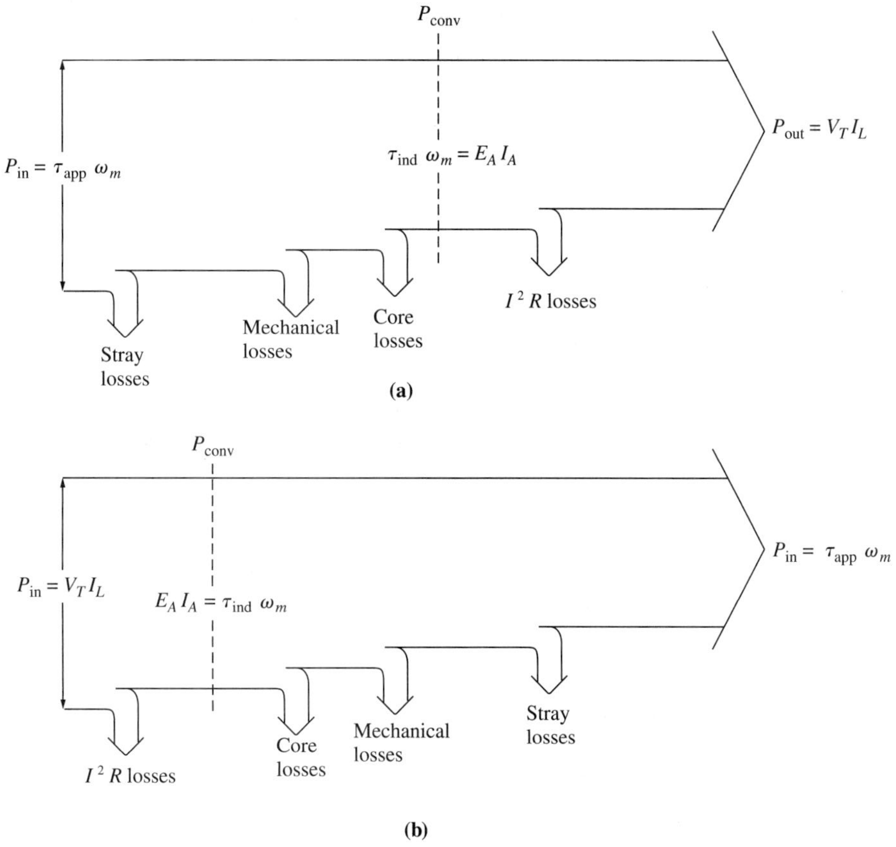

FIGURE 7–39
Power-flow diagrams for dc machine: (a) generator; (b) motor.

and then the stray losses, mechanical losses, and core losses are subtracted. After they have been subtracted, the remaining power is ideally converted from mechanical to electrical form at the point labeled P_{conv}. The mechanical power that is converted is given by

$$\boxed{P_{\text{conv}} = \tau_{\text{ind}}\omega_m} \tag{7–55}$$

and the resulting electric power produced is given by

$$\boxed{P_{\text{conv}} = E_A I_A} \tag{7–56}$$

However, this is not the power that appears at the machine's terminals. Before the terminals are reached, the electrical I^2R losses and the brush losses must be subtracted.

In the case of dc motors, this power-flow diagram is simply reversed. The power-flow diagram for a motor is shown in Figure 7–39b.

Example problems involving the calculation of motor and generator efficiencies will be given in Chapters 8 and 9.

7.8 SUMMARY

DC machines convert mechanical power to dc electric power, and vice versa. In this chapter, the basic principles of dc machine operation were explained first by looking at a simple linear machine and then by looking at a machine consisting of a single rotating loop.

The concept of commutation as a technique for converting the ac voltage in rotor conductors to a dc output was introduced, and its problems were explored. The possible winding arrangements of conductors in a dc rotor (lap and wave windings) were also examined.

Equations were then derived for the induced voltage and torque in a dc machine, and the physical construction of the machines was described. Finally, the types of losses in the dc machine were described and related to its overall operating efficiency.

QUESTIONS

7–1. What is commutation? How can a commutator convert ac voltages on a machine's armature to dc voltages at its terminals?

7–2. Why does curving the pole faces in a dc machine contribute to a smoother dc output voltage from it?

7–3. What is the pitch factor of a coil?

7–4. Explain the concept of electrical degrees. How is the electrical angle of the voltage in a rotor conductor related to the mechanical angle of the machine's shaft?

7–5. What is commutator pitch?

7–6. What is the plex of an armature winding?

7–7. How do lap windings differ from wave windings?

7–8. What are equalizers? Why are they needed on a lap-wound machine but not on a wave-wound machine?

7–9. What is armature reaction? How does it affect the operation of a dc machine?

7–10. Explain the $L\,di/dt$ voltage problem in conductors undergoing commutation.

7–11. How does brush shifting affect the sparking problem in dc machines?

7–12. What are commutating poles? How are they used?

7–13. What are compensating windings? What is their most serious disadvantage?

7–14. Why are laminated poles used in modern dc machine construction?

7–15. What is an insulation class?

7–16. What types of losses are present in a dc machine?

PROBLEMS

7–1. The following information is given about the simple rotating loop shown in Figure 7-6:

$$B = 0.4\ \text{T} \qquad V_B = 48\ \text{V}$$
$$l = 0.5\ \text{m} \qquad R = 0.4\ \Omega$$
$$r = 0.25\ \text{m} \qquad \omega = 500\ \text{rad/s}$$

(a) Is this machine operating as a motor or a generator? Explain.

(b) What is the current i flowing into or out of the machine? What is the power flowing into or out of the machine?

(c) If the speed of the rotor were changed to 550 rad/s, what would happen to the current flow into or out of the machine?

(d) If the speed of the rotor were changed to 450 rad/s, what would happen to the current flow into or out of the machine?

7–2. Refer to the simple two-pole, eight-coil machine shown in Figure P7–1. The following information is given about this machine:

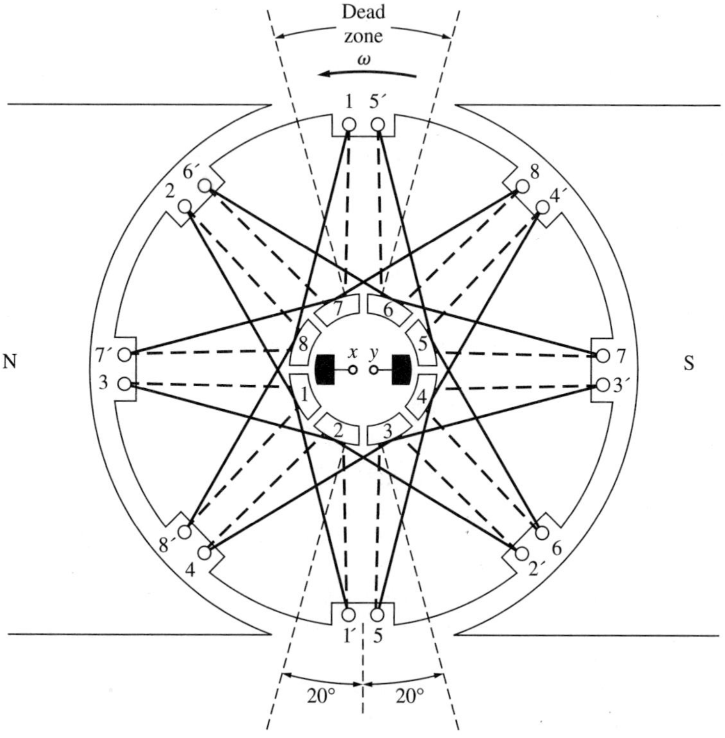

Given: **B** = 1.0 T in the air gap
l = 0.3 m (length of sides)
r = 0.10 m (radius of coils)
n = 1800 r/min

——— Lines on this side of rotor

– – – – Lines on other side of rotor

FIGURE P7–1
The machine in Problem 7–2.

$B = 1.0\text{ T}$ in air gap

$l = 0.3\text{ m}$ (length of coil sides)

$r = 0.10\text{ m}$ (radius of coils)

$n = 1800\text{ r/min}$ CCW

The resistance of each rotor coil is 0.04 Ω.

(a) Is the armature winding shown a progressive or retrogressive winding?

(b) How many current paths are there through the armature of this machine?

(c) What are the magnitude and the polarity of the voltage at the brushes in this machine?

(d) What is the armature resistance R_A of this machine?

(e) If a 5-Ω resistor is connected to the terminals of this machine, how much current flows in the machine? Consider the internal resistance of the machine in determining the current flow.

(f) What are the magnitude and the direction of the resulting induced torque?

(g) Assuming that the speed of rotation and magnetic flux density are constant, plot the terminal voltage of this machine as a function of the current drawn from it.

7–3. Prove that the equation for the induced voltage of a single simple rotating loop

$$e_{\text{ind}} = \frac{2}{\pi}\phi\omega_m \tag{7–6}$$

is just a special case of the general equation for induced voltage in a dc machine

$$E_A = K\phi\omega_m \tag{7–38}$$

7–4. A dc machine has eight poles and a rated current of 120 A. How much current will flow in each path at rated conditions if the armature is (*a*) simplex lap-wound, (*b*) duplex lap-wound, (*c*) simplex wave-wound?

7–5. How many parallel current paths will there be in the armature of a 20-pole machine if the armature is (*a*) simplex lap-wound, (*b*) duplex wave-wound, (*c*) triplex lap-wound, (*d*) quadruplex wave-wound?

7–6. The power converted from one form to another within a dc motor was given by

$$P_{\text{conv}} = E_A I_A = \tau_{\text{ind}}\omega_m$$

Use the equations for E_A and τ_{ind} [Equations (7–38) and (7–49)] to prove that $E_A I_A = \tau_{\text{ind}}\omega_m$; that is, prove that the electric power disappearing at the point of power conversion is exactly equal to the mechanical power appearing at that point.

7–7. An eight-pole, 25-kW, 120-V dc generator has a duplex lap-wound armature which has 64 coils with 10 turns per coil. Its rated speed is 3600 r/min.

(a) How much flux per pole is required to produce the rated voltage in this generator at no-load conditions?

(b) What is the current per path in the armature of this generator at the rated load?

(c) What is the induced torque in this machine at the rated load?

(d) How many brushes must this motor have? How wide must each one be?

(e) If the resistance of this winding is 0.011 Ω per turn, what is the armature resistance R_A of this machine?

7–8. Figure P7–2 shows a small two-pole dc motor with eight rotor coils and 10 turns per coil. The flux per pole in this machine is 0.006 Wb.

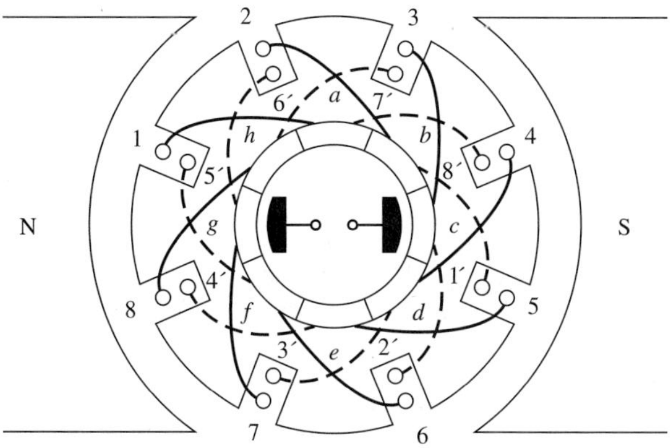

FIGURE P7–2
The machine in Problem 7–8.

(a) If this motor is connected to a 12-V dc car battery, what will the no-load speed of the motor be?

(b) If the positive terminal of the battery is connected to the rightmost brush on the motor, which way will it rotate?

(c) If this motor is loaded down so that it consumes 600 W from the battery, what will the induced torque of the motor be? (Ignore any internal resistance in the motor.)

7–9. Refer to the machine winding shown in Figure P7–3.

(a) How many parallel current paths are there through this armature winding?

(b) Where should the brushes be located on this machine for proper commutation? How wide should they be?

(c) What is the plex of this machine?

(d) If the voltage on any single conductor under the pole faces in this machine is *e*, what is the voltage at the terminals of this machine?

7–10. Describe in detail the winding of the machine shown in Figure P7–4. If a positive voltage is applied to the brush under the north pole face, which way will this motor rotate?

REFERENCES

1. Del Toro, V.: *Electric Machines and Power Systems,* Prentice Hall, Englewood Cliffs, N.J., 1985.
2. Fitzgerald, A. E., C. Kingsley, Jr., and S. D. Umans: *Electric Machinery,* 6th ed., McGraw-Hill, New York, 2003.
3. Hubert, Charles I.: *Preventative Maintenance of Electrical Equipment,* 2d ed., McGraw-Hill, New York, 1969.
4. Kosow, Irving L.: *Electric Machinery and Transformers,* Prentice Hall, Englewood Cliffs, N.J., 1972.
5. National Electrical Manufacturers Association: *Motors and Generators*, Publication No. MG1-2006, NEMA, Washington, D.C., 2006.
6. Siskind, Charles: *Direct Current Machinery,* McGraw-Hill, New York, 1952.
7. Werninck, E. H. (ed.): *Electric Motor Handbook,* McGraw-Hill, London, 1978.

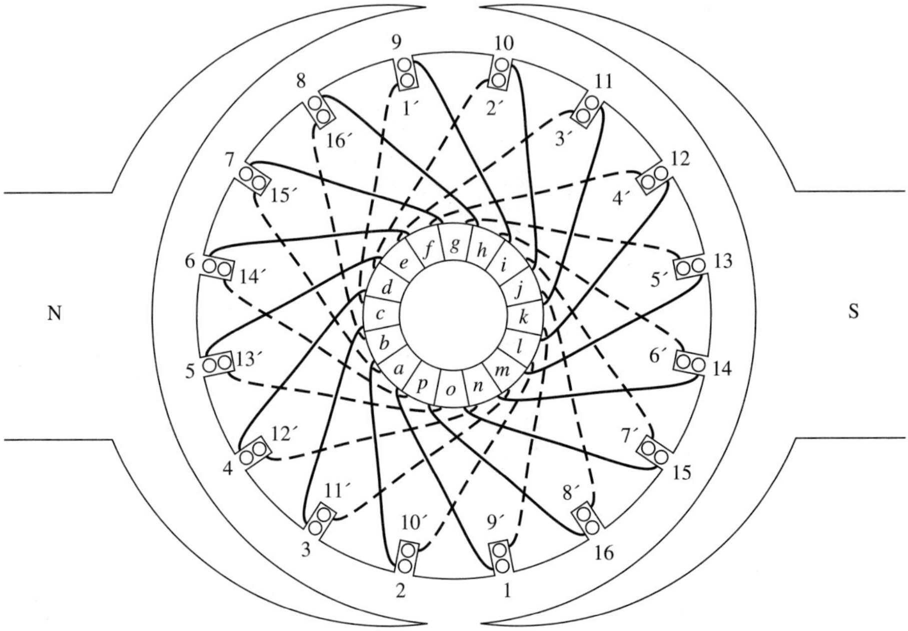

(a)

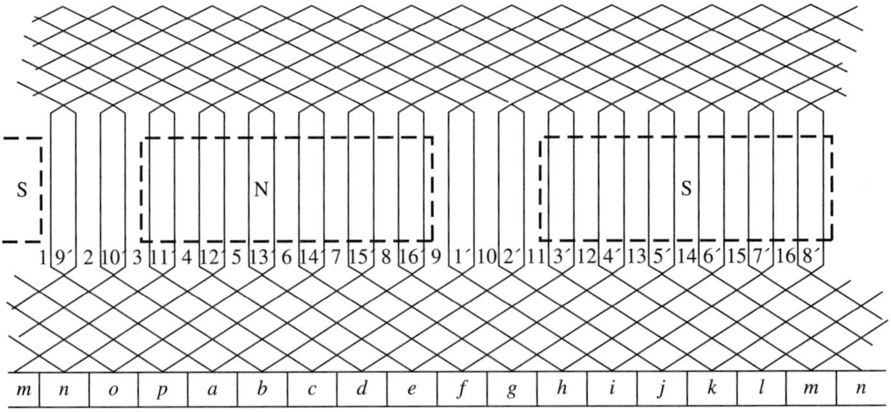

(b)

FIGURE P7–3
(a) The machine in Problem 7–9. (b) The armature winding diagram of this machine.

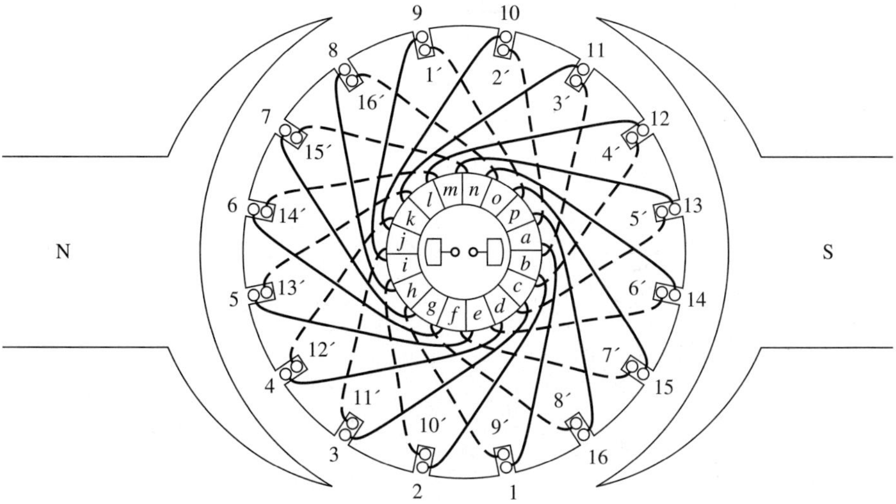

FIGURE P7–4
The machine in Problem 7–10.

CHAPTER 8

DC MOTORS AND GENERATORS

LEARNING OBJECTIVES

- Know the types of dc motors in general use.
- Understand the equivalent circuit of a dc motor.
- Understand how to derive the torque–speed characteristics of separately excited, shunt, series, and compounded dc motors.
- Be able to perform nonlinear analysis of dc motors using the magnetization curve, taking into account armature reaction effects.
- Understand how to control the speed of different types of dc motors.
- Understand the special characteristics of series dc motors, and the applications that they are especially suited for.
- Be able to explain the problems associated with a differentially compounded dc motor.
- Understand the methods of starting dc motors safely.
- Understand the equivalent circuit of a dc generator.
- Understand how a dc generator can start without an external voltage source.
- Understand how to derive the voltage–current characteristics of separately excited, shunt, series, and compounded dc generators.
- Be able to perform nonlinear analysis of dc generators using the magnetization curve, taking into account armature reaction effects.

Dc motors are dc machines used as motors, and dc generators are dc machines used as generators. As noted in Chapter 7, the same physical machine can operate as either a motor or a generator—it is simply a question of the direction of the power flow through it. This chapter will examine the different types of dc motors that can be made and explain the advantages and disadvantages of each. It will include a discussion of dc motor starting and solid-state controls. Finally, the chapter will conclude with a discussion of dc generators.

8.1 INTRODUCTION TO DC MOTORS

The earliest power systems in the United States were dc systems (see Figure 8–1), but by the 1890s ac power systems were clearly winning out over dc systems. Despite this fact, dc motors continued to be a significant fraction of the machinery purchased each year through the 1960s (that fraction has declined in the last 40 years). Why were dc motors so common, when dc power systems themselves were fairly rare?

There were several reasons for the continued popularity of dc motors. One was that dc power systems are still common in cars, trucks, and aircraft. When a vehicle has a dc power system, it makes sense to consider using dc motors. Another application for dc motors was a situation in which wide variations in speed are needed. Before the widespread use of power electronic rectifier-inverters, dc motors were unexcelled in speed control applications. Even if no dc power source were available, solid-state rectifier and chopper circuits were used to create the necessary dc power, and dc motors were used to provide the desired speed control. (Today, induction motors with solid-state drive packages are the preferred choice over dc motors for most speed control applications. However, there are still some applications where dc motors are preferred.)

DC motors are often compared by their speed regulations. The *speed regulation* (SR) of a motor is defined by

$$\boxed{SR = \frac{\omega_{m,nl} - \omega_{m,fl}}{\omega_{m,fl}} \times 100\%} \tag{8–1}$$

$$\boxed{SR = \frac{n_{m,nl} - n_{m,fl}}{n_{m,fl}} \times 100\%} \tag{8–2}$$

It is a rough measure of the shape of a motor's torque–speed characteristic—a positive speed regulation means that a motor's speed drops with increasing load, and a negative speed regulation means a motor's speed increases with increasing load. The magnitude of the speed regulation tells approximately how steep the slope of the torque–speed curve is.

DC motors are, of course, driven from a dc power supply. Unless otherwise specified, *the input voltage to a dc motor is assumed to be constant*, because that assumption simplifies the analysis of motors and the comparison between different types of motors.

(a)

(b)

FIGURE 8–1
Early dc motors. (a) A very early dc motor built by Elihu Thompson in 1886. It was rated at about ½ hp. (*Courtesy of General Electric Company.*) (b) A larger four-pole dc motor from about the turn of the century. Notice the handle for shifting the brushes to the neutral plane. (*Courtesy of General Electric Company.*)

There are five major types of dc motors in general use:

1. The separately excited dc motor
2. The shunt dc motor
3. The permanent-magnet dc motor
4. The series dc motor
5. The compounded dc motor

Each of these types will be examined in turn.

8.2 THE EQUIVALENT CIRCUIT OF A DC MOTOR

The equivalent circuit of a dc motor is shown in Figure 8–2. In this figure, the armature circuit is represented by an ideal voltage source E_A and a resistor R_A. This representation is really the Thevenin equivalent of the entire rotor structure, including rotor coils, interpoles, and compensating windings, if present. The brush voltage drop is represented by a small battery V_{brush} opposing the direction of current flow in the machine. The field coils, which produce the magnetic flux in the generator, are represented by inductor L_F and resistor R_F. The separate resistor R_{adj} represents an external variable resistor used to control the amount of current in the field circuit.

There are a few variations and simplifications of this basic equivalent circuit. The brush drop voltage is often only a very tiny fraction of the generated voltage in a machine. Therefore, in cases where it is not too critical, the brush drop voltage may be left out or approximately included in the value of R_A. Also, the internal resistance of the field coils is sometimes lumped together with the variable resistor, and the total is called R_F (see Figure 8–2b). A third variation is that some generators have more than one field coil, all of which will appear on the equivalent circuit.

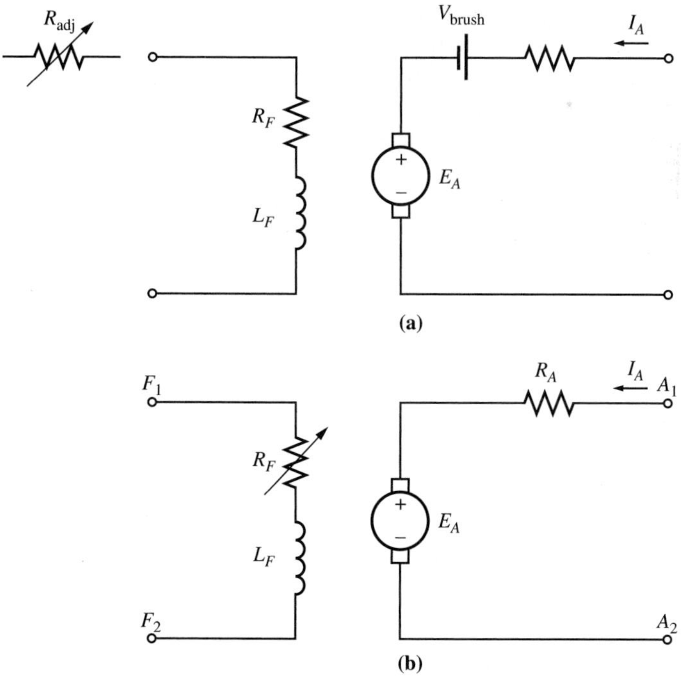

FIGURE 8–2
(a) The equivalent circuit of a dc motor. (b) A simplified equivalent circuit eliminating the brush voltage drop and combining R_{adj} with the field resistance.

The internal generated voltage in this machine is given by the equation

$$\boxed{E_A = K\phi\omega_m} \tag{7–38}$$

and the induced torque developed by the machine is given by

$$\boxed{\tau_{\text{ind}} = K\phi I_A} \tag{7–49}$$

These two equations, the Kirchhoff's voltage law equation of the armature circuit and the machine's magnetization curve, are all the tools necessary to analyze the behavior and performance of a dc motor.

8.3 THE MAGNETIZATION CURVE OF A DC MACHINE

The internal generated voltage E_A of a dc motor or generator is given by Equation (7–38):

$$\boxed{E_A = K\phi\omega_m} \tag{7–38}$$

Therefore, E_A is directly proportional to the flux in the machine and the speed of rotation of the machine. How is the internal generated voltage related to the field current in the machine?

The field current in a dc machine produces a field magnetomotive force given by $\mathscr{F} = N_F I_F$. This magnetomotive force produces a flux in the machine in accordance with its magnetization curve (Figure 8–3). Since the field current is directly proportional to the magnetomotive force and since E_A is directly proportional to the flux, it is customary to present the magnetization curve as a plot of E_A versus field current for a given speed ω_0 (Figure 8–4).

It is worth noting here that, to get the maximum possible power per pound of weight out of a machine, most motors and generators are designed to operate

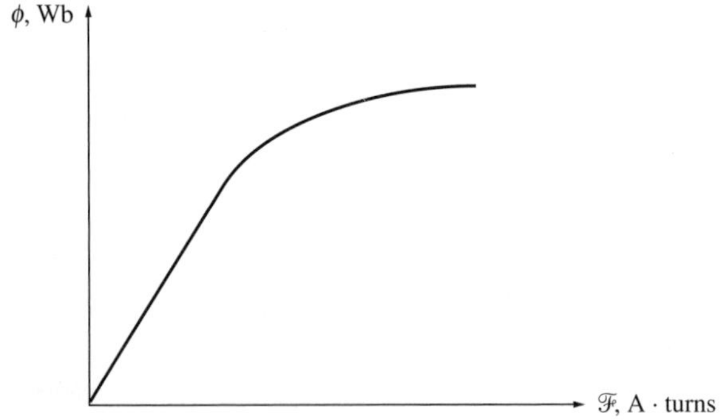

FIGURE 8–3
The magnetization curve of a ferromagnetic material (ϕ versus $\mathscr{F}$).

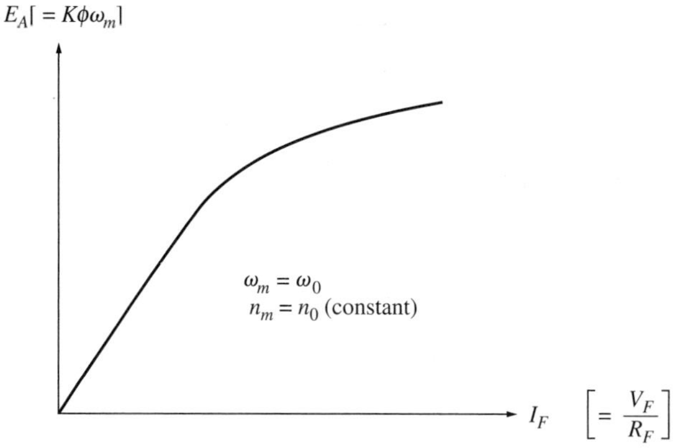

FIGURE 8–4
The magnetization curve of a dc machine expressed as a plot of E_A versus I_F, for a fixed speed ω_o.

near the saturation point on the magnetization curve (at the knee of the curve). This implies that a fairly large increase in field current is often necessary to get a small increase in E_A when operation is near full load.

The dc machine magnetization curves used in this book are also available in electronic form to simplify the solution of problems by MATLAB. Each magnetization curve is stored in a separate MAT file. Each MAT file contains three variables: `if_values`, containing the values of the field current; `ea_values`, containing the corresponding values of E_A; and `n_0`, containing the speed at which the magnetization curve was measured in units of revolutions per minute.

8.4 SEPARATELY EXCITED AND SHUNT DC MOTORS

The equivalent circuit of a separately excited dc motor is shown in Figure 8–5a, and the equivalent circuit of a shunt dc motor is shown in Figure 8–5b. A separately excited dc motor is a motor whose field circuit is supplied from a separate constant-voltage power supply, while a shunt dc motor is a motor whose field circuit gets its power directly across the armature terminals of the motor. When the supply voltage to a motor is assumed constant, there is no practical difference in behavior between these two machines. Unless otherwise specified, whenever the behavior of a shunt motor is described, the separately excited motor is included, too.

The Kirchhoff's voltage law (KVL) equation for the armature circuit of these motors is

$$\boxed{V_T = E_A + I_A R_A} \tag{8–3}$$

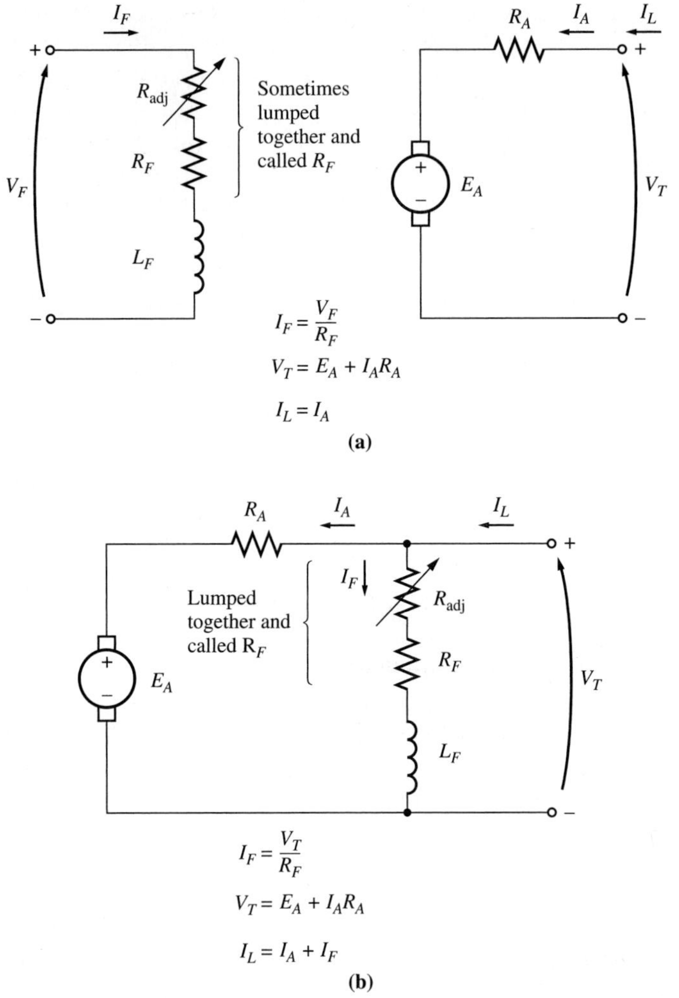

FIGURE 8–5
(a) The equivalent circuit of a separately excited dc motor. (b) The equivalent circuit of a shunt dc motor.

The Terminal Characteristic of a Shunt DC Motor

A terminal characteristic of a machine is a plot of the machine's output quantities versus each other. For a motor, the output quantities are shaft torque and speed, so the terminal characteristic of a motor is a plot of its output *torque versus speed.*

How does a shunt dc motor respond to a load? Suppose that the load on the shaft of a shunt motor is increased. Then the load torque τ_{load} will exceed the induced torque τ_{ind} in the machine, and the motor will start to slow down. When the motor slows down, its internal generated voltage drops ($E_A = K\phi\omega_m\downarrow$), so the armature current in the motor $I_A = (V_T - E_A\downarrow)/R_A$ increases. As the armature current rises, the induced torque in the motor increases ($\tau_{\text{ind}} = K\phi I_A\uparrow$), and finally the induced torque will equal the load torque at a lower mechanical speed of rotation ω_m.

The output characteristic of a shunt dc motor can be derived from the induced voltage and torque equations of the motor plus Kirchhoff's voltage law (KVL). The KVL equation for a shunt motor is

$$V_T = E_A + I_A R_A \tag{8–3}$$

The induced voltage $E_A = K\phi\omega_m$, so

$$V_T = K\phi\omega_m + I_A R_A \tag{8–4}$$

Since $\tau_{\text{ind}} = K\phi I_A$, current I_A can be expressed as

$$I_A = \frac{\tau_{\text{ind}}}{K\phi} \tag{8–5}$$

Combining Equations (8–4) and (8–5) produces

$$V_T = K\phi\omega_m + \frac{\tau_{\text{ind}}}{K\phi} R_A \tag{8–6}$$

Finally, solving for the motor's speed yields

$$\boxed{\omega_m = \frac{V_T}{K\phi} - \frac{R_A}{(K\phi)^2}\tau_{\text{ind}}} \tag{8–7}$$

This equation is just a straight line with a negative slope. The resulting torque–speed characteristic of a shunt dc motor is shown in Figure 8–6a.

It is important to realize that, in order for the speed of the motor to vary linearly with torque, the other terms in this expression must be constant as the load changes. The terminal voltage supplied by the dc power source is assumed to be constant—if it is not constant, then the voltage variations will affect the shape of the torque–speed curve.

Another effect *internal to the motor* that can also affect the shape of the torque–speed curve is armature reaction. If a motor has armature reaction, then as its load increases, the flux-weakening effects *reduce* its flux. As Equation (8–7) shows, the effect of a reduction in flux is to increase the motor's speed at any given load over the speed it would run at without armature reaction. The torque–speed characteristic of a shunt motor with armature reaction is shown in

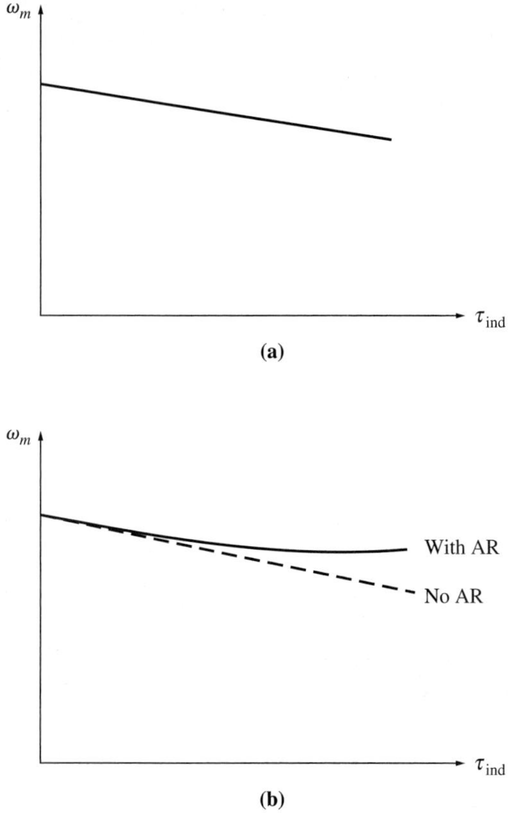

FIGURE 8–6
(a) Torque–speed characteristic of a shunt or separately excited dc motor with compensating windings to eliminate armature reaction. (b) Torque–speed characteristic of the motor with armature reaction present.

Figure 8–6b. If a motor has compensating windings, of course there will be no flux-weakening problems in the machine, and the flux in the machine will be constant.

If a shunt dc motor has compensating windings so that *its flux is constant regardless of load*, and the motor's speed and armature current are known at any one value of load, then it is possible to calculate its speed at any other value of load, as long as the armature current at that load is known or can be determined. Example 8–1 illustrates this calculation.

Example 8–1. A 50-hp, 250-V, 1200 r/min dc shunt motor with compensating windings has an armature resistance (including the brushes, compensating windings, and

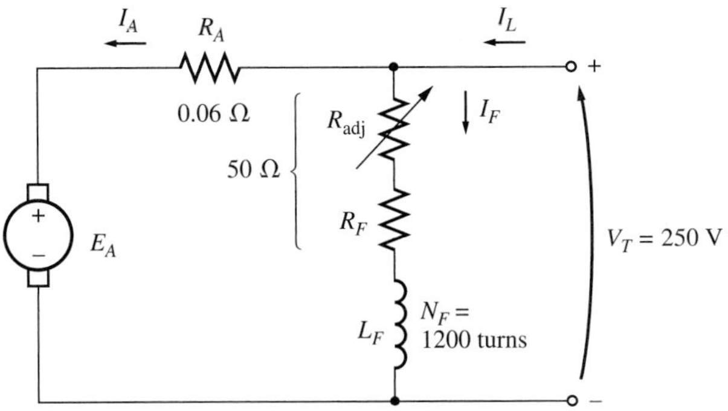

FIGURE 8–7
The shunt motor in Example 8–1.

interpoles) of 0.06 Ω. Its field circuit has a total resistance $R_{adj} + R_F$ of 50 Ω, which produces a *no-load* speed of 1200 r/min. There are 1200 turns per pole on the shunt field winding (see Figure 8–7).

(a) Find the speed of this motor when its input current is 100 A.
(b) Find the speed of this motor when its input current is 200 A.
(c) Find the speed of this motor when its input current is 300 A.
(d) Plot the torque–speed characteristic of this motor.

Solution

The internal generated voltage of a dc machine with its speed expressed in revolutions per minute is given by

$$E_A = K'\phi n_m \tag{7–41}$$

Since the field current in the machine is constant (because V_T and the field resistance are both constant), and since there are no armature reaction effects, *the flux in this motor is constant.* The relationship between the speeds and internal generated voltages of the motor at two different load conditions is thus

$$\frac{E_{A2}}{E_{A1}} = \frac{K'\phi n_{m2}}{K'\phi n_{m1}} \tag{8–8}$$

The constant K' cancels, since it is a constant for any given machine, and the flux ϕ cancels as described above. Therefore,

$$n_{m2} = \frac{E_{A2}}{E_{A1}} n_{m1} \tag{8–9}$$

At no load, the armature current is zero, so $E_{A1} = V_T = 250$ V, while the speed $n_{m1} =$ 1200 r/min. If we can calculate the internal generated voltage at any other load, it will be possible to determine the motor speed at that load from Equation (8–9).

(a) If $I_L = 100$ A, then the armature current in the motor is

$$I_A = I_L - I_F = I_L - \frac{V_T}{R_F}$$

$$= 100 \text{ A} - \frac{250 \text{ V}}{50 \ \Omega} = 95 \text{ A}$$

Therefore, E_A at this load will be

$$E_A = V_T - I_A R_A$$

$$= 250 \text{ V} - (95 \text{ A})(0.06 \ \Omega) = 244.3 \text{ V}$$

The resulting speed of the motor is

$$n_{m2} = \frac{E_{A2}}{E_{A1}} n_{m1} = \frac{244.3 \text{ V}}{250 \text{ V}} 1200 \text{ r/min} = 1173 \text{ r/min}$$

(b) If $I_L = 200$ A, then the armature current in the motor is

$$I_A = 200 \text{ A} - \frac{250 \text{ V}}{50 \ \Omega} = 195 \text{ A}$$

Therefore, E_A at this load will be

$$E_A = V_T - I_A R_A$$

$$= 250 \text{ V} - (195 \text{ A})(0.06 \ \Omega) = 238.3 \text{ V}$$

The resulting speed of the motor is

$$n_{m2} = \frac{E_{A2}}{E_{A1}} n_{m1} = \frac{238.3 \text{ V}}{250 \text{ V}} 1200 \text{ r/min} = 1144 \text{ r/min}$$

(c) If $I_L = 300$ A, then the armature current in the motor is

$$I_A = I_L - I_F = I_L - \frac{V_T}{R_F}$$

$$= 300 \text{ A} - \frac{250 \text{ V}}{50 \ \Omega} = 295 \text{ A}$$

Therefore, E_A at this load will be

$$E_A = V_T - I_A R_A$$

$$= 250 \text{ V} - (295 \text{ A})(0.06 \ \Omega) = 232.3 \text{ V}$$

The resulting speed of the motor is

$$n_{m2} = \frac{E_{A2}}{E_{A1}} n_{m1} = \frac{232.3 \text{ V}}{250 \text{ V}} 1200 \text{ r/min} = 1115 \text{ r/min}$$

(d) To plot the output characteristic of this motor, it is necessary to find the torque corresponding to each value of speed. At no load, the induced torque τ_{ind} is clearly zero. The induced torque for any other load can be found from the fact that power converted in a dc motor is

$$\boxed{P_{\text{conv}} = E_A I_A = \tau_{\text{ind}} \omega_m} \qquad (7\text{–}55, 7\text{–}56)$$

From this equation, the induced torque in a motor is

$$\tau_{\text{ind}} = \frac{E_A I_A}{\omega_m} \tag{8–10}$$

Therefore, the induced torque when $I_L = 100$ A is

$$\tau_{\text{ind}} = \frac{(244.3 \text{ V})(95 \text{ A})}{(1173 \text{ r/min})(1 \text{ min/60s})(2\pi \text{ rad/r})} = 190 \text{ N} \cdot \text{m}$$

The induced torque when $I_L = 200$ A is

$$\tau_{\text{ind}} = \frac{(238.3 \text{ V})(95 \text{ A})}{(1144 \text{ r/min})(1 \text{ min/60s})(2\pi \text{ rad/r})} = 388 \text{ N} \cdot \text{m}$$

The induced torque when $I_L = 300$ A is

$$\tau_{\text{ind}} = \frac{(232.3 \text{ V})(295 \text{ A})}{(1115 \text{ r/min})(1 \text{ min/60s})(2\pi \text{ rad/r})} = 587 \text{ N} \cdot \text{m}$$

The resulting torque–speed characteristic for this motor is plotted in Figure 8–8.

Nonlinear Analysis of a Shunt DC Motor

The flux ϕ and hence the internal generated voltage E_A of a dc machine is a *nonlinear* function of its magnetomotive force. Therefore, anything that changes the

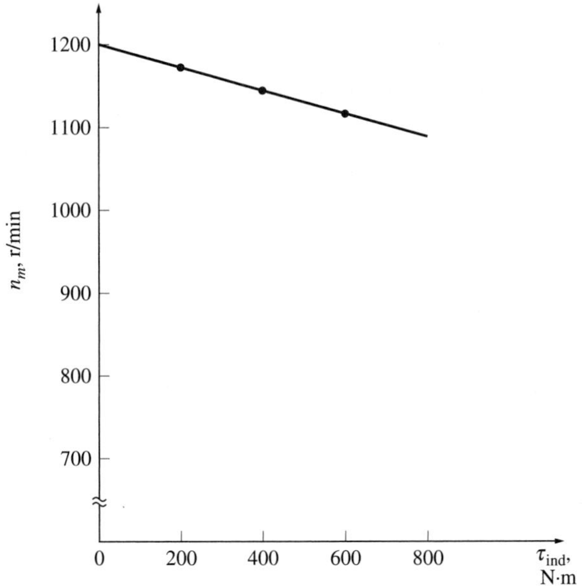

FIGURE 8–8
The torque–speed characteristic of the motor in Example 8–1.

magnetomotive force in a machine will have a nonlinear effect on the internal generated voltage of the machine. Since the change in E_A cannot be calculated analytically, the magnetization curve of the machine must be used to accurately determine its E_A for a given magnetomotive force. The two principal contributors to the magnetomotive force in the machine are its field current and its armature reaction, if present.

Since the magnetization curve is a direct plot of E_A versus I_F for a given speed ω_o, the effect of changing a machine's field current can be determined directly from its magnetization curve.

If a machine has armature reaction, its flux will be reduced with each increase in load. The total magnetomotive force in a shunt dc motor is the field circuit magnetomotive force less the magnetomotive force due to armature reaction (AR):

$$\mathcal{F}_{\text{net}} = N_F I_F - \mathcal{F}_{\text{AR}} \tag{8–11}$$

Since magnetization curves are expressed as plots of E_A versus field current, it is customary to define an *equivalent field current* that would produce the same output voltage as the combination of all the magnetomotive forces in the machine. The resulting voltage E_A can then be determined by locating that equivalent field current on the magnetization curve. The equivalent field current of a shunt dc motor is given by

$$I_F^* = I_F - \frac{\mathcal{F}_{\text{AR}}}{N_F} \tag{8–12}$$

One other effect must be considered when nonlinear analysis is used to determine the internal generated voltage of a dc motor. The magnetization curves for a machine are drawn for a particular speed, usually the rated speed of the machine. How can the effects of a given field current be determined if the motor is turning at other than rated speed?

The equation for the induced voltage in a dc machine when speed is expressed in revolutions per minute is

$$E_A = K'\phi n_m \tag{7–41}$$

For a given effective field current, the flux in a machine is fixed, so the internal generated voltage is related to speed by

$$\boxed{\frac{E_A}{E_{A0}} = \frac{n_m}{n_0}} \tag{8–13}$$

where E_{A0} and n_0 represent the reference values of voltage and speed, respectively. If the reference conditions are known from the magnetization curve and the actual E_A is known from Kirchhoff's voltage law, then it is possible to determine the actual speed n from Equation (8–13). The use of the magnetization curve and Equations (8–12) and (8–13) is illustrated in the following example, which analyzes a dc motor with armature reaction.

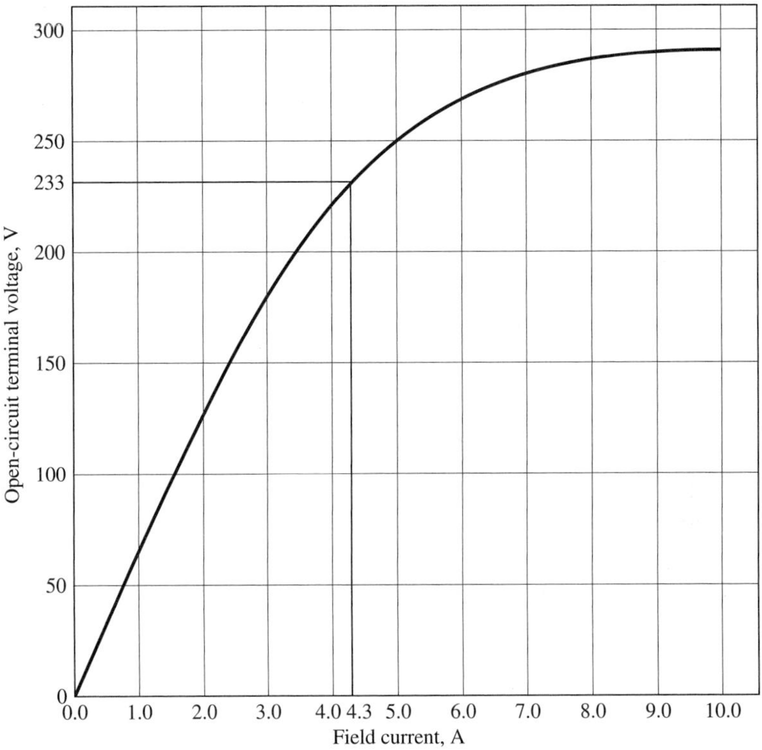

FIGURE 8–9
The magnetization curve of a typical 250-V dc motor, taken at a speed of 1200 r/min.

Example 8–2. A 50-hp, 250-V, 1200 r/min dc shunt motor *without* compensating windings has an armature resistance (including the brushes and interpoles) of 0.06 Ω. Its field circuit has a total resistance $R_F + R_{adj}$ of 50 Ω, which produces a no-load speed of 1200 r/min. There are 1200 turns per pole on the shunt field winding, and the armature reaction produces a demagnetizing magnetomotive force of 840 A • turns at a load current of 200 A. The magnetization curve of this machine is shown in Figure 8–9.

(*a*) Find the speed of this motor when its input current is 200 A.
(*b*) This motor is essentially identical to the one in Example 8–1 except for the absence of compensating windings. How does its speed compare to that of the previous motor at a load current of 200 A?
(*c*) Calculate and plot the torque–speed characteristic for this motor.

Solution

(*a*) If $I_L = 200$ A, then the armature current of the motor is

$$I_A = I_L - I_F = I_L - \frac{V_T}{R_F}$$

$$= 200 \text{ A} - \frac{250 \text{ V}}{50 \ \Omega} = 195 \text{ A}$$

Therefore, the internal generated voltage of the machine is

$$\begin{aligned} E_A &= V_T - I_A R_A \\ &= 250 \text{ V} - (195 \text{ A})(0.06\ \Omega) = 238.3 \text{ V} \end{aligned}$$

At $I_L = 200$ A, the demagnetizing magnetomotive force due to armature reaction is 840 A • turns, so the effective shunt field current of the motor is

$$I_F^* = I_F - \frac{\mathscr{F}_{\text{AR}}}{N_F} \tag{8–12}$$

$$= 5.0 \text{ A} - \frac{840 \text{ A} \bullet \text{turns}}{1200 \text{ turns}} = 4.3 \text{ A}$$

From the magnetization curve, this effective field current would produce an internal generated voltage E_{A0} of 233 V at a speed n_0 of 1200 r/min.

We know that the internal generated voltage E_{A0} would be 233 V at a speed of 1200 r/min. Since the actual internal generated voltage E_A is 238.3 V, the actual operating speed of the motor must be

$$\frac{E_A}{E_{A0}} = \frac{n_m}{n_0} \tag{8–13}$$

$$n_m = \frac{E_A}{E_{A0}} n_0 = \frac{238.3 \text{ V}}{233 \text{ V}} (1200 \text{ r/min}) = 1227 \text{ r/min}$$

(b) At 200 A of load in Example 8–1, the motor's speed was $n_m = 1144$ r/min. In this example, the motor's speed is 1227 r/min. *Notice that the speed of the motor with armature reaction is higher than the speed of the motor with no armature reaction.* This relative increase in speed is due to the flux weakening in the machine with armature reaction.

(c) To derive the torque–speed characteristic of this motor, we must calculate the torque and speed for many different conditions of load. Unfortunately, the demagnetizing armature reaction magnetomotive force is only given for one condition of load (200 A). Since no additional information is available, we will assume that the strength of $\mathscr{F}_{\text{AR}}$ varies linearly with load current.

A MATLAB M-file which automates this calculation and plots the resulting torque–speed characteristic is shown below. It performs the same steps as part *a* to determine the speed for each load current, and then calculates the induced torque at that speed. Note that it reads the magnetization curve from a file called `fig8_9.mat`. This file and the other magnetization curves in this chapter are available for download from the book's World Wide Web site (see Preface for details).

```
% M-file: shunt_ts_curve.m
% M-file create a plot of the torque-speed curve of the
%   the shunt dc motor with armature reaction in
%   Example 8-2.

% Get the magnetization curve.  This file contains the
% three variables if_value, ea_value, and n_0.
load fig8_9.mat.

% First, initialize the values needed in this program.
v_t = 250;                % Terminal voltage (V)
```

```
r_f = 50;                   % Field resistance (ohms)
r_a = 0.06;                 % Armature resistance (ohms)
i_l = 10:10:300;            % Line currents (A)
n_f = 1200;                 % Number of turns on field
f_ar0 = 840;                % Armature reaction @ 200 A (A-t/m)

% Calculate the armature current for each load.
i_a = i_l - v_t / r_f;

% Now calculate the internal generated voltage for
% each armature current.
e_a = v_t - i_a * r_a;

% Calculate the armature reaction MMF for each armature
% current.
f_ar = (i_a / 200) * f_ar0;

% Calculate the effective field current.
i_f = v_t / r_f - f_ar / n_f;

% Calculate the resulting internal generated voltage at
% 1200 r/min by interpolating the motor's magnetization
% curve.
e_a0 = interp1(if_values,ea_values,i_f,'spline');

% Calculate the resulting speed from Equation (8-13).
n = ( e_a ./ e_a0 ) * n_0;

% Calculate the induced torque corresponding to each
% speed from Equations (7-55) and (7-56).
t_ind = e_a .* i_a ./ (n * 2 * pi / 60);

% Plot the torque-speed curve
plot(t_ind,n,'k-','LineWidth',2.0);
hold on;
xlabel('\bf\tau_{ind} (N-m)');
ylabel('\bf\itn_{m} (r/min)');
title ('\bfShunt DC motor torque-speed characteristic');
axis([ 0 600 1100 1300]);
grid on;
hold off;
```

The resulting torque–speed characteristic is shown in Figure 8–10. Note that for any given load, the speed of the motor with armature reaction is higher than the speed of the motor without armature reaction.

Speed Control of Shunt DC Motors

How can the speed of a shunt dc motor be controlled? There are two common methods and one less common method in use. The common methods have already been seen in the simple linear machine in Chapter 1 and the simple rotating loop in Chapter 7. The two common ways in which the speed of a shunt dc machine can be controlled are by

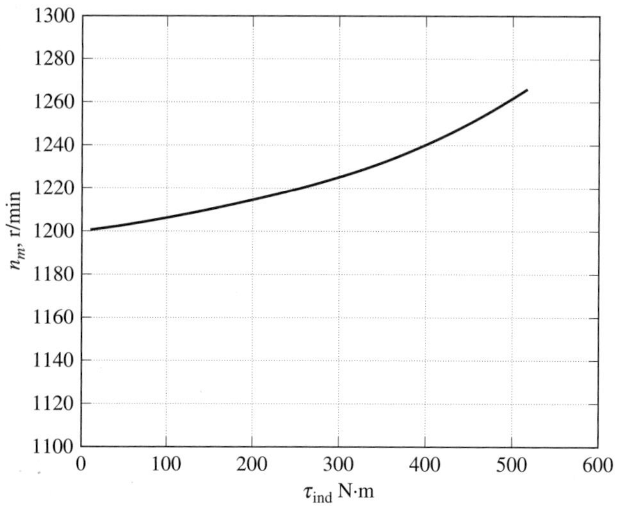

FIGURE 8–10
The torque–speed characteristic of the motor with armature reaction in Example 8–2.

1. Adjusting the field resistance R_F (and thus the field flux)
2. Adjusting the terminal voltage applied to the armature.

The less common method of speed control is by

3. Inserting a resistor in series with the armature circuit.

Each of these methods is described in detail below.

CHANGING THE FIELD RESISTANCE. To understand what happens when the field resistor of a dc motor is changed, assume that the field resistor increases and observe the response. If the field resistance increases, then the field current decreases ($I_F = V_T / R_F \uparrow$), and as the field current decreases, the flux ϕ decreases with it. A decrease in flux causes an instantaneous decrease in the internal generated voltage $E_A (= K\phi\downarrow\omega_m)$, which causes a large increase in the machine's armature current, since

$$I_A\uparrow = \frac{V_T - E_A\downarrow}{R_A}$$

The induced torque in a motor is given by $\tau_{\text{ind}} = K\phi I_A$. Since the flux ϕ in this machine decreases while the current I_A increases, which way does the induced torque change? The easiest way to answer this question is to look at an example. Figure 8–11 shows a shunt dc motor with an internal resistance of 0.25 Ω. It is currently operating with a terminal voltage of 250 V and an internal generated voltage of 245 V. Therefore, the armature current flow is $I_A = (250\text{ V} -$

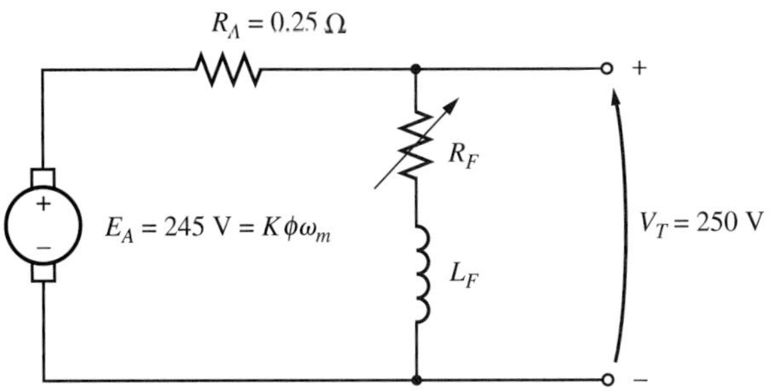

FIGURE 8–11
A 250-V shunt dc motor with typical values of E_A and R_A.

245 V)/0.25 Ω = 20 A. What happens in this motor *if there is a 1 percent decrease in flux?* If the flux decreases by 1 percent, then E_A must decrease by 1 percent too, because $E_A = K\phi\omega_m$. Therefore, E_A will drop to

$$E_{A2} = 0.99\ E_{A1} = 0.99(245\ \text{V}) = 242.55\ \text{V}$$

The armature current must then rise to

$$I_A = \frac{250\ \text{V} - 242.55\ \text{V}}{0.25\ \Omega} = 29.8\ \text{A}$$

Thus a 1 percent decrease in flux produced a 49 percent increase in armature current.

So to get back to the original discussion, the increase in current predominates over the decrease in flux, and the induced torque rises:

$$\tau_{\text{ind}} = K\overset{\downarrow}{\phi}\overset{\Uparrow}{I_A}$$

Since $\tau_{\text{ind}} > \tau_{\text{load}}$, the motor speeds up.

However, as the motor speeds up, the internal generated voltage E_A rises, causing I_A to fall. As I_A falls, the induced torque τ_{ind} falls too, and finally τ_{ind} again equals τ_{load} at a higher steady-state speed than originally.

To summarize the cause-and-effect behavior involved in this method of speed control:

1. Increasing R_F causes $I_F (= V_T / R_F \uparrow)$ to decrease.
2. Decreasing I_F decreases ϕ.
3. Decreasing ϕ lowers E_A $(= K\phi\downarrow\omega_m)$.
4. Decreasing E_A increases $I_A (= V_T - E_A \downarrow)/R_A$.
5. Increasing I_A increases $\tau_{\text{ind}} (= K\phi\downarrow I_A \Uparrow)$, with the change in I_A dominant over the change in flux).
6. Increasing τ_{ind} makes $\tau_{\text{ind}} > \tau_{\text{load}}$, and the speed ω_m increases.
7. Increasing ω_m increases $E_A = K\phi\omega_m\uparrow$ again.

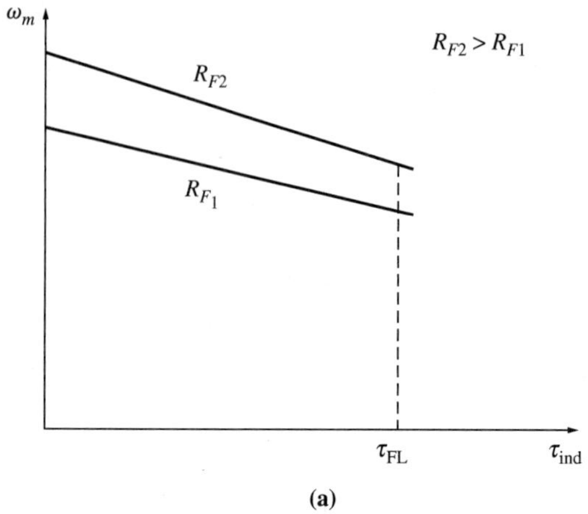

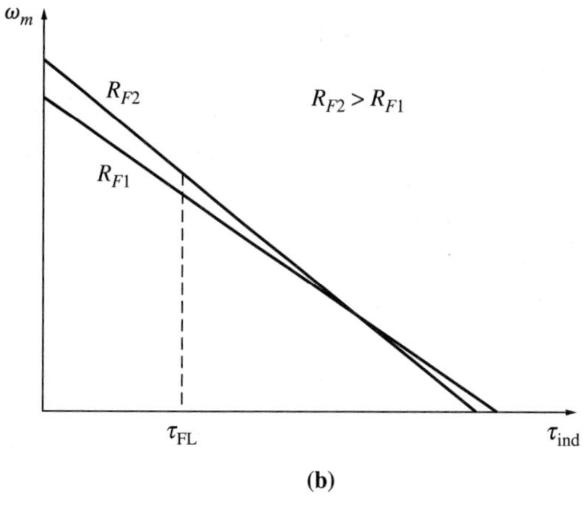

FIGURE 8–12
The effect of field resistance speed control on a shunt motor's torque–speed characteristic: (a) over the motor's normal operating range; (b) over the entire range from no-load to stall conditions.

8. Increasing E_A decreases I_A.
9. Decreasing I_A decreases τ_{ind} until $\tau_{ind} = \tau_{load}$ at a higher speed ω_m.

The effect of increasing the field resistance on the output characteristic of a shunt motor is shown in Figure 8–12a. Notice that as the flux in the machine decreases, the no-load speed of the motor increases, while the slope of the torque–speed curve becomes steeper. Naturally, decreasing R_F would reverse the whole process, and the speed of the motor would drop.

A WARNING ABOUT FIELD RESISTANCE SPEED CONTROL. The effect of increasing the field resistance on the output characteristic of a shunt dc motor is shown in Figure 8–12. Notice that as the flux in the machine decreases, the no-load speed of the motor increases, while the slope of the torque–speed curve becomes steeper. This shape is a consequence of Equation (8–7), which describes the terminal characteristic of the motor. In Equation (8–7), the no-load speed is

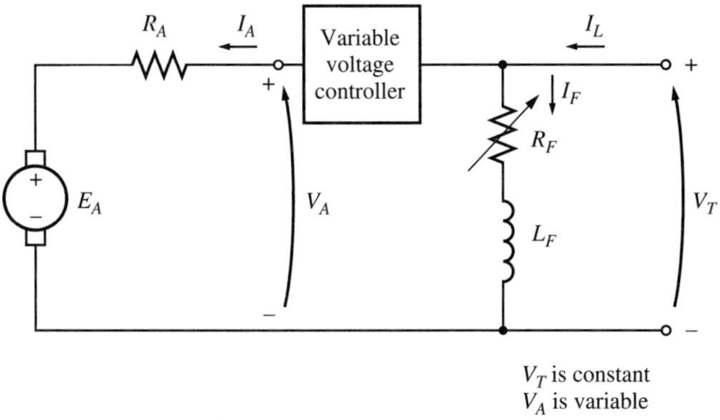

FIGURE 8–13
Armature voltage control of a shunt (or separately excited) dc motor.

proportional to the reciprocal of the flux in the motor, while the slope of the curve is proportional to the reciprocal of the flux squared. Therefore, a decrease in flux causes the slope of the torque–speed curve to become steeper.

Figure 8–12a shows the terminal characteristic of the motor over the range from no-load to full-load conditions. Over this range, an increase in field resistance increases the motor's speed, as previously described in this section. For motors operating between no-load and full-load conditions, an increase in R_F may reliably be expected to increase operating speed.

Now examine Figure 8–12b. This figure shows the terminal characteristic of the motor over the full range from no-load to stall conditions. It is apparent from the figure that at *very slow* speeds an increase in field resistance will actually *decrease* the speed of the motor. This effect occurs because, at very low speeds, the increase in armature current caused by the decrease in E_A is no longer large enough to compensate for the decrease in flux in the induced torque equation. With the flux decrease actually larger than the armature current increase, the induced torque decreases, and the motor slows down.

Some small dc motors used for control purposes actually operate at speeds close to stall conditions. For these motors, an increase in field resistance might have no effect, or it might even decrease the speed of the motor. Since the results are not predictable, field resistance speed control should not be used in these types of dc motors. Instead, the armature voltage method of speed control should be employed.

CHANGING THE ARMATURE VOLTAGE. The second form of speed control involves changing the voltage applied to the armature of the motor *without changing the voltage applied to the field.* A connection similar to that in Figure 8–13 is necessary for this type of control. In effect, the motor must be *separately excited* to use armature voltage control.

If the voltage V_A is increased, then the armature current in the motor must rise $[I_A = (V_A \uparrow - E_A)/R_A]$. As I_A increases, the induced torque $\tau_{\text{ind}} = K\phi I_A\uparrow$ increases, making $\tau_{\text{ind}} > \tau_{\text{load}}$, and the speed ω_m of the motor increases.

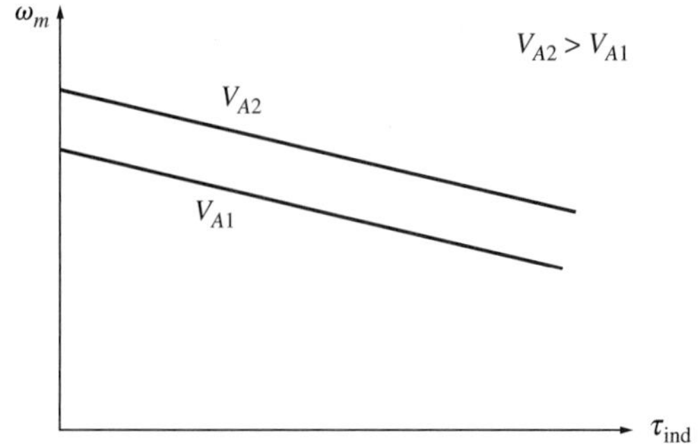

FIGURE 8–14
The effect of armature voltage speed control on a shunt motor's torque–speed characteristic.

But as the speed ω_m increases, the internal generated voltage $E_A(= K\phi\omega_m\uparrow)$ increases, causing the armature current to decrease. This decrease in I_A decreases the induced torque, causing τ_{ind} to equal τ_{load} at a higher rotational speed ω.

To summarize the cause-and-effect behavior in this method of speed control:

1. An increase in V_A increases I_A $[= (V_A\uparrow - E_A)/R_A]$.
2. Increasing I_A increases τ_{ind} $(= K\phi I_A\uparrow)$.
3. Increasing τ_{ind} makes $\tau_{\text{ind}} > \tau_{\text{load}}$ increasing ω_m.
4. Increasing ω_m increases $E_A(= K\phi\omega_m\uparrow)$.
5. Increasing E_A decreases I_A $[= (V_A\uparrow - E_A)/R_A]$.
6. Decreasing I_A decreases τ_{ind} until $\tau_{\text{ind}} = \tau_{\text{load}}$ at a higher ω_m.

The effect of an increase in V_A on the torque–speed characteristic of a separately excited motor is shown in Figure 8–14. Notice that the no-load speed of the motor is shifted by this method of speed control, but the slope of the curve remains constant.

INSERTING A RESISTOR IN SERIES WITH THE ARMATURE CIRCUIT. If a resistor is inserted in series with the armature circuit, the effect is to drastically increase the slope of the motor's torque–speed characteristic, making it operate more slowly if loaded (Figure 8–15). This fact can easily be seen from Equation (8–7). The insertion of a resistor is a very wasteful method of speed control, since the losses in the inserted resistor are very large. For this reason, it is rarely used. It will be found only in applications in which the motor spends almost all its time operating at full speed or in applications too inexpensive to justify a better form of speed control.

The two most common methods of shunt motor speed control—field resistance variation and armature voltage variation—have different safe ranges of operation.

In field resistance control, the lower the field current in a shunt (or separately excited) dc motor, the faster it turns; and the higher the field current, the

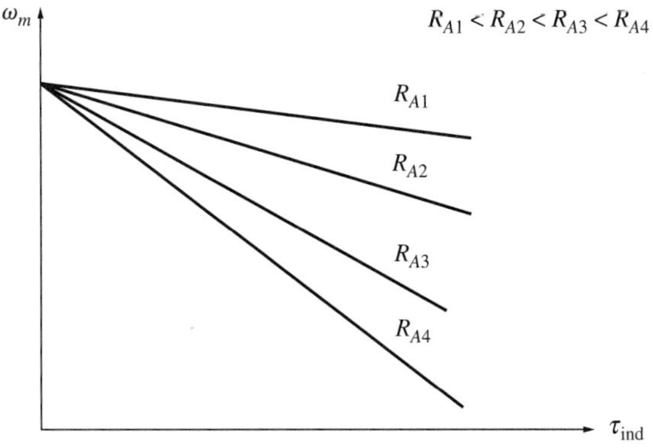

FIGURE 8–15
The effect of armature resistance speed control on a shunt motor's torque–speed characteristic.

slower it turns. Since an increase in field current causes a decrease in speed, there is always a minimum achievable speed by field circuit control. This minimum speed occurs when the motor's field circuit has the maximum permissible current flowing through it.

If a motor is operating at its rated terminal voltage, power, and field current, then it will be running at rated speed, also known as *base speed*. Field resistance control can control the speed of the motor for speeds above base speed but not for speeds below base speed. To achieve a speed slower than base speed by field circuit control would require excessive field current, possibly burning up the field windings.

In armature voltage control, the lower the armature voltage on a separately excited dc motor, the slower it turns; and the higher the armature voltage, the faster it turns. Since an increase in armature voltage causes an increase in speed, there is always a maximum achievable speed by armature voltage control. This maximum speed occurs when the motor's armature voltage reaches its maximum permissible level.

If the motor is operating at its rated voltage, field current, and power, it will be turning at base speed. Armature voltage control can control the speed of the motor for speeds below base speed but not for speeds above base speed. To achieve a speed faster than base speed by armature voltage control would require excessive armature voltage, possibly damaging the armature circuit.

These two techniques of speed control are obviously complementary. Armature voltage control works well for speeds below base speed, and field resistance or field current control works well for speeds above base speed. By combining the two speed-control techniques in the same motor, it is possible to get a range of speed variations of up to 40 to 1 or more. Shunt and separately excited dc motors have excellent speed control characteristics.

There is a significant difference in the torque and power limits on the machine under these two types of speed control. The limiting factor in either case is the heating of the armature conductors, which places an upper limit on the magnitude of the armature current I_A.

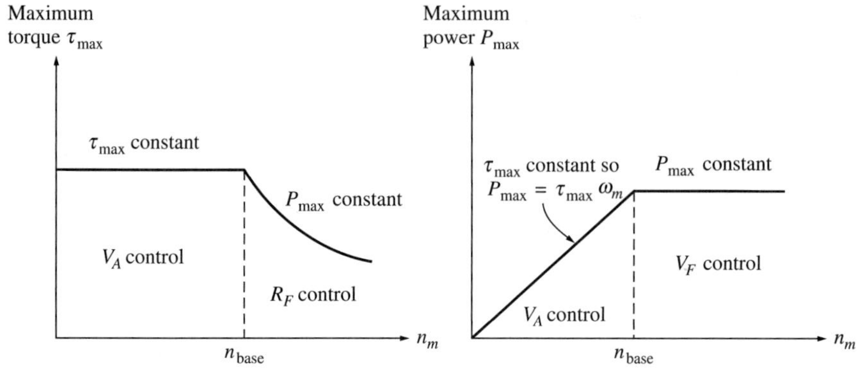

FIGURE 8–16
Power and torque limits as a function of speed for a shunt motor under armature volt and field resistance control.

For armature voltage control, *the flux in the motor is constant*, so the maximum torque in the motor is

$$\tau_{max} = K\phi I_{A,max} \tag{8–14}$$

This maximum torque is constant, regardless of the speed of the rotation of the motor. Since the power out of the motor is given by $P = \tau\omega$, the maximum power of the motor at any speed under armature voltage control is

$$P_{max} = \tau_{max}\omega_m \tag{8–15}$$

Thus *the maximum power out of the motor is directly proportional to its operating speed* under armature voltage control.

On the other hand, when field resistance control is used, the flux does change. In this form of control, a speed increase is caused by a decrease in the machine's flux. In order for the armature current limit not to be exceeded, the induced torque limit must decrease as the speed of the motor increases. Since the power out of the motor is given by $P = \tau\omega$, and the torque limit decreases as the speed of the motor increases, *the maximum power out of a dc motor under field current control is constant*, while *the maximum torque varies as the reciprocal of the motor's speed.*

These shunt dc motor power and torque limitations for safe operation as a function of speed are shown in Figure 8–16.

The following examples illustrate how to find the new speed of a dc motor if it is varied by field resistance or armature voltage control methods.

Example 8–3. Figure 8–17a shows a 100-hp, 250-V, 1200 r/min shunt dc motor with an armature resistance of 0.03 Ω and a field resistance of 41.67 Ω. The motor has compensating windings, so armature reaction can be ignored. Mechanical and core losses may be assumed to be negligible for the purposes of this problem. The motor is assumed to be driving

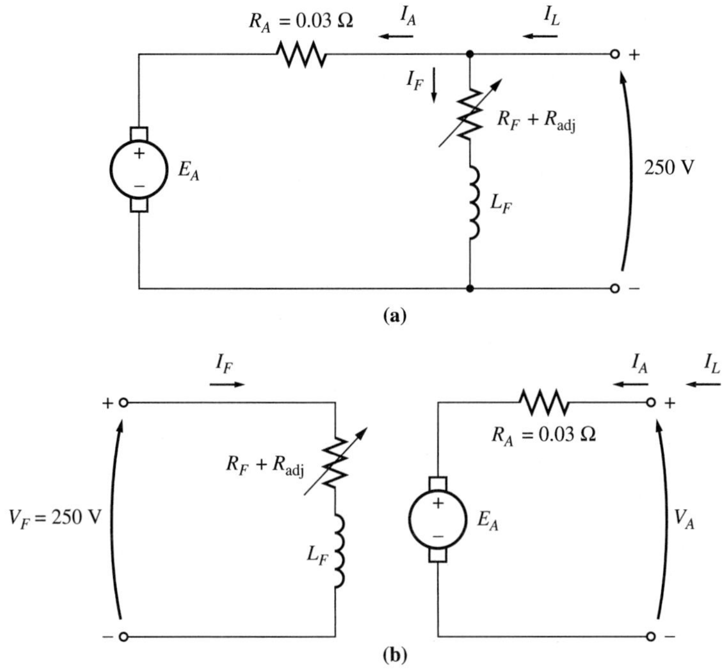

FIGURE 8–17
(a) The shunt motor in Example 8–3. (b) The separately excited dc motor in Example 8–4.

a load with a line current of 126 A and an initial speed of 1103 r/min. To simplify the problem, assume that the amount of armature current drawn by the motor remains constant.

(a) If the machine's magnetization curve is shown in Figure 8–9, what is the motor's speed if the field resistance is raised to 50 Ω?
(b) Calculate and plot the speed of this motor as a function of the field resistance R_F assuming a constant-current load.

Solution
(a) The motor has an initial line current of 126 A, so the initial armature current is

$$I_{A1} = I_{L1} - I_{F1} = 126\text{ A} - \frac{150\text{ V}}{41.67\ \Omega} = 120\text{ A}$$

Therefore, the internal generated voltage is

$$\begin{aligned} E_{A1} &= V_T - I_{A1}R_A = 250\text{ V} - (120\text{ A})(0.03\ \Omega) \\ &= 246.4\text{ V} \end{aligned}$$

After the field resistance is increased to 50 Ω, the field current will become

$$I_{F2} = \frac{V_T}{R_F} = \frac{250\text{ V}}{50\ \Omega} = 5\text{ A}$$

The ratio of the internal generated voltage at one speed to the internal generated voltage at another speed is given by the ratio of Equation (7–41) at the two speeds:

$$\frac{E_{A2}}{E_{A1}} = \frac{K'\phi_2 n_{m2}}{K'\phi_1 n_{m1}} \tag{8–16}$$

Because the armature current is assumed constant, $E_{A1} = E_{A2}$, and this equation reduces to

$$1 = \frac{\phi_2 n_{m2}}{\phi_1 n_{m1}}$$

or

$$n_{m2} = \frac{\phi_1}{\phi_2} n_{m1} \tag{8–17}$$

A magnetization curve is a plot of E_A versus I_F for a given speed. Since the values of E_A on the curve are directly proportional to the flux, the ratio of the internal generated voltages read off the curve is equal to the ratio of the fluxes within the machine. At $I_F = 5$ A, $E_{A0} = 250$ V, while at $I_F = 6$ A, $E_{A0} = 268$ V. Therefore, the ratio of fluxes is given by

$$\frac{\phi_1}{\phi_2} = \frac{268\text{ V}}{250\text{ V}} = 1.076$$

and the new speed of the motor is

$$n_{m2} = \frac{\phi_1}{\phi_2} n_{m1} = (1.076)(1103\text{ r/min}) = 1187\text{ r/min}$$

(b) A MATLAB M-file that calculates the speed of the motor as a function of R_F follows:

```
% M-file: rf_speed_control.m
% M-file create a plot of the speed of a shunt dc
%   motor as a function of field resistance, assuming
%   a constant armature current (Example 8-3).

% Get the magnetization curve.  This file contains the
% three variables if_value, ea_value, and n_0.
load fig8_9.mat

% First, initialize the values needed in this program.
v_t = 250;                 % Terminal voltage (V)
r_f = 40:1:70;             % Field resistance (ohms)
r_a = 0.03;                % Armature resistance (ohms)
i_a = 120;                 % Armature currents (A)

% The approach here is to calculate the e_a0 at the
% reference field current, and then to calculate the
% e_a0 for every field current.  The reference speed is
% 1103 r/min, so by knowing the e_a0 and reference
% speed, we will be able to calculate the speed at the
% other field current.

% Calculate the internal generated voltage at 1200 r/min
% for the reference field current (5 A) by interpolating
```

```
% the motor's magnetization curve.  The reference speed
% corresponding to this field current is 1103 r/min.
e_a0_ref = interp1(if_values,ea_values,5,'spline');
n_ref = 1103;

% Calculate the field current for each value of field
% resistance.
i_f = v_t ./ r_f;

% Calculate the E_a0 for each field current by
% interpolating the motor's magnetization curve.
e_a0 = interp1(if_values,ea_values,i_f,'spline');

% Calculate the resulting speed from Equation (8-17):
% n2 = (phi1 / phi2) * n1 = (e_a0_1 / e_a0_2 ) * n1
n2 = ( e_a0_ref ./ e_a0 ) * n_ref;

% Plot the speed versus r_f curve.
plot(r_f,n2,'k-','LineWidth',2.0);
hold on;
xlabel('\bfField resistance, \Omega');
ylabel('\bf\itn_{m} \rm\bf(r/min)');
title ('\bfSpeed vs \itR_{F} \rm\bffor a Shunt DC Motor');
axis([40 70 0 1400]);
grid on;
hold off;
```

The resulting plot is shown in Figure 8–18.

Note that the assumption of a constant armature current as R_F changes is not a very good one for real loads. The current in the armature will vary with speed in a fashion dependent on the torque required by the type of load attached to the

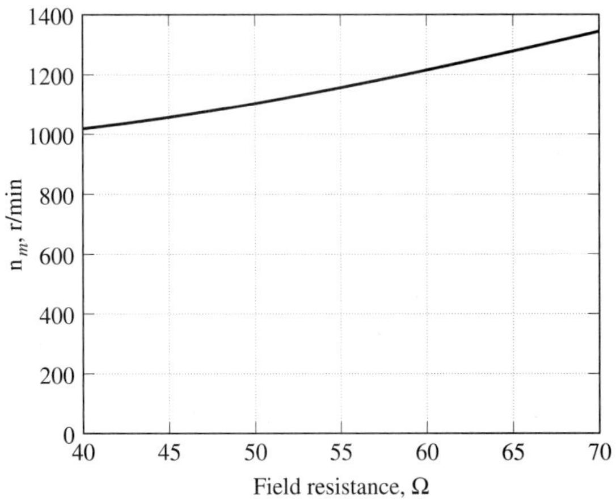

FIGURE 8–18
Plot of speed versus field resistance for the shunt dc motor of Example 8–3.

motor. These differences will cause a motor's speed-versus-R_F curve to be slightly different than the one shown in Figure 8–18, but it will have a similar shape.

Example 8–4. The motor in Example 8–3 is now connected separately excited, as shown in Figure 8–17b. The motor is initially running with $V_A = 250$ V, $I_A = 120$ A, and $n = 1103$ r/min, while supplying a constant-torque load. What will the speed of this motor be if V_A is reduced to 200 V?

Solution

The motor has an initial line current of 120 A and an armature voltage V_A of 250 V, so the internal generated voltage E_A is

$$E_A = V_T - I_A R_A = 250\text{ V} - (120\text{ A})(0.03\ \Omega) = 246.4\text{ V}$$

By applying Equation (8–16) and realizing that the flux ϕ is constant, the motor's speed can be expressed as

$$\frac{E_{A2}}{E_{A1}} = \frac{K'\phi_2 n_{m2}}{K'\phi_1 n_{m1}} \tag{8–16}$$

$$= \frac{n_{m2}}{n_{m1}}$$

$$n_{m2} = \frac{E_{A2}}{E_{A1}} n_{m1}$$

To find E_{A2} use Kirchhoff's voltage law:

$$E_{A2} = V_T - I_{A2} R_A$$

Since the torque is constant and the flux is constant, I_A is constant. This yields a voltage of

$$E_{A2} = 200\text{ V} - (120\text{ A})(0.03\ \Omega) = 196.4\text{ V}$$

The final speed of the motor is thus

$$n_{m2} = \frac{E_{A2}}{E_{A1}} n_{m1} = \frac{196.4\text{ V}}{246.4\text{ V}}\ 1103\text{ r/min} = 879\text{ r/min}$$

The Effect of an Open Field Circuit

The previous section of this chapter contained a discussion of speed control by varying the field resistance of a shunt motor. As the field resistance increased, the speed of the motor increased with it. What would happen if this effect were taken to the extreme, if the field resistor *really* increased? What would happen if the field circuit actually opened while the motor was running? From the previous discussion, the flux in the machine would drop drastically, all the way down to ϕ_{res}, and $E_A (= K\phi\omega_m)$ would drop with it. This would cause an enormous increase in the armature current, and the resulting induced torque would be quite a bit higher than the load torque on the motor. Therefore, the motor's speed would start to rise and would just keep going up.

The results of an open field circuit can be quite spectacular. When the author was an undergraduate "in Electrical Engineering at Louisiana State University," his laboratory group once made a mistake of this sort. The group was working with a small motor-generator set being driven by a 3-hp shunt dc motor. The motor was connected and ready to go, but there was just *one* little mistake—when the field circuit was connected, it was fused with a 0.3-A fuse instead of the 3-A fuse that was supposed to be used.

When the motor was started, it ran normally for about 3 s, and then suddenly there was a flash from the fuse. Immediately, the motor's speed skyrocketed. Someone turned the main circuit breaker off within a few seconds, but by that time the tachometer attached to the motor had pegged at 4000 r/min. The motor itself was only rated for 800 r/min.

Needless to say, that experience scared everyone present very badly and taught them to be *most* careful about field circuit protection. In dc motor starting and protection circuits, a *field loss relay* is normally included to disconnect the motor from the line in the event of a loss of field current.

A similar effect can occur in ordinary shunt dc motors operating with light fields if their armature reaction effects are severe enough. If the armature reaction on a dc motor is severe, an increase in load can weaken its flux enough to actually cause the motor's speed to rise. However, most loads have torque–speed curves whose torque *increases* with speed, so the increased speed of the motor increases its load, which increases its armature reaction, weakening its flux again. The weaker flux causes a further increase in speed, further increasing load, etc., until the motor overspeeds. This condition is known as *runaway*.

In motors operating with very severe load changes and duty cycles, this flux-weakening problem can be solved by installing compensating windings. Unfortunately, compensating windings are too expensive for use on ordinary run-of-the-mill motors. The solution to the runaway problem employed for less-expensive, less-severe duty motors is to provide a turn or two of cumulative compounding to the motor's poles. As the load increases, the magnetomotive force from the series turns increases, which counteracts the demagnetizing magnetomotive force of the armature reaction. A shunt motor equipped with just a few series turns like this is called a *stabilized shunt* motor.

8.5 THE PERMANENT-MAGNET DC MOTOR

A permanent-magnet dc (PMDC) *motor* is a dc motor whose poles are made of permanent magnets. Permanent-magnet dc motors offer a number of benefits compared with shunt dc motors in some applications. Since these motors do not require an external field circuit, they do not have the field circuit copper losses associated with shunt dc motors. Because no field windings are required, they can be smaller than corresponding shunt dc motors. PMDC motors can be commonly found in sizes up to about 10 hp, and in recent years some motors have been built in sizes up to 100 hp. However, they are especially common in smaller fractional- and sub fractional-horsepower sizes, where the expense and space of a separate field circuit cannot be justified.

PMDC motors are generally less expensive, smaller in size, simpler, and higher efficiency than corresponding dc motors with separate electromagnetic fields. This makes them a good choice in many dc motor applications. The armatures of PMDC motors are essentially identical to the armatures of motors with separate field circuits, so their costs are similar too. However, the elimination of separate electromagnets on the stator reduces the size of the stator, the cost of the stator, and the losses in the field circuits.

PMDC motors also have disadvantages. Permanent magnets cannot produce as high a flux density as an externally supplied shunt field, so a PMDC motor will have a lower induced torque τ_{ind} per ampere of armature current I_A than a shunt motor of the same size and construction. In addition, PMDC motors run the risk of demagnetization. As mentioned in Chapter 7, the armature current I_A in a dc machine produces an armature magnetic field of its own. The armature mmf subtracts from the mmf of the poles under some portions of the pole faces and adds to the mmf of the poles under other portions of the pole faces (see Figures 8–23 and 8–25), reducing the overall net flux in the machine. This is the *armature reaction* effect. In a PMDC machine, the pole flux is just the residual flux in the permanent magnets. If the armature current becomes very large, there is some risk that the armature mmf may demagnetize the poles, permanently reducing and reorienting the residual flux in them. Demagnetization may also be caused by the excessive heating which can occur due to shock (dropping the motor) or during prolonged periods of overload. In addition, PMDC materials are weaker physically than most normal steels, so stators constructed out of them can be limited by the physical torque requirements of the motor.

Figure 8–19a shows a magnetization curve for a typical ferromagnetic material. It is a plot of flux density **B** versus magnetizing intensity **H** (or equivalently, a plot of flux ϕ versus mmf $\mathcal{F}$). When a strong external magnetomotive force is applied to this material and then removed, a residual flux $\mathbf{B}_{\text{res}}$ will remain in the material. To force the residual flux to zero, it is necessary to apply a coercive magnetizing intensity $\mathbf{H}_C$ with a polarity opposite to the polarity of the magnetizing intensity **H** that originally established the magnetic field. For normal machine applications such as rotors and stators, a ferromagnetic material should be picked which has the smallest $\mathbf{B}_{\text{res}}$ and $\mathbf{H}_C$ possible, since such a material will have low hysteresis losses.

On the other hand, a good material for the poles of a PMDC motor should have *the largest residual flux density* $\mathbf{B}_{\text{res}}$ *possible,* while simultaneously having *the largest coercive magnetizing intensity* $\mathbf{H}_C$ *possible.* The magnetization curve of such a material is shown in Figure 8–19b. The large $\mathbf{B}_{\text{res}}$ produces a large flux in the machine, while the large $\mathbf{H}_C$ means that a very large current would be required to demagnetize the poles.

In the last 40 years, a number of new magnetic materials have been developed which have desirable characteristics for making permanent magnets. The major types of materials are the ceramic (ferrite) magnetic materials and the rare-earth magnetic materials. Figure 8–19c shows the second quadrant of the magnetization curves of some typical ceramic and rare-earth magnets, compared to the magnetization curve of a conventional ferromagnetic alloy (Alnico 5). It is obvious from the comparison that the best rare-earth magnets can produce the same residual flux as the best conventional ferromagnetic alloys, while simultaneously being largely immune to demagnetization problems due to armature reaction.

A permanent-magnet dc motor is basically the same machine as a shunt dc motor, except that *the flux of a PMDC motor is fixed.* Therefore, it is not possible to control the speed of a PMDC motor by varying the field current or flux. The only methods of speed control available for a PMDC motor are armature voltage control and armature resistance control.

The techniques to analyze a PMDC motor are basically the same as the techniques to analyze a shunt dc motor with the field current held constant.

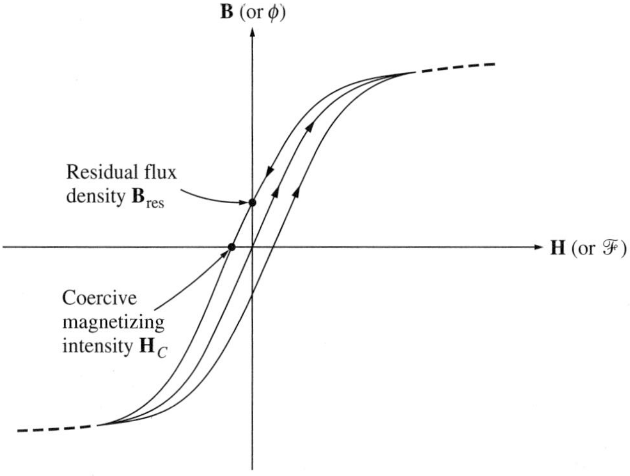

FIGURE 8–19
(a) The magnetization curve of a typical ferromagnetic material. Note the hysteresis loop. After a large magnetizing intensity **H** is applied to the core and then removed, a residual flux density $\mathbf{B}_{res}$ remains behind in the core. This flux can be brought to zero if a coercive magnetizing intensity $\mathbf{H}_C$ is applied to the core with the opposite polarity. In this case, a relatively small value of it will demagnetize the core.

For more information about PMDC motors, see References 4 and 10.

8.6 THE SERIES DC MOTOR

A series dc motor is a dc motor whose field windings consist of a relatively few turns connected in series with the armature circuit. The equivalent circuit of a series dc motor is shown in Figure 8–20. In a series motor, the armature current, field current, and line current are all the same. The Kirchhoff's voltage law equation for this motor is

$$V_T = E_A + I_A(R_A + R_S) \quad (8–18)$$

Induced Torque in a Series DC Motor

The terminal characteristic of a series dc motor is very different from that of the shunt motor previously studied. The basic behavior of a series dc motor is due to the fact that *the flux is directly proportional to the armature current*, at least until saturation is reached. As the load on the motor increases, its flux increases too. As seen earlier, an increase in flux in the motor causes a decrease in its speed. The result is that a series motor has a sharply drooping torque–speed characteristic.

The induced torque in this machine is given by Equation (7–49):

$$\tau_{ind} = K\phi I_A \quad (7–49)$$

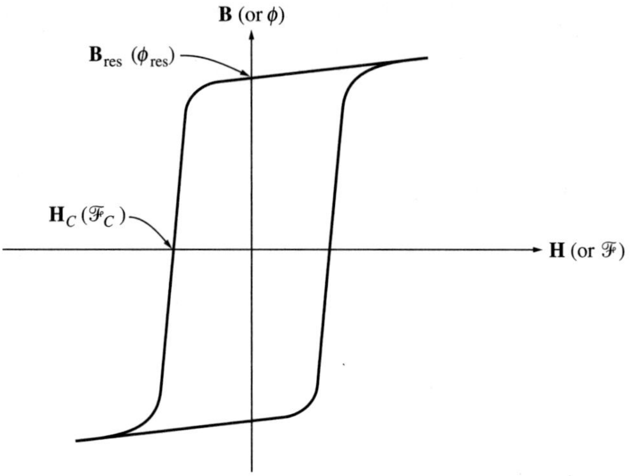

(b)

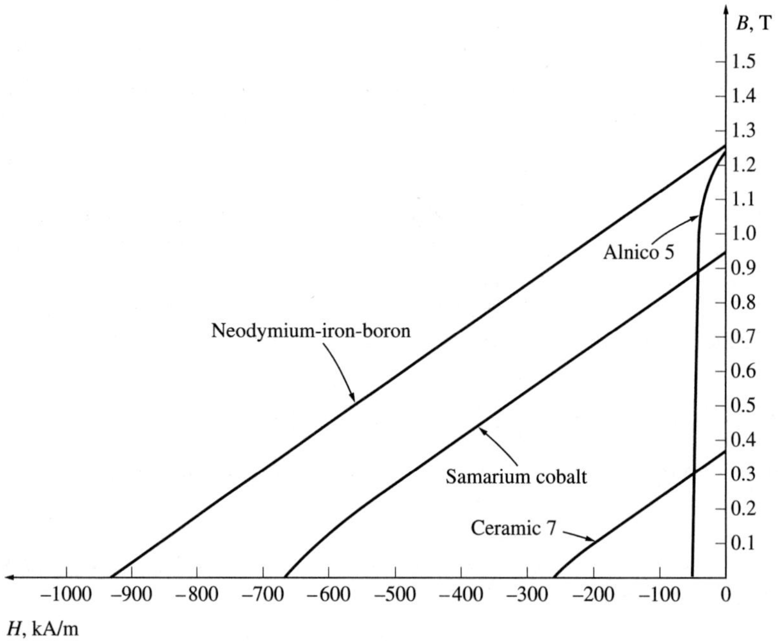

(c)

FIGURE 8–19 (concluded)
(b) The magnetization curve of a ferromagnetic material suitable for use in permanent magnets. Note the high residual flux density $\mathbf{B}_{res}$ and the relatively large coercive magnetizing intensity $\mathbf{H}_C$. (c) The second quadrant of the magnetization curves of some typical magnetic materials. Note that the rare-earth magnets combine a high residual flux with a high coercive magnetizing intensity.

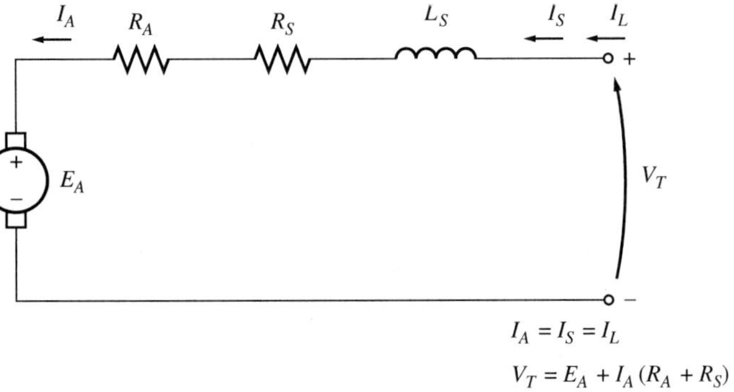

FIGURE 8–20
The equivalent circuit of a series dc motor.

The flux in this machine is directly proportional to its armature current (at least until the metal saturates). Therefore, the flux in the machine can be given by

$$\phi = cI_A \tag{8–19}$$

where c is a constant of proportionality. The induced torque in this machine is thus given by

$$\tau_{\text{ind}} = K\phi I_A = KcI_A^2 \tag{8–20}$$

In other words, the torque in the motor is proportional to the square of its armature current. As a result of this relationship, it is easy to see that a series motor gives more torque per ampere than any other dc motor. It is therefore used in applications requiring very high torques. Examples of such applications are the starter motors in cars, elevator motors, and tractor motors in locomotives.

The Terminal Characteristic of a Series DC Motor

To determine the terminal characteristic of a series dc motor, an analysis will be based on the assumption of a linear magnetization curve, and then the effects of saturation will be considered in a graphical analysis.

The assumption of a linear magnetization curve implies that the flux in the motor will be given by Equation (8–19):

$$\phi = cI_A \tag{8–19}$$

This equation will be used to derive the torque–speed characteristic curve for the series motor.

The derivation of a series motor's torque–speed characteristic starts with Kirchhoff's voltage law:

$$V_T = E_A + I_A(R_A + R_S) \tag{8–18}$$

From Equation (8–20), the armature current can be expressed as

$$I_A = \sqrt{\frac{\tau_{\text{ind}}}{Kc}}$$

Also, $E_A = K\phi\omega_m$. Substituting these expressions in Equation (8–18) yields

$$V_T = K\phi\omega_m + \sqrt{\frac{\tau_{\text{ind}}}{Kc}}(R_A + R_S) \tag{8–21}$$

If the flux can be eliminated from this expression, it will directly relate the torque of a motor to its speed. To eliminate the flux from the expression, notice that

$$I_A = \frac{\phi}{c}$$

and the induced torque equation can be rewritten as

$$\tau_{\text{ind}} = \frac{K}{c}\phi^2$$

Therefore, the flux in the motor can be rewritten as

$$\phi = \sqrt{\frac{c}{K}}\sqrt{\tau_{\text{ind}}} \tag{8–22}$$

Substituting Equation (8–22) into Equation (8–21) and solving for speed yields

$$V_T = K\sqrt{\frac{c}{K}}\sqrt{\tau_{\text{ind}}}\,\omega_m + \sqrt{\frac{\tau_{\text{ind}}}{Kc}}\,(R_A + R_S)$$

$$\sqrt{Kc}\,\sqrt{\tau_{\text{ind}}}\,\omega_m = V_T - \frac{R_A + R_S}{\sqrt{Kc}}\sqrt{\tau_{\text{ind}}}$$

$$\omega_m = \frac{V_T}{\sqrt{Kc}\,\sqrt{\tau_{\text{ind}}}} - \frac{R_A + R_S}{Kc}$$

The resulting torque–speed relationship is

$$\boxed{\omega_m = \frac{V_T}{\sqrt{Kc}}\frac{1}{\sqrt{\tau_{\text{ind}}}} - \frac{R_A + R_S}{Kc}} \tag{8–23}$$

Notice that for an unsaturated series motor, the speed of the motor varies as the reciprocal of the square root of the torque. That is quite an unusual relationship! This ideal torque–speed characteristic is plotted in Figure 8–21.

One disadvantage of series motors can be seen immediately from this equation. When the torque on this motor goes to zero, its speed goes to infinity. In practice, the torque can never go entirely to zero because of the mechanical, core, and stray losses that must be overcome. However, if no other load is connected to the motor, it can turn fast enough to seriously damage itself. *Never* completely unload a series motor, and never connect one to a load by a belt or other mechanism that could break. If that were to happen and the motor were to become unloaded while running, the results could be serious.

The nonlinear analysis of a series dc motor with magnetic saturation effects, but ignoring armature reaction, is illustrated in Example 8–5.

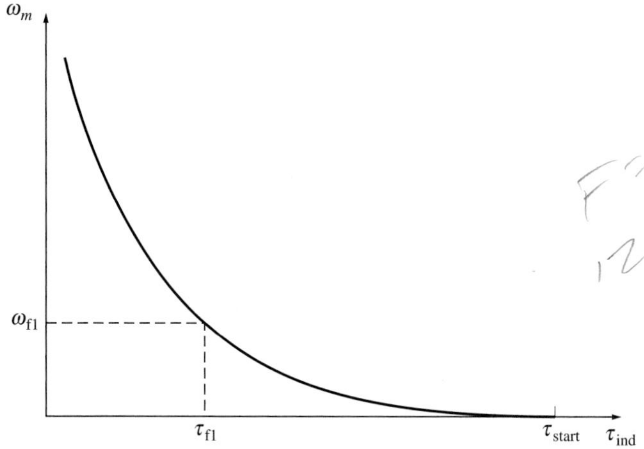

FIGURE 8–21
The torque–speed characteristic of a series dc motor.

Example 8–5. Figure 8–20 shows a 250-V series dc motor with compensating windings, and a total series resistance $R_A + R_S$ of 0.08 Ω. The series field consists of 25 turns per pole, with the magnetization curve shown in Figure 8–22.

(a) Find the speed and induced torque of this motor for when its armature current is 50 A.

(b) Calculate and plot the torque–speed characteristic for this motor.

Solution

(a) To analyze the behavior of a series motor with saturation, pick points along the operating curve and find the torque and speed for each point. Notice that the magnetization curve is given in units of magnetomotive force (ampere-turns) versus E_A for a speed of 1200 r/min, so calculated E_A values must be compared to the equivalent values at 1200 r/min to determine the actual motor speed.

For $I_A = 50$ A,

$$E_A = V_T - I_A(R_A + R_S) = 250 \text{ V} - (50\text{A})(0.08\ \Omega) = 246 \text{ V}$$

Since $I_A = I_F = 50$ A, the magnetomotive force is

$$\mathscr{F} = NI = (25 \text{ turns})(50 \text{ A}) = 1250 \text{ A} \bullet \text{turns}$$

From the magnetization curve at $\mathscr{F} = 1250$ A • turns, $E_{A0} = 80$ V. To get the correct speed of the motor, remember that, from Equation (8–13),

$$n_m = \frac{E_A}{E_{A0}} n_0$$

$$= \frac{246 \text{ V}}{80 \text{ V}} 120 \text{ r/min} = 3690 \text{ r/min}$$

To find the induced torque supplied by the motor at that speed, recall that $P_{\text{conv}} = E_A I_A = \tau_{\text{ind}} \omega_m$. Therefore,

$$\tau_{\text{ind}} = \frac{E_A I_A}{\omega_m}$$

$$= \frac{(246\text{ V})(50\text{ A})}{(3690\text{ r/min})(1\text{ min}/60\text{ s})(2\pi\text{ rad/r})} = 31.8\text{ N} \bullet \text{m}$$

(*b*) To calculate the complete torque–speed characteristic, we must repeat the steps in *a* for many values of armature current. A MATLAB M-file that calculates the torque–speed characteristics of the series dc motor is shown below. Note that the magnetization curve used by this program works in terms of field magnetomotive force instead of effective field current.

```
% M-file: series_ts_curve.m
% M-file create a plot of the torque-speed curve of the
%   the series dc motor with armature reaction in
%   Example 8-5.

% Get the magnetization curve.  This file contains the
% three variables mmf_values, ea_values, and n_0.
load fig8_22.mat
```

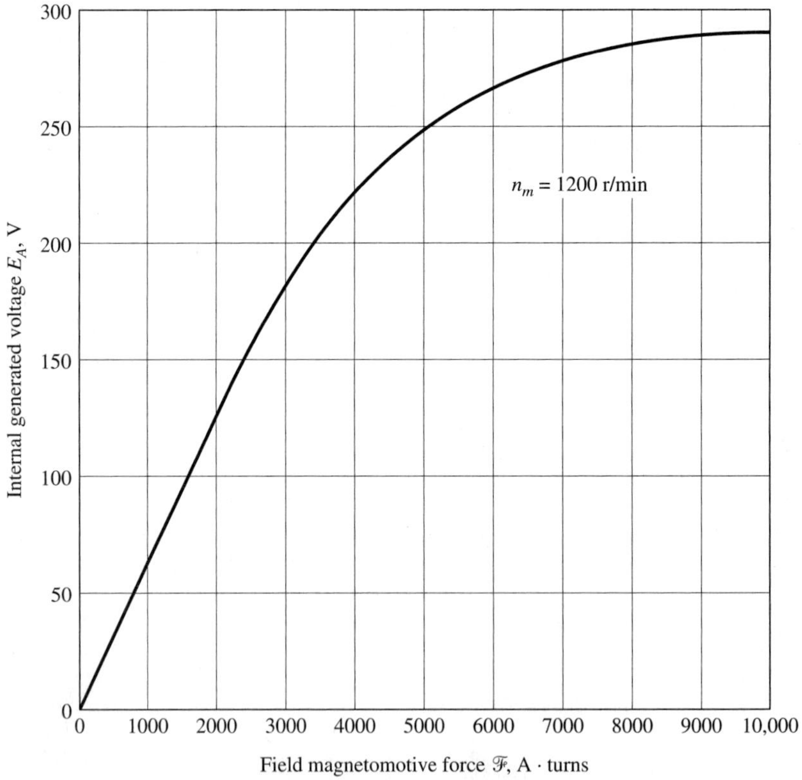

FIGURE 8–22
The magnetization curve of the motor in Example 8–5. This curve was taken at speed n_m = 1200 r/min.

```
% First, initialize the values needed in this program.
v_t = 250;                  % Terminal voltage (V)
r_a = 0.08;                 % Armature + field resistance (ohms)
i_a = 10:10:300;            % Armature (line) currents (A)
n_s = 25;                   % Number of series turns on field

% Calculate the MMF for each load
f = n_s * i_a;

% Calculate the internal generated voltage e_a.
e_a = v_t - i_a * r_a;

% Calculate the resulting internal generated voltage at
% 1200 r/min by interpolating the motor's magnetization
% curve.
e_a0 = interp1(mmf_values,ea_values,f,'spline');

% Calculate the motor's speed from Equation (8-13).
n = (e_a ./ e_a0) * n_0;

% Calculate the induced torque corresponding to each
% speed from Equations (7-55) and (7-56).
t_ind = e_a .* i_a ./ (n * 2 * pi / 60);

% Plot the torque-speed curve
plot(t_ind,n,'Color','k','LineWidth',2.0);
hold on;
xlabel('\bf\tau_{ind} (N-m)');
ylabel('\bf\itn_{m} \rm\bf(r/min)');
title ('\bfSeries DC Motor Torque-Speed Characteristic');
axis([ 0 700 0 5000]);
grid on;
hold off;
```

The resulting motor torque–speed characteristic is shown in Figure 8–23. Notice the severe overspeeding at very small torques.

Speed Control of Series DC Motors

Unlike with the shunt dc motor, there is only one efficient way to change the speed of a series dc motor. That method is to change the terminal voltage of the motor. If the terminal voltage is increased, the first term in Equation (8–23) is increased, resulting in a *higher speed for any given torque*.

The speed of series dc motors can also be controlled by the insertion of a series resistor into the motor circuit, but this technique is very wasteful of power and is used only for intermittent periods during the start-up of some motors.

Until the last 40 years or so, there was no convenient way to change V_T, so the only method of speed control available was the wasteful series resistance method. That has all changed today with the introduction of solid-state control circuits.

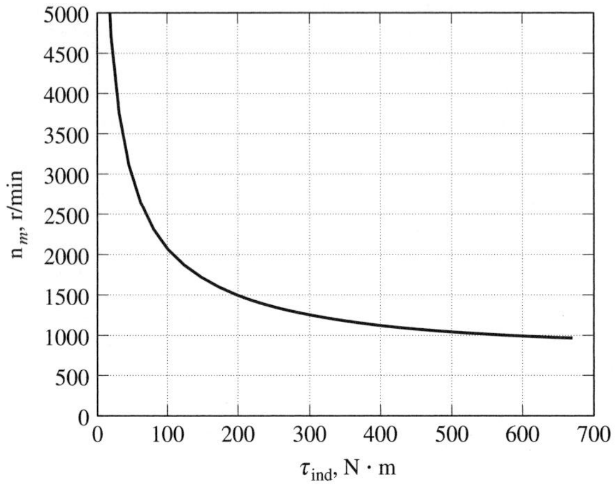

FIGURE 8–23
The torque–speed characteristic of the series dc motor in Example 8–5.

8.7 THE COMPOUNDED DC MOTOR

A compounded dc motor is a motor with *both a shunt and a series field.* Such a motor is shown in Figure 8–24. The dots that appear on the two field coils have the same meaning as the dots on a transformer: *Current flowing into a dot produces a positive magnetomotive force.* If current flows into the dots on both field coils, the resulting magnetomotive forces combine to produce a larger total magnetomotive force. This situation is known as *cumulative compounding.* If current flows into the dot on one field coil and out of the dot on the other field coil, the resulting magnetomotive forces subtract. In Figure 8–24 the round dots correspond to cumulative compounding of the motor, and the squares correspond to differential compounding.

The Kirchhoff's voltage law equation for a compounded dc motor is

$$V_T = E_A + I_A(R_A + R_S) \tag{8–24}$$

The currents in the compounded motor are related by

$$I_A = I_L - I_F \tag{8–25}$$

$$I_F = \frac{V_T}{R_F} \tag{8–26}$$

The net magnetomotive force and the effective shunt field current in the compounded motor are given by

$$\boxed{\mathcal{F}_{\text{net}} = \mathcal{F}_F \pm \mathcal{F}_{\text{SE}} - \mathcal{F}_{\text{AR}}} \tag{8–27}$$

and

$$\boxed{I_F^* = I_F \pm \frac{N_{SE}}{N_F} I_A - \frac{\mathcal{F}_{\text{AR}}}{N_F}} \tag{8–28}$$

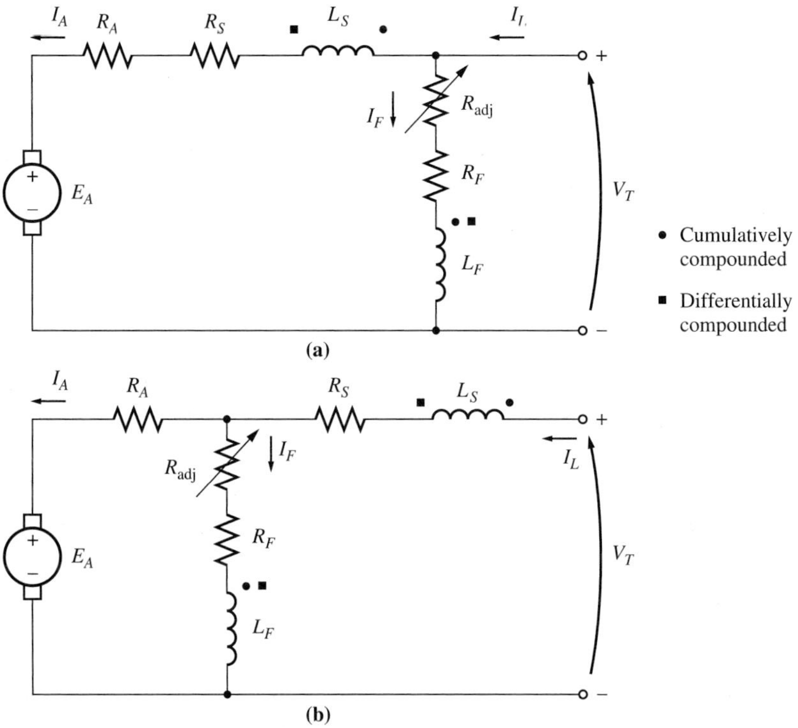

FIGURE 8–24
The equivalent circuit of compounded dc motors: (a) long-shunt connection; (b) short-shunt connection.

where the positive sign in the equations is associated with a cumulatively compounded motor and the negative sign is associated with a differentially compounded motor.

The Torque–Speed Characteristic of a Cumulatively Compounded DC Motor

In the cumulatively compounded dc motor, there is a component of flux which is constant and another component which is proportional to its armature current (and thus to its load). Therefore, the cumulatively compounded motor has a higher starting torque than a shunt motor (whose flux is constant), but a lower starting torque than a series motor (whose entire flux is proportional to armature current).

In a sense, the cumulatively compounded dc motor combines the best features of both the shunt and the series motors. Like a series motor, it has extra torque for starting; like a shunt motor, it does not overspeed at no load.

At light loads, the series field has a very small effect, so the motor behaves approximately as a shunt dc motor. As the load gets very large, the series flux

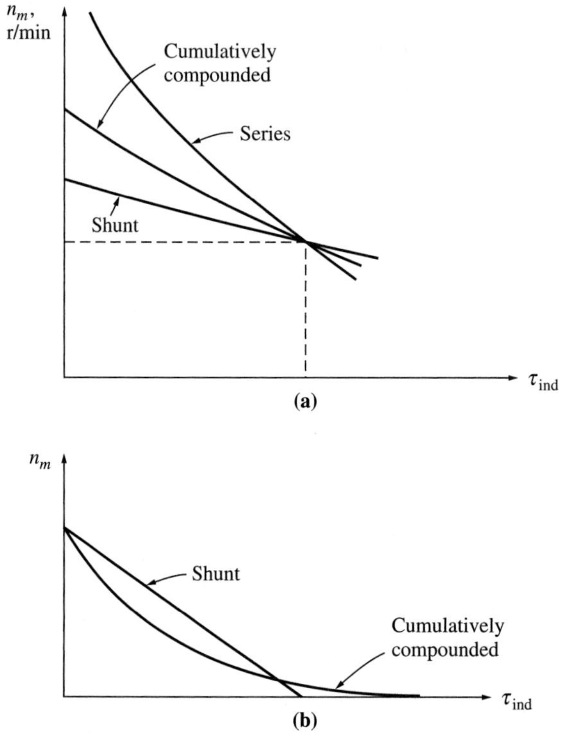

FIGURE 8–25
(a) The torque–speed characteristic of a cumulatively compounded dc motor compared to series and shunt motors with the same full-load rating. (b) The torque–speed characteristic of a cumulatively compounded dc motor compared to a shunt motor with the same no-load speed.

becomes quite important and the torque–speed curve begins to look like a series motor's characteristic. A comparison of the torque–speed characteristics of each of these types of machines is shown in Figure 8–25.

To determine the characteristic curve of a cumulatively compounded dc motor by nonlinear analysis, the approach is similar to that for the shunt and series motors seen before. Such an analysis will be illustrated in a later example.

The Torque–Speed Characteristic of a Differentially Compounded DC Motor

In a differentially compounded dc motor, *the shunt magnetomotive force and series magnetomotive force subtract from each other*. This means that as the load on the motor increases, I_A increases and *the flux in the motor decreases*. But as the flux decreases, the speed of the motor increases. This speed increase causes another increase in load, which further increases I_A, further decreasing the flux, and increasing the speed again. The result is that a differentially compounded motor is

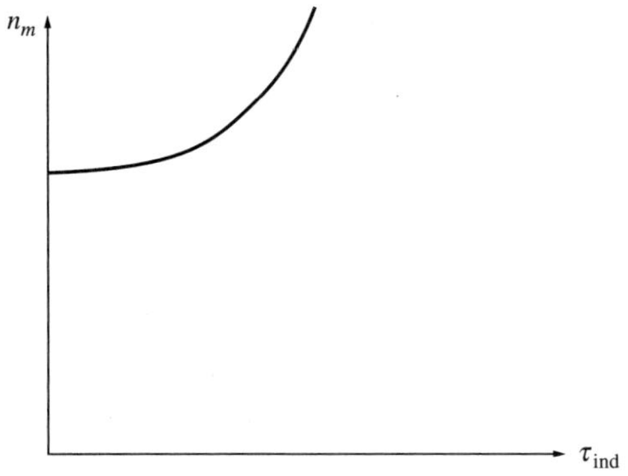

FIGURE 8–26
The torque–speed characteristic of a differentially compounded dc motor.

unstable and tends to run away. This instability is *much* worse than that of a shunt motor with armature reaction. It is so bad that a differentially compounded motor is unsuitable for any application.

To make matters worse, it is impossible to start such a motor. At starting conditions the armature current and the series field current are very high. Since the series flux subtracts from the shunt flux, the series field can actually reverse the magnetic polarity of the machine's poles. The motor will typically remain still or turn slowly in the wrong direction while burning up, because of the excessive armature current. When this type of motor is to be started, its series field must be short-circuited, so that it behaves as an ordinary shunt motor during the starting period.

Because of the stability problems of the differentially compounded dc motor, it is almost never *intentionally* used. However, a differentially compounded motor can result if the direction of power flow reverses in a cumulatively compounded generator. For that reason, if cumulatively compounded dc generators are used to supply power to a system, they will have a reverse-power trip circuit to disconnect them from the line if the power flow reverses. No motor–generator set in which power is expected to flow in both directions can use a differentially compounded motor, and therefore it cannot use a cumulatively compounded generator.

A typical terminal characteristic for a differentially compounded dc motor is shown in Figure 8–26.

The Nonlinear Analysis of Compounded DC Motors

The determination of the torque and speed of a compounded dc motor is illustrated in Example 8–6.

Example 8–6. A 100-hp, 250-V compounded dc motor with compensating windings has an internal resistance, including the series winding, of 0.04 Ω. There are 1000 turns per pole on the shunt field and 3 turns per pole on the series winding. The machine is shown in Figure 8–27, and its magnetization curve is shown in Figure 8–9. At no load, the field resistor has been adjusted to make the motor run at 1200 r/min. The core, mechanical, and stray losses may be neglected.

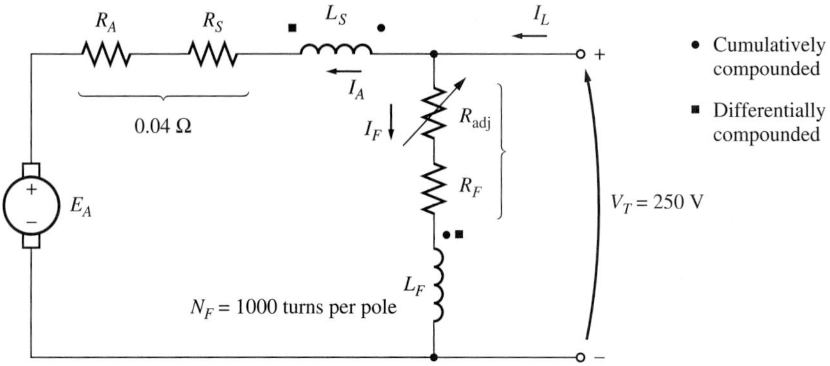

FIGURE 8–27
The compounded dc motor in Example 8–6.

(*a*) What is the shunt field current in this machine at no load?
(*b*) If the motor is cumulatively compounded, find its speed when $I_A = 200$ A.
(*c*) If the motor is differentially compounded, find its speed when $I_A = 200$ A.

Solution

(*a*) At no load, the armature current is zero, so the internal generated voltage of the motor must equal V_T, which means that it must be 250 V. From the magnetization curve, a field current of 5 A will produce a voltage E_A of 250 V at 1200 r/min. Therefore, the shunt field current must be 5 A.

(*b*) When an armature current of 200 A flows in the motor, the machine's internal generated voltage is

$$E_A = V_T - I_A(R_A + R_S)$$
$$= 250 \text{ V} - (200 \text{ A})(0.04\ \Omega) = 242 \text{ V}$$

The effective field current of this cumulatively compounded motor is

$$I_F^* = I_F + \frac{N_{SE}}{N_F} I_A - \frac{\mathscr{F}_{AR}}{N_F} \tag{8–28}$$
$$= 5 \text{ A} + \frac{3}{1000} 200 \text{ A} = 5.6 \text{ A}$$

From the magnetization curve, $E_{A0} = 262$ V at speed $n_0 = 1200$ r/min. Therefore, the motor's speed will be

$$n_m = \frac{E_A}{E_{A0}} n_0$$
$$= \frac{242 \text{ V}}{262 \text{ V}} 1200 \text{ r/min} = 1108 \text{ r/min}$$

(*c*) If the machine is differentially compounded, the effective field current is

$$I_F^* = I_F - \frac{N_{SE}}{N_F} I_A - \frac{\mathscr{F}_{AR}}{N_F} \tag{8–28}$$
$$= 5 \text{ A} - \frac{3}{1000} 200 \text{ A} = 4.4 \text{ A}$$

From the magnetization curve, $E_{A0} = 236$ V at speed $n_0 = 1200$ r/min. Therefore, the motor's speed will be

$$n_m = \frac{E_A}{E_{A0}} n_0$$

$$= \frac{242 \text{ V}}{236 \text{ V}} 1200 \text{ r/min} = 1230 \text{ r/min}$$

Notice that the speed of the cumulatively compounded motor decreases with load, while the speed of the differentially compounded motor increases with load.

Speed Control in the Cumulatively Compounded DC Motor

The techniques available for the control of speed in a cumulatively compounded dc motor are the same as those available for a shunt motor:

1. Change the field resistance R_F.
2. Change the armature voltage V_A.
3. Change the armature resistance R_A.

The arguments describing the effects of changing R_F or V_A are very similar to the arguments given earlier for the shunt motor.

Theoretically, the differentially compounded dc motor could be controlled in a similar manner. Since the differentially compounded motor is almost never used, that fact hardly matters.

8.8 DC MOTOR STARTERS

In order for a dc motor to function properly on the job, it must have some special control and protection equipment associated with it. The purposes of this equipment are

1. To protect the motor against damage due to short circuits in the equipment
2. To protect the motor against damage from long-term overloads
3. To protect the motor against damage from excessive starting currents
4. To provide a convenient manner in which to control the operating speed of the motor

The first three functions will be discussed in this section, and the fourth function will be considered in Section 8.9.

DC Motor Problems on Starting

In order for a dc motor to function properly, it must be protected from physical damage during the starting period. At starting conditions, the motor is not turning, and so $E_A = 0$ V. Since the internal resistance of a normal dc motor is very low

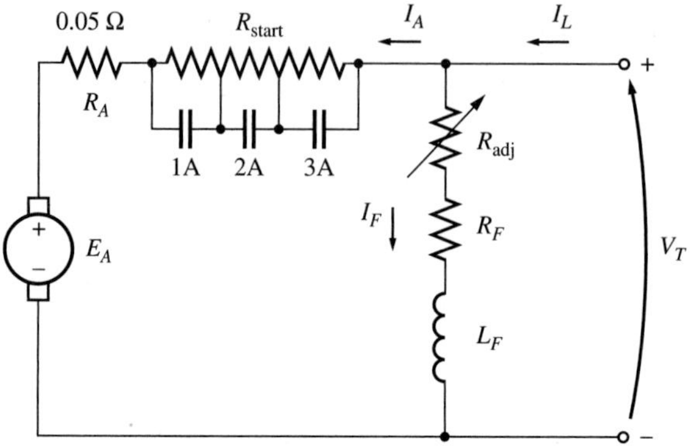

FIGURE 8–28
A shunt motor with a starting resistor in series with its armature. Contacts 1A, 2A, and 3A short-circuit portions of the starting resistor when they close.

compared to its size (3 to 6 percent per unit for medium-size motors), a *very* high current flows.

Consider, for example, the 50-hp, 250-V motor in Example 8–1. This motor has an armature resistance R_A of 0.06 Ω, and a full-load current less than 200 A, but the current on starting is

$$I_A = \frac{V_T - E_A}{R_A}$$

$$= \frac{250\text{ V} - 0\text{ V}}{0.06\ \Omega} = 4167\text{ A}$$

This current is over 20 times the motor's rated full-load current. It is possible for a motor to be severely damaged by such currents, even if they last for only a moment.

A solution to the problem of excess current during starting is to insert a *starting resistor* in series with the armature to limit the current flow until E_A can build up to do the limiting. This resistor must not be in the circuit permanently, because it would result in excessive losses and would cause the motor's torque–speed characteristic to drop off excessively with an increase in load.

Therefore, a resistor must be inserted into the armature circuit to limit current flow at starting, and it must be removed again as the speed of the motor builds up. In modern practice, a starting resistor is made up of a series of pieces, each of which is removed from the motor circuit in succession as the motor speeds up, in order to limit the current in the motor to a safe value while never reducing it to too low a value for rapid acceleration.

Figure 8–28 shows a shunt motor with an extra starting resistor that can be cut out of the circuit in segments by the closing of the 1A, 2A, and 3A contacts. Two actions are necessary in order to make a working motor starter. The first is to pick the size and number of resistor segments necessary in order to limit the starting current to its desired bounds. The second is to design a control circuit that

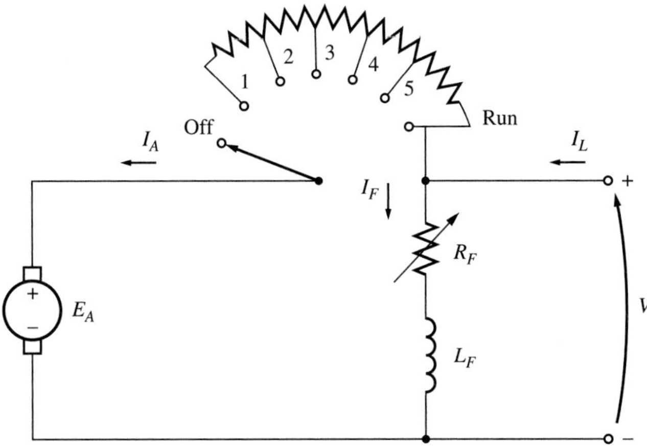

FIGURE 8–29
A manual dc motor starter.

shuts the resistor bypass contacts at the proper time to remove those parts of the resistor from the circuit.

Some older dc motor starters used a continuous starting resistor which was gradually cut out of the circuit by a person moving its handle (Figure 8–29). This type of starter had problems, as it largely depended on the person starting the motor not to move its handle too quickly or too slowly. If the resistance were cut out too quickly (before the motor could speed up enough), the resulting current flow would be too large. On the other hand, if the resistance were cut out too slowly, the starting resistor could burn up. Since they depended on a person for their correct operation, these motor starters were subject to the problem of human error. They have almost entirely been displaced in new installations by automatic starter circuits.

Example 8–7 illustrates the selection of the size and number of resistor segments needed by an automatic starter circuit. The question of the timing required to cut the resistor segments out of the armature circuit will be examined later.

Example 8–7. Figure 8–28 shows a 100-hp, 250-V, 350-A shunt dc motor with an armature resistance of 0.05 Ω. We wish to design a starter circuit for this motor which will limit the maximum starting current to *twice* its rated value and which will switch out sections of resistance as the armature current falls to its rated value.

(*a*) How many stages of starting resistance will be required to limit the current to the range specified?

(*b*) What must the value of each segment of the resistor be? At what voltage should each stage of the starting resistance be cut out?

Solution

(*a*) The starting resistor must be selected so that the current flow equals twice the rated current of the motor when it is first connected to the line. As the motor starts to speed up, an internal generated voltage E_A will be produced in the motor. Since this voltage opposes the terminal voltage of the motor, the increasing internal generated voltage decreases the current flow in the motor. When the current flowing in the motor falls to rated current, a section of the starting resistor must be taken out to increase the starting current back up to 200 percent of rated current. As the motor continues to speed up, E_A continues to rise and the armature current continues to fall. When the current flowing in the motor falls to rated current again, another section of the starting resistor must be taken out. This process repeats until the starting resistance to be removed at a given stage is less than the resistance of the motor's armature circuit. At that point, the motor's armature resistance will limit the current to a safe value all by itself.

How many steps are required to accomplish the current limiting? To find out, define R_{tot} as the original resistance in the starting circuit. So R_{tot} is the sum of the resistance of each stage of the starting resistor together with the resistance of the armature circuit of the motor:

$$R_{tot} = R_1 + R_2 + \cdots + R_A \tag{8–29}$$

Now define $R_{tot,i}$ as the total resistance left in the starting circuit after stages 1 to i have been shorted out. The resistance left in the circuit after removing stages 1 through i is

$$R_{tot,i} = R_{i+1} + \cdots + R_A \tag{8–30}$$

Note also that the initial starting resistance must be

$$R_{tot} = \frac{V_T}{I_{max}}$$

In the first stage of the starter circuit, resistance R_1 must be switched out of the circuit when the current I_A falls to

$$I_A = \frac{V_T - E_A}{R_{tot}} = I_{min}$$

After switching that part of the resistance out, the armature current must jump to

$$I_A = \frac{V_T - E_A}{R_{tot,1}} = I_{max}$$

Since E_A $(= K\phi\omega)$ is directly proportional to the speed of the motor, which cannot change instantaneously, the quantity $V_T - E_A$ must be constant at the instant the resistance is switched out. Therefore,

$$I_{min}R_{tot} = V_T - E_A = I_{max}R_{tot,1}$$

or the resistance left in the circuit after the first stage is switched out is

$$R_{tot,1} = \frac{I_{min}}{I_{max}} R_{tot} \tag{8–31}$$

By direct extension, the resistance left in the circuit after the nth stage is switched out is

$$R_{tot,n} = \left(\frac{I_{min}}{I_{max}}\right)^n R_{tot} \tag{8–32}$$

The starting process is completed when $R_{tot,n}$ for stage n is less than or equal to the internal armature resistance R_A of the motor. At that point, R_A can limit the current to the desired value all by itself. At the boundary where $R_A = R_{tot,n}$

$$R_A = R_{tot,n} = \left(\frac{I_{min}}{I_{max}}\right)^n R_{tot} \tag{8–33}$$

$$\frac{R_A}{R_{tot}} = \left(\frac{I_{min}}{I_{max}}\right)^n \tag{8–34}$$

Solving for n yields

$$n = \frac{\log (R_A/R_{tot})}{\log (I_{min}/I_{max})} \tag{8–35}$$

where n must be rounded up to the next integer value, since it is not possible to have a fractional number of starting stages. If n has a fractional part, then when the final stage of starting resistance is removed, the armature current of the motor will jump up to a value smaller than I_{max}.

In this particular problem, the ratio $I_{min}/I_{max} = 0.5$, and R_{tot} is

$$R_{tot} = \frac{V_T}{I_{max}} = \frac{250\text{ V}}{700\text{ A}} = 0.357\ \Omega$$

so

$$n = \frac{\log (R_A/R_{tot})}{\log (I_{min}/I_{max})} = \frac{\log (0.05\ \Omega/0.357\ \Omega)}{\log (350\text{ A}/700\text{ A})} = 2.84$$

The number of stages required will be three.

(*b*) The armature circuit will contain the armature resistor R_A and three starting resistors R_1, R_2, and R_3. This arrangement is shown in Figure 8–28.

At first, $E_A = 0$ V and $I_A = 700$ A, so

$$I_A = \frac{V_T}{R_A + R_1 + R_2 + R_3} = 700\text{ A}$$

Therefore, the total resistance must be

$$R_A + R_1 + R_2 + R_3 = \frac{250\text{ V}}{700\text{ A}} = 0.357\ \Omega \tag{8–36}$$

This total resistance will be placed in the circuit until the current falls to 350 A. This occurs when

$$E_A = V_T - I_A R_{tot} = 250\text{ V} - (350\text{ A})(0.357\ \Omega) = 125\text{ V}$$

When $E_A = 125$ V, I_A has fallen to 350 A and it is time to cut out the first starting resistor R_1. When it is cut out, the current should jump back to 700 A. Therefore,

$$R_A + R_2 + R_3 = \frac{V_T - E_A}{I_{max}} = \frac{250\text{ V} - 125\text{ V}}{700\text{ A}} = 0.1786\ \Omega \tag{8–37}$$

This total resistance will be in the circuit until I_A again falls to 350 A. This occurs when E_A reaches

$$E_A = V_T - I_A R_{tot} = 250\text{ V} - (350\text{ A})(0.1786\ \Omega) = 187.5\text{ V}$$

When $E_A = 187.5$ V, I_A has fallen to 350 A and it is time to cut out the second starting resistor R_2. When it is cut out, the current should jump back to 700 A. Therefore,

$$R_A + R_3 = \frac{V_T - E_A}{I_{\max}} = \frac{250 \text{ V} - 187.5 \text{ V}}{700 \text{ A}} = 0.0893 \ \Omega \tag{8–38}$$

This total resistance will be in the circuit until I_A again falls to 350 A. This occurs when E_A reaches

$$E_A = V_T - I_A R_{\text{tot}} = 250 \text{ V} - (350 \text{ A})(0.0893 \ \Omega) = 218.75 \text{ V}$$

When $E_A = 218.75$ V, I_A has fallen to 350 A and it is time to cut out the third starting resistor R_3. When it is cut out, only the internal resistance of the motor is left. By now, though, R_A alone can limit the motor's current to

$$I_A = \frac{V_T - E_A}{R_A} = \frac{250 \text{ V} - 218.75 \text{ V}}{0.05 \ \Omega}$$
$$= 625 \text{ A} \qquad \text{(less than allowed maximum)}$$

From this point on, the motor can speed up by itself.

From Equations (8–34) to (8–36), the required resistor values can be calculated:

$$R_3 = R_{\text{tot},3} - R_A = 0.0893 \ \Omega - 0.05 \ \Omega = 0.0393 \ \Omega$$
$$R_2 = R_{\text{tot},2} - R_3 - R_A = 0.1786 \ \Omega - 0.0393 \ \Omega - 0.05 \ \Omega = 0.0893 \ \Omega$$
$$R_1 = R_{\text{tot},1} - R_2 - R_3 - R_A = 0.357 \ \Omega - 0.1786 \ \Omega - 0.0393 \ \Omega - 0.05 \ \Omega = 0.1786 \ \Omega$$

And R_1, R_2, and R_3 are cut out when E_A reaches 125, 187.5, and 218.75 V, respectively.

DC Motor Starting Circuits

Once the starting resistances have been selected, how can their shorting contacts be controlled to ensure that they shut at exactly the correct moment? Several different schemes are used to accomplish this switching, and two of the most common approaches will be examined in this section. Before that is done, though, it is necessary to introduce some of the components used in motor-starting circuits.

Figure 8–30 illustrates some of the devices commonly used in motor-control circuits. The devices illustrated are fuses, push button switches, relays, time delay relays, and overloads.

Figure 8–30a shows a symbol for a fuse. The fuses in a motor-control circuit serve to protect the motor against the danger of short circuits. They are placed in the power supply lines leading to motors. If a motor develops a short circuit, the fuses in the line leading to it will burn out, opening the circuit before any damage has been done to the motor itself.

Figure 8–30b shows spring-type push button switches. There are two basic types of such switches—normally open and normally shut. *Normally open* contacts are open when the button is resting and closed when the button has been

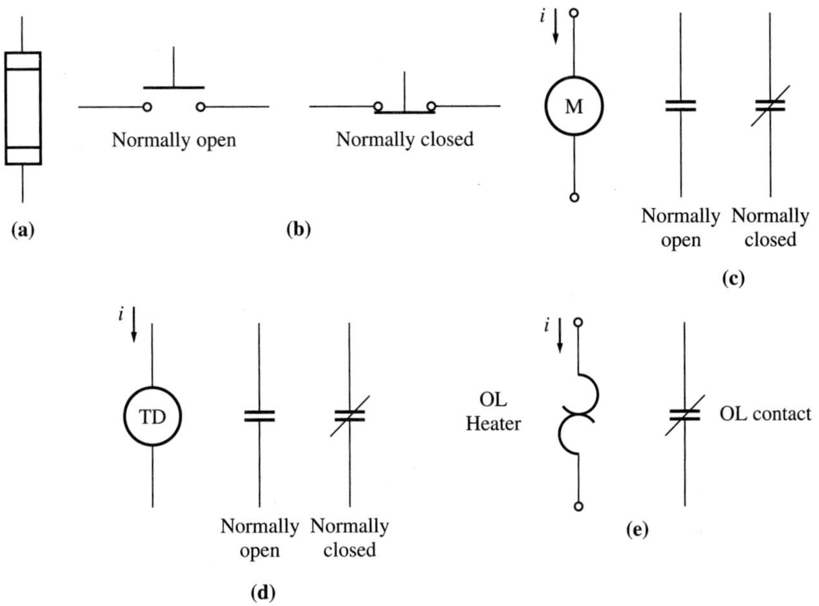

FIGURE 8–30
(a) A fuse. (b) Normally open and normally closed push button switches. (c) A relay coil and contacts. (d) A time delay relay and contacts. (e) An overload and its normally closed contacts.

pushed, while *normally closed* contacts are closed when the button is resting and open when the button has been pushed.

A relay is shown in Figure 8–30c. It consists of a main coil and a number of contacts. The main coil is symbolized by a circle, and the contacts are shown as parallel lines. The contacts are of two types—normally open and normally closed. *A normally open* contact is one which is open when the relay is deenergized, and a *normally closed* contact is one which is closed when the relay is deenergized. When electric power is applied to the relay (the relay is energized), its contacts change state: The normally open contacts close, and the normally closed contacts open.

A time delay relay is shown in Figure 8–30d. It behaves exactly like an ordinary relay except that when it is energized there is an adjustable time delay before its contacts change state.

An overload is shown in Figure 8–30e. It consists of a heater coil and some normally shut contacts. The current flowing to a motor passes through the heater coils. If the load on a motor becomes too large, then the current flowing to the motor will heat up the heater coils, which will cause the normally shut contacts of the overload to open. These contacts can in turn activate some types of motor protection circuitry.

One common motor-starting circuit using these components is shown in Figure 8–31. In this circuit, a series of time delay relays shut contacts which remove each section of the starting resistor at approximately the correct time after

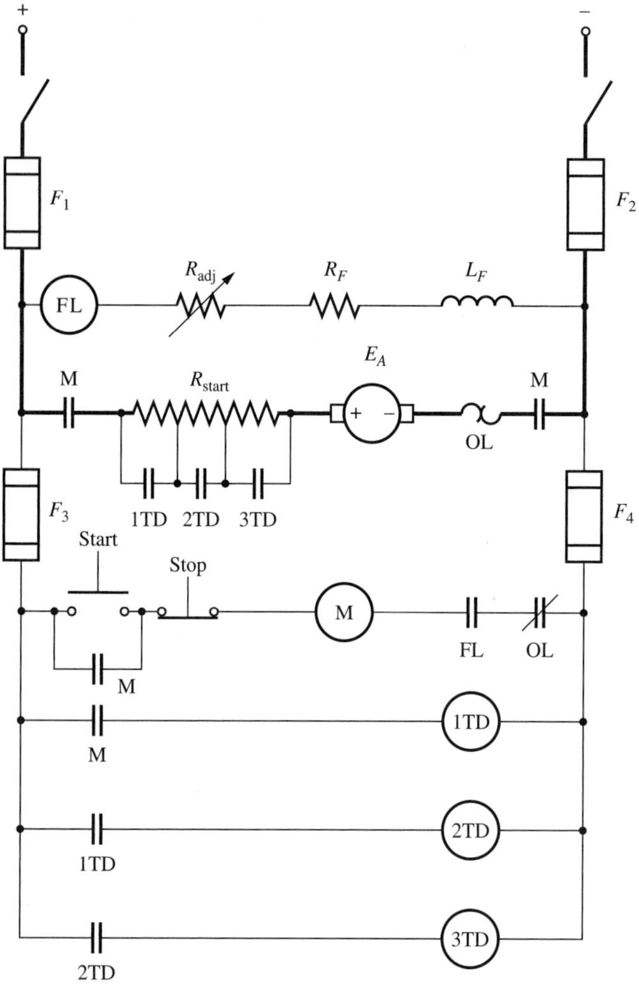

FIGURE 8–31
A dc motor starting circuit rising time delay relays to cut out the starting resistor.

power is applied to the motor. When the start button is pushed in this circuit, the motor's armature circuit is connected to its power supply, and the machine starts with all resistance in the circuit. However, relay 1TD energizes at the same time as the motor starts, so after some delay the 1TD contacts will shut and remove part of the starting resistance from the circuit. Simultaneously, relay 2TD is energized, so after another time delay the 2TD contacts will shut and remove the second part of the timing resistor. When the 2TD contacts shut, the 3TD relay is energized, so the process repeats again, and finally the motor runs at full speed with no starting resistance present in its circuit. If the time delays are picked properly, the starting resistors can be cut out at just the right times to limit the motor's current to its design values.

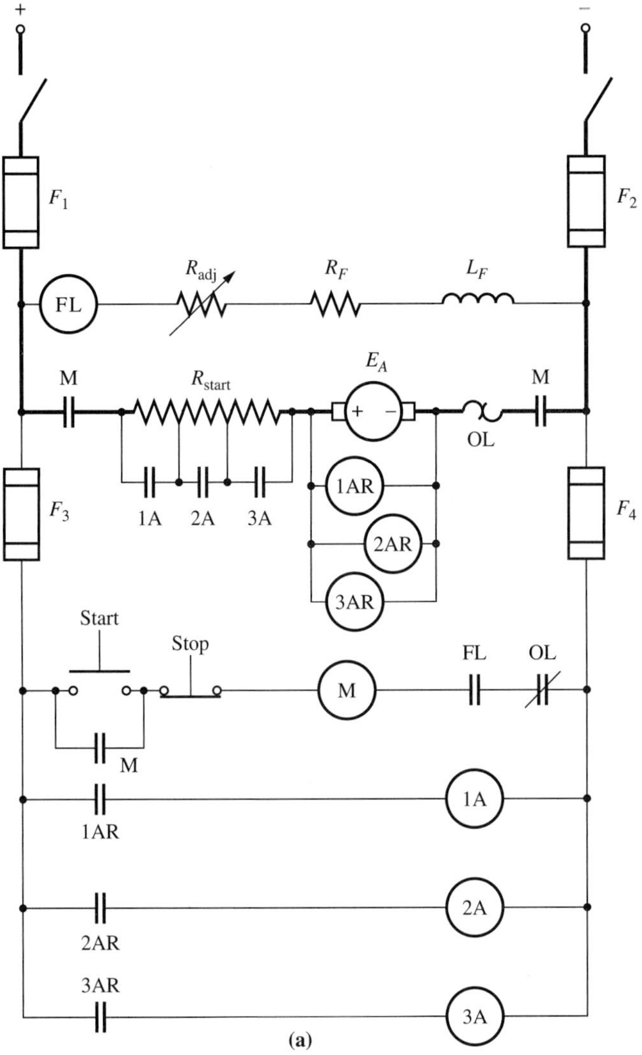

FIGURE 8–32
(a) A dc motor starting circuit using countervoltage-sensing relays to cut out the starting resistor.

Another type of motor starter is shown in Figure 8–32. Here, a series of relays sense the value of E_A in the motor and cut out the starting resistance as E_A rises to preset levels. This type of starter is better than the previous one, since if the motor is loaded heavily and starts more slowly than normal, its armature resistance is still cut out when its current falls to the proper value.

Notice that both starter circuits have a relay in the field circuit labeled FL. This is a *field loss relay*. If the field current is lost for any reason, the field loss

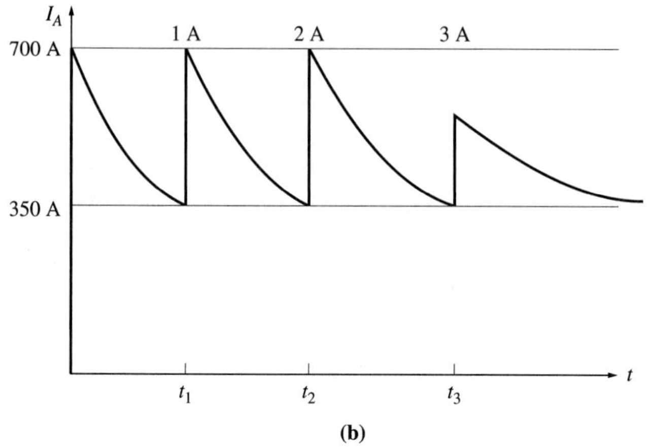

FIGURE 8–32 (concluded)
(b) The armature current in a dc motor during starting.

relay is deenergized, which turns off power to the M relay. When the M relay deenergizes, its normally open contacts open and disconnect the motor from the power supply. This relay prevents the motor from running away if its field current is lost.

Notice also that there is an overload in each motor-starter circuit. If the power drawn from the motor becomes excessive, these overloads will heat up and open the OL normally shut contacts, thus turning off the M relay. When the M relay deenergizes, its normally open contacts open and disconnect the motor from the power supply, so the motor is protected against damage from prolonged excessive loads.

8.9 THE WARD-LEONARD SYSTEM AND SOLID-STATE SPEED CONTROLLERS

The speed of a separately excited, shunt, or compounded dc motor can be varied in one of three ways: by changing the field resistance, changing the armature voltage, or changing the armature resistance. Of these methods, perhaps the most useful is armature voltage control, since it permits wide speed variations without affecting the motor's maximum torque.

A number of motor-control systems have been developed over the years to take advantage of the high torques and variable speeds available from the armature voltage control of dc motors. In the days before solid-state electronic components became available, it was difficult to produce a varying dc voltage. In fact, the normal way to vary the armature voltage of a dc motor was to provide it with its own separate dc generator.

An armature voltage control system of this sort is shown in Figure 8–33. This figure shows an ac motor serving as a prime mover for a dc generator, which

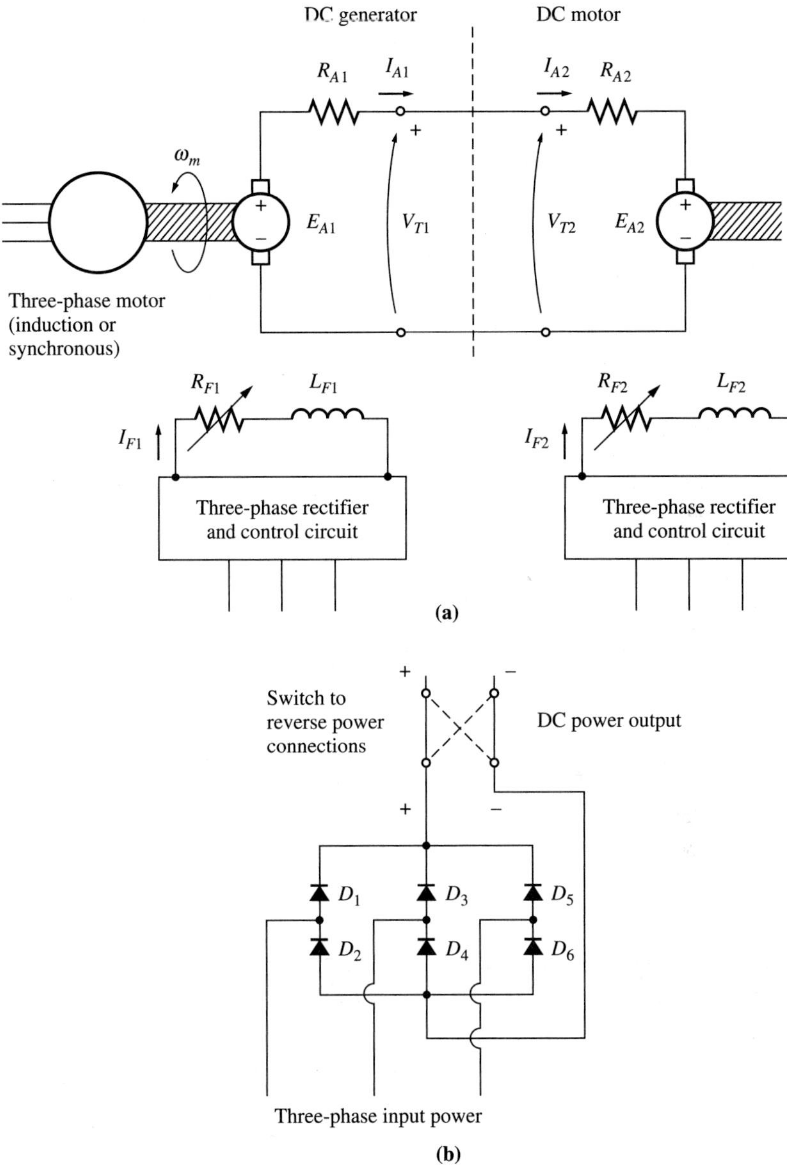

FIGURE 8–33
(a) A Ward-Leonard system for dc motor speed control. (b) The circuit for producing field current in the dc generator and dc motor.

in turn is used to supply a dc voltage to a dc motor. Such a system of machines is called a *Ward-Leonard system*, and it is extremely versatile.

In such a motor-control system, the armature voltage of the motor can be controlled by varying the field current of the dc generator. This armature voltage

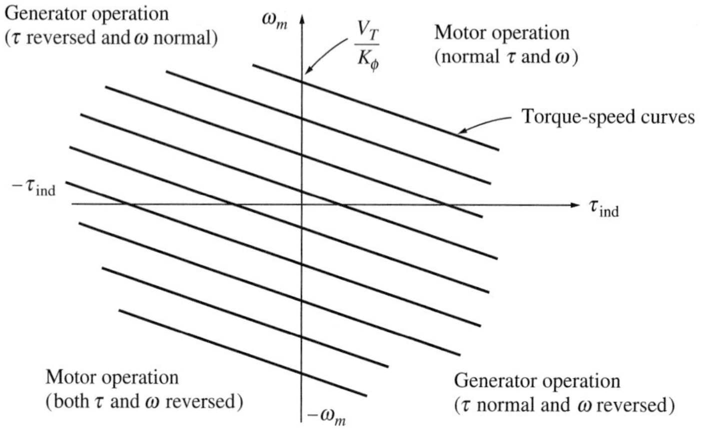

FIGURE 8–34
The operating range of a Ward-Leonard motor-control system. The motor can operate as a motor in either the forward (quadrant 1) or reverse (quadrant 3) direction and it can also regenerate in quadrants 2 and 4.

allows the motor's speed to be smoothly varied between a very small value and the base speed. The speed of the motor can be adjusted above the base speed by reducing the motor's field current. With such a flexible arrangement, total motor speed control is possible.

Furthermore, if the field current of the generator is reversed, then the polarity of the generator's armature voltage will be reversed, too. This will reverse the motor's direction of rotation. Therefore, it is possible to get a very wide range of speed variations *in either direction of rotation* out of a Ward-Leonard dc motor-control system.

Another advantage of the Ward-Leonard system is that it can "regenerate," or return the machine's energy of motion to the supply lines. If a heavy load is first raised and then lowered by the dc motor of a Ward-Leonard system, when the load is falling, the dc motor acts as a generator, supplying power back to the power system. In this fashion, much of the energy required to lift the load in the first place can be recovered, reducing the machine's overall operating costs.

The possible modes of operation of the dc machine are shown in the torque–speed diagram in Figure 8–34. When this motor is rotating in its normal direction and supplying a torque in the direction of rotation, it is operating in the first quadrant of this figure. If the generator's field current is reversed, that will reverse the terminal voltage of the generator, in turn reversing the motor's armature voltage. When the armature voltage reverses with the motor field current remaining unchanged, both the torque and the speed of the motor are reversed, and the machine is operating as a motor in the third quadrant of the diagram. If the torque or the speed alone of the motor reverses while the other quantity does not, then the machine serves as a generator, returning power to the dc power system. Because

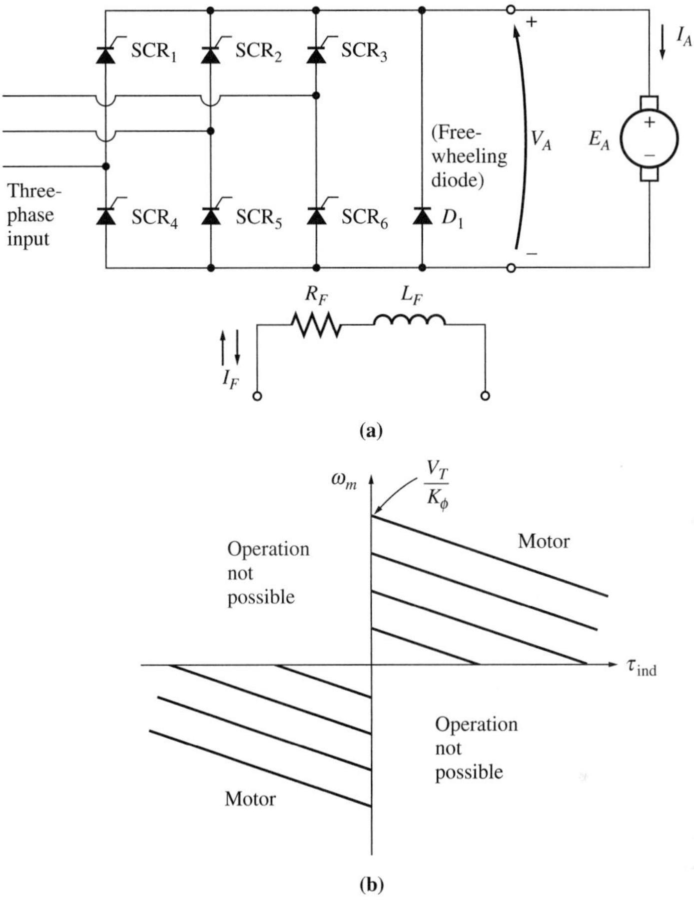

FIGURE 8–35
(a) A two-quadrant solid-state dc motor controller. Since current cannot flow out of the positive terminals of the armature, this motor cannot act as a generator, returning power to the system. (b) The possible operating quadrants of this motor controller.

a Ward-Leonard system permits rotation and regeneration in either direction, it is called a *four-quadrant control system.*

The disadvantages of a Ward-Leonard system should be obvious. One is that the user is forced to buy *three* full machines of essentially equal ratings, which is quite expensive. Another is that three machines will be much less efficient than one. Because of its expense and relatively low efficiency, the Ward-Leonard system has been replaced in new applications by thyristor-based controller circuits.

A simple dc armature voltage controller circuit is shown in Figure 8–35. The average voltage applied to the armature of the motor, and therefore the average speed of the motor, depends on the fraction of the time the supply voltage is applied to the armature. This in turn depends on the relative phase at which the

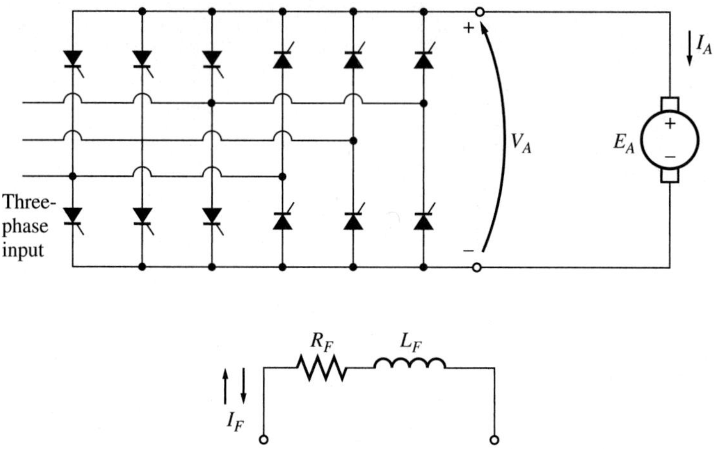

(a)

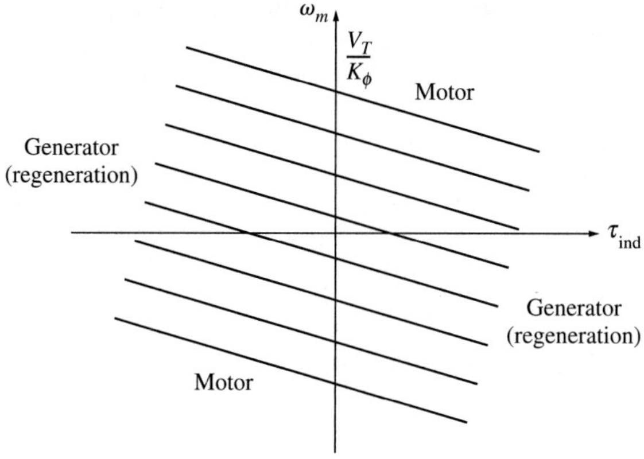

(b)

FIGURE 8–36
(a) A four-quadrant solid-state dc motor controller. (b) The possible operating quadrants of this motor controller.

thyristors in the rectifier circuit are triggered. This particular circuit is only capable of supplying an armature voltage with one polarity, so the motor can only be reversed by switching the polarity of its field connection. Notice that it is not possible for an armature current to flow out the positive terminal of this motor, since current cannot flow backward through a thyristor. Therefore, this motor *cannot* regenerate, and any energy supplied to the motor cannot be recovered. This type of control circuit is a two-quadrant controller, as shown in Figure 8–35b.

A more advanced circuit capable of supplying an armature voltage with either polarity is shown in Figure 8–36. This armature voltage control circuit can

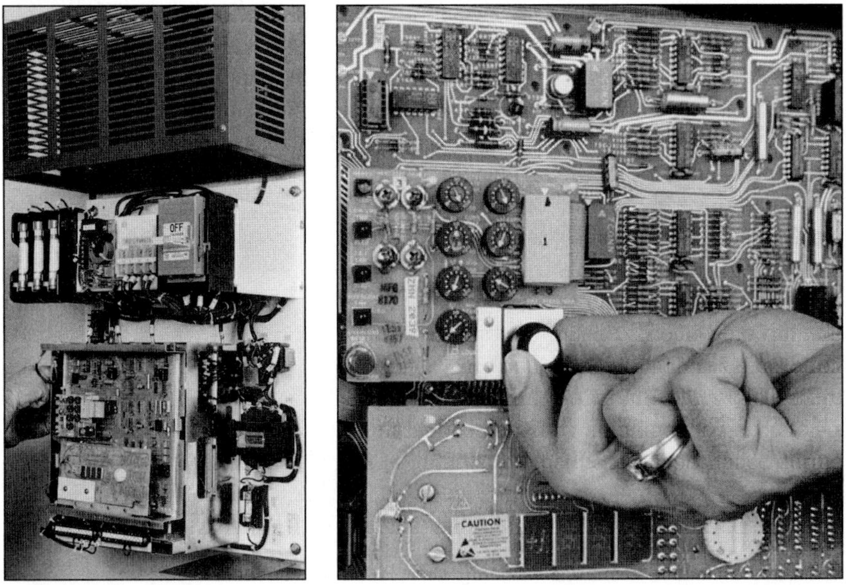

(a) (b)

FIGURE 8–37
(a) A typical solid-state shunt dc motor drive. (*Courtesy of MagneTek, Inc.*) (b) A close-up view of the low-power electronics circuit board, showing the adjustments for current limits, acceleration rate, deceleration rate, minimum speed, and maximum speed. (*Courtesy of MagneTek, Inc.*)

permit a current flow out of the positive terminals of the generator, so a motor with this type of controller can regenerate. If the polarity of the motor field circuit can be switched as well, then the solid-state circuit is a full four-quadrant controller like the Ward-Leonard system.

A two-quadrant or a full four-quadrant controller built with thyristors is cheaper than the two extra complete machines needed for the Ward-Leonard system, so solid-state speed-control systems have largely displaced Ward-Leonard systems in new applications.

A typical two-quadrant shunt dc motor drive with armature voltage speed control is shown in Figure 8–37, and a simplified block diagram of the drive is shown in Figure 8–38. This drive has a constant field voltage supplied by a three-phase full-wave rectifier, and a variable armature terminal voltage supplied by six thyristors arranged as a three-phase full-wave rectifier. The voltage supplied to the armature of the motor is controlled by adjusting the firing angle of the thyristors in the bridge. Since this motor controller has a fixed field voltage and a variable armature voltage, it is only able to control the speed of the motor at speeds less than or equal to the base speed (see "Changing the Armature Voltage" in Section 8.4). The controller circuit is identical with that shown in Figure 8–35, except that all of the control electronics and feedback circuits are shown.

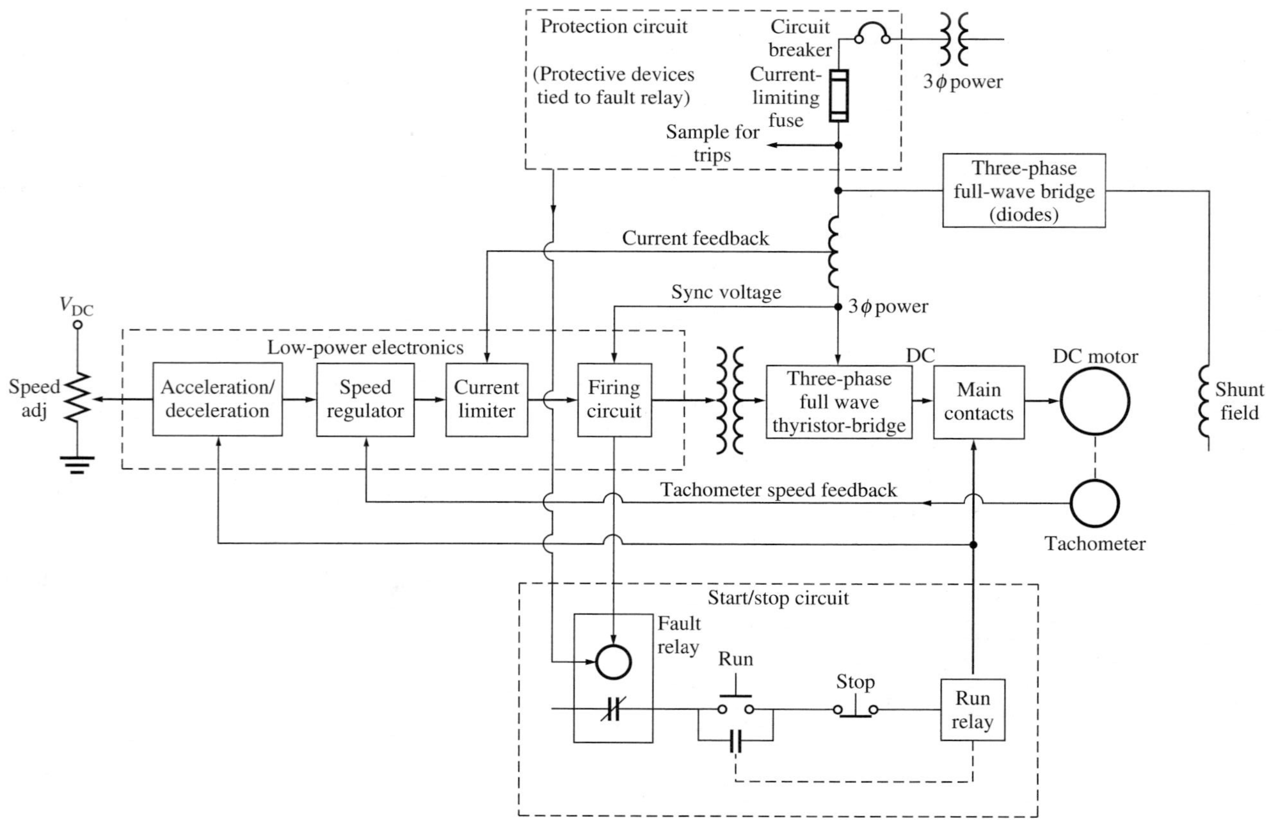

FIGURE 8–38
A simplified block diagram of the typical solid-state shunt dc motor drive shown in Figure 8–37. (*Simplified from a block diagram provided by MagneTek, Inc.*)

The major sections of this dc motor drive include:

1. A protection circuit section to protect the motor from excessive armature currents, low terminal voltage, and loss of field current.
2. A start/stop circuit to connect and disconnect the motor from the line.
3. A high-power electronics section to convert three-phase ac power to dc power for the motor's armature and field circuits.
4. A low-power electronics section to provide firing pulses to the thyristors which supply the armature voltage to the motor. This section contains several major subsections, which will be described below.

Protection Circuit Section

The protection circuit section combines several different devices which together ensure the safe operation of the motor. Some typical safety devices included in this type of drive are

1. *Current-limiting fuses*, to disconnect the motor quickly and safely from the power line in the event of a short circuit within the motor. Current-limiting fuses can interrupt currents of up to several hundred thousand amperes.
2. An *instantaneous static trip*, which shuts down the motor if the armature current exceeds 300 percent of its rated value. If the armature current exceeds the maximum allowed value, the trip circuit activates the fault relay, which deenergizes the run relay, opening the main contactors and disconnecting the motor from the line.
3. An *inverse-time overload trip*, which guards against sustained overcurrent conditions not great enough to trigger the instantaneous static trip but large enough to damage the motor if allowed to continue indefinitely. The term *inverse time* implies that the higher the overcurrent flowing in the motor, the faster the overload acts (Figure 8–39). For example, an inverse-time trip might take a full minute to trip if the current flow were 150 percent of the rated current of the motor, but take 10 seconds to trip if the current flow were 200 percent of the rated current of the motor.
4. An *undervoltage trip*, which shuts down the motor if the line voltage supplying the motor drops by more than 20 percent.
5. A *field loss trip*, which shuts down the motor if the field circuit is lost.
6. An *overtemperature trip*, which shuts down the motor if it is in danger of overheating.

Start/Stop Circuit Section

The start/stop circuit section contains the controls needed to start and stop the motor by opening or closing the main contacts connecting the motor to the line. The motor is started by pushing the run button, and it is stopped either by pushing the

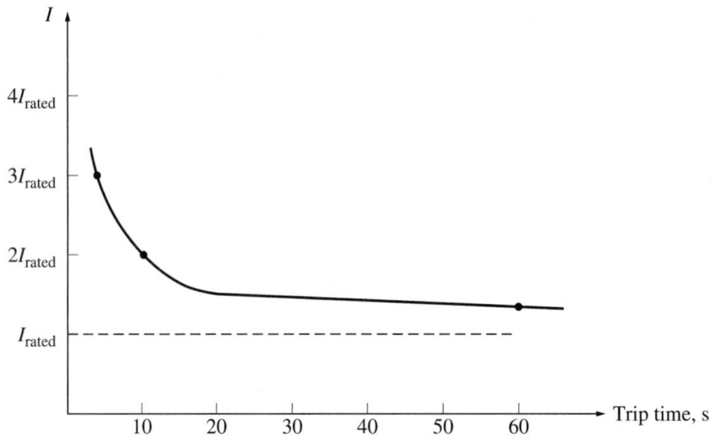

FIGURE 8–39
An inverse-time trip characteristic.

stop button or by energizing the fault relay. In either case, the run relay is deenergized, and the main contacts connecting the motor to the line are opened.

High-Power Electronics Section

The high-power electronics section contains a three-phase full-wave diode rectifier to provide a constant voltage to the field circuit of the motor and a three-phase full-wave thyristor rectifier to provide a variable voltage to the armature circuit of the motor.

Low-Power Electronics Section

The low-power electronics section provides firing pulses to the thyristors, which supply the armature voltage to the motor. By adjusting the firing time of the thyristors, the low-power electronics section adjusts the motor's average armature voltage. The low-power electronics section contains the following subsystems:

1. *Speed regulation circuit.* This circuit measures the speed of the motor with a tachometer, compares that speed with the desired speed (a reference voltage level), and increases or decreases the armature voltage as necessary to keep the speed constant at the desired value. For example, suppose that the load on the shaft of the motor is increased. If the load is increased, then the motor will slow down. The decrease in speed will reduce the voltage generated by the tachometer, which is fed into the speed regulation circuit. Because the voltage level corresponding to the speed of the motor has fallen below the reference voltage, the speed regulator circuit will advance the firing time of the thyristors, producing a higher armature voltage. The higher armature voltage will tend to increase the speed of the motor back to the desired level (see Figure 8–40).

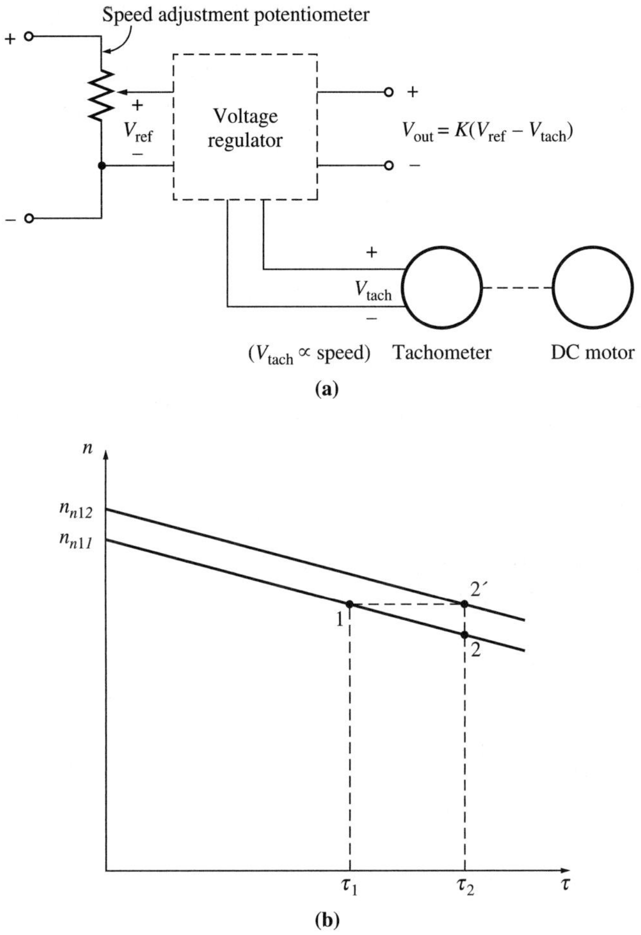

FIGURE 8–40
(a) The speed regulator circuit produces an output voltage which is proportional to the difference between the desired speed of the motor (set by V_{ref}) and the actual speed of the motor (measured by V_{tach}). This output voltage is applied to the firing circuit in such a way that the larger the output voltage becomes, the earlier the thyristors in the drive turn on and the higher the average terminal voltage becomes. (b) The effect of increasing load on a shunt dc motor with a speed regulator. The load in the motor is increased. If no regulator were present, the motor would slow down and operate at point 2. When the speed regulator is present, it detects the decrease in speed and boosts the armature voltage of the motor to compensate. This raises the whole torque–speed characteristic curve of the motor, resulting in operation at point 2′.

With proper design, a circuit of this type can provide speed regulations of 0.1 percent between no-load and full-load conditions.

The desired operating speed of the motor is controlled by changing the reference voltage level. The reference voltage level can be adjusted with a small potentiometer, as shown in Figure 8–40.

2. *Current-limiting circuit.* This circuit measures the steady-state current flowing to the motor, compares that current with the desired maximum current (set by a reference voltage level), and decreases the armature voltage as necessary to keep the current from exceeding the desired maximum value. The desired maximum current can be adjusted over a wide range, say from 0 to 200 percent or more of the motor's rated current. This current limit should typically be set at greater than rated current, so that the motor can accelerate under full-load conditions.
3. *Acceleration/deceleration circuit.* This circuit limits the acceleration and deceleration of the motor to a safe value. Whenever a dramatic speed change is commanded, this circuit intervenes to ensure that the transition from the original speed to the new speed is smooth and does not cause an excessive armature current transient in the motor.

The acceleration/deceleration circuit completely eliminates the need for a starting resistor, since starting the motor is just another kind of large speed change, and the acceleration/deceleration circuit acts to cause a smooth increase in speed over time. This gradual smooth increase in speed limits the current flowing in the machine's armature to a safe value.

8.10 DC MOTOR EFFICIENCY CALCULATIONS

To calculate the efficiency of a dc motor, the following losses must be determined:

1. Copper losses
2. Brush drop losses
3. Mechanical losses
4. Core losses
5. Stray losses

The copper losses in the motor are the I^2R losses in the armature and field circuits of the motor. These losses can be found from a knowledge of the currents in the machine and the two resistances. To determine the resistance of the armature circuit in a machine, block its rotor so that it cannot turn and apply a *small* dc voltage to the armature terminals. Adjust that voltage until the current flowing in the armature is equal to the rated armature current of the machine. The ratio of the applied voltage to the resulting armature current flow is R_A. The reason that the current should be about equal to full-load value when this test is done is that R_A varies with temperature, and at the full-load value of the current, the armature windings will be near their normal operating temperature.

The resulting resistance will not be entirely accurate, because

1. The cooling that normally occurs when the motor is spinning will not be present.

2. Since there is an ac voltage in the rotor conductors during normal operation, they suffer from some amount of skin effect, which further raises armature resistance.

IEEE Standard 113 (Reference 5) deals with test procedures for dc machines. It gives a more accurate procedure for determining R_A, which can be used if needed.

The field resistance is determined by supplying the full-rated field voltage to the field circuit and measuring the resulting field current. The field resistance R_F is just the ratio of the field voltage to the field current.

Brush drop losses are often approximately lumped together with copper losses. If they are treated separately, they can be determined from a plot of contact potential versus current for the particular type of brush being used. The brush drop losses are just the product of the brush voltage drop V_{BD} and the armature current I_A.

The core and mechanical losses are usually determined together. If a motor is allowed to turn freely at no load and at rated speed, then there is no output power from the machine. Since the motor is at no load, I_A is very small and the armature copper losses are negligible. Therefore, if the field copper losses are subtracted from the input power of the motor, the remaining input power must consist of the mechanical and core losses of the machine at that speed. These losses are called the *no-load rotational losses* of the motor. As long as the motor's speed remains nearly the same as it was when the losses were measured, the no-load rotational losses are a good estimate of mechanical and core losses under load in the machine.

Example 8–8 illustrates the determination of a motor's efficiency.

Example 8–8. A 50-hp, 250-V, 1200 r/min shunt dc motor has a rated armature current of 170 A and a rated field current of 5 A. When its rotor is blocked, an armature voltage of 10.2 V (exclusive of brushes) produces 170 A of current flow, and a field voltage of 250 V produces a field current flow of 5 A. The brush voltage drop is assumed to be 2 V. At no load with the terminal voltage equal to 240 V, the armature current is equal to 13.2 A, the field current is 4.8 A, and the motor's speed is 1150 r/min.

(a) How much power is output from this motor at rated conditions?
(b) What is the motor's efficiency?

Solution

The armature resistance of this machine is approximately

$$R_A = \frac{10.2\text{ V}}{170\text{ A}} = 0.06\ \Omega$$

and the field resistance is

$$R_F = \frac{250\text{ V}}{5\text{ A}} = 50\ \Omega$$

Therefore, at full load the armature I^2R losses are

$$P_A = (170\text{ A})^2(0.06\ \Omega) = 1734\text{ W}$$

and the field circuit I^2R losses are

$$P_F = (5 \text{ A})^2(50 \ \Omega) = 1250 \text{ W}$$

The brush losses at full load are given by

$$P_{\text{brush}} = V_{\text{BD}}I_A = (2 \text{ V})(170 \text{ A}) = 340 \text{ W}$$

The rotational losses at full load are essentially equivalent to the rotational losses at no load, since the no-load and full-load speeds of the motor do not differ too greatly. These losses may be ascertained by determining the input power to the armature circuit at no load and assuming that the armature copper and brush drop losses are negligible, meaning that the no-load armature input power is equal to the rotational losses:

$$P_{\text{tot}} = P_{\text{core}} + P_{\text{mech}} = (240 \text{ V})(13.2 \text{ A}) = 3168 \text{ W}$$

(a) The input power of this motor at the rated load is given by

$$P_{\text{in}} = V_T I_L = (250 \text{ V})(175 \text{ A}) = 43{,}750 \text{ W}$$

Its output power is given by

$$\begin{aligned} P_{\text{out}} &= P_{\text{in}} - P_{\text{brush}} - P_{\text{cu}} - P_{\text{core}} - P_{\text{mech}} - P_{\text{stray}} \\ &= 43{,}750 \text{ W} - 340 \text{ W} - 1734 \text{ W} - 1250 \text{ W} - 3168 \text{ W} - (0.01)(43{,}750 \text{ W}) \\ &= 36{,}820 \text{ W} \end{aligned}$$

where the stray losses are taken to be 1 percent of the input power.

(b) The efficiency of this motor at full load is

$$\begin{aligned} \eta &= \frac{P_{\text{out}}}{P_{\text{out}}} \times 100\% \\ &= \frac{36{,}820 \text{ W}}{43{,}750 \text{ W}} \times 100\% = 84.2\% \end{aligned}$$

8.11 INTRODUCTION TO DC GENERATORS

DC generators are dc machines used as generators. As previously noted, there is no real difference between a generator and a motor except for the direction of power flow. There are five major types of dc generators, classified according to the manner in which their field flux is produced:

1. *Separately excited generator.* In a separately excited generator, the field flux is derived from a separate power source independent of the generator itself.
2. *Shunt generator.* In a shunt generator, the field flux is derived by connecting the field circuit directly across the terminals of the generator.
3. *Series generator.* In a series generator, the field flux is produced by connecting the field circuit in series with the armature of the generator.
4. *Cumulatively compounded generator.* In a cumulatively compounded generator, both a shunt and a series field are present, and their effects are additive.
5. *Differentially compounded generator.* In a differentially compounded generator, both a shunt and a series field are present, but their effects are subtractive.

These various types of dc generators differ in their terminal (voltage–current) characteristics, and therefore in the applications to which they are suited.

FIGURE 8–41
The first practical dc generator. This is an exact duplicate of the "long-legged Mary Ann," Thomas Edison's first commercial dc generator, which was built in 1879. It was rated at 5 kW, 100 V, and 1200 r/min. (*Courtesy of General Electric Company.*)

DC generators are compared by their voltages, power ratings, efficiencies, and voltage regulations. *Voltage regulation* (VR) is defined by the equation

$$\boxed{\mathrm{VR} = \frac{V_{\mathrm{nl}} - V_{\mathrm{fl}}}{V_{\mathrm{fl}}} \times 100\%} \tag{8–39}$$

where V_{nl} is the no-load terminal voltage of the generator and V_{fl} is the full-load terminal voltage of the generator. It is a rough measure of the shape of the generator's voltage–current characteristic—a positive voltage regulation means a drooping characteristic, and a negative voltage regulation means a rising characteristic.

All generators are driven by a source of mechanical power, which is usually called the *prime mover* of the generator. A prime mover for a dc generator may be a steam turbine, a diesel engine, or even an electric motor. Since the speed of the prime mover affects the output voltage of a generator, and since prime movers can vary widely in their speed characteristics, it is customary to compare the voltage regulation and output characteristics of different generators, *assuming constant-speed prime movers*. Throughout this chapter, a generator's speed will be assumed to be constant unless a specific statement is made to the contrary.

Dc generators are quite rare in modern power systems. Even dc power systems such as those in automobiles now use ac generators plus rectifiers to produce dc power. However, they have had a limited renaissance in the last few years as power sources for standalone cellular telephone towers.

The equivalent circuit of a dc generator is shown in Figure 8–42, and a simplified version of the equivalent circuit is shown in Figure 8–43. They look similar to the equivalent circuits of a dc motor, except that the direction of current flow and the brush loss are reversed.

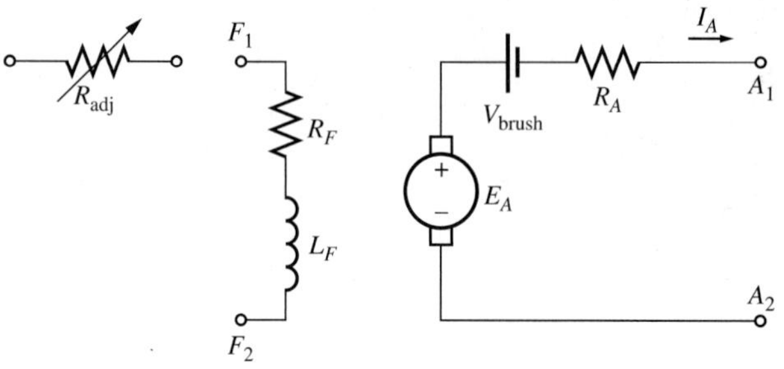

FIGURE 8–42
The equivalent circuit of a dc generator.

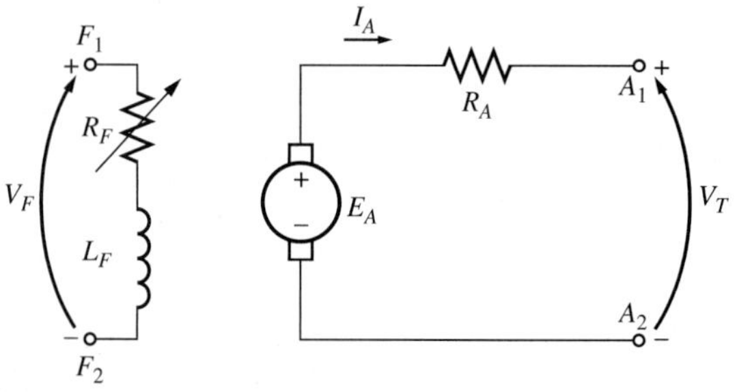

FIGURE 8–43
A simplified equivalent circuit of a dc generator, with R_F combining the resistances of the field coils and the variable control resistor.

8.12 THE SEPARATELY EXCITED GENERATOR

A separately excited dc generator is a generator whose field current is supplied by a separate external dc voltage source. The equivalent circuit of such a machine is shown in Figure 8–44. In this circuit, the voltage V_T represents the actual voltage measured at the terminals of the generator, and the current I_L represents the current flowing in the lines connected to the terminals. The internal generated voltage is E_A, and the armature current is I_A. It is clear that the armature current is equal to the line current in a separately excited generator:

$$I_A = I_L \tag{8–40}$$

The Terminal Characteristic of a Separately Excited DC Generator

The *terminal characteristic* of a device is a plot of the output quantities of the device versus each other. For a dc generator, the output quantities are its terminal voltage and line current. The terminal characteristic of a separately excited generator is

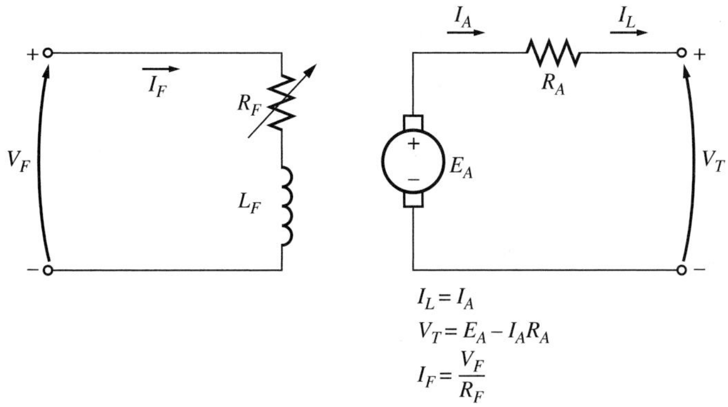

FIGURE 8–44
A separately excited dc generator.

thus a plot of V_T versus I_L for a constant speed ω. By Kirchhoff's voltage law, the terminal voltage is

$$\boxed{V_T = E_A - I_A R_A} \tag{8–41}$$

Since the internal generated voltage is independent of I_A, the terminal characteristic of the separately excited generator is a straight line, as shown in Figure 8–45a.

What happens in a generator of this sort when the load is increased? When the load supplied by the generator is increased, I_L (and therefore I_A) increases. As the armature current increases, the $I_A R_A$ drop increases, so the terminal voltage of the generator falls.

This terminal characteristic is not always entirely accurate. In generators without compensating windings, an increase in I_A causes an increase in armature reaction, and armature reaction causes flux weakening. This flux weakening causes a decrease in $E_A = K\phi\downarrow\omega_m$ which further decreases the terminal voltage of the generator. The resulting terminal characteristic is shown in Figure 8–45b. In all future plots, the generators will be assumed to have compensating windings unless stated otherwise. However, it is important to realize that armature reaction can modify the characteristics if compensating windings are not present.

Control of Terminal Voltage

The terminal voltage of a separately excited dc generator can be controlled by changing the internal generated voltage E_A of the machine. By Kirchhoff's voltage law $V_T = E_A - I_A R_A$, so if E_A increases, V_T will increase, and if E_A decreases, V_T will decrease. Since the internal generated voltage E_A is given by the equation $E_A = K\phi\omega_m$, there are two possible ways to control the voltage of this generator:

1. *Change the speed of rotation.* If ω increases, then $E_A = K\phi\omega_m\uparrow$ increases, so $V_T = E_A\uparrow - I_A R_A$ increases as well.

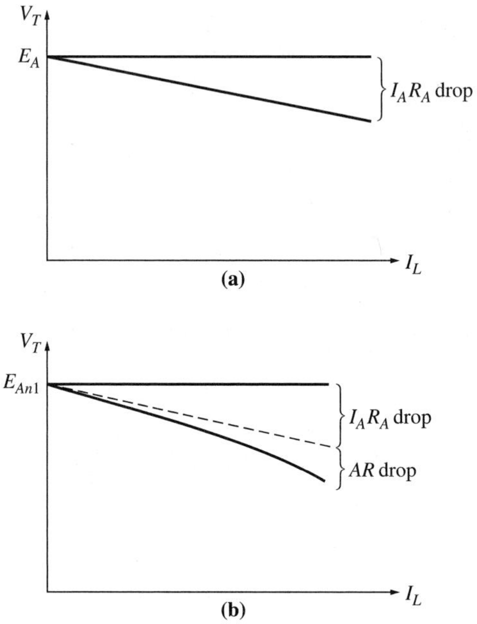

FIGURE 8–45
The terminal characteristic of a separately excited dc generator (a) with and (b) without compensating windings.

2. *Change the field current.* If R_F is decreased, then the field current increases ($I_F = V_F / R_F\downarrow$). Therefore, the flux ϕ in the machine increases. As the flux rises, $E_A = K\phi\uparrow\omega_m$ must rise too, so $V_T = E_A\uparrow - I_AR_A$ increases.

In many applications, the speed range of the prime mover is quite limited, so the terminal voltage is most commonly controlled by changing the field current. A separately excited generator driving a resistive load is shown in Figure 8–46a. Figure 8–46b shows the effect of a decrease in field resistance on the terminal voltage of the generator when it is operating under a load.

Nonlinear Analysis of a Separately Excited DC Generator

Because the internal generated voltage of a generator is a nonlinear function of its magnetomotive force, it is not possible to calculate simply the value of E_A to be expected from a given field current. The magnetization curve of the generator must be used to accurately calculate its output voltage for a given input voltage.

In addition, if a machine has armature reaction, its flux will be reduced with each increase in load, causing E_A to decrease. The only way to accurately determine the output voltage in a machine with armature reaction is to use graphical analysis.

The total magnetomotive force in a separately excited generator is the field circuit magnetomotive force less the magnetomotive force due to armature reaction (AR):

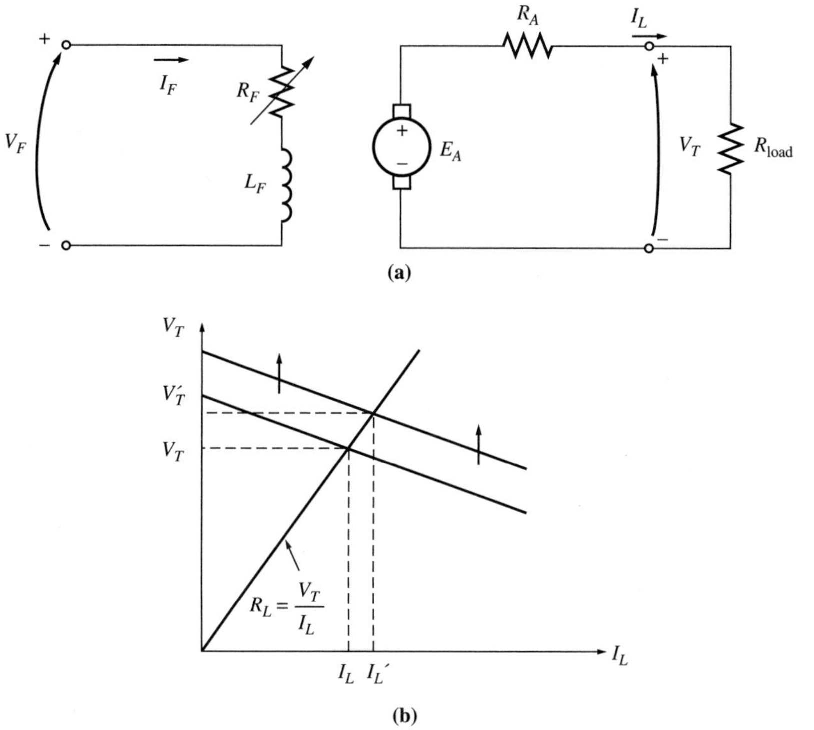

FIGURE 8–46
(a) A separately excited dc generator with a resistive load. (b) The effect of a decrease in field resistance on the output voltage of the generator.

$$\mathscr{F}_{\text{net}} = N_F I_F - \mathscr{F}_{\text{AR}} \tag{8–42}$$

As with dc motors, it is customary to define an *equivalent field current* that would produce the same output voltage as the combination of all the magnetomotive forces in the machine. The resulting voltage E_{A0} can then be determined by locating that equivalent field current on the magnetization curve. The equivalent field current of a separately excited dc generator is given by

$$I_F^* = I_F - \frac{\mathscr{F}_{\text{AR}}}{N_F} \tag{8–43}$$

Also, the difference between the speed of the magnetization curve and the real speed of the generator must be taken into account using Equation (8–13):

$$\frac{E_A}{E_{A0}} = \frac{n_m}{n_0} \tag{8–13}$$

The following example illustrates the analysis of a separately excited dc generator.

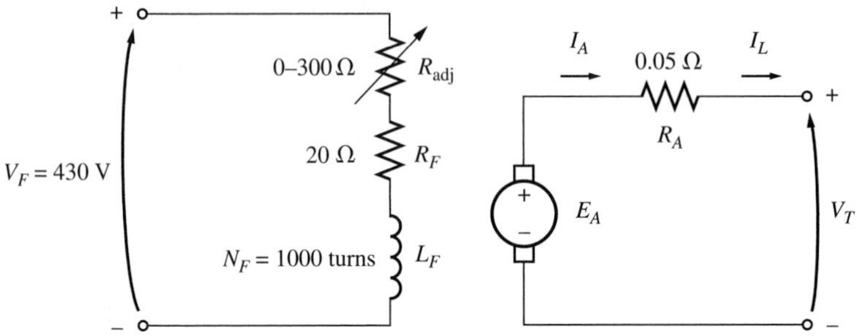

FIGURE 8–47
The separately excited dc generator in Example 8–9.

Example 8–9. A separately excited dc generator is rated at 172 kW, 430 V, 400 A, and 1800 r/min. It is shown in Figure 8–47, and its magnetization curve is shown in Figure 8–48. This machine has the following characteristics:

$$R_A = 0.05\ \Omega \qquad V_F = 430\text{ V}$$
$$R_F = 20\ \Omega \qquad N_F = 1000\text{ turns per pole}$$
$$R_{adj} = 0\text{ to }300\ \Omega$$

(a) If the variable resistor R_{adj} in this generator's field circuit is adjusted to 63 Ω and the generator's prime mover is driving it at 1600 r/min, what is this generator's no-load terminal voltage?

(b) What would its voltage be if a 360-A load were connected to its terminals? Assume that the generator has compensating windings.

(c) What would its voltage be if a 360-A load were connected to its terminals but the generator does not have compensating windings? Assume that its armature reaction at this load is 450 A • turns.

(d) What adjustment could be made to the generator to restore its terminal voltage to the value found in part *a*?

(e) How much field current would be needed to restore the terminal voltage to its no-load value? (Assume that the machine has compensating windings.) What is the required value for the resistor R_{adj} to accomplish this?

Solution

(a) If the generator's total field circuit resistance is

$$R_F + R_{adj} = 83\ \Omega$$

then the field current in the machine is

$$I_F = \frac{V_F}{R_F} = \frac{430\text{ V}}{83\ \Omega} = 5.2\text{ A}$$

From the machine's magnetization curve, this much current would produce a voltage $E_{A0} = 430$ V at a speed of 1800 r/min. Since this generator is actually turning at $n_m = 1600$ r/min, its internal generated voltage E_A will be

$$\frac{E_A}{E_{A0}} = \frac{n_m}{n_0} \tag{8–13}$$

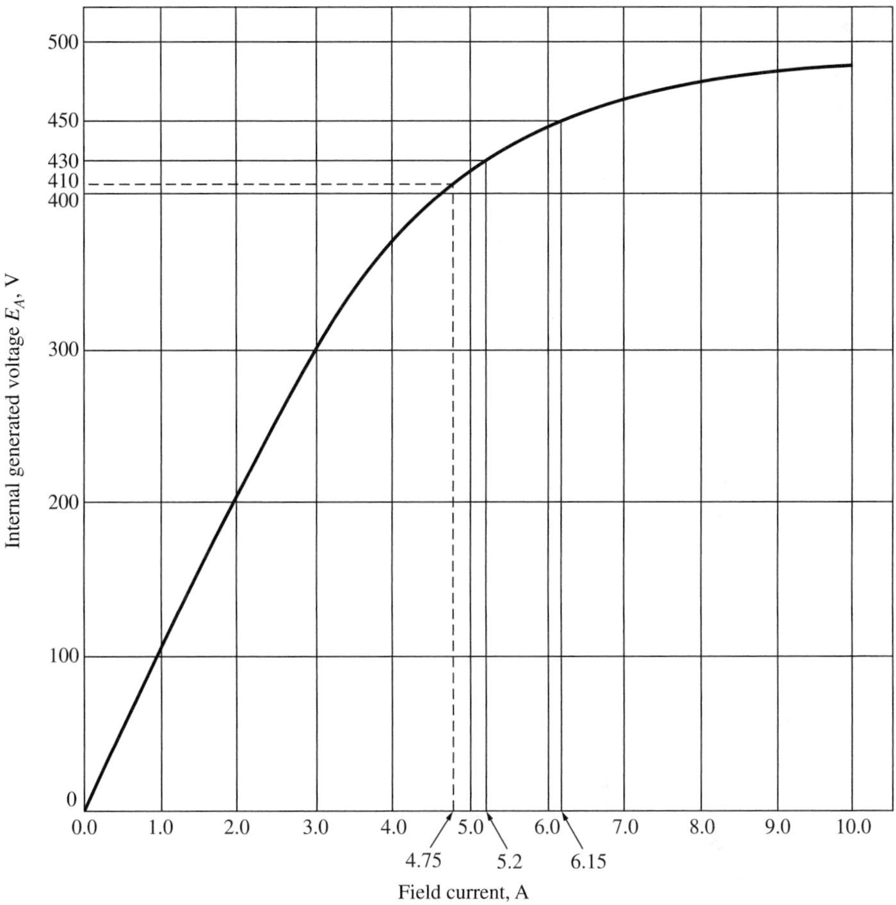

Note: When the field current is zero, E_A is about 3 V.

FIGURE 8–48
The magnetization curve for the generator in Example 8–9.

$$E_A = \frac{1600 \text{ r/min}}{1800 \text{ r/min}} 430 \text{ V} = 382 \text{ V}$$

Since $V_T = E_A$ at no-load conditions, the output voltage of the generator is $V_T =$ 382 V.

(b) If a 360-A load were connected to this generator's terminals, the terminal voltage of the generator would be

$$V_T = E_A - I_A R_A = 382 \text{ V} - (360 \text{ A})(0.05 \ \Omega) = 364 \text{ V}$$

(c) If a 360-A load were connected to this generator's terminals and the generator had 450 A • turns of armature reaction, the effective field current would be

$$I_F^* = I_F - \frac{\mathscr{F}_{\text{AR}}}{N_F} = 5.2 \text{ A} - \frac{450 \text{ A} \bullet \text{turns}}{1000 \text{ turns}} = 4.75 \text{ A}$$

From the magnetization curve, $E_{A0} = 410$ V, so the internal generated voltage at 1600 r/min would be

$$\frac{E_A}{E_{A0}} = \frac{n}{n_0} \qquad (8\text{–}13)$$

$$E_A = \frac{1600 \text{ r/min}}{1800 \text{ r/min}} 410 \text{ V} = 364 \text{ V}$$

Therefore, the terminal voltage of the generator would be

$$V_T = E_A - I_A R_A = 364 \text{ V} - (360 \text{ A})(0.05 \ \Omega) = 346 \text{ V}$$

It is lower than before due to the armature reaction.

(d) The voltage at the terminals of the generator has fallen, so to restore it to its original value, the voltage of the generator must be increased. This requires an increase in E_A, which implies that R_{adj} must be decreased to increase the field current of the generator.

(e) For the terminal voltage to go back up to 382 V, the required value of E_A is

$$E_A = V_T + I_A R_A = 382 \text{ V} + (360 \text{ A})(0.05 \ \Omega) = 400 \text{ V}$$

To get a voltage E_A of 400 V at $n_m = 1600$ r/min, the equivalent voltage at 1800 r/min would be

$$\frac{E_A}{E_{A0}} = \frac{n_m}{n_0} \qquad (8\text{–}13)$$

$$E_{A0} = \frac{1800 \text{ r/min}}{1600 \text{ r/min}} 400 \text{ V} = 450 \text{ V}$$

From the magnetization curve, this voltage would require a field current of $I_F =$ 6.15 A. The field circuit resistance would have to be

$$R_F + R_{\text{adj}} = \frac{V_F}{I_F}$$

$$20 \ \Omega + R_{\text{adj}} = \frac{430 \text{ V}}{6.15 \text{ A}} = 69.9 \ \Omega$$

$$R_{\text{adj}} = 49.9 \ \Omega \approx 50 \ \Omega$$

Notice that, for the same field current and load current, the generator with armature reaction had a lower output voltage than the generator without armature reaction. The armature reaction in this generator is exaggerated to illustrate its effects—it is a good deal smaller in well-designed modern machines.

8.13 THE SHUNT DC GENERATOR

A shunt dc generator is a dc generator that supplies its own field current by having its field connected directly across the terminals of the machine. The equivalent circuit of a shunt dc generator is shown in Figure 8–49. In this circuit, the armature current of the machine supplies both the field circuit and the load attached to the machine:

$$\boxed{I_A = I_F + I_L} \qquad (8\text{–}44)$$

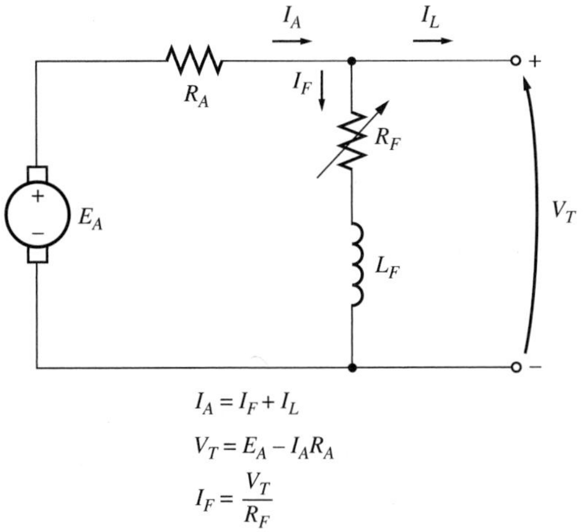

FIGURE 8–49
The equivalent circuit of a shunt dc generator.

The Kirchhoff's voltage law equation for the armature circuit of this machine is

$$\boxed{V_T = E_A - I_A R_A} \tag{8–45}$$

This type of generator has a distinct advantage over the separately excited dc generator in that no external power supply is required for the field circuit. But that leaves an important question unanswered: If the generator supplies its own field current, how does it get the initial field flux to start when it is first turned on?

Voltage Buildup in a Shunt Generator

Assume that the generator in Figure 8–49 has no load connected to it and that the prime mover starts to turn the shaft of the generator. How does an initial voltage appear at the terminals of the machine?

The voltage buildup in a dc generator depends on the presence of a *residual flux* in the poles of the generator. When a generator first starts to turn, an internal voltage will be generated, which is given by

$$E_A = K\phi_{\text{res}}\omega_m$$

This voltage appears at the terminals of the generator (it may only be a volt or two). But when that voltage appears at the terminals, it causes a current to flow in the generator's field coil ($I_F = V_T\uparrow/R_F$). This field current produces a magnetomotive force in the poles, which increases the flux in them. The increase in flux causes an increase in $E_A = K\phi\uparrow\omega_m$, which increases the terminal voltage V_T. When V_T rises, I_F increases further, increasing the flux ϕ more, which increases E_A, etc.

This voltage buildup behavior is shown in Figure 8–50. Notice that it is the effect of magnetic saturation in the pole faces which eventually limits the terminal voltage of the generator.

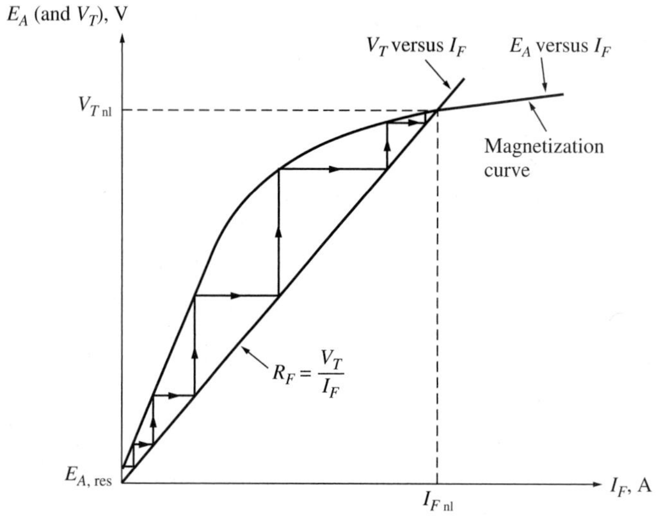

FIGURE 8–50
Voltage buildup on starting in a shunt dc generator.

Figure 8–50 shows the voltage buildup as though it occurred in discrete steps. These steps are drawn in to make obvious the positive feedback between the generator's internal voltage and its field current. In a real generator, the voltage does not build up in discrete steps: Instead, both E_A and I_F increase simultaneously until steady-state conditions are reached.

What if a shunt generator is started and no voltage builds up? What could be wrong? There are several possible causes for the voltage to fail to build up during starting. Among them are

1. *There may be no residual magnetic flux* in the generator to start the process going. If the residual flux $\phi_{\text{res}} = 0$, then $E_A = 0$, and the voltage never builds up. If this problem occurs, disconnect the field from the armature circuit and connect it directly to an external dc source, such as a battery. The current flow from this external dc source will leave a residual flux in the poles, which will then allow normal starting. This procedure is known as "flashing the field."
2. *The direction of rotation of the generator may have been reversed,* or the connections of the field may have been reversed. In either case, the residual flux produces an internal generated voltage E_A. The voltage E_A produces a field current which produces a flux *opposing* the residual flux, instead of adding to it. Under these circumstances, the flux actually *decreases* below ϕ_{res} and no voltage can ever build up.

   If this problem occurs, it can be fixed by reversing the direction of rotation, by reversing the field connections, or by flashing the field with the opposite magnetic polarity.

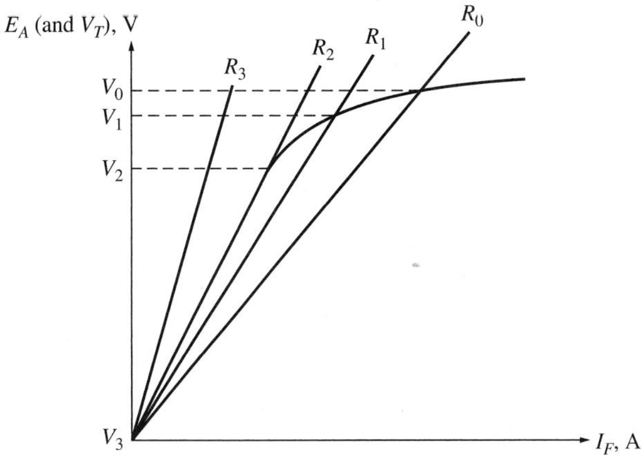

FIGURE 8–51
The effect of shunt field resistance on no-load terminal voltage in a dc generator. If $R_F > R_2$ (the critical resistance), then the generator's voltage will never build up.

3. *The field resistance may be adjusted to a value greater than the critical resistance.* To understand this problem, refer to Figure 8–51. Normally, the shunt generator will build up to the point where the magnetization curve intersects the field resistance line. If the field resistance has the value shown at R_2 in the figure, its line is nearly parallel to the magnetization curve. At that point, the voltage of the generator can fluctuate very widely with only tiny changes in R_F or I_A. This value of the resistance is called the *critical resistance*. If R_F exceeds the critical resistance (as at R_3 in the figure), then the steady-state operating voltage is essentially at the residual level, and it never builds up. The solution to this problem is to reduce R_F.

Since the voltage of the magnetization curve varies as a function of shaft speed, the critical resistance also varies with speed. In general, the lower the shaft speed, the lower the critical resistance.

The Terminal Characteristic of a Shunt DC Generator

The terminal characteristic of a shunt dc generator differs from that of a separately excited dc generator, because the amount of field current in the machine depends on its terminal voltage. To understand the terminal characteristic of a shunt generator, start with the machine unloaded and add loads, observing what happens.

As the load on the generator is increased, I_L increases and so $I_A = I_F + I_L \uparrow$ also increases. An increase in I_A increases the armature resistance voltage drop $I_A R_A$, causing $V_T = E_A - I_A \uparrow R_A$ to decrease. This is precisely the same behavior observed in a separately excited generator. However, when V_T decreases, the field current in the machine decreases with it. This causes the flux in the ma-

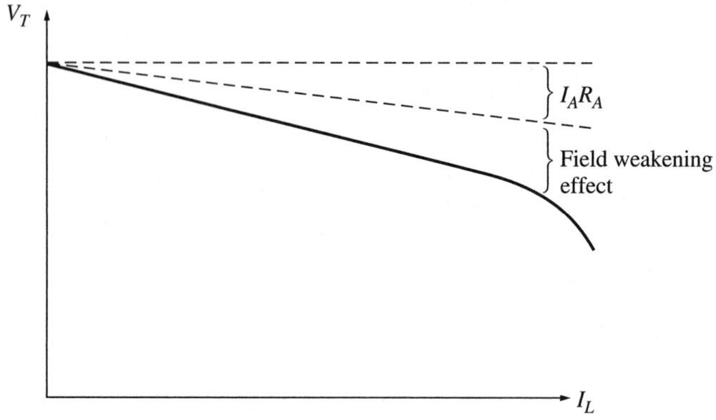

FIGURE 8–52
The terminal characteristic of a shunt dc generator.

chine to decrease, decreasing E_A. Decreasing E_A causes a further decrease in the terminal voltage $V_T = E_A\downarrow - I_A R_A$. The resulting terminal characteristic is shown in Figure 8–52. Notice that the voltage drop-off is steeper than just the $I_A R_A$ drop in a separately excited generator. In other words, the voltage regulation of this generator is worse than the voltage regulation of the same piece of equipment connected separately excited.

Voltage Control for a Shunt DC Generator

As with the separately excited generator, there are two ways to control the voltage of a shunt generator:

1. Change the shaft speed ω_m of the generator.
2. Change the field resistor of the generator, thus changing the field current.

Changing the field resistor is the principal method used to control terminal voltage in real shunt generators. If the field resistor R_F is decreased, then the field current $I_F = V_T / R_F\downarrow$ increases. When I_F increases, the machine's flux ϕ increases, causing the internal generated voltage E_A to increase. The increase in E_A causes the terminal voltage of the generator to increase as well.

The Analysis of Shunt DC Generators

The analysis of a shunt dc generator is somewhat more complicated than the analysis of a separately excited generator, because the field current in the machine depends directly on the machine's own output voltage. First the analysis of shunt generators is studied for machines with no armature reaction, and afterward the effects of armature reaction are included.

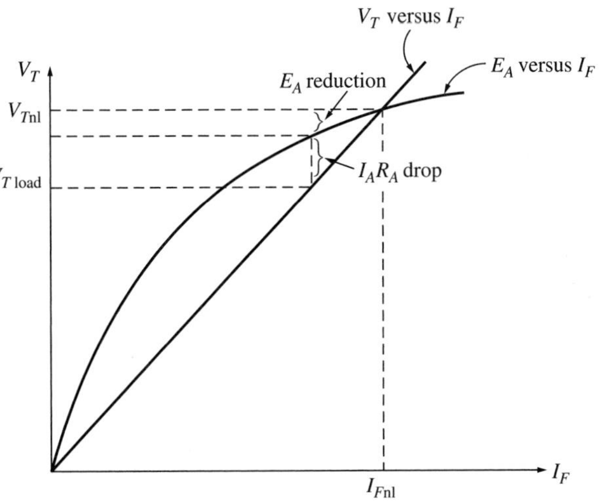

FIGURE 8–53
Graphical analysis of a shunt dc generator with compensating windings.

Figure 8–53 shows a magnetization curve for a shunt dc generator drawn at the actual operating speed of the machine. The field resistance R_F, which is just equal to V_T/I_F, is shown by a straight line laid over the magnetization curve. *At no load,* $V_T = E_A$ and the generator operates at the voltage where the magnetization curve intersects the field resistance line.

The key to understanding the graphical analysis of shunt generators is to remember Kirchhoff's voltage law (KVL):

$$V_T = E_A - I_A R_A \tag{8–45}$$

or

$$\boxed{E_A - V_T = I_A R_A} \tag{8–46}$$

The difference between the internal generated voltage and the terminal voltage is just the I_AR_A drop in the machine. The line of all possible values of E_A is the magnetization curve, and the line of all possible terminal voltages is the resistor line ($I_F = V_T/R_F$). Therefore, to find the terminal voltage for a given load, just determine the I_AR_A drop and locate the place on the graph where that drop fits *exactly* between the E_A line and the V_T line. There are at most two places on the curve where the I_AR_A drop will fit exactly. If there are two possible positions, the one nearer the no-load voltage will represent a normal operating point.

A detailed plot showing several different points on a shunt generator's characteristic is shown in Figure 8–54. Note the dashed line in Figure 8–54b. This line is the terminal characteristic when the load is being reduced. The reason that it does not coincide with the line of increasing load is the hysteresis in the stator poles of the generator.

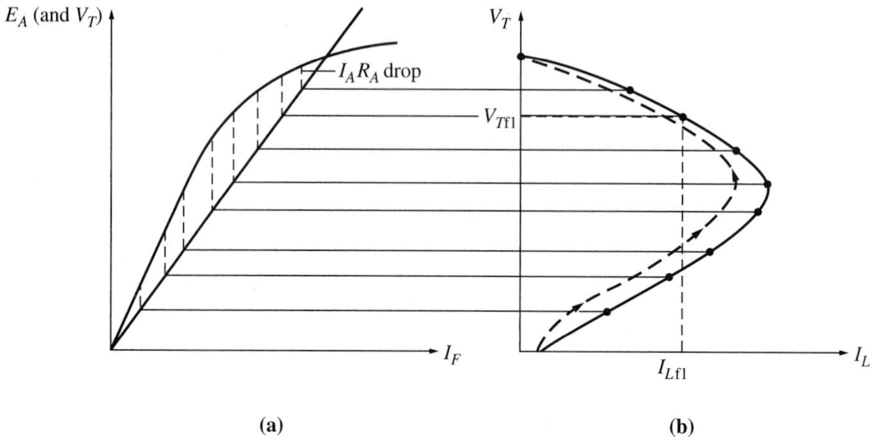

FIGURE 8–54
Graphical derivation of the terminal characteristic of a shunt dc generator.

If armature reaction is present in a shunt generator, this process becomes a little more complicated. The armature reaction produces a demagnetizing magnetomotive force in the generator at the same time that the I_AR_A drop occurs in the machine.

To analyze a generator with armature reaction present, assume that its armature current is known. Then the resistive voltage drop I_AR_A is known, and the demagnetizing magnetomotive force of the armature current is known. The terminal voltage of this generator must be large enough to supply the generator's flux *after the demagnetizing effects of armature reaction have been subtracted.* To meet this requirement both the armature reaction magnetomotive force and the I_AR_A drop must fit between the E_A line and the V_T line. To determine the output voltage for a given magnetomotive force, simply locate the place under the magnetization curve where the triangle formed by the armature reaction and I_AR_A effects *exactly fits* between the line of possible V_T values and the line of possible E_A values (see Figure 8–55).

8.14 THE SERIES DC GENERATOR

A series dc generator is a generator whose field is connected in series with its armature. Since the armature has a *much* higher current than a shunt field, the series field in a generator of this sort will have only a very few turns of wire, and the wire used will be much thicker than the wire in a shunt field. Because magnetomotive force is given by the equation $\mathscr{F} = NI$, exactly the same magnetomotive force can be produced from a few turns with high current as can be produced from many turns with low current. Since the full-load current flows through it, a series field is designed to have the lowest possible resistance. The equivalent circuit of a series dc generator is shown in Figure 8–56. Here, the armature current, field

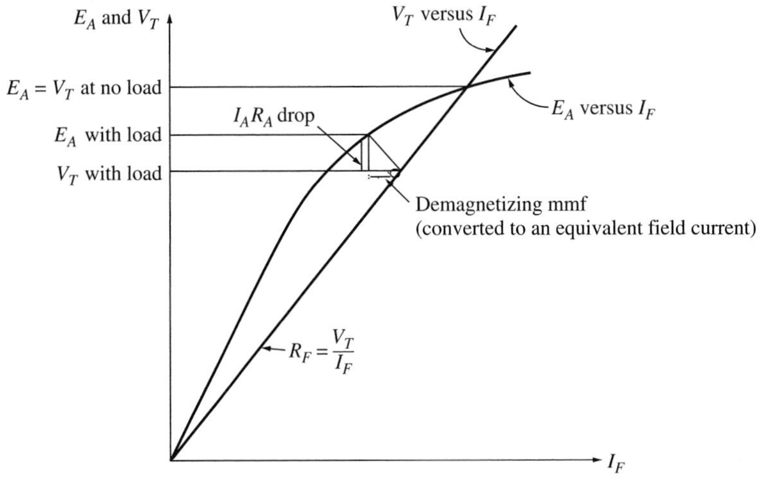

FIGURE 8–55
Graphical analysis of a shunt dc generator with armature reaction.

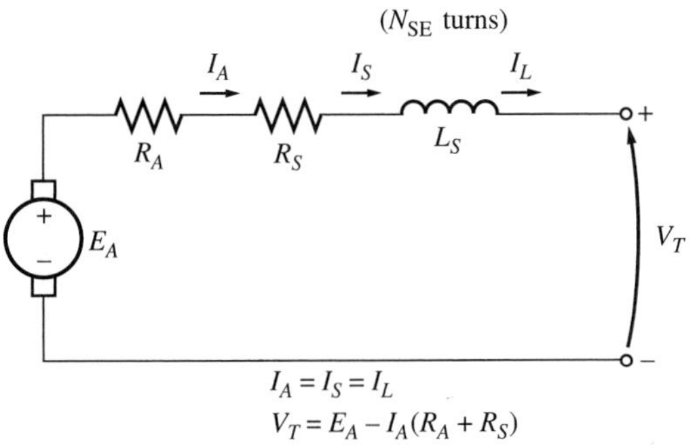

FIGURE 8–56
The equivalent circuit of a series dc generator.

current, and line current all have the same value. The Kirchhoff's voltage law equation for this machine is

$$\boxed{V_T = E_A - I_A(R_A + R_S)} \tag{8–47}$$

The Terminal Characteristic of a Series Generator

The magnetization curve of a series dc generator looks very much like the magnetization curve of any other generator. At no load, however, there is no field current, so V_T is reduced to a small level given by the residual flux in the machine. As the load increases, the field current rises, so E_A rises rapidly. The $I_A(R_A + R_S)$ drop goes up too, but at first the increase in E_A goes up more rapidly than the $I_A(R_A + R_S)$ drop rises, so V_T increases. After a while, the machine approaches saturation, and E_A

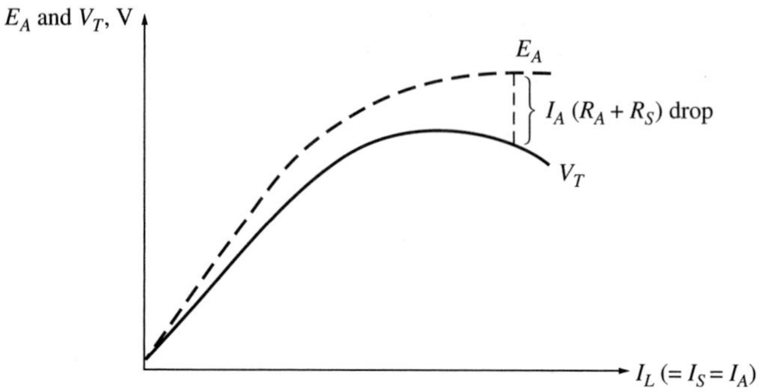

FIGURE 8–57
Derivation of the terminal characteristic for a series dc generator.

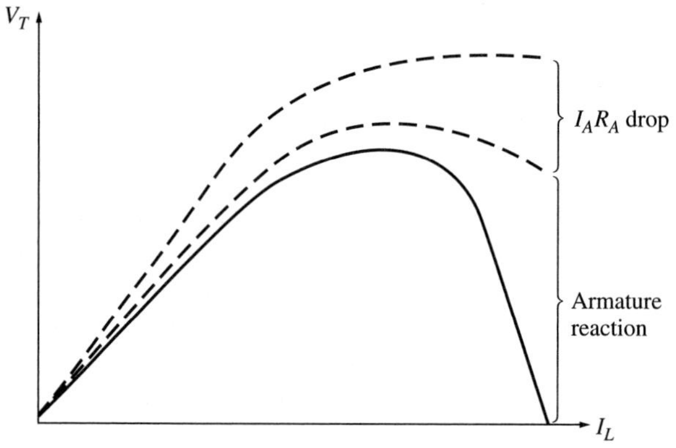

FIGURE 8–58
A series generator terminal characteristic with large armature reaction effects, suitable for electric welders.

becomes almost constant. At that point, the resistive drop is the predominant effect, and V_T starts to fall.

This type of characteristic is shown in Figure 8–57. It is obvious that this machine would make a bad constant-voltage source. In fact, its voltage regulation is a large negative number.

Series generators are used only in a few specialized applications, where the steep voltage characteristic of the device can be exploited. One such application is arc welding. Series generators used in arc welding are deliberately designed to have a large armature reaction, which gives them a terminal characteristic like the one shown in Figure 8–58. Notice that when the welding electrodes make contact with each other before welding commences, a very large current flows. As the operator separates the welding electrodes, there is a very steep rise in the generator's voltage while the current remains high. This voltage ensures that a welding arc is maintained through the air between the electrodes.

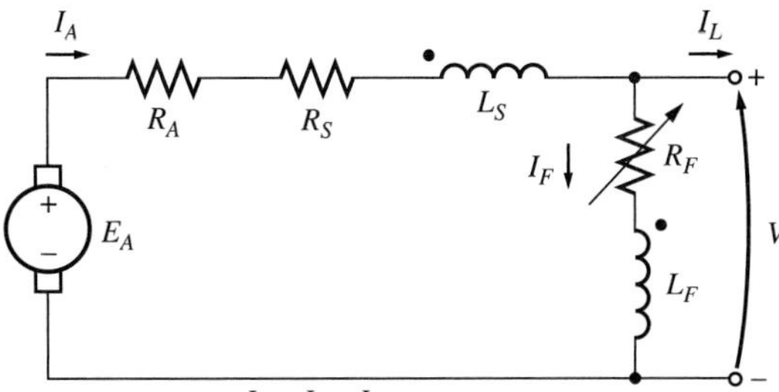

$$I_A = I_L + I_F$$
$$V_T = E_A - I_A(R_A + R_S)$$
$$I_F = \frac{V_T}{R_F}$$
$$\mathscr{F}_{\text{net}} = N_F I_F + N_{\text{SE}} I_A - \mathscr{F}_{\text{AR}}$$

FIGURE 8–59
The equivalent circuit of a cumulatively compounded dc generator with a long-shunt connection.

8.15 THE CUMULATIVELY COMPOUNDED DC GENERATOR

A cumulatively compounded dc generator is a dc generator with *both series and shunt fields*, connected so that the magnetomotive forces from the two fields are additive. Figure 8–59 shows the equivalent circuit of a cumulatively compounded dc generator in the "long-shunt" connection. The dots that appear on the two field coils have the same meaning as the dots on a transformer: *Current flowing into a dot produces a positive magnetomotive force.* Notice that the armature current flows into the dotted end of the series field coil and that the shunt current I_F flows into the dotted end of the shunt field coil. Therefore, the total magnetomotive force on this machine is given by

$$\boxed{\mathscr{F}_{\text{net}} = \mathscr{F}_F + \mathscr{F}_{\text{SE}} - \mathscr{F}_{\text{AR}}} \tag{8–48}$$

where $\mathscr{F}_F$ is the shunt field magnetomotive force, $\mathscr{F}_{\text{SE}}$ is the series field magnetomotive force, and $\mathscr{F}_{\text{AR}}$ is the armature reaction magnetomotive force. The equivalent effective shunt field current for this machine is given by

$$N_F I_F^* = N_F I_F + N_{\text{SE}} I_A - \mathscr{F}_{\text{AR}}$$

$$\boxed{I_F^* = I_F + \frac{N_{\text{SE}}}{N_F} I_A - \frac{\mathscr{F}_{\text{AR}}}{N_F}} \tag{8–49}$$

The other voltage and current relationships for this generator are

$$\boxed{I_A = I_F + I_L} \tag{8–50}$$

$$\boxed{V_T = E_A - I_A(R_A + R_S)} \tag{8–51}$$

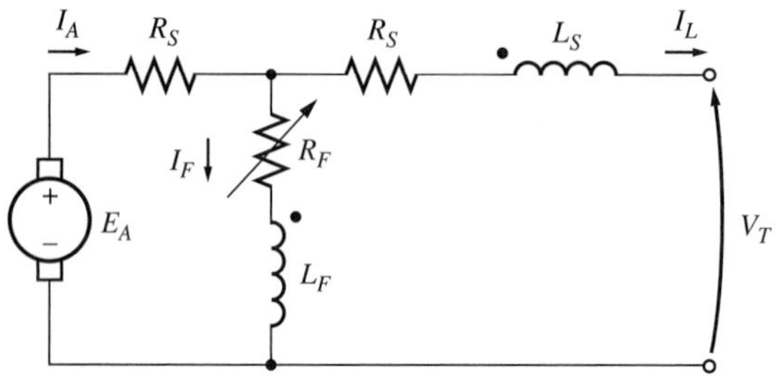

FIGURE 8–60
The equivalent circuit of a cumulatively compounded dc generator with a short-shunt connection.

$$\boxed{I_F = \frac{V_T}{R_F}} \tag{8–52}$$

There is another way to hook up a cumulatively compounded generator. It is the "short-shunt" connection, where the series field is outside the shunt field circuit and has current I_L flowing through it instead of I_A. A short-shunt cumulatively compounded dc generator is shown in Figure 8–60.

The Terminal Characteristic of a Cumulatively Compounded DC Generator

To understand the terminal characteristic of a cumulatively compounded dc generator, it is necessary to understand the competing effects that occur within the machine.

Suppose that the load on the generator is increased. Then as the load increases, the load current I_L increases. Since $I_A = I_F + I_L\uparrow$, the armature current I_A increases too. At this point two effects occur in the generator:

1. As I_A increases, the $I_A(R_A + R_S)$ voltage drop increases as well. This tends to cause a decrease in the terminal voltage $V_T = E_A - I_A\uparrow(R_A + R_S)$.
2. As I_A increases, the series field magnetomotive force $\mathscr{F}_{SE} = N_{SE}I_A$ increases too. This increases the total magnetomotive force $\mathscr{F}_{tot} = N_F I_F + N_{SE}I_A\uparrow$ which increases the flux in the generator. The increased flux in the generator increases E_A, which in turn tends to make $V_T = E_A\uparrow - I_A(R_A + R_S)$ rise.

These two effects oppose each other, with one tending to *increase* V_T and the other tending to *decrease* V_T. Which effect predominates in a given machine? It all depends on just how many series turns were placed on the poles of the machine. The question can be answered by taking several individual cases:

1. *Few series turns* (N_{SE} *small*). If there are only a few series turns, the resistive voltage drop effect wins hands down. The voltage falls off just as in a shunt

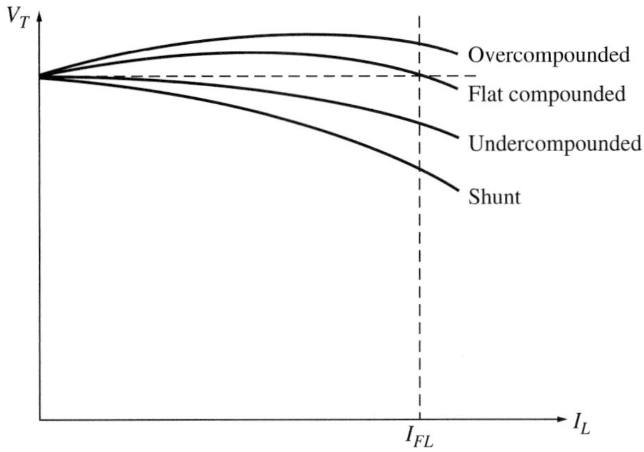

FIGURE 8–61
Terminal characteristics of cumulatively compounded dc generators.

generator, but not quite as steeply (Figure 8–61). This type of construction, where the full-load terminal voltage is less than the no-load terminal voltage, is called *undercompounded.*

2. *More series turns* (N_{SE} *larger*). If there are a few more series turns of wire on the poles, then at first the flux-strengthening effect wins, and the terminal voltage rises with the load. However, as the load continues to increase, magnetic saturation sets in, and the resistive drop becomes stronger than the flux increase effect. In such a machine, *the terminal voltage first rises and then falls as the load increases.* If V_T at no load is equal to V_T at full load, the generator is called *flat-compounded.*
3. *Even more series turns are added* (N_{SE} *large*). If even more series turns are added to the generator, the flux-strengthening effect predominates for a longer time before the resistive drop takes over. The result is a characteristic with the full-load terminal voltage actually higher than the no-load terminal voltage. If V_T at a full load exceeds V_T at no load, the generator is called *overcompounded.*

All these possibilities are illustrated in Figure 8–61.

It is also possible to realize all these voltage characteristics in a *single generator* if a diverter resistor is used. Figure 8–62 shows a cumulatively compounded dc generator with a relatively large number of series turns N_{SE}. A diverter resistor is connected around the series field. If the resistor R_{div} is adjusted to a large value, most of the armature current flows through the series field coil, and the generator is overcompounded. On the other hand, if the resistor R_{div} is adjusted to a small value, most of the current flows around the series field through R_{div}, and the generator is undercompounded. It can be smoothly adjusted with the resistor to have any desired amount of compounding.

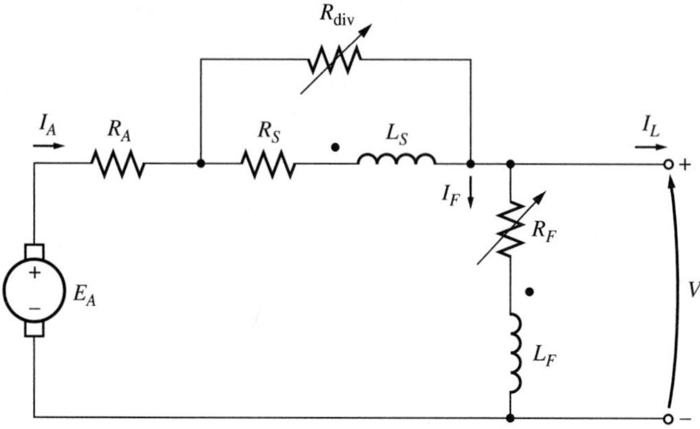

FIGURE 8–62
A cumulatively compounded dc generator with a series diverter resistor.

Voltage Control of Cumulatively Compounded DC Generators

The techniques available for controlling the terminal voltage of a cumulatively compounded dc generator are exactly the same as the techniques for controlling the voltage of a shunt dc generator:

1. Change the speed of rotation. An increase in ω causes $E_A = K\phi\omega_m\uparrow$ to increase, increasing the terminal voltage $V_T = E_A\uparrow - I_A(R_A + R_S)$.
2. Change the field current. A decrease in R_F causes $I_F = V_T/R_F\downarrow$ to increase, which increases the total magnetomotive force in the generator. As $\mathcal{F}_{tot}$ increases, the flux ϕ in the machine increases, and $E_A = K\phi\uparrow\omega_m$ increases. Finally, an increase in E_A raises V_T.

Analysis of Cumulatively Compounded DC Generators

Equations (8–53) and (8–54) are the key to describing the terminal characteristics of a cumulatively compounded dc generator. The equivalent shunt field current I_{eq} due to the effects of the series field and armature reaction is given by

$$\boxed{I_{eq} = \frac{N_{SE}}{N_F} I_A - \frac{\mathcal{F}_{AR}}{N_F}} \tag{8–53}$$

Therefore, the total effective shunt field current in the machine is

$$I_F^* = I_F + I_{eq} \tag{8–54}$$

This equivalent current I_{eq} represents a horizontal distance to the left or the right of the field resistance line ($R_F = V_T/R_F$) along the axes of the magnetization curve.

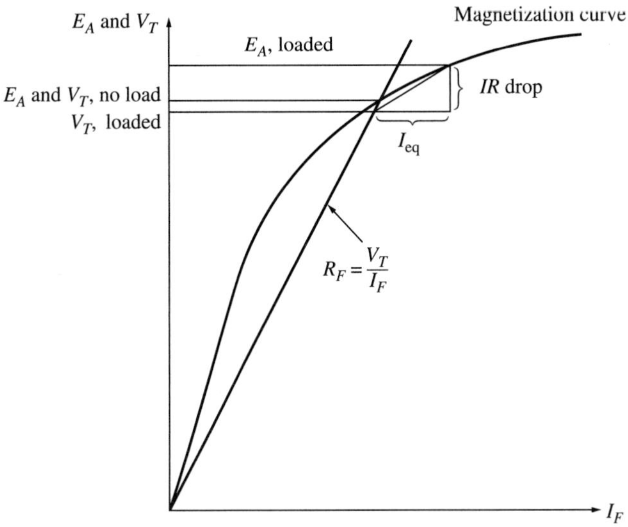

FIGURE 8–63
Graphical analysis of a cumulatively compounded dc generator.

The resistive drop in the generator is given by $I_A(R_A + R_S)$, which is a length along the vertical axis on the magnetization curve. Both the equivalent current I_{eq} and the resistive voltage drop $I_A(R_A + R_S)$ depend on the strength of the armature current I_A. Therefore, they form the two sides of a triangle whose magnitude is a function of I_A. To find the output voltage for a given load, determine the size of the triangle and find the one point where it *exactly* fits between the field current line and the magnetization curve.

This idea is illustrated in Figure 8–63. The terminal voltage at no-load conditions will be the point at which the resistor line and the magnetization curve intersect, as before. As load is added to the generator, the series field magnetomotive force increases, increasing the equivalent shunt field current I_{eq} and the resistive voltage drop $I_A(R_A + R_S)$ in the machine. To find the new output voltage in this generator, slide the leftmost edge of the resulting triangle along the shunt field current line until the upper tip of the triangle touches the magnetization curve. The upper tip of the triangle then represents the internal generated voltage in the machine, while the lower line represents the terminal voltage of the machine.

Figure 8–64 shows this process repeated several times to construct a complete terminal characteristic for the generator.

8.16 THE DIFFERENTIALLY COMPOUNDED DC GENERATOR

A differentially compounded dc generator is a generator with both shunt and series fields, but this time *their magnetomotive forces subtract from each other.*

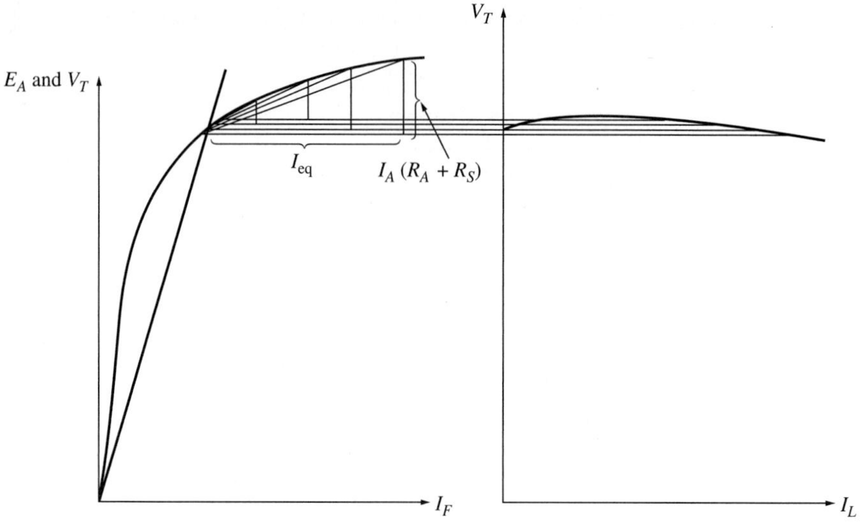

FIGURE 8–64
Graphical derivation of the terminal characteristic of a cumulatively compounded dc generator.

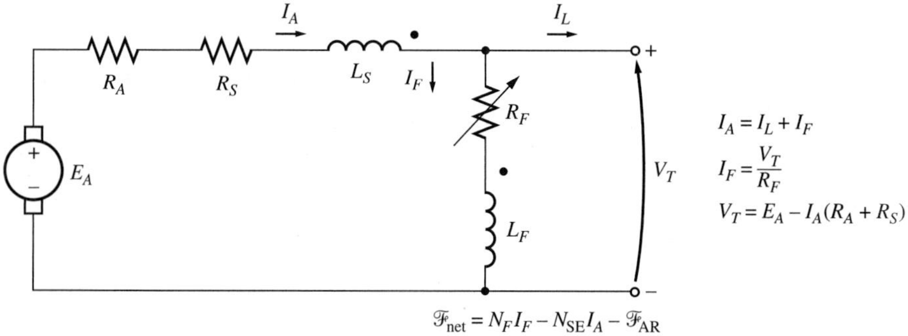

FIGURE 8–65
The equivalent circuit of a differentially compounded dc generator with a long-shunt connection.

The equivalent circuit of a differentially compounded dc generator is shown in Figure 8–65. Notice that the armature current is now flowing *out of* a dotted coil end, while the shunt field current is flowing *into* a dotted coil end. In this machine, the net magnetomotive force is

$$\boxed{\mathscr{F}_{\text{net}} = \mathscr{F}_F - \mathscr{F}_{\text{SE}} - \mathscr{F}_{\text{AR}}} \tag{8–55}$$

$$\boxed{\mathscr{F}_{\text{net}} = N_F I_F - N_{\text{SE}} I_A - \mathscr{F}_{\text{AR}}} \tag{8–56}$$

and the equivalent shunt field current due to the series field and armature reaction is given by

$$\boxed{I_{eq} = -\frac{N_{SE}}{N_F} I_A - \frac{\mathcal{F}_{AR}}{N_F}} \tag{8–57}$$

The total effective shunt field current in this machine is

$$I_F^* = I_F + I_{eq} \tag{8–58a}$$

or

$$\boxed{I_F^* = I_F - \frac{N_{SE}}{N_F} I_A - \frac{\mathcal{F}_{AR}}{N_F}} \tag{8–58b}$$

Like the cumulatively compounded generator, the differentially compounded generator can be connected in either long-shunt or short-shunt fashion.

The Terminal Characteristic of a Differentially Compounded DC Generator

In the differentially compounded dc generator, the same two effects occur that were present in the cumulatively compounded dc generator. This time, though, the effects both act in the same direction. They are

1. As I_A increases, the $I_A(R_A + R_S)$ voltage drop increases as well. This increase tends to cause the terminal voltage to decrease $V_T = E_A - I_A\uparrow (R_A + R_S)$.
2. As I_A increases, the series field magnetomotive force $\mathcal{F}_{SE} = N_{SE} I_A$ increases too. This increase in series field magnetomotive force *reduces* the net magnetomotive force on the generator ($\mathcal{F}_{tot} = N_F I_F - N_{SE} I_A\uparrow$), which in turn reduces the net flux in the generator. A decrease in flux decreases E_A, which in turn decreases V_T.

Since both these effects tend to *decrease* V_T, the voltage drops drastically as the load is increased on the generator. A typical terminal characteristic for a differentially compounded dc generator is shown in Figure 8–66.

Voltage Control of Differentially Compounded DC Generators

Even though the voltage drop characteristics of a differentially compounded dc generator are quite bad, it is still possible to adjust the terminal voltage at any given load setting. The techniques available for adjusting terminal voltage are exactly the same as those for shunt and cumulatively compounded dc generators:

1. Change the speed of rotation ω_m.
2. Change the field current I_F.

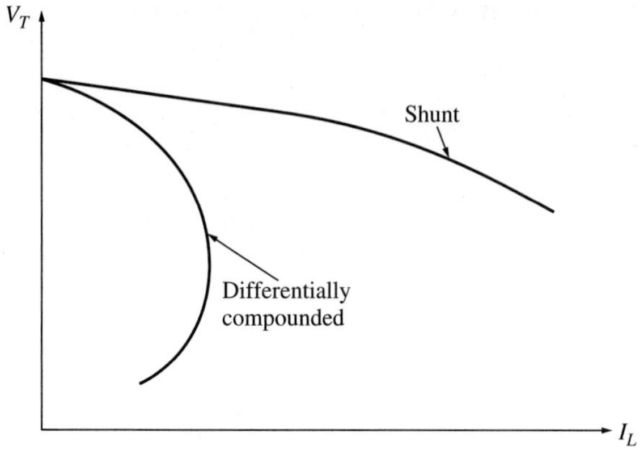

FIGURE 8–66
The terminal characteristic of a differentially compounded dc generator.

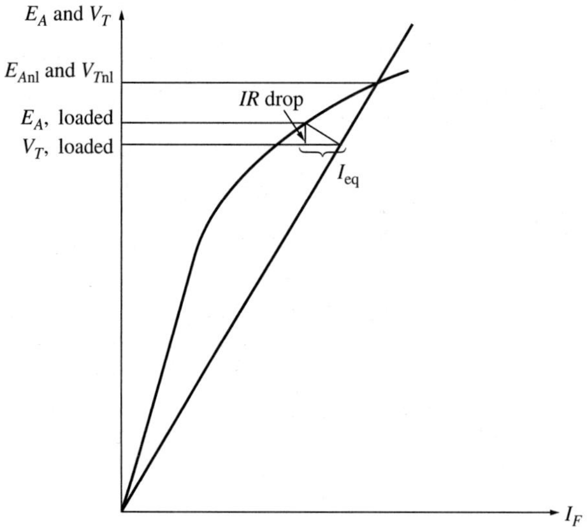

FIGURE 8–67
Graphical analysis of a differentially compounded dc generator.

Graphical Analysis of a Differentially Compounded DC Generator

The voltage characteristic of a differentially compounded dc generator is graphically determined in precisely the same manner as that used for the cumulatively compounded dc generator. To find the terminal characteristic of the machine, refer to Figure 8–67.

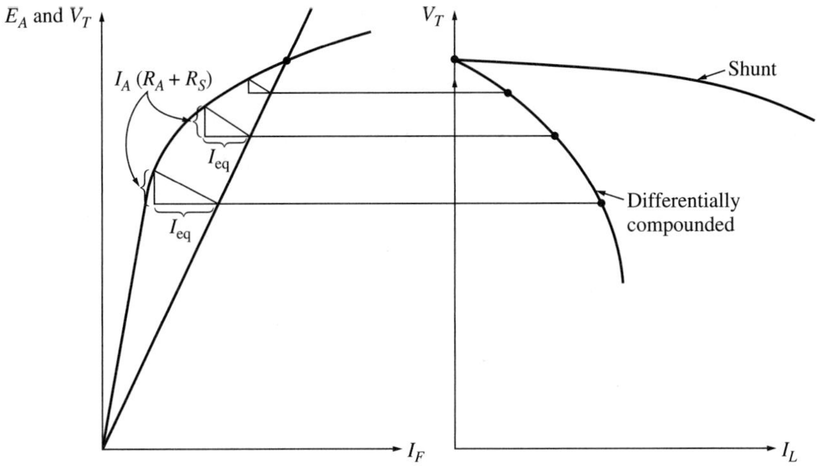

FIGURE 8–68
Graphical derivation of the terminal characteristic of a differentially compounded dc generator.

The portion of the effective shunt field current due to the actual shunt field is always equal to V_T/R_F, since that much current is present in the shunt field. The remainder of the effective field current is given by I_{eq} and is the sum of the series field and armature reaction effects. This equivalent current I_{eq} represents a *negative* horizontal distance along the axes of the magnetization curve, since both the series field and the armature reaction are subtractive.

The resistive drop in the generator is given by $I_A(R_A + R_S)$, which is a length along the vertical axis on the magnetization curve. To find the output voltage for a given load, determine the size of the triangle formed by the resistive voltage drop and I_{eq}, and find the one point where it *exactly* fits between the field current line and the magnetization curve.

Figure 8–68 shows this process repeated several times to construct a complete terminal characteristic for the generator.

8.17 SUMMARY

There are several types of dc motors, differing in the manner in which their field fluxes are derived. These types of motors are separately excited, shunt, permanent-magnet, series, and compounded. The manner in which the flux is derived affects the way it varies with the load, which in turn affects the motor's overall torque–speed characteristic.

A shunt or separately excited dc motor has a torque–speed characteristic whose speed drops linearly with increasing torque. Its speed can be controlled by changing its field current, its armature voltage, or its armature resistance.

A permanent-magnet dc motor is the same basic machine except that its flux is derived from permanent magnets. Its speed can be controlled by any of the above methods except varying the field current.

A series motor has the highest starting torque of any dc motor but tends to overspeed at no load. It is used for very high-torque applications where speed regulation is not important, such as a car starter.

A cumulatively compounded dc motor is a compromise between the series and the shunt motor, having some of the best characteristics of each. On the other hand, a differentially compounded dc motor is a complete disaster. It is unstable and tends to overspeed as load is added to it.

DC generators are dc machines used as generators. There are several different types of dc generators, differing in the manner in which their field fluxes are derived. These methods affect the output characteristics of the different types of generators. The common dc generator types are separately excited, shunt, series, cumulatively compounded, and differentially compounded.

The shunt and compounded dc generators depend on the nonlinearity of their magnetization curves for stable output voltages. If the magnetization curve of a dc machine were a straight line, then the magnetization curve and the terminal voltage line of the generator would never intersect. There would thus be no stable no-load voltage for the generator. Since nonlinear effects are at the heart of the generator's operation, the output voltages of dc generators can only be determined graphically or numerically by using a computer.

Today, dc generators have been replaced in many applications by ac power sources and solid-state electronic components. This is true even in the automobile, which is one of the most common users of dc power.

QUESTIONS

8–1. What is the speed regulation of a dc motor?

8–2. How can the speed of a shunt dc motor be controlled? Explain in detail.

8–3. What is the practical difference between a separately excited and a shunt dc motor?

8–4. What effect does armature reaction have on the torque–speed characteristic of a shunt dc motor? Can the effects of armature reaction be serious? What can be done to remedy this problem?

8–5. What are the desirable characteristics of the permanent magnets in PMDC machines?

8–6. What are the principal characteristics of a series dc motor? What are its uses?

8–7. What are the characteristics of a cumulatively compounded dc motor?

8–8. What are the problems associated with a differentially compounded dc motor?

8–9. What happens in a shunt dc motor if its field circuit opens while it is running?

8–10. Why is a starting resistor used in dc motor circuits?

8–11. How can a dc starting resistor be cut out of a motor's armature circuit at just the right time during starting?

8–12. What is the Ward-Leonard motor control system? What are its advantages and disadvantages?

8–13. What is regeneration?

8–14. What are the advantages and disadvantages of solid-state motor drives compared to the Ward-Leonard system?

8–15. What is the purpose of a field loss relay?

8–16. What types of protective features are included in typical solid-state dc motor drives? How do they work?

8–17. How can the direction of rotation of a separately excited dc motor be reversed?

8–18. How can the direction of rotation of a shunt dc motor be reversed?

8–19. How can the direction of rotation of a series dc motor be reversed?

8–20. Name and describe the features of the five types of generators covered in this chapter.

8–21. How does the voltage buildup occur in a shunt dc generator during starting?

8–22. What could cause voltage buildup on starting to fail to occur? How can this problem be remedied?

8–23. How does armature reaction affect the output voltage in a separately excited dc generator?

8–24. What causes the extraordinarily fast voltage drop with increasing load in a differentially compounded dc generator?

PROBLEMS

Problems 8–1 to 8–12 refer to the following dc motor:

$$P_{\text{rated}} = 30 \text{ hp} \qquad I_{L,\text{rated}} = 110 \text{ A}$$

$$V_T = 240 \text{ V} \qquad N_F = 2700 \text{ turns per pole}$$

$$n_{\text{rated}} = 1800 \text{ r/min} \qquad N_{\text{SE}} = 14 \text{ turns per pole}$$

$$R_A = 0.19\ \Omega \qquad R_F = 75\ \Omega$$

$$R_S = 0.02\ \Omega \qquad R_{\text{adj}} = 100 \text{ to } 400\ \Omega$$

Rotational losses = 3550 W at full load.

Magnetization curve is as shown in Figure P8–1.

In Problems 8–1 through 8–7, assume that the motor can be connected in shunt. The equivalent circuit of the shunt motor is shown in Figure P8–2.

8–1. If the resistor R_{adj} is adjusted to 175 Ω what is the rotational speed of the motor at no-load conditions?

8–2. Assuming no armature reaction, what is the speed of the motor at full load? What is the speed regulation of the motor?

8–3. If the motor is operating at full load and if its variable resistance R_{adj} is increased to 250 Ω, what is the new speed of the motor? Compare the full-load speed of the motor with $R_{\text{adj}} = 175\ \Omega$ to the full-load speed with $R_{\text{adj}} = 250\ \Omega$. (Assume no armature reaction, as in the previous problem.)

8–4. Assume that the motor is operating at full load and that the variable resistor R_{adj} is again 175 Ω. If the armature reaction is 2000 A • turns at full load, what is the speed of the motor? How does it compare to the result for Problem 8–2?

8–5. If R_{adj} can be adjusted from 100 to 400 Ω, what are the maximum and minimum no-load speeds possible with this motor?

8–6. What is the starting current of this machine if it is started by connecting it directly to the power supply V_T? How does this starting current compare to the full-load current of the motor?

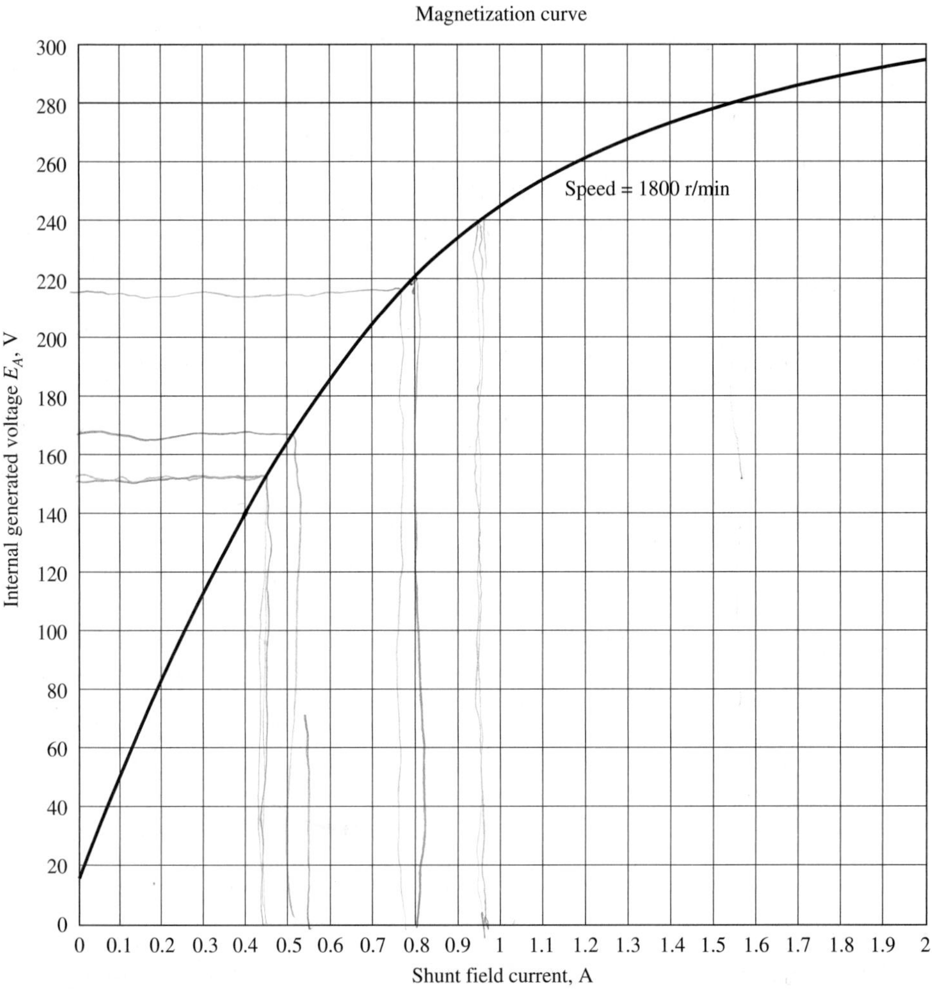

FIGURE P8–1
The magnetization curve for the dc motor in Problems 8–1 to 8–12. This curve was made at a constant speed of 1800 r/min.

8–7. Plot the torque–speed characteristic of this motor assuming no armature reaction, and again assuming a full-load armature reaction of 1200 A • turns. (Assume that the armature reaction increases linearly with increases in armature current.)

For Problems 8–8 and 8–9, the shunt dc motor is reconnected separately excited, as shown in Figure P8–3. It has a fixed field voltage V_F of 240 V and an armature voltage V_A that can be varied from 120 to 240 V.

8–8. What is the no-load speed of this separately excited motor when $R_{adj} = 175\ \Omega$ and (*a*) $V_A = 120$ V, (*b*) $V_A = 180$ V, (*c*) $V_A = 240$ V?

8–9. For the separately excited motor of Problem 8–8:
(*a*) What is the maximum no-load speed attainable by varying both V_A and R_{adj}?

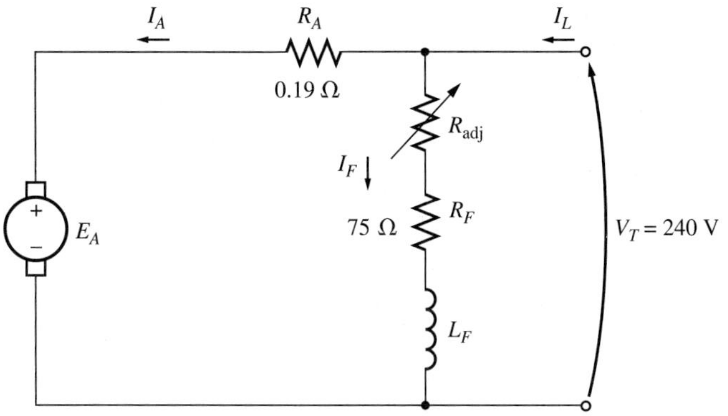

FIGURE P8–2
The equivalent circuit of the shunt motor in Problems 8–1 to 8–7.

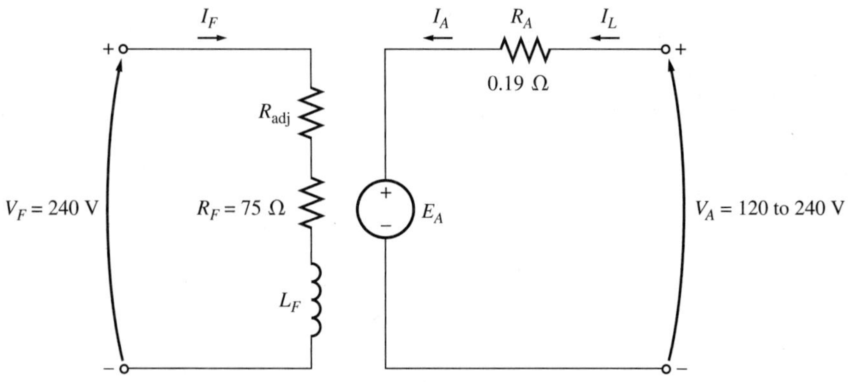

FIGURE P8–3
The equivalent circuit of the separately excited motor in Problems 8–8 and 8–9.

(b) What is the minimum no-load speed attainable by varying both V_A and R_{adj}?
(c) What is the motor's efficiency at rated conditions? [*Note:* Assume that (1) the brush voltage drop is 2 V; (2) the core loss is to be determined at an armature voltage equal to the armature voltage under full load; and (3) stray load losses are 1 percent of full load.]

For Problems 8–10 to 8–11, the motor is connected cumulatively compounded as shown in Figure P8–4

8–10. If the motor is connected cumulatively compounded with $R_{adj} = 175\ \Omega$:
(a) What is its no-load speed of the motor?
(b) What is its full-load speed of the motor?
(c) What is its speed regulation?
(d) Calculate and plot the torque–speed characteristic for this motor. (Neglect armature effects in this problem.)

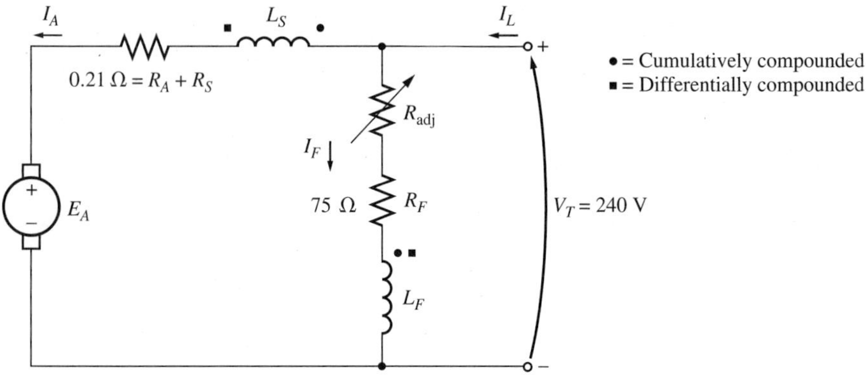

FIGURE P8–4
The equivalent circuit of the compounded motor in Problems 8–10 to 8–12.

8–11. The motor is connected cumulatively compounded and is operating at full load. What will the new speed of the motor be if R_{adj} is increased to 250 Ω? How does the new speed compare to the full-load speed calculated in Problem 8–10?

For Problem 8–12, the motor is now connected differentially compounded as shown in Figure P8–4.

8–12. The motor is now connected differentially compounded.

(a) If $R_{adj} = 175\ \Omega$, what is the no-load speed of the motor?

(b) What is the motor's speed when the armature current reaches 20 A? 40 A? 60 A?

(c) Calculate and plot the torque–speed characteristic curve of this motor.

8–13. A 15-hp, 120-V series dc motor has an armature resistance of 0.1 Ω and a series field resistance of 0.08 Ω. At full load, the current input is 115 A, and the rated speed is 1050 r/min. Its magnetization curve is shown in Figure P8–5. The core losses are 420 W, and the mechanical losses are 460 W at full load. Assume that the mechanical losses vary as the cube of the speed of the motor and that the core losses are constant.

(a) What is the efficiency of the motor at full load?

(b) What are the speed and efficiency of the motor if it is operating at an armature current of 70 A?

(c) Plot the torque–speed characteristic for this motor.

8–14. A 20-hp, 240-V, 76-A, 900 r/min series motor has a field winding of 33 turns per pole. Its armature resistance is 0.09 Ω, and its field resistance is 0.06 Ω. The magnetization curve expressed in terms of magnetomotive force versus E_A at 900 r/min is given by the following table:

E_A, V	95	150	188	212	229	243
$\mathscr{F}$, A • turns	500	1000	1500	2000	2500	3000

Armature reaction is negligible in this machine.

(a) Compute the motor's torque, speed, and output power at 33, 67, 100, and 133 percent of full-load armature current. (Neglect rotational losses.)

(b) Plot the terminal characteristic of this machine.

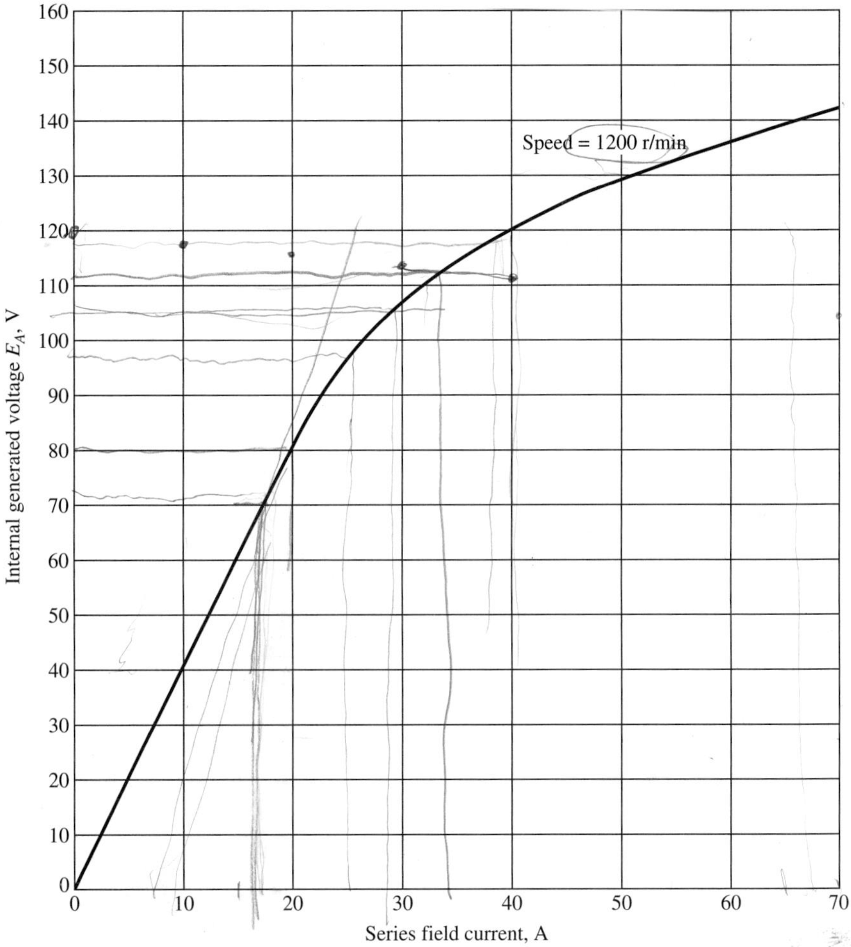

FIGURE P8–5
The magnetization curve for the series motor in Problem 8–13. This curve was taken at a constant speed of 1200 r/min.

8–15. A 300-hp, 440-V, 560-A, 863 r/min shunt dc motor has been tested, and the following data were taken:

Blocked-rotor test:

$V_A = 14.9$ V exclusive of brushes $\qquad V_F = 440$ V

$I_A = 500$ A $\qquad I_F = 7.52$ A

No-load operation:

$V_A = 440$ V including brushes $\qquad I_F = 7.50$ A

$I_A = 23.1$ A $\qquad n = 863$ r/min

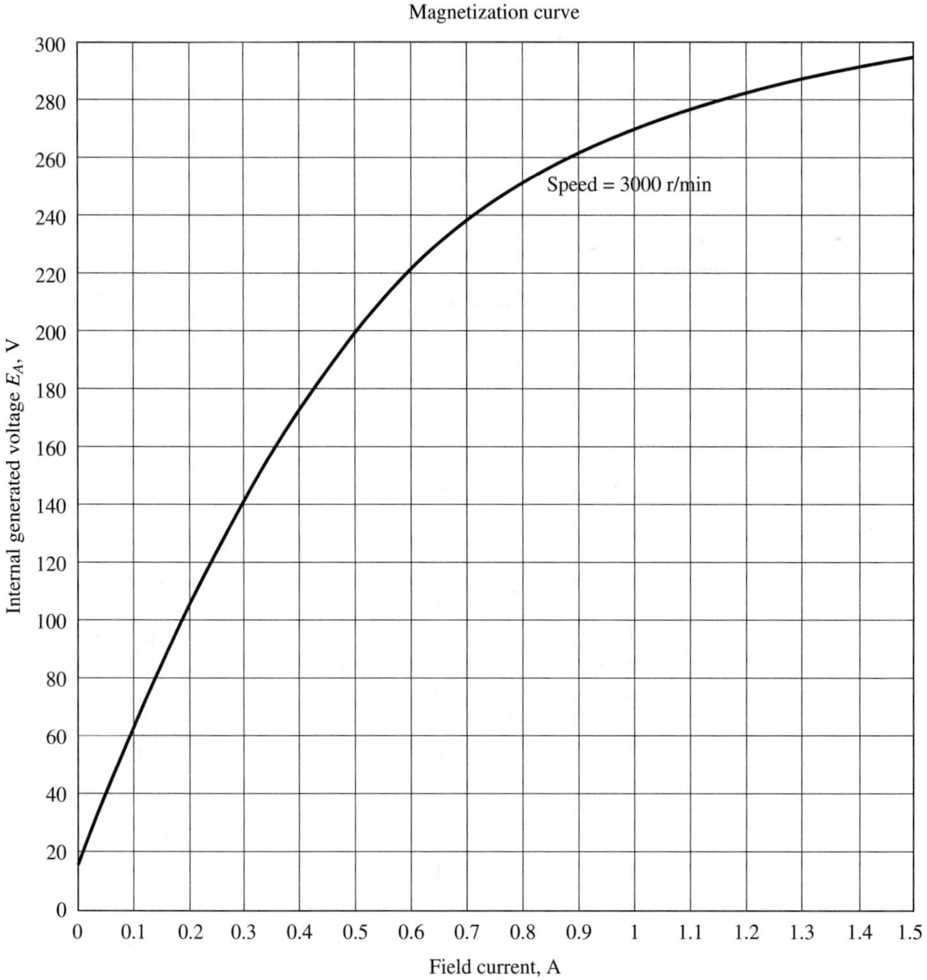

FIGURE P8–6
The magnetization curve for the dc motor in Problems 8–16 to 8–19. This curve was made at a constant speed of 3000 r/min.

What is this motor's efficiency at the rated conditions? [*Note:* Assume that (1) the brush voltage drop is 2 V; (2) the core loss is to be determined at an armature voltage equal to the armature voltage under full load; and (3) stray load losses are 1 percent of full load.]

Problems 8–16 to 8–19 refer to a 240-V, 100-A dc motor which has both shunt and series windings. Its characteristics are

$R_A = 0.14\ \Omega$ $\qquad N_F = 1500$ turns

$R_S = 0.05\ \Omega$ $\qquad N_{SE} = 15$ turns

$R_F = 200\ \Omega$ $\qquad n_m = 3000$ r/min

$R_{adj} = 0$ to $300\ \Omega$, currently set to $120\ \Omega$

This motor has compensating windings and interpoles. The magnetization curve for this motor at 3000 r/min is shown in Figure P8–6.

8–16. The motor described above is connected in *shunt*.

(*a*) What is the no-load speed of this motor when $R_{adj} = 120\ \Omega$?

(*b*) What is its full-load speed?

(*c*) What is its speed regulation?

(*d*) Plot the torque–speed characteristic for this motor.

(*e*) Under no-load conditions, what range of possible speeds can be achieved by adjusting R_{adj}?

8–17. This machine is now connected as a cumulatively compounded dc motor with $R_{adj} = 120\ \Omega$.

(*a*) What is the no-load speed of this motor?

(*b*) What is its full-load speed?

(*c*) What is its speed regulation?

(*d*) Plot the torque–speed characteristic for this motor.

8–18. The motor is reconnected differentially compounded with $R_{adj} = 120\ \Omega$. Derive the shape of its torque–speed characteristic.

8–19. A series motor is now constructed from this machine by leaving the shunt field out entirely. Derive the torque–speed characteristic of the resulting motor.

8–20. An automatic starter circuit is to be designed for a shunt motor rated at 20 hp, 240 V, and 75 A. The armature resistance of the motor is 0.12 Ω, and the shunt field resistance is 40 Ω. The motor is to start with no more than 250 percent of its rated armature current, and as soon as the current falls to rated value, a starting resistor stage is to be cut out. How many stages of starting resistance are needed, and how big should each one be?

8–21. A 10-hp, 120-V, 1000 r/min shunt dc motor has a full-load armature current of 70 A when operating at rated conditions. The armature resistance of the motor is $R_A = 0.12\ \Omega$, and the field resistance R_F is 40 Ω. The adjustable resistance in the field circuit R_{adj} may be varied over the range from 0 to 200 Ω and is currently set to 100 Ω. Armature reaction may be ignored in this machine. The magnetization curve for this motor, taken at a speed of 1000 r/min, is given in the following table:

E_A, V	5	78	95	112	118	126
I_F, A	0.00	0.80	1.00	1.28	1.44	2.88

(*a*) What is the speed of this motor when it is running at the rated conditions specified above?

(*b*) The output power from the motor is 10 hp at rated conditions. What is the output torque of the motor?

(*c*) What are the copper losses and rotational losses in the motor at full load (ignore stray losses)?

(*d*) What is the efficiency of the motor at full load?

(*e*) If the motor is now unloaded with no changes in terminal voltage or R_{adj}, what is the no-load speed of the motor?

(*f*) Suppose that the motor is running at the no-load conditions described in part (*e*). What would happen to the motor if its field circuit were to open? Ignoring armature reaction, what would the final steady-state speed of the motor be under those conditions?

(*g*) What range of no-load speeds is possible in this motor, given the range of field resistance adjustments available with R_{adj}?

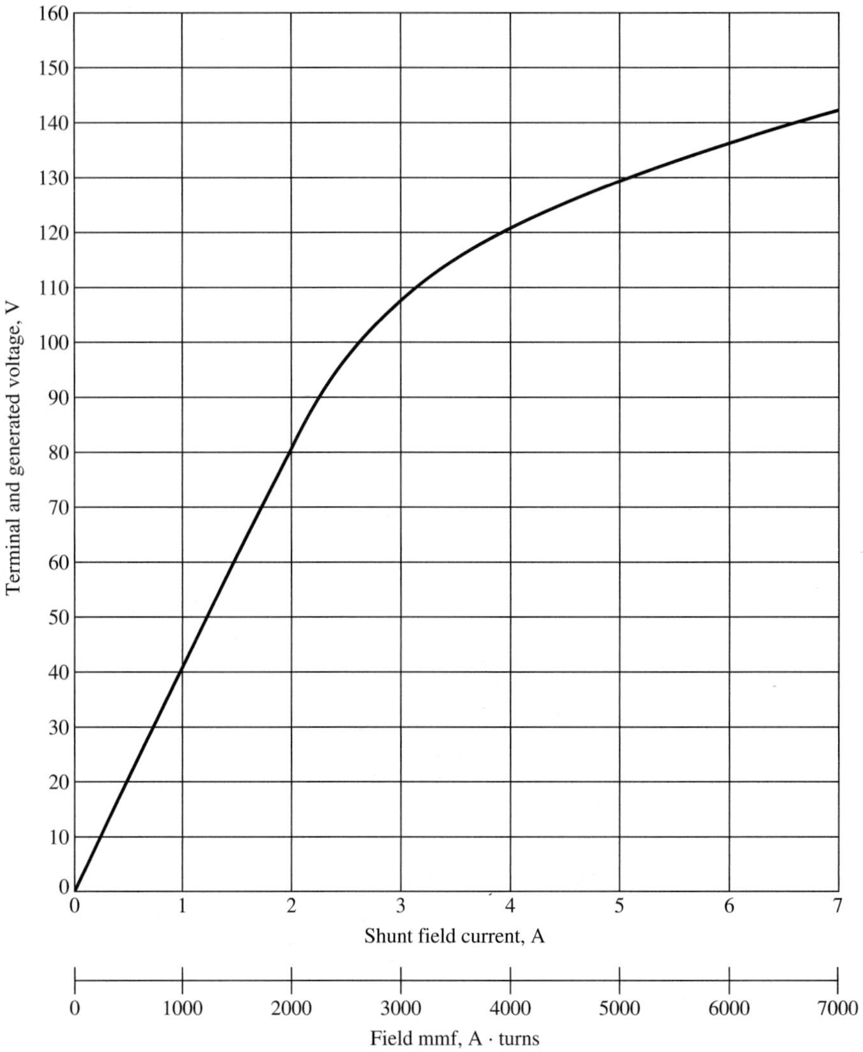

FIGURE P8–7
The magnetization curve for Problems 8–22 to 8–28. This curve was taken at a speed of 1800 r/min.

8–22. The magnetization curve for a separately excited dc generator is shown in Figure P8–7. The generator is rated at 6 kW, 120 V, 50 A, and 1800 r/min and is shown in Figure P8–8. Its field circuit is rated at 5A. The following data are known about the machine:

$R_A = 0.18\ \Omega$ $\qquad V_F = 120$ V

$R_{adj} = 0$ to $40\ \Omega$ $\qquad R_F = 20\ \Omega$

$N_F = 1000$ turns per pole

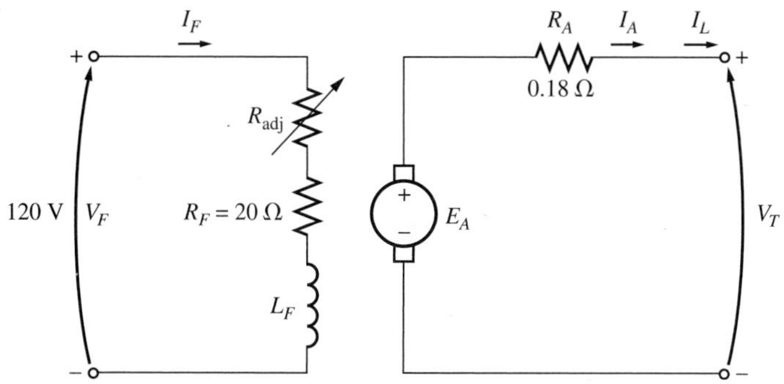

FIGURE P8–8
The separately excited dc generator in Problems 8–22 to 8–24.

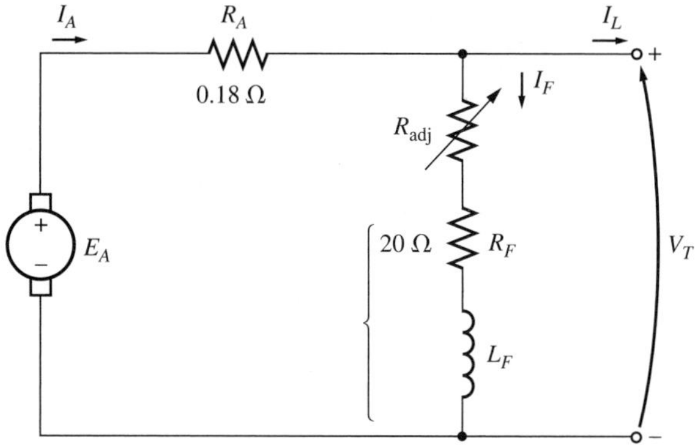

FIGURE P8–9
The shunt dc generator in Problems 8–25 and 8–26.

Answer the following questions about this generator, assuming no armature reaction.

(a) If this generator is operating at no load, what is the range of voltage adjustments that can be achieved by changing R_{adj}?

(b) If the field rheostat is allowed to vary from 0 to 30 Ω and the generator's speed is allowed to vary from 1500 to 2000 r/min, what are the maximum and minimum no-load voltages in the generator?

8–23. If the armature current of the generator in Problem 8–22 is 50 A, the speed of the generator is 1700 r/min, and the terminal voltage is 106 V, how much field current must be flowing in the generator?

8–24. Assuming that the generator in Problem 8–22 has an armature reaction at full load equivalent to 400 A • turns of magnetomotive force, what will the terminal voltage of the generator be when $I_F = 5$ A, $n_m = 1700$ r/min, and $I_A = 50$ A?

8–25. The machine in Problem 8–22 is reconnected as a shunt generator and is shown in Figure P8–9. The shunt field resistor R_{adj} is adjusted to 10 Ω, and the generator's speed is 1800 r/min.

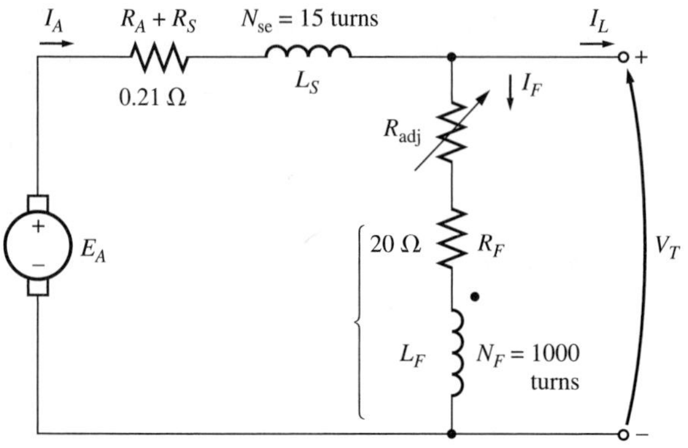

FIGURE P8–10
The compounded dc generator in Problems 8–27 and 8–28.

(*a*) What is the no-load terminal voltage of the generator?
(*b*) Assuming no armature reaction, what is the terminal voltage of the generator with an armature current of 20 A? 40 A?
(*c*) Assuming an armature reaction equal to 300 A • turns at full load, what is the terminal voltage of the generator with an armature current of 20 A? 40 A?
(*d*) Calculate and plot the terminal characteristics of this generator with and without armature reaction.

8–26. If the machine in Problem 8–25 is running at 1800 r/min with a field resistance $R_{adj} = 10\ \Omega$ and an armature current of 25 A, what will the resulting terminal voltage be? If the field resistor decreases to 5 Ω while the armature current remains 25 A, what will the new terminal voltage be? (Assume no armature reaction.)

8–27. A 120-V, 50-A, cumulatively compounded dc generator has the following characteristics:

$$R_A + R_S = 0.21\ \Omega \qquad N_F = 1000 \text{ turns}$$
$$R_F = 20\ \Omega \qquad N_{SE} = 25 \text{ turns}$$
$$R_{adj} = 0 \text{ to } 30\ \Omega, \text{ set to } 10\ \Omega \qquad n_m = 1800 \text{ r/min}$$

The machine has the magnetization curve shown in Figure P8–7. Its equivalent circuit is shown in Figure P8–10. Answer the following questions about this machine, assuming no armature reaction.

(*a*) If the generator is operating at no load, what is its terminal voltage?
(*b*) If the generator has an armature current of 20 A, what is its terminal voltage?
(*c*) If the generator has an armature current of 40 A, what is its terminal voltage?
(*d*) Calculate and plot the terminal characteristic of this machine.

8–28. If the machine described in Problem 8–27 is reconnected as a differentially compounded dc generator, what will its terminal characteristic look like? Derive it in the same fashion as in Problem 8–27.

8–29. A cumulatively compounded dc generator is operating properly as a flat-compounded dc generator. The machine is then shut down, and its shunt field connections are reversed.

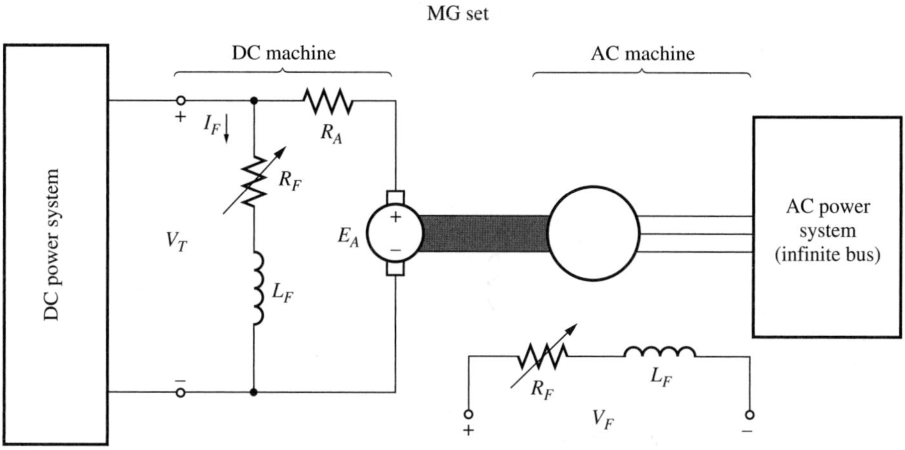

FIGURE P8–11
The motor–generator set in Problem 8–30.

(a) If this generator is turned in the same direction as before, will an output voltage be built up at its terminals? Why or why not?

(b) Will the voltage build up for rotation in the opposite direction? Why or why not?

(c) For the direction of rotation in which a voltage builds up, will the generator be cumulatively or differentially compounded?

8–30. A three-phase synchronous machine is mechanically connected to a shunt dc machine, forming a motor–generator set, as shown in Figure P8–11. The dc machine is connected to a dc power system supplying a constant 240 V, and the ac machine is connected to a 480-V, 60-Hz infinite bus.

The dc machine has four poles and is rated at 50 kW and 240 V. It has a per-unit armature resistance of 0.03. The ac machine has four poles and is Y-connected. It is rated at 50 kVA, 480 V, and 0.8 PF, and its saturated synchronous reactance is 3.0 Ω per phase.

All losses except the dc machine's armature resistance may be neglected in this problem. Assume that the magnetization curves of both machines are linear.

(a) Initially, the ac machine is supplying 50 kVA at 0.8 PF lagging to the ac power system.

1. How much power is being supplied to the dc motor from the dc power system?

2. How large is the internal generated voltage E_A of the dc machine?

3. How large is the internal generated voltage $\mathbf{E}_A$ of the ac machine?

(b) The field current in the ac machine is now decreased by 5 percent. What effect does this change have on the real power supplied by the motor–generator set? On the reactive power supplied by the motor–generator set? Calculate the real and reactive power supplied or consumed by the ac machine under these conditions. Sketch the ac machine's phasor diagram before and after the change in field current.

(c) Starting from the conditions in part *(b)*, the field current in the dc machine is now decreased by 1 percent. What effect does this change have on the real power supplied by the motor–generator set? On the reactive power supplied by the motor–generator set? Calculate the real and reactive power supplied or consumed by the ac machine under these conditions. Sketch the ac machine's phasor diagram before and after the change in the dc machine's field current.

(d) From the preceding results, answer the following questions:

1. How can the real power flow through an ac-dc motor–generator set be controlled?

2. How can the reactive power supplied or consumed by the ac machine be controlled without affecting the real power flow?

REFERENCES

1. Chaston, A. N.: *Electric Machinery*, Reston Publications, Reston, Va., 1986.
2. Fitzgerald, A. E., and C. Kingsley, Jr.: *Electric Machinery*, McGraw-Hill, New York, 1952.
3. Fitzgerald, A. E., C. Kingsley, Jr., and S. D. Umans: *Electric Machinery*, 6th ed., McGraw-Hill, New York, 2003.
4. Heck, C.: *Magnetic Materials and Their Applications*, Butterworth & Co., London, 1974.
5. IEEE Standard 113-1985, *Guide on Test Procedures for DC Machines*, IEEE, Piscataway, N.J., 1985. (Note that this standard has been officially withdrawn but is still available.)
6. Kloeffler, S. M., R. M. Kerchner, and J. L. Brenneman: *Direct Current Machinery*, rev. ed. Macmillan, New York, 1948.
7. Kosow, Irving L.: *Electric Machinery and Transformers*, Prentice Hall, Englewood Cliffs, N.J., 1972.
8. McPherson, George: *An Introduction to Electrical Machines and Transformers*, Wiley, New York, 1981.
9. Siskind, Charles S.: *Direct-Current Machinery*, McGraw-Hill, New York, 1952.
10. Slemon, G. R., and A. Straughen. *Electric Machines*, Addison-Wesley, Reading, Mass., 1980.
11. Werninck, E. H. (ed.): *Electric Motor Handbook*, McGraw-Hill, London, 1978.

CHAPTER 9

SINGLE-PHASE AND SPECIAL-PURPOSE MOTORS

LEARNING OBJECTIVES

- Understand why a universal motor is called "universal."
- Understand how it is possible to develop unidirectional torque from a pulsing magnetic field in a single-phase induction motor.
- Understand how to start single-phase induction motors.
- Understand the characteristics of the different single-phase induction motor classes: split-phase, capacitor-type, and shaded pole.
- Be able to calculate induced torque in a single-phase induction motor.
- Understand the basic operation of reluctance and hysteresis motors.
- Understand the operation of a stepper motor.
- Understand the operation of a brushless dc motor.

Chapters 3 through 6 were devoted to the operation of the two major classes of ac machines (synchronous and induction) on *three-phase* power systems. Motors and generators of these types are by far the most common ones in larger commercial and industrial settings. However, most homes and small businesses do not have three-phase power available. For such locations, all motors must run from single-phase power sources. This chapter deals with the theory and operation of two major types of single-phase motors: the universal motor and the single-phase induction motor. The universal motor, which is a straightforward extension of the series dc motor, is described in Section 9.1.

The single-phase induction motor is described in Sections 9.2 to 9.5. The major problem associated with the design of single-phase induction motors is that, unlike three-phase power sources, a single-phase source does *not* produce a

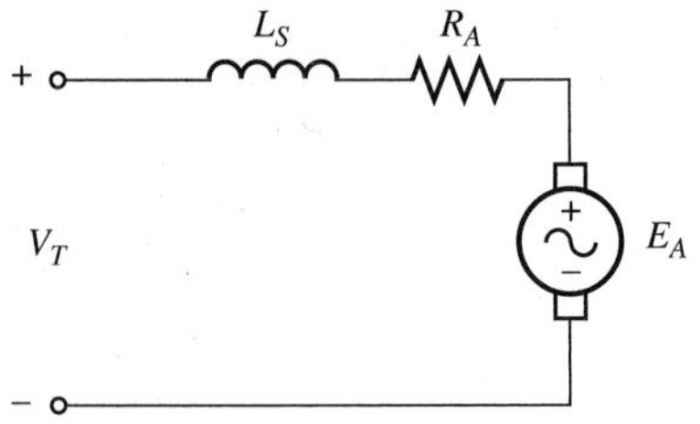

FIGURE 9–1
Equivalent circuit of a universal motor.

rotating magnetic field. Instead, the magnetic field produced by a single-phase source remains stationary in position and *pulses* with time. Since there is no net rotating magnetic field, conventional induction motors cannot function, and special designs are necessary.

In addition, there are a number of special-purpose motors which have not been previously covered. These include reluctance motors, hysteresis motors, stepper motors, and brushless dc motors. They are included in Section 9.6.

9.1 THE UNIVERSAL MOTOR

Perhaps the simplest approach to the design of a motor that will operate on a single-phase ac power source is to take a dc machine and run it from an ac supply. Recall from Chapter 7 that the induced torque of a dc motor is given by

$$\tau_{\text{ind}} = K\phi I_A \tag{7–49}$$

If the polarity of the voltage applied to a shunt or series dc motor is reversed, *both* the direction of the field flux *and* the direction of the armature current reverse, and the resulting induced torque continues in the same direction as before. Therefore, it should be possible to achieve a pulsating but unidirectional torque from a dc motor connected to an ac power supply.

Such a design is practical only for the series dc motor (see Figure 9–1), since the armature current and the field current in the machine must reverse at exactly the same time. For shunt dc motors, the very high field inductance tends to delay the reversal of the field current and thus to unacceptably reduce the average induced torque of the motor.

In order for a series dc motor to function effectively on ac, its field poles and stator frame must be completely laminated. If they were not completely laminated, their core losses would be enormous. When the poles and stator are laminated, this motor is often called a *universal motor*, since it can run from either an ac or a dc source.

When the motor is running from an ac source, the commutation will be much poorer than it would be with a dc source. The extra sparking at the brushes is caused by transformer action inducing voltages in the coils undergoing commutation. These sparks significantly shorten brush life and can be a source of radio-frequency interference in certain environments.

A typical torque–speed characteristic of a universal motor is shown in Figure 9–2. It differs from the torque–speed characteristic of the same machine operating from a dc voltage source for two reasons:

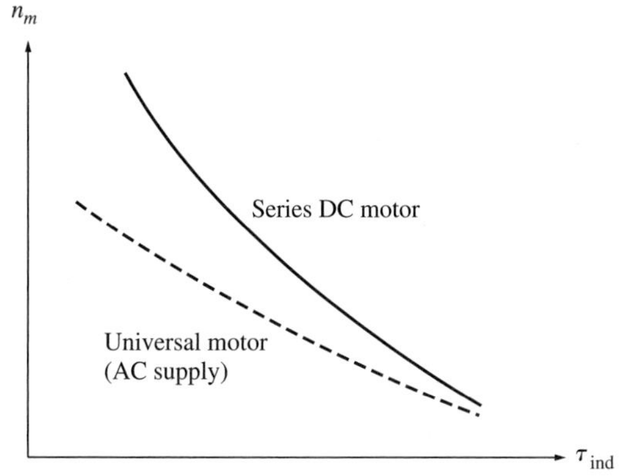

FIGURE 9–2
Comparison of the torque–speed characteristic of a universal motor when operating from ac and dc power supplies.

1. The armature and field windings have quite a large reactance at 50 or 60 Hz. A significant part of the input voltage is dropped across these reactances, and therefore E_A is *smaller* for a given input voltage during ac operation than it is during dc operation. Since $E_A = K\phi\omega_m$, the motor is *slower* for a given armature current and induced torque on alternating current than it would be on direct current.
2. In addition, the peak voltage of an ac system is $\sqrt{2}$ times its rms value, so magnetic saturation could occur near the peak current in the machine. This saturation could significantly lower the rms flux of the motor for a given current level, tending to reduce the machine's induced torque. Recall that a decrease in flux increases the speed of a dc machine, so this effect may partially offset the speed decrease caused by the first effect.

Applications of Universal Motors

The universal motor has the sharply drooping torque–speed characteristic of a dc series motor, so it is not suitable for constant-speed applications. However, it is compact and gives more torque per ampere than any other single-phase motor. It is therefore used where light weight and high torque are important.

Typical applications for this motor are vacuum cleaners, drills, similar portable tools, and kitchen appliances.

Speed Control of Universal Motors

As with dc series motors, the best way to control the speed of a universal motor is to vary its rms input voltage. The higher the rms input voltage, the greater the resulting speed of the motor. Typical torque–speed characteristics of a universal motor as a function of voltage are shown in Figure 9–3.

In practice, the average voltage applied to such a motor is varied with a solid-state speed control circuit. Two such speed control circuits are shown in Figure 9–4. The variable resistors shown in these figures are the speed adjustment knobs of the motors (e.g., such a resistor would be the trigger of a variable-speed drill).

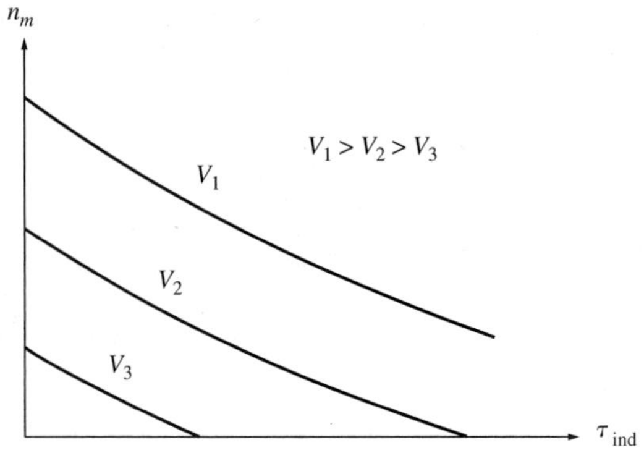

FIGURE 9–3
The effect of changing terminal voltage on the torque–speed characteristic of a universal motor.

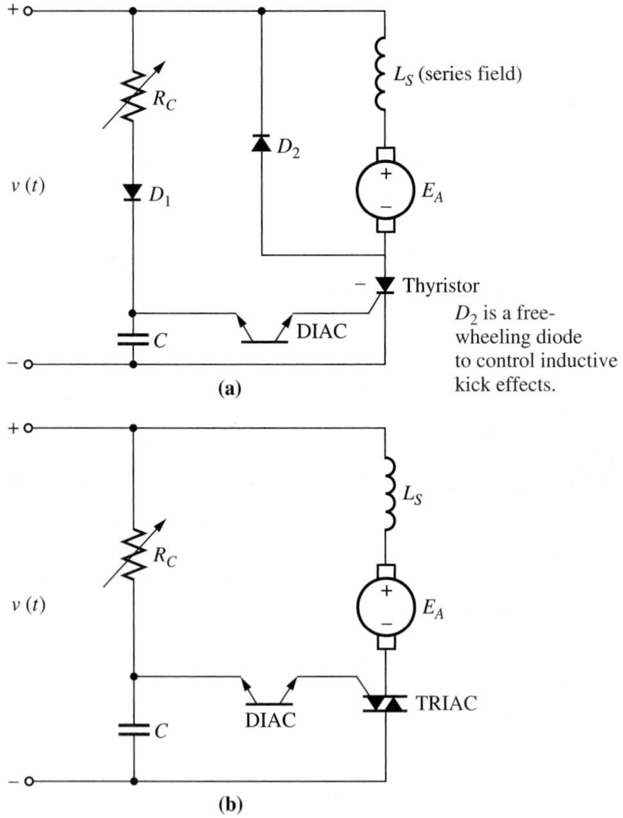

FIGURE 9–4
Sample universal motor speed-control circuits. (a) Half-wave; (b) full-wave.

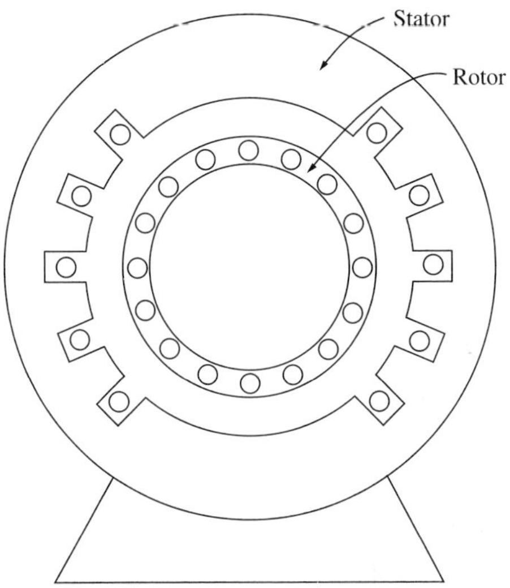

FIGURE 9–5
Construction of a single-phase induction motor. The rotor is the same as in a three-phase induction motor, but the stator has only a single distributed phase.

9.2 INTRODUCTION TO SINGLE-PHASE INDUCTION MOTORS

Another common single-phase motor is the single-phase version of the induction motor. An induction motor with a squirrel-cage rotor and a single-phase stator is shown in Figure 9–5.

Single-phase induction motors suffer from a severe handicap. Since there is only one phase on the stator winding, the magnetic field in a single-phase induction motor does not rotate. Instead, it *pulses*, getting first larger and then smaller, but always remaining in the same direction. Because there is no rotating stator magnetic field, a single-phase induction motor has *no starting torque*.

This fact is easy to see from an examination of the motor when its rotor is stationary. The stator flux of the machine first increases and then decreases, but it always points in the same direction. Since the stator magnetic field does not rotate, there is *no relative motion* between the stator field and the bars of the rotor. Therefore, there is no induced voltage due to relative motion in the rotor, no rotor current flow due to relative motion, and no induced torque. Actually, a voltage is induced in the rotor bars by transformer action ($d\phi/dt$), and since the bars are short-circuited, current flows in the rotor. However, this magnetic field is lined up with the stator magnetic field, and it produces no net torque on the rotor,

$$\begin{aligned}\tau_{\text{ind}} &= k\mathbf{B}_R \times \mathbf{B}_S \qquad (3\text{–}58)\\ &= kB_RB_S \sin \gamma \\ &= kB_RB_S \sin 180° = 0\end{aligned}$$

At stall conditions, the motor looks like a transformer with a short-circuited secondary winding (see Figure 9–6).

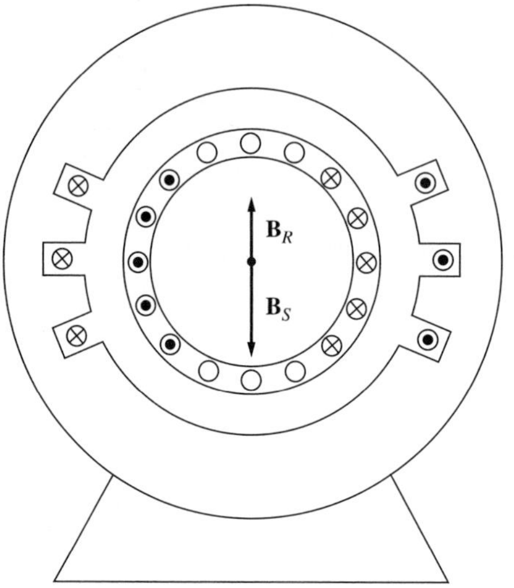

FIGURE 9–6
The single-phase induction motor at starting conditions. The stator winding induces opposing voltages and currents into the rotor circuit, resulting in a rotor magnetic field *lined up* with the stator magnetic field. $\tau_{\text{ind}} = 0$.

The fact that single-phase induction motors have no intrinsic starting torque was a serious impediment to early development of the induction motor. When induction motors were first being developed in the late 1880s and early 1890s, the first available ac power systems were 133-Hz, single-phase. With the materials and techniques then available, it was impossible to build a motor that worked well. The induction motor did not become an off-the-shelf working product until three-phase, 25-Hz power systems were developed in the mid-1890s.

However, *once the rotor begins to turn, an induced torque will be produced in it*. There are two basic theories which explain why a torque is produced in the rotor once it is turning. One is called the *double-revolving-field theory* of single-phase induction motors, and the other is called the *cross-field theory* of single-phase induction motors. Each of these approaches will be described below.

The Double-Revolving-Field Theory of Single-Phase Induction Motors

The double-revolving-field theory of single-phase induction motors basically states that a stationary pulsating magnetic field can be resolved into two *rotating* magnetic fields, each of equal magnitude but rotating in opposite directions. The induction motor responds to each magnetic field separately, and the net torque in the machine will be the sum of the torques due to each of the two magnetic fields.

Figure 9–7 shows how a stationary pulsating magnetic field can be resolved into two equal and oppositely rotating magnetic fields. The flux density of the stationary magnetic field is given by

$$\mathbf{B}_S(t) = (\mathbf{B}_{\max} \cos \omega t)\,\hat{\mathbf{j}} \tag{9–1}$$

A clockwise-rotating magnetic field can be expressed as

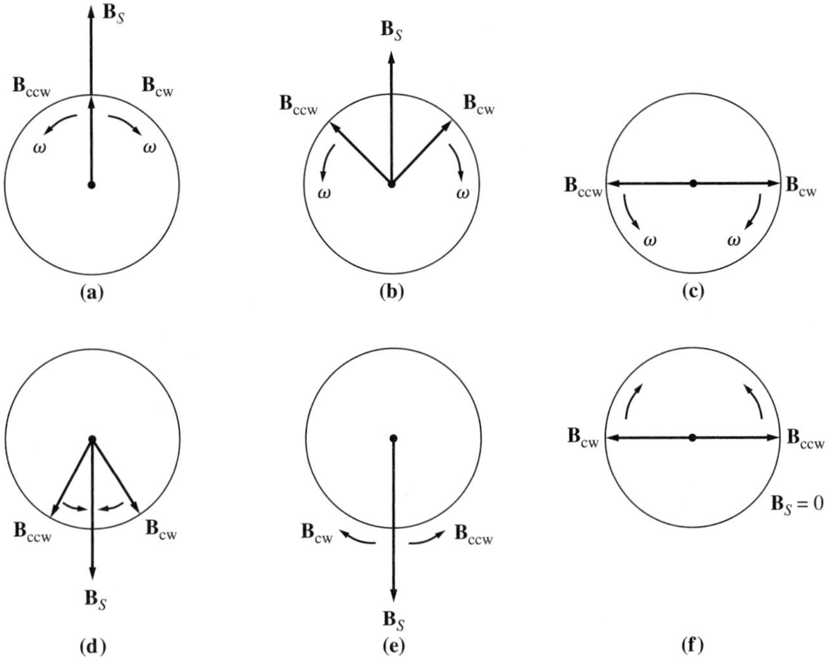

FIGURE 9–7
The resolution of a single pulsating magnetic field into two magnetic fields of equal magnitude by rotation in opposite directions. Notice that at all times the vector sum of the two magnetic fields lies in the vertical plane.

$$\mathbf{B}_{\mathrm{CW}}(t) = \left(\frac{1}{2}B_{\max}\cos\omega t\right)\hat{\mathbf{i}} - \left(\frac{1}{2}B_{\max}\sin\omega t\right)\hat{\mathbf{j}} \tag{9–2}$$

and a counterclockwise-rotating magnetic field can be expressed as

$$\mathbf{B}_{\mathrm{CCW}}(t) = \left(\frac{1}{2}B_{\max}\cos\omega t\right)\hat{\mathbf{i}} + \left(\frac{1}{2}B_{\max}\sin\omega t\right)\hat{\mathbf{j}} \tag{9–3}$$

Notice that the sum of the clockwise and counterclockwise magnetic fields is equal to the stationary pulsating magnetic field $\mathbf{B}_S$:

$$\mathbf{B}_S(t) = \mathbf{B}_{\mathrm{CW}}(t) + \mathbf{B}_{\mathrm{CCW}}(t) \tag{9–4}$$

The torque–speed characteristic of a three-phase induction motor in response to its single rotating magnetic field is shown in Figure 9–8a. A single-phase induction motor responds to each of the two magnetic fields present within it, so the net induced torque in the motor is the *difference* between the two torque–speed curves. This net torque is shown in Figure 9–8b. Notice that there is no net torque at zero speed, so this motor has no starting torque.

The torque–speed characteristic shown in Figure 9–8b is not quite an accurate description of the torque in a single-phase motor. It was formed by the

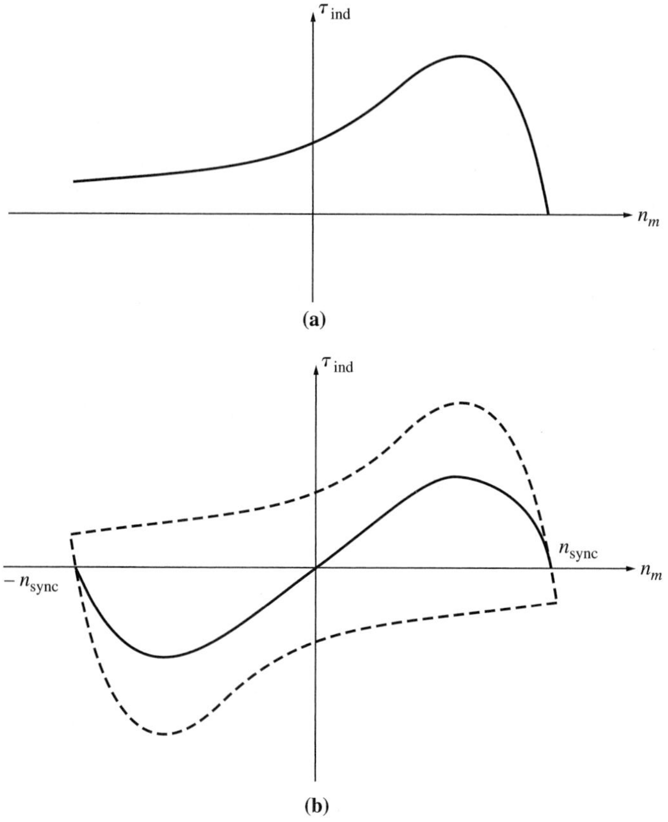

FIGURE 9–8
(a) The torque–speed characteristic of a three-phase induction motor. (b) The torque–speed characteristic curves of the two equal and oppositely rotating stator magnetic fields.

superposition of two three-phase characteristics and ignored the fact that both magnetic fields are present *simultaneously* in the single-phase motor.

If power is applied to a three-phase motor while it is forced to turn backward, its rotor currents will be very high (see Figure 9–9a). However, the rotor frequency is also very high, making the rotor's reactance much larger than its resistance. Since the rotor's reactance is so very high, the rotor current lags the rotor voltage by almost 90°, producing a magnetic field that is nearly 180° from the stator magnetic field (see Figure 9–10). The induced torque in the motor is proportional to the sine of the angle between the two fields, and the sine of an angle near 180° is a very small number. The motor's torque would be very small, except that the extremely high rotor currents partially offset the effect of the magnetic field angles (see Figure 9–9b).

On the other hand, in a single-phase motor, both the forward and the reverse magnetic fields are present and both are produced by the *same* current. The forward

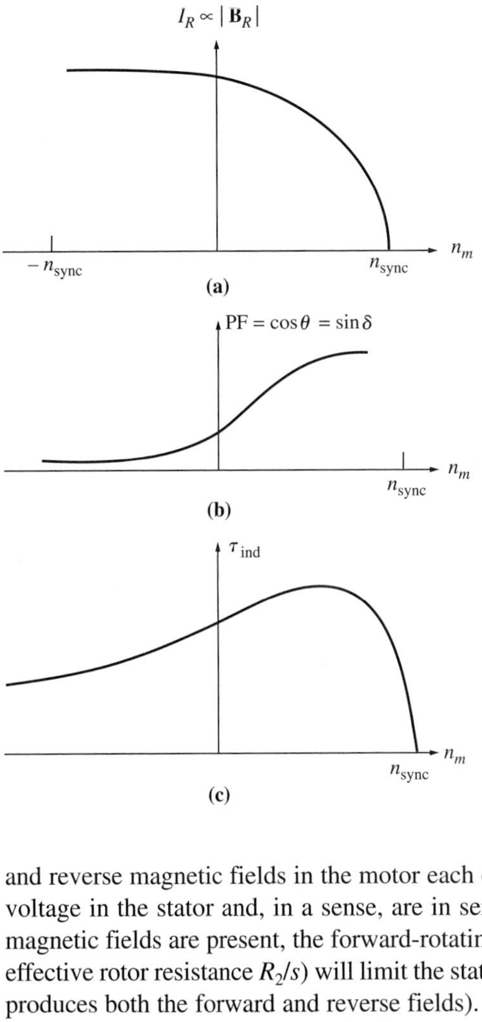

FIGURE 9–9
The torque–speed characteristic of a three-phase induction motor is proportional to both the strength of the rotor magnetic field and the sine of the angle between the fields. When the rotor is turned backward, I_R and I_S are very high, but the angle between the fields is very large, and that angle limits the torque of the motor.

and reverse magnetic fields in the motor each contribute a component to the total voltage in the stator and, in a sense, are in series with each other. Because both magnetic fields are present, the forward-rotating magnetic field (which has a high effective rotor resistance R_2/s) will limit the stator current flow in the motor (which produces both the forward and reverse fields). Since the current supplying the reverse stator magnetic field is limited to a small value and since the reverse rotor magnetic field is at a very large angle with respect to the reverse stator magnetic field, the torque due to the reverse magnetic fields is *very* small near synchronous speed. A more accurate torque–speed characteristic for the single-phase induction motor is shown in Figure 9–11.

In addition to the average net torque shown in Figure 9–11, there are torque pulsations at twice the stator frequency. These torque pulsations are caused when the forward and reverse magnetic fields cross each other twice each cycle. Although these torque pulsations produce no average torque, they do increase vibration, and they make single-phase induction motors noisier than three-phase motors of the same size. There is no way to eliminate these pulsations, since instantaneous power always comes in pulses in a single-phase circuit. A motor designer must allow for this inherent vibration in the mechanical design of single-phase motors.

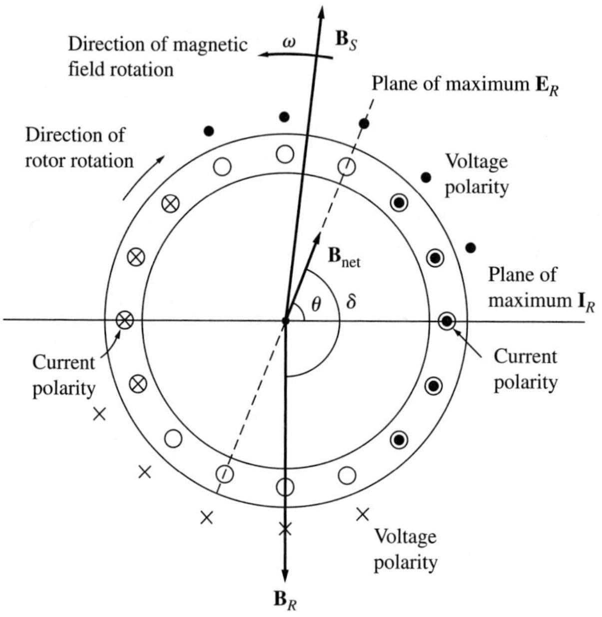

FIGURE 9–10
When the rotor of the motor is forced to turn backward, the angle γ between $\mathbf{B}_R$ and $\mathbf{B}_S$ approaches 180°.

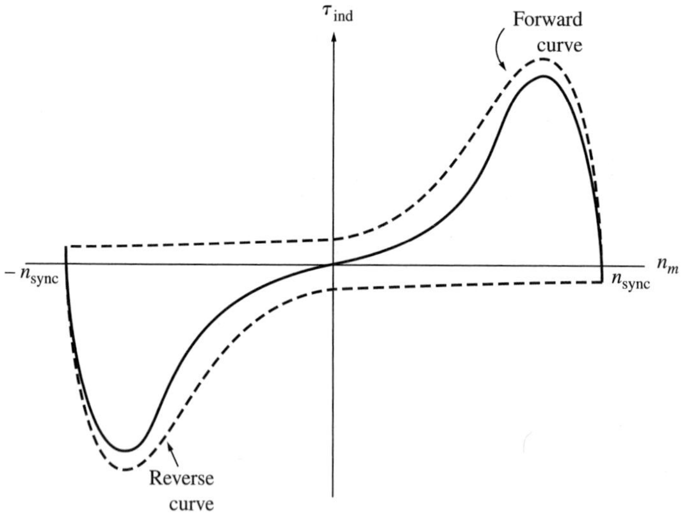

FIGURE 9–11
The torque–speed characteristic of a single-phase induction motor, taking into account the current limitation on the backward-rotating magnetic field caused by the presence of the forward-rotating magnetic field.

The Cross-Field Theory of Single-Phase Induction Motors

The cross-field theory of single-phase induction motors looks at the induction motor from a totally different point of view. This theory is concerned with the voltages and currents that the stationary stator magnetic field can induce in the bars of the rotor when the rotor is moving.

Consider a single-phase induction motor with a rotor which has been brought up to speed by some external method. Such a motor is shown in Figure 9–12a. Voltages are induced in the bars of this rotor, with the peak voltage occurring in the windings passing directly under the stator windings. These rotor voltages produce a current flow in the rotor, but because of the rotor's high reactance, the current lags the voltage by almost 90°. Since the rotor is rotating at nearly synchronous speed, that 90° time lag in current produces an almost 90° *angular* shift between the plane of peak rotor voltage and the plane of peak current. The resulting rotor magnetic field is shown in Figure 9–12b.

The rotor magnetic field is somewhat smaller than the stator magnetic field, because of the losses in the rotor, but they differ by nearly 90° in *both space and*

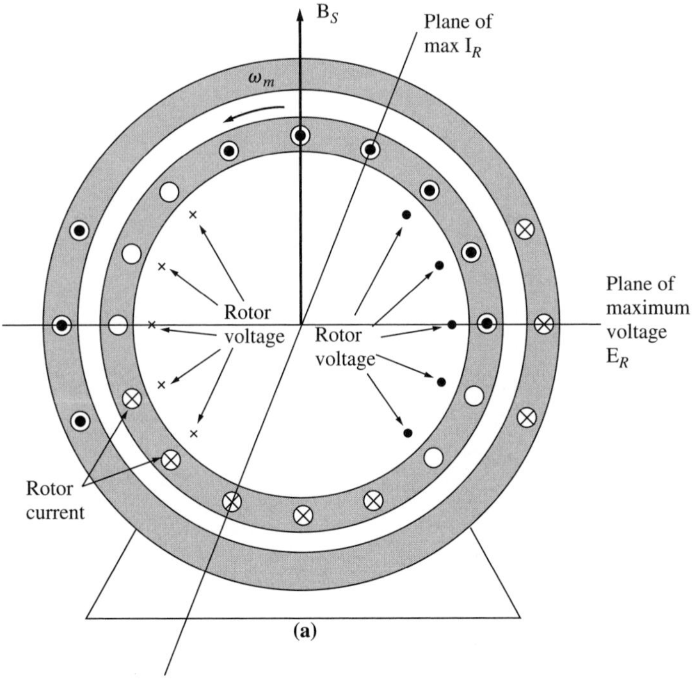

FIGURE 9–12
(a) The development of induced torque in a single-phase induction motor, as explained by the cross-field theory. If the stator field is pulsing, it will induce voltages in the rotor bars, as shown by the marks inside the rotor. However, the rotor current is delayed by nearly 90° behind the rotor voltage, and if the rotor is turning, the rotor current will peak at an angle different from that of the rotor voltage.

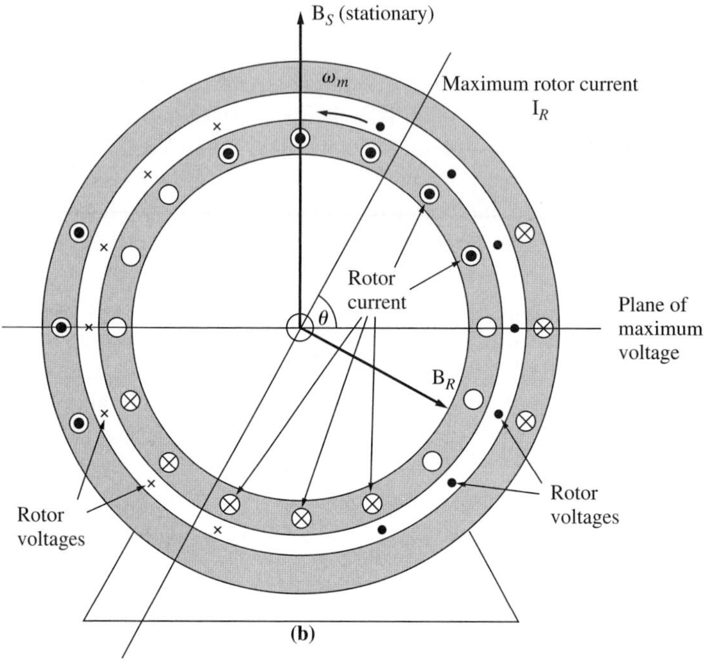

FIGURE 9–12 (concluded)
(b) This delayed rotor current produces a rotor magnetic field at an angle different from the angle of the stator magnetic field.

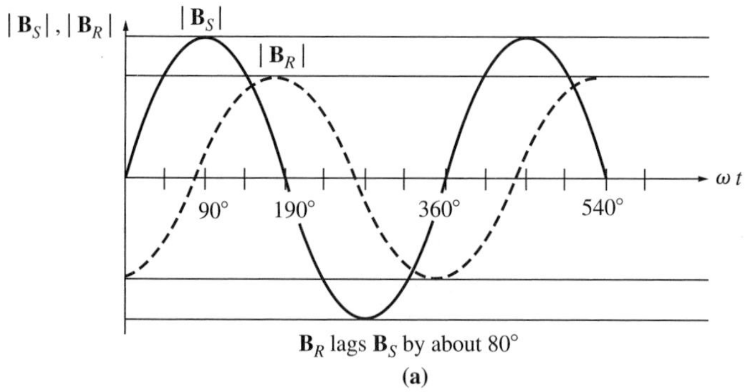

FIGURE 9–13
(a) The *magnitudes* of the magnetic fields as a function of time.

time. If these two magnetic fields are added at different times, one sees that the total magnetic field in the motor is rotating in a counterclockwise direction (see Figure 9–13). With a rotating magnetic field present in the motor, the induction

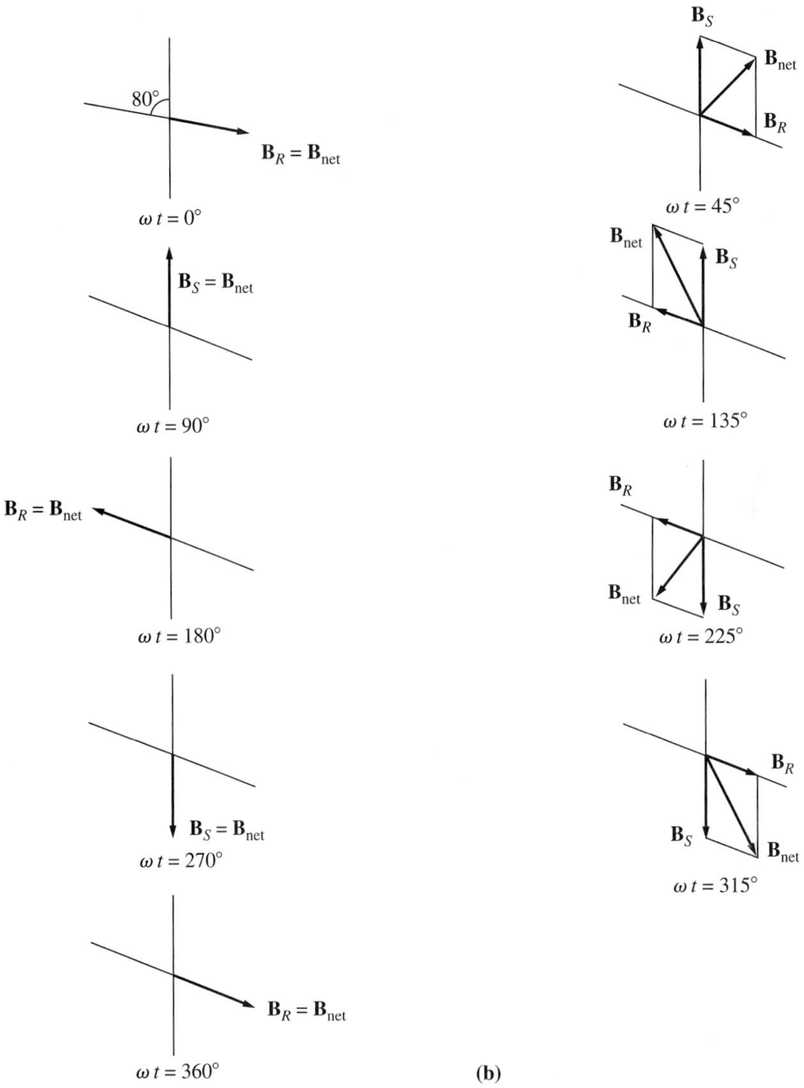

FIGURE 9–13 (concluded)
(b) The vector sum of the rotor and stator magnetic fields at various times, showing a net magnetic field which rotates in a counterclockwise direction.

motor will develop a net torque in the direction of motion, and that torque will keep the rotor turning.

If the motor's rotor had originally been turned in a clockwise direction, the resulting torque would be clockwise and would again keep the rotor turning.

9.3 STARTING SINGLE-PHASE INDUCTION MOTORS

As previously explained, a single-phase induction motor has no intrinsic starting torque. There are three techniques commonly used to start these motors, and single-phase induction motors are classified according to the methods used to produce their starting torque. These starting techniques differ in cost and in the amount of starting torque produced, and an engineer normally uses the least expensive technique that meets the torque requirements in any given application. The three major starting techniques are

1. Split-phase windings
2. Capacitor-type windings
3. Shaded stator poles

All three starting techniques are methods of making one of the two revolving magnetic fields in the motor stronger than the other and so giving the motor an initial nudge in one direction or the other.

Split-Phase Windings

A split-phase motor is a single-phase induction motor with two stator windings, a main stator winding (M) and an auxiliary starting winding (A) (see Figure 9–14). These two windings are set 90 electrical degrees apart along the stator of the motor, and the auxiliary winding is designed to be switched out of the circuit at some set speed by a centrifugal switch. The auxiliary winding is designed to have

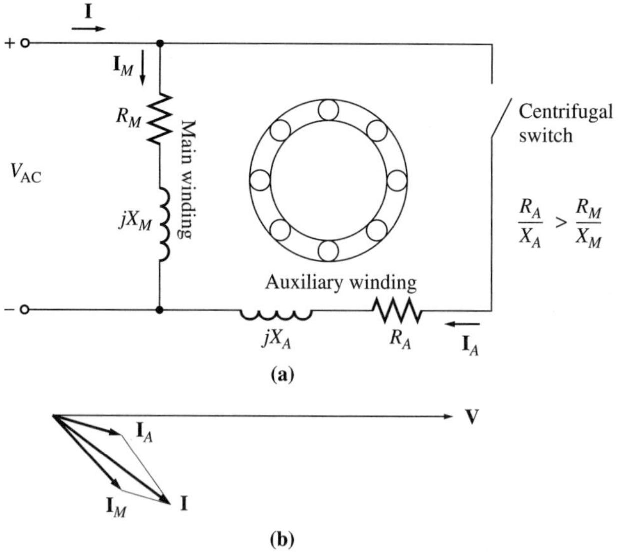

FIGURE 9–14
(a) A split-phase induction motor. (b) The currents in the motor at starting conditions.

a higher resistance/reactance ratio than the main winding, so that the current in the auxiliary winding *leads* the current in the main winding. This higher *R/X* ratio is usually accomplished by using smaller wire for the auxiliary winding. Smaller wire is permissible in the auxiliary winding because it is used only for starting and therefore does not have to take full current continuously.

To understand the function of the auxiliary winding, refer to Figure 9–15. Since the current in the auxiliary winding leads the current in the main winding,

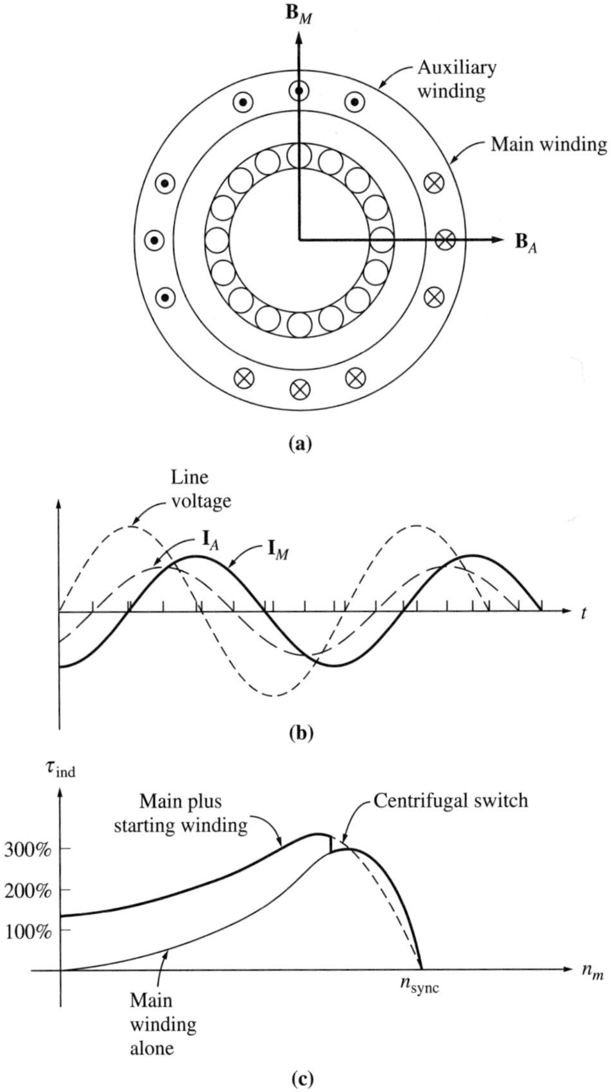

FIGURE 9–15
(a) Relationship of main and auxiliary magnetic fields. (b) $\mathbf{I}_A$ peaks before $\mathbf{I}_M$, producing a net counterclockwise rotation of the magnetic fields. (c) The resulting torque–speed characteristic.

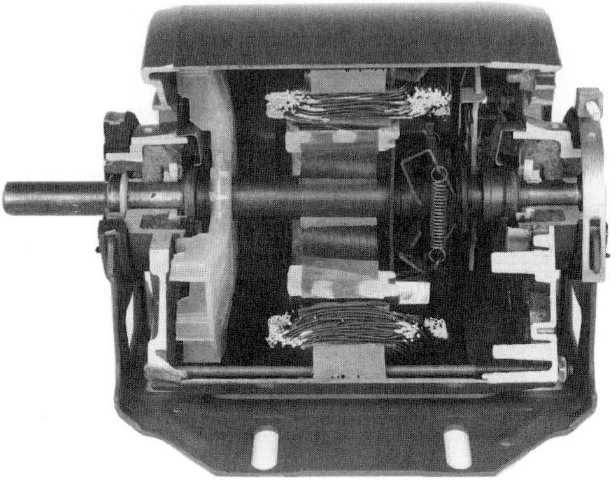

FIGURE 9–16
Cutaway view of a split-phase motor, showing the main and auxiliary windings and the centrifugal switch. (*Courtesy of Westinghouse Electric Corporation.*)

the magnetic field $\mathbf{B}_A$ peaks before the main magnetic field $\mathbf{B}_M$. Since $\mathbf{B}_A$ peaks first and then $\mathbf{B}_M$, there is a net counterclockwise rotation in the magnetic field. In other words, the auxiliary winding makes one of the oppositely rotating stator magnetic fields larger than the other one and provides a net starting torque for the motor. A typical torque–speed characteristic is shown in Figure 9–15c.

A cutaway diagram of a split-phase motor is shown in Figure 9–16. It is easy to see the main and auxiliary windings (the auxiliary windings are the smaller-diameter wires) and the centrifugal switch that cuts the auxiliary windings out of the circuit when the motor approaches operating speed.

Split-phase motors have a moderate starting torque with a fairly low starting current. They are used for applications which do not require very high starting torques, such as fans, blowers, and centrifugal pumps. They are available for sizes in the fractional-horsepower range and are quite inexpensive.

In a split-phase induction motor, the current in the auxiliary windings always peaks before the current in the main winding, and therefore the magnetic field from the auxiliary winding always peaks before the magnetic field from the main winding. The direction of rotation of the motor is determined by whether the space angle of the magnetic field from the auxiliary winding is 90° ahead or 90° behind the angle of the main winding. Since that angle can be changed from 90° ahead to 90° behind just by switching the connections on the auxiliary winding, *the direction of rotation of the motor can be reversed by switching the connections of the auxiliary winding* while leaving the main winding's connections unchanged.

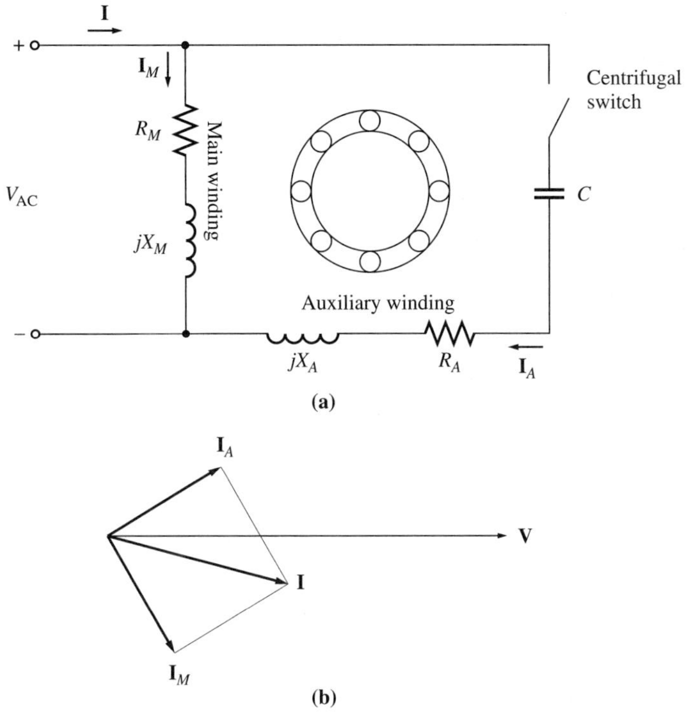

FIGURE 9–17
(a) A capacitor-start induction motor. (b) Current angles at starting in this motor.

Capacitor-Start Motors

For some applications, the starting torque supplied by a split-phase motor is insufficient to start the load on a motor's shaft. In those cases, capacitor-start motors may be used (Figure 9–17). In a capacitor-start motor, a capacitor is placed in series with the auxiliary winding of the motor. By proper selection of capacitor size, the magnetomotive force of the starting current in the auxiliary winding can be adjusted to be equal to the magnetomotive force of the current in the main winding, and the phase angle of the current in the auxiliary winding can be made to lead the current in the main winding by 90°. Since the two windings are physically separated by 90°, a 90° phase difference in current will yield a single uniform rotating stator magnetic field, and the motor will behave just as though it were starting from a three-phase power source. In this case, the starting torque of the motor can be more than 300 percent of its rated value (see Figure 9–18).

Capacitor-start motors are more expensive than split-phase motors, and they are used in applications where a high starting torque is absolutely required. Typical applications for such motors are compressors, pumps, air conditioners, and other pieces of equipment that must start under a load (See Figure 9–19).

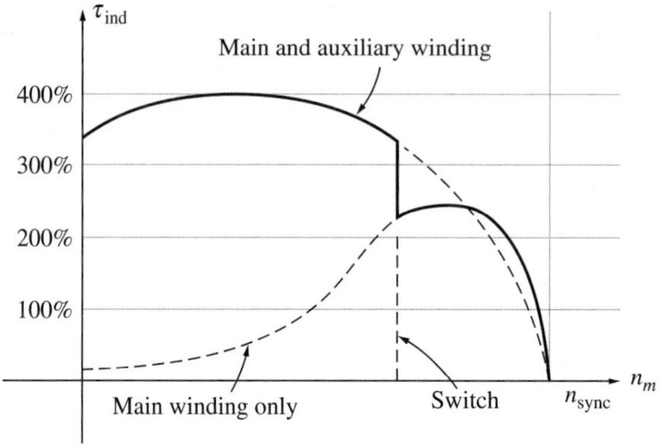

FIGURE 9–18
Torque–speed characteristic of a capacitor-start induction motor.

Permanent Split-Capacitor and Capacitor-Start, Capacitor-Run Motors

The starting capacitor does such a good job of improving the torque–speed characteristic of an induction motor that an auxiliary winding with a smaller capacitor is sometimes left permanently in the motor circuit. If the capacitor's value is chosen correctly, such a motor will have a perfectly uniform rotating magnetic field at some specific load, and it will behave just like a three-phase induction motor at that point. Such a design is called a *permanent split-capacitor* or *capacitor-start-and-run* motor (Figure 9–20). Permanent split-capacitor motors are simpler than capacitor-start motors, since the starting switch is not needed. At normal loads, they are more efficient and have a higher power factor and a smoother torque than ordinary single-phase induction motors.

However, permanent split-capacitor motors have a *lower starting torque* than capacitor-start motors, since the capacitor must be sized to balance the currents in the main and auxiliary windings at normal-load conditions. Since the starting current is much greater than the normal-load current, a capacitor that balances the phases under normal loads leaves them very unbalanced under starting conditions.

If both the largest possible starting torque and the best running conditions are needed, two capacitors can be used with the auxiliary winding. Motors with two capacitors are called *capacitor-start, capacitor-run*, or *two-value capacitor* motors (Figure 9–21). The larger capacitor is present in the circuit only during starting, when it ensures that the currents in the main and auxiliary windings are roughly balanced, yielding very high starting torques. When the motor gets up to speed, the centrifugal switch opens, and the permanent capacitor is left by itself in the auxiliary winding circuit. The permanent capacitor is just large enough to balance the currents at normal motor loads, so the motor again operates efficiently with a high torque and power factor. The permanent capacitor in such a motor is typically about 10 to 20 percent of the size of the starting capacitor.

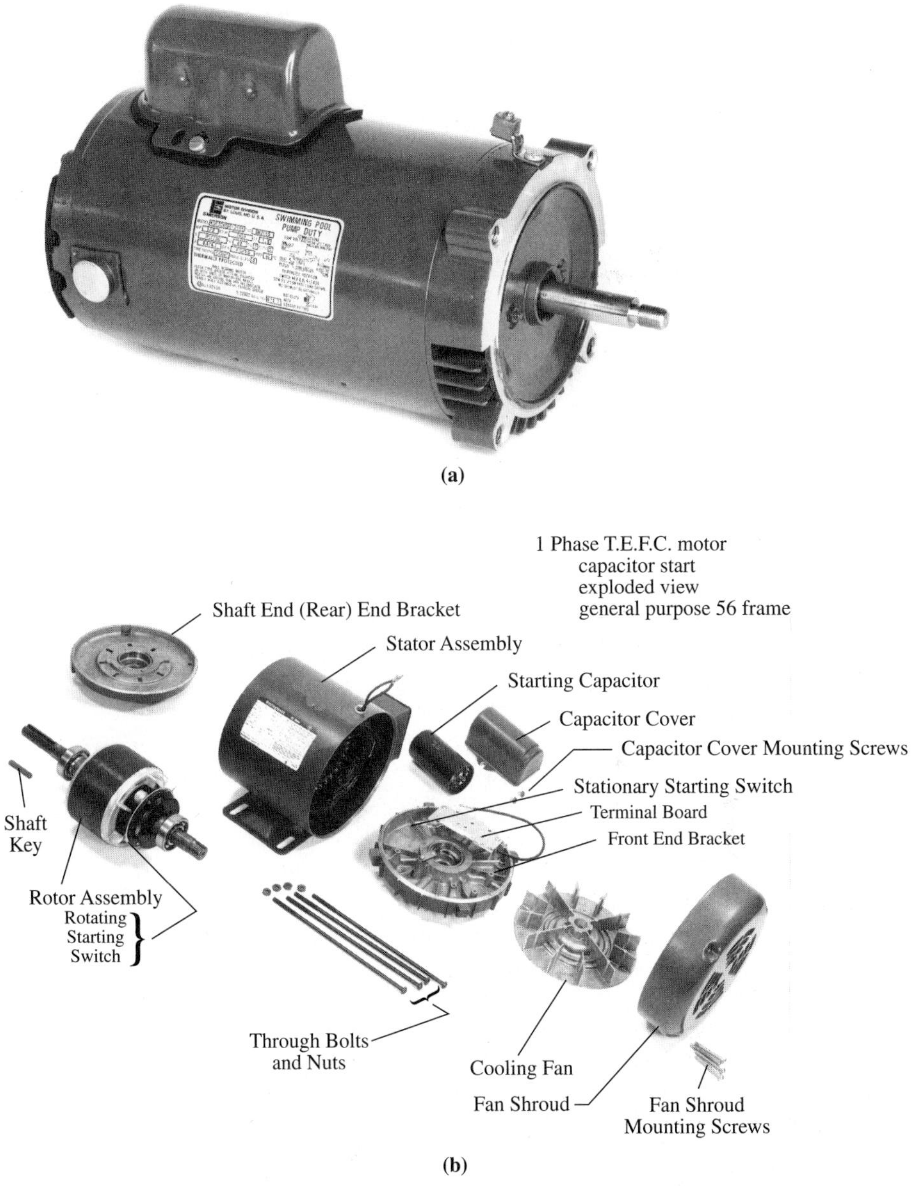

FIGURE 9–19
(a) A capacitor-start induction motor. (*Courtesy of Emerson Electric Company.*) (b) Exploded view of a capacitor-start induction motor. (*Courtesy of Westinghouse Electric Corporation.*)

The direction of rotation of any capacitor-type motor may be reversed by switching the connections of its auxiliary windings.

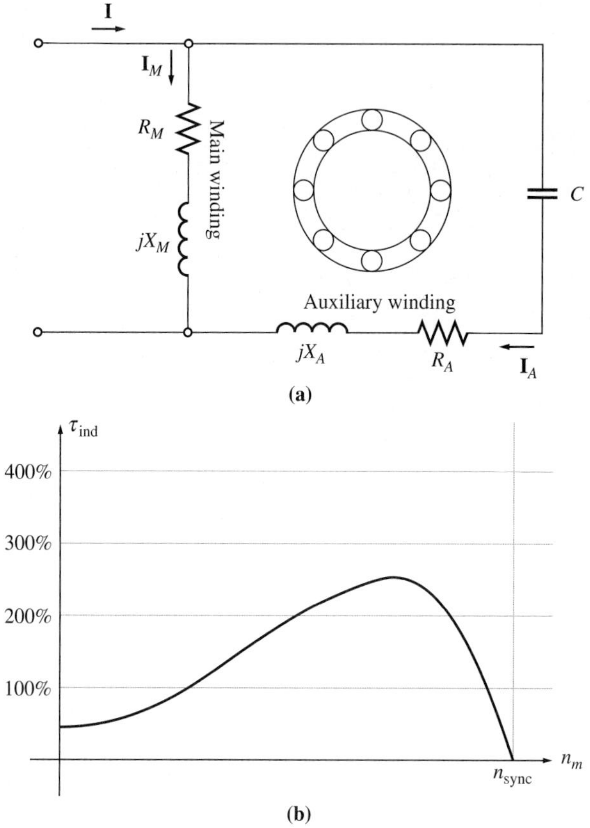

FIGURE 9–20
(a) A permanent split-capacitor induction motor. (b) Torque–speed characteristic of this motor.

Shaded-Pole Motors

A shaded-pole induction motor is an induction motor with only a main winding. Instead of having an auxiliary winding, it has salient poles, and one portion of each pole is surrounded by a short-circuited coil called a *shading coil* (see Figure 9–22a). A time-varying flux is induced in the poles by the main winding. When the pole flux varies, it induces a voltage and a current in the shading coil which *opposes* the original change in flux. This opposition *retards* the flux changes under the shaded portions of the coils and therefore produces a slight imbalance between the two oppositely rotating stator magnetic fields. The net rotation is in the direction from the unshaded to the shaded portion of the pole face. The torque–speed characteristic of a shaded-pole motor is shown in Figure 9–22b.

Shaded poles produce less starting torque than any other type of induction motor starting system. They are much less efficient and have a much higher slip

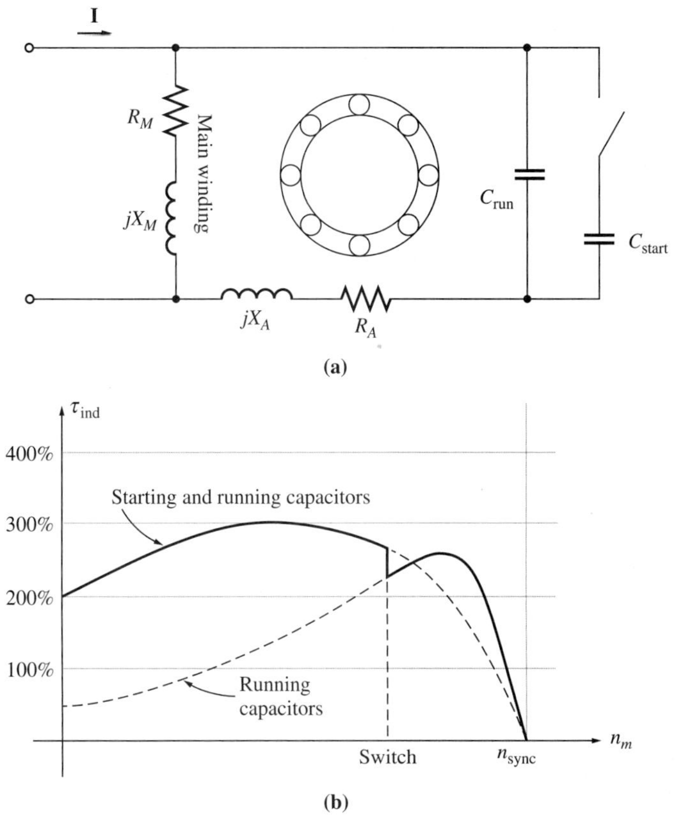

FIGURE 9–21
(a) A capacitor-start, capacitor-run induction motor. (b) Torque–speed characteristic of this motor.

than other types of single-phase induction motors. Such poles are used only in very small motors ($\frac{1}{20}$ hp and less) with very low starting torque requirements. Where it is possible to use them, shaded-pole motors are the cheapest design available.

Because shaded-pole motors rely on a shading coil for their starting torque, there is no easy way to reverse the direction of rotation of such a motor. To achieve reversal, it is necessary to install two shading coils on each pole face and to selectively short one or the other of them. See Figures 9–23 and 9–24.

Comparison of Single-Phase Induction Motors

Single-phase induction motors may be ranked from best to worst in terms of their starting and running characteristics:

1. Capacitor-start, capacitor-run motor
2. Capacitor-start motor

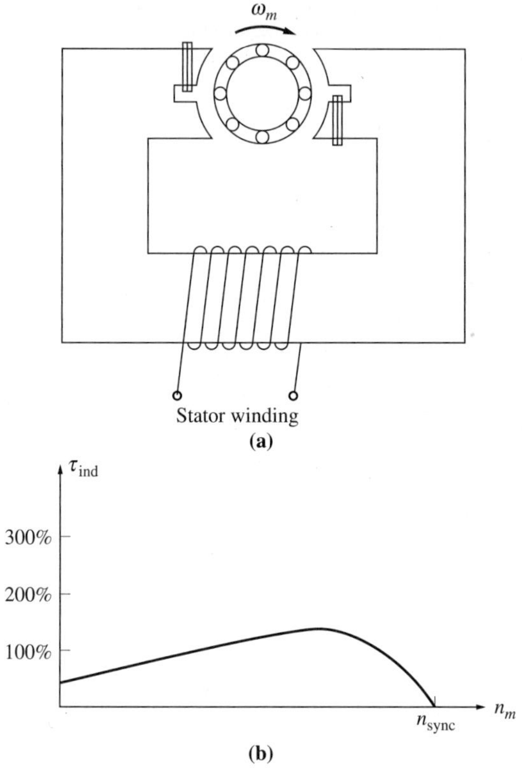

FIGURE 9–22
(a) A basic shaded-pole induction motor. (b) The resulting torque–speed characteristic.

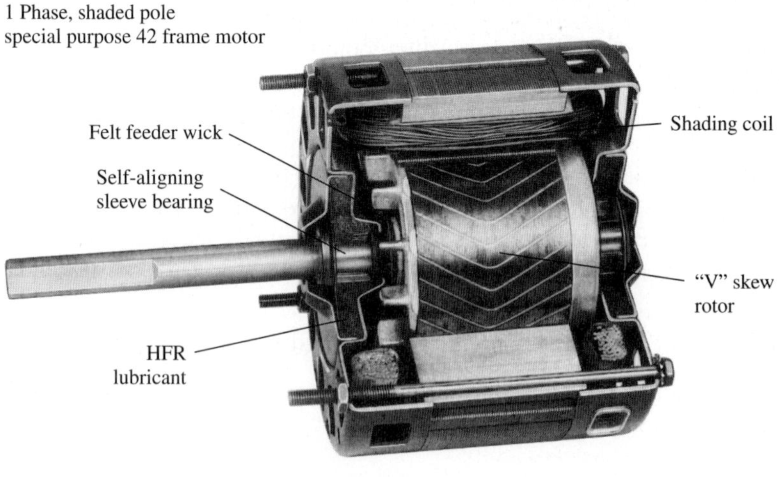

FIGURE 9–23
Cutaway view of a shaded-pole induction motor. (*Courtesy of Westinghouse Electric Corporation.*)

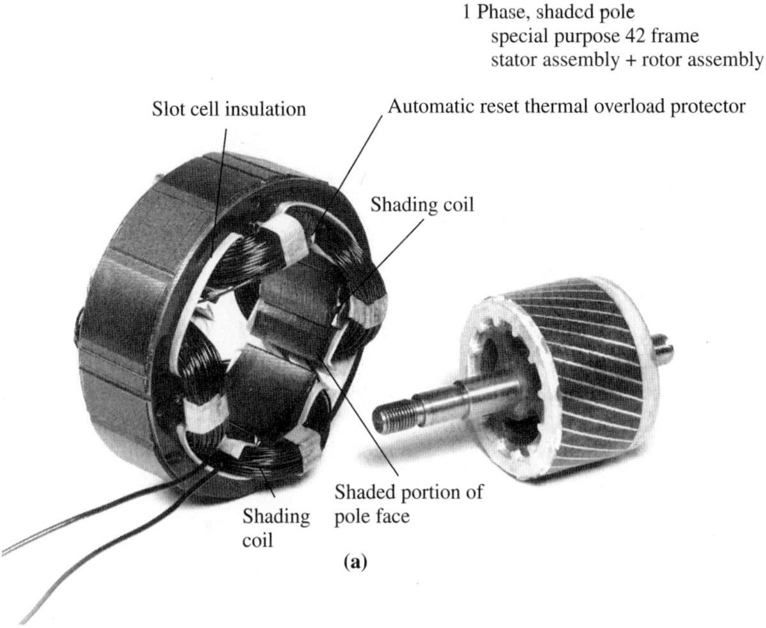

(a)

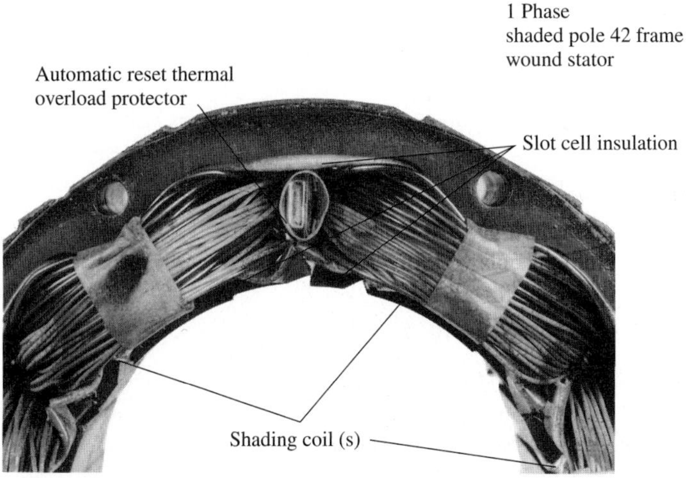

(b)

FIGURE 9–24
Close-up views of the construction of a four-pole shaded-pole induction motor. (*Courtesy of Westinghouse Electric Corporation.*)

3. Permanent split-capacitor motor
4. Split-phase motor
5. Shaded-pole motor

Naturally, the best motor is also the most expensive, and the worst motor is the least expensive. Also, not all these starting techniques are available in all motor size ranges. It is up to the design engineer to select the cheapest available motor for any given application that will do the job.

9.4 SPEED CONTROL OF SINGLE-PHASE INDUCTION MOTORS

In general, the speed of single-phase induction motors may be controlled in the same manner as the speed of polyphase induction motors. For squirrel-cage rotor motors, the following techniques are available:

1. Vary the stator frequency.
2. Change the number of poles.
3. Change the applied terminal voltage V_T.

In practical designs involving fairly high-slip motors, the usual approach to speed control is to vary the terminal voltage of the motor. The voltage applied to a motor may be varied in one of three ways:

1. An autotransformer may be used to continually adjust the line voltage. This is the most expensive method of voltage speed control and is used only when very smooth speed control is needed.
2. A solid-state controller circuit may be used to reduce the rms voltage applied to the motor by ac phase control. Solid-state control circuits are considerably cheaper than autotransformers and so are becoming more and more common.
3. A resistor may be inserted in series with the motor's stator circuit. This is the cheapest method of voltage control, but it has the disadvantage that considerable power is lost in the resistor, reducing the overall power conversion efficiency.

Another technique is also used with very high-slip motors such as shaded-pole motors. Instead of using a separate autotransformer to vary the voltage applied to the stator of the motor, *the stator winding itself* can be used as an autotransformer. Figure 9–25 shows a schematic representation of a main stator winding, with a number of taps along its length. Since the stator winding is wrapped about an iron core, it behaves as an autotransformer.

When the full line voltage V is applied across the entire main winding, then the induction motor operates normally. Suppose instead that the full line voltage is applied to tap 2, the center tap of the winding. Then an identical voltage will be induced in the upper half of the winding by transformer action, and the total winding voltage will be twice the applied line voltage. The total voltage applied to the winding has effectively been doubled.

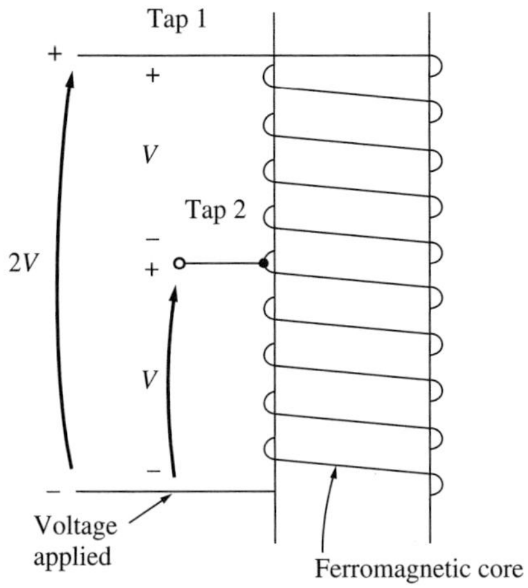

FIGURE 9–25
The use of a stator winding as an autotransformer. If voltage V is applied to the winding at the center tap, the total winding voltage will be $2V$.

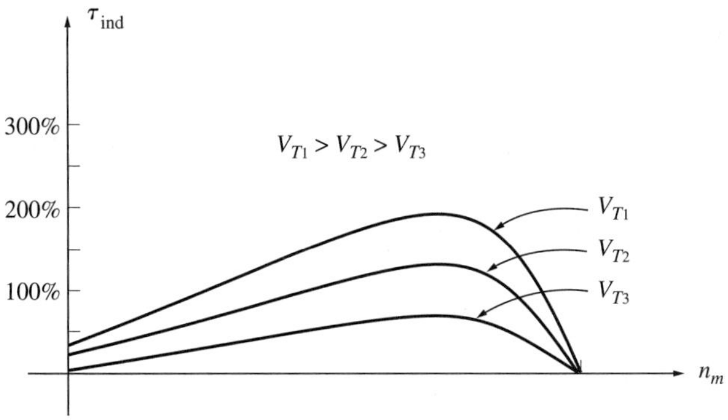

FIGURE 9–26
The torque–speed characteristic of a shaded-pole induction motor as the terminal voltage is changed. Increases in V_T may be accomplished either by actually raising the voltage across the whole winding or by switching to a lower tap on the stator winding.

Therefore, the smaller the fraction of the total coil that the line voltage is applied across, the greater the total voltage will be across the whole winding, and the higher the speed of the motor will be for a given load (see Figure 9–26).

This is the standard approach used to control the speed of single-phase motors in many fan and blower applications. Such speed control has the advantage that it is quite inexpensive, since the only components necessary are taps on the main motor winding and an ordinary multiposition switch. It also has the advantage that the autotransformer effect does not consume power the way series resistors would.

9.5 THE CIRCUIT MODEL OF A SINGLE-PHASE INDUCTION MOTOR

As previously described, an understanding of the induced torque in a single-phase induction motor can be achieved through either the double-revolving-field theory or the cross-field theory of single-phase motors. Either approach can lead to an equivalent circuit of the motor, and the torque–speed characteristic can be derived through either method.

This section is restricted to an examination of an equivalent circuit based on the double-revolving-field theory—in fact, to only a special case of that theory. We will develop an equivalent circuit of the *main winding* of a single-phase induction motor when it is operating alone. The technique of symmetrical components is necessary to analyze a single-phase motor with both main and auxiliary windings present, and since symmetrical components are beyond the scope of this book, that case will not be discussed. For a more detailed analysis of single-phase motors, see Reference 4.

The best way to begin the analysis of a single-phase induction motor is to consider the motor when it is stalled. At that time, the motor appears to be just a single-phase transformer with its secondary circuit shorted out, and so its equivalent circuit is that of a transformer. This equivalent circuit is shown in Figure 9–27a. In this figure, R_1 and X_1 are the resistance and reactance of the stator winding, X_M is the magnetizing reactance, and R_2 and X_2 are the referred values of the rotor's resistance and reactance. The core losses of the machine are not shown and will be lumped together with the mechanical and stray losses as a part of the motor's rotational losses.

Now recall that the pulsating air-gap flux in the motor at stall conditions can be resolved into two equal and opposite magnetic fields within the motor. Since these fields are of equal size, each one contributes an equal share to the resistive and reactive voltage drops in the rotor circuit. It is possible to split the rotor equivalent circuit into two sections, each one corresponding to the effects of one of the magnetic fields. The motor equivalent circuit with the effects of the forward and reverse magnetic fields separated is shown in Figure 9–27b.

Now suppose that the motor's rotor begins to turn with the help of an auxiliary winding and that the winding is switched out again after the motor comes up to speed. As shown in Chapter 8, the effective rotor resistance of an induction motor depends on the amount of relative motion between the rotor and the stator magnetic fields. However, there are two magnetic fields in this motor, and the amount of relative motion differs for each of them.

For the *forward* magnetic field, the per-unit difference between the rotor speed and the speed of the magnetic field is the slip s, where slip is defined in the same manner as it was for three-phase induction motors. The rotor resistance in the part of the circuit associated with the forward magnetic field is thus $0.5R_2/s$.

The forward magnetic field rotates at speed n_{sync} and the reverse magnetic field rotates at speed $-n_{\text{sync}}$. Therefore, the total per-unit difference in speed (on a base of n_{sync}) between the forward and reverse magnetic fields is 2. Since the rotor

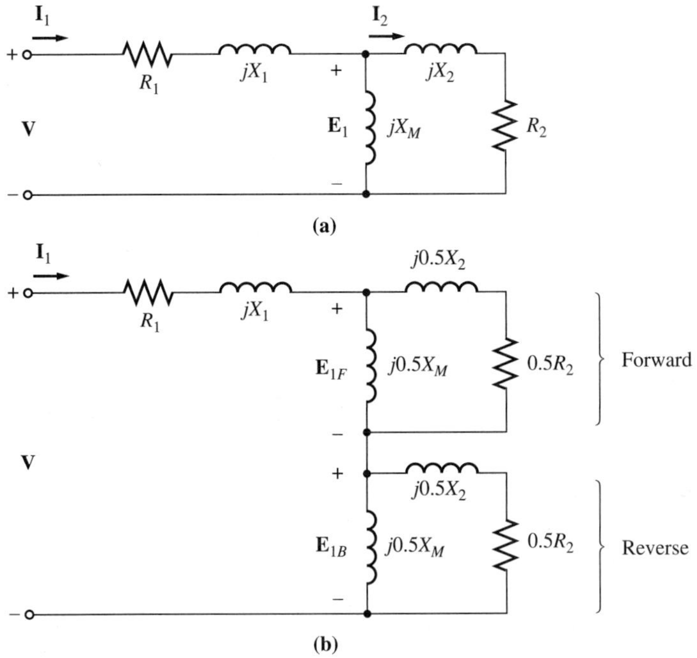

FIGURE 9–27
(a) The equivalent circuit of a single-phase induction motor at standstill. Only its main windings are energized. (b) The equivalent circuit with the effects of the forward and reverse magnetic fields separated.

is turning at a speed s slower than the forward magnetic field, the total per-unit difference in speed between the rotor and the reverse magnetic field is $2 - s$. Therefore, the effective rotor resistance in the part of the circuit associated with the reverse magnetic field is $0.5R_2/(2 - s)$.

The final induction motor equivalent circuit is shown in Figure 9–28.

Circuit Analysis with the Single-Phase Induction Motor Equivalent Circuit

The single-phase induction motor equivalent circuit in Figure 9–28 is similar to the three-phase equivalent circuit, except that there are both forward and backward components of power and torque present. The same general power and torque relationships that applied for three-phase motors also apply for either the forward or the backward components of the single-phase motor, and the net power and torque in the machine is the *difference* between the forward and reverse components.

The power-flow diagram of an induction motor is repeated in Figure 9–29 for easy reference.

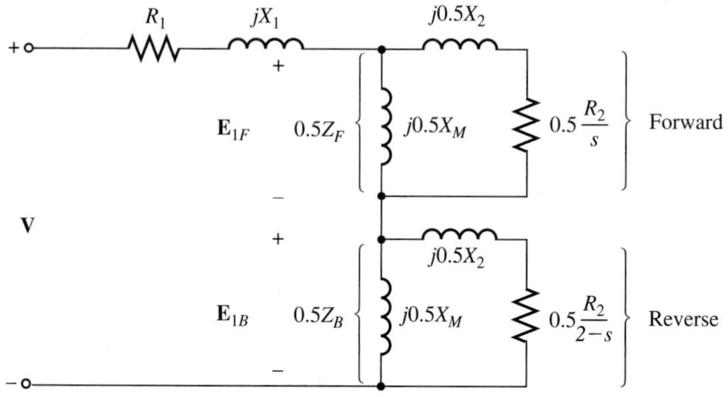

FIGURE 9–28
The equivalent circuit of a single-phase induction motor running at speed with only its main windings energized.

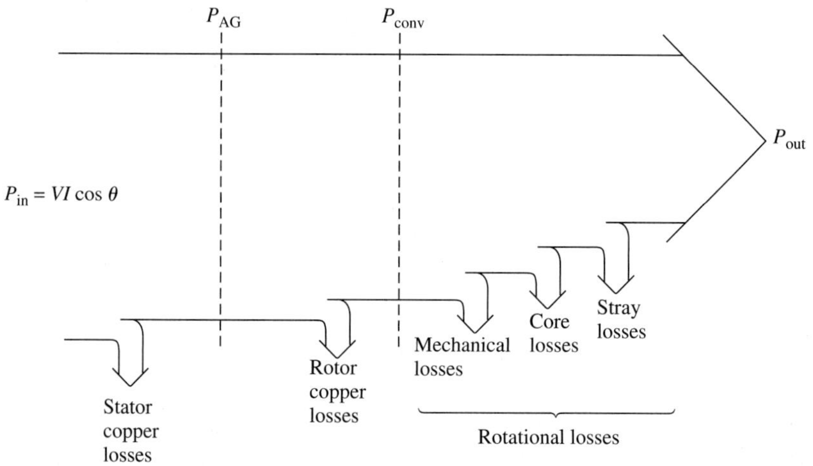

FIGURE 9–29
The power-flow diagram of a single-phase induction motor.

To make the calculation of the input current flow into the motor simpler, it is customary to define impedances Z_F and Z_B, where Z_F is a single impedance equivalent to all the forward magnetic field impedance elements and Z_B is a single impedance equivalent to all the backward magnetic field impedance elements (see Figure 9–30). These impedances are given by

$$Z_F = R_F + jX_F = \frac{(R_2/s + jX_2)(jX_M)}{(R_2/s + jX_2) + jX_M} \tag{9–5}$$

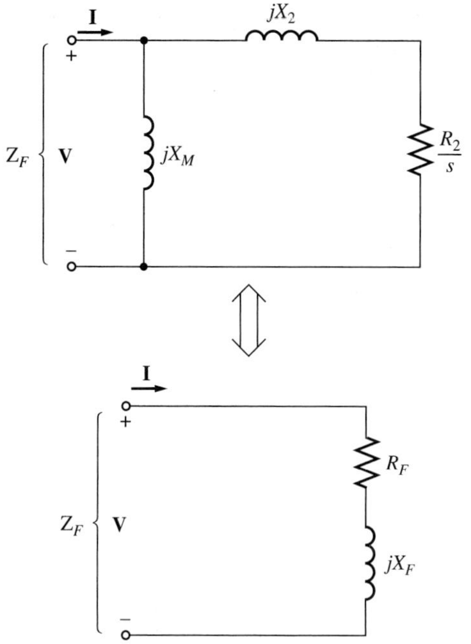

FIGURE 9–30
A series combination of R_F and jX_F is the Thevenin equivalent of the forward-field impedance elements, and therefore R_F must consume the same power from a given current as R_2/s would.

$$Z_B = R_B + jX_B = \frac{[R_2/(2 - s) + jX_2](jX_M)}{[R_2/(2 - s) + jX_2] + jX_M} \tag{9–6}$$

In terms of Z_F and Z_B, the current flowing in the induction motor's stator winding is

$$\mathbf{I}_1 = \frac{\mathbf{V}}{R_1 + jX_1 + 0.5Z_F + 0.5Z_B} \tag{9–7}$$

The per-phase air-gap power of a three-phase induction motor is the power consumed in the rotor circuit resistance $0.5R_2/s$. Similarly, the forward air-gap power of a single-phase induction motor is the power consumed by $0.5R_2/s$, and the reverse air-gap power of the motor is the power consumed by $0.5R_2/(2 - s)$. Therefore, the air-gap power of the motor could be calculated by determining the power in the forward resistor $0.5R_2/s$, determining the power in the reverse resistor $0.5R_2/(2 - s)$, and subtracting one from the other.

The most difficult part of this calculation is the determination of the separate currents flowing in the two resistors. Fortunately, a simplification of this calculation is possible. Notice that the *only* resistor within the circuit elements composing the equivalent impedance Z_F is the resistor R_2/s. Since Z_F is equivalent to that circuit, any power consumed by Z_F must also be consumed by the original circuit, and since R_2/s is the only resistor in the original circuit, its power consumption must equal that of impedance Z_F. Therefore, the air-gap power for the forward magnetic field can be expressed as

$$P_{\text{AG},F} = I_1^2(0.5\ R_F) \tag{9–8}$$

Similarly, the air-gap power for the reverse magnetic field can be expressed as

$$P_{\text{AG},B} = I_1^2(0.5\ R_B) \tag{9–9}$$

The advantage of these two equations is that only the one current I_1 needs to be calculated to determine both powers.

The total air-gap power in a single-phase induction motor is thus

$$P_{\text{AG}} = P_{\text{AG},F} - P_{\text{AG},B} \tag{9–10}$$

The induced torque in a three-phase induction motor can be found from the equation

$$\tau_{\text{ind}} = \frac{P_{\text{AG}}}{\omega_{\text{sync}}} \tag{9–11}$$

where P_{AG} is the net air-gap power given by Equation (9–10).

The rotor copper losses can be found as the sum of the rotor copper losses due to the forward field and the rotor copper losses due to the reverse field.

$$P_{\text{RCL}} = P_{\text{RCL},F} + P_{\text{RCL},B} \tag{9–12}$$

The rotor copper losses in a three-phase induction motor were equal to the per-unit relative motion between the rotor and the stator field (the slip) times the air-gap power of the machine. Similarly, the forward rotor copper losses of a single-phase induction motor are given by

$$P_{\text{RCL},F} = sP_{\text{AG},F} \tag{9–13}$$

and the reverse rotor copper losses of the motor are given by

$$P_{\text{RCL},B} = sP_{\text{AG},B} \tag{9–14}$$

Since these two power losses in the rotor are at different frequencies, the total rotor power loss is just their sum.

The power converted from electrical to mechanical form in a single-phase induction motor is given by the same equation as P_{conv} for three-phase induction motors. This equation is

$$P_{\text{conv}} = \tau_{\text{ind}}\omega_m \tag{9–15}$$

Since $\omega_m = (1 - s)\omega_{\text{sync}}$, this equation can be reexpressed as

$$P_{\text{conv}} = \tau_{\text{ind}}(1 - s)\omega_m \tag{9–16}$$

From Equation (9–11), $P_{\text{AG}} = \tau_{\text{ind}}\omega_{\text{sync}}$, so P_{conv} can also be expressed as

$$P_{\text{conv}} = (1 - s)\ P_{\text{AG}} \tag{9–17}$$

As in the three-phase induction motor, the shaft output power is not equal to P_{conv}, since the rotational losses must still be subtracted. In the single-phase induction motor model used here, the core losses, mechanical losses, and stray losses must be subtracted from P_{conv} in order to get P_{out}.

Example 9–1. A ⅓-hp, 110-V, 60-Hz, six-pole, split-phase induction motor has the following impedances:

$$R_1 = 1.52\ \Omega \qquad X_1 = 2.10\ \Omega \qquad X_M = 58.2\ \Omega$$
$$R_2 = 3.13\ \Omega \qquad X_2 = 1.56\ \Omega$$

The core losses of this motor are 35 W, and the friction, windage, and stray losses are 16 W. The motor is operating at the rated voltage and frequency with its starting winding open, and the motor's slip is 5 percent. Find the following quantities in the motor at these conditions:

(*a*) Speed in revolutions per minute
(*b*) Stator current in amperes
(*c*) Stator power factor
(*d*) P_{in}
(*e*) P_{AG}
(*f*) P_{conv}
(*g*) τ_{ind}
(*h*) P_{out}
(*i*) τ_{load}
(*j*) Efficiency

Solution

The forward and reverse impedances of this motor at a slip of 5 percent are

$$Z_F = R_F + jX_F = \frac{(R_2/s + jX_2)(jX_M)}{(R_2/s + jX_2) + jX_M} \tag{9–5}$$

$$= \frac{(3.13\ \Omega/0.05 + j1.56\ \Omega)(j58.2\ \Omega)}{(3.13\ \Omega/0.05 + j1.56\ \Omega) + j58.2\ \Omega}$$

$$= \frac{(62.6\angle 1.43°\ \Omega)(j58.2\ \Omega)}{(62.6\ \Omega + j1.56\ \Omega) + j58.2\ \Omega}$$

$$= 39.9\angle 50.5°\ \Omega = 25.4 + j30.7\ \Omega$$

$$Z_B = R_B + jX_B = \frac{[R_2/(2 - s) + jX_2](jX_M)}{[R_2/(2 - s) + jX_2] + jX_M} \tag{9–6}$$

$$= \frac{(3.13\ \Omega/1.95 + j1.56\ \Omega)(j58.2\ \Omega)}{(3.13\ \Omega/1.95 + j1.56\ \Omega) + j58.2\ \Omega}$$

$$= \frac{(2.24\angle 44.2°\ \Omega)(j58.2\ \Omega)}{(1.61\ \Omega + j1.56\ \Omega) + j58.2\ \Omega}$$

$$= 2.18\angle 45.9°\ \Omega = 1.51 + j1.56\ \Omega$$

These values will be used to determine the motor current, power, and torque.

(*a*) The synchronous speed of this motor is

$$n_{\text{sync}} = \frac{120 f_{se}}{P} = \frac{120(60\text{ Hz})}{6\text{ pole}} = 1200\text{ r/min}$$

Since the motor is operating at 5 percent slip, its mechanical speed is

$$n_m = (1 - s)n_{\text{sync}}$$

$$n_m = (1 - 0.05)(1200 \text{ r/min}) = 1140 \text{ r/min}$$

(b) The stator current in this motor is

$$\mathbf{I}_1 = \frac{\mathbf{V}}{R_1 + jX_1 + 0.5Z_F + 0.5Z_B} \tag{9–7}$$

$$= \frac{110\angle 0° \text{ V}}{1.52\ \Omega + j2.10\ \Omega + 0.5(25.4\ \Omega + j30.7\ \Omega) + 0.5(1.51\ \Omega + j1.56\ \Omega)}$$

$$= \frac{110\angle 0° \text{ V}}{14.98\ \Omega + j18.23\ \Omega} = \frac{110\angle 0° \text{ V}}{23.6\angle 50.6°\ \Omega} = 4.66\angle -50.6° \text{ A}$$

(c) The stator power factor of this motor is

$$\text{PF} = \cos(-50.6°) = 0.635 \text{ lagging}$$

(d) The input power to this motor is

$$P_{\text{in}} = VI\cos\theta$$
$$= (110 \text{ V})(4.66 \text{ A})(0.635) = 325 \text{ W}$$

(e) The forward-wave air-gap power is

$$P_{\text{AG},F} = I_1^2(0.5\,R_F) \tag{9–8}$$
$$= (4.66 \text{ A})^2(12.7\ \Omega) = 275.8 \text{ W}$$

and the reverse-wave air-gap power is

$$P_{\text{AG},B} = I_1^2(0.5\,R_B) \tag{9–9}$$
$$= (4.66 \text{ A})2(0.755 \text{ V}) = 16.4 \text{ W}$$

Therefore, the total air-gap power of this motor is

$$P_{\text{AG}} = P_{\text{AG},F} - P_{\text{AG},B} \tag{9–10}$$
$$= 275.8 \text{ W} - 16.4 \text{ W} = 259.4 \text{ W}$$

(f) The power converted from electrical to mechanical form is

$$P_{\text{conv}} = (1 - s)\,P_{\text{AG}} \tag{9–17}$$
$$= (1 - 0.05)(259.4 \text{ W}) = 246 \text{ W}$$

(g) The induced torque in the motor is given by

$$\tau_{\text{ind}} = \frac{P_{\text{AG}}}{\omega_{\text{sync}}} \tag{9–11}$$

$$= \frac{259.4 \text{ W}}{(1200 \text{ r/min})(1 \text{ min}/60 \text{ s})(2\pi \text{ rad/r})} = 2.06 \text{ N} \bullet \text{m}$$

(h) The output power is given by

$$P_{\text{out}} = P_{\text{conv}} - P_{\text{rot}} = P_{\text{conv}} - P_{\text{core}} - P_{\text{mech}} - P_{\text{stray}}$$
$$= 246 \text{ W} - 35 \text{ W} - 16 \text{ W} = 195 \text{ W}$$

(i) The load torque of the motor is given by

$$\tau_{\text{load}} = \frac{P_{\text{out}}}{\omega_m}$$

$$= \frac{195 \text{ W}}{(1140 \text{ r/min})(1 \text{ min}/60 \text{ s})(2\pi \text{ rad/r})} = 1.63 \text{ N} \bullet \text{m}$$

(*j*) Finally, the efficiency of the motor at these conditions is

$$\eta = \frac{P_{\text{out}}}{P_{\text{in}}} \times 100\% = \frac{195 \text{ W}}{325 \text{ W}} \times 100\% = 60\%$$

9.6 OTHER TYPES OF MOTORS

Two other types of motors—reluctance motors and hysteresis motors—are used in certain special-purpose applications. These motors differ in rotor construction from the ones previously described, but use the same stator design. Like induction motors, they can be built with either single- or three-phase stators. A third type of special-purpose motor is the stepper motor. A stepper motor requires a polyphase stator, but it does not require a three-phase power supply. The final special-purpose motor discussed is the brushless dc motor, which, as the name suggests, runs on a dc power supply.

Reluctance Motors

A *reluctance motor* is a motor which depends on reluctance torque for its operation. Reluctance torque is the torque induced in an iron object (such as a pin) in the presence of an external magnetic field, which causes the object to line up with the external magnetic field. This torque occurs because the external field induces an internal magnetic field in the iron of the object, and a torque appears between the two fields, twisting the object around to line up with the external field. In order for a reluctance torque to be produced in an object, it must be elongated along axes at angles corresponding to the angles between adjacent poles of the external magnetic field.

A simple schematic of a two-pole reluctance motor is shown in Figure 9–31. It can be shown that the torque applied to the rotor of this motor is proportional to $\sin 2\delta$, where δ is the electrical angle between the rotor and the stator magnetic fields. Therefore, the reluctance torque of a motor is maximum when the angle between the rotor and the stator magnetic fields is 45°.

A simple reluctance motor of the sort shown in Figure 9–31 is a *synchronous motor*, since the rotor will be locked into the stator magnetic fields as long as the pullout torque of the motor is not exceeded. Like a normal synchronous motor, it has no starting torque and will not start by itself.

A *self-starting reluctance motor* that will operate at synchronous speed until its maximum reluctance torque is exceeded can be built by modifying the rotor of an induction motor as shown in Figure 9–32. In this figure, the rotor has salient poles for steady-state operation as a reluctance motor and also has cage or amortisseur windings for starting. The stator of such a motor may be of either single- or three-phase construction. The torque–speed characteristic of this motor, which is sometimes called a *synchronous induction motor*, is shown in Figure 9–33.

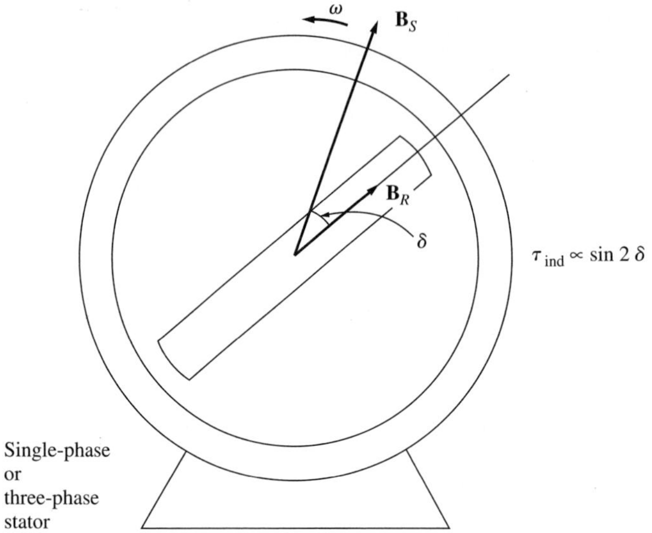

FIGURE 9–31
The basic concept of a reluctance motor.

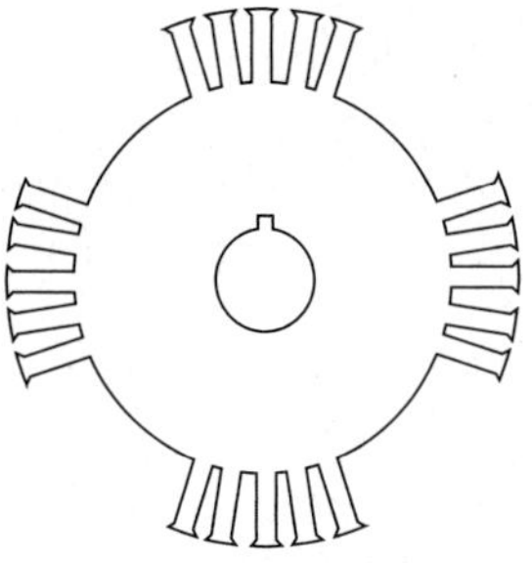

FIGURE 9–32
The rotor design of a "synchronous induction" or self-starting reluctance motor.

An interesting variation on the idea of the reluctance motor is the Syncrospeed motor, which is manufactured in the United States by MagneTek, Inc. The rotor of this motor is shown in Figure 9–34. It uses "flux guides" to increase the coupling between adjacent pole faces and therefore to increase the maximum-reluctance torque of the motor. With these flux guides, the maximum-reluctance torque is increased to about 150 percent of the rated torque, as compared to just over 100 percent of the rated torque for a conventional reluctance motor.

Hysteresis Motors

Another special-purpose motor employs the phenomenon of hysteresis to produce a mechanical torque. The rotor of a hysteresis motor is a smooth cylinder of magnetic material with no teeth, protrusions, or windings. The stator of the motor can be either single- or three-phase; but if it is single-phase, a permanent capacitor

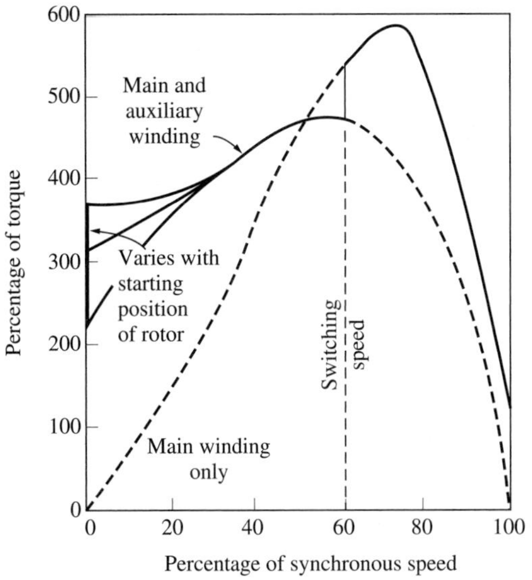

FIGURE 9–33
The torque–speed characteristic of a single-phase self-starting reluctance motor.

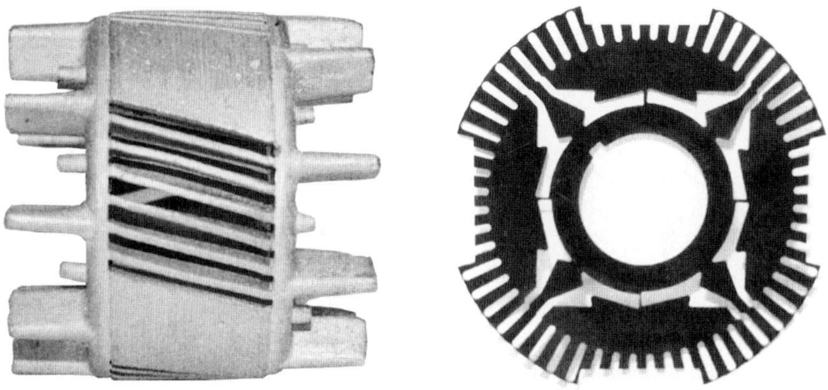

(a) (b)

FIGURE 9–34
(a) The aluminum casting of a Synchrospeed motor rotor. (b) A rotor lamination from the motor. Notice the flux guides connecting the adjacent poles. These guides increase the reluctance torque of the motor. (*Courtesy of MagneTek, Inc.*)

should be used with an auxiliary winding to provide as smooth a magnetic field as possible, since this greatly reduces the losses of the motor.

Figure 9–35 shows the basic operation of a hysteresis motor. When a three-phase (or single-phase with auxiliary winding) current is applied to the stator of the motor, a rotating magnetic field appears within the machine. This rotating magnetic field magnetizes the metal of the rotor and induces poles within it.

When the motor is operating below synchronous speed, there are two sources of torque within it. Most of the torque is produced by hysteresis. When the

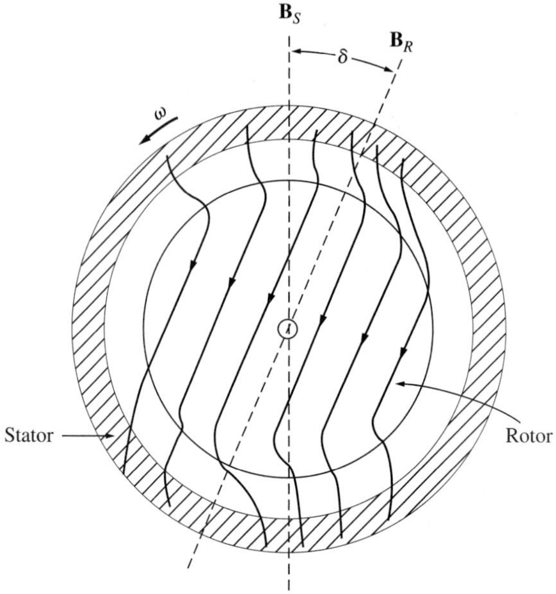

FIGURE 9–35
The construction of a hysteresis motor. The main component of torque in this motor is proportional to the angle between the rotor and stator magnetic fields.

magnetic field of the stator sweeps around the surface of the rotor, the rotor flux cannot follow it exactly, because the metal of the rotor has a large hysteresis loss. The greater the intrinsic hysteresis loss of the rotor material, the greater the angle by which the rotor magnetic field lags the stator magnetic field. Since the rotor and stator magnetic fields are at different angles, a finite torque will be produced in the motor. In addition, the stator magnetic field will produce eddy currents in the rotor, and these eddy currents produce a magnetic field of their own, further increasing the torque on the rotor. The greater the relative motion between the rotor and the stator magnetic field, the greater the eddy currents and eddy-current torques.

When the motor reaches synchronous speed, the stator flux ceases to sweep across the rotor, and the rotor acts like a permanent magnet. The induced torque in the motor is then proportional to the angle between the rotor and the stator magnetic field, up to a maximum angle set by the hysteresis in the rotor.

The torque–speed characteristic of a hysteresis motor is shown in Figure 9–36. Since the amount of hysteresis within a particular rotor is a function of only the stator flux density and the material from which it is made, the hysteresis torque of the motor is approximately constant for any speed from zero to n_{sync}. The eddy-current torque is roughly proportional to the slip of the motor. These two facts taken together account for the shape of the hysteresis motor's torque–speed characteristic.

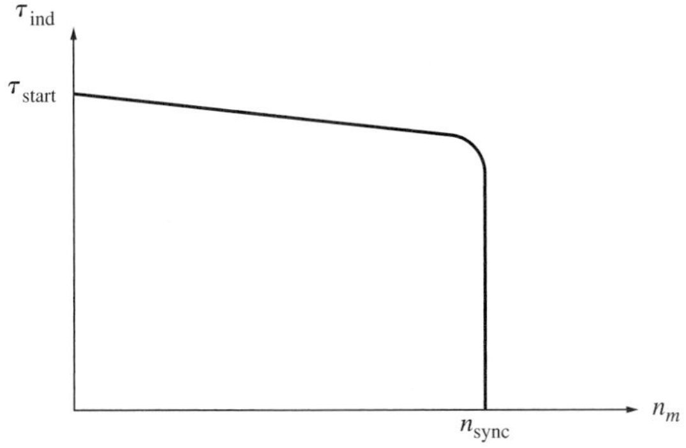

FIGURE 9–36
The torque–speed characteristic of a motor.

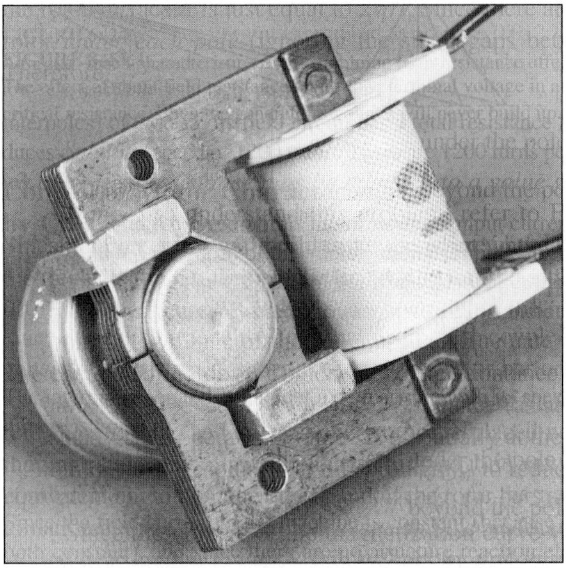

FIGURE 9–37
A small hysteresis motor with a shaded-pole stator, suitable for running an electric clock. Note the shaded stator poles. (*Stephen J. Chapman*)

Since the torque of a hysteresis motor at any subsynchronous speed is greater than its maximum synchronous torque, a hysteresis motor can accelerate any load that it can carry during normal operation.

A very small hysteresis motor can be built with shaded-pole stator construction to create a tiny self-starting low-power synchronous motor. Such a motor is shown in Figure 9–37. It is commonly used as the driving mechanism in electric clocks. An electric clock is therefore synchronized to the line frequency of the power system, and the resulting clock is just as accurate (or as inaccurate) as the frequency of the power system to which it is tied.

Stepper Motors

A *stepper motor* is a special type of synchronous motor which is designed to rotate a specific number of degrees for every electric pulse received by its control unit. Typical steps are 7.5° or 15° per pulse. These motors are used in many control systems, since the position of a shaft or other piece of machinery can be controlled precisely with them.

A simple stepper motor and its associated control unit are shown in Figure 9–38. To understand the operation of the stepper motor, examine Figure 9–39. This figure shows a two-pole, three-phase stator with a permanent-magnet rotor. If a dc voltage is applied to phase a of the stator and no voltage is applied to phases b and c, then a torque will be induced in the rotor which causes it to line up with the stator magnetic field $\mathbf{B}_S$, as shown in Figure 9–39b.

Now assume that phase a is turned off and that a negative dc voltage is applied to phase c. The new stator magnetic field is rotated 60° with respect to the previous magnetic field, and the rotor of the motor follows it around. By continuing this pattern, it is possible to construct a table showing the rotor position as a function of the voltage applied to the stator of the motor. If the voltage produced by the control unit changes with each input pulse in the order shown in Table 9–1, then the stepper motor will advance by 60° with each input pulse.

It is easy to build a stepper motor with finer step size by increasing the number of poles on the motor. From Equation (3–31) the number of mechanical degrees corresponding to a given number of electrical degrees is

$$\theta_m = \frac{2}{P}\,\theta_e \tag{9–18}$$

Since each step in Table 9–1 corresponds to 60 electrical degrees, the number of mechanical degrees moved per step decreases with increasing numbers of poles. For example, if the stepper motor has eight poles, then the mechanical angle of the motor's shaft will change by 15° per step.

The speed of a stepper motor can be related to the number of pulses into its control unit per unit time by using Equation (9–18). Equation (9–18) gives the mechanical angle of a stepper motor as a function of the electrical angle. If both sides of this equation are differentiated with respect to time, then we have a relationship between the electrical and mechanical rotational speeds of the motor:

$$\omega_m = \frac{2}{P}\,\omega_e \tag{9–19a}$$

or

$$n_m = \frac{2}{P}\,n_e \tag{9–19b}$$

Since there are six input pulses per electrical revolution, the relationship between the speed of the motor in revolutions per minute and the number of pulses per minute becomes

$$\boxed{n_m = \frac{1}{3P}\,n_{\text{pulses}}} \tag{9–20}$$

where n_{pulses} is the number of pulses per minute.

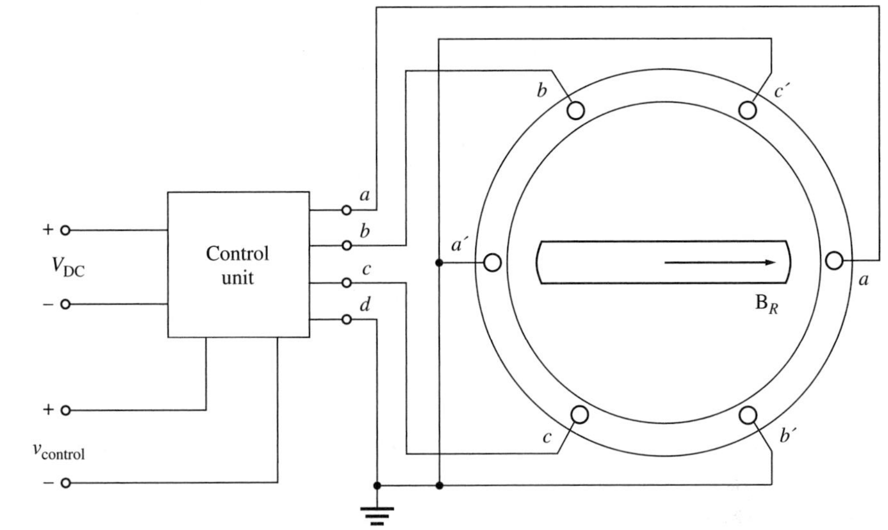

(a)

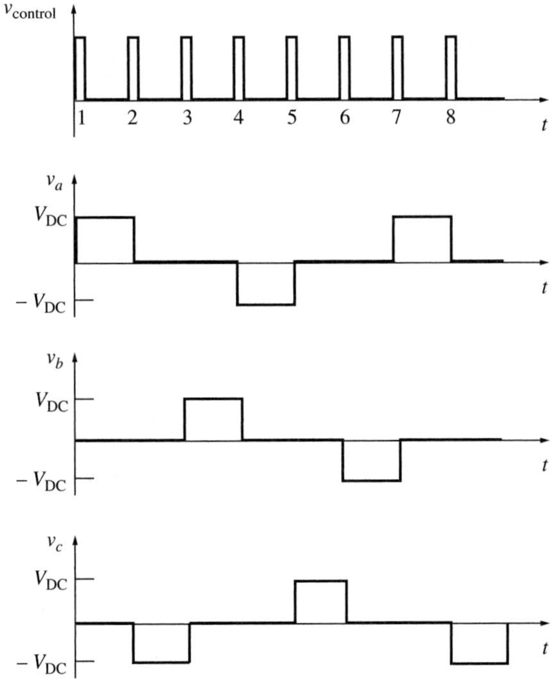

(b)

Pulse number	Phase voltages, V		
	v_a	v_b	v_c
1	V_{DC}	0	0
2	0	0	$-V_{DC}$
3	0	V_{DC}	0
4	$-V_{DC}$	0	0
5	0	0	V_{DC}
6	0	$-V_{DC}$	0

(c)

FIGURE 9–38
(a) A simple three-phase stepper motor and its associated control unit. The inputs to the control unit consist of a dc power source and a control signal consisting of a train of pulses. (b) A sketch of the output voltage from the control unit as a series of control pulses are input. (c) A table showing the output voltage from the control unit as a function of pulse number.

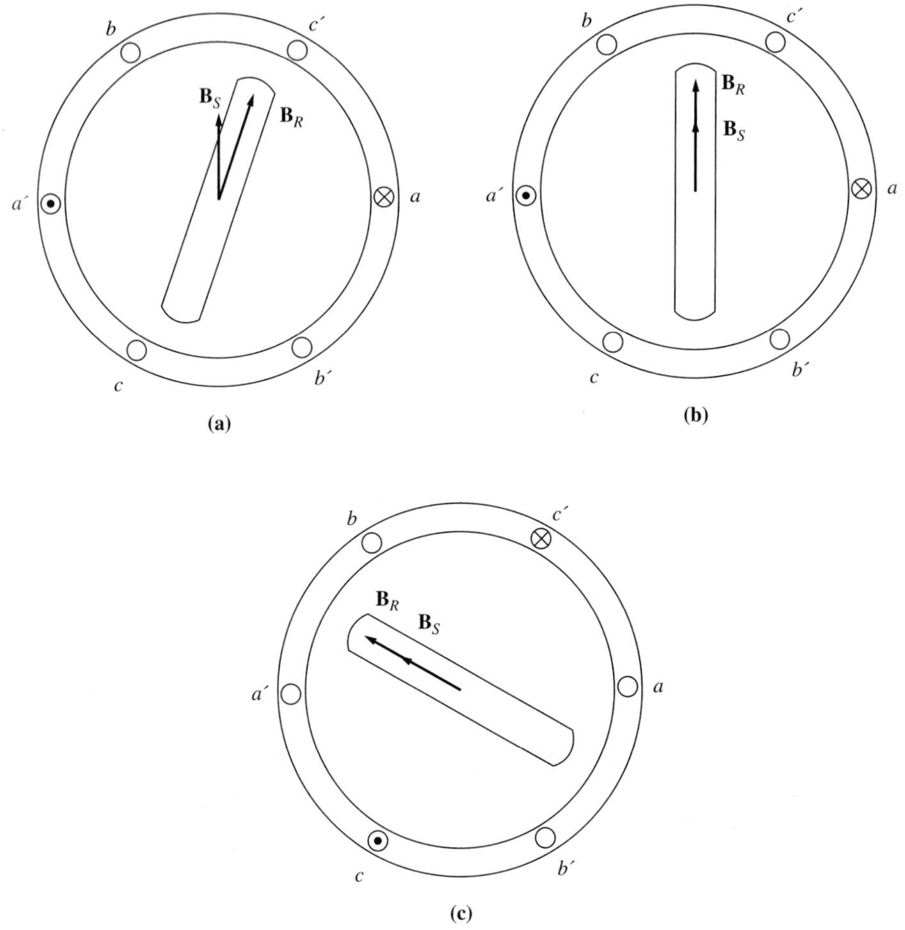

FIGURE 9–39
Operation of a stepper motor. (a) A voltage V is applied to phase a of the stator, causing a current to flow in phase a and producing a stator magnetic field $\mathbf{B}_S$. The interaction of $\mathbf{B}_R$ and $\mathbf{B}_S$ produces a counterclockwise torque on the rotor. (b) When the rotor lines up with the stator magnetic field, the net torque falls to zero. (c) A voltage $-V$ is applied to phase c of the stator, causing a current to flow in phase c and producing a stator magnetic field $\mathbf{B}_S$. The interaction of $\mathbf{B}_R$ and $\mathbf{B}_S$ produces a counterclockwise torque on the rotor, causing the rotor to line up with the new position of the magnetic field.

There are two basic types of stepper motors, differing only in rotor construction: *permanent-magnet type* and *reluctance type*. The permanent-magnet type of stepper motor has a permanent-magnet rotor, while the reluctance-type stepper motor has a ferromagnetic rotor which is not a permanent magnet. (The rotor of the reluctance motor described previously in this section is the reluctance type.) In general, the permanent-magnet stepper motor can produce more torque

TABLE 9–1
Rotor position as a function of voltage in a two-pole stepper motor

	Phase voltages			
Input pulse number	***a***	***b***	***c***	**Rotor position**
1	V	0	0	0°
2	0	0	$-V$	60°
3	0	V	0	120°
4	$-V$	0	0	180°
5	0	0	V	240°
6	0	$-V$	0	300°

than the reluctance type, since the permanent-magnet stepper motor has torque from both the permanent rotor magnetic field and reluctance effects.

Reluctance-type stepper motors are often built with a four-phase stator winding instead of the three-phase stator winding described previously. A four-phase stator winding reduces the steps between pulses from 60 electrical degrees to 45 electrical degrees. As mentioned earlier, the torque in a reluctance motor varies as sin 2δ, so the reluctance torque between steps will be maximum for an angle of 45°. Therefore, a given reluctance-type stepper motor can produce more torque with a four-phase stator winding than with a three-phase stator winding.

Equation (9–20) can be generalized to apply to all stepper motors, regardless of the number of phases on their stator windings. In general, if a stator has N phases, it takes $2N$ pulses per electrical revolution in that motor. Therefore, the relationship between the speed of the motor in revolutions per minute and the number of pulses per minute becomes

$$n_m = \frac{1}{NP} n_{\text{pulses}} \tag{9–21}$$

Stepper motors are very useful in control and positioning systems because the computer doing the controlling can know both the exact *speed* and *position* of the stepper motor without needing feedback information from the shaft of the motor. For example, if a control system sends 1200 pulses per minute to the two-pole stepper motor shown in Figure 9–38, then the speed of the motor will be exactly

$$\begin{aligned} n_m &= \frac{1}{3P} n_{\text{pulses}} \qquad (9–20) \\ &= \frac{1}{3(2 \text{ poles})}(1200 \text{ pulses/min}) \\ &= 200 \text{ r/min} \end{aligned}$$

Furthermore, if the initial position of the shaft is known, then the computer can determine the exact angle of the rotor shaft at any future time by simply counting the total number of pulses which it has sent to the control unit of the stepper motor.

Example 9–2. A three-phase permanent-magnet stepper motor required for one particular application must be capable of controlling the position of a shaft in steps of 7.5°, and it must be capable of running at speeds of up to 300 r/min.

(a) How many poles must this motor have?
(b) At what rate must control pulses be received in the motor's control unit if it is to be driven at 300 r/min?

Solution

(a) In a three-phase stepper motor, each pulse advances the rotor's position by 60 electrical degrees. This advance must correspond to 7.5 mechanical degrees. Solving Equation (9–18) for P yields

$$P = 2\frac{\theta_e}{\theta_m} = 2\left(\frac{60^\circ}{7.5^\circ}\right) = 16 \text{ poles}$$

(b) Solving Equation (9–21) for n_{pulses} yields

$$\begin{aligned} n_{\text{pulses}} &= NPn_m \\ &= (3 \text{ phases})(16 \text{ poles})(300 \text{ r/min}) \\ &= 240 \text{ pulses/s} \end{aligned}$$

Brushless DC Motors

Conventional dc motors have traditionally been used in applications where dc power sources are available, such as on aircraft and automobiles. However, small dc motors of these types have a number of disadvantages. The principal disadvantage is excessive sparking and brush wear. Small, fast dc motors are too small to use compensating windings and interpoles, so armature reaction and $L\,di/dt$ effects tend to produce sparking on their commutator brushes. In addition, the high rotational speed of these motors causes increased brush wear and requires regular maintenance every few thousand hours. If the motors must work in a low-pressure environment (such as at high altitudes in an aircraft), brush wear can be so bad that the brushes require replacement after less than an hour of operation!

In some applications, the regular maintenance required by the brushes of these dc motors may be unacceptable. Consider for example a dc motor in an artificial heart—regular maintenance would require opening the patient's chest. In other applications, the sparks at the brushes may create an explosion danger, or unacceptable RF noise. For all of these cases, there is a need for a small, fast dc motor that is highly reliable and has low noise and long life.

Such motors have been developed in the last 25 years by combining a small motor much like a permanent magnetic stepper motor with a rotor position sensor and a solid-state electronic switching circuit. These motors are called *brushless dc motors* because they run from a dc power source but do not have commutators and brushes. A sketch of a small brushless dc motor is shown in Figure 9–40, and a photograph of a typical brushless dc motor is shown in Figure 9–41. The rotor is similar to that of a permanent magnet stepper motor, except that it is nonsalient. The stator can have three or more phases (there are four phases in the example shown in Figure 9-40).

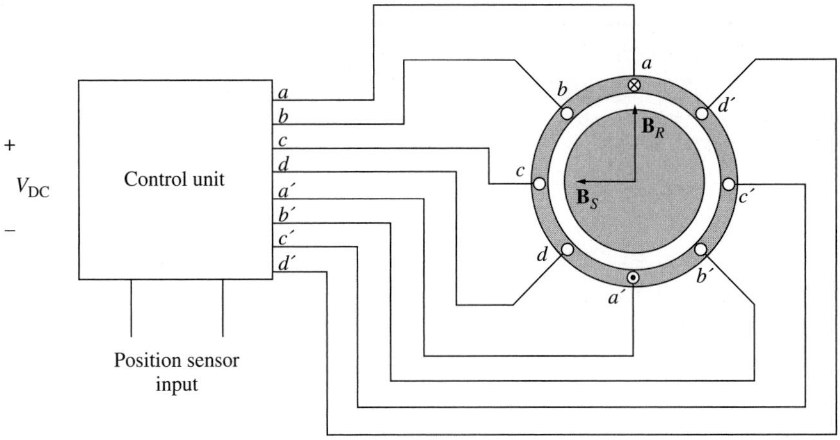

(a)

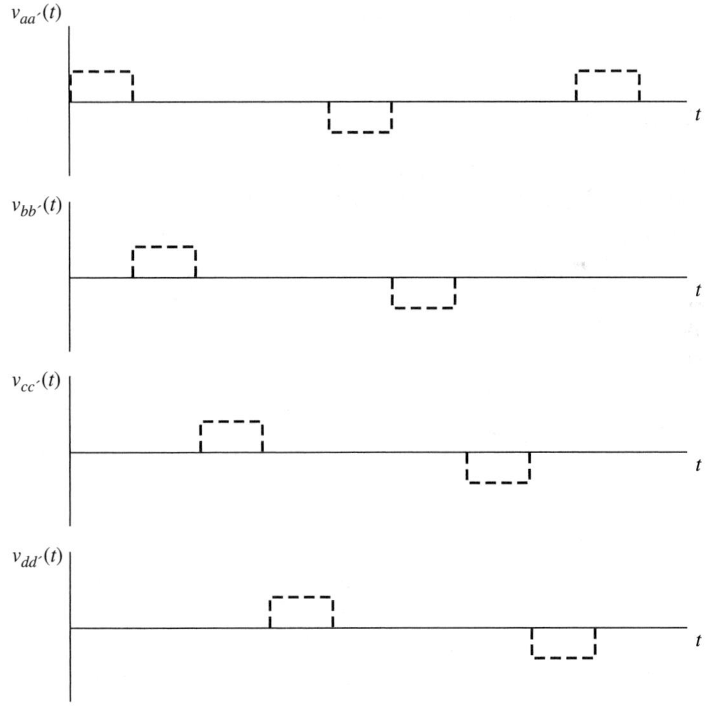

(b)

FIGURE 9–40
(a) A simple brushless dc motor and its associated control unit. The inputs to the control unit consist of a dc power source and a signal proportional to the current rotor position. (b) The voltages applied to the stator coils.

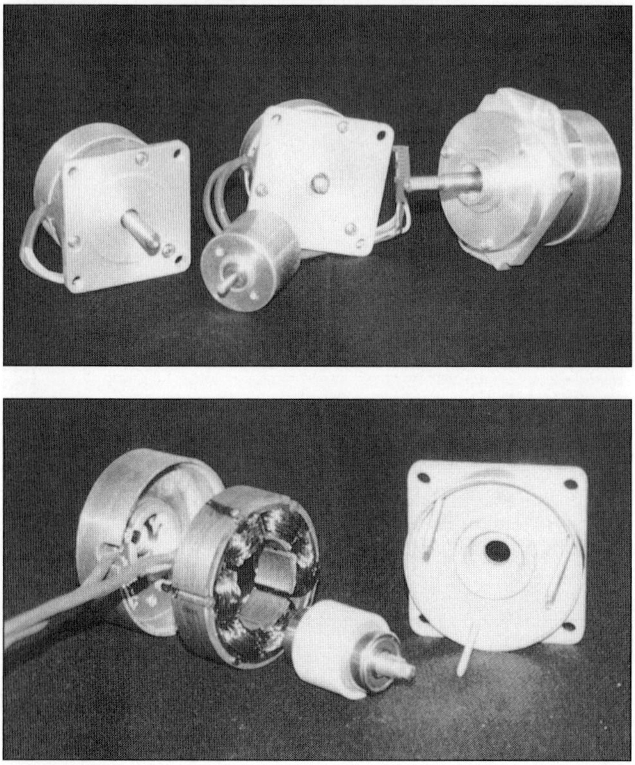

(a)

(b)

FIGURE 9–41
(a) Typical brushless dc motors. (b) Exploded view showing the permanent magnet rotor and a three-phase (6-pole) stator. (*Courtesy of Carson Technologies, Inc.*)

The basic components of a brushless dc motor are

1. A permanent magnet rotor
2. A stator with a three-, four-, or more phase winding
3. A rotor position sensor
4. An electronic circuit to control the phases of the rotor winding

A brushless dc motor functions by energizing one stator coil at a time with a constant dc voltage. When a coil is turned on, it produces a stator magnetic field $\mathbf{B}_S$, and a torque is produced on the rotor given by

$$\tau_{\text{ind}} = k\mathbf{B}_R \times \mathbf{B}_S$$

which tends to align the rotor with the stator magnetic field. At the time shown in Figure 9–40a, the stator magnetic field $\mathbf{B}_S$ points to the left while the permanent magnet rotor magnetic field $\mathbf{B}_R$ points up, producing a counterclockwise torque on the rotor. As a result the rotor will turn to the left.

If coil a remained energized all of thc timc, the rotor would turn until the two magnetic fields were aligned, and then it would stop, just like a stepper motor. The key to the operation of a brushless dc motor is that it includes a *position sensor*, so that the control circuit will know when the rotor is almost aligned with the stator magnetic field. At that time coil a will be turned off and coil b will be turned on, causing the rotor to again experience a counterclockwise torque, and to continue rotating. This process continues indefinitely with the coils turned on in the order a, b, c, d, $-a$, $-b$, $-c$, $-d$, etc., so that the motor turns continuously.

The electronics of the control circuit can be used to control both the speed and direction of the motor. The net effect of this design is a motor that runs from a dc power source, with full control over both the speed and the direction of rotation.

Brushless dc motors are available only in small sizes, up to 20 W or so, but they have many advantages in the size range over which they are available. Some of the major advantages include:

1. Relatively high efficiency
2. Long life and high reliability
3. Little or no maintenance
4. Very little RF noise compared to a dc motor with brushes
5. Very high speeds are possible (greater than 50,000 r/min)

The principal disadvantage is that a brushless dc motor is more expensive than a comparable brush dc motor.

9.7 SUMMARY

The ac motors described in previous chapters required three-phase power to function. Since most residences and small businesses have only single-phase power sources, these motors cannot be used. A series of motors capable of running from a single-phase power source was described in this chapter.

The first motor described was the universal motor. A universal motor is a series dc motor adapted to run from an ac supply, and its torque–speed characteristic is similar to that of a series dc motor. The universal motor has a very high torque, but its speed regulation is very poor.

Single-phase induction motors have no intrinsic starting torque, but once they are brought up to speed, their torque–speed characteristics are almost as good as those of three-phase motors of comparable size. Starting is accomplished by the addition of an auxiliary winding with a current whose phase angle differs from that of the main winding or by shading portions of the stator poles.

The starting torque of a single-phase induction motor depends on the phase angle between the current in the primary winding and the current in the auxiliary winding, with maximum torque occurring when that angle reaches 90°. Since the split-phase construction provides only a small phase difference between the main and auxiliary windings, its starting torque is modest. Capacitor-start motors have auxiliary windings with an approximately 90° phase shift, so they have large starting torques. Permanent split-capacitor motors, which have smaller capacitors, have starting torques intermediate between those of the split-phase motor and the capacitor-start motor. Shaded-pole motors have a very small effective phase shift and therefore a small starting torque.

Reluctance motors and hysteresis motors are special-purpose ac motors which can operate at synchronous speed without the rotor field windings required by synchronous motors and which can accelerate up to synchronous speed by themselves. These motors can have either single- or three-phase stators.

Stepper motors are motors used to advance the position of a shaft or other mechanical device by a fixed amount each time a control pulse is received. They are used extensively in control systems for positioning objects.

Brushless dc motors are similar to stepper motors with permanent magnet rotors, except that they include a position sensor. The position sensor is used to switch the energized stator coil whenever the rotor is almost aligned with it, keeping the rotor rotating a speed set by the control electronics. Brushless dc motors are more expensive than ordinary dc motors, but require low maintenance and have high reliability, long life, and low RF noise. They are available only in small sizes (20 W and under).

QUESTIONS

9–1. What changes are necessary in a series dc motor to adapt it for operation from an ac power source?

9–2. Why is the torque–speed characteristic of a universal motor on an ac source different from the torque–speed characteristic of the same motor on a dc source?

9–3. Why is a single-phase induction motor unable to start itself without special auxiliary windings?

9–4. How is induced torque developed in a single-phase induction motor (*a*) according to the double revolving-field theory and (*b*) according to the cross-field theory?

9–5. How does an auxiliary winding provide a starting torque for single-phase induction motors?

9–6. How is the current phase shift accomplished in the auxiliary winding of a split-phase induction motor?

9–7. How is the current phase shift accomplished in the auxiliary winding of a capacitor-start induction motor?

9–8. How does the starting torque of a permanent split-capacitor motor compare to that of a capacitor-start motor of the same size?

9–9. How can the direction of rotation of a split-phase or capacitor-start induction motor be reversed?

9–10. How is starting torque produced in a shaded-pole motor?

9–11. How does a reluctance motor start?

9–12. How can a reluctance motor run at synchronous speed?

9–13. What mechanisms produce the starting torque in a hysteresis motor?

9–14. What mechanism produces the synchronous torque in a hysteresis motor?

9–15. Explain the operation of a stepper motor.

9–16. What is the difference between a permanent-magnet type of stepper motor and a reluctance-type stepper motor?

9–17. What is the optimal spacing between phases for a reluctance-type stepper motor? Why?

9–18. What are the advantages and disadvantages of brushless dc motors compared to ordinary brush dc motors?

PROBLEMS

9–1. A 120-V, ¼-hp, 60-Hz, four-pole, split-phase induction motor has the following impedances:

$$R_1 = 2.00\ \Omega \qquad X_1 = 2.56\ \Omega \qquad X_M = 60.5\ \Omega$$
$$R_2 = 2.80\ \Omega \qquad X_2 = 2.56\ \Omega$$

At a slip of 0.05, the motor's rotational losses are 51 W. The rotational losses may be assumed constant over the normal operating range of the motor. If the slip is 0.05, find the following quantities for this motor:

(a) Input power
(b) Air-gap power
(c) P_{conv}
(d) P_{out}
(e) τ_{ind}
(f) τ_{load}
(g) Overall motor efficiency
(h) Stator power factor

9–2. Repeat Problem 9–1 for a rotor slip of 0.025.

9–3. Suppose that the motor in Problem 9–1 is started and the auxiliary winding fails open while the rotor is accelerating through 400 r/min. How much induced torque will the motor be able to produce on its main winding alone? Assuming that the rotational losses are still 51 W, will this motor continue accelerating or will it slow down again? Prove your answer.

9–4. Use MATLAB to calculate and plot the torque–speed characteristic of the motor in Problem 9–1, ignoring the starting winding.

9–5. A 220-V, 1.5-hp, 50-Hz, six-pole, capacitor-start induction motor has the following main-winding impedances:

$$R_1 = 1.30\ \Omega \qquad X_1 = 2.01\ \Omega \qquad X_M = 105\ \Omega$$
$$R_2 = 1.73\ \Omega \qquad X_2 = 2.01\ \Omega$$

At a slip of 0.05, the motor's rotational losses are 291 W. The rotational losses may be assumed constant over the normal operating range of the motor. Find the following quantities for this motor at 5 percent slip:

(a) Stator current
(b) Stator power factor
(c) Input power
(d) P_{AG}
(e) P_{conv}

(*f*) P_{out}
(*g*) τ_{ind}
(*h*) τ_{load}
(*i*) Efficiency

9–6. Find the induced torque in the motor in Problem 9-5 if it is operating at 5 percent slip and its terminal voltage is (*a*) 190 V, (*b*) 208 V, (*c*) 230 V.

9–7. What type of motor would you select to perform each of the following jobs? Why?
(*a*) Vacuum cleaner
(*b*) Refrigerator
(*c*) Air conditioner compressor
(*d*) Air conditioner fan
(*e*) Variable-speed sewing machine
(*f*) Clock
(*g*) Electric drill

9–8. For a particular application, a three-phase stepper motor must be capable of stepping in 10° increments. How many poles must it have?

9–9. How many pulses per second must be supplied to the control unit of the motor in Problem 9–8 to achieve a rotational speed of 600 r/min?

9–10. Construct a table showing step size versus number of poles for three-phase and four-phase stepper motors.

REFERENCES

1. Fitzgerald, A. E., and C. Kingsley, Jr.: *Electric Machinery*, McGraw-Hill, New York, 1952.
2. National Electrical Manufacturers Association, *Motors and Generators*, Publication No. MG1-1993, NEMA, Washington, 1993.
3. Werninck, E. H. (ed.): *Electric Motor Handbook*, McGraw-Hill, London, 1978.
4. Veinott, G. C.: *Fractional and Subfractional Horsepower Electric Motors*, McGraw-Hill, New York, 1970.

APPENDIX A

THREE-PHASE CIRCUITS

Almost all electric power generation and most of the power transmission in the world today is in the form of three-phase ac circuits. A three-phase ac power system consists of three-phase generators, transmission lines, and loads. Ac power systems have a great advantage over dc systems in that their voltage levels can be changed with transformers to reduce transmission losses, as described in Chapter 2. *Three-phase* ac power systems have two major advantages over single-phase ac power systems: (1) it is possible to get more power per kilogram of metal from a three-phase machine and (2) the power delivered to a three-phase load is constant at all times, instead of pulsing as it does in single-phase systems. Three-phase systems also make the use of induction motors easier by allowing them to start without special auxiliary starting windings.

A.1 GENERATION OF THREE-PHASE VOLTAGES AND CURRENTS

A three-phase generator consists of three single-phase generators, with voltages equal in magnitude but differing in phase angle from the others by 120°. Each of these three generators could be connected to one of three identical loads by a pair of wires, and the resulting power system would be as shown in Figure A–1c. Such a system consists of three single-phase circuits that happen to differ in phase angle by 120°. The current flowing to each load can be found from the equation

$$\mathbf{I} = \frac{\mathbf{V}}{\mathbf{Z}} \tag{A–1}$$

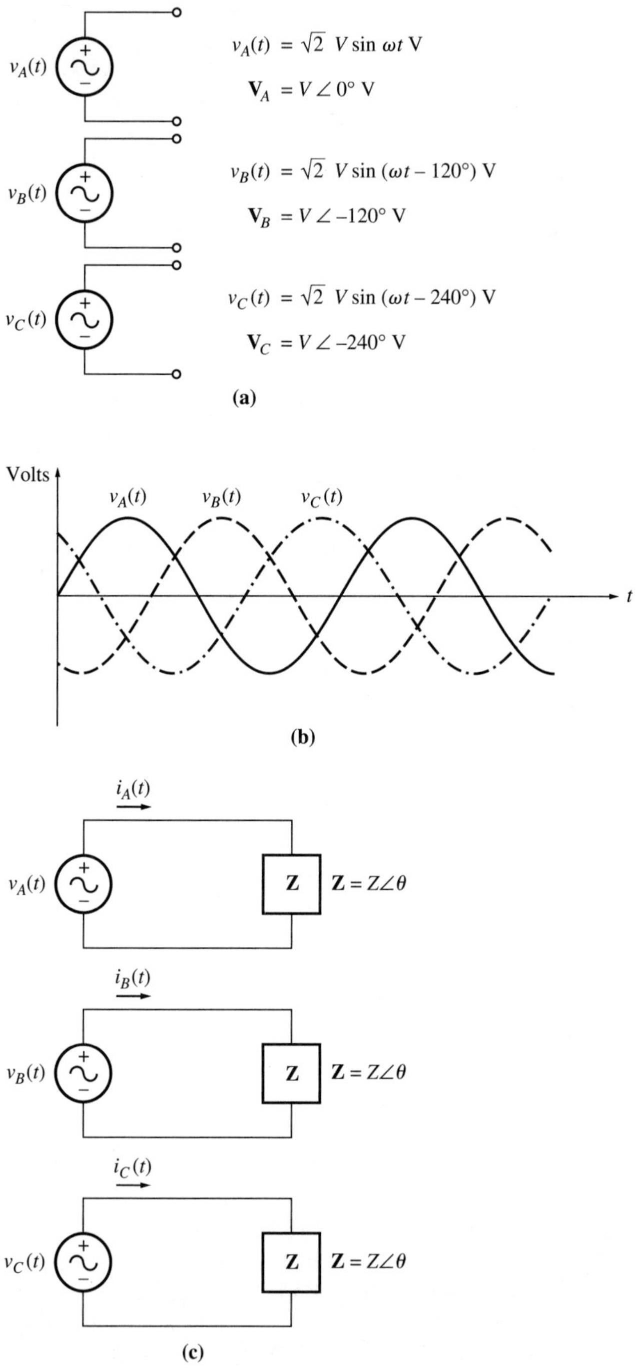

FIGURE A–1
(a) A three-phase generator, consisting of three single-phase sources equal in magnitude and 120° apart in phase. (b) The voltages in each phase of the generator. (c) The three phases of the generator connected to three identical loads.

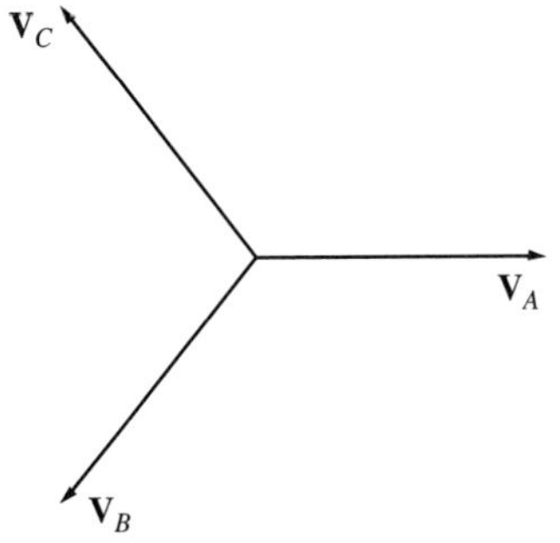

FIGURE A–1 (concluded)
(d) Phasor diagram showing the voltages in each phase.

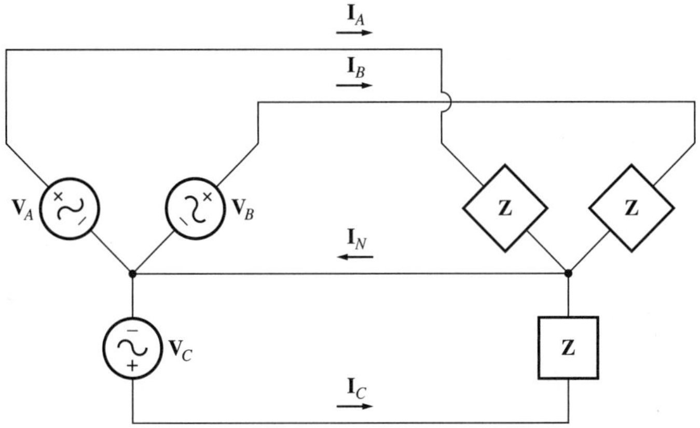

FIGURE A–2
The three circuits connected together with a common neutral.

Therefore, the currents flowing in the three phases are

$$\mathbf{I}_A = \frac{V\angle 0°}{Z\angle\theta} = I\angle -\theta \tag{A–2}$$

$$\mathbf{I}_B = \frac{V\angle -120°}{Z\angle\theta} = I\angle -120° - \theta \tag{A–3}$$

$$\mathbf{I}_C = \frac{V\angle -240°}{Z\angle\theta} = I\angle -240° - \theta \tag{A–4}$$

It is possible to connect the negative ends of these three single-phase generators and loads together, so that they share a common return line (called the *neutral*). The resulting system is shown in Figure A–2; note that now only *four* wires are required to supply power from the three generators to the three loads.

How much current is flowing in the single neutral wire shown in Figure A–2? The return current will be the sum of the currents flowing to each individual load in the power system. This current is given by

$$\begin{aligned}
\mathbf{I}_N &= \mathbf{I}_A + \mathbf{I}_B + \mathbf{I}_C \qquad\qquad (A\text{–}5)\\
&= I\angle -\theta + I\angle -\theta - 120^\circ + I\angle -\theta - 240^\circ\\
&= I\cos(-\theta) + jI\sin(-\theta)\\
&\quad + I\cos(-\theta - 120^\circ) + jI\sin(-\theta - 120^\circ)\\
&\quad + I\cos(-\theta - 240^\circ) + jI\sin(-\theta - 240^\circ)\\
&= I\,[\cos(-\theta) + \cos(-\theta - 120^\circ) + \cos(-\theta - 240^\circ)]\\
&\quad + jI\,[\sin(-\theta) + \sin(-\theta - 120^\circ) + \sin(-\theta - 240^\circ)]
\end{aligned}$$

Recall the elementary trigonometric identities:

$$\cos(\alpha - \beta) = \cos\alpha\cos\beta + \sin\alpha\sin\beta \qquad (A\text{–}6)$$

$$\sin(\alpha - \beta) = \sin\alpha\cos\beta - \cos\alpha\sin\beta \qquad (A\text{–}7)$$

Applying these trigonometric identities yields

$$\begin{aligned}
\mathbf{I}_N = {} & I[\cos(-\theta) + \cos(-\theta)\cos 120^\circ + \sin(-\theta)\sin 120^\circ + \cos(-\theta)\cos 240^\circ\\
& + \sin(-\theta)\sin 240^\circ]\\
& + jI[\sin(-\theta) + \sin(-\theta)\cos 120^\circ - \cos(-\theta)\sin 120^\circ\\
& + \sin(-\theta)\cos 240^\circ - \cos(-\theta)\sin 240^\circ]
\end{aligned}$$

$$\begin{aligned}
\mathbf{I}_N = {} & I\left[\cos(-\theta) - \frac{1}{2}\cos(-\theta) + \frac{\sqrt{3}}{2}\sin(-\theta) - \frac{1}{2}\cos(-\theta) - \frac{\sqrt{3}}{2}\sin(-\theta)\right]\\
& + jI\left[\sin(-\theta) - \frac{1}{2}\sin(-\theta) - \frac{\sqrt{3}}{2}\cos(-\theta) - \frac{1}{2}\sin(-\theta) + \frac{\sqrt{3}}{2}\cos(-\theta)\right]
\end{aligned}$$

$$\mathbf{I}_N = 0\text{ A}$$

As long as the three loads are equal, the return current in the neutral is zero! A three-phase power system in which the three generators have voltages that are exactly equal in magnitude and 120° different in phase, and in which all three loads are identical, is called a *balanced three-phase system.* In such a system, the neutral is actually unnecessary, and we could get by with only *three* wires instead of the original six.

PHASE SEQUENCE. The *phase sequence* of a three-phase power system is the order in which the voltages in the individual phases peak. The three-phase power system illustrated in Figure A–1 is said to have phase sequence *abc*, since the voltages in the three phases peak in the order *a*, *b*, *c* (see Figure A–1b). The phasor diagram of a power system with an *abc* phase sequence is shown in Figure A–3a.

It is also possible to connect the three phases of a power system so that the voltages in the phases peak in the order *a*, *c*, *b*. This type of power system is said to have phase sequence *acb*. The phasor diagram of a power system with an *acb* phase sequence is shown in Figure A–3b.

The result derived above is equally valid for both *abc* and *acb* phase sequences. In either case, if the power system is balanced, the current flowing in the neutral will be 0.

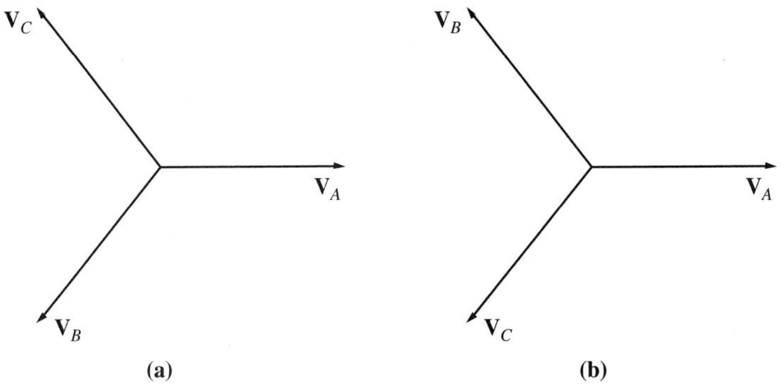

FIGURE A–3
(a) The phase voltages in a power system with an *abc* phase sequence. (b) The phase voltages in a power system with an *acb* phase sequence.

A.2 VOLTAGES AND CURRENTS IN A THREE-PHASE CIRCUIT

A connection of the sort shown in Figure A–2 is called a wye (Y) connection because it looks like the letter Y. Another possible connection is the delta (Δ) connection, in which the three generators are connected head to tail. The Δ connection is possible because the sum of the three voltages $\mathbf{V}_A + \mathbf{V}_B + \mathbf{V}_C = 0$, so that no short-circuit currents will flow when the three sources are connected head to tail.

Each generator and each load in a three-phase power system may be either Y- or Δ-connected. Any number of Y- and Δ-connected generators and loads may be mixed on a power system.

Figure A–4 shows three-phase generators connected in Y and in Δ. The voltages and currents in a given phase are called *phase quantities,* and the voltages between lines and currents in the lines connected to the generators are called *line quantities.* The relationship between the line quantities and phase quantities for a given generator or load depends on the type of connection used for that generator or load. These relationships will now be explored for each of the Y and Δ connections.

Voltages and Currents in the Wye (Y) Connection

A Y-connected three-phase generator with an *abc* phase sequence connected to a resistive load is shown in Figure A–5. The phase voltages in this generator are given by

$$\begin{aligned} \mathbf{V}_{an} &= V_\phi \angle 0° \\ \mathbf{V}_{bn} &= V_\phi \angle -120° \\ \mathbf{V}_{cn} &= V_\phi \angle -240° \end{aligned} \tag{A–8}$$

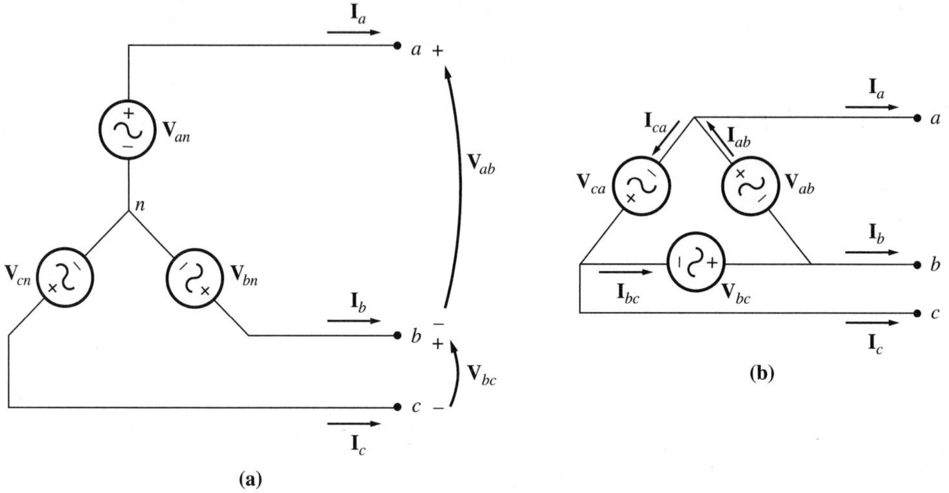

FIGURE A–4
(a) Y connection. (b) Δ connection.

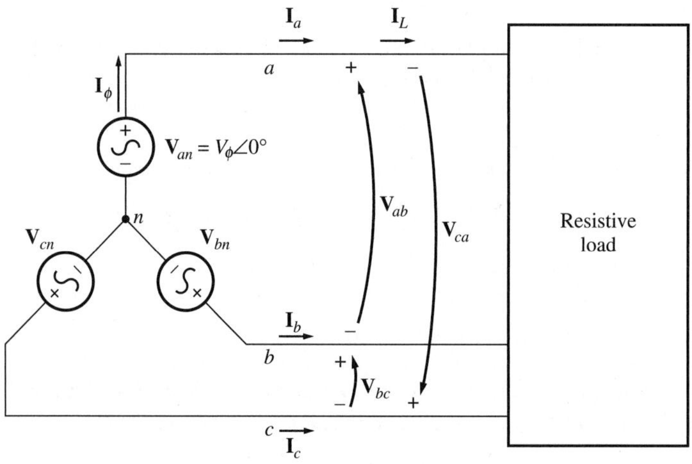

FIGURE A–5
Y-connected generator with a resistive load.

Since the load connected to this generator is assumed to be resistive, the current in each phase of the generator will be at the same angle as the voltage. Therefore, the current in each phase will be given by

$$\begin{aligned} \mathbf{I}_a &= I_\phi\angle 0^\circ \\ \mathbf{I}_b &= I_\phi\angle -120^\circ \\ \mathbf{I}_c &= I_\phi\angle -240^\circ \end{aligned} \tag{A–9}$$

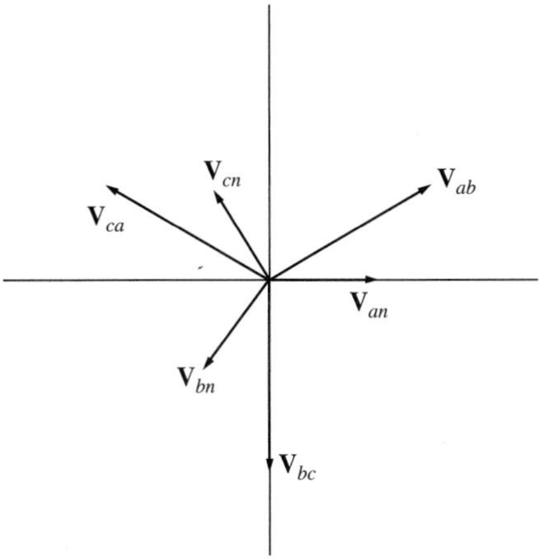

FIGURE A–6
Line-to-line and phase (line-to-neutral) voltages for the Y connection in Figure A–5.

From Figure A–5, it is obvious that the current in any line is the same as the current in the corresponding phase. Therefore, for a Y connection,

$$\boxed{I_L = I_\phi \qquad \text{Y connection}} \tag{A–10}$$

The relationship between line voltage and phase voltage is a bit more complex. By Kirchhoff's voltage law, the line-to-line voltage $\mathbf{V}_{ab}$ is given by

$$\begin{aligned}
\mathbf{V}_{ab} &= \mathbf{V}_a - \mathbf{V}_b \\
&= V_\phi \angle 0° - V_\phi \angle -120° \\
&= V_\phi - \left(-\frac{1}{2}V_\phi - j\frac{\sqrt{3}}{2}V_\phi\right) = \frac{3}{2}V_\phi + j\frac{\sqrt{3}}{2}V_\phi \\
&= \sqrt{3}V_\phi\left(\frac{\sqrt{3}}{2} + j\frac{1}{2}\right) \\
&= \sqrt{3}V_\phi \angle 30°
\end{aligned}$$

Therefore, the relationship between the magnitudes of the line-to-line voltage and the line-to-neutral (phase) voltage in a Y-connected generator or load is

$$\boxed{V_{LL} = \sqrt{3}V_\phi \qquad \text{Y connection}} \tag{A–11}$$

In addition, the line voltages are shifted 30° with respect to the phase voltages. A phasor diagram of the line and phase voltages for the Y connection in Figure A–5 is shown in Figure A–6.

Note that for Y connections with the *abc* phase sequence such as the one in Figure A–5, the voltage of a line *leads* the corresponding phase voltage by 30°. For Y connections with the *acb* phase sequence, the voltage of a line *lags* the corresponding phase voltage by 30°, as you will be asked to demonstrate in a problem at the end of the appendix.

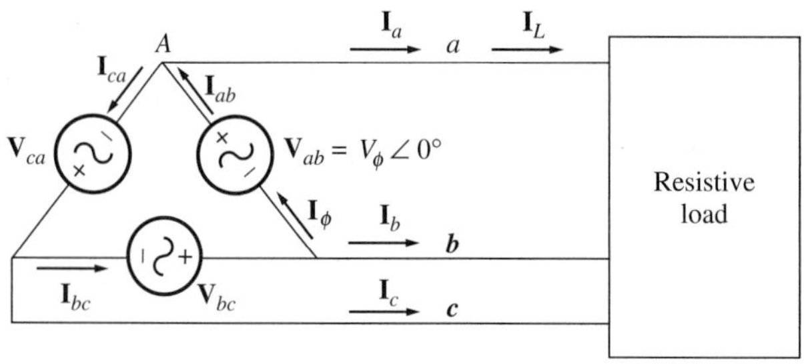

FIGURE A–7
Δ-connected generator with a resistive load.

Although the relationships between line and phase voltages and currents for the Y connection were derived for the assumption of a unity power factor, they are in fact valid for any power factor. The assumption of unity-power-factor loads simply made the mathematics slightly easier in this development.

Voltages and Currents in the Delta (Δ) Connection

A Δ-connected three-phase generator connected to a resistive load is shown in Figure A–7. The phase voltages in this generator are given by

$$\begin{aligned} \mathbf{V}_{ab} &= V_\phi \angle 0° \\ \mathbf{V}_{bc} &= V_\phi \angle -120° \\ \mathbf{V}_{ca} &= V_\phi \angle -240° \end{aligned} \tag{A–12}$$

Because the load is resistive, the phase currents are given by

$$\begin{aligned} \mathbf{I}_{ab} &= I_\phi \angle 0° \\ \mathbf{I}_{bc} &= I_\phi \angle -120° \\ \mathbf{I}_{ca} &= I_\phi \angle -240° \end{aligned} \tag{A–13}$$

In the case of the Δ connection, it is obvious that the line-to-line voltage between any two lines will be the same as the voltage in the corresponding phase. *In a Δ connection,*

$$\boxed{V_{LL} = V_\phi \qquad \Delta \text{ connection}} \tag{A–14}$$

The relationship between line current and phase current is more complex. It can be found by applying Kirchhoff's current law at a node of the Δ. Applying Kirchhoff's current law to node A yields the equation

$$\begin{aligned} \mathbf{I}_a &= \mathbf{I}_{ab} - \mathbf{I}_{ca} \\ &= I_\phi \angle 0° - I_\phi \angle -240° \\ &= I_\phi - \left(-\frac{1}{2} I_\phi + j\frac{\sqrt{3}}{2} I_\phi\right) = \frac{3}{2} I_\phi - j\frac{\sqrt{3}}{2} I_\phi \end{aligned}$$

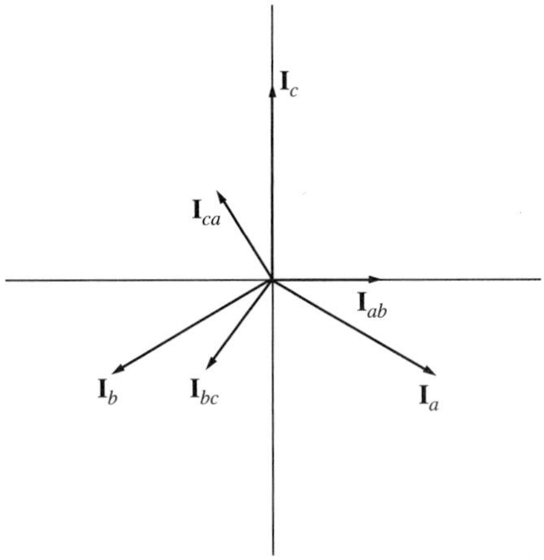

FIGURE A–8
Line and phase currents for the Δ connection in Figure A–7.

Table A–1
Summary of relationships in Y and Δ connections

	Y connection	Δ connection
Voltage magnitudes	$V_{LL} = \sqrt{3}\,V_\phi$	$V_{LL} = V_\phi$
Current magnitudes	$I_L = I_\phi$	$I_L = \sqrt{3}\,I_\phi$
abc phase sequence	$\mathbf{V}_{ab}$ leads $\mathbf{V}_a$ by 30°	$\mathbf{I}_a$ lags $\mathbf{I}_{ab}$ by 30°
acb phase sequence	$\mathbf{V}_{ab}$ lags $\mathbf{V}_a$ by 30°	$\mathbf{I}_a$ leads $\mathbf{I}_{ab}$ by 30°

$$= \sqrt{3}I_\phi\left(\frac{\sqrt{3}}{2} - j\frac{1}{2}\right)$$
$$= \sqrt{3}I_\phi\angle -30^\circ$$

Therefore, the relationship between the magnitudes of the line and phase currents in a Δ-connected generator or load is

$$\boxed{I_L = \sqrt{3}I_\phi \qquad \Delta \text{ connection}} \tag{A–15}$$

and the line currents are shifted 30° relative to the corresponding phase currents.

Note that for Δ connections with the *abc* phase sequence such as the one shown in Figure A–7, the current of a line *lags* the corresponding phase current by 30° (see Figure A–8). For Δ connections with the *acb* phase sequence, the current of a line *leads* the corresponding phase current by 30°.

The voltage and current relationships for Y- and Δ-connected sources and loads are summarized in Table A–1.

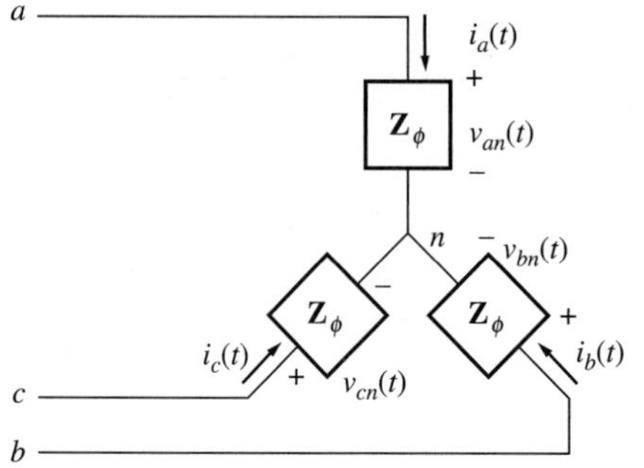

FIGURE A–9
A balanced Y-connected load.

A.3 POWER RELATIONSHIPS IN THREE-PHASE CIRCUITS

Figure A–9 shows a balanced Y-connected load whose phase impedance is $\mathbf{Z}_\phi = Z\angle\theta°$. If the three-phase voltages applied to this load are given by

$$\begin{aligned} v_{an}(t) &= \sqrt{2}V \sin \omega t \\ v_{bn}(t) &= \sqrt{2}V \sin(\omega t - 120°) \\ v_{cn}(t) &= \sqrt{2}V \sin(\omega t - 240°) \end{aligned} \tag{A–16}$$

then the three-phase currents flowing in the load are given by

$$\begin{aligned} i_a(t) &= \sqrt{2}I \sin(\omega t - \theta) \\ i_b(t) &= \sqrt{2}I \sin(\omega t - 120° - \theta) \\ i_c(t) &= \sqrt{2}I \sin(\omega t - 240° - \theta) \end{aligned} \tag{A–17}$$

where $I = V/Z$. How much power is being supplied to this load from the source?

The instantaneous power supplied to one phase of the load is given by the equation

$$\boxed{p(t) = v(t)i(t)} \tag{A–18}$$

Therefore, the instantaneous power supplied to each of the three phases is

$$\begin{aligned} p_a(t) &= v_{an}(t)i_a(t) = 2VI \sin(\omega t) \sin(\omega t - \theta) \\ p_b(t) &= v_{bn}(t)i_b(t) = 2VI \sin(\omega t - 120°) \sin(\omega t - 120° - \theta) \\ p_c(t) &= v_{cn}(t)i_c(t) = 2VI \sin(\omega t - 240°) \sin(\omega t - 240° - \theta) \end{aligned} \tag{A–19}$$

A trigonometric identity states that

$$\sin \alpha \sin \beta = \frac{1}{2}[\cos(\alpha - \beta) - \cos(\alpha - \beta)] \tag{A–20}$$

Applying this identity to Equations (A–19) yields new expressions for the power in each phase of the load:

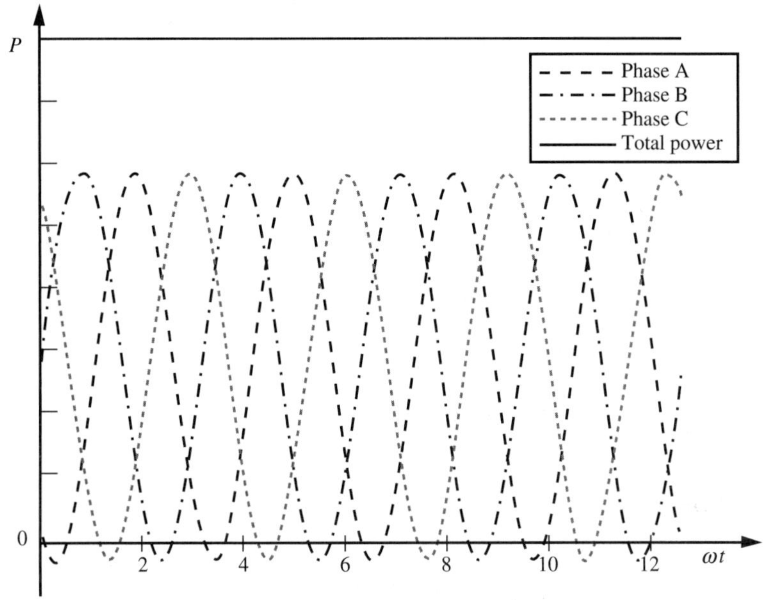

FIGURE A–10
Instantaneous power in phases *a, b,* and *c,* plus the total power supplied to the load.

$$\begin{aligned} p_a(t) &= VI[\cos\theta - \cos(2\omega t - \theta)] \\ p_b(t) &= VI[\cos\theta - \cos(2\omega t - 240^\circ - \theta)] \\ p_c(t) &= VI[\cos\theta - \cos(2\omega t - 480^\circ - \theta)] \end{aligned} \tag{A–21}$$

The total power supplied to the entire three-phase load is the sum of the power supplied to each of the individual phases. The power supplied by each phase consists of a constant component plus a pulsing component. However, *the pulsing components in the three phases cancel each other out since they are 120° out of phase with each other*, and the final power supplied by the three-phase power system is constant. This power is given by the equation:

$$p_{\text{tot}}(t) = p_A(t) + p_B(t) + p_C(t) = 3VI\cos\theta \tag{A–22}$$

The instantaneous power in phases *a*, *b*, and *c* are shown as a function of time in Figure A–10. Note that *the total power supplied to a balanced three-phase load is constant at all times.* The fact that a constant power is supplied by a three-phase power system is one of its major advantages compared to single-phase sources.

Three-Phase Power Equations Involving Phase Quantities

The single-phase power Equations (1–60) to (1–66) apply to *each phase* of a Y- or Δ-connected three-phase load, so the real, reactive, and apparent powers supplied to a balanced three-phase load are given by

$$\boxed{P = 3V_\phi I_\phi \cos\theta} \tag{A–23}$$

$$\boxed{Q = 3V_\phi I_\phi \sin\theta} \tag{A–24}$$

$$\boxed{S = 3V_\phi I_\phi} \tag{A–25}$$

$$\boxed{P = 3I_\phi^2 Z \cos\theta} \tag{A–26}$$

$$\boxed{Q = 3I_\phi^2 Z \sin\theta} \tag{A–27}$$

$$\boxed{S = 3I_\phi^2 Z} \tag{A–28}$$

The angle θ is again the angle between the voltage and the current in any phase of the load (it is the same in all phases), and the power factor of the load is the cosine of the impedance angle θ. The power-triangle relationships apply as well.

Three-Phase Power Equations Involving Line Quantities

It is also possible to derive expressions for the power in a balanced three-phase load in terms of line quantities. This derivation must be done separately for Y- and Δ-connected loads, since the relationships between the line and phase quantities are different for each type of connection.

For a Y-connected load, the power consumed by a load is given by

$$P = 3V_\phi I_\phi \cos\theta \tag{A–23}$$

For this type of load, $I_L = I_\phi$ and $V_{LL} = \sqrt{3}V_\phi$, so the power consumed by the load can also be expressed as

$$P = 3\left(\frac{V_{LL}}{\sqrt{3}}\right) I_L \cos\theta$$

$$\boxed{P = \sqrt{3}V_{LL} I_L \cos\theta} \tag{A–29}$$

For a Δ-connected load, the power consumed by a load is given by

$$P = 3V_\phi I_\phi \cos\theta \tag{A–23}$$

For this type of load, $I_L = \sqrt{3}I_\phi$ and $V_{LL} = V_\phi$, so the power consumed by the load can also be expressed in terms of line quantities as

$$P = 3V_{LL}\left(\frac{I_L}{\sqrt{3}}\right)\cos\theta$$

$$= \sqrt{3}V_{LL}I_L \cos\theta \tag{A–29}$$

This is exactly the same equation that was derived for a Y-connected load, so Equation (A–29) gives the power of a balanced three-phase load in terms of line quantities *regardless of the connection of the load.* The reactive and apparent powers of the load in terms of line quantities are

$$\boxed{Q = \sqrt{3}V_{LL}\, I_L \sin\theta} \tag{A–30}$$

$$\boxed{S = \sqrt{3}V_{LL}\, I_L} \tag{A–31}$$

It is important to realize that the $\cos\theta$ and $\sin\theta$ terms in Equations (A–29) and (A–30) are the cosine and sine of the angle between the *phase* voltage and the *phase* current, not the angle between the line-to-line voltage and the line current. Remember that there is a 30° phase shift between the line-to-line and phase voltage for a Y connection, and between the line and phase current for a Δ connection, so it is important not to take the cosine of the angle between the line-to-line voltage and line current.

A.4 ANALYSIS OF BALANCED THREE-PHASE SYSTEMS

If a three-phase power system is balanced, it is possible to determine the voltages, currents, and powers at various points in the circuit with a *per-phase equivalent circuit.* This idea is illustrated in Figure A–11. Figure A–11a shows a Y-connected generator supplying power to a Y-connected load through a three-phase transmission line.

In such a balanced system, a neutral wire may be inserted with no effect on the system, since no current flows in that wire. This system with the extra wire inserted is shown in Figure A–11b. Also, notice that each of the three phases is *identical* except for a 120° shift in phase angle. Therefore, it is possible to analyze a circuit consisting of *one phase and the neutral,* and the results of that analysis will be valid for the other two phases as well if the 120° phase shift is included. Such a per-phase circuit is shown in Figure A–11c.

There is one problem associated with this approach, however. It requires that a neutral line be available (at least conceptually) to provide a return path for current flow from the loads to the generator. This is fine for Y-connected sources and loads, but no neutral can be connected to Δ-connected sources and loads.

How can Δ-connected sources and loads be included in a power system to be analyzed? The standard approach is to transform the impedances by the Y–Δ transform of elementary circuit theory. For the special case of balanced loads, the Y–Δ transformation states that a Δ-connected load consisting of three equal impedances, each of value Z, is totally equivalent to a Y-connected load consisting of three impedances, each of value $Z/3$ (see Figure A–12). This equivalence means that the voltages, currents, and powers supplied to the two loads cannot be distinguished in any fashion by anything external to the load itself.

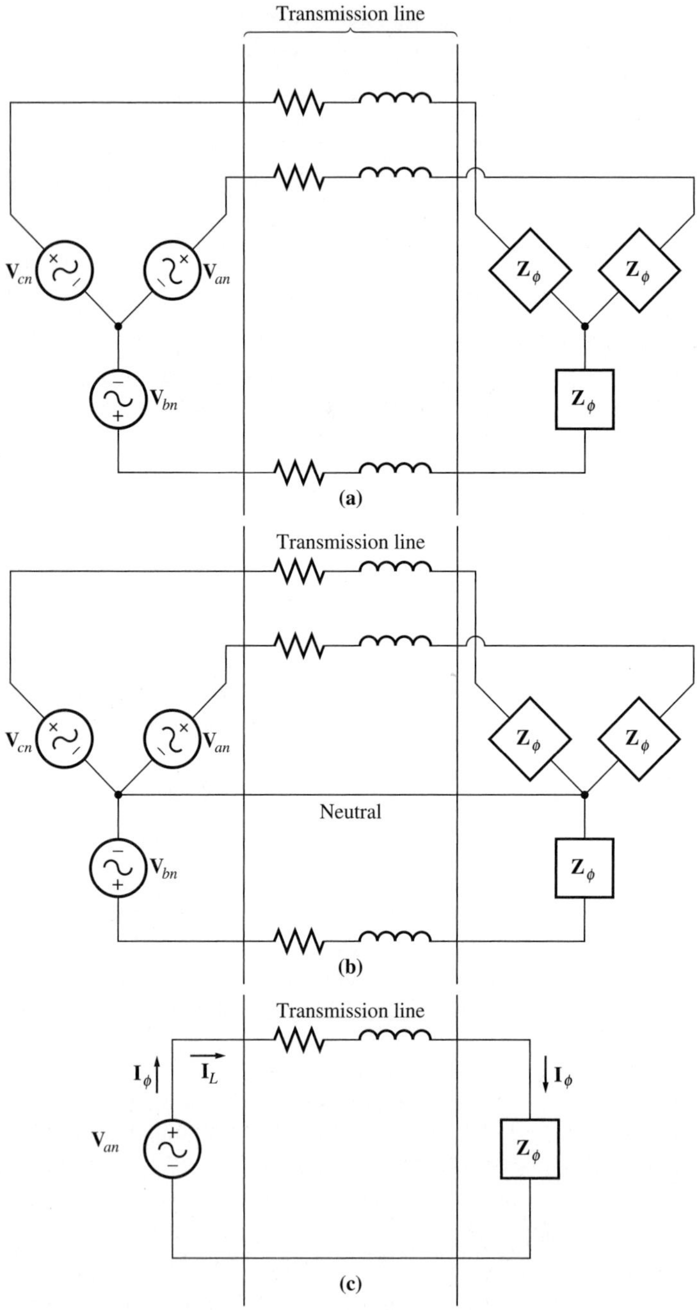

FIGURE A–11
(a) A Y-connected generator and load. (b) System with neutral inserted. (c) The per-phase equivalent circuit.

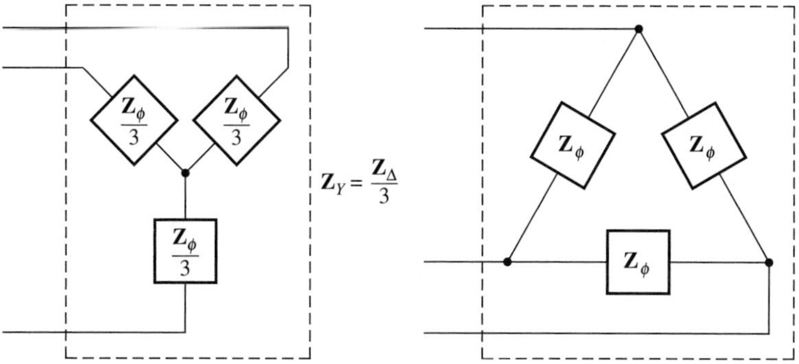

FIGURE A–12
Y-Δ transformation. A Y-connected impedance of Z/3 Ω is totally equivalent to a Δ-connected impedance of Z Ω to any circuit connected to the load's terminals.

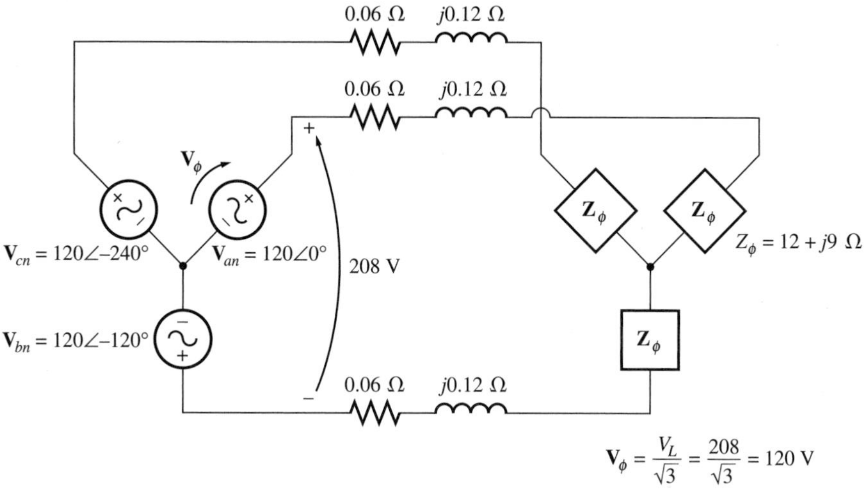

FIGURE A–13
The three-phase circuit of Example A–1.

If Δ-connected sources or loads include voltage sources, then the magnitudes of the voltage sources must be scaled according to Equation (A–11), and the effect of the 30° phase shift must be included as well.

Example A–1. A 208-V three-phase power system is shown in Figure A–13. It consists of an ideal 208-V Y-connected three-phase generator connected through a three-phase transmission line to a Y-connected load. The transmission line has an impedance of $0.06 + j0.12\ \Omega$ per phase, and the load has an impedance of $12 + j9\ \Omega$ per phase. For this simple power system, find

(a) The magnitude of the line current I_L
(b) The magnitude of the load's line and phase voltages V_{LL} and $V_{\phi L}$

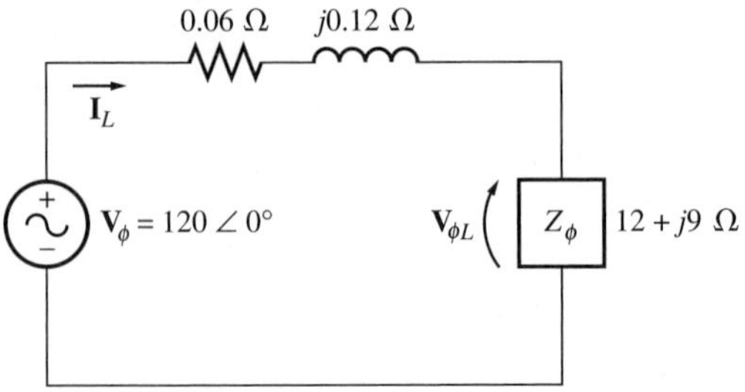

FIGURE A–14
Per-phase circuit in Example A–1.

(c) The real, reactive, and apparent powers consumed by the load
(d) The power factor of the load
(e) The real, reactive, and apparent powers consumed by the transmission line
(f) The real, reactive, and apparent powers supplied by the generator
(g) The generator's power factor

Solution

Since both the generator and the load on this power system are Y-connected, it is very simple to construct a per-phase equivalent circuit. This circuit is shown in Figure A–14.

(a) The line current flowing in the per-phase equivalent circuit is given by

$$\begin{aligned}\mathbf{I}_{\text{line}} &= \frac{\mathbf{V}}{\mathbf{Z}_{\text{line}} + \mathbf{Z}_{\text{load}}} \\ &= \frac{120\,\angle 0°\text{ V}}{(0.06 + j0.12\ \Omega) + (12 + j9\Omega)} \\ &= \frac{120\,\angle 0°}{12.06 + j9.12} = \frac{120\,\angle 0°}{15.12\angle 37.1°} \\ &= 7.94\angle -37.1°\text{ A}\end{aligned}$$

The magnitude of the line current is thus 7.94 A.

(b) The phase voltage on the load is the voltage across one phase of the load. This voltage is the product of the phase impedance and the phase current of the load:

$$\begin{aligned}\text{V}_{\phi L} &= \mathbf{I}_{\phi L}\mathbf{Z}_{\phi L} \\ &= (7.94\angle -37.1°\text{ A})(12 + j9\ \Omega) \\ &= (7.94\angle -37.1°\text{ A})(15\angle 36.9°\ \Omega) \\ &= 119.1\angle -0.2°\text{ V}\end{aligned}$$

Therefore, the magnitude of the load's phase voltage is

$$\text{V}_{\phi L} = 119.1\text{ V}$$

and the magnitude of the load's line voltage is

$$\text{V}_{LL} = \sqrt{3}\text{V}_{\phi L} = 206.3\text{ V}$$

(c) The real power consumed by the load is

$$\begin{aligned}P_{\text{load}} &= 3\text{V}_{\phi}I_{\phi}\cos\theta \\ &= 3(119.1\text{ V})(7.94\text{ A})\cos 36.9° \\ &= 2270\text{ W}\end{aligned}$$

The reactive power consumed by the load is

$$\begin{aligned} Q_{\text{load}} &= 3V_\phi I_\phi \sin\theta \\ &= 3(119.1 \text{ V})(7.94 \text{ A}) \sin 36.9^\circ \\ &= 1702 \text{ var} \end{aligned}$$

The apparent power consumed by the load is

$$\begin{aligned} S_{\text{load}} &= 3V_\phi I_\phi \\ &= 3(119.1 \text{ V})(7.94 \text{ A}) \\ &= 2839 \text{ VA} \end{aligned}$$

(d) The load power factor is

$$\text{PF}_{\text{load}} = \cos\theta = \cos 36.9^\circ = 0.8 \text{ lagging}$$

(e) The current in the transmission line is $7.94\angle -37.1$ A, and the impedance of the line is $0.06 + j0.12\ \Omega$ or $0.134\angle 63.4^\circ\ \Omega$ per phase. Therefore, the real, reactive, and apparent powers consumed in the line are

$$\begin{aligned} P_{\text{line}} &= 3I_\phi^2 Z \cos\theta \qquad \text{(A–26)} \\ &= 3(7.94 \text{ A})^2\ (0.134\ \Omega) \cos 63.4^\circ \\ &= 11.3 \text{ W} \end{aligned}$$

$$\begin{aligned} Q_{\text{line}} &= 3I_\phi^2 Z \sin\theta \qquad \text{(A–27)} \\ &= 3(7.94 \text{ A})^2\ (0.134\ \Omega) \sin 63.4^\circ \\ &= 22.7 \text{ var} \end{aligned}$$

$$\begin{aligned} S_{\text{line}} &= 3I_\phi^2 Z \qquad \text{(A–28)} \\ &= 3(7.94 \text{ A})^2\ (0.134\ \Omega) \\ &= 25.3 \text{ VA} \end{aligned}$$

(f) The real and reactive powers supplied by the generator are the sum of the powers consumed by the line and the load:

$$\begin{aligned} P_{\text{gen}} &= P_{\text{line}} + P_{\text{load}} \\ &= 11.3 \text{ W} + 2270 \text{ W} = 2281 \text{ W} \end{aligned}$$

$$\begin{aligned} Q_{\text{gen}} &= Q_{\text{line}} + Q_{\text{load}} \\ &= 22.7 \text{ var} + 1702 \text{ var} = 1725 \text{ var} \end{aligned}$$

The apparent power of the generator is the square root of the sum of the squares of the real and reactive powers:

$$S_{\text{gen}} = \sqrt{P_{\text{gen}}^2 + Q_{\text{gen}}^2} = 2860 \text{ VA}$$

(g) From the power triangle, the power-factor angle θ is

$$\theta_{\text{gen}} = \tan^{-1}\frac{Q_{\text{gen}}}{P_{\text{gen}}} = \tan^{-1}\frac{1725 \text{ VAR}}{2281 \text{ W}} = 37.1^\circ$$

Therefore, the generator's power factor is

$$\text{PF}_{\text{gen}} = \cos 37.1^\circ = 0.798 \text{ lagging}$$

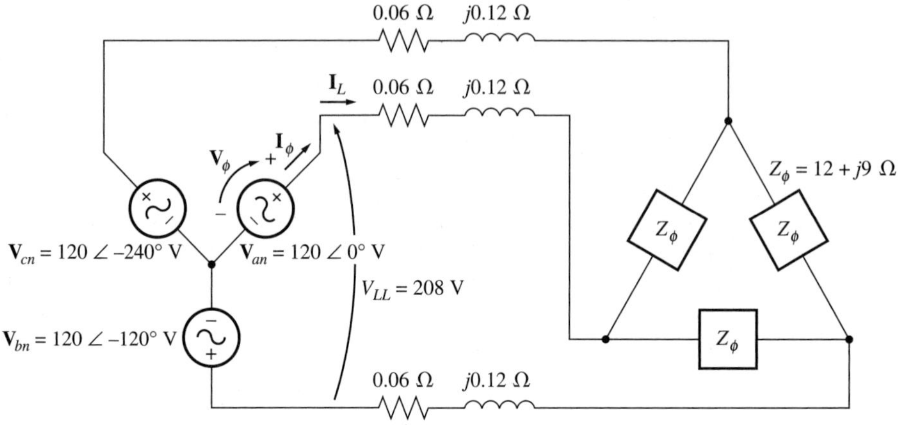

FIGURE A–15
Three-phase circuit in Example A–2.

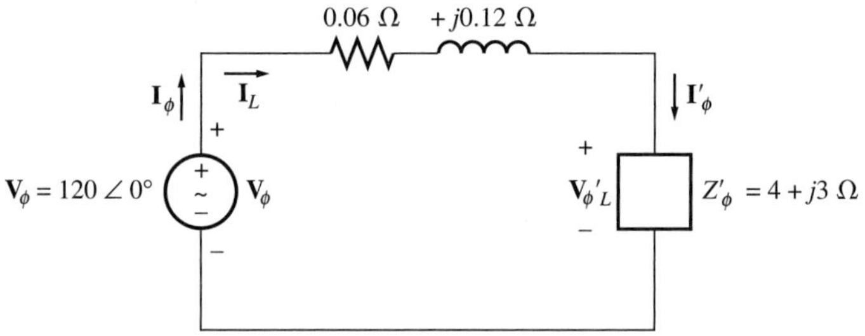

FIGURE A–16
Per-phase circuit in Example A–2.

Example A–2. Repeat Example A–1 for a Δ-connected load, with everything else unchanged.

Solution

This power system is shown in Figure A–15. Since the load on this power system is Δ connected, it must first be converted to an equivalent Y form. The phase impedance of the Δ-connected load is $12 + j9\ \Omega$ so the equivalent phase impedance of the corresponding Y form is

$$Z_Y = \frac{Z_\Delta}{3} = 4 + j3\ \Omega$$

The resulting per-phase equivalent circuit of this system is shown in Figure A–16.

(*a*) The line current flowing in the per-phase equivalent circuit is given by

$$\mathbf{I}_{\text{line}} = \frac{\mathbf{V}}{\mathbf{Z}_{\text{line}} + \mathbf{Z}_{\text{load}}}$$

$$= \frac{120\angle 0° \text{ V}}{(0.06 + j0.12\ \Omega) + (4 + j3\ \Omega)}$$
$$= \frac{120\angle 0°}{4.06 + j3.12} = \frac{120\angle 0°}{5.12\angle 37.5°}$$
$$= 23.4\angle -37.5° \text{ A}$$

The magnitude of the line current is thus 23.4 A.

(b) The phase voltage on the equivalent Y load is the voltage across one phase of the load. This voltage is the product of the phase impedance and the phase current of the load:

$$\mathbf{V}'_{\phi L} = \mathbf{I}'_{\phi L}\mathbf{Z}'_{\phi L}$$
$$= (23.4\angle -37.5° \text{ A})(4 + j3\ \Omega)$$
$$= (23.4\angle -37.5° \text{ A})(5\angle 36.9°\ \Omega) = 117\angle -0.6° \text{ V}$$

The original load was Δ connected, so the phase voltage of the original load is

$$V_{\phi L} = \sqrt{3}\,(117 \text{ V}) = 203 \text{ V}$$

and the magnitude of the load's line voltage is

$$V_{LL} = V_{\phi L} = 203 \text{ V}$$

(c) The real power consumed by the equivalent Y load (which is the same as the power in the actual load) is

$$P_{\text{load}} = 3V_\phi I_\phi \cos\theta$$
$$= 3(117 \text{ V})(23.4 \text{ A}) \cos 36.9°$$
$$= 6571 \text{ W}$$

The reactive power consumed by the load is

$$Q_{\text{load}} = 3V_\phi I_\phi \sin\theta$$
$$= 3(117 \text{ V})(23.4 \text{ A}) \sin 36.9°$$
$$= 4928 \text{ var}$$

The apparent power consumed by the load is

$$S_{\text{load}} = 3V_\phi I_\phi$$
$$= 3(117 \text{ V})(23.4 \text{ A})$$
$$= 8213 \text{ VA}$$

(d) The load power factor is

$$\text{PF}_{\text{load}} = \cos\theta = \cos 36.9° = 0.8 \text{ lagging}$$

(e) The current in the transmission is $23.4\angle -37.5°$ A, and the impedance of the line is $0.06 + j0.12\ \Omega$ or $0.134\angle 63.4°\ \Omega$ per phase. Therefore, the real, reactive, and apparent powers consumed in the line are

$$P_{\text{line}} = 3I_\phi^2 Z \cos\theta \tag{A–26}$$
$$= 3(23.4 \text{ A})^2(0.134\ \Omega) \cos 63.4°$$
$$= 98.6 \text{ W}$$

$$Q_{\text{line}} = 3I_\phi^2 Z \sin\theta \tag{A–27}$$
$$= 3(23.4\text{ A})^2(0.134\ \Omega)\sin 63.4°$$
$$= 197\text{ var}$$
$$S_{\text{line}} = 3I_\phi^2 Z \tag{A–28}$$
$$= 3(23.4\text{ A})^2(0.134\ \Omega)$$
$$= 220\text{ VA}$$

(*f*) The real and reactive powers supplied by the generator are the sums of the powers consumed by the line and the load:

$$P_{\text{gen}} = P_{\text{line}} + P_{\text{load}}$$
$$= 98.6\text{ W} + 6571\text{ W} = 6670\text{ W}$$
$$Q_{\text{gen}} = Q_{\text{line}} + Q_{\text{load}}$$
$$= 197\text{ var} + 4928\text{ VAR} = 5125\text{ var}$$

The apparent power of the generator is the square root of the sum of the squares of the real and reactive powers:

$$S_{\text{gen}} = \sqrt{P_{\text{gen}}^2 + Q_{\text{gen}}^2} = 8411\text{ VA}$$

(*g*) From the power triangle, the power-factor angle θ is

$$\theta_{\text{gen}} = \tan^{-1}\frac{Q_{\text{gen}}}{P_{\text{gen}}} = \tan^{-1}\frac{5125\text{ var}}{6670\text{ W}} = 37.6°$$

Therefore, the generator's power factor is

$$\text{PF}_{\text{gen}} = \cos 37.6° = 0.792\text{ lagging}$$

A.5 ONE-LINE DIAGRAMS

As we have seen in this chapter, a balanced three-phase power system has three lines connecting each source with each load, one for each of the phases in the power system. The three phases are all similar, with voltages and currents equal in amplitude and shifted in phase from each other by 120°. Because the three phases are all basically the same, it is customary to sketch power systems in a simple form with a *single line* representing all three phases of the real power system. These *one-line diagrams* provide a compact way to represent the interconnections of a power system. One-line diagrams typically include all of the major components of a power system, such as generators, transformers, transmission lines, and loads with the transmission lines represented by a single line. The voltages and types of connections of each generator and load are usually shown on the diagram. A simple power system is shown in Figure A–17, together with the corresponding one-line diagram.

A.6 USING THE POWER TRIANGLE

If the transmission lines in a power system can be assumed to have negligible impedance, then an important simplification is possible in the calculation of three-phase

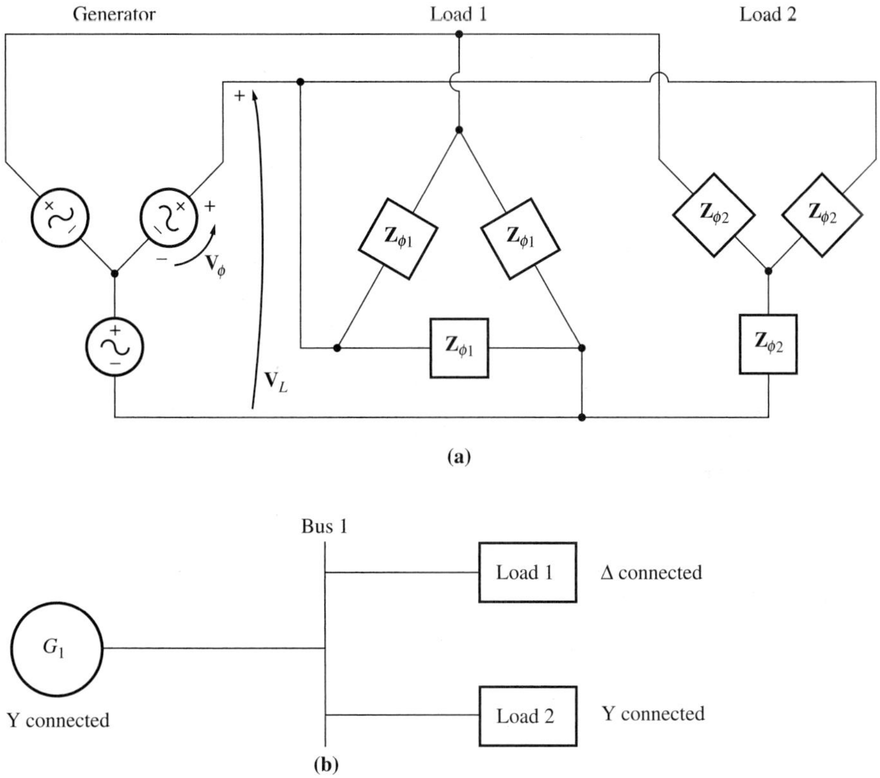

FIGURE A–17
(a) A simple power system with a Y-connected generator, a Δ-connected load, and a Y-connected load. (b) The corresponding one-line diagram.

currents and powers. This simplification depends on the use of the real and reactive powers of each load to determine the currents and power factors at various points in the system.

For example, consider the simple power system shown in Figure A–17. If the transmission line in that power system is assumed to be lossless, the line voltage at the generator will be the same as the line voltage at the loads. If the generator voltage is specified, then we can find the current and power factor at any point in this power system as follows:

1. Determine the line voltage at the generator and the loads. Since the transmission line is assumed to be lossless, these two voltages will be identical.
2. Determine the real and reactive powers of each load on the power system. We can use the known load voltage to perform this calculation.
3. Find the total real and reactive powers supplied to all loads "downstream" from the point being examined.

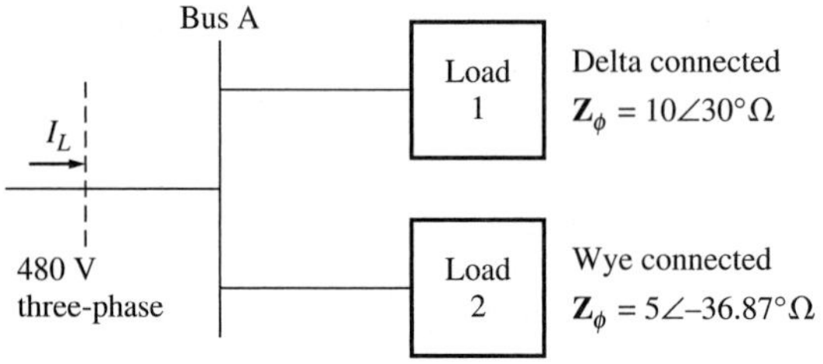

FIGURE A–18
The system in Example A–3.

4. Determine the system power factor at that point, using the power-triangle relationships.
5. Use Equation (A–29) to determine line currents, or Equation (A–23) to determine phase currents, at that point.

This approach is commonly employed by engineers estimating the currents and power flows at various points on distribution systems within an industrial plant. Within a single plant, the lengths of transmission lines will be quite short and their impedances will be relatively small, and so only small errors will occur if the impedances are neglected. An engineer can treat the line voltage as constant, and use the power triangle method to quickly calculate the effect of adding a load on the overall system current and power factor.

Example A–3. Figure A–18 shows a one-line diagram of a small 480-V industrial distribution system. The power system supplies a constant line voltage of 480 V, and the impedance of the distribution lines is negligible. Load 1 is a Δ-connected load with a phase impedance of $10\angle 30°\ \Omega$, and load 2 is a Y-connected load with a phase impedance of $5\angle -36.87°\ \Omega$.

(a) Find the overall power factor of the distribution system.
(b) Find the total line current supplied to the distribution system.

Solution
The lines in this system are assumed impedanceless, so there will be no voltage drops within the system. Since load 1 is Δ connected, its phase voltage will be 480 V. Since load 2 is Y connected, its phase voltage will be $480/\sqrt{3} = 277$ V.

The phase current in load 1 is

$$I_{\phi 1} = \frac{480\text{ V}}{10\ \Omega} = 48\text{ A}$$

Therefore, the real and reactive powers of load 1 are

$$\begin{aligned} P_1 &= 3V_{\phi 1}I_{\phi 1}\cos\theta \\ &= 3(480\text{ V})(48\text{ A})\cos 30° = 59.9\text{ kW} \end{aligned}$$

$$Q_1 = 3V_{\phi 1} I_{\phi 1} \sin \theta$$
$$= 3(480 \text{ V})(48 \text{ A}) \sin 30° = 34.6 \text{ kvar}$$

The phase current in load 2 is

$$I_{\phi 2} = \frac{277 \text{ V}}{5\ \Omega} = 55.4 \text{ A}$$

Therefore, the real and reactive powers of load 2 are

$$P_2 = 3V_{\phi 2} I_{\phi 2} \cos \theta$$
$$= 3(277 \text{ V})(55.4 \text{ A}) \cos(-36.87°) = 36.8 \text{ kW}$$
$$Q_2 = 3V_{\phi 2} I_{\phi 2} \sin \theta$$
$$= 3(277 \text{ V})(55.4 \text{ A}) \sin(-36.87°) = -27.6 \text{ kvar}$$

(a) The total real and reactive powers supplied by the distribution system are

$$P_{\text{tot}} = P_1 + P_2$$
$$= 59.9 \text{ kW} + 36.8 \text{ kW} = 96.7 \text{ kW}$$
$$Q_{\text{tot}} = Q_1 + Q_2$$
$$= 34.6 \text{ kvar} - 27.6 \text{ kvar} = 7.00 \text{ kvar}$$

From the power triangle, the effective impedance angle θ is given by

$$\theta = \tan^{-1} \frac{Q}{P}$$
$$= \tan^{-1} \frac{7.00 \text{ kvar}}{96.7 \text{ kW}} = 4.14°$$

The system power factor is thus

$$\text{PF} = \cos \theta = \cos(4.14°) = 0.997 \text{ lagging}$$

(b) The total line current is given by

$$I_L = \frac{P}{\sqrt{3} V_L \cos \theta}$$
$$I_L = \frac{96.7 \text{ kW}}{\sqrt{3}(480 \text{ V})(0.997)} = 117 \text{ A}$$

QUESTIONS

A–1. What types of connections are possible for three-phase generators and loads?

A–2. What is meant by the term “balanced” in a balanced three-phase system?

A–3. What is the relationship between phase and line voltages and currents for a wye (Y) connection?

A–4. What is the relationship between phase and line voltages and currents for a delta (Δ) connection?

A–5. What is phase sequence?

A–6. Write the equations for real, reactive, and apparent power in three-phase circuits, in terms of both line and phase quantities.

A–7. What is a Y–Δ transform?

PROBLEMS

A–1. Three impedances of $4 + j3\ \Omega$ are Δ connected and tied to a three-phase 208-V power line. Find I_ϕ, I_L, P, Q, S, and the power factor of this load.

A–2. Figure PA–1 shows a three-phase power system with two loads. The Δ-connected generator is producing a line voltage of 480 V, and the line impedance is $0.09 + j0.16\ \Omega$. Load 1 is Y connected, with a phase impedance of $2.5\angle 36.87°\ \Omega$ and load 2 is Δ connected, with a phase impedance of $5\angle -20°\ \Omega$.

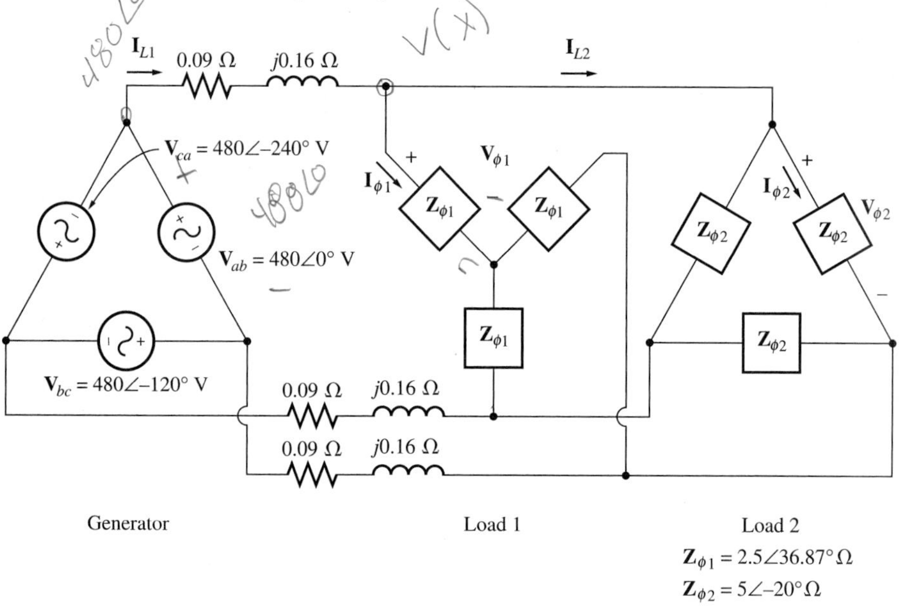

FIGURE PA–1
The system in Problem A–2.

(*a*) What is the line voltage of the two loads?
(*b*) What is the voltage drop on the transmission lines?
(*c*) Find the real and reactive powers supplied to each load.
(*d*) Find the real and reactive power losses in the transmission line.
(*e*) Find the real power, reactive power, and power factor supplied by the generator.

A–3. Figure PA–2 shows a one-line diagram of a simple power system containing a single 480-V generator and three loads. Assume that the transmission lines in this power system are lossless, and answer the following questions.

(*a*) Assume that Load 1 is Y connected. What are the phase voltage and currents in that load?
(*b*) Assume that Load 2 is Δ connected. What are the phase voltage and currents in that load?
(*c*) What real, reactive, and apparent power does the generator supply when the switch is open?
(*d*) What is the total line current I_L when the switch is open?
(*e*) What real, reactive, and apparent power does the generator supply when the switch is closed?

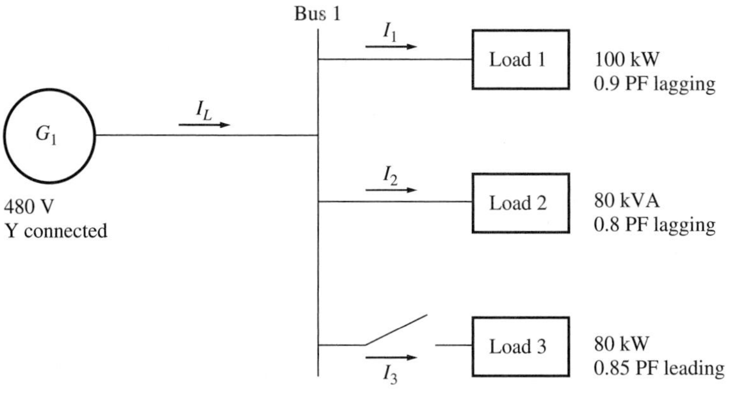

FIGURE PA–2
The power system in Problem A–3.

(*f*) What is the total line current I_L when the switch is closed?

(*g*) How does the total line current I_L compare to the sum of the three individual currents $I_1 + I_2 + I_3$? If they are not equal, why not?

A–4. Prove that the line voltage of a Y-connected generator with an *acb* phase sequence lags the corresponding phase voltage by 30°. Draw a phasor diagram showing the phase and line voltages for this generator.

A–5. Find the magnitudes and angles of each line and phase voltage and current on the load shown in Figure PA–3.

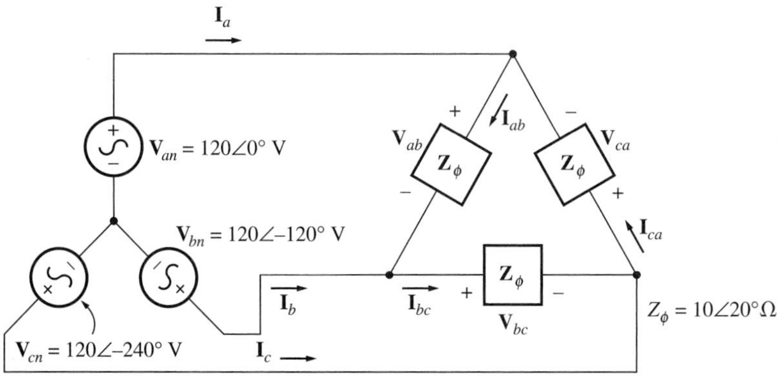

FIGURE PA–3
The system in Problem A–5.

A–6. Figure PA–4 shows a one-line diagram of a small 480-V distribution system in an industrial plant. An engineer working at the plant wishes to calculate the current that will be drawn from the power utility company with and without the capacitor bank switched into the system. For the purposes of this calculation, the engineer will assume that the lines in the system have zero impedance.

(*a*) If the switch shown is open, find the real, reactive, and apparent powers in the system. Find the total current supplied to the distribution system by the utility.

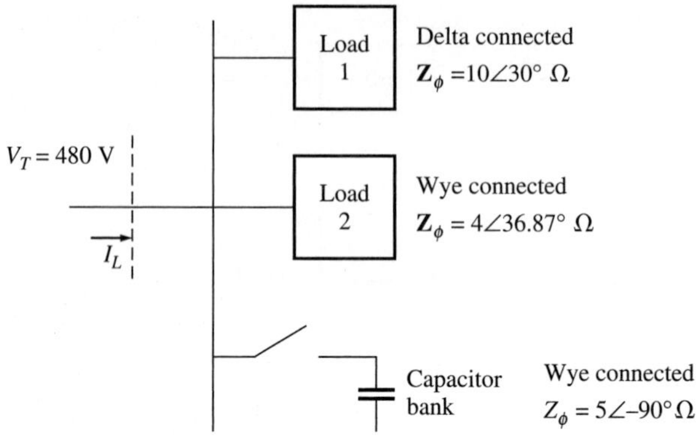

FIGURE PA–4
The system in Problem A–6.

(*b*) Repeat part (*a*) with the switch closed.
(*c*) What happened to the total current supplied by the power system when the switch closed? Why?

REFERENCE

1. Alexander, Charles K., and Matthew N. O. Sadiku: *Fundamentals of Electric Circuits,* McGraw-Hill, 2000.

APPENDIX B

COIL PITCH AND DISTRIBUTED WINDINGS

As mentioned in Chapter 3, the induced voltage in an ac machine is sinusoidal only if the harmonic components of the air-gap flux density are suppressed. This appendix describes two techniques used by machinery designers to suppress harmonics in machines.

B.1 THE EFFECT OF COIL PITCH ON AC MACHINES

In the simple ac machine design of Section 3.4, the output voltages in the stator coils were sinusoidal because the air-gap flux density distribution was sinusoidal. If the air-gap flux density distribution had not been sinusoidal, then the output voltages in the stator would not have been sinusoidal either. They would have had the same nonsinusoidal shape as the flux density distribution.

In general, the air-gap flux density distribution in an ac machine will not be sinusoidal. Machine designers do their best to produce sinusoidal flux distributions, but of course no design is ever perfect. The actual flux distribution will consist of a fundamental sinusoidal component plus harmonics. These harmonic components of flux will generate harmonic components in the stator's voltages and currents.

The harmonic components in the stator voltages and currents are undesirable, so techniques have been developed to suppress the unwanted harmonic components in the output voltages and currents of a machine. One important technique to suppress the harmonics is the use of *fractional-pitch windings*.

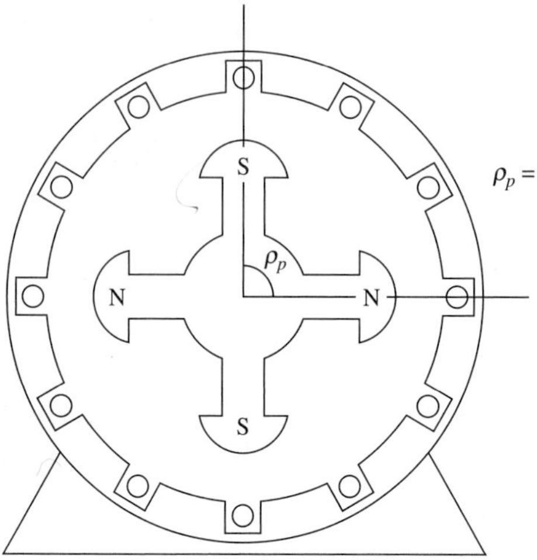

FIGURE B–1
The pole pitch of a four-pole machine is 90 mechanical or 180 electrical degrees.

The Pitch of a Coil

The *pole pitch* is the angular distance between two adjacent poles on a machine. The pole pitch of a machine in *mechanical degrees* is

$$\boxed{\rho_p = \frac{360°}{P}} \tag{B–1}$$

where ρ_p is the pole pitch in *mechanical degrees* and P is the number of poles on the machine. Regardless of the number of poles on the machine, a pole pitch is always 180 *electrical degrees* (see Figure B–1).

If the stator coil stretches across the same angle as the pole pitch, it is called a *full-pitch coil.* If the stator coil stretches across an angle smaller than a pole pitch, it is called a *fractional-pitch coil.* The pitch of a fractional-pitch coil is often expressed as a fraction indicating the portion of the pole pitch it spans. For example, a 5/6-pitch coil spans five-sixths of the distance between two adjacent poles. Alternatively, the pitch of a fractional-pitch coil in electrical degrees is given by Equations (B–2):

$$\boxed{\rho = \frac{\theta_m}{\rho_p} \times 180°} \tag{B–2a}$$

where θ_m is the mechanical angle covered by the coil in degrees and ρ_p is the machine's pole pitch in mechanical degrees, or

$$\boxed{\rho = \frac{\theta_m P}{2} \times 180°} \tag{B–2b}$$

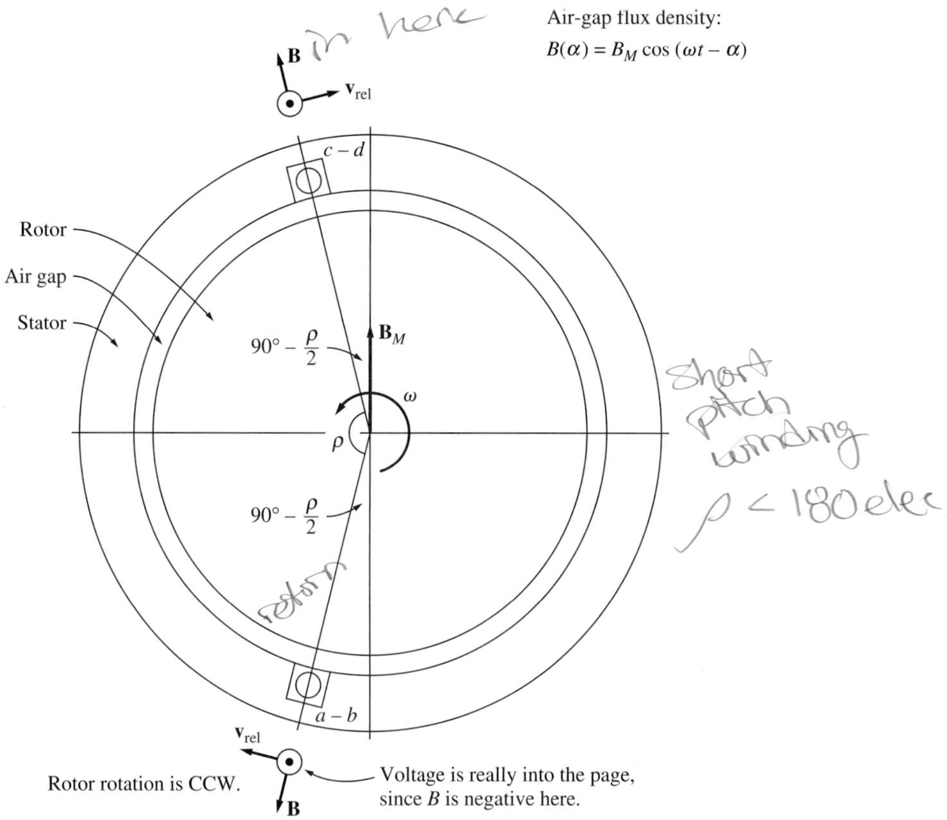

FIGURE B–2
A fractional-pitch winding of pitch ρ. The vector magnetic flux densities and velocities on the sides of the coil. The velocities are from a frame of reference in which the magnetic field is stationary.

where θ_m is the mechanical angle covered by the coil in degrees and P is the number of poles in the machine. Most practical stator coils have a fractional pitch, since a fractional-pitch winding provides some important benefits which will be explained later. Windings employing fractional-pitch coils are known as *chorded windings*.

The Induced Voltage of a Fractional-Pitch Coil

What effect does fractional pitch have on the output voltage of a coil? To find out, examine the simple two-pole machine with a fractional-pitch winding shown in Figure B–2. The pole pitch of this machine is 180°, and the coil pitch is ρ. The voltage induced in this coil by rotating the magnetic field can be found in exactly the same manner as in the previous section, by determining the voltages on each side of the coil. The total voltage will just be the sum of the voltages on the individual sides.

As before, assume that the magnitude of the flux density vector **B** in the air gap between the rotor and the stator varies sinusoidally with mechanical angle, while the direction of **B** is always radially outward. If α is the angle measured from the direction of the peak rotor flux density, then the magnitude of the flux density vector **B** at a point around the rotor is given by

$$B = B_M \cos \alpha \tag{B–3a}$$

Since the rotor is itself rotating within the stator at an angular velocity ω_m, the magnitude of the flux density vector **B** at any angle α around the *stator* is given by

$$\boxed{B = B_M \cos (\omega t - \alpha)} \tag{B–3b}$$

The equation for the induced voltage in a wire is

$$e_{\text{ind}} = (\mathbf{v} \times \mathbf{B}) \cdot \mathbf{l} \tag{1–45}$$

where $\mathbf{v}$ = velocity of the wire relative to the magnetic field
$\mathbf{B}$ = magnetic flux density vector
$\mathbf{l}$ = length of conductor in the magnetic field

This equation can only be used in a frame of reference where the magnetic field appears to be stationary. If we "sit on the magnetic field" so that the field appears to be stationary, the sides of the coil will appear to go by at an apparent velocity $\mathbf{v}_{\text{rel}}$, and the equation can be applied. Figure B–2 shows the vector magnetic field and velocities from the point of view of a stationary magnetic field and a moving wire.

1. *Segment ab*. For segment *ab* of the fractional-pitch coil, $\alpha = 90° + \rho/2$. Assuming that **B** is directed radially outward from the rotor, the angle between **v** and **B** in segment *ab* is 90°, while the quantity **v × B** is in the direction of **l**, so

$$\begin{aligned} e_{ba} &= (\mathbf{v} \times \mathbf{B}) \cdot \mathbf{l} \\ &= vBl \quad \text{directed out of the page} \\ &= -vB_M \cos \left[\omega_m t - \left(90° + \frac{\rho}{2}\right)\right] l \\ &= -vB_M l \cos \left(\omega_m t - 90° - \frac{\rho}{2}\right) \end{aligned} \tag{B–4}$$

   where the negative sign comes from the fact that **B** is really pointing inward when it was assumed to point outward.

2. *Segment bc*. The voltage on segment *bc* is zero, since the vector quantity **v × B** is perpendicular to **l**, so

$$e_{cb} = (\mathbf{v} \times \mathbf{B}) \cdot \mathbf{l} = 0 \tag{B–5}$$

3. *Segment cd*. For segment *cd*, the angle $\alpha = 90° - \rho/2$. Assuming that **B** is directed radially outward from the rotor, the angle between **v** and **B** in segment *cd* is 90°, while the quantity **v** × **B** is in the direction of **l**, so

$$\begin{aligned} e_{dc} &= (\mathbf{v} \times \mathbf{B}) \bullet \mathbf{l} \\ &= vBl \quad \text{directed out of the page} \end{aligned}$$

$$\begin{aligned} e_{ba} &= -vB_M \cos\left[\omega_m t - \left(90° - \frac{\rho}{2}\right)\right] l \\ &= -vB_M l \cos\left(\omega_m t - 90° + \frac{\rho}{2}\right) \end{aligned} \tag{B–6}$$

4. *Segment da*. The voltage on segment *da* is zero, since the vector quantity **v** × **B** is perpendicular to **l**, so

$$e_{ad} = (\mathbf{v} \times \mathbf{B}) \bullet \mathbf{l} = 0 \tag{B–7}$$

Therefore, the total voltage on the coil will be

$$\begin{aligned} e_{\text{ind}} &= e_{ba} + e_{dc} \\ &= -vB_M l \cos\left(\omega_m t - 90° - \frac{\rho}{2}\right) + vB_M l \cos\left(\omega_m t - 90° + \frac{\rho}{2}\right) \end{aligned}$$

By trigonometric identities,

$$\cos\left(\omega_m t - 90° - \frac{\rho}{2}\right) = \cos\,(\omega_m t - 90°) \cos\frac{\rho}{2} + \sin\,(\omega_m t - 90°) \sin\frac{\rho}{2}$$

$$\cos\left(\omega_m t - 90° + \frac{\rho}{2}\right) = \cos\,(\omega_m t - 90°) \cos\frac{\rho}{2} - \sin\,(\omega_m t - 90°) \sin\frac{\rho}{2}$$

$$\sin\,(\omega_m t - 90°) = -\cos\,\omega_m t$$

Therefore, the total resulting voltage is

$$\begin{aligned} e_{\text{ind}} &= vB_M l\left[-\cos\,(\omega_m t - 90°) \cos\frac{\rho}{2} - \sin\,(\omega_m t - 90°) \sin\frac{\rho}{2}\right. \\ &\qquad \left. + \cos\,(\omega_m t - 90°) \cos\frac{\rho}{2} - \sin(\omega_m t - 90°) \sin\frac{\rho}{2}\right] \\ &= -2vB_M l \sin\frac{\rho}{2} \sin\,(\omega_m t - 90°) \\ &= 2vB_M l \sin\frac{\rho}{2} \cos\,\omega_m t \end{aligned}$$

Since $2vB_M l$ is equal to $\phi\omega$, the final expression for the voltage in a single turn is

$$\boxed{e_{\text{ind}} = \phi\omega \sin\frac{\rho}{2} \cos\,\omega_m t} \tag{B–8}$$

This is the same value as the voltage in a full-pitch winding except for the sin $\rho/2$ term. It is customary to define this term as the *pitch factor* k_p of the coil. The pitch factor of a coil is given by

$$\boxed{k_p = \sin\frac{\rho}{2}} \tag{B–9}$$

In terms of the pitch factor, the induced voltage on a single-turn coil is

$$e_{\text{ind}} = k_p \phi \omega_m \cos \omega_m t \tag{B–10}$$

The total voltage in an N-turn fractional-pitch coil is thus

$$e_{\text{ind}} = N_C\, k_p \phi \omega_m \cos \omega_m t \tag{B–11}$$

and its peak voltage is

$$E_{\max} = N_C\, k_p \phi \omega_m \tag{B–12}$$

$$= 2\pi N_C\, k_p \phi f \tag{B–13}$$

Therefore, the rms voltage of any phase of this three-phase stator is

$$E_A = \frac{2\pi}{\sqrt{2}} N_C k_p \phi f \tag{B–14}$$

$$= \sqrt{2}\pi N_C k_p \phi f \tag{B–15}$$

Note that for a full-pitch coil, $\rho = 180°$ and Equation (B–15) reduces to the same result as before.

For machines with more than two poles, Equation (B–9) gives the pitch factor if the coil pitch p is in electrical degrees. If the coil pitch is given in mechanical degrees, then the pitch factor can be given by

$$\boxed{k_p = \sin\frac{\theta_m P}{2}} \tag{B–16}$$

Harmonic Problems and Fractional-Pitch Windings

There is a very good reason for using fractional-pitch windings. It concerns the effect of the nonsinusoidal flux density distribution in real machines. This problem can be understood by examining the machine shown in Figure B–3. This figure shows a salient-pole synchronous machine whose rotor is sweeping across the stator surface. Because the reluctance of the magnetic field path is *much lower* directly under the center of the rotor than it is toward the sides (smaller air gap), the flux is strongly concentrated at that point and the flux density is very high there. The resulting induced voltage in the winding is shown in Figure B–3. *Notice that it is not sinusoidal—it contains many harmonic frequency components.*

Because the resulting voltage waveform is symmetric about the center of the rotor flux, no *even harmonics* are present in the phase voltage. However, all

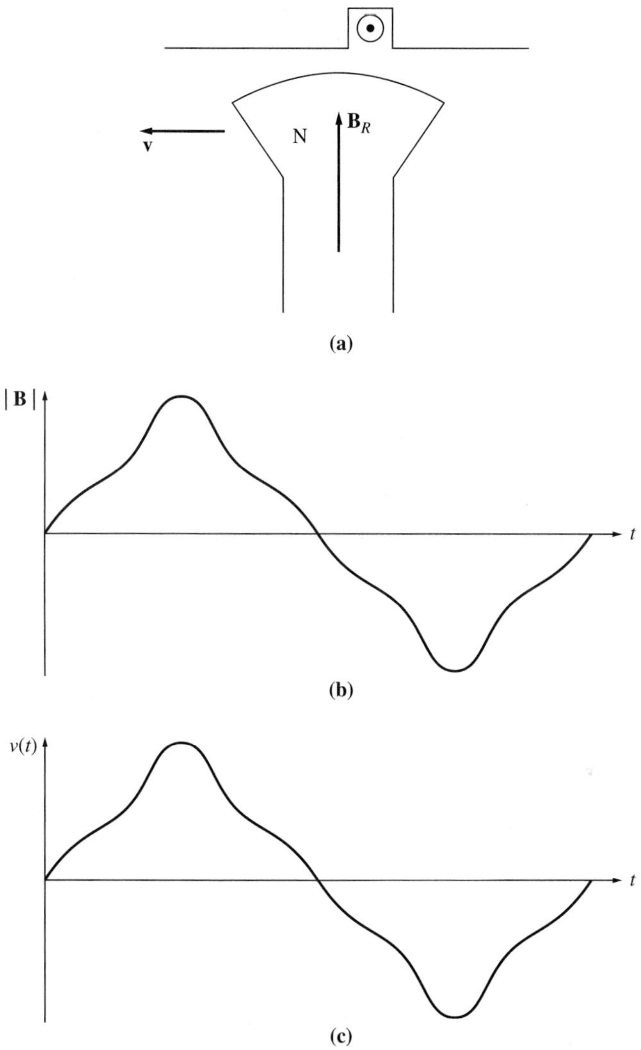

FIGURE B–3
(a) A ferromagnetic rotor sweeping past a stator conductor. (b) The flux density distribution of the magnetic field as a function of time at a point on the stator surface. (c) The resulting induced voltage in the conductor. Note that the voltage is directly proportional to the magnetic flux density at any given time.

the odd harmonics (third, fifth, seventh, ninth, etc.) *are* present in the phase voltage to some extent and need to be dealt with in the design of ac machines. In general, the higher the number of a given harmonic frequency component, the lower its magnitude in the phase output voltage; so beyond a certain point (above the ninth harmonic or so) the effects of higher harmonics may be ignored.

When the three phases are Y or Δ connected, some of the harmonics disappear from the output of the machine as a result of the three-phase connection. The third-harmonic component is one of these. If the fundamental voltages in each of the three phases are given by

$$e_a(t) = E_{M1} \sin \omega t \qquad \text{V} \tag{B–17a}$$

$$e_b(t) = E_{M1} \sin (\omega t - 120°) \qquad \text{V} \tag{B–17b}$$

$$e_c(t) = E_{M1} \sin (\omega t - 240°) \qquad \text{V} \tag{B–17c}$$

then the third-harmonic components of voltage will be given by

$$e_{a3}(t) = E_{M3} \sin 3\omega t \qquad \text{V} \tag{B–18a}$$

$$e_{b3}(t) = E_{M3} \sin (3\omega t - 360°) \qquad \text{V} \tag{B–18b}$$

$$e_{c3}(t) = E_{M3} \sin (3\omega t - 720°) \qquad \text{V} \tag{B–18c}$$

Notice that *the third-harmonic components of voltage are all identical* in each phase. If the synchronous machine is Y-connected, then the third-harmonic voltage *between any two terminals* will be zero (even though there may be a large third-harmonic component of voltage in each phase). If the machine is Δ-connected, then the three third-harmonic components all add and drive a third-harmonic current around inside the Δ-winding of the machine. Since the third-harmonic voltages are dropped across the machine's internal impedances, there is again no significant third-harmonic component of voltage at the terminals.

This result applies not only to third-harmonic components but also to any *multiple* of a third-harmonic component (such as the ninth harmonic). Such special harmonic frequencies are called *triplen harmonics* and are automatically suppressed in three-phase machines.

The remaining harmonic frequencies are the fifth, seventh, eleventh, thirteenth, etc. Since the strength of the harmonic components of voltage decreases with increasing frequency, most of the actual distortion in the sinusoidal output of a synchronous machine is caused by the fifth and seventh harmonic frequencies, sometimes called the *belt harmonics*. If a way could be found to reduce these components, then the machine's output voltage would be essentially a pure sinusoid at the fundamental frequency (50 or 60 Hz).

How can some of the harmonic content of the winding's terminal voltage be eliminated?

One way is to design the rotor itself to distribute the flux in an approximately sinusoidal shape. Although this action will help reduce the harmonic content of the output voltage, it may not go far enough in that direction. An additional step that is used is to design the machine with fractional-pitch windings.

The key to the effect of fractional-pitch windings on the voltage produced in a machine's stator is that the electrical angle of the nth harmonic is n times the electrical angle of the fundamental frequency component. In other words, if a coil spans 150 electrical degrees at its fundamental frequency, it will span 300 electrical degrees at its second-harmonic frequency, 450 electrical degrees at its third-harmonic frequency, and so forth. If ρ represents the electrical angle spanned by the coil at its *fundamental* frequency and ν is the number of the harmonic being examined, then the coil will span $\nu\rho$ electrical degrees at that harmonic frequency. Therefore, the pitch factor of the coil at the harmonic frequency can be expressed as

$$\boxed{k_p = \sin \frac{\nu\rho}{2}} \tag{B–19}$$

The important consideration here is that *the pitch factor of a winding is different for each harmonic frequency*. By a proper choice of coil pitch it is possible to almost eliminate harmonic frequency components in the output of the machine. We can now see how harmonics are suppressed by looking at a simple example problem.

Example B–1. A three-phase, two-pole stator has coils with a 5/6 pitch. What are the pitch factors for the harmonics present in this machine's coils? Does this pitch help suppress the harmonic content of the generated voltage?

Solution

The pole pitch in mechanical degrees of this machine is

$$\rho_p = \frac{360°}{P} = 180° \tag{B–1}$$

Therefore, the mechanical pitch angle of these coils is five-sixths of 180°, or 150°. From Equation (B–2a), the resulting pitch in electrical degrees is

$$\rho = \frac{\theta_m}{\rho_p} \times 180° = \frac{150°}{180°} \times 180° = 150° \tag{B–2a}$$

The mechanical pitch angle is equal to the electrical pitch angle only because this is a two-pole machine. For any other number of poles, they would not be the same.

Therefore, the pitch factors for the fundamental and the higher odd harmonic frequencies (remember, the even harmonics are already gone) are

Fundamental: $k_p = \sin \frac{150°}{2} = 0.966$

Third harmonic: $k_p = \sin \frac{3(150°)}{2} = -0.707$ (This is a triplen harmonic not present in the three-phase output.)

Fifth harmonic: $k_p = \sin \frac{5(150°)}{2} = 0.259$

Seventh harmonic: $k_p = \sin \frac{7(150°)}{2} = 0.259$

Ninth harmonic: $k_p = \sin \frac{9(150°)}{2} = -0.707$ (This is a triplen harmonic not present in the three-phase output.)

The third- and ninth-harmonic components are suppressed only slightly by this coil pitch, but that is unimportant since they do not appear at the machine's terminals anyway. Between the effects of triplen harmonics and the effects of the coil pitch, the *third, fifth, seventh, and ninth harmonics are suppressed relative to the fundamental frequency*. Therefore, employing fractional-pitch windings will drastically reduce the harmonic content of the machine's output voltage while causing only a small decrease in its fundamental voltage.

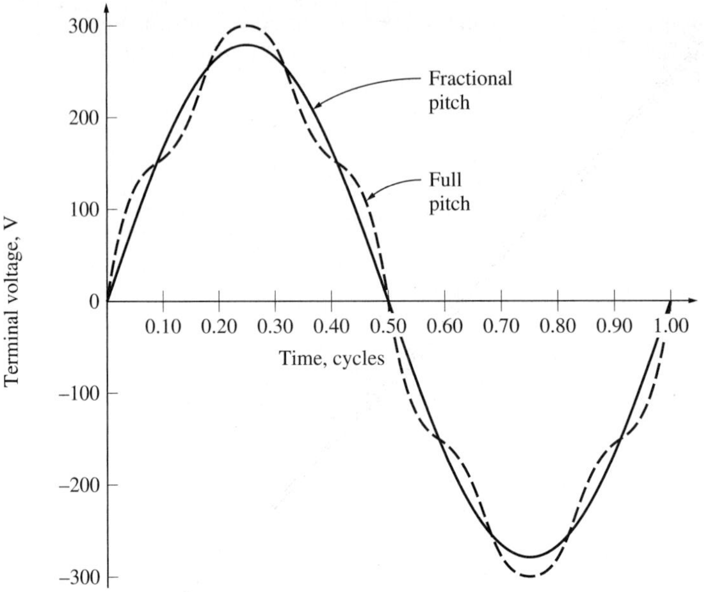

FIGURE B–4
The line voltage out of a three-phase generator with full-pitch and fractional-pitch windings. Although the peak voltage of the fractional-pitch winding is slightly smaller than that of the full-pitch winding, its output voltage is much purer.

The terminal voltage of a synchronous machine is shown in Figure B–4 both for full-pitch windings and for windings with a pitch $\rho = 150°$. Notice that the fractional-pitch windings produce a large visible improvement in waveform quality.

It should be noted that there are certain types of higher-frequency harmonics, called *tooth* or *slot harmonics*, which cannot be suppressed by varying the pitch of stator coils. These slot harmonics will be discussed in conjunction with distributed windings in Section B.2.

B.2 DISTRIBUTED WINDINGS IN AC MACHINES

In the previous section, the windings associated with each phase of an ac machine were implicitly assumed to be concentrated in a single pair of slots on the stator surface. In fact, the windings associated with each phase are almost always distributed among several adjacent pairs of slots, because it is simply impossible to put all the conductors into a single slot.

The construction of the stator windings in real ac machines is quite complicated. Normal ac machine stators consist of several coils in each phase, distributed in slots around the inner surface of the stator. In larger machines, each coil is a preformed unit consisting of a number of turns, each turn insulated from the others and from the side of the stator itself (see Figure B–5). The voltage in any

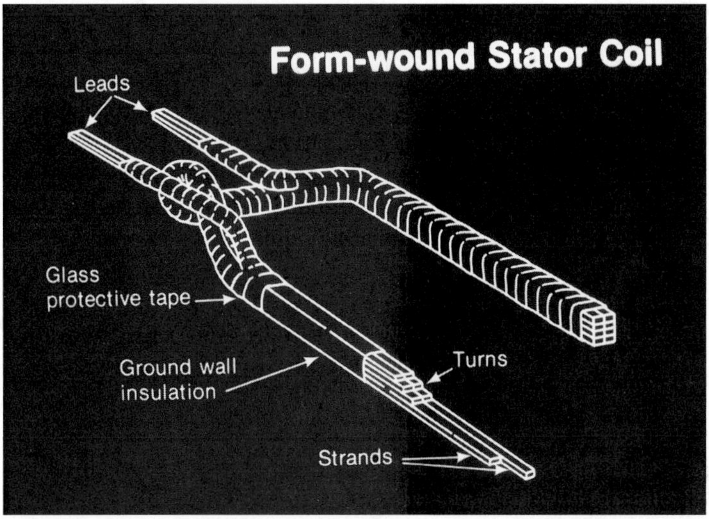

FIGURE B–5
A typical preformed stator coil. (*Courtesy of General Electric Company.*)

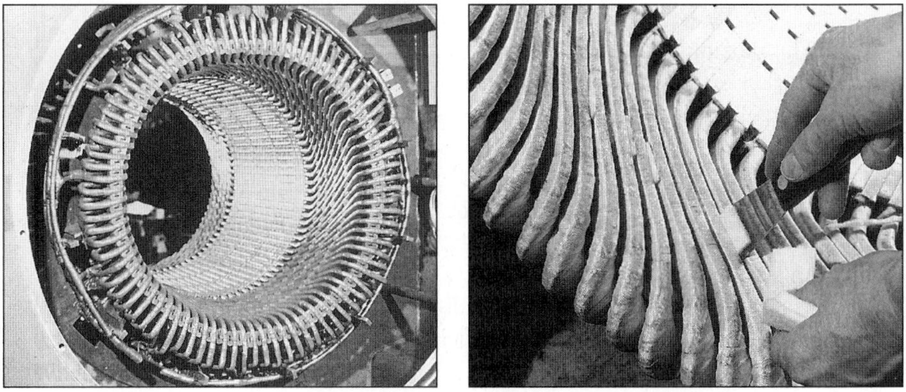

(a) (b)

FIGURE B–6
(a) An ac machine stator with preformed stator coils. (*Courtesy of Westinghouse Electric Company.*) (b) A close-up view of the coil ends on a stator. Note that one side of the coil will be outermost in its slot and the other side will be innermost in its slot. This shape permits a single standard coil form to be used for every slot on the stator. (*Courtesy of General Electric Company.*)

single turn of wire is very small, and it is only by placing many of these turns in series that reasonable voltages can be produced. The large number of turns is normally physically divided among several coils, and the coils are placed in slots equally spaced along the surface of the stator, as shown in Figure B–6.

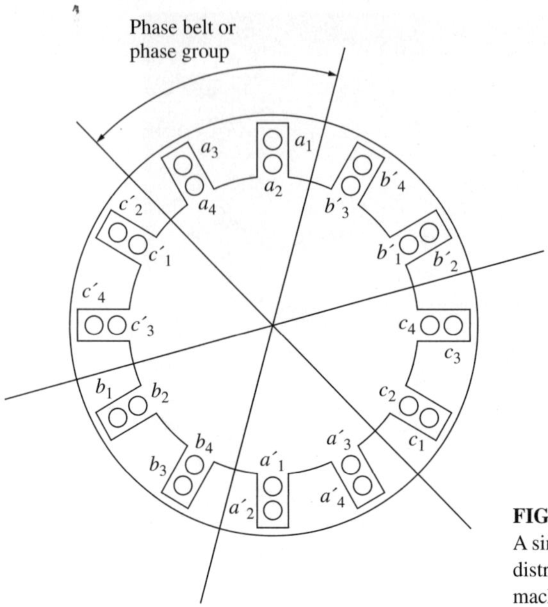

FIGURE B–7
A simple double-layer full-pitch distributed winding for a two-pole ac machine.

The spacing in degrees between adjacent slots on a stator is called the *slot pitch* γ of the stator. The slot pitch can be expressed in either mechanical or electrical degrees.

Except in very small machines, stator coils are normally formed into *double-layer windings*, as shown in Figure B–7. Double-layer windings are usually easier to manufacture (fewer slots for a given number of coils) and have simpler end connections than single-layer windings. They are therefore much less expensive to build.

Figure B–7 shows a distributed full-pitch winding for a two-pole machine. In this winding, there are four coils associated with each phase. All the coil sides of a given phase are placed in adjacent slots, and these sides are known as a *phase belt* or *phase group*. Notice that there are six phase belts on this two-pole stator. In general, there are $3P$ phase belts on a P-pole stator, P of them in each phase.

Figure B–8 shows a distributed winding using fractional-pitch coils. Notice that this winding still has phase belts, but that the phases of coils within an individual slot may be mixed. The pitch of the coils is 5/6 or 150 electrical degrees.

The Breadth or Distribution Factor

Dividing the total required number of turns into separate coils permits more efficient use of the inner surface of the stator, and it provides greater structural strength, since the slots carved in the frame of the stator can be smaller. However, the fact that the turns composing a given phase lie at different angles means that their voltages will be somewhat smaller than would otherwise be expected.

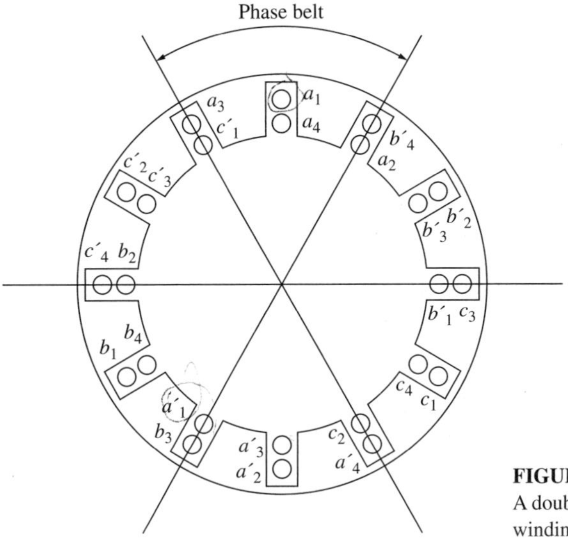

FIGURE B–8
A double-layer fractional-pitch ac winding for a two-pole ac machine.

To illustrate this problem, examine the machine shown in Figure B–9. This machine has a single-layer winding, with the stator winding of each phase (each phase belt) distributed among three slots spaced 20° apart.

If the central coil of phase *a* initially has a voltage given by

$$\mathbf{E}_{a2} = E \angle 0° \text{ V}$$

then the voltages in the other two coils in phase *a* will be

$$\mathbf{E}_{a1} = E \angle -20° \text{ V}$$
$$\mathbf{E}_{a3} = E \angle 20° \text{ V}$$

The total voltage in phase *a* is given by

$$\begin{aligned}\mathbf{E}_a &= \mathbf{E}_{a1} + \mathbf{E}_{a2} + \mathbf{E}_{a3} \\ &= E \angle -20° + E \angle 0° + E \angle 20° \\ &= E \cos(-20°) + jE \sin(-20°) + E + E \cos 20° + jE \sin 20° \\ &= E + 2E \cos 20° = 2.879\, E\end{aligned}$$

This voltage in phase *a* is not quite what would have been expected if the coils in a given phase had all been concentrated in the same slot. Then, the voltage E_a would have been equal to $3E$ instead of $2.879E$. The ratio of the actual voltage in a phase of a distributed winding to its expected value in a concentrated winding with the same number of turns is called the *breadth factor* or *distribution factor* of winding. The distribution factor is defined as

$$k_d = \frac{V_\phi \text{ actual}}{V_\phi \text{ expected with no distribution}} \tag{B–20}$$

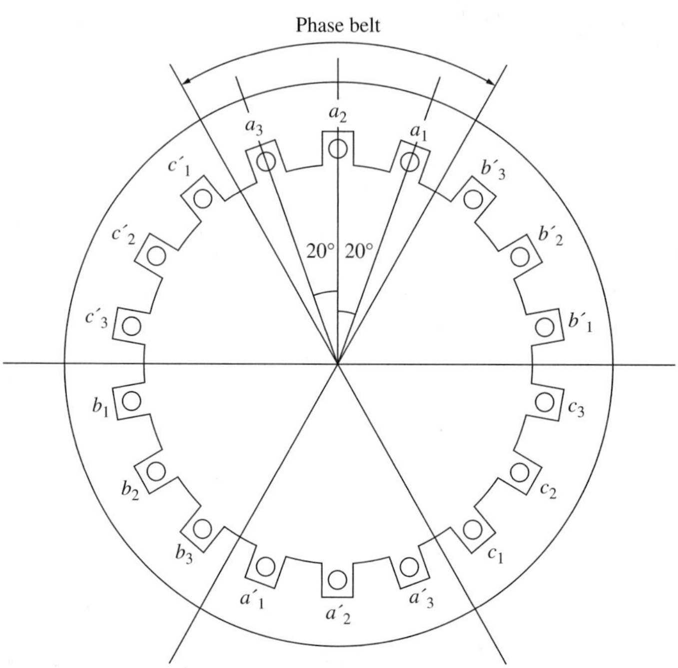

FIGURE B–9
A two-pole stator with a single-layer winding consisting of three coils per phase, each separated by 20°.

The distribution factor for the machine in Figure B–9 is thus

$$k_d = \frac{2.879E}{3E} = 0.960 \tag{B–21}$$

The distribution factor is a convenient way to summarize the decrease in voltage caused by the spatial distribution of the coils in a stator winding.

It can be shown (see Reference 1) that, for a winding with n slots per phase belt spaced γ degrees apart, the distribution factor is given by

$$\boxed{k_d = \frac{\sin(n\gamma/2)}{n \sin(\gamma/2)}} \tag{B–22}$$

Notice that for the previous example with $n = 3$ and $\gamma = 20°$, the distribution factor becomes

$$k_d = \frac{\sin(n\gamma/2)}{n \sin(\gamma/2)} = \frac{\sin[(3)(20°)/2]}{3 \sin(20°/2)} = 0.960 \qquad \text{(B–22)}$$

which is the same result as before.

The Generated Voltage Including Distribution Effects

The rms voltage in a single coil of N_C turns and pitch factor k_p was previously determined to be

$$E_A = \sqrt{2}\pi N_C k_p \phi f \tag{B–15}$$

If a stator phase consists of i coils, each containing N_C turns, then a total of $N_P = iN_C$ turns will be present in the phase. The voltage present across the phase will just be the voltage due to N_P turns all in the same slot times the reduction caused by the distribution factor, so the total phase voltage will become

$$\boxed{E_A = \sqrt{2}\pi N_P k_p k_d \phi f} \tag{B–23}$$

The pitch factor and the distribution factor of a winding are sometimes combined for ease of use into a single *winding factor* k_w. The winding factor of a stator is given by

$$\boxed{k_w = k_p k_d} \tag{B–24}$$

Applying this definition to the equation for the voltage in a phase yields

$$\boxed{E_A = \sqrt{2}\pi N_P k_w \phi f} \tag{B–25}$$

Example B–2. A simple two-pole, three-phase, Y-connected synchronous machine stator is used to make a generator. It has a double-layer coil construction, with four stator coils per phase distributed as shown in Figure B–8. Each coil consists of 10 turns. The windings have an electrical pitch of 150°, as shown. The rotor (and the magnetic field) is rotating at 3000 r/min, and the flux per pole in this machine is 0.019 Wb.

(a) What is the slot pitch of this stator in mechanical degrees? In electrical degrees?
(b) How many slots do the coils of this stator span?
(c) What is the magnitude of the phase voltage of one phase of this machine's stator?
(d) What is the machine's terminal voltage?
(e) How much suppression does the fractional-pitch winding give for the fifth-harmonic component of the voltage relative to the decrease in its fundamental component?

Solution

(a) This stator has 6 phase belts with 2 slots per belt, so it has a total of 12 slots. Since the entire stator spans 360°, the slot pitch of this stator is

$$\gamma = \frac{360°}{12} = 30°$$

This is both its electrical and mechanical pitch, since this is a two-pole machine.

(b) Since there are 12 slots and 2 poles on this stator, there are 6 slots per pole. A coil pitch of 150 electrical degrees is 150°/180° = 5/6, so the coils must span 5 stator slots.

(c) The frequency of this machine is

$$f = \frac{n_m P}{120} = \frac{(3000\ \text{r/min})(2\ \text{poles})}{120} = 50\ \text{Hz}$$

From Equation (B–19), the pitch factor for the fundamental component of the voltage is

$$k_p = \sin\frac{\nu\rho}{2} = \sin\frac{(1)(150°)}{2} = 0.966 \qquad \text{(B–19)}$$

Although the windings in a given phase belt are in three slots, the two outer slots have only one coil each from the phase. Therefore, the winding essentially occupies two complete slots. The winding distribution factor is

$$k_d = \frac{\sin(n\gamma/2)}{n\sin(\gamma/2)} = \frac{\sin[(2)(30°)/2]}{2\sin(30°/2)} = 0.966 \qquad \text{(B–22)}$$

Therefore, the voltage in a single phase of this stator is

$$\begin{aligned} E_A &= \sqrt{2}\,\pi N_P k_p k_d \phi f \\ &= \sqrt{2}\,\pi(40\ \text{turns})(0.966)(0.966)(0.019\ \text{Wb})(50\ \text{Hz}) \\ &= 157\ \text{V} \end{aligned}$$

(d) This machine's terminal voltage is

$$V_T = \sqrt{3}E_A = \sqrt{3}(157\ \text{V}) = 272\ \text{V}$$

(e) The pitch factor for the fifth-harmonic component is

$$k_p = \sin\frac{\nu\rho}{2} = \sin\frac{(5)(150°)}{2} = 0.259 \qquad \text{(B–19)}$$

Since the pitch factor of the fundamental component of the voltage was 0.966 and the pitch factor of the fifth-harmonic component of voltage is 0.259, the fundamental component was decreased 3.4 percent, while the fifth-harmonic component was decreased 74.1 percent. Therefore, the fifth-harmonic component of the voltage is decreased 70.7 percent more than the fundamental component is.

Tooth or Slot Harmonics

Although distributed windings offer advantages over concentrated windings in terms of stator strength, utilization, and ease of construction, the use of distributed windings introduces an additional problem into the machine's design. The presence of uniform slots around the inside of the stator causes regular variations in reluctance and flux along the stator's surface. These regular variations produce harmonic components of voltage called *tooth* or *slot harmonics* (see Figure B–10). Slot harmonics occur at frequencies set by the spacing between adjacent slots and are given by

$$\boxed{\nu_{\text{slot}} = \frac{2MS}{P} \pm 1} \qquad \text{(B–26)}$$

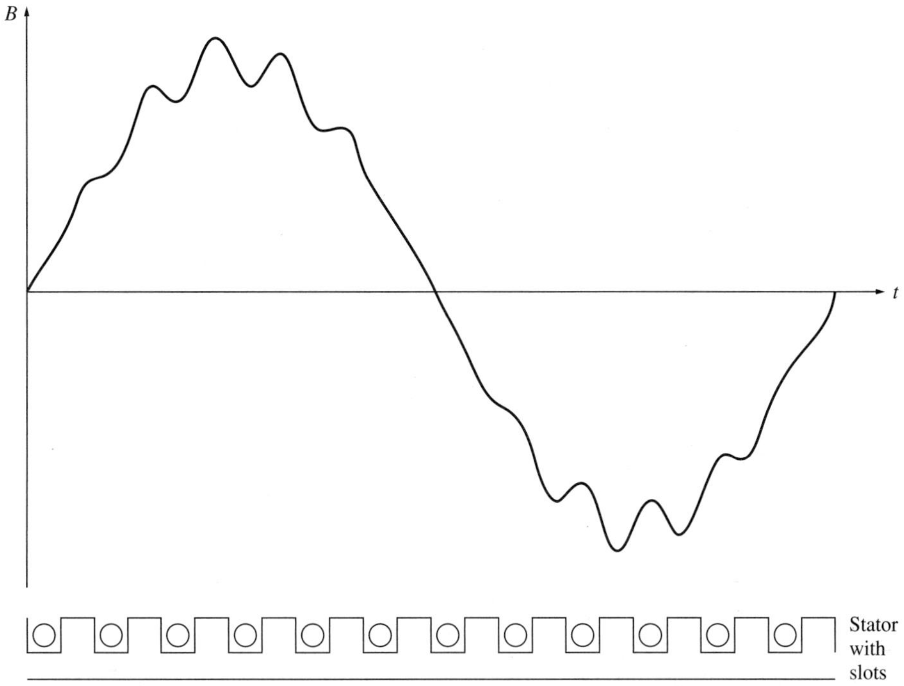

FIGURE B–10
Flux density variations in the air gap due to the tooth or slot harmonics. The reluctance of each slot is higher than the reluctance of the metal surface between the slots, so flux densities are lower directly over the slots.

where v_{slot} = number of the harmonic component
S = number of slots on stator
M = an integer
P = number of poles on machine

The value $M = 1$ yields the lowest-frequency slot harmonics, which are also the most troublesome ones.

Since these harmonic components are set by the spacing *between adjacent coil slots*, variations in coil pitch and distribution cannot reduce these effects. Regardless of a coil's pitch, it *must* begin and end in a slot, and therefore the coil's spacing is an integral multiple of the basic spacing causing slot harmonics in the first place.

For example, consider a 72-slot, six-pole ac machine stator. In such a machine, the two lowest and most troublesome stator harmonics are

$$v_{slot} = \frac{2MS}{P} \pm 1 \tag{B–26}$$

$$= \frac{2(1)(72)}{6} \pm 1 = 23, 25$$

These harmonics are at 1380 and 1500 Hz in a 60-Hz machine.

Slot harmonics cause several problems in ac machines:

1. They induce harmonics in the generated voltage of ac generators.
2. The interaction of stator and rotor slot harmonics produces parasitic torques in induction motors. These torques can seriously affect the shape of the motor's torque–speed curve.
3. They introduce vibration and noise in the machine.
4. They increase core losses by introducing high-frequency components of voltages and currents into the teeth of the stator.

Slot harmonics are especially troublesome in induction motors, where they can induce harmonics of the same frequency into the rotor field circuit, further reinforcing their effects on the machine's torque.

Two common approaches are taken in reducing slot harmonics. They are *fractional-slot windings* and *skewed rotor conductors*.

Fractional-slot windings involve using a fractional number of slots per rotor pole. All previous examples of distributed windings have been integral-slot windings; i.e., they have had 2, 3, 4, or some other integral number of slots per pole. On the other hand, a fractional-slot stator might be constructed with 2½ slots per pole. The offset between adjacent poles provided by fractional-slot windings helps to reduce both belt and slot harmonics. This approach to reducing harmonics may be used on any type of ac machine. Fractional-slot harmonics are explained in detail in References 1 and 2.

The other, much more common, approach to reducing slot harmonics is *skewing* the conductors on the rotor of the machine. This approach is primarily used on induction motors. The conductors on an induction motor rotor are given a slight twist, so that when one end of a conductor is under one stator slot, the other end of the coil is under a neighboring slot. This rotor construction is shown in Figure B–11. Since a single rotor conductor stretches from one coil slot to the next (a distance corresponding to one full electrical cycle of the lowest slot harmonic frequency), the voltage components due to the slot harmonic variations in flux cancel.

B.3 SUMMARY

In real machines, the stator coils are often of fractional pitch, meaning that they do not reach completely from one magnetic pole to the next. Making the stator windings fractional-pitch reduces the magnitude of the output voltage slightly, but at the same time attenuates the harmonic components of voltage drastically, resulting in a much smoother output voltage from the machine. A stator winding using fractional-pitch coils is often called a *chorded winding*.

Certain higher-frequency harmonics, called tooth or slot harmonics, cannot be suppressed with fractional-pitch coils. These harmonics are especially

FIGURE B–11
An induction motor rotor exhibiting conductor skewing. The skew of the rotor conductors is just equal to the distance between one stator slot and the next one. (*Courtesy of MagneTek, Inc.*)

troublesome in induction motors. They can be reduced by employing fractional-slot windings or by skewing the rotor conductors of an induction motor.

Real ac machine stators do not simply have one coil for each phase. In order to get reasonable voltages out of a machine, several coils must be used, each with a large number of turns. This fact requires that the windings be distributed over some range on the stator surface. Distributing the stator windings in a phase reduces the possible output voltage by the distribution factor k_d, but it makes it physically easier to put more windings on the machine.

QUESTIONS

B–1. Why are distributed windings used instead of concentrated windings in ac machine stators?

B–2. (*a*) What is the distribution factor of a stator winding? (*b*) What is the value of the distribution factor in a concentrated stator winding?

B–3. What are chorded windings? Why are they used in an ac stator winding?

B–4. What is pitch? What is the pitch factor? How are they related to each other?

B–5. Why are third-harmonic components of voltage not found in three-phase ac machine outputs?

B–6. What are triplen harmonics?

B–7. What are slot harmonics? How can they be reduced?

B–8. How can the magnetomotive force (and flux) distribution in an ac machine be made more nearly sinusoidal?

PROBLEMS

B–1. A two-slot three-phase stator armature is wound for two-pole operation. If fractional-pitch windings are to be used, what is the best possible choice for winding pitch if it is desired to eliminate the fifth-harmonic component of voltage?

B–2. Derive the relationship for the winding distribution factor k_d in Equation (B–22).

B–3. A three-phase four-pole synchronous machine has 96 stator slots. The slots contain a double-layer winding (two coils per slot) with four turns per coil. The coil pitch is 19/24.

(a) Find the slot and coil pitch in electrical degrees.

(b) Find the pitch, distribution, and winding factors for this machine.

(c) How well will this winding suppress third, fifth, seventh, ninth, and eleventh harmonics? Be sure to consider the effects of both coil pitch and winding distribution in your answer.

B–4. A three-phase four-pole winding of the double-layer type is to be installed on a 48-slot stator. The pitch of the stator windings is 5/6, and there are 10 turns per coil in the windings. All coils in each phase are connected in series, and the three phases are connected in Δ. The flux per pole in the machine is 0.054 Wb, and the speed of rotation of the magnetic field is 1800 r/min.

(a) What is the pitch factor of this winding?

(b) What is the distribution factor of this winding?

(c) What is the frequency of the voltage produced in this winding?

(d) What are the resulting phase and terminal voltages of this stator?

B–5. A three-phase, Y-connected, six-pole synchronous generator has six slots per pole on its stator winding. The winding itself is a chorded (fractional-pitch) double-layer winding with eight turns per coil. The distribution factor $k_d = 0.956$, and the pitch factor $k_p = 0.981$. The flux in the generator is 0.02 Wb per pole, and the speed of rotation is 1200 r/min. What is the line voltage produced by this generator at these conditions?

B–6. A three-phase, Y-connected, 50-Hz, two-pole synchronous machine has a stator with 18 slots. Its coils form a double-layer chorded winding (two coils per slot), and each coil has 60 turns. The pitch of the stator coils is 8/9.

(a) What rotor flux would be required to produce a terminal (line-to-line) voltage of 6 kV?

(b) How effective are coils of this pitch at reducing the fifth-harmonic component of voltage? The seventh-harmonic component of voltage?

B–7. What coil pitch could be used to completely eliminate the seventh-harmonic component of voltage in ac machine armature (stator)? What is the *minimum* number of slots needed on an eight-pole winding to exactly achieve this pitch? What would this pitch do to the fifth-harmonic component of voltage?

B–8. A 13.8-kV, Y-connected, 60-Hz, 12-pole, three-phase synchronous generator has 180 stator slots with a double-layer winding and eight turns per coil. The coil pitch on the stator is 12 slots. The conductors from all phase belts (or groups) in a given phase are connected in series.

(a) What flux per pole would be required to give a no-load terminal (line) voltage of 13.8 kV?

(b) What is this machine's winding factor k_w?

REFERENCES

1. Fitzgerald, A. E., and Charles Kingsley. *Electric Machinery*. New York: McGraw-Hill, 1952.
2. Liwschitz-Garik, Michael, and Clyde Whipple. *Alternating-Current Machinery.* Princeton, N.J.: Van Nostrand, 1961.
3. Werninck. E. H. (ed.). *Electric Motor Handbook.* London: McGraw-Hill, 1978.

APPENDIX C

SALIENT-POLE THEORY OF SYNCHRONOUS MACHINES

The equivalent circuit for a synchronous generator derived in Chapter 4 is in fact valid only for machines built with cylindrical rotors, and not for machines built with salient-pole rotors. Likewise, the expression for the relationship between the torque angle δ and the power supplied by the generator [Equation (4–20)] is valid only for cylindrical rotors. In Chapter 4, we ignored any effects due to the saliency of rotors and assumed that the simple cylindrical theory applied. This assumption is in fact not too bad for steady-state work, but it is quite poor for examining the transient behavior of generators and motors.

The problem with the simple equivalent circuit of induction motors is that it ignores the effect of the *reluctance torque* on the generator. To understand the idea of reluctance torque, refer to Figure C–1. This figure shows a salient-pole rotor with no windings inside a three-phase stator. If a stator magnetic field is produced as shown in the figure, it will induce a magnetic field in the rotor. Since it is *much* easier to produce a flux along the axis of the rotor than it is to produce a flux across the axis, the flux induced in the rotor will line up with the axis of the rotor. Since there is an angle between the stator magnetic field and the rotor magnetic field, a torque will be induced in the rotor which will tend to line up the rotor with the stator field. The magnitude of this torque is proportional to the sine of twice the angle between the two magnetic fields ($\sin 2\delta$).

Since the cylindrical rotor theory of synchronous machines ignores the fact that it is easier to establish a magnetic field in some directions than in others (i.e., ignores the effect of reluctance torques), it is inaccurate when salient-pole rotors are involved.

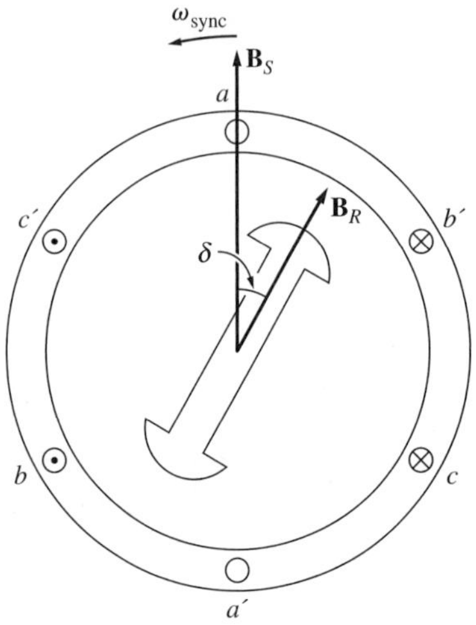

FIGURE C–1
A salient-pole rotor, illustrating the idea of reluctance torque. A magnetic field is induced in the rotor by the stator magnetic field, and a torque is produced on the rotor that is proportional to the sine of twice the angle between the two fields.

C.1 DEVELOPMENT OF THE EQUIVALENT CIRCUIT OF A SALIENT-POLE SYNCHRONOUS GENERATOR

As was the case for the cylindrical rotor theory, there are four elements in the equivalent circuit of a synchronous generator:

1. The internal generated voltage of the generator $\mathbf{E}_A$
2. The armature reaction of the synchronous generator
3. The stator winding's self-inductance
4. The stator winding's resistance

The first, third, and fourth elements are unchanged in the salient-pole theory of synchronous generators, but the armature-reaction effect must be modified to explain the fact that it is easier to establish a flux in some directions than in others.

This modification of the armature-reaction effects is accomplished as explained below. Figure C–2 shows a two-pole salient-pole rotor rotating counterclockwise within a two-pole stator. The rotor flux of this rotor is called $\mathbf{B}_R$, and it points upward. By the equation for the induced voltage on a moving conductor in the presence of a magnetic field,

$$e_{\text{ind}} = (\mathbf{v} \times \mathbf{B}) \bullet \mathbf{l} \tag{1–45}$$

the voltage in the conductors in the upper part of the stator will be positive out of the page, and the voltage in the conductors in the lower part of the stator will be into the page. The plane of maximum induced voltage will lie directly under the rotor pole at any given time.

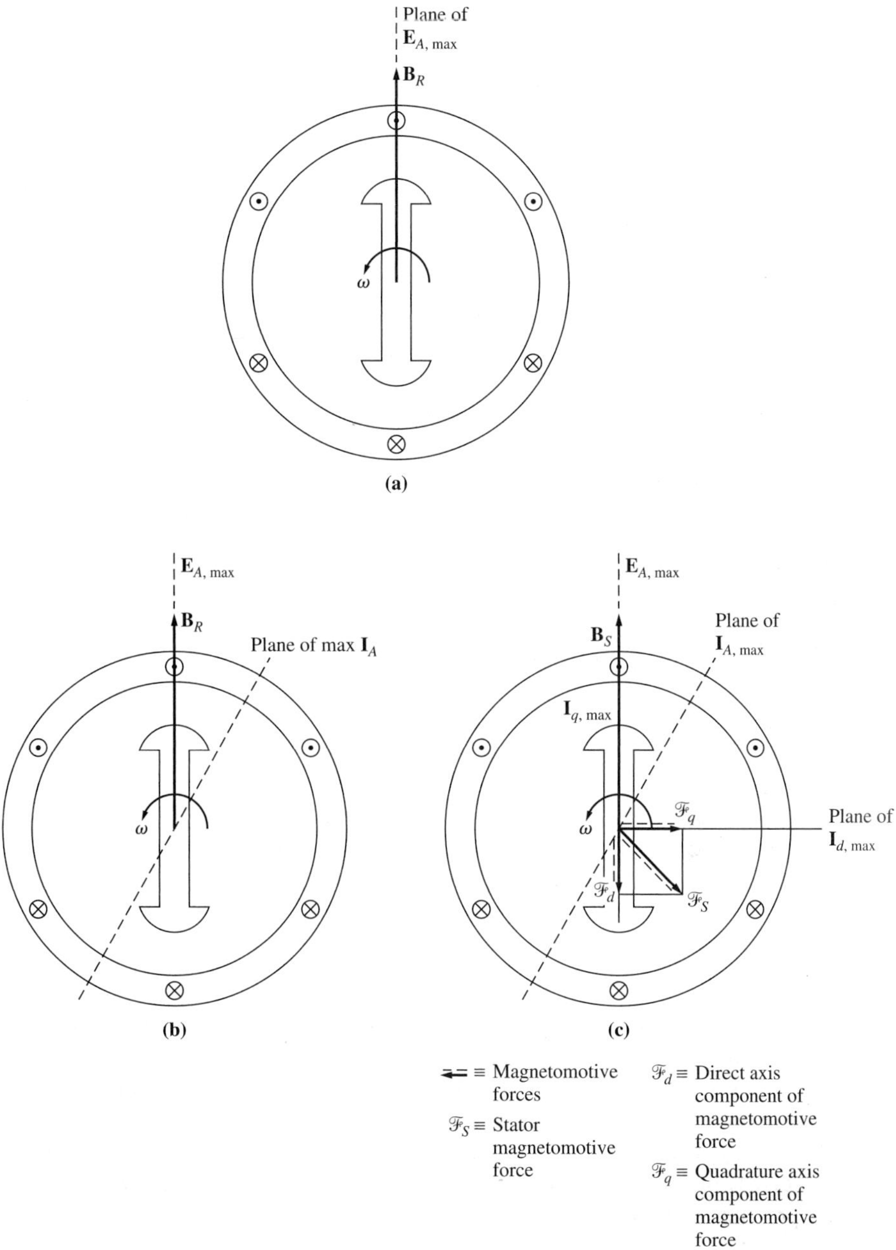

FIGURE C–2
The effects of armature reaction in a salient-pole synchronous generator. (a) The rotor magnetic field induces a voltage in the stator which peaks in the wires directly under the pole faces. (b) If a lagging load is connected to the generator, a stator current will flow that peaks at an angle behind $\mathbf{E}_A$. (c) This stator current $\mathbf{I}_A$ produces a stator magnetomotive force in the machine.

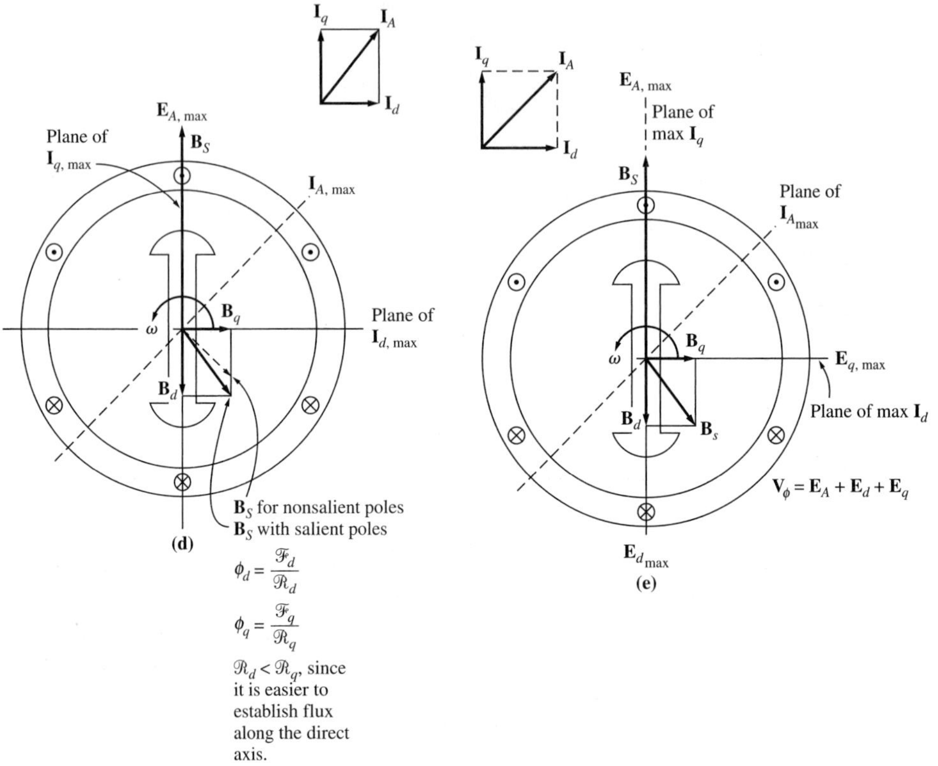

FIGURE C–2 (concluded)
(d) The stator magnetomotive force produces a stator flux $\mathbf{B}_S$. However, the direct-axis component of magnetomotive force produces more flux per ampere-turn than the quadrature-axis component does, since the reluctance of the direct-axis flux path is lower than the reluctance of the quadrature-axis flux path. (e) The direct- and quadrature-axis stator fluxes produce armature reaction voltages in the stator of the machine.

If a lagging load is now connected to the terminals of this generator, then a current will flow whose peak is delayed behind the peak voltage. This current is shown in Figure C–2b.

The stator current flow produces a magnetomotive force that lags 90° behind the plane of peak stator current, as shown in Figure C–2c. In the cylindrical theory, this magnetomotive force then produces a stator magnetic field $\mathbf{B}_S$ that lines up with the stator magnetomotive force. However, it is actually easier to produce a magnetic field in the direction of the rotor than it is to produce one in the direction perpendicular to the rotor. Therefore, we will break down the stator magnetomotive force into components parallel to and perpendicular to the rotor's axis. Each of these magnetomotive forces produces a magnetic field, but more flux is produced per ampere-turn along the axis than is produced perpendicular (*in quadrature*) to the axis.

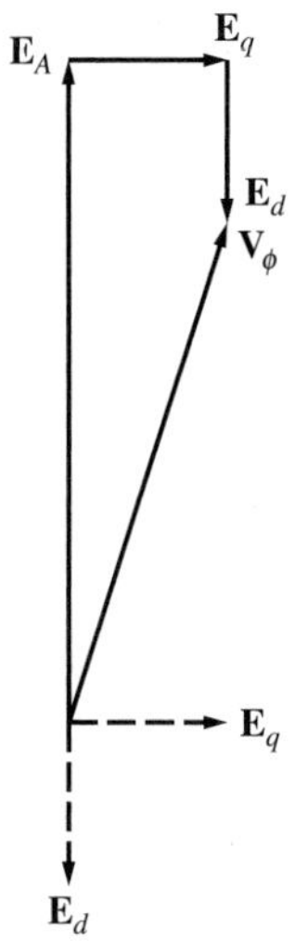

FIGURE C–3
The phase voltage of the generator is just the sum of its internal generated voltage and its armature reaction voltages.

The resulting stator magnetic field is shown in Figure C–2d, compared to the field predicted by the cylindrical rotor theory.

Now, each component of the stator magnetic field produces a voltage of its own in the stator winding by armature reaction. These armature-reaction voltages are shown in Figure C–2e.

The total voltage in the stator is thus

$$\mathbf{V}_\phi = \mathbf{E}_A + \mathbf{E}_d + \mathbf{E}_q \tag{C–1}$$

where $\mathbf{E}_d$ is the direct-axis component of the armature-reaction voltage and $\mathbf{E}_q$ is the quadrature-axis component of armature reaction voltage (see Figure C–3). As in the case of the cylindrical rotor theory, each armature-reaction voltage is *directly proportional to its stator current* and *delayed 90°* behind the stator current. Therefore, each armature-reaction voltage can be modeled by

$$\mathbf{E}_d = -jx_d\mathbf{I}_d \tag{C–2}$$

$$\mathbf{E}_q = -jx_q\mathbf{I}_q \tag{C–3}$$

and the total stator voltage becomes

$$\mathbf{V}_\phi = \mathbf{E}_A - jx_d\mathbf{I}_d - jx_q\mathbf{I}_q \tag{C–4}$$

The armature resistance and self-reactance must now be included. Since the armature self-reactance X_A is independent of the rotor angle, it is normally added to the direct and quadrature armature-reaction reactances to produce the *direct synchronous reactance* and the *quadrature synchronous reactance* of the generator:

$$\boxed{X_d = x_d + X_A} \tag{C–5}$$

$$\boxed{X_q = x_q + X_A} \tag{C–6}$$

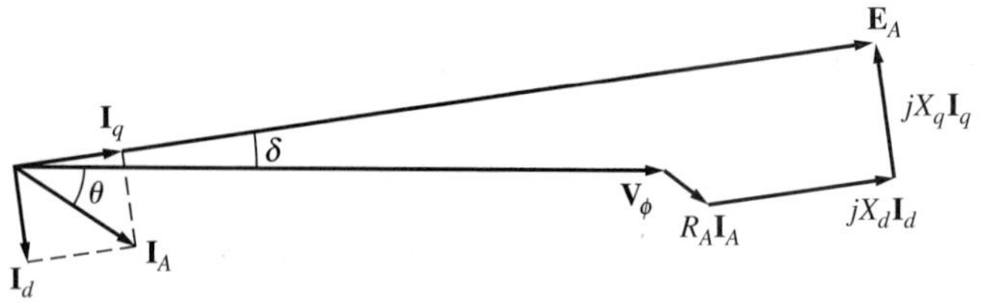

FIGURE C–4
The phasor diagram of a salient-pole synchronous generator.

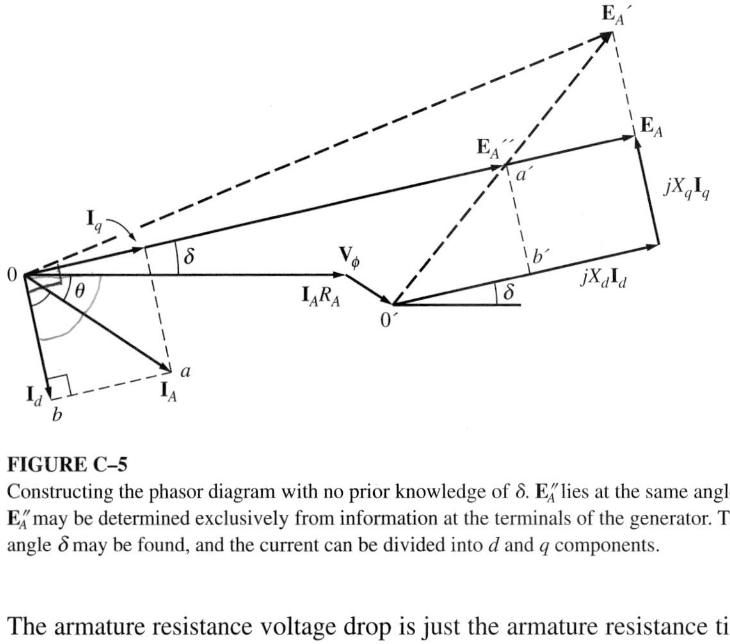

FIGURE C–5
Constructing the phasor diagram with no prior knowledge of δ. $\mathbf{E}_A''$ lies at the same angle as $\mathbf{E}_A$, and $\mathbf{E}_A''$ may be determined exclusively from information at the terminals of the generator. Therefore, the angle δ may be found, and the current can be divided into d and q components.

The armature resistance voltage drop is just the armature resistance times the armature current $\mathbf{I}_A$.

Therefore, the final expression for the phase voltage of a salient-pole synchronous motor is

$$\boxed{\mathbf{V}_\phi = \mathbf{E}_A - jX_d\mathbf{I}_d - jX_q\mathbf{I}_q - R_A\mathbf{I}_A} \tag{C–7}$$

and the resulting phasor diagram is shown in Figure C–4.

Note that this phasor diagram requires that the armature current be resolved into components in parallel with $\mathbf{E}_A$ and in quadrature with $\mathbf{E}_A$. However, the *angle* between $\mathbf{E}_A$ and $\mathbf{I}_A$ is $\delta + \theta$, which is *not usually known* before the diagram is constructed. Normally, only the power-factor angle θ is known in advance.

It is possible to construct the phasor diagram without advance knowledge of the angle δ, as shown in Figure C–5. The solid lines in Figure C–5 are the same as the lines shown in Figure C–4, while the dotted lines present the phasor diagram as though the machine had a cylindrical rotor with synchronous reactance X_d.

The angle δ of $\mathbf{E}_A$ can be found by using information known at the terminals of the generator. Notice that the phasor $\mathbf{E}_A''$, which is given by

$$\boxed{\mathbf{E}_A'' = \mathbf{V}_\phi + R_A\mathbf{I}_A + jX_q\mathbf{I}_A} \tag{C–8}$$

is collinear with the internal generated voltage $\mathbf{E}_A$. Since $\mathbf{E}_A''$ is determined by the current at the terminals of the generator, the angle δ can be determined with a knowledge of the armature current. Once the angle δ is known, the armature current can be broken down into direct and quadrature components, and the internal generated voltage can be determined.

Example C–1. A 480-V, 60-Hz, Δ-connected, four-pole synchronous generator has a direct-axis reactance of 0.1 Ω, and a quadrature-axis reactance of 0.075 Ω. Its armature resistance may be neglected. At full load, this generator supplies 1200 A at a power factor of 0.8 lagging.

(a) Find the internal generated voltage $\mathbf{E}_A$ of this generator at full load, assuming that it has a cylindrical rotor of reactance X_d.
(b) Find the internal generated voltage $\mathbf{E}_A$ of this generator at full load, assuming it has a salient-pole rotor.

Solution

(a) Since this generator is Δ-connected, the armature current at full load is

$$I_A = \frac{1200 \text{ A}}{\sqrt{3}} = 693 \text{ A}$$

The power factor of the current is 0.8 lagging, so the impedance angle θ of the load is

$$\theta = \cos^{-1} 0.8 = 36.87°$$

Therefore, the internal generated voltage is

$$\begin{aligned}\mathbf{E}_A &= \mathbf{V}_\phi + jX_S\mathbf{I}_A \\ &= 480 \angle 0° \text{ V} + j(0.1\ \Omega)(693 \angle -36.87° \text{ A}) \\ &= 480 \angle 0° + 69.3 \angle 53.13° = 524.5 \angle 6.1° \text{ V}\end{aligned}$$

Notice that the torque angle δ is 6.1°.

(b) Assume that the rotor is salient. To break down the current into direct- and quadrature-axis components, it is necessary to know the *direction* of $\mathbf{E}_A$. This direction may be determined from Equation (C–8):

$$\mathbf{E}_A'' = \mathbf{V}_\phi + R_A\mathbf{I}_A + jX_q\mathbf{I}_A \tag{C–8}$$

$$= 480\angle 0° \text{ V} + 0 \text{ V} + j(0.075\ \Omega)(693\angle -36.87° \text{ A})$$

$$= 480\angle 0° + 52\angle 53.13° = 513\angle 4.65° \text{ V}$$

The direction of $\mathbf{E}_A$ is $\delta = 4.65°$. The magnitude of the direct-axis component of current is thus

$$\begin{aligned}I_d &= I_A \sin(\theta + \delta) \\ &= (693 \text{ A}) \sin(36.87 + 4.65) = 459 \text{ A}\end{aligned}$$

and the magnitude of the quadrature-axis component of current is

$$I_q = I_A \cos(\theta + \delta)$$
$$= (693 \text{ A}) \cos(36.87 + 4.65) = 519 \text{ A}$$

Combining magnitudes and angles yields

$$\mathbf{I}_d = 459 \angle -85.35° \text{ A}$$
$$\mathbf{I}_q = 519 \angle 4.65° \text{ A}$$

The resulting internal generated voltage is

$$\mathbf{E}_A = \mathbf{V}_\phi + R_A\mathbf{I}_A + jX_d\mathbf{I}_d + jX_q\mathbf{I}_q$$
$$= 480 \angle 0° \text{ V} + 0 \text{ V} + j(0.1\ \Omega)(459 \angle -85.35° \text{ A}) + j(0.075\ \Omega)(519 \angle 4.65° \text{ A})$$
$$= 524.3 \angle 4.65° \text{ V}$$

Notice that the *magnitude of* $\mathbf{E}_A$ is not much affected by the salient poles, but the *angle* of $\mathbf{E}_A$ is considerably different with salient poles than it is without salient poles.

C.2 TORQUE AND POWER EQUATIONS OF A SALIENT-POLE MACHINES

The power output of a synchronous generator with a cylindrical rotor as a function of the torque angle was given in Chapter 4 as

$$\boxed{P = \frac{3V_\phi E_A \sin\delta}{X_S}} \qquad (4\text{–}20)$$

This equation assumed that the armature resistance was negligible. Making the same assumption, what is the output power of a salient-pole generator as a function of torque angle? To find out, refer to Figure C–6. The power out of a synchronous generator is the sum of the power due to the direct-axis current and the power due to the quadrature-axis current:

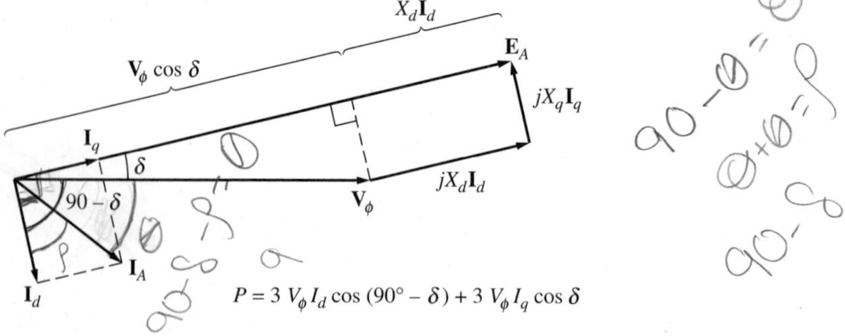

FIGURE C–6
Determining the power output of a salient-pole synchronous generator. Both $\mathbf{I}_d$ and $\mathbf{I}_q$ contribute to the output power, as shown.

$$
\begin{aligned}
P &= P_d + P_q \\
&= 3V_\phi I_d \cos(90° - \delta) + 3V_\phi I_q \cos \delta \\
&= 3V_\phi I_d \sin \delta + 3V_\phi I_q \cos \delta
\end{aligned} \tag{C–9}
$$

From Figure C–6, the direct-axis current is given by

$$I_d = \frac{E_A - V_\phi \cos \delta}{X_d} \tag{C–10}$$

and the quadrature-axis current is given by

$$I_q = \frac{V_\phi \sin \delta}{X_q} \tag{C–11}$$

Substituting Equations (C–10) and (C–11) into Equation (C–9) yields

$$
\begin{aligned}
P &= 3V_\phi\left(\frac{E_A - V_\phi \cos \delta}{X_d}\right) \sin \delta + 3V_\phi\left(\frac{V_\phi \sin \delta}{X_q}\right) \cos \delta \\
&= \frac{3V_\phi E_A}{X_d} \sin \delta + 3V_\phi^2\left(\frac{1}{X_q} - \frac{1}{X_d}\right) \sin \delta \cos \delta
\end{aligned}
$$

Since $\sin \delta \cos \delta = \frac{1}{2} \sin 2\delta$, this expression reduces to

$$\boxed{P = \frac{3V_\phi E_A}{X_d} \sin \delta + \frac{3V_\phi^2}{2}\left(\frac{X_d - X_q}{X_d X_q}\right) \sin 2\delta} \tag{C–12}$$

The first term of this expression is the same as the power in a cylindrical rotor machine, and the second term is the additional power due to the reluctance torque in the machine.

Since the induced torque in the generator is given by $\tau_{\text{ind}} = P_{\text{conv}}/\omega_m$, the induced torque in the motor can be expressed as

$$\boxed{\tau_{\text{ind}} = \frac{3V_\phi E_A}{\omega_m X_d} \sin \delta + \frac{3V_\phi^2}{2\omega_m}\left(\frac{X_d - X_q}{X_d X_q}\right) \sin 2\delta} \tag{C–13}$$

The induced torque out of a salient-pole generator as a function of the torque angle δ is plotted in Figure C–7.

PROBLEMS

C–1. A 2300-V, 1000-kVA, 0.8-PF-lagging, 60-Hz, four-pole, Y-connected synchronous generator has a direct-axis reactance of 1.1 Ω, a quadrature-axis reactance of 0.8 Ω, and an armature resistance of 0.15 Ω. Friction, windage, and stray losses may be assumed negligible. The generator's open-circuit characteristic is given by Figure P4–1.

(a) How much field current is required to make V_T equal to 2300 V when the generator is running at no load?

(b) What is the internal generated voltage of this machine when it is operating at rated conditions? How does this value of $\mathbf{E}_A$ compare to that of Problem 4–2b?

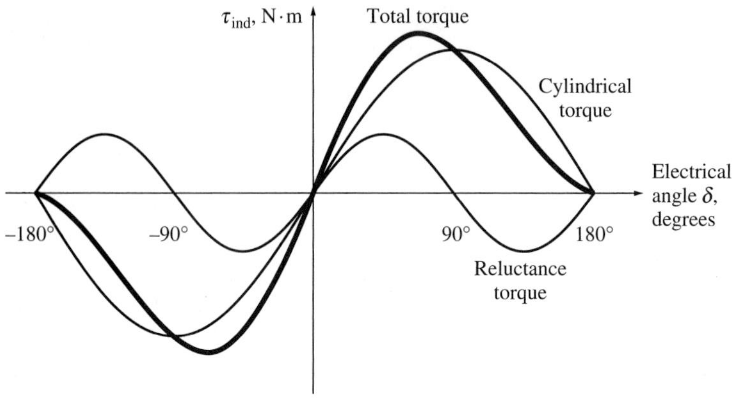

FIGURE C–7
Plot of torque versus torque angle for a salient-pole synchronous generator. Note the component of torque due to rotor reluctance.

(c) What fraction of this generator's full-load power is due to the reluctance torque of the rotor?

C–2. A 14-pole, Y-connected, three-phase, water-turbine-driven generator is rated at 120 MVA, 13.2 kV, 0.8 PF lagging, and 60 Hz. Its direct-axis reactance is 0.62 Ω and its quadrature-axis reactance is 0.40 Ω. All rotational losses may be neglected.

(a) What internal generated voltage would be required for this generator to operate at the rated conditions?

(b) What is the voltage regulation of this generator at the rated conditions?

(c) Sketch the power-versus-torque-angle curve for this generator. At what angle δ is the power of the generator maximum?

(d) How does the maximum power out of this generator compare to the maximum power available if it were of cylindrical rotor construction?

C–3. Suppose that a salient-pole machine is to be used as a motor.

(a) Sketch the phasor diagram of a salient-pole synchronous machine used as a motor.

(b) Write the equations describing the voltages and currents in this motor.

(c) Prove that the torque angle δ between $\mathbf{E}_A$ and $\mathbf{V}_\phi$ on this motor is given by

$$\delta = \tan^{-1}\frac{I_A X_q \cos\theta - I_A R_A \sin\theta}{V_\phi + I_A X_q \sin\theta + I_A R_A \cos\theta}$$

C–4. If the machine in Problem C–1 is running as a *motor* at the rated conditions, what is the maximum torque that can be drawn from its shaft without it slipping poles *when the field current is zero?*

APPENDIX D

TABLES OF CONSTANTS AND CONVERSION FACTORS

Constants	
Charge of the electron	$e = -1.6 \times 10^{-19}$ C
Permeability of free space	$\mu_0 = 4\pi \times 10^{-7}$ H/m
Permittivity of free space	$\epsilon_0 = 8.854 \times 10^{-12}$ F/m

Conversion factors		
Length	1 meter (m)	= 3.281 ft
		= 39.37 in
Mass	1 kilogram (kg)	= 0.0685 slug
		= 2.205 lb mass (lbm)
Force	1 newton (N)	= 0.2248 lb force (lb • f)
		= 7.233 poundals
		= 0.102 kg (force)
Torque	1 newton-meter (N • m)	= 0.738 pound-feet (lb • ft)
Energy	1 joule (J)	= 0.738 foot-pounds (ft • lb)
		= 3.725×10^{-7} horsepower-hour (hp • h)
		= 2.778×10^{-7} kilowatt-hour (kWh)
Power	1 watt (W)	= 1.341×10^{-3} hp
		= 0.7376 ft • lbf / s
	1 horsepower	= 746 W
Magnetic flux	1 weber (Wb)	= 10^8 maxwells (lines)
Magnetic flux density	1 tesla (T)	= 1 Wb / m^2
		= 10,000 gauss (G)
		= 64.5 kilolines/in^2
Magnetizing intensity	1 ampere • turn/m	= 0.0254 A • turns/in
		= 0.0126 oersted (Oe)

INDEX

A

Acceleration, 4, 6
Acceleration/deceleration circuits, 524
Acceleration ramps, 378
ac circuits, power in, 46–52
ac current, 2, 26–28, 67
ac machines
  distributed windings, 648–56
  effects of coil pitch, 639–48
  induced torque, 160–69, 178–81
  induced voltage, 172–78
  magnetomotive force and flux distribution, 169–72
  modeling with simple wire loop, 153–60
  overview, 152–53
  power flows and losses, 182–86
  winding insulation, 182, 183
Across-the-line starting, 357, 359–61
Air-gap flux density distribution, 639
Air-gap line, 209
Air-gap power, 321, 324, 333, 593–94
Air gaps, 14, 449
Alternators, 191. *See also* Synchronous generators
Amortisseur windings, 293–97
Ampere's law, 8–9
Angular acceleration, 4, 7
Angular position, 3
Angular velocity, 3–4
Apparent impedance, 72–73. *See also* Impedance
Apparent power
  autotransformers, 112–15
  in ideal transformer, 72
  open-delta transformer connection, 127–29
  in single-phase ac circuits, 49–50, 52
  synchronous generators, 252–53, 255
  in three-phase circuits, 625
  transformer ratings, 138–39
Approximate transformer models, 89
Arc welding generators, 542
Armature reaction
  cylindrical rotor synchronous generators, 198, 199–201
  dc machines, 433–36, 439
  permanent-magnet dc motors, 492
  salient-pole synchronous generators, 660, 663–64
  shunt dc motors, 471–72, 476
Armatures. *See* Rotors
Armature voltage controls
  basic principles, 483–84
  use with field resistance controls, 485–86
  Ward-Lenard system, 514–17, 519
Armature windings
  coils and, 421–23
  connection to commutator segments, 423–24
  copper losses, 455
  defined, 192, 451
  maximum acceptable current, 252–53, 260
Autotransformers
  apparent power rating advantage, 112–15
  basic features, 109–11
  internal impedance, 115–16
  speed control with, 588
  voltage and current relationships, 111–12
Auxiliary winding, 578–80
Average flux per turn, 78

B

Base speed, 367, 368–69, 370
Belt harmonics, 646
Blocked-rotor condition, 317
Boldface type, 3
Breadth factor, 650–52
Breakdown torque, 336. *See also* Pullout torque
Brush drop losses, 525
Brush drop voltage, 467
Brushes
  basic features, 193–94, 410, 420
  effects of commutation problems, 435, 438, 454
  in real dc machines, 421, 452–54
  voltage at, 418
Brushless dc motors, 606–9
Brushless exciters, 194–95, 196, 292
Brush losses, 456
Brush shifting, 439

C

Cage rotors
  basic features, 308, 309–10
  effects of design on induction motor characteristics, 345–48
  starting codes, 357, 358
Capability diagrams, 254–59
Capacitive loads, 51, 52
Capacitors
  synchronous motors as, 289, 290
  voltage versus current in, 389, 390
Capacitor-start, capacitor-run motors, 582–83
Capacitor-start motors, 581, 582, 583
Chamfered pole faces, 451
Chorded windings, 422, 641
Classification. *See also* Ratings
  insulation, 182, 261, 454–55
  rotor construction, 345–50
Clock motors, 600
Code letters, 357, 358
Coercive magnetizing intensity, 492, 494
Coercive magnetomotive force, 27
Coils. *See also* Windings
  construction in dc machine rotors, 421–23
  harmonics and, 639, 644–48
  pitch, 640–41 (*See also* Fractional-pitch coils)

Common current, 109
Common voltage, 109
Common windings, 109
Commutating poles, 439–42
Commutation
  frog-leg windings, 432, 433
  lap windings, 424–27
  problem remedies, 439–44
  problems in real dc machines, 433–38
  in rotating wire loop, 410
  rotor coils, 421–23
  simple four-loop dc machine, 416–21
  wave windings, 427–31
  winding connection types, 423–24
Commutator construction, 452
Commutator pitch, 423, 427
Commutator segments, 410, 420–21
Compensating windings, 443–44
Complex power, 50–51
Compounded dc motors, 500–505
Condensers, synchronous motors as, 289, 290
Conductors, 34–35, 421
Consequent-poles method, 363–64
Constant-horsepower connections, 366
Constants, 669
Constant-torque connections, 366
Conversion factors, 669
Copper losses
  ac machines, 184
  dc machines, 455–56, 524
  impact on transformer efficiency, 102
  induction motors, 321, 322, 324, 325–26, 594
  in no-load test, 380
  in real transformer behavior, 86
Core-form transformers, 67, 68
Core-loss current, 81, 83, 87
Core losses
  ac machines, 184
  basic principles, 26–28, 31
  dc machines, 456, 525
  induction motors, 322, 324
  synchronous generators, 205, 206
Countervoltage-sensing relays, 513
Critical resistance, 537
Cross-field theory, 575–77
Cumulative compounding, 500
Cumulatively compounded dc generators, 543–47
Cumulatively compounded dc motors, 501–2
Current
  in delta connections, 620–21
  induction motor limits, 394
  induction motor starting codes, 357–59
  production of magnetic flux, 11
  in single-phase ac circuits, 48
  starting problems, 42–43
  three-phase generation, 613–16
  in wye (Y) connections, 617–20
Current inrush, 139–40
Current-limiting circuits, 524
Current-limiting fuses, 521
Current paths in rotor windings, 424–27
Current ratio, 83–86
Current transformers, 69, 141–42

D

Damper windings, 248, 293–97
dc current
  behavior in ferromagnetic materials, 21–24
  early power systems based on, 66–67
  starting problems, 42–43
dc generators
  basic features, 527
  cumulatively compounded, 543–47
  differentially compounded, 547–51
  equivalent circuits, 528
  main types, 526
  separately excited, 528–34
  series, 540–42
  shunt, 534–40
  Ward-Lenard system, 516–17
dc machines
  basic linear model, 36–37
  commutation in simple four-loop machine, 416–21
  commutation problem remedies, 439–44
  commutation problems, 433–38
  construction, 449–55
  defined, 404
  frog-leg windings, 432, 433
  as generators, 41–42
  induced torque, 40–41, 447–48
  internal generated voltage, 445–47
  lap windings in, 424–28
  magnetization curve, 468–69
  modeling with simple wire loop, 404–13
  as motors, 39–41
  power flows and losses, 455–57
  rotor coils, 421–23
  starting, 37–39, 42–43
  wave windings in, 428–31, 432
dc motors
  ac power supply for, 566
  basic principles, 40, 465–66
  brushless, 606–9
  common applications, 465
  compounded, 500–505
  efficiency, 524–26
  equivalent circuits, 467–68
  permanent-magnet, 491–93
  separately excited, 469, 470
  series, 493–99, 500, 566
  shunt motor nonlinear analysis, 475–76
  shunt motor open field circuits, 490–91
  shunt motor terminal characteristics, 470–72
  shunt type defined, 470
  speed control, 479–90, 514–24
  starting, 505–14
dc test, 382–83
Deceleration ramps, 378
Deep-bar rotors, 347–48
Delta-delta connections, 123
Delta-wye connections, 121–22
Delta connections, 620–21, 624
Demagnetization, 492
Derating, 135, 367
Design classes. *See* Classification
Developed mechanical power, 324
Developed torque, 325
Differentially compounded dc generators, 547–51
Differentially compounded dc motors, 502–3
Direct synchronous reactance, 663
Distributed windings, 648–56
Distribution factor, 650–52
Distribution transformers, 69
Diverter resistors, 545, 546
Domains in ferromagnetic materials, 27–28
Dot convention, 70–71, 83–84, 85, 500
Double-cage rotors, 348

Double-layer windings, 650
Double-revolving-field theory, 570–73, 590
Drooping frequency-power characteristics, 229, 242–44
Duplex windings, 423
Dynamic stability limit, 245–46

E

Eccentric pole faces, 451
Eddy current losses
- ac machines, 184
- defined, 28
- impact on transformer efficiency, 102
- induction motors, 321
- in real transformer behavior, 86
- reducing, 31–32

Edison, Thomas, 66
Efficiency
- ac machines, 184
- dc motors, 524–26
- induction motors, 354–57
- relation to power factor in synchronous motors, 288
- transformers, 102–7

Electrical losses, 184, 455–56. *See also* Copper losses
Electrical machines, 1–2
Electric clocks, 600
Electric power grid, generator behavior in, 233–37
Energy conversion factors, 669
Energy losses
- ac machines, 182–85
- compensating for, 83
- dc machines, 455–58, 524–26
- in early power generating systems, 66
- in ferromagnetic cores, 26–28, 31
- induction motors, 321–22, 324–26, 380
- major types in real transformers, 86, 102
- reducing, 31–32, 354–55
- single-phase induction motors, 592, 594
- synchronous generators, 194, 205, 206, 253
- synchronous motors, 288
- transformer functions and, 67

Energy recovery, 391–92
English units, 2–3, 669
Equalizers, 425–27
Equivalent circuits
- compounded dc motors, 501
- cumulatively compounded dc generator, 543
- dc generators, 527, 528
- dc motors, 467–68
- differentially compounded dc generators, 548
- induction motors, 315–21, 380–87
- real transformers, 86–94
- salient-pole synchronous generator, 660–66
- separately excited dc generators, 528, 529
- series dc generator, 541
- series dc motor, 495
- shunt dc generators, 535
- single-phase induction motors, 590–97
- synchronous generators, 198–202, 203
- synchronous motors, 272–73

Equivalent field currents, 476, 531
Excitation current
- modeling for transformers, 87
- as percentage of full load, 89
- real single-phase transformers, 83, 84

Exciters, brushless, 194–95, 196, 292

F

Fan torque connections, 366
Fan torque patterns, 379
Faraday's law, 28–32, 77, 80, 367
Faults in synchronous generators, 246–50
Ferromagnetic materials
- defined, 9
- energy losses, 26–28
- magnetic behavior, 21–24
- permeability, 10, 13–14, 21–24, 25

Field current
- compounded dc motor, 500
- cumulatively compounded dc generator, 543
- dc machines, 468–69
- differentially compounded dc generators, 549
- generator ratings, 253, 255
- impact of change on synchronous motors, 280–85
- increasing in parallel synchronous generators, 237, 238, 242
- relation to field resistance, 216
- relation to flux in synchronous generators, 208–9
- relation to internal generated voltage, 198, 204
- separately excited dc generators, 528, 530
- short-circuit ratio, 212
- shunt dc motors, 476
- sources in ac machines, 152, 194–95, 196

Field loss relays, 491, 513–14
Field loss trips, 521
Field resistance
- dc motors, 525
- effect on dc generator terminal voltage, 537
- speed control by changing, 480–83

Field windings. *See also* Stator windings
- copper losses, 455
- defined, 192, 451
- disconnecting for motor startup, 294, 296
- maximum acceptable current, 253, 260

Flashing the field, 536
Flashover, 435
Flat-compounded generators, 545
Flux
- average per turn, 78
- behavior in ferromagnetic materials, 21–24, 26–28
- calculating, 10–11, 13–14, 16–17
- conversion factors, 669
- distribution in ac machines, 169–72
- Faraday's law, 28–32
- induction motor core, 367
- leakage, 14, 79–80, 86–87
- maximum in transformers, 139–40
- relation to field intensity, 9–10
- relation to magnetomotive force, 12–13
- series dc motor, 493, 495

Flux density of magnetic fields
- in ac machines, 163–64, 169–70
- basic principles, 9–11
- conversion factors, 669
- in PMDC motor poles, 492, 494

Flux guides, 598
Flux linkage, 31, 77–78

Flux per pole, 446, 447
Flux weakening, 435–36, 439, 442, 443
Force conversion factors, 669
Forward air-gap power, 593–94
Four-loop dc machines, 416–21
Four-phase stator windings, 605
Four-quadrant control systems, 517, 518–19
Fractional-pitch coils
 defined, 422, 640
 describing, 640–41
 with distributed windings, 650, 651
 harmonics suppression using, 639, 644–48
 induced voltage, 641–44
Fractional-slot windings, 656
Frame (dc machine), 449
Frequency
 reducing for motor startup, 291–92
 relation to power in synchronous generators, 230
 speed control by changing, 367–70, 372–78
 synchronous generator ratings, 251–52
 transformer ratings, 134–35
Frequency adjustment, 373–74
Friction losses
 ac machines, 184
 dc machines, 456
 induction motors, 321
Fringing effects, 14
Frog-leg windings, 432, 433
Full-load voltage regulation, 100
Full-pitch coils, 421, 640
Fuses, 510, 511, 521

G

Generator action, 35
Generators. *See also specific types of generators*
 basic principles, 1–2, 8, 35
 induction, 388–93
 linear dc machines as, 41–42
 power-flow diagrams, 185
 three-phase, 613–16
 voltage regulation, 186, 215–16
 Ward-Lenard system, 516–17
Ghost phase, 127
Governors
 function on prime movers, 229–30
 impact on ac generators operating together, 238, 242
 with infinite-bus systems, 236–37

H

Harmonic components of flux, 118–20, 639, 644–48, 654–56
Heat. *See also* Overheating
 ac generator ratings and, 252–53
 effects on machine insulation, 139, 182, 183, 454
 effects on transformer insulation, 139
High-efficiency induction motor nameplates, 393–94
High-power electronics section, 522
High-slip region, 332
High-starting-torque pattern, 377, 378
House diagrams, 234
Hysteresis, 26–27, 77, 78
Hysteresis loss
 ac machines, 184
 defined, 28
 effect of magnetomotive force excursions, 28, 29
 impact on transformer efficiency, 102
 induction motors, 321
 in real transformer behavior, 86
Hysteresis motors, 598–601

I

Ideal transformers
 analyzing circuits containing, 73
 basic features, 69–71
 magnetization curve, 85
 power in, 71–72
 real single-phase transformers as, 85–86
IEEE Standard 112, 355, 380
IEEE Standard 113, 525
Impedance
 autotransformers, 115–16
 defined, 72
 in ideal transformer, 72–73
 ignoring in power triangle approach, 632, 634
 induction motor circuit model, 382, 384, 592–93
 internal machine, 210
 rotor circuit model, 318–19
 Thevenin equivalent, 335
Impedance angle
 induction motor circuit model, 384
 in single-phase ac circuits, 48, 49, 50–52
Induced force on wires, 33–34
Induced torque
 ac machines, 158, 159, 178–81
 dc machines, 40–41, 447–48, 468
 effects of changing field resistance, 480–81, 483
 induction motors, 325, 329–36, 594
 in rotating wire loop, 156–60, 411–13
 salient-pole synchronous generators, 667, 668
 series dc motor, 493–95
 shunt dc motors, 471
 synchronous generators, 207–8, 244–45, 274
 synchronous motors, 274–75, 277, 290, 294
Induced voltage
 ac machines, 172–78
 amortisseur windings, 294
 conductor moving in magnetic field, 34–35
 dc machines, 445–47
 Faraday's law, 28–32
 fractional-pitch coils, 641–44
 induction motors, 311–12, 317–20
 in rotating wire loop, 153–56, 404–9
Inductance leakage, 86, 87
Induction generators, 388–93
Induction machines defined, 152, 307
Induction motors
 basic principles, 311–15
 construction, 309–10, 311
 deriving torque–speed characteristics, 328–43
 design evolution, 353–57
 equivalent circuit, 315–21, 380–87
 as generators, 388–93
 power and torque, 321–28
 ratings, 393–94
 single-phase (*see* Single-phase induction motors)
 slot harmonics in, 656
 speed control, 363–70, 371
 starting, 343–44, 357–62

Induction motors—*Cont.*
variations in torque–speed characteristics, 343–53
Inductive loads, 51
Inductors, 359
Infinite bus, 233–37, 238
Input impedance, 382
Input windings, 66
Instantaneous power in single-phase ac circuits, 47, 48
Instantaneous static trips, 521
Instrument transformers, 140–42
Insulation
ac machine windings, 182, 183, 260
in autotransformers, 114–15
dc machine windings, 454–55
overheating's effects on, 139, 182, 183, 454
Insulation system classes, 182, 261, 454–55
Integral-slot windings, 656
Intensity of magnetic fields, 9–10
Internal angle, 199, 206
Internal generated voltage
dc machines, 445–47, 468, 476
synchronous generators, 197–98
Internal impedance in autotransformers, 115–16
Internal machine impedance, 210
Interpoles, 439–42, 444, 452
Inverse-time overload trips, 521
Iron permeability, 10, 27

K

Kirchhoff's voltage law
compounded dc motor, 500
equivalent circuit of synchronous generator, 200–201
equivalent circuit of synchronous motor, 272–73, 289
linear dc machines, 37–38
separately excited and shunt dc motors, 469, 471
separately excited dc generators, 529
series dc generator, 541
series dc motor, 493
shunt dc generators, 535, 539
transformers, 101
Knee of saturation curve, 21

L

Lagging power factor, 101
Laminations, 31
Lap windings, 424–28, 432
L di/dt voltages, 436–38
Leading loads, 288
Leading power factors, 298
Leakage flux, 14, 79–80, 86–87
Leakage inductance, 86, 87
Leakage reactance, 344–46
Length conversion factors, 669
Lenz's law, 29–30
Linear dc machines
basic model, 36–37
as generators, 41–42
as motors, 39–41
starting, 37–39, 42–43
Line frequency. *See* Frequency
Line quantities, 617, 624–25
Line voltage. *See* Voltage
Load changes
cumulatively compounded dc generators, 544–45
induction generators, 391
shunt dc generators, 537–38
synchronous generators in parallel, 230–31
synchronous generators operating alone, 214–16
synchronous motors, 277–80
Load imbalances, 118–19
Locked-rotor condition, 317
Locked-rotor power factor, 384
Locked-rotor test, 382, 383–85
Long-legged Mary Ann, 527
Loop models. *See* Rotating wire loop
Losses. *See* Energy losses
Low-power electronics section, 522–24
Low-slip region, 332

M

Magnetic circuit principles, 11–14
Magnetic field intensity, 9–10
Magnetic fields
basic principles, 8–11
circuit model, 11–14
effects on current-carrying wires, 33–35, 36–37
Faraday's law, 28–32
induction motors, 330–31, 570–73, 575–77, 590–91
rotation in ac machines, 160–69
Magnetic flux density, 9–11
Magnetic induction motor starter circuits, 359–62
Magnetic neutral plane, 433–34
Magnetic permeability. *See* Permeability
Magnetization current, 81–83
Magnetization curves
basic features, 21, 22–23
dc machines, 468–69
ideal transformers, 85
induction machines, 316, 317, 389, 390
shunt dc generators, 539
synchronous generators, 198
typical ferromagnetic material, 492, 493
Magnetizing current, 389
Magnetizing intensity, 21–23, 669
Magnetomotive force
ac machines, 169–72
behavior in ferromagnetic materials, 21
calculating, 11–12
coercive, 27
compounded dc motor, 500
cumulatively compounded dc generator, 543
dc machines, 468, 475–76
differentially compounded dc generators, 547–49
real single-phase transformers, 83–85
salient-pole synchronous generators, 662
separately excited dc generators, 530–31
Main transformers, 131
Mass conversion factors, 669
MATLAB programs
core flux, 16–17
linear dc machine speed, 46
magnetization curves, 469
rotating magnetic field, 168–69
series dc motor torque–speed curve, 498–99
shunt dc motor speed control, 488–89
shunt dc motor torque–speed curve, 478–79
synchronous generator terminal characteristic, 223–24
synchronous motor V curve, 284–85
torque–speed characteristic of induction motor, 342–43, 351–53
transformer magnetization curve, 136–38
voltage regulation as function of load, 106–7

Maximum power in synchronous generators, 245
Maximum torque in synchronous generators, 245
Measurement. *See* Units of measure
Mechanical losses
ac machines, 184–85
dc machines, 456
dc motors, 525
synchronous generators, 205, 206
Mechanical power, 457
Method of consequent poles, 363–64
Metric units, 2–3, 669
Miscellaneous losses. *See* Stray losses
Moderate-slip region, 332
Moment of inertia, 7
Motors. *See also specific types of motors*
basic principles, 1–2, 8, 34
linear dc machines as, 39–41, 42
power-flow diagrams, 185
speed regulation, 186, 465
Multiple stator windings, 364–67
Multiplex windings, 424, 431–32
Mutual flux, 79, 80

N

Nameplates
induction motors, 393–94
synchronous motors, 298
transformer, 140, 141
NEMA design class A, 345, 346, 349
NEMA design class B, 346, 347, 348, 349
NEMA design class C, 346, 347, 348, 349
NEMA design class D, 346, 347, 350
NEMA design class F, 351
NEMA insulation classes, 182, 261, 454–55
NEMA nominal efficiency, 355, 356
NEMA Standard MG1-1993, 455
Neutral connections
three-phase power systems, 615–16, 625
three-phase transformers, 119
Neutral-plane shift, 433–35
Newton's law, 6–7, 37
No-load frequency, 234–35
No-load rotational losses, 185, 525
No-load test, 380–82
Nonsalient-pole rotors, 169, 170, 192
Normally closed contacts, 511
Normally open contacts, 510–11
Notation, 3

O

Ohm's law, 11
Oncoming generators, 226
One-dimensional linear velocity, 3
One-line diagrams, 632, 633
Open-circuit characteristics
shunt dc motors, 490–91
synchronous generators, 198, 208–9
Open-circuit power factor, 91
Open-circuit test, 90–91, 92, 208–9
Open-delta connections, 126–29
Open-wye-open-delta connections, 130–31
Output windings, 66
Overcompounded generators, 545
Overexcited synchronous motors, 282, 288–89
Overheating. *See also* Heat
ac machines, 182, 183
dc machines, 454–55
synchronous generators, 252–53, 260
transformers, 139
Overload protection
induction motors, 361
solid-state variable-frequency drives, 378
symbols for, 511
Overtemperature trips, 521

P

Parallel operation of ac generators
advantages, 224–26
with large power systems, 233–37
power characteristics, 229–33
procedures, 228–29
requirements, 226–28
for same-sized generators, 237–44
Permanent-magnet dc motors, 491–93
Permanent-magnet rotors, 608–9
Permanent magnets, 28
Permanent-magnet stepper motors, 604–5
Permanent split-capacitor motors, 582–83, 584
Permeability
basic principles, 10, 21
behavior in ferromagnetic materials, 21–24, 25
variation with existing flux, 14
Permeance, 13
Per-phase air-gap power, 593
Per-phase equivalent circuit
synchronous generators, 202, 204
system analysis with, 625, 626
Per-unit system of measure, 94–99, 123–26
Phase belts, 650
Phase groups, 650
Phase quantities, 617, 623–24
Phase sequence, 226–28, 616
Phase shifts, 120
Phasor diagrams
power flows in synchronous machines, 297
salient-pole synchronous generators, 664
synchronous generators, 202–4, 205
synchronous motors, 277, 278
transformers, 100–102
for two-phase sequences, 617
Phasors, representing in print, 3
Pilot exciters, 195, 196
Pitch, 423, 427
Pitch factor, 422, 644, 647
Plugging, 337
Polarity
in ideal transformer, 70–71
Lenz's law, 29–31
magnetomotive force, 12
real single-phase transformers, 83
Pole changing, 363–67
Pole faces, 449, 451
Pole pieces, 449
Pole pitch, 640
Pole shoes, 449
Position sensors, 606, 609
Potential transformers, 69, 141
Power
ac machine losses, 182–86
basic principles, 7–8
conversion factors, 669
in ideal transformer, 71–72
induction motor limits, 394
induction motors, 321–28, 369

Power—*Cont.*
in linear dc machines, 39–42
locked-rotor test, 384
open-delta transformer connection, 127–29
in per-unit system, 95
salient-pole synchronous generators, 666–67
in single-phase ac circuits, 46–52
synchronous generators, 205–8, 229–33, 245, 254–55
three-phase circuits, 613, 622–25
Power factor
induction motor rotor, 331–32
locked-rotor test, 384
in model transformer, 91–92, 101–2
in single-phase ac circuits, 52
Power-factor angle, 91, 331–32
Power-factor correction, 285–89
Power-flow diagrams
ac machines, 185–86
dc machines, 456–57
induction motors, 321–22, 592
Power grid, generator behavior in, 233–37
Power losses. *See* Energy losses
Power transformers, 67–69. *See also* Transformers
Power triangle, 51–52, 632–35
Preformed rotor coils, 421, 422
Preformed stator coils, 648–49
Primary windings
defined, 66
flux leakage, 79
in ideal transformer, 69
in transformer construction, 67
Prime movers
for dc generators, 527
external types for motor startup, 292
for synchronous generators, 192, 205, 229–30, 236–37
Progressive windings, 423, 431
Protection circuits, 521
Pullout torque
induction motors, 330, 332, 336, 338–39
synchronous motors, 276–77
Pulse-width modulation, 372–73, 374, 375
Push-button switches, 510–11
Pushover torque, 388

Q

Quadrature synchronous reactance, 663

R

Rare-earth magnets, 492
Ratings. *See also* Classification
induction motors, 393–94
synchronous generators, 251–60
synchronous motors, 298
transformers, 134–39
Reactance
induction motors, 316, 318, 344–46
locked-rotor test, 385
modeling for transformers, 87, 98
rotor versus stator, 385
salient-pole synchronous generators, 660, 663–64
synchronous generators, 201, 210–11, 250
Thevenin equivalent, 335
universal motors, 567
Reactive load, 391
Reactive power
expressing for synchronous generators, 206, 255
in ideal transformer, 72
induction generator requirements, 388, 389
open-delta transformer connection, 129
with parallel operation of ac generators, 238, 242
relation to terminal voltage in synchronous generators, 230–31, 242
in single-phase ac circuits, 49, 50, 52
versus terminal voltage in infinite bus, 233–34, 236–37
in three-phase circuits, 625
variation in synchronous machines, 297
Real power
ideal transformer, 71–72
single-phase ac circuits, 48–49, 50, 52
synchronous generators, 205–6, 254
variation in synchronous machines, 297
Referring, 73
Relative permeability, 10
Relays
field loss, 491, 513–14
symbols for, 511
time-delay, 361–62, 511–12
Reluctance, 12, 13
Reluctance motors, 597–98
Reluctance torque, 597, 659, 660
Reluctance-type stepper motors, 604, 605
Residual flux
defined, 27
in PMDC machines, 492
in shunt dc generators, 535, 536
Resistance
dc motor speed control based on, 480–83, 484–86
dc test, 382–83
induction generators, 392–93
induction motors, 316, 318
locked-rotor test, 385
modeling for transformers, 87, 98
speed control by changing, 370, 371
synchronous generators, 201, 211–12
Thevenin equivalent, 335
Resistors
adding to control shunt dc motor speed, 484–86
for dc motor starting, 506–10
diverter, 545, 546
for induction motor speed control, 588
for induction motor starting, 359, 361
Retrogressive windings, 423, 431
Reverse air-gap power, 593, 594
Reverse-power trips, 235
Reversing magnetic field rotation, 167–68
Right-hand rule, 12, 161
Rotating magnetic fields, 160–69
Rotating wire loop
dc voltage from, 409–10
torque induced in, 156–60, 411–13
voltage induced in, 153–56, 404–9
Rotational losses, 322, 380–82
Rotational motion principles, 3–8
Rotor coils, 421–23
Rotor copper losses
induction motors, 321, 324, 325–26, 594
synchronous ac machines, 184

Rotor frequency, 314
Rotor resistance
  dc test, 382–83
  effects on induction generators, 392–93
  locked-rotor test, 385
  speed control by changing, 370
Rotors
  basic features in synchronous generators, 192
  construction in dc machines, 452
  copper losses, 184, 321, 324, 325–26, 594
  damper windings, 293–97
  design developments, 355
  effects of design on induction motor characteristics, 345–48, 392–93
  induction motor circuit model, 317–20
  induction motor types, 309–10, 311
  NEMA standard designs, 346, 347, 349–50
  salient-pole (*See* Salient-pole rotors)
  skewed conductors, 656, 657
Rotor slip. *See* Slip
Rotor windings. *See* Armature windings
Runaway, 491

S

Salient-pole rotors
  cylindrical versus, 659
  defined, 169, 192
  equivalent circuit of machine, 660–66
  illustrated, 170, 193
  torque and power of machine, 666–67, 668
Salient poles in dc machines, 452
Saturation curves, 21, 22–23, 28
Scalars, 3
Scott-T connection, 131, 132
Secondary windings
  defined, 66
  flux leakage, 79–80
  in ideal transformer, 69
  in transformer construction, 67
Self-inductance, 201
Self-starting reluctance motors, 597, 598, 599
Separately excited dc generators, 528–34
Separately excited dc motors, 469, 470, 483
Series current (autotransformer), 109
Series dc generators, 540–42
Series dc motors, 493–99, 500, 566
Series voltage (autotransformer), 109
Series windings (autotransformer), 109
Service factors, 260
Shaded-pole motors, 584–85, 586, 587
Shading coils, 584, 585, 587
Shell-form transformers, 67, 68
Short-circuit characteristics, 209–10
Short-circuit protection, 361
Short-circuit ratio, 212
Short circuits in synchronous generators, 246–50
Short-circuit test, 91–92, 209–10
Shorting rings, 293, 310
"Short-shunt" connections, 544
Short-time operation limits, synchronous generators, 260
Shunt dc generators, 534–40
Shunt dc motors
  nonlinear analysis, 475–76
  open field circuits, 490–91
  speed control, 479–90
  with starting resistor, 506
  terminal characteristics, 470–72
Simplex windings, 423, 431
Single-phase ac circuits, 46–52
Single-phase induction motors
  capacitor-start, 581, 582, 583
  capacitor-start, capacitor-run, 582–83, 585
  circuit model, 590–97
  cross-field theory, 575–77
  double-revolving-field theory, 570–73, 590
  permanent split-capacitor, 582–83, 584
  shaded-pole, 584–85, 586, 587
  speed control, 588–89
  split-phase, 578–80
  starting problems, 569–70
  types compared, 585–88
Single-phase transformers, 116, 117
Singly excited machines, 316. *See also* Induction motors
SI units, 2–3, 669
Skewed rotor conductors, 656, 657
Slip (induction motors)
  basic principles, 313–14
  relation to rotor resistance, 339
  relation to voltage, 317–18, 320
  variations on torque–speed curve, 332
Slipping poles, 277
Slip rings, 193–94
Slip speed, 313
Slot harmonics, 648, 654–56
Slot pitch, 650
Soft-start induction motors, 350, 378
Solid-state controllers
  dc motor speed control, 517–24
  induction motor speed control, 370, 372–78, 379
  for starting synchronous motors, 292
Special-purpose motors
  brushless, 606–9
  hysteresis motors, 598–601
  reluctance motors, 597–98
  stepper motors, 597, 602–6
Speed, representing, 4
Speed control
  approaches for induction motors, 363–70, 371
  cumulatively compounded dc motors, 505
  dc motor advantages, 465
  series dc motor, 499
  shunt dc motors, 479–90
  single-phase induction motors, 588–89
  solid-state drives for, 370, 372–78, 379
  synchronous motors, 276
  universal motors, 567
  Ward-Lenard system, 514–17, 519
Speed droop, 229, 242–44
Speed ratings, 251–52
Speed regulation, 186, 465
Speed regulation circuits, 522–23
Split-phase induction motors, 578–80
Spring-type switches, 510–11
Stabilized shunt motors, 491
Standard units, 2–3, 669

Starting
  current inrush in transformers, 139–40
  dc motors, 505–14
  induction motors, 343–44, 357–62
  linear dc machines, 37–39, 42–43
  single-phase induction motors, 578–88
  synchronous motors, 290–97
Starting resistors, 506–10
Start/stop circuits, 521–22
Static stability limit, 207, 246
Stator copper losses
  induction motors, 321, 322, 324
  synchronous ac machines, 184
Stator resistance, 382–83
Stators
  copper losses, 184, 321, 322, 324
  design developments, 355
  induction motors, 308
  peak voltage, 177
  rotating magnetic field in, 160–69
Stator windings
  as autotransformer, 588–89
  defined, 192
  distributed, 648–56
  induction motors, 308, 321
  reluctance-type stepper motors, 605
  speed control approaches, 363–67
  typical construction, 648–50
Steady-state current during faults, 249
Steady-state period, 248
Step-down autotransformers, 109, 110
Step-down transformers, 77
Stepper motors, 597, 602–6
Step-up autotransformers, 109, 110
Step-up transformers, 77
Stray losses
  ac machines, 185
  dc machines, 456
  synchronous generators, 205, 206
Substation transformers, 69
Subtransient current during faults, 248–49
Subtransient period, 248
Subtransient reactance during faults, 250
Switches, 510–11
Synchronous capacitors, 289, 290
Synchronous generators
  amortisseur windings, 296
  behavior in single operation, 213–24
  construction, 192–95
  cylindrical versus salient-pole rotors, 659
  defined, 191
  equivalent circuit determination, 198–202
  internal generated voltage, 197–98
  model parameters, 208–13
  parallel operation overview, 224–26
  parallel operation power characteristics, 229–33
  parallel operation procedures, 228–29
  parallel operation requirements, 226–28
  parallel operation with large power systems, 233–37
  parallel operation with others of same size, 237–44
  phasor diagrams, 202–4, 205
  power and torque, 205–8
  ratings, 251–60
  rotation speed, 197
  salient-pole equivalent circuit, 660–66
  salient-pole torque and power equations, 666–67
  synchronous motors and, 297–98
  transients, 244–50
Synchronous machines, 152, 192
Synchronous motors
  basic principles, 271–75
  effect of field current changes, 280–85
  effect of load changes, 277–80
  power-factor correction, 285–89
  ratings, 298
  starting, 290–97
  stepper motors, 602–6
  synchronous generators versus, 297–98
  torque-speed curves, 275–77
Synchronous motor V curve, 281
Synchronous reactance, 201, 210–11
Synchronous reluctance motors, 597
Synchronous speed, 363
Synchroscopes, 228–29
Syncrospeed motor, 598, 599
Système International (SI), 2–3, 669

T

Tap changing under load transformers, 109
Taps, 108–9, 131
TCUL transformers, 109
Teaser transformers, 131
Terminal characteristic
  cumulatively compounded dc generators, 544–45, 546–47, 548
  differentially compounded dc generators, 549, 550, 551
  separately excited dc generators, 528–29, 530
  series dc generators, 541–42
  series dc motors, 495–96
  shunt dc generators, 537–38, 539–40
  shunt dc motors, 470–72
Terminal voltage
  control in induction generators, 388, 389, 390
  correcting variation in synchronous generators, 216
  cumulatively compounded dc generators, 546
  determining in synchronous generators, 201, 204
  differentially compounded dc generators, 549
  effects of load changes in induction generators, 391
  effects of load changes in synchronous generators, 215, 230–31
  reducing for induction motor speed control, 367, 588
  reducing for motor startup, 358–59
  relation to reactive power in synchronous generators, 231, 242
  separately excited dc generators, 529–30
  shunt dc generators, 538, 540
  simple four-loop dc machine, 420

wye- versus delta-connected ac machines, 177
Tertiary windings, 66
Tesla, Nicola, 353
Test Procedure for Polyphase Induction Motors and Generators, 355
Thermal damage curves, 260, 261
Thevenin's theorem, 333–35
Third-harmonic voltages, 118–20, 646
Three-light-bulb method, 227, 228
Three-phase circuits
analyzing, 625–32
delta connection voltages and currents, 620–21
power generation, 613–16
power relationships, 622–25
wye (Y) connection voltages and currents, 617–20
Three-phase power systems
advantages, 613
analyzing, 625–32
diagramming, 632, 633
power triangle approach, 632–35
Three-phase T connection, 131, 133
Three-phase transformers
approaches to creating, 116, 117
per-unit measurement, 123–26
primary-secondary connection types, 118–23
two-transformer connection methods, 126–31, 132, 133
Thyristor-based controller circuits, 517–19, 588
Time-delay relays, 361–62, 511–12
Tooth harmonics, 648, 654–56
Torque
from amortisseur windings, 294–96
basic principles, 5–6, 7
conversion factors, 669
deriving for induction motors, 329–39
induced in dc machines, 40–41, 447–48, 468
induced in series dc motor, 493–95
induced in shunt dc motors, 471
induction motors, 311–13, 321–28, 594
linear dc machines, 40–41
produced in ac machines, 160–69, 178–81
reluctance, 597, 659, 660
reluctance-type stepper motors, 605
in rotating wire loop, 156–60, 411–13
salient-pole synchronous generators, 666–67
synchronous generators, 207–8, 244–45, 274
synchronous motors, 274–77
Torque–speed characteristics
ac machines, 186
capacitor-start motor, 582
changing in induction motors, 310
compounded dc motors, 501–3
deriving for induction motors, 328–43
effect of resistance changes, 371, 482–83, 485–86
effect of voltage controls in shunt motors, 484
effects of varying stator connections, 366
generator region of induction machine, 388
hysteresis motors, 600, 601
of increased rotor resistance in induction machines, 392–93
prime movers, 236–37
self-starting reluctance motors, 599
series dc motor, 496
shaded-pole motor, 586
shunt dc motors, 470–72
single-phase induction motors, 571–73, 574, 579
speed regulation as rough measure, 465
synchronous motors, 275–77
universal motors, 566–67, 568
variations in induction motors, 343–53
for varying frequencies, 368–70, 375–78
Ward-Lenard system, 516
Torque angle, 199, 206, 207, 208
Transformers
autotransformers, 109–16, 588
basic principles, 2, 8, 66–67
current inrush, 139–40
efficiency, 102–7
equivalent circuit determination, 86–94
Faraday's law, 28–32
ideal devices, 69–77
induction motors as, 316–17
nameplates, 140, 141
per-unit measurement, 94–99, 123–26
ratings, 134–39
single-phase operation, 77–86
special-purpose, 140–42
taps, 108–9
three-phase connection types, 118–23, 126–31
three-phase overview, 116
types and construction, 67–69
voltage regulation, 99–102, 108–9
Transient current during faults, 249
Transient period, 248
Transient reactance during faults, 250
Transients, 244–50
Transmission losses, 66–67
Trip circuits, 521
Triplen harmonics, 646
Triplex windings, 424
Turns ratio
in ideal transformer, 69, 70, 320
induction motors, 317, 320
taps and, 108–9
Two-layer windings, 422
Two-pole stator, 173–76, 363, 364
Two-pole synchronous motor, 271–72
Two-quadrant control systems, 517, 518
Two-value capacitor motors, 582–83

U

Unbalanced loads, 118–19
Undercompounded generators, 545
Underexcited synchronous motors, 282
Undervoltage protection, 361, 378, 521
Units of measure
acceleration, 6
angular velocity and angular acceleration, 4
apparent power, 50
conversion factors, 669
flux density, 10
flux linkage, 31
force, 6
magnetic field, 9, 10
magnetic field intensity, 10

Units of measure—*Cont.*
 magnetomotive force, 11
 overview, 2–3
 permeability, 10
 per-unit system, 94–99
 power, 7
 reactive power, 49
 real power, 31, 48
 reluctance, 12
 torque, 6
 work, 7
Unit transformers, 68
Universal motors, 566–67, 568
Unsaturated regions, 21
Unsaturated synchronous reactance, 211

V

Variable autotransformers, 115
Variable-frequency drives
 for induction motor speed control, 367, 370, 372–78, 379
 for synchronous motor startup, 292
Vectors, 3
Velocity, 5
Voltage. *See also* Terminal voltage
 brush drop, 467
 at brushes of simple dc machine, 418
 buildup in induction generator, 389–91
 buildup in shunt generators, 535–37
 changing to control shunt dc motor speed, 483–84, 485
 in delta connections, 620–21
 Faraday's law, 28–32
 induced in ac machines, 172–78
 induced in fractional-pitch coils, 641–44
 induced on conductor in magnetic field, 34–35, 36
 induction motor limits, 394
 internal generated, 197–98, 445–47, 468, 476
 Ohm's law, 11
 in per-unit system, 95
 for power transmission, 67, 76–77
 produced in induction motors, 311–12
 reducing for motor startup, 358–59
 in rotating wire loop, 153–56, 404–9
 in single-phase ac circuits, 48
 speed control by changing, 370, 371, 372–78
 synchronous generator ratings, 251–52
 Thevenin equivalent, 333–35
 three-phase generation, 613–16
 transformer functions, 2, 67
 transformer ratings, 134–35
 in wye (Y) connections, 617–20
Voltage ratio
 across transformer, 78–81
 three-phase transformers, 118, 120, 122, 123
Voltage regulation
 dc generators, 527
 generators, 186, 215–16
 transformers, 99–102, 108–9
Voltage regulators, 109

W

Ward-Lenard system, 514–17, 519
Wave windings, 428–31, 432
Windage losses
 ac machines, 184–85
 dc machines, 456
 induction motors, 321
Winding factor, 653
Windings
 amortisseur, 293–97
 compensating, 443–44
 connection to commutator segments, 423–24
 copper losses, 455–56
 distributed, 648–56
 fractional-pitch, 422, 639–48
 insulation, 454–55
 main types in dc machines, 451
 rotor coils and, 421–23
 split-phase, 578–80
 of synchronous generators, 192, 252–53, 260
Windmills, 392
Wire loop models. *See* Rotating wire loop
Work, basic principles, 7
Wound rotors, 310, 311, 344, 392–93
Wye (Y) connections, 617–20, 624
Wye-delta connections
 induction motor starters, 358, 359
 three-phase transformers, 120–21
Wye-wye (Y-Y) connections, 118–20

Y

Yokes, 452